ALGEBRA
AND TRIGONOMETRY:
A Functions Approach
THIRD EDITION

M. L. Keedy
PURDUE UNIVERSITY

Marvin L. Bittinger
INDIANA UNIVERSITY—
PURDUE UNIVERSITY AT INDIANAPOLIS

Pg 324 tan — Pg380
Pg 325 cotan —
rive —
cos

326 → sec
csc

Graphs of the trig functions

ADDISON-WESLEY PUBLISHING COMPANY
Reading, Massachusetts ▪ Menlo Park, California
London ▪ Amsterdam ▪ Don Mills, Ontario ▪ Sydney

Sponsoring Editor: *Patricia Mallion*
Production Editor: *Martha Morong*
Designer: *Vanessa Piñeiro*
Illustrator: *VAP International Communications Ltd.*
Art Coordinator: *Joseph Vetere*
Cover Design: *Vanessa Piñeiro*
Cover Illustrator: *Bob Trevor*

Library of Congress Cataloging in Publication Data

Keedy, Mervin Laverne.
  Algebra and trigonometry.

  Includes index.
  1. Algebra. 2. Trigonometry. I. Bittinger,
Marvin L. II. Title.
QA154.2.K43 1982   512′.13   81-14953
ISBN 0-201-13404-7   AACR2

ISBN 0-201-13404-7
ABCDEFGHIJK-MU-8987654321

PREFACE

There are substantial differences between the second edition of this text and the present third edition. Following is a list of features that characterize this edition.

1. *Completeness.* Topics have been added in an effort to satisfy the needs of various users. As a result, this book has become "topic-wise" very complete. It is hoped that it will now meet the needs of any user, besides being a valuable reference source.

   Naturally, no class can cover this entire book in an ordinary freshman course. Topics must be chosen judiciously and some topics will have to be omitted. The following feature is therefore very important.

2. *Flexibility.* There are many ways in which this book can be used. Many topics are optional, and there are numerous paths that one can take through various topics, teaching them in various orders. Some of the possibilities are detailed on pp. iv–vii.

3. *Exercises.* This edition contains many new exercises. In response to comments from users, we have added exercises that require something of the student other than an understanding of the immediate objectives of the lesson at hand, yet are not necessarily highly challenging. The challenge exercises of the second edition have been augmented here. Thus the first exercises in an exercise set are very much like the examples in the text for that section. These exercises are graded in difficulty and are paired. That is, each even-numbered exercise is very much like the one that precedes it. The next exercises (marked ☆) require the student to go beyond the immediate objectives. They may, for example, ask the student to synthesize objectives in the section, or those from preceding chapters with those of this section. For the first two types of exercises,

answers to the odd-numbered ones are given at the back of the book. The instructor can therefore easily make an assignment that is varied in terms of availability of answers. The challenge exercises are marked ★, and some of them are quite difficult. Answers to the challenge exercises are not in the text, but are given in the answer booklet.

4. *Keying of objectives.* For each section, the objectives (listed in the margins) are identified as [i], [ii], [iii], and so on. The text material in which the objectives are developed is marked similarly: [i], [ii], [iii], and so on. In the exercise sets, those exercises that pertain to a specific objective are also marked in like fashion. The questions in the tests and reviews are keyed to the section and subsection to which they apply. For example, a question pertaining to objective [iii] in Section 4.2 will be marked [4.2, iii].

   A student having difficulty with an exercise set or a test/review can use the keying system to find text material, worked examples, and margin exercises that pertain to the particular area of difficulty.

5. *Calculators.* Many exercises, as well as some parts of the development, are designed with the calculator in mind. Although it is perfectly feasible to use this text without a calculator, we have indicated those exercises, examples, or sections in which the use of a calculator is recommended by the symbol ▦. A calculator will be found useful in many other places as well.

   Most of the calculator exercises in the second edition were much like the other exercises, except that the numbers were more complicated. In this edition the use of the calculator has been made much more comprehensive.

6. *Readability and understandability.* Although readability and understandability are related, they are actually separate features of a text. It is easy to write prose that is easy to read but impossible to understand. Therefore, we discuss these items separately.

   With respect to *readability*, we have striven to say what we feel needs to be said, but without excess verbiage. The goal was to make the reading level of this text quite low, without sounding condescending. This book is written *to the student*. Theorems, principles, and procedures are stated for maximum student understanding, yet the tone of the book is still mature.

   With respect to *understandability*, the goal was to produce a sequence in which each topic is developed, step by carefully-described step. At each appropriate point, examples are given, sufficient in coverage that the routine part of the homework exercises is thoroughly covered. *Cautions* are frequently given to the student: for example, "Don't make the mistake of thinking that $\sqrt{a^2 + b^2} = a + b$."

7. *Functions and transformations.* This text applies the concepts of function, relation, and transformation quite thoroughly. The idea of transformations makes Chapter 3 unique and sets up the study of later material.

Chapter 3 may be a bit long. Users have given us valuable feedback with respect to this chapter. Some have recommended that it be broken up, with some of the topics imbedded in other chapters, whereas others feel quite positive about teaching the entire chapter. After careful consideration, we have decided to leave the chapter intact, but with some rearrangement of topics. Although the topics can be taught in various orders, there is no *best* order. Moreover, for reference purposes, it is desirable to leave all the material together in its own chapter. Some possible reorderings of the topics are given below.

The concept of transformation helps to simplify topics related to quadratic functions and to graphing. It provides a new approach to solving inequalities with absolute value, and many of the trigonometric identities are rendered very simple. Instructors tell us of cases in which students are afraid that they may not be understanding the material because they grasp the ideas so readily.

8. *Supplementary materials.* The following supplementary materials are available.

- An Answer Booklet containing the answers to the even-numbered exercises and to the more challenging exercises.

- A Student Solutions Booklet containing worked-out solutions to selected odd-numbered exercises.

- A Test Booklet containing five classroom-ready tests of various types for each chapter and five final examinations.

TOPIC SELECTION AND ORDERING

There are numerous topics that can be omitted, depending on the nature of a particular class and/or course. Chapter 6 can easily be omitted entirely, for example, as can several of the later chapters. We will not attempt to list all the topics that can be omitted. Such a list would be unduly long and probably not very helpful. Rather, we shall point out some possibilities that might not be so easy to see on cursory inspection of the book.

Functions and Transformations

Some users consider Chapter 3 a bit long. The following may help such users see how to break it up or omit parts of it.

Sections 3.1, 3.2, and 3.3, on Cartesian products, graphs, and functions, respectively, are the parts of the chapter that are essential in sequence. They should be taught before moving on.

Sections 3.4 and 3.5, on symmetry and transformations, can be postponed until Chapter 4, prior to the material on quadratic functions. Section 3.4 can be partly or entirely omitted without causing difficulty at this point.

Section 3.7, on inverses of relations, is needed for the material on logarithmic and exponential functions in Chapter 6, and can be postponed until the beginning of that chapter.

Section 3.6, on special classes of functions (including periodic functions), is needed for Chapter 8 on the circular functions and can be taught along with that chapter.

Trigonometry

The trigonometry occurs in successive chapters, but need not necessarily be taught in a block. There are now several different possible tracks through the trigonometry. We detail three such tracks, one with analytic emphasis, the others with triangle emphasis.

Track I (Analytic emphasis)

Proceed through Chapters 8-11 as written, omitting Section 9.6. This track gives a thorough grounding in trigonometry with an initial emphasis on the analytic aspects.

Section 8.1, which is an introduction to triangle trigonometry, could be postponed until the beginning of Chapter 9. Certain topics, such as vectors and polar coordinates, can be omitted.

Track II (Triangle emphasis)

Section 8.1	
Chapter 9 (omitting 9.6)	This is triangle trigonometry.
Chapter 11	
Remainder of Chapter 8	This is analytic trigonometry.
Chapter 10	

Track III (Minimal course, triangle emphasis)

Section 8.1	
Chapter 9 (omitting 9.6)	
Chapter 11 (omitting certain optional sections if desired)	This is triangle trigonometry.
Section 9.6	
Perhaps Section 8.7	This is analytic trigonometry.
Chapter 10	

Note that in the minimal course, most of Chapter 8 is omitted. There are other variations possible. One could teach Chapters 8, 9, 11, and then 10, for example.

Among the salient features of the text that have been retained are the following.

1. *Use of margins.* The margins contain objectives for each lesson for student reference. They also contain developmental exercises, which have proved to be extremely effective. The text refers the student to these exercises at appropriate places. When students come to these exercises, they are to stop reading and do them. Then the answers can be checked at the back of the text. Thus students receive reinforcement, guidance, and practice before continuing with the text development. The exercises in the first part of the exercise sets are very similar to these developmental exercises in the margins. The margin materials constitute a built-in study guide.

2. *Flexibility of teaching method.* There are many ways in which to use this book. The instructor who wishes to use it in a traditional way should simply ignore the margins and have students ignore them. The instructor who wishes to use the lecture method primarily, but also wants to introduce some student-centered activity into the class, can easily do so by interrupting the lecture and having students work the exercises in the margins at appropriate times. On the other hand, the book is well suited for use in math labs or other systems of individualized instruction, or in any approach that is essentially self-study. The book can be used with minimal instructor guidance, so it is particularly effective for use in large classes.

The authors wish to thank the following reviewers, who helped with the development of this new edition: Ronald D. Ferguson, San Antonio College; James E. Keisler, Louisiana State University at Baton Rouge; Robert H. Lohman, Kent State University; Gordon L. Nipp, California State College, Bakersfield; and Samuel B. Thompson, United States Air Force Academy.

January 1982   M. L. K.
M. L. B.

CONTENTS

ix

1
BASIC
CONCEPTS
OF ALGEBRA

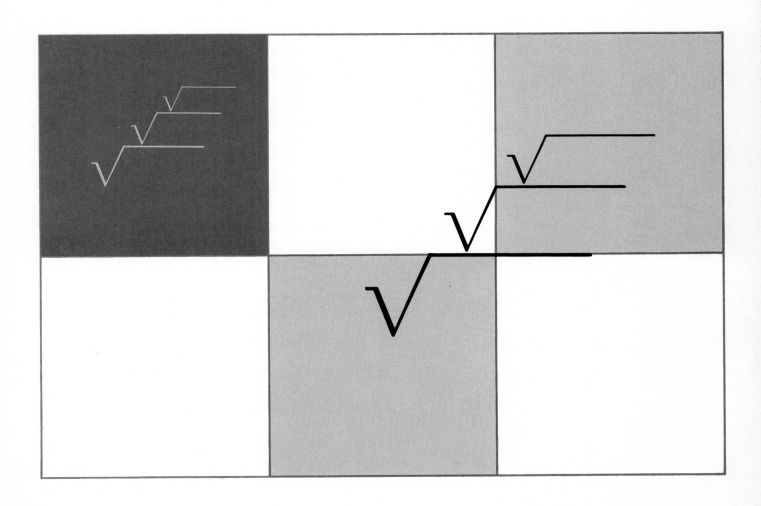

OBJECTIVES

You should be able to:

[i] Identify various kinds of real numbers.
[ii] Find the absolute value of a real number.
[iii] Add, subtract, multiply, and divide positive and negative real numbers.
[iv] Find the additive inverse of a real number.

Consider the numbers

$$1, \frac{3}{4}, -6, 0, 19, -\frac{8}{7}.$$

1. Which are natural numbers?

2. Which are whole numbers?

3. Which are integers?

4. Which are rational numbers?

1.1 THE REAL-NUMBER SYSTEM

[i] Real Numbers

There are various kinds of numbers. Those most used in elementary algebra are the so-called *real numbers*. Later we consider a more comprehensive system of numbers called the *complex numbers*. The real numbers are often shown in one-to-one correspondence with the points of a line, as follows.

The positive numbers are shown to the right of 0 and the negative numbers to the left. Zero itself is neither positive nor negative.

There are several subsystems of the real-number system. They are as follows.

Natural Numbers. **Those numbers used for counting: 1, 2, 3, ...**

Whole Numbers. **The natural numbers and 0: that is, 0, 1, 2, 3, ...**

Integers. **The whole numbers and their additive inverses: 0, 1, −1, 2, −2, 3, −3, ...**

Rational Numbers. **The integers and all quotients of integers (excluding division by 0): for example,** $\frac{4}{5}, \frac{-4}{7}, \frac{9}{1}, 6, -4, 0, \frac{78}{-5}, -\frac{2}{3}$ **(can also be named** $\frac{-2}{3}$**, or** $\frac{2}{-3}$**).**

DO EXERCISES 1–4 (in the margin).

Any real number that is not rational is called *irrational*. The rational numbers and the irrational numbers can be described in several ways.

The *rational numbers* **are:**

1. **Those numbers that can be named with fractional notation** *a/b*, **where** *a* **and** *b* **are integers and** $b \neq 0$ **(definition);**
2. **Those numbers for which decimal notation either ends or repeats.**

Examples All of these are rational.

1. $\frac{5}{16} = 0.3125$ Ending (terminating) decimal

2. $-\frac{8}{7} = -1.142857142857\ldots = -1.\overline{142857}$ Repeating decimal. The bar indicates the repeating part.

3. $\frac{3}{11} = 0.2727\ldots = 0.\overline{27}$ Repeating decimal

The *irrational numbers* are:

1. Those real numbers that are not rational (definition);
2. Those real numbers that cannot be named with fractional notation a/b, where a and b are integers and $b \neq 0$;
3. Those real numbers for which decimal notation does not end and does not repeat.

There are many irrational numbers. For example, $\sqrt{2}$ is irrational. We can find rational numbers a/b for which $(a/b) \cdot (a/b)$ is close to 2, but we cannot find such a number a/b for which $(a/b) \cdot (a/b)$ is *exactly* 2.

Unless a whole number is a perfect square its square root is irrational. For example, $\sqrt{9}$ and $\sqrt{25}$ are rational, but all of the following are irrational:

$$\sqrt{3}, \quad -\sqrt{14}, \quad \sqrt{45}.$$

There are also many irrational numbers that cannot be obtained by taking square roots. The number π is an example.* Decimal notation for π does not end and does not repeat.

Examples All of these are irrational.

4. $\pi = 3.1415926535\ldots$ Numeral does not repeat.
5. $-1.101001000100001000001\ldots$ Numeral does not repeat.
6. $\sqrt[3]{2} = 1.25992105\ldots$ Numeral does not repeat.

In Example 5 there is a pattern, but it is not a pattern formed by a repeating block of digits.

DO EXERCISES 5–10.

In arithmetic we use numbers, performing calculations to obtain certain answers. In algebra, we use arithmetic symbolism, but in addition we use symbols to represent unknown numbers. We do calculations and manipulations of symbols, based on the properties of numbers, which we review now. Algebra is thus an extension of arithmetic and a more powerful tool for solving problems.

[ii] Absolute Value

Before considering addition we need to recall that the absolute value of a number is its distance from 0 on a number line. We shall define absolute value more precisely later. The absolute value of a number a is denoted $|a|$.

Examples Simplify.

7. $|-7|$ The distance of -7 from 0 is 7, so $|-7| = 7$.

---

*$\frac{22}{7}$ and 3.14 are only rational approximations to the irrational number π.

Which of the following are rational? Which are irrational?

5. $\dfrac{-4}{5}$

6. $\dfrac{59}{37}$

7. 7.42

8. 0.47474747...
   (Numeral repeats)

9. 2.57340046631...
   (Numeral does not repeat)

10. $\sqrt{7}$

4

Simplify.

11. $|2|$ **12.** $|\sqrt{3}|$

13. $|-11.3|$ **14.** $\left|-\dfrac{3}{4}\right|$

Add.

15. $-5 + (-7)$

16. $-1.2 + (-3.5)$

17. $-\dfrac{6}{5} + \dfrac{2}{5}$

18. $0.5 + (-0.7)$

19. $8 + (-3)$

20. $\dfrac{14}{3} + \left(-\dfrac{14}{3}\right)$

8. $|5|$ The distance of 5 from 0 is 5, so $|5| = 5$.

9. $|0|$ The distance of 0 from 0 is 0, so $|0| = 0$.

To get the absolute value of a negative number, change its sign (make it positive). The absolute value of a nonnegative number is that number itself. Thus, $\left|\frac{2}{3}\right| = \frac{2}{3}$ and $|-9.7| = 9.7$.

DO EXERCISES 11–14.

[iii] Operations on the Real Numbers

Addition

Assuming that addition of nonnegative real numbers poses no problem, let us review how the definition of addition is extended to include the negative numbers.

> 1. **To add two negative numbers, add their absolute values (the sum is negative).**
> 2. **To add a negative number and a positive number, find the difference of their absolute values. The result will have the sign of the summand with the larger absolute value. If the absolute values are the same, the sum is 0.**

Examples Add.

$$-5 + (-6) = -11,$$
$$8.6 + (-4.2) = 4.4,$$
$$-5 + 3 = -2,$$
$$-\frac{3}{4} + \left(-\frac{7}{8}\right) = -\frac{13}{8},$$
$$\pi + (-\pi) = 0,$$
$$-\sqrt{3} + (-4\sqrt{3}) = -5\sqrt{3},$$
$$8 + (-5) = 3,$$
$$-\frac{9}{5} + \frac{3}{5} = -\frac{6}{5}$$

DO EXERCISES 15–20.

Properties of Real Numbers Under Addition

We now list, for review, the fundamental properties of real numbers under addition. These are properties on which algebraic manipulations are based, especially when symbols for unknown numbers are used.

> **COMMUTATIVITY.** For any real numbers a and b,
>
> $$a + b = b + a.$$
>
> (The *order* in which numbers are added does not affect the sum.)
>
> **ASSOCIATIVITY.** For any real numbers a, b, and c,
>
> $$a + (b + c) = (a + b) + c.$$
>
> (When *only* additions are involved, parentheses for grouping purposes may be placed as we please without affecting the sum.)
>
> **IDENTITY.** There exists a unique real number 0 such that for any real number a,
>
> $$a + 0 = 0 + a = a.$$
>
> (Adding 0 to any number gives that same number.)
>
> **INVERSES.** For every real number a, there exists a unique number, denoted $-a$, called the additive inverse of a, for which
>
> $$-a + a = a + (-a) = 0.$$

[iv] Additive inverses

Concerning additive inverses, we caution the reader about one of the most misunderstood, or confusing, ideas in elementary algebra: It is common to read an expression such as $-x$ as "negative x." This can be confusing, because $-x$ may be positive, negative, or zero, depending on the value of x. The symbol $-$, used in this way, indicates an *additive inverse;* somewhat unfortunately the same symbol may also indicate a negative number, as in -5, or it may indicate subtraction, as in $3 - x$.

> **CAUTION!** An initial $-$ sign, as in $-x$ or $-(x^2 + 3x - 2)$, should always be interpreted as meaning "the additive inverse of." The entire expression may be positive, negative, or zero, depending on the value of the part of the expression that follows the $-$ sign.*

Examples Find the additive inverse.

10. $-x$, when x = 5 $-(5) = -5$ (negative 5)

11. $-x$, when x = -3 $-(-3) = 3$ The inverse of negative 3 is 3.

12. $-x$, when x = 0 $-(0) = 0$ The inverse of 0 is 0.

13. $-(x^2 + 4x + 2)$, when x = 1 $-(1^2 + 4 \cdot 1 + 2) = -7$
$-(x^2 + 4x + 2)$, when x = -1 $-((-1)^2 + 4(-1) + 2) = 1$

It is easily shown that $-1 \cdot x = -x$ for any number x. That is, multiplying a number x by negative 1 produces the additive inverse of x. We list this as a theorem.

---

*Taking the additive inverse is sometimes called "changing the sign."

6

Find $-x$ and $-1 \cdot x$ when:

21. $x = 6$.

22. $x = -8$.

23. $x = -3.4$.

24. Find the additive inverse $-(x^2 - 3x)$ when $x = 2$ and $x = -1$.

Multiply.

25. $4 \cdot (-6)$

26. $-\dfrac{7}{5} \cdot \left(-\dfrac{3}{5}\right)$

27. $(-2) \cdot (-3) \cdot (-4) \cdot (-6)$

THEOREM 1

> **For any real number x, $-1 \cdot x = -x$. (Multiplying a number by -1 produces its additive inverse.)**

DO EXERCISES 21–24.

Multiplication

Assuming that multiplication of nonnegative real numbers poses no problem, let us review how the definition of multiplication is extended to include the negative numbers.

> 1. **To multiply two negative numbers, multiply their absolute values (the product is positive).**
> 2. **To multiply a positive number and a negative number, multiply their absolute values and take the additive inverse of the result (the product is negative).**

Examples Multiply.

14. $3 \cdot (-4) = -12$

15. $1.5 \cdot (-3.8) = -5.7$

16. $-5 \cdot (-4) = 20$

17. $-\dfrac{2}{3} \cdot \left(-\dfrac{4}{5}\right) = \dfrac{8}{15}$

18. $-3 \cdot (-2) \cdot (-4) = -24$

DO EXERCISES 25–27.

We now list the properties of the real numbers under multiplication.

> **COMMUTATIVITY.** For any real numbers a and b,
>
> $$ab = ba.$$
>
> (The *order* in which numbers are multiplied does not affect the product.)
>
> **ASSOCIATIVITY.** For any real numbers a, b, and c,
>
> $$a(bc) = (ab)c.$$
>
> (When *only* multiplications are involved, parentheses for grouping purposes may be placed as we please without affecting the product.)
>
> **IDENTITY.** There exists a unique real number 1 such that for any real number a,
>
> $$a \cdot 1 = 1 \cdot a = a.$$
>
> (Multiplying any number by 1 gives that same number.)
>
> **INVERSES.** For each nonzero real number a, there exists a unique number, denoted $\dfrac{1}{a}$ or a^{-1}, called the *multiplicative inverse* or *reciprocal,* for which
>
> $$a\left(\frac{1}{a}\right) = \frac{1}{a}(a) = 1.$$

Examples

19. The multiplicative inverse, or reciprocal, of 2 is $\frac{1}{2}$.

20. The multiplicative inverse of $-\frac{2}{3}$ is $-\frac{3}{2}$.

21. The reciprocal of 0.16 is 6.25.

There is a very special property that connects addition and multiplication, as follows.

DISTRIBUTIVITY. **For any real numbers a, b, and c,**

$$a(b + c) = ab + ac.^*$$

(This is also called the distributive law of multiplication over addition.)

Any number system having the preceding properties for addition and multiplication is called a *field*. Thus we refer to this list of properties as the *field properties*.

Many other properties important in algebraic manipulations can be proved from the field properties. We list some of these as theorems.

Subtraction

Subtraction is the operation opposite to addition, as given in the following definition.

DEFINITION

Subtraction. **For any real numbers a and b, $a - b = c$ if and only if $b + c = a$ ($a - b$ is the number which when added to b gives a).**

In any field, we actually subtract by adding an inverse, according to the following theorem.

THEOREM 2†

For any real numbers a and b, $a - b = a + (-b)$.

Theorem 2 follows immediately from the definitions of subtraction and additive inverse. It says that to subtract a number we can add its additive inverse.

---

*The expression $ab + ac$ means $(a \cdot b) + (a \cdot c)$. By agreement, we can omit parentheses around multiplications. According to this agreement, multiplications are to be done before additions or subtractions.

†Theorem 2 is often used as a *definition* of subtraction. The definition of subtraction used here is more general, since it does not depend on the existence of inverses. Our definition is valid in the system of natural numbers—for example, where Theorem 2 would not even make sense since additive inverses do not exist.

Subtract.

28. $2.5 - 1.2$

29. $12 - (-5)$

30. $-\dfrac{8}{5} - \dfrac{3}{5}$

31. $-20 - (-7)$

Divide.

32. $\dfrac{-20}{-5}$

33. $\dfrac{4.5}{-1.5}$

34. $-\dfrac{4}{5} \div \dfrac{3}{10}$

35. $-\dfrac{5}{6} \div \left(-\dfrac{5}{12}\right)$

Examples Subtract.

$$8 - 5 = 8 + (-5) = 3, \qquad\qquad 3 - 7 = 3 + (-7) = -4,$$
$$8.6 - (-2.3) = 8.6 + 2.3 = 10.9, \qquad -15 - (-5) = -15 + 5 = -10,$$
$$10 - (-4) = 10 + 4 = 14, \qquad\qquad \frac{5}{9} - \frac{2}{9} = \frac{5}{9} + \left(-\frac{2}{9}\right) = \frac{1}{3}$$

DO EXERCISES 28–31.

Multiplication is distributive over subtraction in the real-number system, as the following theorem states.

THEOREM 3

Distributivity. **For any real numbers** a, b, **and** c,

$$a(b - c) = ab - ac.$$

(This is the distributive law of multiplication over subtraction.)

Theorem 3 follows easily from Theorem 2 and the other distributive laws.

Division

Division is the operation opposite to multiplication, as given in the following definition.

DEFINITION

Division. **For any number** a **and any nonzero number** b, $a \div b = c$ **if and only if** $b \cdot c = a$ **(**$a \div b$ **is the number which when multiplied by** b **gives** a**).**

In any field, we usually divide by multiplying by a reciprocal, according to the following theorem.

THEOREM 4

For any real number a **and nonzero number** b, $a \div b = a\left(\dfrac{1}{b}\right)$.

The definition of division parallels the one for subtraction, and Theorems 2 and 4 are also parallels.

Examples Divide by multiplying by a reciprocal.

22. $\dfrac{3}{4} \div \left(-\dfrac{2}{3}\right) = \dfrac{3}{4}\left(-\dfrac{3}{2}\right) = -\dfrac{9}{8}$

23. $-\dfrac{6}{7} \div \left(-\dfrac{3}{5}\right) = -\dfrac{6}{7}\left(-\dfrac{5}{3}\right) = \dfrac{10}{7}$

From Theorem 4 it follows easily that the quotient of two negative numbers is positive and that the quotient of a positive number and a negative number is negative.

DO EXERCISES 32–35.

Order

The order of the real numbers is shown intuitively by a number line. If a number a is pictured to the left of a number b, then a is less than b ($a < b$). In this event, if we subtract a from b, the answer will be a positive number. This idea motivates the definition of $<$.

DEFINITION

For any real numbers a and b, $a < b$ if and only if $b - a$ is positive.

Examples Verify each inequality using the definition of $<$.

24. $2 < 8$ $8 - 2 = 6$ and 6 is positive.

25. $-4 < 9$ $9 - (-4) = 13$ and 13 is positive.

26. $-7 < -5$ $-5 - (-7) = 2$ and 2 is positive.

The symbol for "greater than" ($>$) is defined in terms of $<$, as follows.

DEFINITION

$a > b$ means $b < a$.

Thus to show that $a > b$ we can show that $a - b$ is positive.

The Use of Calculators

In the exercise sets there are problems designed for the use of a calculator. They are indicated by the symbol ▦. Certain examples and discussions are similarly marked. A calculator will be useful (but optional) in many other places as well.

Keep in mind that there can be differences in answers found on the calculator because of rounding-error differences. For example, suppose you were asked to approximate $\sqrt{18}$, precise to seven decimal places. On a certain kind of calculator with an eight-digit readout the display would show

$$\sqrt{18} = 4.2426406$$

On another kind of calculator with a ten-digit readout the display would show

$$\sqrt{18} = 4.242640687$$

The answer would be found by rounding back to the seventh decimal place and would be given as

$$\sqrt{18} = 4.2426407$$

Thus there can be variance in the seventh decimal place.

EXERCISE SET 1.1

[i] Consider the numbers

$$-6, \quad 0, \quad 3, \quad -\frac{1}{2}, \quad \sqrt{3}, \quad -2, \quad -\sqrt{7}, \quad \sqrt[3]{2}, \quad \frac{5}{8}, \quad 14, \quad -\frac{9}{4}, \quad 8.53, \quad 9\frac{1}{2}.$$

1. Which are natural numbers?   **2.** Which are whole numbers?

3. Which are irrational numbers?   **4.** Which are rational numbers?

5. Which are integers?   **6.** Which are real numbers?

Which of the following are rational? Which are irrational?

7. $-\dfrac{6}{5}$ **8.** $-\dfrac{3}{7}$ **9.** -9.032 **10.** 3.14

11. $4.516\overline{516}$ (Numeral repeats)   **12.** $-7.323\overline{232}$ (Numeral repeats)

13. $4.303003000300003\ldots$ (Numeral does not repeat)

14. $6.414114111411114\ldots$ (Numeral does not repeat)

15. $\sqrt{6}$ **16.** $\sqrt{7}$ **17.** $-\sqrt{14}$ **18.** $-\sqrt{12}$

19. $\sqrt{49}$ **20.** $-\sqrt{16}$ **21.** $\sqrt[3]{5}$ **22.** $\sqrt[4]{10}$

[ii] Simplify.

23. $|12|$ **24.** $|-2.56|$ **25.** $|-47|$ **26.** $|0|$

[iv] Find $-x$ and $-1 \cdot x$, when:

27. $x = -7.$ **28.** $x = -\dfrac{10}{3}.$ **29.** $x = 57.$ **30.** $x = \dfrac{13}{14}.$

Find $-(x^2 - 5x + 3)$, when: Find $-(7 - y)$, when:

31. $x = 12.$ **32.** $x = -8.$ **33.** $y = -9.$ **34.** $y = 19.$

[iii] Compute.

35. $-3.1 + (-7.2)$ **36.** $-735 + 319$ **37.** $\dfrac{9}{2} + \left(-\dfrac{3}{5}\right)$ **38.** $-6 + (-4) + (-10)$

39. $-7(-4)$ **40.** $-\dfrac{8}{3}\left(-\dfrac{9}{2}\right)$ **41.** $(-8.2) \times 6$ **42.** $-6(-2)(-4)$

43. $-7(-2)(-3)(-5)$ **44.** $(-7.1)(-2.3)$ **45.** $-\dfrac{14}{3}\left(-\dfrac{17}{5}\right)\left(-\dfrac{21}{2}\right)$ **46.** $-\dfrac{13}{4}\left(-\dfrac{16}{5}\right)\left(\dfrac{23}{2}\right)$

47. $\dfrac{-20}{-4}$ **48.** $\dfrac{49}{-7}$ **49.** $\dfrac{-10}{70}$ **50.** $\dfrac{-40}{8}$

51. $\dfrac{2}{7} \div \left(-\dfrac{14}{3}\right)$ **52.** $-\dfrac{3}{5} \div \left(-\dfrac{6}{7}\right)$ **53.** $-\dfrac{10}{3} \div \left(-\dfrac{2}{15}\right)$ **54.** $-\dfrac{12}{5} \div (-0.3)$

55. $11 - 15$ **56.** $-12 - 17$ **57.** $12 - (-6)$ **58.** $-13 - (-4)$

59. $15.8 - 27.4$ **60.** $-19.04 - 15.76$ **61.** $-\dfrac{21}{4} - \left(-\dfrac{7}{4}\right)$ **62.** $\dfrac{10}{3} - \left(-\dfrac{17}{3}\right)$

☆ ───

Calculate. Round to six decimal places. The symbol indicates an exercise meant to be done with a calculator.

63. a) $(1.4)^2$
              $(1.41)^2$
              $(1.414)^2$
              $(1.4142)^2$
              $(1.41421)^2$

b) What number does the sequence of numbers 1.4, 1.41, 1.414, and so on, seem to approach as a limit?

64. ▦ a) $(2.1)^3$
    $(2.15)^3$
    $(2.154)^3$
    $(2.1544)^3$
    $(2.15443)^3$

b) What number does the sequence of numbers 2.1, 2.15, 2.154, and so on, seem to approach as a limit?

What property is illustrated by each sentence?

65. $k + 0 = k$

66. $ax = xa$

67. $-1(x + y) = (-1x) + (-1y)$

68. $4 + (t + 6) = (4 + t) + 6$

69. $c + d = d + c$

70. $-67 \cdot 1 = -67$

71. $4(xy) = (4x)y$

72. $5(a + t) = 5a + 5t$

73. $y\left(\dfrac{1}{y}\right) = 1, \quad y \neq 0$

74. $-x + x = 0$

75. Show that subtraction is not commutative. That is, find real numbers a and b such that $a - b \neq b - a$.

76. Show that division is not commutative.

77. Show that division is not associative.

78. Show that subtraction is not associative.

★ ───────────────────────────────────

79. Prove that for any real numbers $(b + c)a = ba + ca$.

80. Prove that any positive number is greater than 0.

1.2 EXPONENTIAL NOTATION

[i] Integers as Exponents

When an integer greater than 1 is used as an exponent, the integer gives the number of times the base is used as a factor. For example, 5^3 means $5 \cdot 5 \cdot 5$. An exponent of 1 does not change the meaning of an expression. For example, $(-3)^1 = -3$. When 0 occurs as the exponent of a nonzero expression, we agree that the expression is equal to 1. For example, $37^0 = 1$. When a minus sign occurs with exponential notation, a certain caution is in order. For example, $(-4)^2$ means that -4 is to be raised to the second power. Hence $(-4)^2 = (-4)(-4) = 16$. On the other hand, -4^2 represents the additive inverse of 4^2. Thus $-4^2 = -16$. It may help to think of $-x^2$ as $-1 \cdot x^2$, according to Theorem 1.

DO EXERCISES 1–10.

Negative integers as exponents have meaning as follows.

DEFINITION

If n is any positive integer, then a^{-n} means $1/a^n$ for $a \neq 0$. In other words, a^n and a^{-n} are reciprocals of each other.

Examples

1. $\dfrac{1}{5^2} = 5^{-2}$

2. $7^{-3} = \dfrac{1}{7^3}$

OBJECTIVES

You should be able to:

[i] Simplify expressions with integer exponents.

[ii] Convert from decimal notation to scientific notation, and vice versa.

Rename with exponents.

1. $8 \cdot 8 \cdot 8 \cdot 8$

2. xxx

3. $4y \cdot 4y \cdot 4y \cdot 4y$

Rename without exponents.

4. 3^4

5. $(5x)^4$

6. $(-5)^4$

7. -5^4

8. $(3x)^0$

Simplify.

9. $(5y)^2$

10. $(-2x)^3$

11. Rename $\frac{1}{4^3}$ using a negative exponent.

12. Rename 10^{-4} without a negative exponent.

13. Write three other symbols for 4^{-3}.

Multiply and simplify.

14. $8^{-3} \cdot 8^7$

15. $y^7 y^{-2}$

16. $(9x^4)(-2x^7)$

17. $(-3x^{-4})(25x^{-10})$

18. $(5x^{-3}y^4)(-2x^{-9}y^{-2})$

19. $(4x^{-2}y^4)(15x^2y^{-3})$

3. $5^{-4} = \dfrac{1}{5^4} = \dfrac{1}{5 \cdot 5 \cdot 5 \cdot 5} = \dfrac{1}{625}$

DO EXERCISES 11–13.

Properties of Exponents

Multiplication

Let us consider an example involving multiplication:

$$b^5 \cdot b^{-2} = (b \cdot b \cdot b \cdot b \cdot b) \cdot \frac{1}{b \cdot b} = \frac{b \cdot b}{b \cdot b} \cdot (b \cdot b \cdot b)$$
$$= 1 \cdot (b \cdot b \cdot b) = b^3.$$

Note that the result can be obtained by adding the exponents. This is true in general.

THEOREM 5

For any number a and any integers m and n, $a^m \cdot a^n = a^{m+n}$.

Examples Multiply and simplify.

4. $x^4 \cdot x^{-2} = x^{4+(-2)} = x^2$

5. $5^4 \cdot 5^6 = 5^{10}$

6. $c^{-3} \cdot c^{-2} = c^{-5}$

7. $a^4 \cdot a^3 = a^7$

DO EXERCISES 14–19.

Division

Let us consider an example involving division:

$$\frac{8^5}{8^3} = 8^5 \cdot \frac{1}{8^3} = 8^5 \cdot 8^{-3} = 8^{5-3} = 8^2.$$

Note that the result could also be obtained by subtracting the exponents. Here is another example:

$$\frac{7^{-2}}{7^3} = 7^{-2} \cdot 7^{-3} = 7^{-2-3} = 7^{-5}.$$

Again, the result could be obtained by subtracting the exponents. This is true in general.

THEOREM 6

For any nonzero a and any integers m and n, $a^m/a^n = a^{m-n}$.

Examples Divide and simplify.

8. $\dfrac{9^{-2}}{9^5} = 9^{-2-5} = 9^{-7}$

9. $\dfrac{7^{-4}}{7^{-5}} = 7^{-4-(-5)} = 7^1 = 7$

This property can be used to show why a^0 is not defined when $a = 0$. Consider the following:

$$a^0 = a^{3-3} = \dfrac{a^3}{a^3}.$$

If a were 0, we would then have $0/0$, which is meaningless.

DO EXERCISES 20–26.

Raising a Power to a Power

Consider this example:

$$(5^2)^4 = 5^2 \cdot 5^2 \cdot 5^2 \cdot 5^2 = 5^8.$$

The result can be obtained by multiplying the exponents. This is true in general.

THEOREM 7

> For any number a and any integers m and n, $(a^m)^n = a^{mn}$.

Examples Simplify.

10. $(8^{-2})^3 = 8^{-2} \cdot 8^{-2} \cdot 8^{-2} = 8^{-6}$

11. $(x^{-5})^4 = x^{-20}$

12. $(3x^2y^{-2})^3 = 3^3(x^2)^3(y^{-2})^3 = 3^3x^6y^{-6} = 27x^6y^{-6}$

13. $(5x^3y^{-5}z^2)^4 = 5^4x^{12}y^{-20}z^8 = 625x^{12}y^{-20}z^8$

14. $(4x^2y^{-3})^{-4} = 4^{-4}x^{-8}y^{12} = \frac{1}{256}x^{-8}y^{12}$

> **CAUTION!** When raising a product such as $8x^2y^{-2}$ to a power, don't forget to raise *all* the factors to the power. For example,
>
> $$(8x^2y^{-2})^3 = 8^3(x^2)^3(y^{-2})^3.$$

DO EXERCISES 27–33.

[ii] Scientific Notation

Scientific notation is particularly useful for naming very large or very small numbers. It also has uses for our later study of logarithms. The following are examples of scientific notation:

$$7.8 \times 10^{13},$$
$$5.64 \times 10^{-8}.$$

Divide and simplify.

20. $\dfrac{4^8}{4^5}$

21. $\dfrac{5^4}{5^{-2}}$

22. $\dfrac{10^{-5}}{10^9}$

23. $\dfrac{9^{-8}}{9^{-2}}$

24. $\dfrac{y^6}{y^{-5}}$

25. $\dfrac{10y^2}{2y^3}$

26. $\dfrac{42x^7y^6}{-21y^{-3}x^{10}}$

Simplify.

27. $(3^7)^7$

28. $(8^2)^{-7}$

29. $(y^4)^{-7}$

30. $(2xy)^3$

31. $(4x^{-2}y^7)^2$

32. $(3x^4y^2)^{-3}$

33. $(10x^{-4}y^7z^{-2})^3$

14

BASIC CONCEPTS OF ALGEBRA

Convert to scientific notation.

34. 465,000

35. 3789

Convert to scientific notation.

36. 0.000145

37. 0.00000000067

Convert to decimal notation.

38. 4.67×10^{-5}

39. 7.894×10^{12}

DEFINITION

Scientific notation for a number consists of exponential notation for a power of 10, and, if needed, decimal notation for a number *a* between 1 and 10 and a multiplication sign. The following are both scientific notation: $a \times 10^{b}$; 10^{b}.

We can convert to scientific notation by multiplying by 1, choosing an appropriate symbol $10^{k}/10^{k}$ for 1.

Example 15 Convert to scientific notation.

$96,000 = 96,000 \times \dfrac{10^4}{10^4}$ We use 10^4 in order to "move the decimal point" between 9 and 6.

$ = \dfrac{96,000}{10^4} \times 10^4$

$ = 9.6 \times 10^4$

With practice such conversions can be done mentally, and you should try to do this as much as possible.

DO EXERCISES 34 AND 35.

Example 16 Convert to scientific notation.

$0.00000478 = 0.00000478 \times \dfrac{10^6}{10^6}$ We use 10^6 in order to "move the decimal point" between 4 and 7.

$ = (0.00000478 \times 10^6) \times 10^{-6}$

$ = 4.78 \times 10^{-6}$

Again you should try to make conversions mentally as much as possible.

Examples Convert to decimal notation.

17. $6.043 \times 10^5 = 604,300$

18. $4.7 \times 10^{-8} = 0.000000047$

DO EXERCISES 36–39.

EXERCISE SET 1.2

[i] Simplify.

1. $2^3 \cdot 2^{-4}$

2. $3^4 \cdot 3^{-5}$

3. $b^2 \cdot b^{-2}$

4. $c^3 \cdot c^{-3}$

5. $4^2 \cdot 4^{-5} \cdot 4^6$

6. $5^2 \cdot 5^{-4} \cdot 5^5$

7. $2x^3 \cdot 3x^2$

8. $3y^4 \cdot 4y^3$

9. $(5a^2b)(3a^{-3}b^4)$

10. $(4xy^2)(3x^{-4}y^5)$

11. $(2x)^3(3x)^2$

12. $(4y)^2(3y)^3$

13. $(6x^5y^{-2}z^3)(-3x^2y^3z^{-2})$

14. $(5x^4y^{-3}z^2)(-2x^2y^4z^{-1})$

15. $\dfrac{b^{40}}{b^{37}}$

16. $\dfrac{a^{39}}{a^{32}}$

17. $\dfrac{x^2y^{-2}}{x^{-1}y}$

18. $\dfrac{x^3y^{-3}}{x^{-1}y^2}$

19. $\dfrac{9a^2}{(-3a)^2}$

20. $\dfrac{16y^2}{(-4y)^2}$

21. $\dfrac{24a^5b^3}{8a^4b}$ **22.** $\dfrac{30x^6y^4}{5x^3y^2}$ **23.** $\dfrac{12x^2y^3z^{-2}}{21xy^2z^3}$ **24.** $\dfrac{15x^3y^4z^{-3}}{45xyz^5}$

25. $(2ab^2)^3$ **26.** $(4xy^3)^2$ **27.** $(-2x^3)^4$ **28.** $(-3x^2)^4$

29. $-(2x^3)^4$ **30.** $-(3x^2)^4$ **31.** $(6a^2b^3c)^2$ **32.** $(5x^3y^2z)^2$

33. $(-5c^{-1}d^{-2})^{-2}$ **34.** $(-4x^{-1}z^{-2})^{-2}$ **35.** $\dfrac{4^{-2}+2^{-4}}{8^{-1}}$ **36.** $\dfrac{3^{-2}+2^{-3}}{7^{-1}}$

37. $\dfrac{(-2)^4+(-4)^2}{(-1)^8}$ **38.** $\dfrac{(-3)^2+(-2)^4}{(-1)^6}$ **39.** $\dfrac{(3a^2b^{-2}c^4)^3}{(2a^{-1}b^2c^{-3})^2}$ **40.** $\dfrac{(2a^3b^{-3}c^3)^3}{(3a^{-1}b^{-3}c^{-5})^2}$

41. $\dfrac{6^{-2}x^{-3}y^2}{3^{-3}x^{-4}y}$ **42.** $\dfrac{5^{-2}x^{-4}y^3}{2^{-3}x^{-5}y}$

[ii] Convert to scientific notation.

43. 58,000,000   **44.** 27,000   **45.** 365,000   **46.** 3645

47. 0.0000027   **48.** 0.0000658   **49.** 0.027   **50.** 0.0038

51. $910,000,000,000 (a recent national debt)

52. 93,000,000 (the distance, in miles, from the earth to the sun)

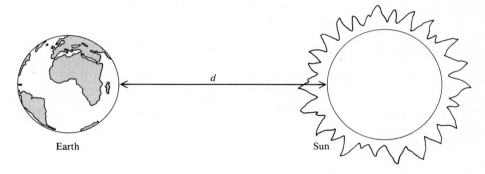

Convert to decimal notation.

53. 4×10^5 **54.** 5×10^{-4} **55.** 6.2×10^{-3} **56.** 7.8×10^6

57. 7.69×10^{12} **58.** 8.54×10^{-7}

59. 9.46×10^{12} (the distance, in kilometers, that light travels in one year)

60. 1.7×10^{-24} (the mass, in grams, of a hydrogen atom)

Find $-x^2$ and $(-x)^2$, when:

61. $x = 5$.   **62.** $x = -7$.   **63.** $x = -1.08$.   **64.** $x = \sqrt{3}$.

☆ ───

The formula

$$M = P\left[\dfrac{\dfrac{i}{12}\left(1+\dfrac{i}{12}\right)^n}{\left(1+\dfrac{i}{12}\right)^n - 1}\right]$$

gives the monthly mortgage payment M on a home loan of P dollars at interest rate i and n is the total number of payments (12 times the number of years).

65. ▦ The cost of a house is $92,000. The down payment is $14,000. The interest rate is $15\frac{3}{4}\%$. The loan period is 25 years. What is the monthly payment?

66. ▦ Repeat Exercise 65 for loan periods of 20 years and 30 years.

Find the error(s) in each of the following. Explain why each is an error. Then find the correct answer.

67. $x^4(x^3)^2 = x^9$

68. $\dfrac{x^4 y^{-7}}{x^{-2} y^5} = \dfrac{x^2}{y^2}$

69. $(2x^{-4} y^6 z^3)^3 = 6x^{-1} y^3 z^6$

OBJECTIVES

You should be able to:
[i] Determine the degree of each term of a polynomial and the degree of the polynomial.
[ii] Add polynomials.
[iii] Find the additive inverse of a polynomial.
[iv] Subtract polynomials.

1.3 ADDITION AND SUBTRACTION OF ALGEBRAIC EXPRESSIONS

[i] Polynomials

Expressions like the following are called *polynomials in one variable:**

$$-7x + 5, \quad 3y^3 - 5y^2 + 7y - 4, \quad 0, \quad -5t^4, \quad x^5 - 9.$$

DEFINITION

A *polynomial in one variable* is any expression of the type

$$a_n x^n + a_{n-1} x^{n-1} + \cdots + a_2 x^2 + a_1 x + a_0,$$

where n is a nonnegative integer and $a_n, \ldots, a_0$ are real numbers, called *coefficients*. Some or all of the coefficients may be 0. Each of the parts separated by plus signs is called a *term*.

The question might arise whether an expression such as

$$8x^3 - 6x^2 + 7x - 5$$

is a polynomial. It is indeed since it can be renamed

$$8x^2 + (-6x^2) + 7x + (-5).$$

Note that coefficients can be negative.

Expressions like the following are called *polynomials in several variables:*

$$5x^2 y^3 + 17x^2 y - 2, \quad 14a^2 b, \quad \pi r^2 + 2\pi rh.$$

Example 1 The polynomial $5x^3 y - 7xy^2 + 2$ has three terms. They are

$$5x^3 y, \quad -7xy^2, \quad \text{and} \quad 2.$$

The coefficients of the terms are 5, -7, and 2.

The *degree of a term* is the sum of the exponents of the variables. The *degree* of a nonzero polynomial is the degree of the term of highest degree. The polynomial consisting only of the number 0 has no degree.

---

*A *variable* is a symbol that can represent different numbers. Letters are usually used for variables. Letters used to represent numbers are not always variables, however. For example, if we choose to represent the distance to the moon by the letter d, then in that context d is not a variable, but a *constant*.

Example 2 In the polynomial $5x^3y - 7xy^2 + 2$, the degrees of the terms are 4, 3, and 0. The polynomial is of degree 4.

A polynomial with just one term is called a *monomial*. If there are just two terms, a polynomial is called a *binomial*. If there are just three terms, it is called a *trinomial*.

DO EXERCISES 1 AND 2.

> CAUTION! **In many applications, lower-case and capital letters are used to represent different numbers, as in**
>
> $$R + r.$$
>
> **In copying expressions, do *not* use a capital letter if a lower-case letter is given.**

[ii] Addition

Much of the algebraic manipulation we do with polynomials can also be done with expressions that are not polynomials. Here are some examples of expressions that are not polynomials.

Examples

3. $3\sqrt{x} + 4y$

4. $\dfrac{3x^2 + 2}{x - 1}$

5. $4x^{1/2} - 5y^{3/2}$

If two terms of an expression have the same letters raised to the same powers, the terms are called *similar*. Similar terms can be "combined" using the distributive laws.

Examples

6. $3x^2 - 4y + 2x^2 = 3x^2 + 2x^2 - 4y$ Rearranging using the commutative and associative laws

$\qquad = (3 + 2)x^2 - 4y$ Using a distributive law
$\qquad = 5x^2 - 4y$

7. $4x^{1/2}y + 7x^{1/2}y = 11x^{1/2}y$

8. $-2x^2\sqrt{y^3} + 5x^2\sqrt{y^3} = 3x^2\sqrt{y^3}$

DO EXERCISES 3–5.

The sum of two polynomials can be found by writing a plus sign between them and then combining similar terms. Ordinarily this can be done mentally.

Determine the degree of each term and the degree of the polynomial.

1. $x^8 - 7x^6 + 2x^4 - 3x^9 + 2$

2. $8y^4 - 7xy^3 + 6x^2y^3 - 9x^5y - 1$

Combine similar terms.

3. $5x^3y^2 - 2x^2y^3 + 4x^3y^2$

4. $3xy^2 - 4x^2y + 4xy^2 + 2x^2y$

5. $5x^4\sqrt{y} - 2x^4\sqrt{y} + 2$

Add.

6. $3x^3 + 4x^2 - 7x - 2$ and
$-7x^3 - 2x^2 + 3x + \frac{1}{2}$

7. $5p^2q^4 - 2p^2q^2 - 3q$ and
$-6pq^2 + 3p^2q^2 + 5$

8. Find the additive inverse.
$5x^2t^2 - 4xy^2t - 3xt + 6x - 5$

9. Remove the parentheses.
$-(-3x^2y + 5xy - 7x + 4y + 2)$

Example 9 Add $-3x^3 + 2x - 4$ and $4x^3 + 3x^2 + 2$.

$$(-3x^3 + 2x - 4) + (4x^3 + 3x^2 + 2) = x^3 + 3x^2 + 2x - 2$$

DO EXERCISES 6 AND 7.

[iii] Additive Inverses

The additive inverse of a polynomial can be found as described in the following theorem.

THEOREM 8

The additive inverse of a polynomial can be found by replacing every term by its additive inverse.

Example 10 The additive inverse of $-3xy^2 + 4x^2y - 5x - 3$ is

$$3xy^2 - 4x^2y + 5x + 3.$$

Example 11 The additive inverse of $7xy^2 - 6xy - 4y + 3$ can be symbolized as

$$-(7xy^2 - 6xy - 4y + 3).$$

Thus,

$$-(7xy^2 - 6xy - 4y + 3) = -7xy^2 + 6xy + 4y - 3.$$

The preceding example may bring to mind a rule that says: To remove parentheses preceded by an additive inverse sign, change the sign of every term inside the parentheses.

DO EXERCISES 8 AND 9.

[iv] Subtraction

By Theorem 2 we can subtract by adding an inverse. Thus to subtract one polynomial from another, we add its additive inverse. We change the sign of each term of the polynomial to be subtracted and then add. In simple cases this can be done mentally.

Example 12 Subtract.

$$(-9x^5 - x^3 + 2x^2 + 4) - (2x^5 - x^4 + 4x^3 - 3x^2)$$
$$= (-9x^5 - x^3 + 2x^2 + 4) + [-(2x^5 - x^4 + 4x^3 - 3x^2)]$$
$$= (-9x^5 - x^3 + 2x^2 + 4) + (-2x^5 + x^4 - 4x^3 + 3x^2)$$
$$= -11x^5 + x^4 - 5x^3 + 5x^2 + 4$$

On occasion, it may be helpful to write polynomials to be subtracted with similar terms in columns.

Example 13 Subtract the second polynomial from the first.

$$4x^2y - 6x^3y^2 \qquad\qquad + x^2y^2 - 5y$$
$$\underline{4x^2y + \ x^3y^2 + 3x^2y^3 \qquad\qquad + \ 6y} \qquad \text{Mentally, change signs and add.}$$
$$\quad\ -7x^3y^2 - 3x^2y^3 + x^2y^2 - 11y$$

DO EXERCISES 10 AND 11.

10. Subtract.
$$(5xy^4 - 7xy^2 + 4x^2 - 3)$$
$$\ - (-3xy^4 + 2xy^2 - 2y + 4)$$

11. Subtract.
$$5x^2y - 7x^3y^2 \qquad\ - x^2y^2 + 4y$$
$$\underline{-2x^2y + 2x^3y^2 - 5x^2y^3 \qquad\quad - 5y}$$

EXERCISE SET 1.3

[i] Determine the degree of each term and the degree of the polynomial.

1. $-11x^4 - x^3 + x^2 + 3x - 9$

2. $t^3 - 3t^2 + t + 1$

3. $y^3 + 2y^6 + x^2y^4 - 8$

4. $u^2 + 3v^5 - u^3v^4 - 7$

5. $a^5 + 4a^2b^4 + 6ab + 4a - 3$

6. $8p^6 + 2p^4t^4 - 7p^3t + 5p^2 - 14$

[ii] Add.

7. $5x^2y - 2xy^2 + 3xy - 5$ and
$-2x^2y - 3xy^2 + 4xy + 7$

8. $6x^2y - 3xy^2 + 5xy - 3$ and
$-4x^2y - 4xy^2 + 3xy + 8$

9. $-3pq^2 - 5p^2q + 4pq + 3$ and
$-7pq^2 + 3pq - 4p + 2q$

10. $-5pq^2 - 3p^2q + 6pq + 5$ and
$-4pq^2 + 5pq - 6p + 4q$

11. $2x + 3y + z - 7$ and
$4x - 2y - z + 8$ and
$-3x + y - 2z - 4$

12. $2x^2 + 12xy - 11$ and
$6x^2 - 2x + 4$ and
$-x^2 - y - 2$

13. $7x\sqrt{y} - 3y\sqrt{x} + \dfrac{1}{5}$ and

$-2x\sqrt{y} - y\sqrt{x} - \dfrac{3}{5}$

14. $10x\sqrt{y} - 4y\sqrt{x} + \dfrac{4}{3}$ and

$-3x\sqrt{y} - y\sqrt{x} - \dfrac{1}{3}$

[iii] Rename each additive inverse without parentheses.

15. $-(5x^3 - 7x^2 + 3x - 6)$

16. $-(-4y^4 + 7y^2 - 2y - 1)$

[iv] Subtract.

17. $(3x^2 - 2x - x^3 + 2)$
$\ - (5x^2 - 8x - x^3 + 4)$

18. $(5x^2 + 4xy - 3y^2 + 2)$
$\ - (9x^2 - 4xy + 2y^2 - 1)$

19. $(4a - 2b - c + 3d)$
$\ - (-2a + 3b + c - d)$

20. $(5a - 3b - c + 4d)$
$\ - (-3a + 5b + c - 2d)$

21. $(x^4 - 3x^2 + 4x)$
$\ - (3x^3 + x^2 - 5x + 3)$

22. $(2x^4 - 5x^2 + 7x)$
$\ - (5x^3 + 2x^2 - 3x + 5)$

23. $(7x\sqrt{y} - 4y\sqrt{x} + 7.5)$
$\ - (-2x\sqrt{y} - y\sqrt{x} - 1.6)$

24. $\left(10x\sqrt{y} - 4y\sqrt{x} + \dfrac{4}{3}\right) - \left(-3x\sqrt{y} + y\sqrt{x} - \dfrac{1}{3}\right)$

☆ ──────────────────────────────

[v] Simplify.

25. ▦ $(0.565p^2q - 2.167pq^2 + 16.02pq - 17.1)$
$\quad + (-1.612p^2q - 0.312pq^2 - 7.141pq - 87.044)$

26. ▦ $(5003.2xy^{-2} + 3102.4\sqrt{xy} - 5280)$
$\quad - (2143.6xy^{-2} + 6153.8xy - 4141\sqrt{xy} + 4979.12)$

OBJECTIVE

You should be able to:

[i] Multiply any two polynomials, striving for speed and accuracy. Whenever possible you should write only the answer. In particular, you should be able to:
  a) Square a binomial, writing only the answer.
  b) Multiply the sum and difference of the same two terms, writing only the answer.
  c) Multiply any two binomials, writing only the answer.
  d) Cube a binomial.

Multiply.

1. $3x^2y - 2xy + 3y$ and $xy + 2y$

2. $p^2q + 2pq + 2q$ and $2p^2q - pq + q$

Multiply.

3. $(2xy + 3x)(x^2 - 2)$

4. $(3x - 2y)(5x + 3y)$

5. $(2x + \sqrt{2})(3y - \sqrt{2})$

1.4 MULTIPLICATION OF POLYNOMIALS

[i] Multiplication of Any Two Polynomials

Multiplication of polynomials is based on the distributive laws. To multiply two polynomials, we multiply each term of one by every term of the other and then add the results.

Example 1 Multiply $4x^4y - 7x^2y + 3y$ by $2y - 3x^2y$.

$$
\begin{array}{r}
4x^4y - 7x^2y + 3y \\
2y - 3x^2y \\
\hline
8x^4y^2 - 14x^2y^2 + 6y^2 \\
-12x^6y^2 + 21x^4y^2 - 9x^2y^2 \\
\hline
-12x^6y^2 + 29x^4y^2 - 23x^2y^2 + 6y^2
\end{array}
$$

Multiplying by $2y$
Multiplying by $-3x^2y$
Adding

DO EXERCISES 1 AND 2.

The following methods allow us to multiply binomials more efficiently.

Products of Two Binomials

We can find a product of two binomials mentally. We multiply the first terms, then the outside terms, then the inside terms, then the last terms (this procedure is sometimes abbreviated FOIL), and then add the results. This also works for expressions that are not polynomials.

Examples Multiply.

$$\qquad\qquad\qquad\quad F\qquad O\qquad I\qquad L$$

2. $(3xy + 2x)(x^2 + 2xy^2) = 3x^3y + 6x^2y^3 + 2x^3 + 4x^2y^2$

3. $(x + \sqrt{2})(y - \sqrt{2}) = xy - \sqrt{2}x + \sqrt{2}y - 2$

4. $(2x - \sqrt{3})(y + 2) = 2xy + 4x - \sqrt{3}y - 2\sqrt{3}$

5. $(2x + 3y)(x - 4y) = 2x^2 - 5xy - 12y^2$

DO EXERCISES 3–5.

Squares of Binomials

Multiplying a binomial $a + b$ by itself, we obtain the following:

$$(a + b)^2 = a^2 + 2ab + b^2$$

and

$$(a - b)^2 = a^2 - 2ab + b^2.$$

Thus, to square a binomial we square the first term, add twice the product of the terms, and then add the square of the second term.

Examples Multiply.

6. $(2x + 9y^2)^2 = (2x)^2 + 2(2x)(9y^2) + (9y^2)^2$
$= 4x^2 + 36xy^2 + 81y^4$

7. $(3x^2 - 5xy^2)^2 = (3x^2)^2 - 2(3x^2)(5xy^2) + (5xy^2)^2$ The second term
$= 9x^4 - 30x^3y^2 + 25x^2y^4$ is $-5xy^2$ so
twice the
product of the
terms is
$-2(3x^2)(5xy^2)$.

CAUTION! The square of a sum is *not* the sum of squares; that is,

$$(a + b)^2 \neq a^2 + b^2.$$

DO EXERCISES 6 AND 7.

Products of Sums and Differences

The area of a square with side of length a is a^2.

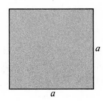

If we increase the length of one side by length b, and decrease the other side by the same length b, the area is found as follows:

$$(a + b)(a - b) = (a + b)a - (a + b)b$$
$$= a^2 + ab - ab - b^2$$
$$= a^2 - b^2.$$

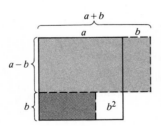

This leads us to a new result to be remembered as follows:

$$(a + b)(a - b) = a^2 - b^2.$$

The product of a sum and a difference of the same two expressions is the difference of their squares. Thus to find such a product, we square the first expression, square the second expression, and write a minus sign between the results.

Multiply.

6. $(4x - 5y)^2$

7. $(2y^2 + 6x^2y)^2$

Multiply.

8. $(4x + 7)(4x - 7)$

9. $(5x^2y + 2y)(5x^2y - 2y)$

10. $(4y^2 + \sqrt{3})(4y^2 - \sqrt{3})$

11. $(2x + 3 + 5y)(2x + 3 - 5y)$

12. $(-2x^3y^2 + 5t)(2x^3y^2 + 5t)$

Multiply.

13. $(x + 1)^3$

14. $(x - 1)^3$

15. $(t^2 - 3b)^3$

16. $(2a^3 - 5b^2)^3$

Examples Multiply.

8. $(y + 5)(y - 5) = y^2 - 25$

9. $(3x + 2)(3x - 2) = (3x)^2 - 2^2$
$$= 9x^2 - 4$$

10. $(2xy^2 + 3x)(2xy^2 - 3x) = (2xy^2)^2 - (3x)^2$
$$= 4x^2y^4 - 9x^2$$

11. $(5x + \sqrt{2})(5x - \sqrt{2}) = (5x)^2 - (\sqrt{2})^2$
$$= 25x^2 - 2$$

12. $(5y + 4 + 3x)(5y + 4 - 3x) = (5y + 4)^2 - (3x)^2$
$$= 25y^2 + 40y + 16 - 9x^2$$

13. $(3xy^2 + 4y)(-3xy^2 + 4y) = -(3xy^2)^2 + (4y)^2$
$$= 16y^2 - 9x^2y^4$$

DO EXERCISES 8–12.

Cubing a Binomial

The following multiplication gives another result to be remembered:

$$(a + b)^3 = (a + b)(a + b)^2$$
$$= (a + b)(a^2 + 2ab + b^2)$$
$$= (a + b)a^2 + (a + b)2ab + (a + b)b^2$$
$$= a^3 + a^2b + 2a^2b + 2ab^2 + ab^2 + b^3$$
$$= a^3 + 3a^2b + 3ab^2 + b^3.$$

The result to be remembered is as follows:

$$\boldsymbol{(a + b)^3 = a^3 + 3a^2b + 3ab^2 + b^3.}$$

Examples Multiply.

14. $(x + 2)^3 = x^3 + 3x^2(2) + 3x(2)^2 + 2^3$
$$= x^3 + 6x^2 + 12x + 8$$

15. $(x - 2)^3 = [x + (-2)]^3$
$$= x^3 + 3x^2(-2) + 3x(-2)^2 + (-2)^3$$
$$= x^3 - 6x^2 + 12x - 8$$

16. $(5m^2 - 4n^3)^3 = (5m^2)^3 + 3(5m^2)^2(-4n^3) + 3(5m^2)(-4n^3)^2 + (-4n^3)^3$
$$= 125m^6 - 300m^4n^3 + 240m^2n^6 - 64n^9$$

Note in Examples 15 and 16 that a separate formula for $(a - b)^3$ need not be memorized. We can think of $(a - b)^3$ as $[a + (-b)]^3$.

DO EXERCISES 13–16.

In the following exercises you should do mentally as much of the calculating as you can. If possible, write only the answer. Work for speed with accuracy.

EXERCISE SET 1.4

[i] Multiply.

1. $2x^2 + 4x + 16$ and $3x - 4$

2. $3y^2 - 3y + 9$ and $2y + 3$

3. $4a^2b - 2ab + 3b^2$ and $ab - 2b + 1$

4. $2x^2 + y^2 - 2xy$ and $x^2 - 2y^2 - xy$

5. $(a - b)(a^2 + ab + b^2)$

6. $(t + 1)(t^2 - t + 1)$

7. $(2x + 3y)(2x + y)$

8. $(2a - 3b)(2a - b)$

9. $\left(4x^2 - \frac{1}{2}y\right)\left(3x + \frac{1}{4}y\right)$

10. $\left(2y^3 + \frac{1}{5}x\right)\left(3y - \frac{1}{4}x\right)$

11. $(\sqrt{2}x^2 - y^2)(\sqrt{2}x - 2y)$

12. $(\sqrt{3}y^2 - 2)(\sqrt{3}y - x)$

13. $(2x + 3y)^2$

14. $(5x + 2y)^2$

15. $(2x^2 - 3y)^2$

16. $(4x^2 - 5y)^2$

17. $(2x^3 + 3y^2)^2$

18. $(5x^3 + 2y^2)^2$

19. $\left(\frac{1}{2}x^2 - \frac{3}{5}y\right)^2$

20. $\left(\frac{1}{4}x^2 - \frac{2}{3}y\right)^2$

21. $(0.5x + 0.7y^2)^2$

22. $(0.3x + 0.8y^2)^2$

23. $(3x - 2y)(3x + 2y)$

24. $(3x + 5y)(3x - 5y)$

25. $(x^2 + yz)(x^2 - yz)$

26. $(2x^2 + 5xy)(2x^2 - 5xy)$

27. $(3x^2 - \sqrt{2})(3x^2 + \sqrt{2})$

28. $(5x^2 - \sqrt{3})(5x^2 + \sqrt{3})$

29. $(2x + 3y + 4)(2x + 3y - 4)$

30. $(5x + 2y + 3)(5x + 2y - 3)$

31. $(x^2 + 3y + y^2)(x^2 + 3y - y^2)$

32. $(2x^2 + y + y^2)(2x^2 + y - y^2)$

33. $(x + 1)(x - 1)(x^2 + 1)$

34. $(y - 2)(y + 2)(y^2 + 4)$

35. $(2x + y)(2x - y)(4x^2 + y^2)$

36. $(5x + y)(5x - y)(25x^2 + y^2)$

37. ▦ $(0.051x + 0.04y)^2$

38. ▦ $(1.032x - 2.512y)^2$

39. ▦ $(37.86x + 1.42)(65.03x - 27.4)$

40. ▦ $(3.601x - 17.5)(47.105x + 31.23)$

41. $(y + 5)^3$

42. $(t - 7)^3$

43. $(m^2 - 2n)^3$

44. $(3t^2 + 4)^3$

☆ ─────────────────────────

Find the error(s) in each of the following. Explain why each is an error. Then find the correct answer.

45. $(3a + b)^2 = 3a^2 + b^2$

46. $(2x - 3y)(2x - 3y) = 4x^2 - 9y^2$

47. $2x(x + 3) + 4(x^2 - 3)$
$= 2x^2 + 3x + 4x^2 - 3$ (1)
$= 6x^2 + x$ (2)

48. $(2a - 3b)(3a + 2b)$
$= 6a^2 - 6b^2$ (1)
$= a^2 - b^2$ (2)

1.5 FACTORING

[i] To factor a polynomial we do the reverse of multiplying; that is, we find an expression that is a product. Facility in factoring is an important algebraic skill.

Terms with Common Factors

When an expression is to be factored, we should always look first for a possible factor that is common to all terms. We then "factor it out" using the distributive laws.

OBJECTIVE

You should be able to:

[i] Determine the kind of factoring to try, when an expression is to be factored. Then factor expressions
a) by removing a common factor.
b) that are differences of squares.
c) that are trinomial squares.
d) that are trinomials not squares.
e) that are sums or differences of cubes.

Factor.

1. $20x^3y + 12x^2y$

2. $(p + q)(x + 2) + (p + q)(x + y)$

3. $4x^2 + 20x - 3x - 15$

Factor.

4. $x^2 - 16$

5. $25y^4 - 16x^2$

6. $2y^4 - 32x^4$

7. $x^2 - 3$

Examples Factor.

1. $4x^2 + 8 = 4(x^2 + 2)$
2. $12x^2y - 20x^3y = 4x^2y(3 - 5x)$
3. $7x\sqrt{y} + 14x^2\sqrt{y} - 21\sqrt{y} = 7\sqrt{y}(x + 2x^2 - 3)$
4. $(a - b)(x + 5) + (a - b)(x - y^2) = (a - b)[(x + 5) + (x - y^2)]$
$$= (a - b)(2x + 5 - y^2)$$

In some polynomials pairs of terms have a common factor that can be removed, as in the following examples. This process is called *factoring by grouping*, and uses the distributive laws repeatedly.

Examples Factor.

5. $y^2 + 3y + 4y + 12 = y(y + 3) + 4(y + 3)$
$$= (y + 4)(y + 3)$$
6. $ax^2 + ay + bx^2 + by = a(x^2 + y) + b(x^2 + y)$
$$= (a + b)(x^2 + y)$$

DO EXERCISES 1–3.

Differences of Squares

Recall that $(a + b)(a - b) = a^2 - b^2$. We can use this equation in reverse to factor an expression that is a difference of two squares.

Examples Factor.

7. $x^2 - 9 = (x + 3)(x - 3)$
8. $y^2 - 2 = (y + \sqrt{2})(y - \sqrt{2})$
9. $9a^2 - 16x^4 = (3a)^2 - (4x^2)^2 = (3a + 4x^2)(3a - 4x^2)$
10. $9y^4 - 9x^4 = 9(y^4 - x^4)$ Remove the common factor first.
$$= 9(y^2 + x^2)(y^2 - x^2)$$
$$= 9(y^2 + x^2)(y + x)(y - x)$$

DO EXERCISES 4–7.

Trinomial Squares

You should recall that

$$a^2 + 2ab + b^2 = (a + b)^2 \quad \text{and} \quad a^2 - 2ab + b^2 = (a - b)^2.$$

We can use these equations to factor trinomials that are squares. To factor a trinomial, you should check to see if it is a square. For this to be the case, two of the terms must be squares and the other term must be twice the product of the square roots, or the additive inverse of that product.

Examples Factor.

11. $x^2 - 10x + 25 = (x - 5)^2$

12. $16y^2 + 56y + 49 = (4y + 7)^2$

13.
$$\begin{aligned}
-4y^2 - 144y^8 + 48y^5 &= -4y^2(1 + 36y^6 - 12y^3) \quad \text{We first} \\
&= -4y^2(1 - 12y^3 + 36y^6) \quad \text{removed the} \\
&= -4y^2(1 - 6y^3)^2 \quad \text{common factor.}
\end{aligned}$$

DO EXERCISES 8–10.

Trinomials That Are Not Squares

Certain trinomials that are not squares can be factored into two binomials. To do this, we factor by trial and error using the equation $acx^2 + (ad + bc)x + bd = (ax + b)(cx + d)$.

Example 14 Factor $x^2 + 7x + 12$.

We look for factors of 12 whose sum is 7. By trial we determine the factorization to be as follows: $(x + 4)(x + 3)$.

Example 15 Factor $2x^2 + 11x + 12$.

We look for binomials $(ax + b)$ and $(cx + d)$. The product of the first terms must be $2x^2$. The product of the last terms must be 12. When we multiply the inside terms, then the outside terms, and add, we must get $11x$. By trial, we determine the factorization to be as follows: $(x + 4)(2x + 3)$.

DO EXERCISES 11–13.

Sums or Differences of Cubes

We can use the following equations to factor a sum or a difference of two cubes:

$$a^3 + b^3 = (a + b)(a^2 - ab + b^2),$$
$$a^3 - b^3 = (a - b)(a^2 + ab + b^2).$$

Check them by multiplying the right-hand sides.

Example 16 Factor $x^3 - 27$.

$$x^3 - 27 = x^3 - 3^3$$

In one set of parentheses we write the cube root of the first expression, x. Then we write the cube root of the second expression, -3. This gives us $x - 3$.

$$(x - 3)(\qquad\qquad)$$

Factor.

8. $9y^2 - 30y + 25$

9. $16x^2 + 72xy + 81y^2$

10. $-12x^4y^2 + 60x^2y^5 - 75y^8$
[*Hint:* First remove a common factor.]

Factor.

11. $x^2 + 5x - 14$

12. $3x^2 + 5x + 2$

13. $6x^4y^6 - 9x^2y^3 - 60$

Factor.

14. $x^3 - 8$

15. $64 - t^3$

Factor.

16. $27x^3 + y^3$

17. $8m^3 + 125t^3$

Factor.

18. $128y^7 - 250x^6y$

To get the next factor we think of $x - 3$ and do the following.

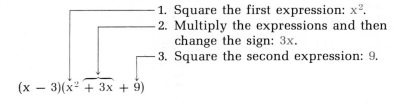

1. Square the first expression: x^2.
2. Multiply the expressions and then change the sign: $3x$.
3. Square the second expression: 9.

$$(x - 3)(x^2 + 3x + 9)$$

Note: We cannot factor $x^2 + 3x + 9$.   (It is not a trinomial square.)

DO EXERCISES 14 AND 15.

Example 17 Factor $125x^3 + y^3$.

$$125x^3 + y^3 = (5x)^3 + y^3$$

In one set of parentheses we write the cube root of the first expression, then a plus sign, and then the cube root of the second expression.

$$(5x + y)(\quad)$$

To get the next factor, we think of $5x + y$ and do the following.

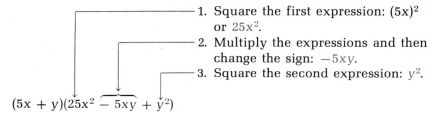

1. Square the first expression: $(5x)^2$ or $25x^2$.
2. Multiply the expressions and then change the sign: $-5xy$.
3. Square the second expression: y^2.

$$(5x + y)(25x^2 - 5xy + y^2)$$

DO EXERCISES 16 AND 17.

Example 18 Factor $16x^7y + 54xy^7$.

We first look for a common factor.

$$2xy(8x^6 + 27y^6) = 2xy[(2x^2)^3 + (3y^2)^3]$$
$$= 2xy(2x^2 + 3y^2)(4x^4 - 6x^2y^2 + 9y^4)$$

DO EXERCISE 18.

Remember the following about factoring sums or differences of squares and cubes.

Sum of cubes:	$a^3 + b^3 = (a + b)(a^2 - ab + b^2)$
Difference of cubes:	$a^3 - b^3 = (a - b)(a^2 + ab + b^2)$
Difference of squares:	$a^2 - b^2 = (a + b)(a - b)$
Sum of squares:	$a^2 + b^2$ **cannot be factored!**

EXERCISE SET 1.5

[i] Factor.

1. $18a^2b - 15ab^2$

2. $4x^2y + 12xy^2$

3. $a(b - 2) + c(b - 2)$

4. $a(x^2 - 3) - 2(x^2 - 3)$

5. $x^2 + 3x + 6x + 18$

6. $3x^3 + x^2 - 18x - 6$

7. $9x^2 - 25$

8. $16x^2 - 9$

9. $4xy^4 - 4xz^2$

10. $5xy^4 - 5xz^4$

11. $y^2 - 6y + 9$

12. $x^2 + 8x + 16$

13. $1 - 8x + 16x^2$

14. $1 + 10x + 25x^2$

15. $4x^2 - 5$

16. $16x^2 - 7$

17. $x^2y^2 - 14xy + 49$

18. $x^2y^2 - 16xy + 64$

19. $4ax^2 + 20ax - 56a$

20. $21x^2y + 2xy - 8y$

21. $a^2 + 2ab + b^2 - c^2$

22. $x^2 - 2xy + y^2 - z^2$

23. $x^2 + 2xy + y^2 - a^2 - 2ab - b^2$

24. $r^2 + 2rs + s^2 - t^2 + 2tv - v^2$

25. $5y^4 - 80x^4$

26. $6y^4 - 96x^4$

27. $x^3 + 8$

28. $y^3 - 64$

29. $3x^3 - \dfrac{3}{8}$

30. $5y^3 + \dfrac{5}{27}$

31. $x^3 + 0.001$

32. $y^3 - 0.125$

33. $3z^3 - 24$

34. $4t^3 + 108$

35. $a^6 - t^6$

36. $64m^6 + y^6$

37. $16a^7b + 54ab^7$

38. $24a^2x^4 - 375a^8x$

39. ▦ $x^2 - 17.6$

40. ▦ $x^2 - 8.03$

41. ▦ $37x^2 - 14.5y^2$
(*Hint:* First remove the common factor 37.)

42. ▦ $1.96x^2 - 17.4y^2$
(*Hint:* First remove the common factor 1.96.)

☆

Factor.

43. $(x + h)^3 - x^3$

44. $(x + 0.01)^2 - x^2$

1.6 SOLVING EQUATIONS AND INEQUALITIES

Solutions and Solution Sets

OBJECTIVES

You should be able to:

[i] Solve simple equations using the addition and multiplication principles and the principle of zero products.

[ii] Solve simple inequalities.

DEFINITION

A *solution* of an equation is any number that makes the equation true when that number is substituted for the variable.

The number 3 is a solution of $5x = 15$ because $5(3) = 15$ is true. The number -4 is *not* a solution of $5x = 15$ because $5(-4) = 15$ is false.

Solve (find the solution set) by trial and error.

1. $x - 2 = 7$

2. $y^2 + y = 0$

DEFINITION

> The set of all solutions of an equation is called its *solution set.* When we find all the solutions of an equation (find its solution set), we say that we have *solved* it.

The number 3 is the only solution of $5x = 15$, so the solution set consists of the number 3 and is denoted $\{3\}$. As another example consider

$$y^2 - y = 0.$$

The number 0 is a solution, and so is the number 1. There are no other solutions, so the solution set is $\{0, 1\}$.

DO EXERCISES 1 AND 2.

[i] Equation-solving Principles

The Addition and Multiplication Principles

Two simple principles* allow us to solve many equations. The first of these is as follows:

The Addition Principle

> For any real numbers a, b, and c, if an equation $a = b$ is true, then $a + c = b + c$ is true.

The second principle is similar to the first.

The Multiplication Principle

> For any real numbers a, b, and c, if an equation $a = b$ is true, then $ac = bc$ is true.

These principles may be used together as needed to solve an equation.

Example 1 Solve $3x + 4 = 15$.

$3x + 4 + (-4) = 15 + (-4)$	Here we used the addition principle, adding -4.
$3x = 11$	Here we simplified.
$\frac{1}{3} \cdot 3x = \frac{1}{3} \cdot 11$	Here we used the multiplication principle, multiplying by $\frac{1}{3}$.
$x = \frac{11}{3}$	Here we simplified.

---

*These "principles" are actually very easy theorems. Suppose $a = b$ is true. Then a and b are the same number. If we add c to this number, the result is $a + c$. It is also $b + c$.

Check:

$$3x + 4 = 15$$

$$\begin{array}{c|c} 3 \cdot \dfrac{11}{3} + 4 & 15 \\[2mm] 11 + 4 & \\[1mm] 15 & \end{array}$$ The only solution is the number $\dfrac{11}{3}$.

The solution set can be indicated as $\{\frac{11}{3}\}$, but for brevity we often omit the braces.

In Example 1 we used the addition and multiplication principles. Thus we know that if $3x + 4 = 15$ is true, then $x = \frac{11}{3}$ is true. This does not guarantee that if $x = \frac{11}{3}$ is true, then $3x + 4 = 15$ is true. Thus it is important to check by substituting $\frac{11}{3}$ in the original equation.

Example 2 Solve $3(7 - 2x) = 14 - 8(x - 1)$.

$$21 - 6x = 14 - 8x + 8 \quad \text{Here we multiplied to remove parentheses.}$$
$$21 - 6x = 22 - 8x \quad \text{Here we simplified.}$$
$$8x - 6x = 22 - 21 \quad \text{Here we added } -21 \text{ and also } 8x.$$
$$2x = 1 \quad \text{Here we combined like terms and simplified.}$$
$$x = \frac{1}{2} \quad \text{Here we multiplied by } \frac{1}{2}.$$

Check:

$$\begin{array}{c|c} 3(7 - 2x) & = 14 - 8(x - 1) \\[2mm] 3\left(7 - 2 \cdot \dfrac{1}{2}\right) & 14 - 8\left(\dfrac{1}{2} - 1\right) \\[3mm] 3(7 - 1) & 14 - 8\left(-\dfrac{1}{2}\right) \\[3mm] 3 \cdot 6 & 14 + 4 \\[1mm] 18 & 18 \end{array}$$

The number $\frac{1}{2}$ checks, so it is a solution.

Example 3 Solve $x + 3 = x$.

$$-x + x + 3 = -x + x \quad \text{Here we added } -x.$$
$$3 = 0 \quad \text{Here we simplified.}$$

In Example 3 we get a false equation. No replacement for x will make the equation true. Thus there are no solutions. The solution set is the *empty set*, denoted $\emptyset$.

DO EXERCISES 3–7.

The Principle of Zero Products

A third principle for solving equations is called the *principle of zero products*. It is as follows.

Solve and check.

3. $9x - 4 = 8$

4. $-4x + 2 + 5x = 3x - 15$

5. $3(y - 1) - 1 = 2 - 5(y + 5)$

6. $x - 7 = x$

7. $2x + 6 = 2x - 4$

Solve.

8. $(x - 7)(2x + 3) = 0$

9. $x^2 - x = 20$

The Principle of Zero Products

> **For any numbers a and b, if $ab = 0$, then $a = 0$ *or* $b = 0$; and if $a = 0$ or $b = 0$, then $ab = 0$.**

To solve an equation using this principle, there must be 0 on one side of the equation and a product on the other. The solutions are then obtained by setting the factors equal to 0 separately.

Example 4 Solve $x^2 + x - 12 = 0$.

$(x + 4)(x - 3) = 0$ Here we factored.

$x + 4 = 0$ *or* $x - 3 = 0$ Here we used the principle of zero products.

$x = -4$ *or* $x = 3$

The solutions are -4 and 3. The solution set is $\{-4, 3\}$.

DO EXERCISES 8 AND 9.

[ii] Solving Inequalities

Principles for solving inequalities are similar to those for solving equations. We can add the same number on both sides of an inequality. We can also multiply on both sides by the same nonzero number, but if that number is negative, we must *reverse* the inequality sign.

Example 5 Solve $3x < 11 - 2x$.

$3x + 2x < 11$ Here we added $2x$.

$5x < 11$ Here we combined similar terms.

$x < \dfrac{11}{5}$ Here we multiplied by $\dfrac{1}{5}$.

Any number less than $\frac{11}{5}$ is a solution. The solution set is the set of all x such that $x < \frac{11}{5}$. We abbreviate this using *set-builder* notation as follows:

$$\left\{ x \mid x < \frac{11}{5} \right\}.$$

For brevity we often write merely $x < \frac{11}{5}$.

Example 6 Solve $16 - 7y \geq 10y - 4$.

$-16 + 16 - 7y \geq -16 + 10y - 4$ We added -16.

$-7y \geq 10y - 20$ We have simplified.

$-17y \geq -20$ We added $-10y$ and simplified.

$y \leq \dfrac{20}{17}$ We multiplied by $-\dfrac{1}{17}$ and reversed the inequality sign.

Any number less than or equal to $\frac{20}{17}$ is a solution. The solution set is $\{y \mid y \leq \frac{20}{17}\}$.

DO EXERCISES 10–15.

Solve.

10. $5x > 12 - 3x$

11. $17 - 5y \leq 8y - 5$

12. $12x - 6 < 10x + 4$

Write set-builder notation for each set.

13. The set of all x such that $x > \frac{5}{2}$.

14. The set of all y such that $y \geq -7$.

15. The set of all x such that $x^2 = 5$.

EXERCISE SET 1.6

[i] Solve.

1. $4x + 12 = 60$

2. $2y - 11 = 37$

3. $4 + \frac{1}{2}x = 1$

4. $4.1 - 0.2y = 1.3$

5. $y + 1 = 2y - 7$

6. $5 - 4x = x - 13$

7. $5x - 2 + 3x = 2x + 6 - 4x$

8. $5x - 17 - 2x = 6x - 1 - x$

9. $1.9x - 7.8 + 5.3x = 3.0 + 1.8x$

10. $2.2y - 5 + 4.5y = 1.7y - 20$

11. $7(3x + 6) = 11 - (x + 2)$

12. $4(5y + 3) = 3(2y - 5)$

13. $2x - (5 + 7x) = 4 - [x - (2x + 3)]$

14. $y - (9y - 8) = [5 - 2y - 3(2y - 3)] + 29$

15. $(2x - 3)(3x - 2) = 0$

16. $(5x - 2)(2x + 3) = 0$

17. $x(x - 1)(x + 2) = 0$

18. $x(x + 2)(x - 3) = 0$

19. $3x^2 + x - 2 = 0$

20. $10x^2 - 16x + 6 = 0$

21. $(x - 1)(x + 1) = 5(x - 1)$

22. $6(y - 3) = (y - 3)(y - 2)$

23. $x[4(x - 2) - 5(x - 1)] = 2$

24. $14[(x - 4) - \frac{1}{14}(x + 2)] = (x + 2)(x - 4)$

25. $(3x^2 - 7x - 20)(2x - 5) = 0$

26. $(8x + 11)(12x^2 - 5x - 2) = 0$

27. $16x^3 = x$

28. $9x^3 = x$

[ii] Solve.

29. $x + 6 < 5x - 6$

30. $3 - x < 4x + 7$

31. $3x - 3 + 2x \geq 1 - 7x - 9$

32. $5y - 5 + y \leq 2 - 6y - 8$

33. $17 - 5y \leq 8y - 5$

34. $12x - 6 < 10x + 4$

35. $-\frac{3}{4}x \geq -\frac{5}{8} + \frac{2}{3}x$

36. $-\frac{5}{6}x \leq \frac{3}{4} + \frac{8}{3}x$

37. $4x(x - 2) < 2(2x - 1)(x - 3)$

38. $(x + 1)(x + 2) > x(x + 1)$

☆

Write set-builder notation for each set.

39. The set of all x such that $x > 2.5$

40. The set of all y such that $y \leq -7$

41. The set of all t such that $t^2 = 5$

42. The set of all m such that $m^3 + 3 = m^2 - 2$

Solve.

43. ▦ $2.905x - 3.214 + 6.789x = 3.012 + 1.805x$

44. ▦ $(13.14x + 17.152)(15.15 - 7.616x) = 0$

45. ▦ $1.52(6.51x + 7.3) < 11.2 - (7.2x + 13.52)$ **46.** ▦ $4.73(5.16y + 3.62) \geq 3.005(2.75y - 6.31)$

★ ──

Solve.

47. $(x + 1)^3 = (x - 1)^3 + 26$ **48.** $(x - 2)^3 = x^3 - 2$

──

OBJECTIVES

You should be able to:

[i] Determine sensible replacements in fractional expressions.
[ii] Simplify fractional expressions.
[iii] Multiply or divide fractional expressions, and simplify.
[iv] Add or subtract fractional expressions, and simplify.
[v] Simplify complex fractional expressions.

Determine the sensible replacements.

1. $\dfrac{x^2 - 9}{x - 3}$

2. $\dfrac{x^3 - xy^2}{x^2 + 7x + 12}$

1.7 FRACTIONAL EXPRESSIONS

[i] Sensible Replacements

Expressions like the following are called *fractional expressions*:

$$\frac{8}{5}, \quad \frac{x^2 - 9}{x - 3},$$

$$\frac{3x^2 + 5\sqrt{x} - 2}{x^2 - y^2}, \qquad \frac{x - 3}{x^2 - x - 2}.$$

Fractional expressions represent division. Certain substitutions are not sensible in such expressions. Since division by 0 is not defined, any number that makes a denominator zero is not a sensible replacement. For example, consider

$$\frac{x - 3}{x^2 - x - 2}.$$

To determine the sensible replacements, we set the denominator equal to 0 and solve:

$$x^2 - x - 2 = 0$$
$$(x + 1)(x - 2) = 0$$
$$x + 1 = 0 \quad \text{or} \quad x - 2 = 0$$
$$x = -1 \quad \text{or} \quad x = 2.$$

Thus -1 and 2 are not sensible replacements. All real numbers except -1 and 2 are sensible replacements.

DO EXERCISES 1 AND 2.

Multiplication and Division

To multiply two fractional expressions, we multiply their numerators and also their denominators. By Theorem 4, when we divide, we multiply by the reciprocal of the divisor. The latter can be obtained by inverting the divisor.

Examples

1. Multiply.

$$\frac{x + 3}{y - 4} \cdot \frac{x^3}{y + 5} = \frac{(x + 3)x^3}{(y - 4)(y + 5)}$$
$$= \frac{x^4 + 3x^3}{y^2 + y - 20}$$

2. Divide.

$$\frac{x-2}{x+1} \div \frac{x+5}{x-3} = \frac{x-2}{x+1} \cdot \frac{x-3}{x+5} \quad \text{Inverting}$$

$$= \frac{(x-2)(x-3)}{(x+1)(x+5)} \quad \text{Multiplying}$$

$$= \frac{x^2 - 5x + 6}{x^2 + 6x + 5}$$

DO EXERCISES 3 AND 4.

[ii] Simplifying

The basis of simplifying fractional expressions rests on the fact that certain expressions have a value of 1 for all sensible replacements. Such expressions have the same numerator and denominator.* Here are some examples:

$$\frac{x-2}{x-2},$$

$$\frac{3x^2 - 4x + 2}{3x^2 - 4x + 2},$$

$$\frac{4x-5}{4x-5}.$$

When we multiply by such an expression we obtain an equivalent expression. This means that the new expression will name the same number as the first for all sensible replacements. The set of sensible replacements may not be the same for the two expressions.

Example 3 Multiply.

$$\frac{y+4}{y-3} \cdot \frac{y-2}{y-2} = \frac{(y+4)(y-2)}{(y-3)(y-2)}$$

$$= \frac{y^2 + 2y - 8}{y^2 - 5y + 6}$$

The expressions $(y+4)/(y-3)$ and $(y^2 + 2y - 8)/(y^2 - 5y + 6)$ are equivalent. That is, they will name the same number for all sensible replacements. The only nonsensible replacement in the first expression is 3. For the second expression the nonsensible replacements are 2 and 3.

DO EXERCISE 5.

Simplification can be accomplished by reversing the procedure in the above example; that is, we try to factor the fractional expression in such a way that one of the factors is equal to 1 and then "remove" that factor.

---

*By Theorem 4, $a \div a = a/a = a(1/a)$, and since a and $1/a$ are reciprocals, their product is 1.

3. Multiply.

$$\frac{x+y}{2x^2 - 1} \cdot \frac{x+y}{7x}$$

4. Divide.

$$\frac{x-2}{x+2} \div \frac{x+2}{x+4}$$

5. Multiply $\frac{x+2}{x-5}$ by $\frac{x+3}{x+3}$ to obtain an equivalent expression. Name the sensible replacements for the two expressions.

Simplify. Name the sensible replacements in the original and the simplified expressions.

6. $\dfrac{6x^2 + 4x}{2x^2 + 4x}$

7. $\dfrac{y^2 + 3y + 2}{y^2 - 1}$

Examples Simplify.

4. $\dfrac{15x^3y^2}{20x^2y} = \dfrac{(5x^2y)3xy}{(5x^2y)4}$ Factoring numerator and denominator

$= \dfrac{5x^2y}{5x^2y} \cdot \dfrac{3xy}{4}$ Factoring the expression

$= \dfrac{3xy}{4}$ "Removing" a factor of 1

Note that in the original expression neither x nor y can be 0. In the simplified expression, however, all replacements are sensible.

5. $\dfrac{x^2 - 1}{2x^2 - x - 1} = \dfrac{(x - 1)(x + 1)}{(2x + 1)(x - 1)}$

$= \dfrac{x - 1}{x - 1} \cdot \dfrac{x + 1}{2x + 1}$

$= \dfrac{x + 1}{2x + 1}$

In the original expression the sensible replacements are all real numbers except 1 and $-\frac{1}{2}$. In the simplified expression all real numbers except $-\frac{1}{2}$ are sensible replacements.

DO EXERCISES 6 AND 7.

Canceling

In Examples 4 and 5 a shortcut called *canceling* saves a step. Canceling, however, gives rise to a great many errors. It should be done cautiously, if at all.

Example 6 Simplify.

$$\dfrac{x^3 - 27}{x^2 + x - 12} = \dfrac{(x - 3)(x^2 + 3x + 9)}{(x + 4)(x - 3)}$$

$$= \dfrac{x^2 + 3x + 9}{x + 4}$$

Note that the canceling is a shortcut for "removing" a factor of 1. When fractional expressions are multiplied or divided, they should be simplified when possible.

CAUTION! The difficulty with canceling is that it is often applied incorrectly in situations such as the following:

$$\dfrac{x + 3}{x} = 3, \quad \dfrac{x + 1}{x + 2} = \dfrac{1}{2}, \quad \dfrac{x^3}{3} = x, \quad \dfrac{15}{54} = \dfrac{1}{4}.$$

Wrong!   **Wrong!**   **Wrong!**   **Wrong!**

In each of these situations the expressions canceled were *not* factors. If you can't factor, you can't cancel! **If you are in doubt, don't cancel.**

[iii] Multiplying, Dividing, and Simplifying

Examples

7. Multiply and simplify.

$$\frac{x + 2}{x - 2} \cdot \frac{x^2 - 4}{x^2 + x - 2} = \frac{(x + 2)(x^2 - 4)}{(x - 2)(x^2 + x - 2)} \quad \text{Multiplying}$$

$$= \frac{(x + 2)(x + 2)(x - 2)}{(x - 2)(x + 2)(x - 1)} \quad \text{Factoring}$$

$$= \frac{(x + 2)(x - 2)}{(x + 2)(x - 2)} \cdot \frac{x + 2}{x - 1} = \frac{x + 2}{x - 1} \quad \text{``Removing'' a factor of 1}$$

8. Divide and simplify.

$$\frac{a^2 - 1}{a + 1} \div \frac{a^2 - 2a + 1}{a + 1} = \frac{a^2 - 1}{a + 1} \cdot \frac{a + 1}{a^2 - 2a + 1}$$

$$= \frac{(a + 1)(a - 1)(a + 1)}{(a + 1)(a - 1)(a - 1)} = \frac{a + 1}{a - 1}$$

DO EXERCISES 8 AND 9.

[iv] Addition and Subtraction

When fractional expressions have the same denominator, we can add or subtract them by adding or subtracting the numerators and retaining the common denominator. If denominators are not the same, we then find equivalent expressions with the same denominator and add. If one denominator is the additive inverse of another, we can find a common denominator by multiplying by $-1/-1$.

Examples Add.

9. $\dfrac{3x^2 + 4x - 8}{x^2 + y^2} + \dfrac{-5x^2 + 5x + 7}{x^2 + y^2} = \dfrac{-2x^2 + 9x - 1}{x^2 + y^2}$

In the following example, one denominator is the additive inverse of the other.

10. $\dfrac{3x^2 + 4}{x - y} + \dfrac{5x^2 - 11}{y - x}$

$$= \frac{3x^2 + 4}{x - y} + \frac{-1}{-1} \cdot \frac{5x^2 - 11}{y - x} \quad \begin{array}{l}\text{We multiply by } -1/-1 \text{ to convert} \\ \text{the second denominator to its addi-} \\ \text{tive inverse.}\end{array}$$

$$= \frac{3x^2 + 4}{x - y} + \frac{-1(5x^2 - 11)}{-1(y - x)}$$

$$= \frac{3x^2 + 4}{x - y} + \frac{11 - 5x^2}{x - y} \quad -1(y - x) = -y + x = x - y$$

$$= \frac{-2x^2 + 15}{x - y}$$

DO EXERCISES 10 AND 11.

8. Multiply and simplify.

$$\frac{x^2 - 2xy + y^2}{x + y} \cdot \frac{3x + 3y}{x^2 - y^2}$$

9. Divide and simplify.

$$\frac{a^2 - b^2}{ab} \div \frac{a^2 - 2ab + b^2}{2a^2b^2}$$

10. Add.

$$\frac{2x^2 + 5x - 9}{x - 5} + \frac{x^2 - x + 11}{x - 5}$$

11. Add.

$$\frac{3x^2 + 4}{x - 5} + \frac{x^2 - 7}{5 - x}$$

12. Add.

$$\frac{x^2 - 4xy + 4y^2}{2x^2 - 3xy + y^2} + \frac{x + 4y}{2x - 2y}$$

When denominators are different, but not additive inverses of each other, we find a common denominator by factoring the denominators. Then we multiply by 1 appropriately to get the common denominator in each expression.

Example 11 Add $\dfrac{1}{2x} + \dfrac{5x}{x^2 - 1} + \dfrac{3}{x + 1}$.

We first find the *Least Common Multiple* (LCM) of the denominators. The denominators, when factored, are

$$2x, \quad (x + 1)(x - 1), \quad x + 1.$$

The LCM is $2x(x + 1)(x - 1)$. Now we multiply each fractional expression by 1 appropriately.

$$\frac{1}{2x} \cdot \frac{(x + 1)(x - 1)}{(x + 1)(x - 1)} + \frac{5x}{(x + 1)(x - 1)} \cdot \frac{2x}{2x} + \frac{3}{(x + 1)} \cdot \frac{2x(x - 1)}{2x(x - 1)}$$

$$= \frac{(x + 1)(x - 1) + 10x^2 + 6x(x - 1)}{2x(x + 1)(x - 1)}$$

$$= \frac{17x^2 - 6x - 1}{2x(x + 1)(x - 1)} \quad \text{or} \quad \frac{17x^2 - 6x - 1}{2x^3 - 2x}$$

DO EXERCISE 12.

13. Subtract.

$$\frac{x}{x^2 + 11x + 30} - \frac{5}{x^2 + 9x + 20}$$

Example 12 Subtract $\dfrac{x}{x^2 + 5x + 6} - \dfrac{2}{x^2 + 3x + 2}$.

$$\frac{x}{x^2 + 5x + 6} - \frac{2}{x^2 + 3x + 2} = \frac{x}{(x + 2)(x + 3)} - \frac{2}{(x + 1)(x + 2)}$$

The LCM is $(x + 1)(x + 2)(x + 3)$.

$$= \frac{x}{(x + 2)(x + 3)} \cdot \frac{x + 1}{x + 1} - \frac{2}{(x + 1)(x + 2)} \cdot \frac{x + 3}{x + 3}$$

$$= \frac{x(x + 1) - [2(x + 3)]}{(x + 1)(x + 2)(x + 3)} \quad \text{We use the colored brackets here to make sure we subtract the } \textit{entire} \text{ numerator, not just part of it.}$$

$$= \frac{x^2 + x - [2x + 6]}{(x + 1)(x + 2)(x + 3)}$$

$$= \frac{x^2 + x - 2x - 6}{(x + 1)(x + 2)(x + 3)}$$

$$= \frac{x^2 - x - 6}{(x + 1)(x + 2)(x + 3)}$$

$$= \frac{(x - 3)(x + 2)}{(x + 1)(x + 2)(x + 3)}$$

$$= \frac{x - 3}{(x + 1)(x + 3)} \quad \text{Always simplify at the end if possible.}$$

DO EXERCISE 13.

> **CAUTION!** When subtracting one fractional expression from another, subtract numerators:
>
> $$\frac{A}{C} - \frac{B}{C} = \frac{A - (B)}{C}.$$
>
> Always be sure to subtract the *entire* numerator B and not just part of it. The use of parentheses or brackets helps. See Example 12.

[v] Complex Fractional Expressions

A complex fractional expression has a fractional expression in its numerator or denominator or both. To simplify such an expression, we can combine as necessary in numerator and denominator in order to obtain a single fractional expression for each. Then we divide the numerator by the denominator.

Example 13 Simplify.

$$\frac{x + \dfrac{1}{5}}{x - \dfrac{1}{3}} = \frac{x \cdot \dfrac{5}{5} + \dfrac{1}{5}}{x \cdot \dfrac{3}{3} - \dfrac{1}{3}}$$

$$= \frac{\dfrac{5x + 1}{5}}{\dfrac{3x - 1}{3}} \quad \text{Now we have a single fractional expression for both numerator and denominator.}$$

$$= \frac{5x + 1}{5} \cdot \frac{3}{3x - 1} \quad \text{Here we divided by multiplying by the reciprocal of the denominator.}$$

$$= \frac{15x + 3}{15x - 5}$$

Example 14 Simplify.

$$\frac{a^{-2} - b^{-2}}{a^{-1} + b^{-1}} = \frac{\dfrac{1}{a^2} - \dfrac{1}{b^2}}{\dfrac{1}{a} + \dfrac{1}{b}}$$

$$= \frac{\dfrac{b^2}{b^2} \cdot \dfrac{1}{a^2} - \dfrac{a^2}{a^2} \cdot \dfrac{1}{b^2}}{\dfrac{b}{b} \cdot \dfrac{1}{a} + \dfrac{a}{a} \cdot \dfrac{1}{b}} = \frac{\dfrac{b^2 - a^2}{a^2 b^2}}{\dfrac{b + a}{ab}}$$

$$= \frac{b^2 - a^2}{a^2 b^2} \cdot \frac{ab}{b + a}$$

$$= \frac{(b - a)(b + a)ab}{(b + a)a^2 b^2}$$

$$= \frac{ab(b + a)}{ab(b + a)} \cdot \frac{b - a}{ab} = \frac{b - a}{ab}$$

DO EXERCISES 14 AND 15.

14. Simplify.

$$\frac{1 + \dfrac{x}{a}}{a - \dfrac{x^2}{a}}$$

15. Simplify.

$$\frac{\dfrac{1}{a} + \dfrac{1}{b}}{\dfrac{1}{a} - \dfrac{1}{b}}$$

EXERCISE SET 1.7

[i] In Exercises 1–3, determine the sensible replacements.

1. $\dfrac{3x - 2}{x(x - 1)}$

2. $\dfrac{(x^2 - 4)(x + 1)}{(x + 2)(x^2 - 1)}$

3. $\dfrac{7y^2 - 2y + 4}{x(x^2 - x - 6)}$

[ii] In Exercises 4–6, simplify. Then determine the replacements that are sensible in the simplified expression.

4. $\dfrac{25x^2y^2}{10xy^2}$

5. $\dfrac{x^2 - 4}{x^2 + 5x + 6}$

6. $\dfrac{x^2 - 3x + 2}{x^2 + x - 2}$

[iii] Multiply or divide, and simplify.

7. $\dfrac{x^2 - y^2}{(x - y)^2} \cdot \dfrac{1}{x + y}$

8. $\dfrac{r - s}{r + s} \cdot \dfrac{r^2 - s^2}{(r - s)^2}$

9. $\dfrac{x^2 - 2x - 35}{2x^3 - 3x^2} \cdot \dfrac{4x^3 - 9x}{7x - 49}$

10. $\dfrac{x^2 + 2x - 35}{3x^3 - 2x^2} \cdot \dfrac{9x^3 - 4x}{7x + 49}$

11. $\dfrac{a^2 - a - 6}{a^2 - 7a + 12} \cdot \dfrac{a^2 - 2a - 8}{a^2 - 3a - 10}$

12. $\dfrac{a^2 - a - 12}{a^2 - 6a + 8} \cdot \dfrac{a^2 + a - 6}{a^2 - 2a - 24}$

13. $\dfrac{m^2 - n^2}{r + s} \div \dfrac{m - n}{r + s}$

14. $\dfrac{a^2 - b^2}{x - y} \div \dfrac{a + b}{x - y}$

15. $\dfrac{3x + 12}{2x - 8} \div \dfrac{(x + 4)^2}{(x - 4)^2}$

16. $\dfrac{a^2 - a - 2}{a^2 - a - 6} \div \dfrac{a^2 - 2a}{2a + a^2}$

17. $\dfrac{x^2 - y^2}{x^3 - y^3} \cdot \dfrac{x^2 + xy + y^2}{x^2 + 2xy + y^2}$

18. $\dfrac{c^3 + 8}{c^2 - 4} \div \dfrac{c^2 - 2c + 4}{c^2 - 4c + 4}$

19. $\dfrac{(x - y)^2 - z^2}{(x + y)^2 - z^2} \div \dfrac{x - y + z}{x + y - z}$

20. $\dfrac{(a + b)^2 - 9}{(a - b)^2 - 9} \cdot \dfrac{a - b - 3}{a + b + 3}$

[iv] Add or subtract, and simplify.

21. $\dfrac{3}{2a + 3} + \dfrac{2a}{2a + 3}$

22. $\dfrac{a - 3b}{a + b} + \dfrac{a + 5b}{a + b}$

23. $\dfrac{y}{y - 1} + \dfrac{2}{1 - y}$

24. $\dfrac{a}{a - b} + \dfrac{b}{b - a}$

25. $\dfrac{x}{2x - 3y} - \dfrac{y}{3y - 2x}$

26. $\dfrac{3a}{3a - 2b} - \dfrac{2a}{2b - 3a}$

27. $\dfrac{3}{x + 2} + \dfrac{2}{x^2 - 4}$

28. $\dfrac{5}{a - 3} - \dfrac{2}{a^2 - 9}$

29. $\dfrac{y}{y^2 - y - 20} + \dfrac{2}{y + 4}$

30. $\dfrac{6}{y^2 + 6y + 9} - \dfrac{5}{y + 3}$

31. $\dfrac{3}{x + y} + \dfrac{x - 5y}{x^2 - y^2}$

32. $\dfrac{a^2 + 1}{a^2 - 1} - \dfrac{a - 1}{a + 1}$

33. $\dfrac{9x + 2}{3x^2 - 2x - 8} + \dfrac{7}{3x^2 + x - 4}$

34. $\dfrac{3y}{y^2 - 7y + 10} - \dfrac{2y}{y^2 - 8y + 15}$

35. $\dfrac{5a}{a - b} + \dfrac{ab}{a^2 - b^2} + \dfrac{4b}{a + b}$

36. $\dfrac{6a}{a - b} - \dfrac{3b}{b - a} + \dfrac{5}{a^2 - b^2}$

37. $\dfrac{7}{x + 2} - \dfrac{x + 8}{4 - x^2} + \dfrac{3x - 2}{4 - 4x + x^2}$

38. $\dfrac{6}{x + 3} - \dfrac{x + 4}{9 - x^2} + \dfrac{2x - 3}{9 - 6x + x^2}$

39. $\dfrac{1}{x + 1} - \dfrac{x}{x - 2} + \dfrac{x^2 + 2}{x^2 - x - 2}$

40. $\dfrac{x - 1}{x - 2} - \dfrac{x + 1}{x + 2} + \dfrac{x - 6}{x^2 - 4}$

[v] Simplify.

41. $\dfrac{\dfrac{x^2 - y^2}{xy}}{\dfrac{x - y}{y}}$

42. $\dfrac{\dfrac{a - b}{b}}{\dfrac{a^2 - b^2}{ab}}$

43. $\dfrac{a - a^{-1}}{a + a^{-1}}$

44. $\dfrac{a - \dfrac{a}{b}}{b - \dfrac{b}{a}}$

45. $\dfrac{c + \dfrac{8}{c^2}}{1 + \dfrac{2}{c}}$

46. $\dfrac{x^{-1} + y^{-1}}{x^{-3} + y^{-3}}$

47. $\dfrac{x^2 + xy + y^2}{\dfrac{x^2}{y} - \dfrac{y^2}{x}}$

48. $\dfrac{\dfrac{a^2}{b} + \dfrac{b^2}{a}}{a^2 - ab + b^2}$

49. $\dfrac{\dfrac{x}{y} - \dfrac{y}{x}}{\dfrac{1}{y} + \dfrac{1}{x}}$

50. $\dfrac{\dfrac{a}{b} - \dfrac{b}{a}}{\dfrac{1}{a} - \dfrac{1}{b}}$

51. $\dfrac{x^2 y^{-2} - y^2 x^{-2}}{xy^{-1} + yx^{-1}}$

52. $\dfrac{a^2 b^{-2} - b^2 a^{-2}}{ab^{-1} - ba^{-1}}$

53. $\dfrac{\dfrac{a}{1 - a} + \dfrac{1 + a}{a}}{\dfrac{1 - a}{a} + \dfrac{a}{1 + a}}$

54. $\dfrac{\dfrac{1 - x}{x} + \dfrac{x}{1 + x}}{\dfrac{1 + x}{x} + \dfrac{x}{1 - x}}$

55. $\dfrac{\dfrac{1}{a^2} + \dfrac{2}{ab} + \dfrac{1}{b^2}}{\dfrac{1}{a^2} - \dfrac{1}{b^2}}$

56. $\dfrac{\dfrac{1}{x^2} - \dfrac{1}{y^2}}{\dfrac{1}{x^2} - \dfrac{2}{xy} + \dfrac{1}{y^2}}$

☆

Simplify.

57. $\dfrac{(x + h)^2 - x^2}{h}$

58. $\dfrac{\dfrac{1}{x + h} - \dfrac{1}{x}}{h}$

59. $\dfrac{(x + h)^3 - x^3}{h}$

60. $\dfrac{\dfrac{1}{(x + h)^2} - \dfrac{1}{x^2}}{h}$

61. $\left[\dfrac{\dfrac{x+1}{x-1}+1}{\dfrac{x+1}{x-1}-1}\right]^5$

63. $1+\dfrac{1}{1+\dfrac{1}{1+\dfrac{1}{1+\dfrac{1}{x}}}}$

Find the error(s) in each of the following. Explain why each is an error. Then find the correct answer.

63. $\dfrac{a}{b}\div\left(\dfrac{a}{3}+\dfrac{b}{4}\right)$

$=\dfrac{a}{b}\cdot\left(\dfrac{3}{a}+\dfrac{4}{b}\right)$ (1)

$=\dfrac{a}{b}\cdot\left(\dfrac{3b+4a}{ab}\right)$ (2)

$=\dfrac{a(3b+4a)}{ab^2}$ (3)

$=\dfrac{4a+3b}{b}$ (4)

64. $\dfrac{5x}{2y}+\dfrac{3y}{4x}$

$=\dfrac{5x^2}{4xy}+\dfrac{3y^2}{4xy}$ (1)

$=\dfrac{5x^2+3y^2}{4xy}$ (2)

$=\dfrac{5x+3y^2}{4y}$ (3)

$=\dfrac{5x+3y}{4}$ (4)

OBJECTIVES

You should be able to:

[i] Determine the sensible replacements in radical expressions.

[ii] Simplify expressions involving absolute value, leaving as little as possible inside absolute value signs.

[iii] Simplify radical expressions.

[iv] Solve applied problems involving radical expressions.

1.8 RADICAL NOTATION AND ABSOLUTE VALUE

The symbol $\sqrt{a}$ denotes the nonnegative square root of the number a. The symbol $\sqrt[3]{a}$ denotes the cube root of a, and $\sqrt[4]{a}$ denotes the nonnegative fourth root of a. In general, $\sqrt[n]{a}$ denotes the nth root of a; that is, a number whose nth power is a. The symbol $\sqrt[n]{}$ is called a *radical* and the symbol under the radical is called the *radicand*. The number n (which is omitted when it is 2) is called the *index*.

Odd and Even Roots

Any positive real number has two square roots, one positive and one negative. The same is true for fourth roots, or roots of any even index. The positive root is called the *principal* root. When a radical such as $\sqrt{4}$ or $\sqrt[4]{18}$ is used, it is understood to represent the principal (nonnegative) root. To denote a nonpositive root we use $-\sqrt{2}$, $-\sqrt[4]{18}$, and so on.

DEFINITION

A radical expression $\sqrt[n]{a}$, where n is even, represents the principal (nonnegative) nth root of a. The nonpositive root is denoted $-\sqrt[n]{a}$.

CAUTION! Again, keep in mind that when the index is an even number E, then $\sqrt[E]{x}$ *never* represents a negative number. For example, $\sqrt{9}$ represents 3, and *not* -3!

[i] Sensible Replacements

Since negative numbers do not have even roots in the system of real numbers, any replacement that makes a radicand negative when the index is even is nonsensible.

Every real number, positive, negative, or zero, has just one cube root, and the same is true for any odd root. Thus $\sqrt[n]{a}$, where n is odd, represents the (only) nth root of a. In this case all real numbers are sensible replacements in the radicand.

Example 1 What are the sensible replacements in $\sqrt{5x - 4}$?

The sensible replacements are those that make the radicand nonnegative; that is, numbers x for which

$$5x - 4 \geq 0$$
$$5x \geq 4$$
$$x \geq \frac{4}{5}.$$

Thus the sensible replacements are any numbers x for which $x \geq \frac{4}{5}$. These form the set $\{x \,|\, x \geq \frac{4}{5}\}$.

DO EXERCISES 1–4.

[ii] Absolute Value

The absolute value of a nonnegative real number is that number itself. The absolute value of a negative number is its additive inverse. We make our definition as follows.

DEFINITION

For any real number x,

$$|x| = x \quad \text{if} \quad x \geq 0,$$

and

$$|x| = -x \quad \text{if} \quad x < 0.$$

This definition may be understood better if it is stated as follows. The absolute value of a number x is

a) the number x itself, if x is not negative;

b) the additive inverse of x, if x is negative.

A common source of confusion arises from interpreting $-x$ as meaning something negative, rather than "the additive inverse of x."

Certain properties of absolute value notation follow at once. For example, the absolute value of a product is the product of absolute values.

What are the sensible replacements in each of the following expressions?

1. $\sqrt{x - 2}$

2. $\sqrt{x + 3}$

3. $\sqrt{x^2}$

4. $\sqrt{x^2 + 1}$

Simplify.

5. $|-6ab|$

$6|ab|$

6. $|x^8|$

x^8

7. $|10m^2n^3|$

$10m^2n^2|n|$

8. $\left|\dfrac{-2x^3}{y^2}\right|$

$\dfrac{|-2x^3|}{|y^2|} = \dfrac{2x^2|x|}{y^2}$

Examples

2. $|-3 \cdot 5| = |-15| = 15$ and $|-3| \cdot |5| = 3 \cdot 5 = 15$,
so $|-3 \cdot 5| = |-3| \cdot |5|$.

Similarly,

3. $|-4 \cdot (-3)| = |12| = 12$ and $|-4| \cdot |-3| = 4 \cdot 3 = 12$,
so $|-4 \cdot (-3)| = |-4| \cdot |-3|$.

The absolute value of a quotient is similarly the quotient of the absolute values.

Example 4

$$\left|\frac{25}{-5}\right| = |-5| = 5 \quad \text{and} \quad \frac{|25|}{|-5|} = \frac{25}{5} = 5, \quad \text{so} \quad \left|\frac{25}{-5}\right| = \frac{|25|}{|-5|}.$$

The absolute value of an even power can be simplified by leaving off the absolute value signs, because no even power can be negative.

The absolute value of the additive inverse of a number is the same as the absolute value of the number. Their distances from 0 are the same on the number line.

Examples

5. $|(-3)^2| = |9| = 9$ and $(-3)^2 = 9$, so $|(-3)^2| = (-3)^2$.

6. $|-3| = 3$ and $|3| = 3$, so $|-3| = |3|$.

THEOREM 9

For any real numbers a and b and any nonzero number c:

1. $|ab| = |a| \cdot |b|$;

2. $\left|\dfrac{a}{c}\right| = \dfrac{|a|}{|c|}$;

3. $|a^n| = a^n$ if n is an even integer;

4. $|-a| = |a|$.

Examples Simplify, leaving as little as possible inside absolute value signs.

7. $|3x| = |3| \cdot |x| = 3|x|$

8. $|x^2| = x^2$

9. $|x^2y^3| = |x^2| \cdot |y^3| = x^2|y^3| = x^2y^2|y|$

10. $\left|\dfrac{x^2}{y}\right| = \dfrac{|x^2|}{|y|} = \dfrac{x^2}{|y|}$

11. $|-3x| = 3|x|$

DO EXERCISES 5–8.

[iii] Simplifying Radical Expressions

Consider the expression $\sqrt{(-3)^2}$. This is equivalent to $\sqrt{9}$, which simplifies to 3. Similarly, $\sqrt{3^2} = 3$. This illustrates an important general principle for simplifying radicals of even index.

THEOREM 10

For any radicand R, $\sqrt{R^2} = |R|$. Similarly, for any even index n, $\sqrt[n]{R^n} = |R|$.

Examples Simplify.

12. $\sqrt{x^2} = |x|$

13. $\sqrt{x^2 - 2ax + a^2} = \sqrt{(x-a)^2} = |x - a|$

14. $\sqrt{x^2 y^6} = \sqrt{(xy^3)^2} = |xy^3| = y^2|xy|$

If an index is odd, no absolute value signs are necessary.

THEOREM 11

For any radicand R and any odd index n, $\sqrt[n]{R^n} = R$.

DO EXERCISES 9–13.

A second property enables us to multiply radicals. We illustrate it with an example.

Example 15 Compare $\sqrt{4} \cdot \sqrt{9}$ and $\sqrt{4 \cdot 9}$.

$\sqrt{4} \cdot \sqrt{9} = 2 \cdot 3 = 6$ and $\sqrt{4 \cdot 9} = \sqrt{36} = 6$, so $\sqrt{4} \cdot \sqrt{9} = \sqrt{4 \cdot 9}$.

THEOREM 12

For any nonnegative real numbers a and b and any index n, $\sqrt[n]{a} \cdot \sqrt[n]{b} = \sqrt[n]{a \cdot b}$.

Examples Multiply.

16. $\sqrt{3} \cdot \sqrt{5} = \sqrt{3 \cdot 5} = \sqrt{15}$

17. $\sqrt{x + 2} \cdot \sqrt{x - 2} = \sqrt{(x+2)(x-2)} = \sqrt{x^2 - 4}$

18. $\sqrt[3]{4} \cdot \sqrt[3]{5} = \sqrt[3]{4 \cdot 5} = \sqrt[3]{20}$

DO EXERCISES 14–16.

Theorem 12 also enables us to simplify radical expressions. The idea is to factor the radicand, obtaining factors that are perfect nth powers.

Examples Simplify.

19. $\sqrt{50} = \sqrt{25 \cdot 2} = \sqrt{25} \cdot \sqrt{2} = 5\sqrt{2}$

20. $\sqrt{5x^2} = \sqrt{x^2 \cdot 5} = \sqrt{x^2} \cdot \sqrt{5} = |x|\sqrt{5}$

Simplify.

9. $\sqrt{(x + 2)^2}$

10. $\sqrt{x^2(y - 2)^2}$

11. $\sqrt[4]{(x + 2)^4}$

12. $\sqrt{x^2 + 8x + 16}$

13. $\sqrt[3]{(-4xy)^3}$

Multiply.

14. $\sqrt{19} \cdot \sqrt{7}$

15. $\sqrt{x + 2y} \cdot \sqrt{x - 2y}$

16. $\sqrt[4]{27} \cdot \sqrt[4]{3}$

Simplify.

17. $\sqrt{300}$

18. $\sqrt{36y^2}$

19. $\sqrt{2x^2 + 4x + 2}$

20. $\sqrt[3]{16}$

21. $\sqrt[3]{(a + b)^4}$

Simplify.

22. $\sqrt{\dfrac{49}{64}}$

23. $\sqrt{\dfrac{25}{y^2}}$

24. $\sqrt{\dfrac{7}{5}}$

25. $\sqrt[3]{\dfrac{7}{125}}$

26. $\dfrac{\sqrt{75}}{\sqrt{3}}$

27. $\dfrac{\sqrt{2x^3}}{\sqrt{50x}}$

28. $\dfrac{\sqrt[3]{24x^3y}}{\sqrt[3]{3y^4}}$

21. $\sqrt[3]{32} = \sqrt[3]{8 \cdot 4} = \sqrt[3]{8} \cdot \sqrt[3]{4} = 2\sqrt[3]{4}$

22. $\sqrt{216x^5y^3} = \sqrt{36 \cdot 6 \cdot x^4 \cdot x \cdot y^2 \cdot y} = |6x^2y|\sqrt{6xy} = 6x^2|y|\sqrt{6xy}$

23. $\sqrt{2x^2 - 4x + 2} = \sqrt{2(x - 1)^2} = |x - 1|\sqrt{2}$

DO EXERCISES 17–21.

A third fundamental property of radicals is as follows.

THEOREM 13

For any nonnegative number a and any positive number b, and any index n,

$$\sqrt[n]{\dfrac{a}{b}} = \dfrac{\sqrt[n]{a}}{\sqrt[n]{b}}.$$

This property can be used to divide and to simplify radical expressions.

Examples Simplify.

24. $\sqrt{16x^3y^{-4}} = \sqrt{\dfrac{16x^3}{y^4}} = \dfrac{\sqrt{16x^3}}{\sqrt{y^4}} = \dfrac{\sqrt{16x^2 \cdot x}}{\sqrt{y^4}} = \dfrac{4|x|\sqrt{x}}{y^2}$

25. $\sqrt[3]{\dfrac{27y^5}{343x^3}} = \dfrac{\sqrt[3]{27y^5}}{\sqrt[3]{343x^3}} = \dfrac{\sqrt[3]{27y^3 \cdot y^2}}{\sqrt[3]{343x^3}} = \dfrac{3y\sqrt[3]{y^2}}{7x}$

Fractional expressions are often considered simpler when the denominator is free of radicals. Thus in simplifying, it is usual to remove the radicals in a denominator. This is called *rationalizing the denominator*, and it can be done by multiplying by 1.

Examples Simplify.

26. $\sqrt{\dfrac{1}{2}} = \sqrt{\dfrac{1}{2} \cdot \dfrac{2}{2}} = \sqrt{\dfrac{2}{4}} = \dfrac{\sqrt{2}}{\sqrt{4}} = \dfrac{\sqrt{2}}{2}$

27. $\sqrt[3]{\dfrac{7}{9}} = \sqrt[3]{\dfrac{7}{9} \cdot \dfrac{3}{3}} = \sqrt[3]{\dfrac{21}{27}} = \dfrac{\sqrt[3]{21}}{\sqrt[3]{27}} = \dfrac{\sqrt[3]{21}}{3}$

Examples Divide and simplify.

28. $\dfrac{18\sqrt{72}}{6\sqrt{6}} = 3\sqrt{\dfrac{72}{6}} = 3\sqrt{12} = 3\sqrt{4 \cdot 3} = 3 \cdot 2\sqrt{3} = 6\sqrt{3}$

29. $\dfrac{\sqrt[3]{32}}{\sqrt[3]{2}} = \sqrt[3]{\dfrac{32}{2}} = \sqrt[3]{16} = \sqrt[3]{8 \cdot 2} = \sqrt[3]{8} \cdot \sqrt[3]{2} = 2\sqrt[3]{2}$

DO EXERCISES 22–28.

A fourth fundamental principle of radicals involves an exponent under the radical. We illustrate with an example.

Example 30 Compare $\sqrt{3^4}$ and $(\sqrt{3})^4$.

$$\sqrt{3^4} = \sqrt{81} = 9,$$
$$(\sqrt{3})^4 = \sqrt{3} \cdot \sqrt{3} \cdot \sqrt{3} \cdot \sqrt{3} = 3 \cdot 3 = 9$$

Thus $\sqrt{3^4} = (\sqrt{3})^4$.

The general principle is given in Theorem 14.

THEOREM 14

For any nonnegative number a and any index n and any natural number m, $\sqrt[n]{a^m} = (\sqrt[n]{a})^m$.

Theorem 14 sometimes facilitates radical simplification.

Examples Simplify.

31. $\sqrt[3]{8^5} = (\sqrt[3]{8})^5 = 2^5 = 32$

32. $(\sqrt{2})^6 = \sqrt{2^6} = 2^3 = 8$

DO EXERCISES 29 AND 30.

[iv] Applied Problems

Example 33 *Speed of a skidding car.* How do police determine the speed of a car that has skidded? The formula

$$r = 2\sqrt{5L}$$

can be used to approximate the speed r, in mph, of a car that has left a skid mark of length L, in feet. What was the speed of a car that left skid marks 50 ft long?

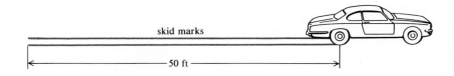

skid marks

50 ft

We substitute 50 for L in the formula $r = 2\sqrt{5L}$:

$$r = 2\sqrt{5(50)} = 2\sqrt{250}$$
$$= 2\sqrt{25 \cdot 10}$$
$$= 2(5)\sqrt{10}$$
$$= 10\sqrt{10}$$
$$\approx 10(3.162) \quad \text{Table 1 or } \boxplus$$
$$\approx 31.62 \text{ mph.}$$

DO EXERCISE 31.

29. $\sqrt[3]{27^{10}}$

30. $(\sqrt{3})^8$

31. What was the speed of a car that left skid marks 70 ft long?

EXERCISE SET 1.8

[i] What are the sensible replacements in each of the following?

1. $\sqrt{x-3}$ **2.** $\sqrt{2x-5}$ **3.** $\sqrt{3-4x}$ **4.** $\sqrt{x^2+3}$

[ii] Simplify.

5. $|9xy|$ **6.** $|y^4|$ **7.** $|3a^2b|$ **8.** $\left|\dfrac{4a}{b^2}\right|$

[iii] Simplify.

9. $\sqrt{(-11)^2}$ **10.** $\sqrt{(-1)^2}$ **11.** $\sqrt{16x^2}$ **12.** $\sqrt{36t^2}$

13. $\sqrt{(b+1)^2}$ **14.** $\sqrt{(2c-3)^2}$ **15.** $\sqrt[3]{-27x^3}$ **16.** $\sqrt[3]{-8y^3}$

17. $\sqrt{x^2-4x+4}$ **18.** $\sqrt{y^2+16y+64}$ **19.** $\sqrt[5]{32}$ **20.** $\sqrt[5]{-32}$

21. $\sqrt{180}$ **22.** $\sqrt{48}$ **23.** $\sqrt[3]{54}$ **24.** $\sqrt[3]{135}$

25. $\sqrt{128c^2d^{-4}}$ **26.** $\sqrt{162c^4d^{-6}}$ **27.** $\sqrt{3}\cdot\sqrt{6}$ **28.** $\sqrt{6}\cdot\sqrt{8}$

[iv] In the following exercises simplify, assuming that all letters represent positive numbers and that all radicands are positive. Thus no absolute value signs will be needed.

29. $\sqrt{2x^3y}\,\sqrt{12xy}$ **30.** $\sqrt{3y^4z}\,\sqrt{20z}$ **31.** $\sqrt[3]{3x^2y}\,\sqrt[3]{36x}$

32. $\sqrt[5]{8x^3y^4}\,\sqrt[5]{4x^4y}$ **33.** $\sqrt[3]{2(x+4)}\,\sqrt[3]{4(x+4)^4}$ **34.** $\sqrt[3]{4(x+1)^2}\,\sqrt[3]{18(x+1)^2}$

35. $\dfrac{\sqrt{21ab^2}}{\sqrt{3ab}}$ **36.** $\dfrac{\sqrt{128ab^2}}{\sqrt{16a^2b}}$ **37.** $\dfrac{\sqrt[3]{40m}}{\sqrt[3]{5m}}$ **38.** $\dfrac{\sqrt{40xy}}{\sqrt{8x}}$

39. $\dfrac{\sqrt[3]{3x^2}}{\sqrt[3]{24x^5}}$ **40.** $\dfrac{\sqrt[3]{40xy^3}}{\sqrt[3]{8x}}$ **41.** $\dfrac{\sqrt{a^2-b^2}}{\sqrt{a-b}}$ **42.** $\dfrac{\sqrt{x^3-y^3}}{\sqrt{x-y}}$

43. $\sqrt{\dfrac{9a^2}{8b}}$ **44.** $\sqrt{\dfrac{5b^2}{12a}}$ **45.** $\sqrt[3]{\dfrac{2x^2 2y^3}{25z^4}}$ **46.** $\sqrt[3]{\dfrac{24x^3y}{3y^4}}$

47. $\dfrac{(\sqrt[3]{32x^4y})^2}{(\sqrt[3]{xy})^2}$ **48.** $\dfrac{(\sqrt[3]{16x^2y})^2}{(\sqrt[3]{xy})^2}$ **49.** $\dfrac{3\sqrt{a^2b^2}\,\sqrt{4xy}}{2\sqrt{a^{-1}b^2}\,\sqrt{9x^{-3}y^{-1}}}$ **50.** $\dfrac{4\sqrt{xy^2}\,\sqrt{9ab}}{3\sqrt{x^{-1}y^{-2}}\,\sqrt{16a^{-5}b^{-1}}}$

51. ▦ *Pendulums.* The *period T* of a pendulum is the time it takes to make a move from one side to the other and back. A formula for the period is

$$T = 2\pi\sqrt{\dfrac{L}{32}},$$

where T is in seconds and L, the length of the pendulum, is in feet. Find the periods of pendulums of lengths 2 ft, 8 ft, 64 ft, and 100 ft. Use 3.14 for π.

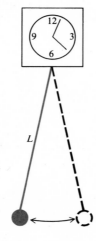

52. A slow-pitch softball diamond is actually a square 60 ft on a side. How far is it from home to second base? (*Hint:* For right triangles, $c = \sqrt{a^2 + b^2}$, where a and b are the lengths of the legs and c is the length of the hypotenuse.)

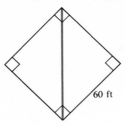

☆

Simplify, assuming that all letters represent positive numbers.

53. ▦ $\sqrt{8.2x^3y}\sqrt{12.5xy}$ **54.** ▦ $\sqrt{0.012y^4z}\sqrt{1.305x}$ **55.** ▦ $\sqrt{\dfrac{6.03a^2}{17.13b}}$ **56.** ▦ $\sqrt{\dfrac{3.2b^2}{82.1a}}$

An *equilateral* triangle is shown at the right.

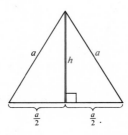

57. Find an expression for its height h in terms of a.   **58.** Find an expression for its area A in terms of a.

★

59. For what integer values of n is $|a^n| = |a|^n$ for all a?   **60.** For what values of x and y is $|x + y| = |x| + |y|$?

1.9 FURTHER CALCULATIONS WITH RADICAL NOTATION

[i] Further Simplifying

Various calculations with radicals can be carried out using the properties of radicals and the properties of numbers, such as the distributive property. The following examples illustrate.

Examples Simplify.

1. $3\sqrt{8} - 5\sqrt{2} = 3\sqrt{4 \cdot 2} - 5\sqrt{2} = 3 \cdot 2\sqrt{2} - 5\sqrt{2}$
$\qquad = (6 - 5)\sqrt{2}$ Here we used a distributive law.
$\qquad = \sqrt{2}$

2. $(4\sqrt{3} + \sqrt{2})(\sqrt{3} - 5\sqrt{2}) = 4(\sqrt{3})^2 - 20\sqrt{3}\sqrt{2} + \sqrt{2}\sqrt{3} - 5(\sqrt{2})^2$
$\qquad = 4 \cdot 3 - 20\sqrt{6} + \sqrt{6} - 5 \cdot 2$
$\qquad = 12 - 19\sqrt{6} - 10$
$\qquad = 2 - 19\sqrt{6}$

DO EXERCISES 1–3.

OBJECTIVES

You should be able to:

[i] Simplify expressions containing radicals, using the distributive property and properties of radicals.

[ii] Rationalize numerators and denominators.

Simplify.

1. $7\sqrt{5} + 3\sqrt{5} - 8\sqrt{20}$

2. $5\sqrt[3]{16y^4} + 7\sqrt[3]{2y}$

3. $(\sqrt{3} - 5\sqrt{2})(2\sqrt{3} + \sqrt{2})$

[ii] Rationalizing Denominators or Numerators

When a fractional symbol contains radicals, we ordinarily rationalize the denominator, but on occasion we prefer to rationalize the numerator. In either case, we can accomplish the rationalization by multiplying by 1, as in the following examples.

Examples Rationalize the denominator.

3. $\dfrac{\sqrt{7}}{\sqrt{5}} = \dfrac{\sqrt{7}}{\sqrt{5}} \cdot \dfrac{\sqrt{5}}{\sqrt{5}} = \dfrac{\sqrt{35}}{\sqrt{25}} = \dfrac{\sqrt{35}}{5}$

4. $\dfrac{\sqrt{2a}}{\sqrt{5b}} = \dfrac{\sqrt{2a}}{\sqrt{5b}} \cdot \dfrac{\sqrt{5b}}{\sqrt{5b}} = \dfrac{\sqrt{10ab}}{\sqrt{(5b)^2}}$

$= \dfrac{\sqrt{10ab}}{|5b|} = \dfrac{\sqrt{10ab}}{5b}$ The absolute value sign in the denominator is not necessary since $\sqrt{5b}$ would not exist at the outset unless $b > 0$.

5. $\dfrac{\sqrt[3]{54x^3}}{\sqrt[3]{4y^5}} = \sqrt[3]{\dfrac{54x^3}{4y^5} \cdot \dfrac{2y}{2y}} = \sqrt[3]{\dfrac{54x^3 \cdot 2y}{8y^6}} = \dfrac{\sqrt[3]{27x^3} \cdot \sqrt[3]{4y}}{\sqrt[3]{8y^6}} = \dfrac{3x \cdot \sqrt[3]{4y}}{2y^2}$

When a numerator or denominator to be rationalized has two terms, we choose the symbol for 1 a little differently. The symbol for 1 will have two terms in its numerator and denominator. The following examples illustrate.

Examples Rationalize the denominator. Assume all letters represent positive numbers.

6. $\dfrac{1}{\sqrt{2} + \sqrt{3}} = \dfrac{1}{\sqrt{2} + \sqrt{3}} \cdot \dfrac{\sqrt{2} - \sqrt{3}}{\sqrt{2} - \sqrt{3}}$

$= \dfrac{\sqrt{2} - \sqrt{3}}{(\sqrt{2} + \sqrt{3})(\sqrt{2} - \sqrt{3})}$

$= \dfrac{\sqrt{2} - \sqrt{3}}{(\sqrt{2})^2 - (\sqrt{3})^2} = \dfrac{\sqrt{2} - \sqrt{3}}{2 - 3}$

$= \dfrac{\sqrt{2} - \sqrt{3}}{-1} = \sqrt{3} - \sqrt{2}$

7. $\dfrac{\sqrt{x} + \sqrt{y}}{\sqrt{x} - \sqrt{y}} = \dfrac{\sqrt{x} + \sqrt{y}}{\sqrt{x} - \sqrt{y}} \cdot \dfrac{\sqrt{x} + \sqrt{y}}{\sqrt{x} + \sqrt{y}} = \dfrac{(\sqrt{x} + \sqrt{y})^2}{(\sqrt{x})^2 - (\sqrt{y})^2}$

$= \dfrac{x + 2\sqrt{x}\sqrt{y} + y}{x - y}$

Examples Rationalize the numerator. Assume all letters represent positive numbers.

8. $\dfrac{1 - \sqrt{2}}{5} = \dfrac{1 - \sqrt{2}}{5} \cdot \dfrac{1 + \sqrt{2}}{1 + \sqrt{2}} = \dfrac{(1 - \sqrt{2})(1 + \sqrt{2})}{5(1 + \sqrt{2})}$

$= \dfrac{1 - 2}{5(1 + \sqrt{2})} = \dfrac{-1}{5 + 5\sqrt{2}}$

9. $\dfrac{\sqrt{x+h}-\sqrt{x}}{h} = \dfrac{\sqrt{x+h}-\sqrt{x}}{h}\cdot\dfrac{\sqrt{x+h}+\sqrt{x}}{\sqrt{x+h}+\sqrt{x}}$

$$= \dfrac{(x+h)-x}{h(\sqrt{x+h}+\sqrt{x})}$$

$$= \dfrac{h}{h(\sqrt{x+h}+\sqrt{x})}$$

$$= \dfrac{1}{\sqrt{x+h}+\sqrt{x}}$$

DO EXERCISES 4–7.

Rationalize the denominator. Assume all letters represent positive numbers.

4. $\dfrac{1}{\sqrt{3}-\sqrt{5}}$

5. $\dfrac{\sqrt{x}-5}{\sqrt{x}+2}$

Rationalize the numerator. Assume all letters represent positive numbers.

6. $\dfrac{\sqrt{a+2}-\sqrt{a}}{2}$

7. $\dfrac{\sqrt{x}-\sqrt{5}}{\sqrt{x}+\sqrt{5}}$

EXERCISE SET 1.9

In this exercise set assume that all letters represent positive numbers and that all radicands are positive. Thus, absolute value signs will not be necessary.

[i] Simplify.

1. $8\sqrt{2}-6\sqrt{20}-5\sqrt{8}$

2. $\sqrt{12}-\sqrt{27}+\sqrt{75}$

3. $2\sqrt[3]{8x^2}+5\sqrt[3]{27x^2}-3\sqrt[3]{x^3}$

4. $5a\sqrt{(a+b)^3}-2ab\sqrt{a+b}-3b\sqrt{(a+b)^3}$

5. $3\sqrt{3y^2}-\dfrac{y\sqrt{48}}{\sqrt{2}}+\sqrt{\dfrac{12}{4y^{-2}}}$

6. $\sqrt[3]{x^5}-\dfrac{2\sqrt[3]{x}}{\sqrt[3]{x^{-1}}}+\sqrt[3]{\dfrac{8}{x^{-5}}}$

7. $(\sqrt{3}-\sqrt{2})(\sqrt{3}+\sqrt{2})$

8. $(\sqrt{8}+2\sqrt{5})(\sqrt{8}-2\sqrt{5})$

9. $(\sqrt{t}-x)^2$

10. $\left(\sqrt{a}+\dfrac{1}{\sqrt{a}}\right)^2$

11. $5\sqrt{7}+\dfrac{35}{\sqrt{7}}$

12. $(\sqrt{a^2b}+3\sqrt{y})(2a\sqrt{b}-\sqrt{y})$

13. $(\sqrt{x+3}-\sqrt{3})(\sqrt{x+3}+\sqrt{3})$

14. $(\sqrt{x+h}-\sqrt{x})(\sqrt{x+h}+\sqrt{x})$

[ii] Rationalize the denominator.

15. $\dfrac{6}{3+\sqrt{5}}$

16. $\dfrac{2}{\sqrt{3}-1}$

17. $\sqrt[3]{\dfrac{16}{9}}$

18. $\dfrac{\sqrt[3]{3}}{\sqrt[3]{6}}$

19. $\dfrac{4\sqrt{x}-3\sqrt{xy}}{2\sqrt{x}+5\sqrt{y}}$

20. $\dfrac{5\sqrt{x}+2\sqrt{xy}}{3\sqrt{x}-2\sqrt{y}}$

[ii] Rationalize the numerator.

21. $\dfrac{\sqrt{2}+\sqrt{5a}}{6}$

22. $\dfrac{\sqrt{3}+\sqrt{5y}}{4}$

23. $\dfrac{\sqrt{x+1}+1}{\sqrt{x+1}-1}$

24. $\dfrac{\sqrt{x+4}-2}{\sqrt{x+4}+2}$

25. $\dfrac{\sqrt{a+3}-\sqrt{3}}{3}$

26. $\dfrac{\sqrt{a+h}-\sqrt{a}}{h}$

☆ ──────────────────

Simplify.

27. $\sqrt{1+x^2}+\dfrac{1}{\sqrt{1+x^2}}$

28. $\sqrt{1-x^2}-\dfrac{x^2}{2\sqrt{1-x^2}}$

29. Show that $\sqrt{a+b} = \sqrt{a} + \sqrt{b}$ is false for positive real numbers a and b by finding two positive numbers a and b for which $\sqrt{a+b} \neq \sqrt{a} + \sqrt{b}$.

30. Show that $(\sqrt{5+\sqrt{24}})^2 = (\sqrt{2}+\sqrt{3})^2$.

Prove the following.

31. For any positive real numbers a and b, there exist positive numbers c and d for which

$$\sqrt{a+b} = \sqrt{c} + \sqrt{d}.$$

Under what conditions does $a = c$ and $b = d$?

OBJECTIVES

You should be able to:

[i] Convert from exponential notation to radical notation.

[ii] Convert from radical notation to exponential notation.

[iii] Simplify expressions using the properties of rational exponents and the arithmetic of rational numbers.

[iv] Simplify certain expressions to expressions containing a single radical.

1.10 RATIONAL EXPONENTS

We are motivated to define fractional exponents so that the same rules, or laws, hold for them as for integer exponents. For example, if the laws of exponents are to hold, we must have

$$a^{1/2} \cdot a^{1/2} = a^{1/2+1/2} = a^1 = a.$$

Thus we are led to define $a^{1/2}$ to mean $\sqrt{a}$. Similarly, $a^{1/n}$ would mean $\sqrt[n]{a}$. Again, if the usual laws of exponents are to hold, we must have

$$(a^{1/n})^m = (a^m)^{1/n} = a^{m/n}.$$

Thus we are led to define $a^{m/n}$ to mean $(\sqrt[n]{a})^m$ or $\sqrt[n]{a^m}$.

DEFINITION

An expression $a^{m/n}$, where a is positive and m and n are natural numbers, is defined to mean $(\sqrt[n]{a})^m$ or $\sqrt[n]{a^m}$. An expression $a^{-m/n}$ is defined to mean $1/a^{m/n}$.

Note that in this definition we require a to be positive. Thus in manipulations with fractional exponents we assume that all letters represent positive numbers and that all radicands are positive. No absolute value signs need be used.

Once the definition of rational exponents is made, the question arises whether the usual laws of exponents actually do hold. We shall not prove it, but the answer is that they do. Thus we can simplify or otherwise manipulate expressions containing rational exponents using those laws and the usual arithmetic of rational numbers.

[i] **Converting to Radical Notation**

Examples Convert to radical notation and simplify, if possible.

1. $m^{2/3} = \sqrt[3]{m^2}$

2. $t^{-1/2} = \dfrac{1}{t^{1/2}} = \dfrac{1}{\sqrt{t}}$

3. $64^{5/2} = (64^{1/2})^5 = (\sqrt{64})^5 = 8^5 = 32{,}768$

4. $16^{-3/4} = \dfrac{1}{16^{3/4}} = \dfrac{1}{(16^{1/4})^3} = \dfrac{1}{(\sqrt[4]{16})^3} = \dfrac{1}{2^3} = \dfrac{1}{8}$

DO EXERCISES 1–4.

[ii] Converting to Exponential Notation

Examples Convert to exponential notation and simplify.

5. $(\sqrt[4]{7xy})^5 = (7xy)^{5/4}$

6. $\sqrt[3]{8^4} = 8^{4/3} = (8^{1/3})^4 = 2^4 = 16$

7. $\sqrt[6]{x^3} = x^{3/6} = x^{1/2}$ (or $\sqrt{x}$)

8. $\sqrt[6]{4} = 4^{1/6} = (4^{1/2})^{1/3} = 2^{1/3}$ (or $\sqrt[3]{2}$)

9. $\sqrt[3]{\sqrt{7}} = \sqrt[3]{7^{1/2}} = (7^{1/2})^{1/3} = 7^{1/6}$ (or $\sqrt[6]{7}$)

10. $\sqrt{6}\,\sqrt[3]{6} = 6^{1/2} \cdot 6^{1/3} = 6^{1/2+1/3} = 6^{5/6}$ (or $\sqrt[6]{6^5}$)

DO EXERCISES 5–9.

[iii] Simplifying Expressions with Rational Exponents

Examples Simplify and then write radical notation.

11. $x^{5/6} \cdot x^{2/3} = x^{5/6+2/3} = x^{9/6} = x^{3/2} = \sqrt{x^3} = x\sqrt{x}$

12. $(a^5)^{-2/3} = a^{-10/3} = \dfrac{1}{a^{10/3}}$

$\qquad = \dfrac{1}{\sqrt[3]{a^{10}}} = \dfrac{1}{a^3 \cdot \sqrt[3]{a}}$

13. $(5^{1/3} - 5^{-5/3}) \cdot 5^{1/3} = 5^{1/3} \cdot 5^{1/3} - 5^{-5/3} \cdot 5^{1/3}$

$\qquad = 5^{2/3} - 5^{-4/3}$

$\qquad = \sqrt[3]{5^2} - \dfrac{1}{\sqrt[3]{5^4}}$

$\qquad = \sqrt[2]{25} - \dfrac{1}{5\sqrt[3]{5}}$

DO EXERCISES 10–12.

[iv] Writing Single Radicals

In certain expressions containing radicals or fractional exponents, it is possible to simplify in such a way that there is a single radical.

Examples Write an expression containing a single radical.

14. $a^{1/2}b^{-1/2}c^{5/6} = a^{3/6}b^{-3/6}c^{5/6} = (a^3b^{-3}c^5)^{1/6} = \sqrt[6]{a^3b^{-3}c^5}$

15. $\dfrac{a^{1/4}b^{3/8}}{a^{1/2}b^{1/8}} = a^{-1/4}b^{1/4} = (a^{-1}b)^{1/4} = \sqrt[4]{a^{-1}b}$

16. $\sqrt[4]{7}\,\sqrt{3} = 7^{1/4} \cdot 3^{1/2} = 7^{1/4} \cdot 3^{2/4} = (7 \cdot 3^2)^{1/4} = \sqrt[4]{63}$

Convert to radical notation and simplify, if possible.

1. $n^{3/2}$

2. $y^{-6/7}$

3. $32^{4/5}$

4. $64^{-2/3}$

Convert to exponential notation and simplify.

5. $\sqrt[3]{(5ab)^4}$

6. $\sqrt[4]{16^3}$

7. $\sqrt[6]{a^4}$

8. $\sqrt{\sqrt[3]{4}}$

9. $\sqrt{5^3}\sqrt[3]{5}$

Simplify and then write radical notation.

10. $a^{3/4} \cdot a^{1/2}$

11. $(x^{-3})^{2/5}$

12. $(2^{1/4} + 2^{-3/4}) \cdot 2^{1/2}$

Write an expression containing a single radical.

13. $\sqrt[3]{5} \cdot \sqrt{2}$

14. $x^{2/3}y^{1/2}z^{5/6}$

15. $\dfrac{\sqrt[4]{(x+y)^3}}{\sqrt{x+y}}$

17. $\dfrac{\sqrt[4]{(x+2)^3}\sqrt[5]{x+2}}{\sqrt{x+2}} = \dfrac{(x+2)^{3/4}(x+2)^{1/5}}{(x+2)^{1/2}}$

$$= (x+2)^{3/4+1/5-1/2}$$

$$= (x+2)^{9/20}$$

$$= \sqrt[20]{(x+2)^9}$$

DO EXERCISES 13–15.

EXERCISE SET 1.10

[i] Convert to radical notation and simplify.

1. $x^{3/4}$ **2.** $y^{2/5}$ **3.** $16^{3/4}$ **4.** $4^{7/2}$

5. $125^{-1/3}$ **6.** $32^{-4/5}$ **7.** $a^{5/4}b^{-3/4}$ **8.** $x^{2/5}y^{-1/5}$

[ii] Convert to exponential notation and simplify.

9. $\sqrt[3]{20^2}$ **10.** $\sqrt[5]{17^3}$ **11.** $(\sqrt[4]{13})^5$ **12.** $(\sqrt[5]{12})^4$

13. $\sqrt[3]{\sqrt{11}}$ **14.** $\sqrt[3]{\sqrt[4]{7}}$ **15.** $\sqrt{5}\sqrt[3]{5}$ **16.** $\sqrt[3]{2}\sqrt{2}$

17. $\sqrt[5]{32^2}$ **18.** $\sqrt[3]{64^{-2}}$ **19.** $\sqrt[3]{8y^6}$ **20.** $\sqrt[5]{32c^{10}d^{15}}$

21. $\sqrt[3]{a^2+b^2}$ **22.** $\sqrt[4]{a^3-b^3}$ **23.** $\sqrt[3]{27a^3b^9}$ **24.** $\sqrt[4]{81x^8y^8}$

25. $\sqrt[6]{\dfrac{m^{12}n^{24}}{64}}$ **26.** $\sqrt[8]{\dfrac{m^{16}n^{24}}{2^8}}$

[iii] Simplify and then write radical notation, unless inappropriate.

27. $(2a^{3/2})(4a^{1/2})$ **28.** $(3a^{5/6})(8a^{2/3})$ **29.** $\left(\dfrac{x^6}{9b^{-4}}\right)^{-1/2}$

30. $\left(\dfrac{x^{2/3}}{4y^{-2}}\right)^{-1/2}$ **31.** $\dfrac{x^{2/3}y^{5/6}}{x^{-1/3}y^{1/2}}$ **32.** $\dfrac{a^{1/2}b^{5/8}}{a^{1/4}b^{3/8}}$

[iv] Write an expression containing a single radical and simplify.

33. $\sqrt[3]{6}\sqrt{2}$ **34.** $\sqrt{2}\sqrt[4]{8}$ **35.** $\sqrt[4]{xy}\sqrt[3]{x^2y}$

36. $\sqrt[3]{ab^2}\sqrt{ab}$ **37.** $\sqrt[3]{a^4}\sqrt{a^3}$ **38.** $\sqrt{a^3}\sqrt[3]{a^2}$

39. $\dfrac{\sqrt{(a+x)^3}\sqrt[3]{(a+x)^2}}{\sqrt[4]{a+x}}$ **40.** $\dfrac{\sqrt[4]{(x+y)^2}\sqrt[3]{(x+y)}}{\sqrt{(x+y)^3}}$

Simplify. (*Note:* Since $x^{1/4} = (x^{1/2})^{1/2}$ you can take a fourth root by taking a square root, and then the square root of the result. Or you can find decimal notation for the exponent, obtaining $x^{0.25}$, and use the power key, x^y. Remember, answers can vary depending on the type and readout of your calculator.) Round to three decimal places.

41. ▦ $(\sqrt[4]{13})^5$ **42.** ▦ $\sqrt[4]{17^3}$ **43.** ▦ $12.3^{3/2}$

44. ▦ $1.345^{5/2}$ **45.** ▦ $105.6^{3/4}$ **46.** ▦ $7.14^{5/4}$

☆ ──

In a psychological study, pavement signs were found to be most readable by a driver when the letters in the sign are of length L, given by,

$$L = \frac{0.000169d^{2.27}}{h},$$

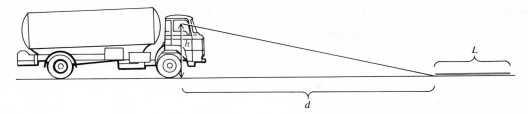

where d is the distance from the car to the lettering and h is the height of the eye above the road. All units are in feet. Find L, given the values of h and d.

47. ▦ $h = 4$ ft, $d = 180$ ft

48. ▦ $h = 4$ ft, $d = 100$ ft

49. ▦ $h = 4$ ft, $d = 200$ ft

50. ▦ $h = 4$ ft, $d = 300$ ft

Simplify.

51. $\left(\sqrt{a^{\sqrt{a}}}\right)^{\sqrt{a}}$

52. $(2a^3b^{5/4}c^{1/7})^4 \div (54a^{-2}b^{2/3}c^{6/5})^{-1/3}$

1.11 HANDLING DIMENSION SYMBOLS

Speed

Speed is often determined by measuring a distance and a time and then dividing the distance by the time (this is *average* speed):

$$\text{Speed} = \frac{\text{Distance}}{\text{Time}}.$$

If a distance is measured in kilometers and the time required to travel that distance is measured in hours, the speed will be computed in *kilometers per hour* (km/h). For example, if a car travels 100 km in 2 h, the average speed is

$$\frac{100 \text{ km}}{2 \text{ h}}, \quad \text{or } 50\frac{\text{km}}{\text{h}}.$$

DO EXERCISES 1 AND 2.

[i] Dimension Symbols

The symbol 100 km/2 h makes it look as if we are dividing 100 km by 2 h. It may be argued that we cannot divide 100 km by 2 h (we can only divide 100 by 2). Nevertheless, it is convenient to treat dimension symbols such as *kilometers, hours, feet, seconds,* and *pounds* as if they were numerals or variables, for the reason that correct results can thus be obtained mechanically. Compare, for example,

$$\frac{100x}{2y} = \frac{100}{2} \cdot \frac{x}{y} = 50\frac{x}{y}$$

with

$$\frac{100 \text{ km}}{2 \text{ h}} = \frac{100}{2} \cdot \frac{\text{km}}{\text{h}} = 50\frac{\text{km}}{\text{h}}.$$

OBJECTIVES

You should be able to:

[i] Perform a given calculation involving dimension symbols and simplify, if possible, without making any unit changes.

[ii] Perform a given change of dimension symbols, using substitution or "multiplying by one."

What is the speed in m/sec?

1. 186,000 m, 10 sec

2. 8 m, 16 sec

Add these measures.

3. 45 ft, 17 ft

4. $\frac{3}{4}$ kg, $\frac{2}{5}$ kg

5. 70 $\frac{cm}{sec}$, 35 $\frac{cm}{sec}$

Perform these calculations and simplify if possible. Do not make any unit changes.

6. 36 ft $\cdot \frac{1\,yd}{3\,ft}$

7. 5 lb $\cdot \frac{16\,oz}{1\,lb}$

8. $\frac{4\,kg}{5\,ft} \cdot \frac{7\,ft}{8\,kg}$

9. $\frac{5\,in. \cdot 9\,lb/hr}{4\,hr}$

10. $\frac{10\,lb}{7\,m} \cdot \frac{14\,lb}{5\,m}$

Perform the following changes of unit. Use substitution.

11. 34 yd, change to in.

12. 11 mi, change to ft.

13. 5 hr, change to sec.

The analogy holds in other situations, as shown in the following examples.

Example 1 Compare

$$3\,ft + 2\,ft = (3 + 2)ft = 5\,ft$$

with

$$3x + 2x = (3 + 2)x = 5x.$$

This looks like a distributive law in use.

DO EXERCISES 3–5.

Example 2 Compare

$$4\,m \cdot 3\,m = 3 \cdot 4 \cdot m \cdot m = 12\,m^2 \text{ (sq m)}$$

with

$$4x \cdot 3x = 4 \cdot 3 \cdot x \cdot x = 12x^2.$$

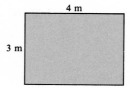

Example 3 Compare

$$5 \text{ men} \cdot 8\,hr = 5 \cdot 8 \cdot \text{man-hr} = 40 \text{ man-hr}$$

with

$$5x \cdot 8y = 5 \cdot 8 \cdot x \cdot y = 40xy.$$

In each of the above examples, dimension symbols are treated as if they were variables or numerals, and as if a symbol such as "3 m" represents a product 3 times m. A symbol like km/h is treated as if it represents a division of km by h (*kilometers* by *hours*). Any two measures can be "multiplied" or "divided."

DO EXERCISES 6–10.

[ii] Changes of Unit

Changes of unit can be accomplished by substitutions.

Example 4 Change to inches: 25 yd.

$$\begin{aligned} 25\,yd &= 25 \cdot 1\,yd \\ &= 25 \cdot 3\,ft \quad \text{Substituting 3 ft for 1 yd} \\ &= 25 \cdot 3 \cdot 1\,ft \\ &= 25 \cdot 3 \cdot 12\,in. \quad \text{Substituting 12 in. for 1 ft} \\ &= 900\,in. \end{aligned}$$

DO EXERCISES 11–13.

The notion of "multiplying by one" can also be used to change units.

Example 5 Change to yd: 7.2 in.

$$7.2 \text{ in.} = 7.2 \text{ in.} \cdot \frac{1 \text{ ft}}{12 \text{ in.}} \cdot \frac{1 \text{ yd}}{3 \text{ ft}}$$

Both of these are equal to 1.

$$= \frac{7.2}{12 \cdot 3} \cdot \frac{\text{in.}}{\text{in.}} \cdot \frac{\text{ft}}{\text{ft}} \text{ yd}$$

$$= 0.2 \text{ yd}$$

In Example 5, we first used the following symbol for 1:

$$\frac{1 \text{ ft}}{12 \text{ in.}}$$

"ft" in the numerator is the unit we are changing *to*.

"in." in the denominator is the unit we are changing *from*.

In the final multiplication we were converting from ft to yd, so 1 yd was in the numerator and 3 ft was in the denominator.

DO EXERCISES 14–16.

Example 6 Change to $\frac{\text{m}}{\text{sec}}$: $60 \frac{\text{km}}{\text{h}}$.

$$60 \frac{\text{km}}{\text{h}} = 60 \frac{\text{km}}{\text{h}} \cdot \frac{1000 \text{ m}}{1 \text{ km}} \cdot \frac{1 \text{ h}}{60 \text{ min}} \cdot \frac{1 \text{ min}}{60 \text{ sec}}$$

$$= \frac{60 \cdot 1000}{60 \cdot 60} \cdot \frac{\text{km}}{\text{km}} \cdot \frac{\text{h}}{\text{h}} \cdot \frac{\text{min}}{\text{min}} \cdot \frac{\text{m}}{\text{sec}}$$

$$= 16.67 \frac{\text{m}}{\text{sec}}$$

$$= 16.67 \text{ m/s.}^*$$

Example 7 Change to $\frac{\text{ft}}{\text{sec}}$: $55 \frac{\text{mi}}{\text{hr}}$.

$$55 \frac{\text{mi}}{\text{hr}} = 55 \frac{\text{mi}}{\text{hr}} \cdot \frac{5280 \text{ ft}}{1 \text{ mi}} \cdot \frac{1 \text{ hr}}{60 \text{ min}} \cdot \frac{1 \text{ min}}{60 \text{ sec}}$$

$$= \frac{55 \cdot 5280}{60 \cdot 60} \cdot \frac{\text{mi}}{\text{mi}} \cdot \frac{\text{hr}}{\text{hr}} \cdot \frac{\text{min}}{\text{min}} \cdot \frac{\text{ft}}{\text{sec}}$$

$$= 80 \frac{2}{3} \frac{\text{ft}}{\text{sec}}$$

DO EXERCISES 17–20.

---

*The standard abbreviation for meters per second is m/s and for kilometers per hour is km/h.

Perform the following changes of unit. Use multiplying by one.

14. 720 in., change to yd.

15. 36,960 m, change to km.

16. 360,000 sec, change to hr.

Perform the following changes of unit. Use multiplying by one.

17. $120 \frac{\text{mi}}{\text{hr}}$, change to $\frac{\text{ft}}{\text{sec}}$.

18. 3600 cm^2, change to m^2.

19. $50 \frac{\text{kg}}{\text{L}}$, change to $\frac{\text{g}}{\text{cm}^3}$.

20. $\frac{\$72}{\text{day}}$, change to $\frac{\cent}{\text{hr}}$.

EXERCISE SET 1.11

[i] Perform these calculations and simplify if possible. Do not make any unit changes.

1. $36\,\text{ft} \cdot \dfrac{1\,\text{yd}}{3\,\text{ft}}$

2. $6\,\text{lb} \cdot \dfrac{16\,\text{oz}}{1\,\text{lb}}$

3. $6\,\text{kg} \cdot 8\dfrac{\text{h}}{\text{kg}}$

4. $9\dfrac{\text{km}}{\text{h}} \cdot 3\,\text{h}$

5. $3\,\text{cm} \cdot \dfrac{2\,\text{g}}{2\,\text{cm}}$

6. $\dfrac{9\,\text{km}}{3\,\text{days}} \cdot 6\,\text{days}$

7. $6\,\text{m} + 2\,\text{m}$

8. $10\,\text{tons} + 6\,\text{tons}$

9. $5\,\text{ft}^3 + 7\,\text{ft}^3$

10. $10\,\text{yd}^3 + 17\,\text{yd}^3$

11. $\dfrac{3\,\text{kg}}{5\,\text{m}} \cdot \dfrac{7\,\text{kg}}{6\,\text{m}}$

12. $3\,\text{acres} \times 60\dfrac{1}{\text{acre}}$

13. $\dfrac{2000\,\text{lb} \cdot (6\,\text{mi/hr})^2}{100\,\text{ft}}$

14. $\dfrac{7\,\text{m} \cdot 8\,\text{kg/sec}}{4\,\text{sec}}$

15. $\dfrac{6\,\text{cm}^2 \cdot 5\,\text{cm/sec}}{2\,\text{sec}^2/\text{cm}^2 \cdot 2\dfrac{1}{\text{kg}}}$

16. $\dfrac{320\,\text{lb} \cdot (5\,\text{ft/sec})^2}{2 \cdot 32\dfrac{\text{ft}}{\text{sec}^2}}$

[ii] Perform the following changes of unit, using substitution or multiplying by one.

17. 72 in., change to ft.

18. 17 hr, change to min.

19. 2 days, change to sec.

20. 360 sec, change to hr.

21. $60\dfrac{\text{kg}}{\text{m}}$, change to g/cm.

22. $44\dfrac{\text{ft}}{\text{sec}}$, change to mi/hr.

23. 216 m², change to cm².

24. $60\dfrac{\text{lb}}{\text{ft}^3}$, change to ton/yd³.

25. $\dfrac{\$36}{\text{day}}$, change to $\dfrac{¢}{\text{hr}}$.

26. 1440 man-hr, change to man-days.

27. $1.73\dfrac{\text{mL}}{\text{sec}}$, change to $\dfrac{\text{L}}{\text{hr}}$.

28. $1800\dfrac{\text{g}}{\text{L}}$, change to $\dfrac{\text{cg}}{\text{mL}}$.

(*Hint:* 1 liter = 1 L = 1000 mL = 1000 milliliters.)

29. $186{,}000\dfrac{\text{mi}}{\text{sec}}$ (speed of light), change to $\dfrac{\text{mi}}{\text{yr}}$. Let 365 days = 1 yr.

30. $1100\dfrac{\text{ft}}{\text{sec}}$ (speed of sound), change to $\dfrac{\text{mi}}{\text{yr}}$. Let 365 days = 1 yr.

Use Table 6 at the back of the book to do the following unit changes.

31. ▦ $89.2\dfrac{\text{ft}}{\text{sec}}$, change to m/min.

32. ▦ 1013 yd³, change to m³.

33. ▦ 640 mi², change to km².

34. ▦ $312.2\dfrac{\text{kg}}{\text{m}}$, change to lb/ft.

35. If a steel rod 2 cm long weighs 5 g, how much does a rod of the same type weigh whose length is 3 cm? 5 m?

36. In Exercise 35, how long is a rod that weighs 4.3 g? 20 cg?

In chemistry, 1 *mole* of a substance is that mass of the substance, in grams, equal to its molecular weight. For example, 1 mole of oxygen is 32 grams because its molecular weight is 32, and 1 mole of neon is 20.2 grams.

Convert to grams.

37. 50 moles of oxygen

38. 44 moles of neon

Convert to moles.

39. 303 grams of neon

40. 377.6 grams of oxygen

CHAPTER 1 REVIEW

[1.1, i] Consider the numbers

$$-43.89, \; 12, \; -3, \; -\frac{1}{5}, \; \sqrt{7}, \; \sqrt[3]{10}, \; -1, \; -\frac{4}{3}, \; 7\frac{2}{3}, \; -19, \; 31, \; 0.$$

1. Which are integers? **2.** Which are natural numbers? **3.** Which are rational numbers?

4. Which are real numbers? **5.** Which are irrational numbers? **6.** Which are whole numbers?

[1.1, iii] Compute.

7. $15 + (-19)$ **8.** $-12 + (-4)$ **9.** $-2.5 + (-2.5)$

10. $22 - (-8)$ **11.** $\dfrac{18}{-3}$ **12.** $(-17)(-9)$

13. $-10(20)(-5)(-3)$ **14.** $-\dfrac{15}{16} + \dfrac{3}{4}$ **15.** $\dfrac{5}{12} - \left(-\dfrac{7}{8}\right)$

[1.2, ii] Convert to decimal notation.

16. 3.261×10^6 **17.** 4.1×10^{-4}

[1.2, ii] Convert to scientific notation.

18. 0.01432 **19.** $43{,}210$

[1.2, i] Simplify.

20. $(7a^2b^4)(-2a^{-4}b^3)$ **21.** $\dfrac{54x^6y^{-4}z^2}{9x^{-3}y^2z^{-4}}$

22. $\sqrt[4]{81}$ **23.** $\sqrt[5]{-32}$

[1.7, v] **24.** $\dfrac{b - a^{-1}}{a - b^{-1}}$ **25.** $\dfrac{\dfrac{x^2}{y} + \dfrac{y^2}{x}}{y^2 - xy + x^2}$

[1.9, i] **26.** $(\sqrt{3} - \sqrt{7})(\sqrt{3} + \sqrt{7})$ **27.** $(5x^2 - \sqrt{2})^2$

[1.9, i] **28.** $8\sqrt{5} + \dfrac{25}{\sqrt{5}}$

[1.4, i] **29.** $(x + t)(x^2 - xt + t^2)$ **30.** $(5a + 4b)^3$

[1.3, iv] **31.** $(5xy^4 - 7xy^2 + 4x^2 - 3) - (-3xy^4 + 2xy^2 - 2y + 4)$

[1.5, i] Factor.

32. $x^3 + 2x^2 - 3x - 6$ **33.** $12a^3 - 27ab^4$

34. $24x + 144 + x^2$ **35.** $9x^3 + 35x^2 - 4x$

36. $8x^3 - 1$ **37.** $27x^6 + 125y^6$

[1.10, iv] Write an expression containing a single radical.

38. $\sqrt{y^5}\,\sqrt[3]{y^2}$ **39.** $\dfrac{\sqrt{(a + b)^3}\,\sqrt[3]{a + b}}{\sqrt[6]{(a + b)^7}}$

[1.10, ii] **40.** Convert to radical notation: $b^{7/5}$.

[1.10, i] **41.** Convert to exponential notation and simplify: $\sqrt[8]{\dfrac{m^{32}n^{16}}{3^8}}$.

[1.6, i] Solve.

42. $y^2 - 3y = 18$ **43.** $3[x - 5(4 + 2x)] = 7x - 10(3x - 2)$

44. $(z^2 - 1) + z = 14 - z$ **45.** $(x - 2)(x + 3) + 4 = 0$

[1.6, ii]   Solve.

46. $14 - 4y < 22$ **47.** $(x - 5)(x + 5) \geq (x - 5)(x + 4)$

[1.7, iii]   **48.** Divide and simplify: $\dfrac{3x^2 - 12}{x^2 + 4x + 4} \div \dfrac{x - 2}{x + 2}$.

[1.7, iv]   **49.** Subtract and simplify: $\dfrac{x}{x^2 + 9x + 20} - \dfrac{4}{x^2 + 7x + 12}$.

[1.9, ii]   **50.** Rationalize the denominator: $\dfrac{\sqrt{x} - \sqrt{y}}{\sqrt{x} + \sqrt{y}}$.

[1.11, ii]   **51.** Change to $\dfrac{m}{\min}$: $10 \dfrac{km}{h}$.

What property is illustrated by each sentence?

52. $t + (-t) = 0$ **53.** $8(a + b) = 8a + 8b$

54. $-3(ab) = (-3a)b$ **55.** $tx = xt$

2

EQUATIONS

x x x

x x x

OBJECTIVES

You should be able to:

[i] Determine whether equations are equivalent.

[ii] Solve equations like $2x^3 - x^2 = 3x$ by factoring and solve fractional equations like

$$\frac{14}{x + 2} - \frac{1}{x - 4} = 1.$$

Determine which pairs of equations are equivalent.

1. $3x = 7$
$-15x = -35$

2. $x = 7$
$5x = 35$

3. $x = -5$
$x^2 = 25$

4. $x = -5$
$x + 1 = -4$

5. $\dfrac{(x - 2)(x + 8)}{x - 2} = x + 8$
$\qquad x + 8 = x + 8$

6. $3x^2 = 5x$
$3x = 5$

7. Explain how you would derive $2x + 5x^2 = 5x^2$ from $2x = 0$.

8. Explain how you would derive $2x = 0$ from $2x + 5x^2 = 5x^2$.

2.1 SOLVING EQUATIONS

[i] Equivalent Equations

DEFINITION

Equations that have the same solution set are called *equivalent equations.*

The three equations in Example 1 are equivalent.

Example 1

$$3x = 6 \qquad 3x + 5 = 11 \qquad -12x = -24$$
Solution set $\{2\}$.   Solution set $\{2\}$.   Solution set $\{2\}$.

The pairs of equations in Examples 2 and 3 are *not* equivalent.

Example 2

$$3x = 4x \qquad\qquad \frac{3}{x} = \frac{4}{x}$$

Solution set $\{0\}$.   The solution set is $\emptyset$, the empty set (no solution, since division by 0 is not defined).

Example 3

$$x = 1 \qquad\qquad x^2 = x$$
Solution set $\{1\}$.   Solution set $\{0, 1\}$.

DO EXERCISES 1–6.

Equation-solving Principles

In Chapter 1 we reviewed some equation-solving principles. Let us consider them in more detail.

The Addition Principle

For any real numbers *a*, *b*, and *c*, if an equation $a = b$ is true, then $a + c = b + c$ is true.

Let us look at a simple example illustrating this principle.

$$x + 5 = 9$$
$$x + 5 + (-5) = 9 + (-5) \quad \text{Adding } -5 \text{ on both sides}$$
$$x = 4$$

In this example the original equation and the last equation had exactly the same solutions. They were equivalent. Whenever the steps in an argument are reversible, such will be the case. When we use the addition principle, such will be the case, unless an expression is added having nonsensible replacements. To see that, note that we start with an equation $a = b$ and obtain $a + c = b + c$. By adding $-c$ we can always obtain $a = b$ again, so the steps are reversible.

DO EXERCISES 7 AND 8.

> The use of the addition principle produces an equation equivalent to the original, unless an expression is added having nonsensible replacements. Checking by substituting is therefore not necessary except to detect errors in solving.

Now let us consider the multiplication principle.

The Multiplication Principle
_____

> For any real numbers a, b, and c, if an equation $a = b$ is true, then $a \cdot c = b \cdot c$ is true.

_____

Does the multiplication principle yield equivalent equations? In other words, are the steps reversible? If the number c by which we multiply is not 0, then we can reverse the step by multiplying by $1/c$, but if we multiply by 0, then $1/c$ does not exist and the step is not reversible.

> The use of the multiplication principle produces an equation equivalent to the original, if we multiply by a nonzero number.

[ii] More Equation Solving

Now let us consider the principle of zero products.

The Principle of Zero Products
_____

> For any real numbers a and b, if $ab = 0$, then $a = 0$ or $b = 0$; and if $a = 0$ or $b = 0$, then $ab = 0$.

_____

According to this principle, if we start with an equation $ab = 0$, we obtain $a = 0$ or $b = 0$. Also, if we start with $a = 0$ or $b = 0$, we can obtain $ab = 0$. Thus when we use this principle, that step is reversible, so we have equivalent equations.

> The use of the principle of zero products yields the solutions of the original equation. Checking by substituting is not necessary except to detect errors in solving.

Example 4 Solve $2x^3 - x^2 = 3x$.

$$2x^3 - x^2 - 3x = 0 \quad \text{Addition principle}$$
$$x(2x^2 - x - 3) = 0 \quad \text{Factoring}$$
$$x(2x - 3)(x + 1) = 0$$
$$x = 0 \quad \text{or} \quad 2x - 3 = 0 \quad \text{or} \quad x + 1 = 0 \quad \text{Principle of zero products}$$
$$x = 0 \quad \text{or} \quad x = \frac{3}{2} \quad \text{or} \quad x = -1$$

The solution set is $\{0, \frac{3}{2}, -1\}$.

DO EXERCISES 9–13.

Solve.

9. $x^2 = 5x$

10. $25x^2 + 10x + 1 = 0$

11. $3x^3 - 11x^2 = 4x$

12. $5x^3 + x^2 - 5x - 1 = 0$
   [*Hint:* $x^2(5x + 1) - 1(5x + 1) = 0$.]

13. $3x^3 + x^2 - 12x - 4 = 0$

Solve.

14. $\dfrac{-5}{x} = \dfrac{7}{x}$

15. $\dfrac{x+4}{x+5} = \dfrac{-1}{x+5}$

16. $\dfrac{2x-7}{x+4} = \dfrac{5}{x+4}$

Fractional Equations

Now let us consider equations containing fractional expressions. These are called *fractional equations*. Solutions of such equations often involve multiplying by expressions with variables, as in the following example.

Example 5 Solve $\dfrac{3}{x} = \dfrac{4}{x}$.

$$\frac{3}{x} = \frac{4}{x}$$

$$\frac{3}{x} \cdot x^2 = \frac{4}{x} \cdot x^2 \quad \text{Here we multiplied by } x^2.$$

$$3x = 4x$$

$$0 = x \quad \text{Here we added } -3x \text{ on both sides.}$$

The number 0 is a solution of the last equation but is not a solution of the first. Hence we did not obtain equivalent equations. What can we do in practice?

> When we use the multiplication principle and multiply by an expression with a variable, we may not obtain equivalent equations. We must check possible solutions by substituting in the original equation.

Example 6 Solve $\dfrac{x-3}{x-7} = \dfrac{4}{x-7}$.

$$(x-7) \cdot \frac{x-3}{x-7} = (x-7) \cdot \frac{4}{x-7} \quad \left\{ \begin{array}{l} \text{Caution! Here we multiplied} \\ \text{by an expression with a} \\ \text{variable. Thus we must check.} \end{array} \right.$$

$$x - 3 = 4$$

$$x = 7$$

The possible solution is 7. We check:

$$\frac{x-3}{x-7} = \frac{4}{x-7}$$

$$\begin{array}{c|c} \dfrac{7-3}{7-7} & \dfrac{4}{7-7} \\[2ex] \dfrac{4}{0} & \dfrac{4}{0} \end{array}$$

Division by 0 is undefined; 7 is not a solution. The equation has no solutions. The solution set is ∅.

DO EXERCISES 14–16.

Example 7 Solve $\dfrac{x^2}{x-3} = \dfrac{9}{x-3}$.

$(x-3)\cdot\dfrac{x^2}{x-3} = (x-3)\cdot\dfrac{9}{x-3}$

> Caution! Here we multiplied by an expression with a variable. Thus we must check.

$$x^2 = 9$$
$$x^2 - 9 = 0$$
$$(x+3)(x-3) = 0$$
$$x + 3 = 0 \quad \text{or} \quad x - 3 = 0$$
$$x = -3 \quad \text{or} \quad x = 3$$

The possible solutions are 3 and -3. We must check since we have multiplied by an expression with a variable.

For 3:

$$\dfrac{x^2}{x-3} = \dfrac{9}{x-3}$$

$\dfrac{3^2}{3-3}$	$\dfrac{9}{3-3}$
$\dfrac{9}{0}$	$\dfrac{9}{0}$

3 does not check.

For -3:

$$\dfrac{x^2}{x-3} = \dfrac{9}{x-3}$$

$\dfrac{(-3)^2}{-3-3}$	$\dfrac{9}{-3-3}$
$-\dfrac{9}{6}$	$-\dfrac{9}{6}$

-3 checks.

Thus the solution set is $\{-3\}$.

DO EXERCISES 17 AND 18.

The usual procedure for solving fractional equations involves multiplying on both sides by the LCM of all the denominators. The procedure is called *clearing of fractions*.

Example 8 Solve $\dfrac{14}{x+2} - \dfrac{1}{x-4} = 1$.

We multiply by the LCM of all the denominators: $(x+2)(x-4)$.

$$(x+2)(x-4)\cdot\dfrac{14}{x+2} - (x+2)(x-4)\cdot\dfrac{1}{x-4} = (x+2)(x-4)\cdot 1$$

$$14(x-4) - (x+2) = (x+2)(x-4)$$
$$14x - 56 - x - 2 = x^2 - 2x - 8$$
$$0 = x^2 - 15x + 50$$
$$0 = (x-10)(x-5)$$
$$x = 10 \quad \text{or} \quad x = 5$$

The possible solutions are 10 and 5. These check, so the solution set is $\{10, 5\}$.

DO EXERCISE 19.

Solve.

17. $\dfrac{y^2}{y+4} = \dfrac{16}{y+4}$

18. $\dfrac{x^2}{x-5} = \dfrac{36}{x-5}$

Solve.

19. $\dfrac{4}{x+5} + \dfrac{1}{x-5} = \dfrac{1}{x^2-25}$

EXERCISE SET 2.1

[i] Determine which pairs of equations are equivalent.

1. $3x + 5 = 12$

   $3x = 7$

2. $x^2 = -7x$

   $x = -7$

.3. $x = 3$

   $x^2 = 9$

4. $2y + 1 = -3$

   $8y + 4 = -12$

5. $\dfrac{(x - 2)(x + 8)}{(x - 2)} = x + 8$

   $x + 8 = x + 8$

6. $x^2 + x - 20 = 0$

   $x^2 - 25 = 0$

[ii] Solve.

7. $2x^2 - 6x = 0$

8. $9x^2 + 18x = 0$

9. $3y^3 - 5y^2 - 2y = 0$

10. $3t^3 - 5t^2 + 2t = 0$

11. $(2x - 3)(3x + 2)(x - 1) = 0$

12. $(y - 4)(4y + 12)(2y + 1) = 0$

13. $(2 - 4y)(y^2 + 3y) = 0$

14. $(y^2 - 9)(y^2 - 36) = 0$

15. $\dfrac{x + 2}{2} + \dfrac{3x + 1}{5} = \dfrac{x - 2}{4}$

16. $\dfrac{2x - 1}{3} - \dfrac{x - 2}{5} = \dfrac{x}{2}$

17. $\dfrac{1}{2} + \dfrac{2}{x} = \dfrac{1}{3} + \dfrac{3}{x}$

18. $\dfrac{1}{t} + \dfrac{1}{2t} + \dfrac{1}{3t} = 5$

19. $\dfrac{4}{x^2 - 1} - \dfrac{2}{x - 1} = \dfrac{3}{x + 1}$

20. $\dfrac{3y + 5}{y^2 + 5y} + \dfrac{y + 4}{y + 5} = \dfrac{y + 1}{y}$

21. $\dfrac{1}{2t} - \dfrac{2}{5t} = \dfrac{1}{10t} - 3$

22. $\dfrac{3}{m + 2} + \dfrac{2}{m - 2} = \dfrac{4m - 4}{m^2 - 4}$

23. $1 - \dfrac{3}{x} = \dfrac{40}{x^2}$

24. $1 - \dfrac{15}{y^2} = \dfrac{2}{y}$

25. $\dfrac{11 - t^2}{3t^2 - 5t + 2} = \dfrac{2t + 3}{3t - 2} - \dfrac{t - 3}{t - 1}$

26. $\dfrac{1}{3y^2 - 10y + 3} = \dfrac{6y}{9y^2 - 1} + \dfrac{2}{1 - 3y}$

27. ▦ $3.12x^2 - 6.715x = 0$

28. ▦ $9.25x^2 + 18.03x = 0$

29. ▦ $\dfrac{2.315}{y} - \dfrac{12.6}{17.4} = \dfrac{6.71}{7} + 0.763$

30. ▦ $\dfrac{6.034}{x} - 43.17 = \dfrac{0.793}{x} + 18.15$

☆ ───────────────────────────────

Solve.

31. $5x^3 + x^2 - 5x - 1 = 0$

   [*Hint:* $x^2(5x + 1) - 1(5x + 1) = 0$.]

32. $3x^3 + x^2 - 12x - 4 = 0$

33. $y^3 + 2y^2 - y - 2 = 0$

34. $t^3 + t^2 - 25t - 25 = 0$

35. Determine whether each equation is equivalent to the one that follows:

$$
\begin{array}{ll}
x^2 - x - 20 = x^2 - 25 & (1) \\
(x - 5)(x + 4) = (x - 5)(x + 5) & (2) \\
x + 4 = x + 5 & (3) \\
4 = 5 & (4)
\end{array}
$$

36. Solve: $x^2 - x - 20 = x^2 - 25$.

Equations that are true for all sensible replacements of the variables are called *identities*. Determine which equations are identities.

37. $\dfrac{x^2 + 6x - 16}{x - 2} = x + 8$

38. $x + 4 = 4 + x$

39. $(x - 1)(x^2 + x + 1) = x^3 - 1$

40. $\dfrac{x^3 + 8}{x^2 - 4} = \dfrac{x^2 - 2x + 4}{x - 2}$

41. $(x + 7)^2 = x^2 + 49$

42. $\sqrt{x^2 - 16} = x - 4$

★ ───────────────────────────────

43. Solve: $\dfrac{x + 3}{x + 2} - \dfrac{x + 4}{x + 3} = \dfrac{x + 5}{x + 4} - \dfrac{x + 6}{x + 5}$.

2.2 FORMULAS AND APPLIED PROBLEMS

[i] Formulas

A formula is a recipe for doing a calculation. An example is $A = \pi rs + \pi r^2$, which gives the area A of a cone in terms of the slant height s and radius of the base r.

Suppose we wanted to find the slant height s when the area A and radius r are known. Our knowledge of equations allows us to get s alone on one side, or as we say, "solve the formula for s."

Example 1 Solve $A = \pi rs + \pi r^2$, for s.

$$A - \pi r^2 = \pi rs$$
$$\frac{A - \pi r^2}{\pi r} = s \quad \frac{A - \pi r^2}{\pi r} \text{ could also be expressed as } \frac{A}{\pi r} - r.$$

Example 2 Solve $\frac{1}{R} = \frac{1}{r_1} + \frac{1}{r_2}$, for R. (This is a formula from electricity.)

Resistance in parallel:
$$\frac{1}{R} = \frac{1}{r_1} + \frac{1}{r_2}$$

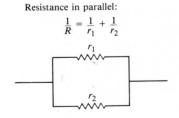

We first multiply by the LCM, which is Rr_1r_2:

$$Rr_1r_2 \cdot \frac{1}{R} = Rr_1r_2 \cdot \left(\frac{1}{r_1} + \frac{1}{r_2}\right)$$

$$Rr_1r_2 \cdot \frac{1}{R} = Rr_1r_2 \cdot \frac{1}{r_1} + Rr_1r_2 \cdot \frac{1}{r_2} \quad \text{Using a distributive law}$$

$$r_1r_2 = Rr_2 + Rr_1 \quad \text{Simplifying}$$

$$r_1r_2 = (r_2 + r_1)R \quad \text{Factoring out the common factor } R$$

$$\frac{r_1r_2}{r_2 + r_1} = R.$$

DO EXERCISES 1 AND 2.

1. Solve $C = \frac{5}{9}(F - 32)$, for F.

(This is a formula for Celsius, or Centigrade, temperature in terms of Fahrenheit temperature F.)

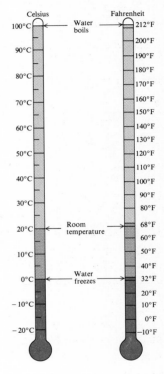

2. Solve $\frac{1}{R} = \frac{1}{r_1} + \frac{1}{r_2}$, for r_2.

[ii] **Applied Problems**

By an *applied problem* we mean a problem in which mathematical techniques are used to answer some question. Problems like this may be posed orally. They may come about in the course of a conversation, or they can be hatched within the mind of one person. Thus to call them "word problems" or "story problems" is misleading.

There is no rule that will enable us to solve applied problems, because they are of many different kinds. We can, however, describe an overall, or general, strategy.

The idea is to translate the problem situation to mathematical language and then calculate to find a solution.

GENERAL STRATEGY FOR SOLVING
APPLIED PROBLEMS

1. **Become familiar with the problem situation. If the problem is presented to you in written words, then of course this means to read carefully.**

   a) **Make a drawing, if it makes sense to do so. It is difficult to overemphasize the importance of this!**

   b) **Make a written list of the known facts and a list of what you wish to find out.**

2. **Translate the problem situation to mathematical language or symbolism. For most of the problems you will encounter in algebra this means to write one or more equations. You will of course also assign certain letters to represent unknown quantities.**

3. **Use your mathematical knowledge to find a possible solution. In algebra this usually means to solve an equation or system of equations.**

4. **Check to see whether your possible solution actually fits the problem situation, and is thus really a solution of the problem.**

Problems stated in textbooks are of necessity somewhat contrived. Problems that you encounter in nonclassroom situations will almost invariably contain insufficient information to obtain a firm, or exact, answer. They also usually contain a good deal of extraneous, useless information. Here is an example that illustrates this point.

Example 3 There are 145 persons in the junior class at Notown College. The prom committee wishes to know how much they will have to charge each person who attends the spring prom, to be held May 19 in the college gymnasium at 8:00 P.M.

The above problem is typical of problems encountered in real life. There is insufficient information to obtain an answer. To get an answer we need to know the following.

1. The cost of holding the prom. This would include such things as cost of music, refreshments, and hall rent.

2. The number of persons who will attend.

The committee will need to find out information about costs, and in the course of finding it, they may revise their formulation of the problem. For example, if a band costs too much, they may have to settle for using records. In any event, they may find that they can determine costs only approximately; in other words, they must *estimate*.

The number 145 may help in estimating the number who will attend, but at best this number will be an estimate. Thus, on the basis of the best information and/or estimates available, the committee will do its arithmetic.

Note that the date, time, and place, while they are interesting information, do not contribute to the solution of the problem. Thus this is extraneous information.

In the following examples we illustrate how a problem situation can be translated to mathematical language. In certain simple situations, the translation is easy because certain words translate directly to mathematical symbols. Note how the word "is" translates to an equals sign, the word "what" translates to a variable, and the word "of" translates to a multiplication sign.

Example 4 What percent of 84 is 11.76?

Translate: $x\%$ $\cdot\ 84 = 11.76$

Solve: $x \cdot 0.01$ $\cdot\ 84 = 11.76$ "%" means "$\cdot\ 0.01$"

      $x \cdot 0.84 = 11.76$

       $x = \dfrac{11.76}{0.84} = 14$

Check: $14\% \cdot 84 = 0.14 \cdot 84 = 11.76$

Example 5 14% of what is 11.76?

Translate: $14\% \ \cdot \quad y \ = 11.76$

Solve: $0.14 \quad \cdot \quad y \ = 11.76$

     $y \ = \dfrac{11.76}{0.14} = 84$

Check: $14\% \cdot 84 = 0.14 \cdot 84 = 11.76$

DO EXERCISES 3–5.

In the remainder of this section we consider applied problems of various types. Although there is no rule for solving applied problems because they can be so different, it does help somewhat to consider a few different types of problems. *The best way to learn to solve applied problems is to solve a lot of them.*

3. What percent of 76 is 13.68?

4. 24% of what is 13.68?

5. 36% of 75 is what?

6. An investment is made at 14%, compounded annually. It grows to $2793 at the end of one year. How much was originally invested?

Compound Interest Problems

Example 6 An investment is made at 16%, compounded annually. It grows to $1740 at the end of one year. How much was originally invested?

There is more than one way to translate the problem situation to mathematical language. The following is one method. We first restate the situation as follows:

The invested amount *plus* the interest is $1740.

Now the interest is 16% of the amount invested, so we have the following, which translates directly:

$$\underbrace{\text{Invested amount}}_{x} \underbrace{\text{plus}}_{+} \underbrace{16\%}_{16\%} \text{of} \underbrace{\text{Invested amount}}_{x} \underbrace{\text{is}}_{=} \underbrace{\$1740}_{1740.}$$

Now we solve the equation:

$$x + 16\%x = 1740$$
$$x + 0.16x = 1740$$
$$(1 + 0.16)x = 1740$$
$$1.16x = 1740$$
$$x = \frac{1740}{1.16} = 1500.$$

The number 1500 checks in the problem situation, so the answer is $1500.

DO EXERCISE 6.

Let us now consider an investment over a period longer than one year. If we invest P dollars at an interest rate i, compounded annually, we will have an amount in the account at the end of a year that we will call A_1. Now $A_1 = P + Pi$, or

$$A_1 = P(1 + i), \quad \text{or}$$
$$A_1 = Pr, \quad \text{where we have let } r = 1 + i.$$

Going into the second year we have Pr dollars. By the end of the second year we will have A_2 dollars, given by

$$A_2 = A_1 \cdot r.$$

But $A_1 = Pr$, so $A_2 = (Pr)r$, or

$$A_2 = Pr^2.$$

Similarly, the amount A_3 in the account at the end of three years is given by

$$A_3 = Pr^3,$$

and so on. In general, the following applies.

THEOREM 1

If principal P is invested at an interest rate i, compounded annually, in t years it will grow to an amount A given by

$$A = P(1 + i)^t.$$

▦ **Example 7** Suppose $1000 is invested at 12%, compounded annually. What amount will be in the account at the end of ten years?

We use the equation $A = P(1 + i)^t$. We get

$$A = 1000(1 + 0.12)^{10} = 1000(1.12)^{10} \quad \text{Use the } x^y \text{ key.}$$
$$\approx 3105.85.$$

The answer is $3105.85.

DO EXERCISE 7.

Interest may be compounded more often than once a year. Suppose it is compounded four times a year, or *quarterly*. The formula derived above can be altered to apply. We consider one-fourth of a year to be an *interest period*. The rate of interest for such a period is then $i/4$. The number of periods will be four times the number of years. This is shown in the following diagram.

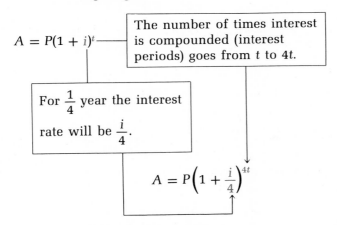

Now suppose the number of interest periods per year is something other than 4, say n. Using the reasoning of the example above, we obtain a general formula.

THEOREM 2

If principal P is invested at an interest rate i, compounded n times per year, in t years it will grow to an amount A given by

$$A = P\left(1 + \frac{i}{n}\right)^{nt}.$$

When problems involving compound interest are translated to mathematical language, the preceding formula is almost always used.

7. ▦ Suppose $1000 is invested at 12.5%, compounded annually. What amount will be in the account at the end of eight years?

8. ▦ Suppose $1000 is invested at 12.5% compounded semiannually ($n = 2$). How much will be in the account at the end of eight years?

9. A rectangular garden is 60 ft by 80 ft. Part of the garden is torn up to install a sidewalk of uniform width around the garden. The new area of the garden is $\frac{1}{6}$ of the old area. How wide is the sidewalk?

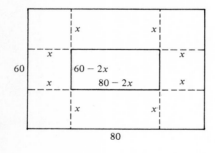

▦ **Example 8** Suppose $1000 is invested at 12%, compounded quarterly. How much will be in the account at the end of ten years?

In this case $n = 4$ and $t = 10$. We substitute into the formula:

$$A = P\left(1 + \frac{i}{n}\right)^{nt} = 1000\left(1 + \frac{0.12}{4}\right)^{4 \cdot 10}$$
$$= 1000(1.03)^{40} \quad \text{Use the } x^y \text{ key.}$$
$$\approx 3262.04.$$

The answer is $3262.04.

DO EXERCISE 8.

Area Problems

Example 9 The radius of a circular swimming pool is 10 ft. A sidewalk of uniform width is constructed around the outside and has an area of 44π ft². How wide is the sidewalk?

First make a drawing:

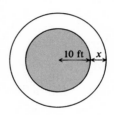

Let x represent the width of the walk. Then, recalling that a formula for the area of a circle is $A = \pi r^2$, we have

    Area of pool $= \pi \cdot 10^2 = 100\pi$;
    Area of sidewalk plus pool $= \pi \cdot (10 + x)^2 = \pi(100 + 20x + x^2)$.

Thus

$$\underbrace{\text{(Area of sidewalk}}_{\pi(100 + 20x + x^2)} - \underbrace{\text{(Area of}}_{100\pi} = \underbrace{\text{Area of}}_{44\pi}$$

$$100 + 20x + x^2 - 100 = 44 \quad \text{Multiplying by } \frac{1}{\pi}$$
$$x^2 + 20x = 44$$
$$x^2 + 20x - 44 = 0$$
$$(x + 22)(x - 2) = 0$$
$$x = -22 \quad \text{or} \quad x = 2.$$

We see that -22 ft is not a solution of the original problem since width has to be positive. The number 2 checks; that is, when the sidewalk is 2 ft wide, the area of the pool is 100π ft². The area of the pool plus sidewalk is $\pi \cdot 12^2$, or 144π ft², so the area of the sidewalk is 44π ft².

DO EXERCISE 9.

Motion Problems

For problems that deal with distance, time, and speed, we almost always need to recall the definition of speed, or something equivalent to it.

Speed = distance/time or $r = d/t$, where r = speed, d = distance, and t = time.

If you memorize $r = d/t$, you can easily obtain either of the two equivalent equations $d = rt$ or $t = d/r$, as needed.

When translating these problems to mathematical language it is often helpful to look for some quantity that is constant in the problem. For example, two cars may travel for the same length of time, or two boats may travel the same distance. Such facts often provide the basis for setting up an equation.

Example 10 A boat travels 246 km downstream in the same time that it takes to travel 180 km upstream. The speed of the current in the stream is 5.5 km/h. Find the speed of the boat in still water.

We first make a drawing and lay out the known facts, and any other information that is pertinent.

$$\overrightarrow{\hspace{5cm}}\text{Downstream}$$

$d_1 = 246$ t_1 unknown r_1 unknown, but equals speed of boat plus speed of current

$$\overleftarrow{\hspace{5cm}}\text{Upstream}$$

$d_2 = 180$ t_2 unknown, but same as t_1 r_2 unknown, but equals speed of boat minus speed of current

In this problem the times are the same, so we have an equation

$$t_1 = t_2.$$

Now, remembering that $t = d/r$, we substitute to obtain

$$\frac{d_1}{r_1} = \frac{d_2}{r_2}.$$

Now d_1, the distance downstream, is 246, and d_2 is 180.

The speed downstream is the speed of the boat *plus* that of the current.

Let's call the boat's speed in still water r.

Then r_1, the speed downstream, is $r + 5.5$, and r_2, the speed upstream, is $r - 5.5$. This gives us the equation

$$\frac{246}{r + 5.5} = \frac{180}{r - 5.5}.$$

10. An airplane flies 1062 km with the wind in the same time that it takes to fly 738 km against the wind. The speed of the plane in still air is 200 km/h. Find the speed of the wind.

11. A train leaves a station and travels north at 75 km/h. Two hours later a second train leaves on a parallel track traveling north at 125 km/h. How far from the station will the second train overtake the first train?

12. A train leaves Oldtown and travels 300 km to Newtown. On the return trip it travels 10 km/h faster. The total time for the round trip was 11 hr. How fast did the train travel on each part of the trip?

Solving for r we get 35.5 km/h. This checks. Thus the speed of the boat in still water is 35.5 km/h.

DO EXERCISE 10.

In the following example, although the distance is the same in two cases, an additional piece of information is given that is the key to the translation.

Example 11 The speed of a boat is still water is 10 mph. It travels 24 miles upstream and 24 miles downstream in a total time of 5 hr. What is the speed of the current?

We first make a drawing and write out pertinent information.

$$\longrightarrow \text{Downstream}$$

| $d_1 = 24$ mi | t_1 unknown | r_1 unknown, but is 10 mph plus speed of current |

$$\longleftarrow \text{Upstream}$$

| $d_2 = 24$ mi | t_2 unknown, but $t_1 + t_2 = 5$ hr | r_2 unknown, but is 10 mph minus speed of current |

This time the basis of our translation is the equation involving times:

$$t_1 + t_2 = 5.$$

This then leads to

$$\frac{d_1}{r_1} + \frac{d_2}{r_2} = 5,$$

and then

$$\frac{24}{10 + r} + \frac{24}{10 - r} = 5,$$

where r is the speed of the current. Solving for r we get $r = -2$ or $r = 2$. Since speed cannot be negative in this problem, -2 cannot be a solution. But 2 checks, so the speed of the current is 2 mph.

DO EXERCISES 11 AND 12.

Work Problems

Suppose an employee can do a certain job in 4 hr. Then the employee can do $\frac{1}{4}$ of it in 1 hr and $\frac{3}{4}$ of it in 3 hr. The basic principle for translating work problems follows.

> If a job can be done in time t, then $1/t$ of it can be done in 1 unit of time.

Example 12 Typist A can do a certain job in 3 hr. Typist B can do the same job in 5 hr. How long would it take both, working together, to do the same job?

a) Make a guess at a solution. Does 4 hr seem reasonable? Sometimes our intuition can fool us. Clearly the answer is less than 3 hr because one can do the job alone in 3 hr.

b) A can do the job in 3 hr. Thus in 1 hr $\frac{1}{3}$ of it can be done. Similarly, B can do $\frac{1}{5}$ of the job in 1 hr. Thus together they can do $\frac{1}{3} + \frac{1}{5}$ of it in 1 hr. Let t represent the amount of time it takes to do the job if they work together. Then together they can do $1/t$ of the job in 1 hr. Thus

$$\frac{1}{3} + \frac{1}{5} = \frac{1}{t}$$

$$5t + 3t = 5 \cdot 3 \quad \text{Multiplying by the LCM, } 3 \cdot 5 \cdot t$$

$$8t = 15$$

$$t = \frac{15}{8}, \quad \text{or} \quad 1\frac{7}{8}\,\text{hr.}$$

And $t = 1\frac{7}{8}$ hr checks. Thus it takes them $1\frac{7}{8}$ hr to do the job if they work together.

DO EXERCISE 13.

Example 13 It takes Red 9 hr longer to build a wall than it takes Mort. If they work together they can build the wall in 20 hr. How long would it take each, working alone, to build the wall?

Let $t =$ the amount of time it takes Mort working alone. Then $t + 9 =$ the amount of time it takes Red working alone.

Then Mort can do $1/t$ of the work in 1 hr and Red can do $1/(t + 9)$ of it in 1 hr. Together they can do $1/t + 1/(t + 9)$ of the work in 1 hr. We also know that they can do $\frac{1}{20}$ of the work in 1 hr. Thus,

$$\frac{1}{t} + \frac{1}{t + 9} = \frac{1}{20}.$$

Solving the equation, we get $t = -5$ or $t = 36$. Since negative time has no meaning in the problem, -5 is not a solution of the original problem. The number 36 checks in the original problem. Thus it would take Mort 36 hr and Red 45 hr.

DO EXERCISE 14.

13. A can mow a lawn in 4 hours. B can mow the same lawn in 5 hours. How long will it take them to mow the lawn if they work together?

14. Fran and Helen work together and get a certain job completed in 4 hr. It would take Helen 6 hr longer, working alone, to do the job than it would Fran. How long would each need to do the job working alone?

EXERCISE SET 2.2

[i]

1. Solve $P = 2l + 2w$, for w.

2. Solve $F = ma$, for a.

3. Solve $E = IR$, for I.

4. Solve $F = \dfrac{km_1 m_2}{d^2}$, for m_2.

5. Solve $\dfrac{P_1 V_1}{T_1} = \dfrac{P_2 V_2}{T_2}$, for T_1.

6. Solve $\dfrac{P_1 V_1}{T_1} = \dfrac{P_2 V_2}{T_2}$, for V_2.

7. Solve $S = \dfrac{H}{m(v_1 - v_2)}$, for v_1.

8. Solve $S = \dfrac{H}{m(v_1 - v_2)}$, for v_2.

9. Solve $\frac{1}{F} = \frac{1}{m} + \frac{1}{p}$, for p.

10. Solve $\frac{1}{F} = \frac{1}{m} + \frac{1}{p}$, for F.

Solve for x.

11. $(x + a)(x - b) = x^2 + 5$

12. $(c + d)x + (c - d)x = c^2$

13. $10(a + x) = 8(a - x)$

14. $4(a + b + x) + 3(a + b - x) = 8a$

[ii] Applied problems

15. 79.2 is what percent of 180?

16. 6% of what number is 480?

17. What percent of 28 is 1.68?

18. What is 7% of 45.3?

19. A person gets an 11% raise that is $1595. What was the old salary? the new salary?

20. A person gets a 12% raise that is $2520. What was the old salary? the new salary?

21. An investment is made at 13%, compounded annually. It grows to $734.50 at the end of one year. How much was originally invested?

22. An investment is made at 14%, compounded annually. It grows to $912 at the end of one year. How much was originally invested?

23. In triangle ABC, angle B is five times as large as angle A. The measure of angle C is 2° less than that of angle A. Find the measures of the angles. (*Hint:* The sum of the angle measures is 180°.)

24. In triangle ABC, angle B is twice as large as angle A. Angle C measures 20° more than angle A. Find the measures of the angles.

25. The perimeter of a rectangle is 322 m. The length is 25 m more than the width. Find the dimensions.

26. The length of a rectangle is twice the width. The perimeter is 39 m. Find the dimensions.

27. A student's scores on three tests are 87%, 64%, and 78%. What must the student score on the fourth test so that the average will be 80%?

28. A student's scores on three tests are 74%, 55%, and 68%. What must the student score on the fourth test so that the average will be 70%?

29. An open box is made from a 10-cm by 20-cm piece of tin by cutting a square from each corner and folding up the edges. The area of the resulting base is 96 cm². What is the length of the sides of the squares?

30. The frame of a picture is 28 cm by 32 cm outside and is of uniform width. What is the width of the frame if 192 cm² of the picture shows?

31. After a 2% increase, the population of a city is 826,200. What was the former population?

32. After a 3% increase, the population of a city is 741,600. What was the former population?

33. A boat goes 50 km downstream in the same time that it takes to go 30 km upstream. The speed of the stream is 3 km/h. Find the speed of the boat in still water.

34. A boat goes 50 km downstream in the same time that it takes to go 30 km upstream. The speed of the boat in still water is 16 km/h. Find the speed of the stream.

35. The speed of train A is 12 mph slower than the speed of train B. Train A travels 230 mi in the same time it takes train B to travel 290 mi. Find the speed of each train.

36. The speed of a passenger train is 14 mph faster than the speed of a freight train. The passenger train travels 400 mi in the same time it takes the freight train to travel 330 mi. Find the speed of each train.

37. An airplane leaves Chicago at a speed of 475 mph. Twenty minutes later a plane leaves Cleveland, which is 350 miles from Chicago, at 500 mph. When they meet, how far are they from Cleveland? (*Hint:* It is usually best to make the units consistent. That is, consider 20 min as $\frac{1}{3}$ hr.)

38. A private airplane leaves an airport and flies due east at 180 km/h. Two hours later a jet leaves the same airport and flies due east at 900 km/h. How far from the airport will the jet overtake the private plane?

39. A can do a certain job in 3 hr, B can do the same job in 5 hr, and C can do the same job in 7 hr. How long would the job take with all working together?

40. Pipe A can fill a tank in 4 hr, pipe B can fill it in 10 hr, and pipe C can fill it in 12 hr. The pipes are connected to the same tank. How long does it take to fill the tank if all three are running together?

41. ▦ A can do a certain job, working alone, in 3.15 hr. A working with B can do the same job in 2.09 hr. How long would it take B, working alone, to do the job?

42. ▦ At a factory, smokestack A pollutes the air 2.13 times as fast as smokestack B. When both stacks operate together they yield a certain amount of pollution in 16.3 hr. Find the amount of time it would take each to yield the same amount of pollution if it operated alone.

43. Suppose $1000 is invested at $13\frac{3}{4}$%. How much is in the account at the end of one year if interest is compounded (a) annually? (b) semiannually? (c) quarterly? (d) daily (use 365 days per yr)? (e) hourly?

44. Suppose $1000 is invested at 15.5%. How much is in the account at the end of five years, if interest is compounded (a) annually? (b) semiannually? (c) quarterly? (d) daily (use 365 days per yr)? (e) hourly?

☆ ————————————————————————————————

45. A car is driven 144 mi. If it had gone 4 mph faster it could have made the trip in $\frac{1}{2}$ hr less time. What was the speed?

46. A car is driven 280 mi. If it had gone 5 mph faster it could have made the trip in 1 hr less time. What was the speed?

★ ————————————————————————————————

47. A commuter drives to work at 45 mph and arrives one minute early. At 40 mph, the commuter would arrive one minute late. How far is it to work?

48. Suppose that b is 20% more than a, c is 25% more than b, and d is k% less than c. Find k such that $a = d$.

2.3 QUADRATIC EQUATIONS

[i] Solving Quadratic Equations

Solving $ax^2 = k$

DEFINITION

A *quadratic equation*, also known as an equation of *second degree*, is any equation equivalent to the following:

$$ax^2 + bx + c = 0, \quad \text{where } a \neq 0.$$

The above is the *standard form* of a quadratic equation. You have solved such equations by factoring. We now consider other methods. Let us first consider $x^2 = 3$. We shall solve this equation two ways.

Example 1 Solve $x^2 = 3$.

Method 1
$$x^2 = 3$$
$$x^2 - 3 = 0$$
$$(x - \sqrt{3})(x + \sqrt{3}) = 0 \quad \text{Factoring}$$
$$x - \sqrt{3} = 0 \quad \text{or} \quad x + \sqrt{3} = 0 \quad \text{Principle of zero products}$$
$$x = \sqrt{3} \quad \text{or} \quad x = -\sqrt{3}$$

The solutions are $\sqrt{3}$ and $-\sqrt{3}$. The solution set is $\{\sqrt{3}, -\sqrt{3}\}$. Remember, you have not solved an equation unless you find *all* the solutions. An answer of $\{\sqrt{3}\}$ is incorrect!

Method 2
$$x^2 = 3$$
$$\sqrt{x^2} = \sqrt{3} \quad \text{Taking principal square root}$$
$$|x| = \sqrt{3}$$
$$x = \sqrt{3} \quad \text{or} \quad -x = \sqrt{3} \quad \text{Definition of absolute value}$$
$$x = \sqrt{3} \quad \text{or} \quad x = -\sqrt{3}$$

The solutions are $\sqrt{3}$ and $-\sqrt{3}$. This is often abbreviated as $\pm\sqrt{3}$.

In general, the equation $x^2 = p$, where p is nonnegative, has two solutions, $\sqrt{p}$ and $-\sqrt{p}$, if $p > 0$. If $p = 0$, there is just one solution. If $p < 0$, there are no real-number solutions.

DO EXERCISES 1–4.

Solve for x.

1. $x^2 = 3$

2. $5x^2 = 0$

3. $3x^2 = \pi$

4. $mx^2 = n$

Solve.

5. $(x + 4)^2 = 7$

6. $(x - 5)^2 = 3$

7. $(x + 5)^2 = 4$

We now use this method in another example.

Example 2 Solve $(x + 5)^2 = 3$.

Taking the principal square root, we have

$$|x + 5| = \sqrt{3}.$$

Then

$$x + 5 = \pm\sqrt{3}$$
$$x = -5 \pm \sqrt{3}.$$

The solution set is $\{-5 - \sqrt{3}, -5 + \sqrt{3}\}$.

DO EXERCISES 5–7.

Completing the Square

If the equation is not in the form $(x - h)^2 = k$, we can put it into that form by *completing the square*.

Example 3 Solve by completing the square: $x^2 - 6x - 12 = 0$.

We consider first the x^2- and x-terms.

$$x^2 - 6x \quad - 12 = 0$$

We construct a trinomial square. First note that $x^2 - 6x + 9$ is a perfect square: $(x - 3)^2$. So if we add 9 to $x^2 - 6x$ we will have a perfect square. But, to get an equivalent equation we have to subtract 9 as well. This is the same as adding $9 - 9$, or 0. Then we proceed as follows:

$$x^2 - 6x + 9 - 9 - 12 = 0$$
$$(x - 3)^2 - 21 = 0$$
$$(x - 3)^2 = 21$$
$$x - 3 = \pm\sqrt{21}$$
$$x = 3 \pm \sqrt{21}.$$

The solution set is $\{3 + \sqrt{21}, 3 - \sqrt{21}\}$, which we often abbreviate as $3 \pm \sqrt{21}$.

The following is a general procedure we use called *completing the square*.

To solve $ax^2 + bx + c = 0$, by *completing the square*:

1. If $a \neq 1$, multiply both sides of the equation by $1/a$ to get the equation in the form $x^2 + Bx + C = 0$.
2. To complete the square on $x^2 + Bx$, take half the coefficient of x and square it. Then add and subtract that number: $(B/2)^2$.
3. Take the square roots and solve for x.

Example 4 Solve by completing the square: $x^2 + 3x - 5 = 0$.

1. Since $a = 1$, no multiplication by $1/a$ is necessary to get the equation in proper form: $x^2 + 3x - 5 = 0$.

2. To complete the square on $x^2 + 3x$, we take half the coefficient of x and square it. Then we add and subtract that number: $(\frac{3}{2})^2$, or $\frac{9}{4}$.

$$x^2 + 3x + \frac{9}{4} - \frac{9}{4} - 5 = 0$$

$$\left(x + \frac{3}{2}\right)^2 - \frac{29}{4} = 0$$

$$\left(x + \frac{3}{2}\right)^2 = \frac{29}{4}$$

3. Take the square roots and solve for x.

$$x + \frac{3}{2} = \pm\sqrt{\frac{29}{4}} = \pm\frac{\sqrt{29}}{2}$$

$$x = -\frac{3}{2} \pm \frac{\sqrt{29}}{2} = \frac{-3 \pm \sqrt{29}}{2}$$

The solutions are $\dfrac{-3 + \sqrt{29}}{2}$ and $\dfrac{-3 - \sqrt{29}}{2}$, or $\dfrac{-3 \pm \sqrt{29}}{2}$.

DO EXERCISES 8–16.

Example 5 Solve by completing the square: $2x^2 - 3x - 1 = 0$.

1. Since $a \neq 1$, we multiply on both sides by $1/a$, which is $\frac{1}{2}$:

$$2x^2 - 3x - 1 = 0$$

$$x^2 - \frac{3}{2}x - \frac{1}{2} = 0. \quad \text{Multiplying by } \frac{1}{2}$$

2. To complete the square on $x^2 - \frac{3}{2}x$, we take half the coefficient of x and square it. Then we add and subtract that number: $(\frac{1}{2} \cdot \frac{3}{2})^2$, or $\frac{9}{16}$.

$$x^2 - \frac{3}{2}x + \frac{9}{16} - \frac{9}{16} - \frac{1}{2} = 0$$

$$\left(x - \frac{3}{4}\right)^2 - \frac{17}{16} = 0$$

$$\left(x - \frac{3}{4}\right)^2 = \frac{17}{16}$$

3. Take the square roots and solve for x:

$$x - \frac{3}{4} = \pm\sqrt{\frac{17}{16}} = \pm\frac{\sqrt{17}}{4}$$

$$x = \frac{3}{4} \pm \frac{\sqrt{17}}{4} = \frac{3 \pm \sqrt{17}}{4}.$$

The solution set is $\left\{\dfrac{3 + \sqrt{17}}{4}, \dfrac{3 - \sqrt{17}}{4}\right\}$.

DO EXERCISES 17 AND 18.

Find the term that completes the square, then fill in the second expression.

8. $x^2 + 4x +$ ___4___ $= (x + 2)^2$

9. $x^2 - 6x +$ ___9___ $= (x - 3)^2$

10. $x^2 + 5x +$ ___$\frac{25}{4}$___ $= (x + \frac{5}{2})^2$

11. $x^2 - 7x +$ ___$\frac{49}{4}$___ $= (x - \frac{7}{2})^2$

12. $x^2 + \frac{3}{4}x +$ ___$\frac{9}{64}$___ $= (x + \frac{3}{8})^2$

13. $x^2 - x +$ ___$\frac{1}{4}$___ $= (x - \frac{1}{2})^2$

Solve by completing the square.

14. $x^2 + 4x - 3 = 0$

15. $x^2 - 6x + 8 = 0$

16. $x^2 - 5x + 6 = 0$

Solve by completing the square.

17. $2x^2 + 2x - 3 = 0$

18. $4x^2 + 3x - 1 = 0$

The Quadratic Formula

We studied completing the square for two reasons. The most important is that it is a useful tool in other places in mathematics. The second reason is that it can be used to derive a general formula for solving quadratic equations, called the quadratic formula. You will be asked to give such a derivation in the exercises (see Exercise 64). In the following space we give another proof of the quadratic formula. We consider the standard form of the quadratic equation, with unspecified coefficients. Solving that equation gives us the formula we seek. We begin with

$$ax^2 + bx + c = 0.$$

We multiply on both sides by $4a$, to obtain

$$4a^2x^2 + 4abx + 4ac = 0.$$

Next we add $b^2 - b^2$ on the left and rearrange, as follows:

$$(4a^2x^2 + 4abx + b^2) - (b^2 - 4ac) = 0,$$

or

$$(2ax + b)^2 - (b^2 - 4ac) = 0.$$

The left side can be factored as a difference of squares, after which we use the principle of zero products.

$$[(2ax + b) - \sqrt{b^2 - 4ac}][(2ax + b) + \sqrt{b^2 - 4ac}] = 0$$

$$2ax + b - \sqrt{b^2 - 4ac} = 0 \quad \text{or} \quad 2ax + b + \sqrt{b^2 - 4ac} = 0$$

$$x = \frac{-b + \sqrt{b^2 - 4ac}}{2a} \quad \text{or} \quad x = \frac{-b - \sqrt{b^2 - 4ac}}{2a}.$$

We can abbreviate the last line using the sign $\pm$. This gives us the quadratic formula.

THEOREM 3

The quadratic formula. If a quadratic equation $ax^2 + bx + c = 0$ has solutions, they are given by

$$x = \frac{-b \pm \sqrt{b^2 - 4ac}}{2a}.$$

When using the quadratic formula it is helpful to first find the standard form so that the coefficients a, b, and c can be determined.

Example 6 Solve $3x^2 + 2x = 7$.

First find the standard form and determine a, b, and c:

$$3x^2 + 2x - 7 = 0,$$
$$a = 3, \quad b = 2, \quad c = -7.$$

Then use the quadratic formula:

$$x = \frac{-b \pm \sqrt{b^2 - 4ac}}{2a} = \frac{-2 \pm \sqrt{2^2 - 4 \cdot 3 \cdot (-7)}}{2 \cdot 3}$$

$$x = \frac{-2 \pm \sqrt{4 + 84}}{6} \qquad \text{To prevent careless mistakes it helps to write out } \textit{all} \text{ the steps.}$$

$$x = \frac{-2 \pm \sqrt{88}}{6} = \frac{-2 \pm \sqrt{4 \cdot 22}}{6}$$

$$= \frac{-2 \pm 2\sqrt{22}}{6} = \frac{2(-1 \pm \sqrt{22})}{2 \cdot 3} = \frac{-1 \pm \sqrt{22}}{3}.$$

Should such irrational solutions arise in an applied problem, we can find approximations using Table 1 at the end of the book or a calculator:

$$\frac{-1 + \sqrt{22}}{3} \approx \frac{-1 + 4.69}{3}$$

$$\approx \frac{3.69}{3} \approx 1.2; \quad \text{Rounded to the nearest tenth}$$

$$\frac{-1 - \sqrt{22}}{2} \approx \frac{-1 - 4.69}{3}$$

$$\approx \frac{-5.69}{3} \approx -1.9. \quad \text{Rounded to the nearest tenth}$$

The solutions of a quadratic equation can *always* be found using the quadratic formula. They are not always easy to find by factoring. A general strategy for solving quadratic equations is as follows.

a) Try factoring.

b) If factoring seems difficult, use the quadratic formula; it *always* works!

DO EXERCISES 19 AND 20.

[ii] The Discriminant

The expression $b^2 - 4ac$, called the *discriminant*, determines the nature of the solutions. If x_1 and x_2 represent the solutions, then

$$x_1 = \frac{-b + \sqrt{b^2 - 4ac}}{2a}, \quad x_2 = \frac{-b - \sqrt{b^2 - 4ac}}{2a}.$$

Note that if $b^2 - 4ac = 0$, then x_1 and x_2 are the same, $-b/2a$. If $b^2 - 4ac > 0$, then the expression $\sqrt{b^2 - 4ac}$ represents a positive real number that is to be added to or subtracted from $-b$. In this case the solutions are different. If $b^2 - 4ac < 0$, the expression $\sqrt{b^2 - 4ac}$ does not represent a real number. Thus there are no real-number solutions. In a later chapter we will consider a number system in which such solutions exist.

19. a) Solve $2x^2 + 7x = 4$, by factoring.
    b) Solve the equation in (a) using the quadratic formula. Compare your answers.

20. Solve $5x^2 - 8x = 3$ using the quadratic formula.

Determine the nature of the solutions of each equation. Do not solve.

21. $9x^2 - 12x + 4 = 0$

22. $x^2 + 5x + 8 = 0$

23. $x^2 + 5x + 6 = 0$

For each pair of numbers, write a quadratic equation having those numbers as solutions.

24. $-4, \dfrac{5}{3}$

25. $-7, 8$

26. m, n

THEOREM 4

If $b^2 - 4ac = 0$, $ax^2 + bx + c = 0$ has just one real-number solution.
If $b^2 - 4ac > 0$, $ax^2 + bx + c = 0$ has two real-number solutions.
If $b^2 - 4ac < 0$, $ax^2 + bx + c = 0$ has no real-number solution.

DO EXERCISES 21–23.

[iii] Writing Equations from Solutions

We can use the principle of zero products to write a quadratic equation whose solutions are known.

Example 7 Write a quadratic equation whose solutions are 3 and $-\frac{2}{5}$.

$$x = 3 \quad \text{or} \quad x = -\frac{2}{5}$$

$$x - 3 = 0 \quad \text{or} \quad x + \frac{2}{5} = 0$$

$$(x - 3)\left(x + \frac{2}{5}\right) = 0 \quad \text{Multiplying}$$

$$x^2 - \frac{13}{5}x - \frac{6}{5} = 0 \quad \text{or} \quad 5x^2 - 13x - 6 = 0$$

DO EXERCISES 24–26.

EXERCISE SET 2.3

[i] Solve for x.

1. $2x^2 = 14$ **2.** $6x^2 = 36$ **3.** $9x^2 = 5$ **4.** $4x^2 = 3$

5. $ax^2 = b$ **6.** $\pi x^2 = k$ **7.** $(x - 7)^2 = 5$ **8.** $(x + 3)^2 = 2$

9. $(x - h)^2 = a$ **10.** $y = a(x - h)^2 + k$

[i] Solve by completing the square. (It is important to practice this method since we will need it later.)

11. $x^2 + 6x + 4 = 0$ **12.** $x^2 - 6x - 4 = 0$ **13.** $y^2 + 7y - 30 = 0$ **14.** $y^2 - 7y - 30 = 0$

15. $5x^2 - 4x - 2 = 0$ **16.** $12y^2 - 14y + 3 = 0$ **17.** $2x^2 + 7x - 15 = 0$ **18.** $9x^2 - 30x = -25$

[i] Solve, using the quadratic formula.

19. $x^2 + 4x - 5 = 0$ **20.** $x^2 - 2x = 15$ **21.** $2y^2 - 3y - 2 = 0$ **22.** $5m^2 + 3m - 2 = 0$

23. $3t^2 + 8t + 5 = 0$ **24.** $3u^2 = 18u - 6$ **25.** $u^2 = 12u - 3$ **26.** $5p^2 + 30p + 40 = 0$

[ii] Determine the nature of the solutions of each equation. Do not solve.

27. $x^2 - 2x + 10 = 0$ **28.** $4x^2 - 4\sqrt{3}x + 3 = 0$ **29.** $9x^2 - 6x = 0$ **30.** $2x^2 - 12x + 1 = 0$

[iii] Write quadratic equations whose solutions are as follows.

31. $-11, 9$ **32.** $-4, 4$ **33.** 7, only solution   **34.** $-\dfrac{2}{3}$, only solution

35. $-\frac{2}{5}, \frac{6}{5}$ **36.** $-\frac{1}{4}, -\frac{1}{2}$ **37.** $\frac{c}{2}, \frac{d}{2}$ **38.** $\frac{k}{3}, \frac{m}{4}$

39. $\sqrt{2}, 3\sqrt{2}$ **40.** $-\sqrt{3}, 2\sqrt{3}$

☆ ───────────────────────────────────────

Solve.

41. ▦ $x^2 - 0.75x - 0.5 = 0$ **42.** ▦ $5.33x^2 - 8.23x - 3.24 = 0$

Solve, using any method. (In general, try factoring first. Then use the quadratic formula if factoring is not possible.)

43. $2x^2 - x - 6 = 0$ **44.** $3x^2 + 5x + 2 = 0$ **45.** $x^2 + 3x = 8$ **46.** $x^2 - 6x = 4$

47. $x + \frac{1}{x} = \frac{13}{6}$ **48.** $\frac{3}{x} + \frac{x}{3} = \frac{5}{2}$ **49.** $t^2 + 0.2t - 0.3 = 0$ **50.** $p^2 + 0.3p - 0.2 = 0$

51. $x^2 + x - \sqrt{2} = 0$ **52.** $x^2 - x - \sqrt{3} = 0$

53. $x^2 + \sqrt{5}x - \sqrt{3} = 0$ **54.** $2x^2 + \sqrt{3}x - \pi = 0$

55. $\sqrt{2}x^2 - \sqrt{3}x - \sqrt{5} = 0$ **56.** $\sqrt{2}x^2 + 5x + \sqrt{2} = 0$

57. $(2t - 3)^2 + 17t = 15$ **58.** $2y^2 - (y + 2)(y - 3) = 12$

59. $(x + 3)(x - 2) = 2(x + 11)$ **60.** $9t(t + 2) - 3t(t - 2) = 2(t + 4)(t + 6)$

61. $2x^2 + (x - 4)^2 = 5x(x - 4) + 24$ **62.** $(c + 2)^2 + (c - 2)(c + 2) = 44 + (c - 2)^2$

★ ───────────────────────────────────────

63. Prove that the solutions of $ax^2 + bx + c = 0$ are the reciprocals of the solutions of the equation

$$cx^2 + bx + a = 0, \quad c \neq 0, a \neq 0.$$

64. Derive the quadratic formula by completing the square.

2.4 FORMULAS AND APPLIED PROBLEMS

[i] Formulas

To solve a formula for a certain letter, we use the principles of equation solving we have developed until we have an equation with the letter alone on one side. In most formulas the letters represent non-negative numbers, so we usually do not need to use absolute values when taking principal square roots.

Example 1 Solve $V = \pi r^2 h$, for r.

$$V = \pi r^2 h$$

$$\frac{V}{\pi h} = r^2$$

$$\sqrt{\frac{V}{\pi h}} = r$$

DO EXERCISE 1.

1. Solve $V = \frac{1}{3}\pi r^2 h$, for r.

2. Solve $S = 16t^2 + v_0 t$, for t.

Example 2 Solve $A = \pi rs + \pi r^2$, for r.

$$\pi r^2 + \pi rs - A = 0$$

Then $a = \pi$, $b = \pi s$, $c = -A$, and we use the quadratic formula:

$$r = \frac{-b \pm \sqrt{b^2 - 4ac}}{2a}$$

$$= \frac{-\pi s \pm \sqrt{(\pi s)^2 - 4 \cdot \pi \cdot (-A)}}{2\pi}$$

$$= \frac{-\pi s \pm \sqrt{\pi^2 s^2 + 4\pi A}}{2\pi}$$

or just

$$r = \frac{-\pi s + \sqrt{\pi^2 s^2 + 4\pi A}}{2\pi},$$

since the negative square root would result in a negative solution.

DO EXERCISE 2.

[ii] Applied Problems

Example 3 (*An interest problem*). $2560 is invested at interest rate r, compounded annually. In 2 years it grows to $3240. What is the interest rate?

We substitute 2560 for P, 3240 for A, and 2 for t in the formula $A = P(1 + i)^t$, and solve for i.

$$A = P(1 + i)^t$$
$$3240 = 2560(1 + i)^2$$

$$\frac{3240}{2560} = (1 + i)^2$$

$$\sqrt{\frac{324}{256}} = |1 + i| \quad \text{Taking the principal square root}$$

$$\pm \frac{18}{16} = 1 + i$$

$$-1 + \frac{18}{16} = i \quad \text{or} \quad -1 - \frac{18}{16} = i$$

$$\frac{2}{16} = i \quad \text{or} \quad -\frac{34}{16} = i$$

Since the interest rate cannot be negative, $i = \frac{2}{16} = \frac{1}{8} = 0.125 = 12.5\%$.

DO EXERCISE 3.

3. $2560 is invested at interest rate i, compounded annually. In 2 years it grows to $3610. What is the interest rate?

Example 4 A ladder 10 ft long leans against a wall. The bottom of the ladder is 6 ft from the wall. The bottom of the ladder is then pulled out 3 ft farther. How much does the top end move down the wall?

The first thing to do is to make a drawing and label it, with the known and unknown data. The dotted line shows the ladder in its original position, with the lower end 6 ft from the wall.

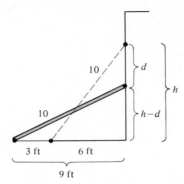

In the figure there are right triangles. This is a clue that we may wish to use the Pythagorean theorem. We use that theorem and begin to write equations. From the taller triangle, we get

$$h^2 + 6^2 = 10^2.$$

We solve this for h and find that $h = 8$ ft.

From the other triangle, we get

$$9^2 + (h - d)^2 = 10^2$$

but we know that $h = 8$, so we have

$$9^2 + (8 - d)^2 = 10^2$$

or

$$d^2 - 16d + 45 = 0.$$

Using the quadratic formula we get

$$d = \frac{16 \pm \sqrt{76}}{2} = \frac{16 \pm 2\sqrt{19}}{2} = 8 \pm \sqrt{19}.$$

The length $8 + \sqrt{19}$ is not a solution since it exceeds the original length. Thus the solution is $8 - \sqrt{19} \approx 8 - 4.359 = 3.641$. Therefore the top moves down 3.641 ft when the bottom is moved out 3 ft.

DO EXERCISE 4.

When an object is dropped or thrown downward, the distance, in meters, that it falls in t seconds is given by the following formula:

$$s = 4.9t^2 + v_0 t.$$

In this formula v_0 is the initial velocity.

▓ **Example 5**

a) An object is dropped from the top of the Gateway Arch in St. Louis, which is 195 meters high. How long does it take to reach the ground?

4. A 13-ft ladder leans against a wall. The bottom of the ladder is 5 ft from the wall. The bottom is then pulled out 4 ft farther. How much does the top end move down the wall?

5. a) An object is dropped from the top of the Statue of Liberty, which is 92 m tall. How long does it take to reach the ground?

b) An object is thrown downward from the statue at an initial velocity of 40 m/s. How long does it take to reach the ground?

c) How far will an object fall in 1 sec, thrown downward from the statue at an initial velocity of 40 m/s?

Since the object was *dropped* its initial velocity was 0. So we substitute 0 for v_0 and 195 for s and then solve for t:

$$195 = 4.9t^2 + 0 \cdot t$$
$$195 = 4.9t^2$$
$$t^2 = 39.8$$
$$t = \sqrt{39.8} \approx 6.31. \quad \text{Use the square root key.}$$

Thus it takes about 6.31 seconds to reach the ground.

b) An object is thrown downward from the arch at an initial velocity of 16 m/sec. How long does it take to reach the ground?

We substitute 195 for s and 16 for v_0 and solve for t:

$$195 = 4.9t^2 + 16t$$
$$0 = 4.9t^2 + 16t - 195.$$

By the quadratic formula and a calculator we obtain

$$t = -8.15 \quad \text{or} \quad t = 4.88.$$

The negative answer is meaningless in this problem, so the answer is about 4.88 sec.

c) How far will an object fall in 3 seconds if it is thrown downward from the arch at an initial velocity of 16 m/s?

We substitute 16 for v_0 and 3 for t and solve for s:

$$s = 4.9t^2 + v_0 t = 4.9(3)^2 + 16 \cdot 3 = 92.1.$$

Thus the object falls 92.1 meters in 3 seconds.

DO EXERCISE 5.

EXERCISE SET 2.4

[i] Solve each formula for the given letter. Assume that all letters represent positive numbers.

1. $F = \dfrac{kM_1 M_2}{d^2}$, for d

2. $E = mc^2$, for c

3. $S = \dfrac{1}{2}at^2$, for t

4. $V = 4\pi r^2$, for r

5. $s = -16t^2 + v_0 t$, for t

6. $A = 2\pi r^2 + 3\pi rh$, for r

7. $d = \dfrac{n^2 - 3n}{2}$, for n

8. $\sqrt{2}t^2 + 3k = \pi t$, for t

9. $A = P(1 + i)^2$, for i

10. $A = P\left(1 + \dfrac{i}{2}\right)^2$, for i

[ii] Applied problems

What is the interest rate if interest is compounded annually?

11. $2560 grows to $3610 in 2 years

12. $1000 grows to $1210 in 2 years

13. $8000 grows to $9856.80 in 2 years

14. $1000 grows to $1271.26 in 2 years

The number of diagonals, d, of a polygon of n sides is given by

$$d = \frac{n^2 - 3n}{2}.$$

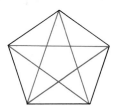

15. A polygon has 27 diagonals. How many sides does it have?

16. A polygon has 44 diagonals. How many sides does it have?

17. A ladder 10 ft long leans against a wall. The bottom of the ladder is 6 ft from the wall. How much would the lower end of the ladder have to be pulled away so that the top end would be pulled down the same amount?

18. A ladder 13 ft long leans against a wall. The bottom of the ladder is 5 ft from the wall. How much would the lower end of the ladder have to be pulled away so that the top end would be pulled down the same amount?

19. The area of a triangle is 18 cm². The base is 3 cm longer than the height. Find the height.

20. A baseball diamond is a square 90 ft on a side. How far is it directly from second base to home?

21. Trains A and B leave the same city at right angles at the same time. Train B travels 5 mph faster than train A. After 2 hr they are 50 mi apart. Find the speed of each train.

22. Trains A and B leave the same city at right angles at the same time. Train A travels 14 km/h faster than train B. After 5 hr they are 130 km apart. Find the speed of each train.

For Exercises 23 and 24 use the formula $s = 4.9t^2 + v_0 t$.

23. a) An object is dropped 75 m from an airplane. How long does it take to reach the ground?

  b) An object is thrown downward from the plane at an initial velocity of 30 m/s. How long does it take to reach the ground?

  c) How far will an object fall in 2 sec, thrown downward at an initial velocity of 30 m/s?

24. a) An object is dropped 500 m from an airplane. How long does it take to reach the ground?

  b) An object is thrown downward from the plane at an initial velocity of 30 m/s. How long does it take to reach the ground?

  c) How far will an object fall in 5 sec, thrown downward at an initial velocity of 30 m/s?

25. The diagonal of a square is 1.341 cm longer than a side. Find the length of the side.

26. The hypotenuse of a right triangle is 8.312 cm long. The sum of the lengths of the legs is 10.23 cm. Find the lengths of the legs.

☆ ──

The following formula will be helpful in Exercises 27 through 30:

$$T = c \cdot N.$$

Total cost = (Cost per item) · (Number of items)
Total cost = (Cost per person) · (Number of persons)

27. A group of students share equally in the $140 cost of a boat. At the last minute 3 students drop out and this raises the share of each remaining student $15. How many students were in the group at the outset?

28. An investor bought a group of lots for $8400. All but 4 of them were sold for $8400. The selling price for each lot was $350 greater than the cost. How many lots were bought?

29. An investor buys some stock for $720. If each share had cost $15 less 4 more shares could have been bought for the same $720. How many shares of stock were bought?

30. A sorority is going to spend $112 for a party. When 14 new pledges join the sorority, this reduces each student's cost by $4. How much did it cost each student before?

Solve for x.

31. $kx^2 + (3 - 2k)x - 6 = 0$

32. $x^2 - 2x + kx + 1 = kx^2$

33. $(m + n)^2x^2 + (m + n)x = 2$

34. Solve $x^2 - 3xy - 4y^2 = 0$
(a) for x; (b) for y.

35. ▦ For interest compounded annually, what is the interest rate when \$9826 grows to \$13,704 in 3 years?

★ ─────────────────────────────────────

36. A rectangle of 12-cm² area is inscribed in the right triangle ABC as shown in the drawing. What are its dimensions?

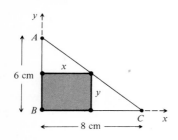

37. The world record for free-fall without a parachute by a woman is 175 ft and is held by Kitty O'Neill. Approximately how long did the fall take?

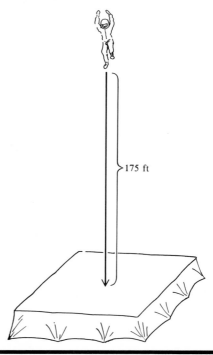

175 ft

OBJECTIVE

You should be able to:

[i] Solve radical equations using the principle of powers.

2.5 RADICAL EQUATIONS

[i] A *radical equation* is an equation in which variables occur in one or more radicands. For example, $\sqrt[3]{x} + \sqrt[3]{4x - 7} = 2$. To solve such equations we need a new principle.

THEOREM 5

─────────────────────────────────────

The principle of powers. **For any number** n, **if an equation** $a = b$ **is true, then** $a^n = b^n$ **is true.**

─────────────────────────────────────

This principle does *not* always yield equivalent equations. For example, the solution of $x = 3$ is 3. If we square both sides, we get $x^2 = 9$, which has solutions -3 and 3. As another example, consider $\sqrt{x} = -3$. At the outset we should note that this equation has no solution because, by definition, $\sqrt{x}$ must be a *nonnegative* number. Suppose, though, that we should try to solve by squaring both sides. We would get $(\sqrt{x})^2 = (-3)^2$, or $x = 9$. The number 9 does not check.

> **CAUTION!** It is imperative when using the principle of powers to check possible solutions in the original equation.

Example 1 Solve $x - 5 = \sqrt{x + 7}$.

$(x - 5)^2 = (\sqrt{x + 7})^2$ Principle of powers; squaring both sides
$x^2 - 10x + 25 = x + 7$
$x^2 - 11x + 18 = 0$
$(x - 9)(x - 2) = 0$
$x = 9$ or $x = 2$

The possible solutions are 9 and 2. We check:

For 9:
$$\frac{x - 5 \;=\; \sqrt{x + 7}}{9 - 5 \;\bigm|\; \sqrt{9 + 7}}$$
$$4 \;\bigm|\; 4$$

For 2:
$$\frac{x - 5 \;=\; \sqrt{x + 7}}{2 - 5 \;\bigm|\; \sqrt{2 + 7}}$$
$$-3 \;\bigm|\; 3$$

Since 9 checks but 2 does not, the only solution is 9.

DO EXERCISES 1 AND 2.

Example 2 Solve $\sqrt[3]{4x^2 + 1} = 5$.

$(\sqrt[3]{4x^2 + 1})^3 = 5^3$ Principle of powers; cubing both sides
$4x^2 + 1 = 125$
$4x^2 = 124$
$x^2 = 31$
$x = \pm\sqrt{31}$

Both $\sqrt{31}$ and $-\sqrt{31}$ check. These are the solutions.

DO EXERCISE 3.

Equations with Two Radical Terms

A general strategy for solving equations with two radical terms is as follows.

> 1. **Isolate one of the radical terms.**
> 2. **Use the principle of powers.**
> 3. **If a radical remains, perform steps 1 and 2 again.**
> 4. **Check possible solutions.**

Solve.

1. $\sqrt{2x} = -5$

2. $x - 1 = \sqrt{x + 5}$

Solve.

3. $\sqrt[4]{3x - 1} = 2$

4. Solve: $\sqrt{3x + 1} - \sqrt{x + 4} = 1$.

Example 3 Solve $\sqrt{2x - 5} - \sqrt{x - 3} = 1$.

$$\sqrt{2x - 5} = 1 + \sqrt{x - 3} \quad \text{Adding } \sqrt{x - 3} \text{ to isolate one of the radical terms}$$

$$(\sqrt{2x - 5})^2 = (1 + \sqrt{x - 3})^2 \quad \text{Principle of powers; squaring both sides}$$

$$2x - 5 = 1 + 2\sqrt{x - 3} + (x - 3) \quad \text{On the right we squared a binomial.}$$

$$x - 3 = 2\sqrt{x - 3} \quad \text{Isolating the remaining radical term}$$

$$(x - 3)^2 = (2\sqrt{x - 3})^2 \quad \text{Squaring both sides}$$

$$x^2 - 6x + 9 = 4(x - 3)$$

$$x^2 - 6x + 9 = 4x - 12$$

$$x^2 - 10x + 21 = 0$$

$$(x - 7)(x - 3) = 0$$

$$x = 7 \quad \text{or} \quad x = 3$$

The numbers 7 and 3 check. Thus the solutions are 7 and 3.

DO EXERCISE 4.

5. Solve $A = \sqrt{1 + \dfrac{a^2}{b^2}}$, for b.

Assume the variables represent nonnegative numbers.

Example 4 Solve $A = \sqrt{1 + \dfrac{a^2}{b^2}}$, for a. Assume the variables represent nonnegative numbers.

$$A^2 = 1 + \frac{a^2}{b^2}$$

$$b^2 A^2 = b^2 + a^2$$

$$b^2 A^2 - b^2 = a^2$$

$$\sqrt{b^2 A^2 - b^2} = a$$

$$\sqrt{b^2(A^2 - 1)} = a$$

$$b\sqrt{A^2 - 1} = a$$

DO EXERCISE 5.

EXERCISE SET 2.5

[i] Solve.

1. $\sqrt{3x - 4} = 1$

2. $\sqrt[3]{2x + 1} = -5$

3. $\sqrt[4]{x^2 - 1} = 1$

4. $\sqrt{m + 1} - 5 = 8$

5. $\sqrt{y - 1} + 4 = 0$

6. $5 + \sqrt{3x^2 + \pi} = 0$

7. $\sqrt{x - 3} + \sqrt{x + 5} = 4$

8. $\sqrt{x} - \sqrt{x - 5} = 1$

9. $\sqrt{3x - 5} + \sqrt{2x + 3} + 1 = 0$

10. $\sqrt{2m - 3} = \sqrt{m + 7} - 2$

11. $\sqrt[3]{6x + 9} + 8 = 5$

12. $\sqrt[5]{3x + 4} = 2$

13. $\sqrt{6x + 7} = x + 2$

14. $\sqrt{6x + 7} - \sqrt{3x + 3} = 1$

15. $\sqrt{20 - x} = \sqrt{9 - x} + 3$

16. $\sqrt{n + 2} + \sqrt{3n + 4} = 2$

17. ▦ $\sqrt{7.35x + 8.051} = 0.345x + 0.067$

18. ▦ $\sqrt{1.213x + 9.333} = 5.343x + 2.312$

For Exercises 19 and 20 assume that the variables represent nonnegative numbers.

19. Solve $T = 2\pi\sqrt{L/g}$, for L; for g.

20. Solve $H = \sqrt{c^2 + d^2}$, for c.

☆ ───

The formula $V = 1.2\sqrt{h}$ can be used to approximate the distance V, in miles, that a person can see to the horizon from a height h, in feet.

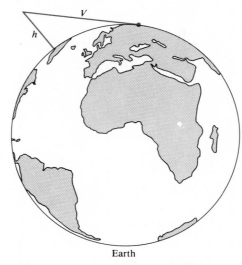

Earth

21. ▦ How far can you see to the horizon through an airplane window at a height of 30,000 feet?

22. ▦ How far can a sailor see to the horizon from the top of a 72-ft mast?

23. ▦ A person can see 144 miles to the horizon from an airplane window. How high is the airplane?

24. ▦ A sailor can see 11 miles to the horizon from the top of a mast. How high is the mast?

Solve.

25. $(x - 5)^{2/3} = 2$

26. $(x - 3)^{2/3} = 2$

27. $\dfrac{x + \sqrt{x + 1}}{x - \sqrt{x + 1}} = \dfrac{5}{11}$

28. $\sqrt{\sqrt{x + 25}} - \sqrt{x} = 5$

29. $\sqrt{x + 2} - \sqrt{x - 2} = \sqrt{2x}$

30. $2\sqrt{x + 3} = \sqrt{x} + \sqrt{x + 8}$

31. $\sqrt[4]{x + 2} = \sqrt{3x + 1}$

32. $\sqrt[3]{2x - 1} = \sqrt[6]{x + 1}$

───

2.6 EQUATIONS REDUCIBLE TO QUADRATIC

[i] Certain equations that are not really quadratic can be thought of in such a way that they can be solved as quadratic. For example,

$$x \quad + 3\sqrt{x} - 10 = 0$$
$$\downarrow \qquad \downarrow \qquad \downarrow$$
$$(\sqrt{x})^2 + 3\sqrt{x} - 10 = 0 \quad \text{Thinking of } x \text{ as } (\sqrt{x})^2$$
$$\downarrow \qquad \downarrow \qquad \downarrow$$
$$u^2 \quad + \quad 3u \quad - 10 = 0 \quad \text{To make this clearer,}$$
$$\text{write } u \text{ instead of } \sqrt{x}.$$

The equation $u^2 + 3u - 10 = 0$ can be solved by factoring or the quadratic formula. After that, we can find x by remembering that $\sqrt{x} = u$. Equations that can be solved like this are said to be *reducible to quadratic*.

OBJECTIVES

You should be able to:

[i] Solve equations that are reducible to quadratic.

[ii] Solve applied problems involving equations reducible to quadratic.

1. a) Solve $x + \sqrt{x} - 12 = 0$.

> To solve equations reducible to quadratic we first make a substitution, solve for the new variable, and then solve for the original variable.

Example 1 Solve $x + 3\sqrt{x} - 10 = 0$.

Let $u = \sqrt{x}$. Then we solve the equation resulting from substituting u for $\sqrt{x}$:

$$u^2 + 3u - 10 = 0$$
$$(u + 5)(u - 2) = 0$$
$$u = -5 \quad \text{or} \quad u = 2$$

Now we substitute $\sqrt{x}$ for u and solve these equations:

$$\sqrt{x} = -5 \quad \text{or} \quad \sqrt{x} = 2$$
$$\text{(no solution)} \quad \text{or} \quad x = 4$$

b) Can you think of another procedure for solving this equation? See the previous section. Which seemed easier?

Check:

$$\begin{array}{c|c} x + 3\sqrt{x} - 10 = 0 & \\ \hline 4 + 3\sqrt{4} - 10 & 0 \\ 4 + \quad 6 - 10 & \\ 0 & \end{array}$$

The solution is 4.

DO EXERCISE 1.

Example 2 Solve $x^4 - 6x^2 + 7 = 0$.

Let $u = x^2$. Then we solve the equation resulting from substituting u for x^2:

$$u^2 - 6u + 7 = 0$$
$$a = 1, \; b = -6, \; c = 7$$
$$u = \frac{-b \pm \sqrt{b^2 - 4ac}}{2a} = \frac{-(-6) \pm \sqrt{(-6)^2 - 4 \cdot 1 \cdot 7}}{2 \cdot 1}$$
$$u = \frac{6 \pm \sqrt{8}}{2} = \frac{2 \cdot 3 \pm 2\sqrt{2}}{2 \cdot 1} = 3 \pm \sqrt{2}.$$

2. Solve $2x^4 - 10x^2 + 11 = 0$.

Now we substitute x^2 for u and solve for x:

$$x^2 = 3 + \sqrt{2} \quad \text{or} \quad x^2 = 3 - \sqrt{2}$$
$$x = \pm\sqrt{3 + \sqrt{2}} \quad \text{or} \quad x = \pm\sqrt{3 - \sqrt{2}}.$$

Thus we have the four solutions: $\sqrt{3 + \sqrt{2}}, \; -\sqrt{3 + \sqrt{2}}, \; \sqrt{3 - \sqrt{2}}$, and $-\sqrt{3 - \sqrt{2}}$.

DO EXERCISE 2.

Example 3 Solve $(x^2 - x)^2 - 14(x^2 - x) + 24 = 0$.

Let $u = x^2 - x$. Then we solve the equation resulting from substituting u for $x^2 - x$:

$$u^2 - 14u + 24 = 0$$
$$(u - 12)(u - 2) = 0$$
$$u = 12 \quad \text{or} \quad u = 2.$$

Now we substitute $x^2 - x$ for u and solve:

$$x^2 - x = 12 \quad \text{or} \quad x^2 - x = 2$$
$$x^2 - x - 12 = 0 \quad \text{or} \quad x^2 - x - 2 = 0$$
$$(x - 4)(x + 3) = 0 \quad \text{or} \quad (x - 2)(x + 1) = 0$$
$$x = 4 \quad \text{or} \quad x = -3 \quad \text{or} \quad x = 2 \quad \text{or} \quad x = -1.$$

The solutions are $4, -3, 2, -1$.

DO EXERCISE 3.

Example 4 Solve $t^{2/5} - t^{1/5} - 2 = 0$.

Let $u = t^{1/5}$. Then we solve the equation resulting from substituting u for $t^{1/5}$:

$$u^2 - u - 2 = 0$$
$$(u - 2)(u + 1) = 0$$
$$u = 2 \quad \text{or} \quad u = -1.$$

Now we substitute $t^{1/5}$ for u and solve:

$$t^{1/5} = 2 \quad \text{or} \quad t^{1/5} = -1$$
$$t = 32 \quad \text{or} \quad t = -1 \qquad \text{Principles of powers;}$$
$$\text{raising to the 5th power}$$

The solutions are 32 and -1.

DO EXERCISE 4.

[ii] **An Applied Problem**

Example 5 (*A well problem*). An object is dropped into a well. Two seconds later the sound of the splash is heard at the top. The speed of sound is 1100 ft/sec. How deep is the well?

a) Let $s = $ the depth of the well. The formula for falling objects (p. 83) when the distance is in feet becomes

$$s = 16t^2 + v_0 t.$$

Since the object is dropped, $v_0 = 0$. The time t_1 that it takes for the object to reach the bottom of the well can be found as follows:

$$s = 16t_1^2, \quad \text{or} \quad t_1 = \frac{\sqrt{s}}{4}.$$

b) Now how long does it take the sound of the splash to reach the top of the well? We can use the formula $d = rt$, and let $d = s$, $r = 1100$, and $t = t_2$, the time it takes the sound to get to the top. Then

$$s = 1100 \cdot t_2, \quad \text{or} \quad t_2 = \frac{s}{1100}.$$

3. Solve $(x^2 - 1)^2 - (x^2 - 1) - 2 = 0$.

4. Solve $t^{2/3} - 3t^{1/3} - 10 = 0$.

5. An object is dropped into a well. In 5 sec the sound of the splash reaches the top of the well. If we assume that the speed of sound is 1100 ft/sec, how deep is the well?

c) The total time for the object to fall and the sound to get back to the top is given by

$$t_1 + t_2 = 2, \quad \text{or} \quad \frac{\sqrt{s}}{4} + \frac{s}{1100} = 2.$$

Then multiplying by 1100, we get

$$275\sqrt{s} + s = 2200, \quad \text{or} \quad s + 275\sqrt{s} - 2200 = 0.$$

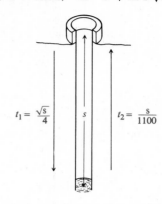

This equation is reducible to quadratic with $u = \sqrt{s}$. Substituting, we get

$$u^2 + 275u - 2200 = 0.$$

Using the quadratic formula we can solve for u:

$$u = \frac{-275 + \sqrt{275^2 + 8800}}{2} \quad \text{We want the positive solution.}$$

$$= \frac{-275 + \sqrt{84,425}}{2} \approx \frac{-275 + \sqrt{840 \cdot 100}}{2}$$

$$= \frac{-275 + 10\sqrt{840}}{2}$$

$$\approx \frac{-275 + 10 \cdot 29}{2}$$

$$u = \frac{15}{2} = 7.5.$$

Thus $u = 7.5 = \sqrt{s}$, so $s = 56.25$; that is, the well is about 56.25 ft deep.

DO EXERCISE 5.

EXERCISE SET 2.6

[i] Solve.

1. $x - 10\sqrt{x} + 9 = 0$

2. $2x - 9\sqrt{x} + 4 = 0$

3. $x^4 - 10x^2 + 25 = 0$

4. $x^4 - 3x^2 + 2 = 0$

5. $t^{2/3} + t^{1/3} - 6 = 0$

6. $w^{2/3} - 2w^{1/3} - 8 = 0$

7 $z^{1/2} = z^{1/4} + 2$

8. $6 = m^{1/3} - m^{1/6}$

9. $(x^2 - 6x)^2 - 2(x^2 - 6x) - 35 = 0$

10. $(1 + \sqrt{x})^2 + (1 + \sqrt{x}) - 6 = 0$

11. $(y^2 - 5y)^2 + (y^2 - 5y) - 12 = 0$

12. $(2t^2 + t)^2 - 4(2t^2 + t) + 3 = 0$

13. $w^4 - 4w^2 - 2 = 0$

14. $t^4 - 5t^2 + 5 = 0$

15. $x^{-2} - x^{-1} - 6 = 0$

16. $4x^{-2} - x^{-1} - 5 = 0$

17. $2x^{-2} + x^{-1} = 1$

18. $10 - 9m^{-1} = m^{-2}$

19. $\left(\dfrac{x^2 - 2}{x}\right)^2 - 7\left(\dfrac{x^2 - 2}{x}\right) - 18 = 0$

20. $\left(\dfrac{x^2 + 1}{x}\right)^2 - 8\left(\dfrac{x^2 + 1}{x}\right) + 15 = 0$

21. $\dfrac{x}{x - 1} - 6\sqrt{\dfrac{x}{x - 1}} - 40 = 0$

22. $\dfrac{2x + 1}{x} + 30 = 7\sqrt{\dfrac{2x + 1}{x}}$

23. $5\left(\dfrac{x + 2}{x - 2}\right)^2 = 3\left(\dfrac{x + 2}{x - 2}\right) + 2$

24. $\left(\dfrac{x + 1}{x + 3}\right)^2 + \left(\dfrac{x + 1}{x + 3}\right) - 6 = 0$

[ii] Applied problems

25. A stone is dropped from a cliff. In 3 sec the sound of the stone striking the ground reaches the top of the cliff. Assuming the speed of sound is 1100 ft/sec, how high is the cliff?

26. A stone is dropped from a cliff. In 4 sec the sound of the stone striking the ground reaches the top of the cliff. Assuming the speed of sound is 1100 ft/sec, how high is the cliff?

☆ ——————————————————————

Solve. Check possible solutions by substituting into the original equation.

27. $6.75x = \sqrt{35x} + 5.36$

28. $\pi x^4 - \sqrt{99.3} = \pi^2 x^2$

Solve.

29. $9x^{3/2} - 8 = x^3$

30. $\sqrt[3]{2x + 3} = \sqrt[6]{2x + 3}$

★ ——————————————————————

Solve.

31. $\dfrac{2x + 1}{x} = 3 + 7\sqrt{\dfrac{2x + 1}{x}}$

2.7 VARIATION

[i] Types of Variation

Direct Variation

There are many situations that yield linear equations like $y = kx$, where k is some positive constant. Note that as x increases, y increases. In such a situation we say we have *direct variation*, and k is called the *variation constant*. Usually only positive values of x and y are considered.

DEFINITION

If two variables x and y are related as in the equation $y = kx$, where k is a positive constant, we say that "y varies directly as x," or that "y is directly proportional to x."

For example, the circumference C of a circle varies directly as its diameter D: $C = \pi D$. A car moves at a constant speed of 65 mph. The distance d it travels varies directly as the time t: $d = 65t$.

OBJECTIVE

You should be able to:

[i] Find equations of variation and solve applied problems involving variation.

1. Find an equation of variation where y varies directly as x, and $y = 32$ when $x = 0.2$.

Example 1 Find an equation of variation where y varies directly as x, and $y = 5.6$ when $x = 8$.

We know that $y = kx$, so $5.6 = k \cdot 8$ and $0.7 = k$. Thus $y = 0.7x$.

Suppose y varies directly as x. Then we have $y = kx$. When (x_1, y_1) and (x_2, y_2) are solutions of the equation, we have $y_1 = kx_1$ and $y_2 = kx_2$, and

$$\frac{y_2}{y_1} = \frac{kx_2}{kx_1} = \frac{k}{k} \cdot \frac{x_2}{x_1} = \frac{x_2}{x_1}.$$

The equation $y_2/y_1 = x_2/x_1$ is called a *proportion*. Such equations are helpful in solving applied problems.

DO EXERCISE 1.

2. Under the conditions of Example 2, what weight would be required to stretch the spring 60 cm?

Example 2 (*A spring problem*). *Hooke's law* states that the distance d an elastic object such as a spring is stretched by placing a certain weight on it varies directly as the weight w of the object. If the distance is 40 cm when the weight is 3 kg, what is the distance stretched when a 2-kg weight is attached?

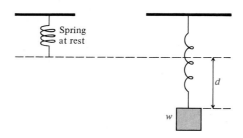

3. *Ohm's law* states that the voltage V, in an electric circuit, varies directly as the number of amperes I of electric current in the circuit. If the voltage is 10 volts when the current is 3 amperes, what is the voltage when the current is 15 amperes?

Method 1. First find k using the first set of data. Then solve for d_2 using the second set of data.

$$d_1 = kw_1 \qquad d_2 = \frac{40}{3} \cdot w_2$$

$$40 = k \cdot 3 \qquad d_2 = \frac{40}{3} \cdot 2$$

$$\frac{40}{3} = k \qquad d_2 = \frac{80}{3}, \quad \text{or } 26.7 \text{ cm}$$

Method 2. Use a proportion to solve for d_2 without first finding the variation constant.

4. The amount of garbage G produced in the United States varies directly as the number of people N who produce the garbage. It is known that 50 tons of garbage is produced by 200 people in 1 yr. The population of San Francisco is 705,000. How much garbage is produced by San Francisco in 1 yr?

$$\frac{d_2}{d_1} = \frac{w_2}{w_1}$$

$$\frac{d_2}{40} = \frac{2}{3}$$

$$3 \cdot d_2 = 40 \cdot 2$$

$$d_2 = 26.7 \text{ cm}$$

DO EXERCISES 2–4.

Inverse Variation

There are also situations that yield equations of the type $y = k/x$, where k is a positive constant. Note that as positive values of x increase, y decreases. In such a situation, we say we have *inverse variation*, and k is called the *variation constant*. In applications we are usually considering only positive values of x and y.

DEFINITION

If two variables x and y are related as in the equation $y = k/x$, where k is a positive constant, we say that "y varies inversely as x," or that "y is inversely proportional to x."

DO EXERCISE 5.

Suppose y varies inversely as x. Then we have $y = k/x$. When (x_1, y_1) and (x_2, y_2) are solutions of the equation, we have $y_1 = k/x_1$ and $y_2 = k/x_2$. Then

$$\frac{y_2}{y_1} = \frac{\dfrac{k}{x_2}}{\dfrac{k}{x_1}} = \frac{k}{x_2} \cdot \frac{x_1}{k} = \frac{k}{k} \cdot \frac{x_1}{x_2} = \frac{x_1}{x_2}.$$

The equation $y_2/y_1 = x_1/x_2$ is a proportion that also can be helpful in solving applied problems. Compare it to the proportion for direct variation. Note how the variables are interchanged.

Example 3 (*Stocks and gold*). Certain economists theorize that stock prices are inversely proportional to the price of gold. That is, when the price of gold goes up, the prices of stock go down; and when the price of gold goes down, the prices of stock go up. Let us assume that the Dow-Jones Industrial Average, D, an index of the overall price of stock, is inversely proportional to the price of gold, G, in dollars per ounce. One day the Dow-Jones Industrial Average was 846 and the price of gold was $528 per ounce. What will the Dow-Jones average be if the price of gold drops to $480 per ounce?

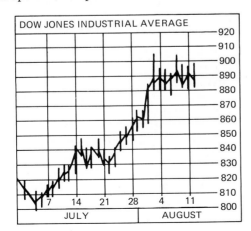

5. Find an equation of variation where y varies inversely as x, and $y = 32$ when $x = 0.2$.

6. The time t required to drive a fixed distance varies inversely as the speed r. It takes 5 hr at 60 km/h to drive a fixed distance; how long would it take to drive the fixed distance at 40 km/h?

7. Find a proportion involving $\dfrac{A_2}{A_1}$, where A_1 and A_2 are areas of circles.

8. Find an equation of variation where y varies directly as the square of x, and $y = 12$ when $x = 2$.

9. Find a proportion involving $\dfrac{W_2}{W_1}$.

10. Find an equation of variation where y varies inversely as the square of x, and $y = \frac{1}{4}$ when $x = 6$.

Method 1. Find k using the first set of data. Then solve for D_2.

$$D_1 = \frac{k}{G_1}$$

$$846 = \frac{k}{528}$$

$$k = 846(528)$$
$$= 446{,}688$$

$$D_2 = \frac{k}{G_2}$$

$$= \frac{446{,}688}{480}$$

$$= 930.6$$

Method 2. Use a proportion to solve for D_2 without first finding the variation constant.

$$\frac{D_2}{D_1} = \frac{G_1}{G_2}$$

$$\frac{D_2}{846} = \frac{528}{480}$$

$$480(D_2) = 846(528)$$
$$D_2 = 930.6$$

Warning! Do not put too much "stock" in the equation of Example 3. It is meant to give us an idea of economic relationships. An equation to predict the stock market accurately has not been found!

DO EXERCISE 6.

Other Kinds of Variation

There are many other types of variation.

DEFINITION

> y **varies directly as the square of** x if there is some positive constant k such that $y = kx^2$.

For example, consider the equation for the area of a circle: $A = \pi r^2$.

DO EXERCISES 7 AND 8.

DEFINITION

> y **varies inversely as the square of** x if there is some positive constant k such that $y = k/x^2$.

For example, the weight W of a body varies inversely as the square of the distance d from the center of the earth: $W = k/d^2$.

DO EXERCISES 9 AND 10.

DEFINITION

y varies *jointly* as *x* and *z* if there is some positive constant *k* such that
$y = kxz$.

For example, consider the equation for the area of a triangle: $A = \frac{1}{2}bh$. The area varies jointly as *b* and *h*. The variation constant is $\frac{1}{2}$.

DO EXERCISES 11 AND 12.

Several kinds of variation can occur together. For example, if $y = k \cdot xz^3/w^2$, then *y* varies jointly as *x* and the cube of *z* and inversely as the square of *w*.

DO EXERCISE 13.

Applied Problems

Example 4 (*The volume of a tree*). The volume of wood *V* in a tree varies jointly as the height *h* and the square of the girth *g* (girth is distance around). If the volume is 7750 ft³ when the height is 100 ft and the girth is 5 ft, what is the height when the volume is 37,975 ft³ and the girth is 7 ft?

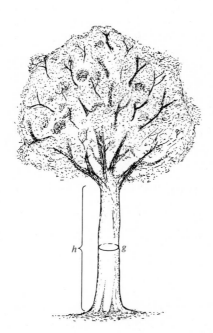

Method 1. First find *k* using the first set of data. Then solve for h_2 using the second set of data.

$$V_1 = k \cdot h_1 \cdot g_1^2$$
$$7750 = k \cdot 100 \cdot 5^2$$
$$3.1 = k$$

11. Find a proportion involving $\dfrac{A_2}{A_1}$.

12. Find an equation of variation where *y* varies jointly as *x* and *z*, and $y = 42$ when $x = 2$ and $z = 3$.

13. Find an equation of variation where *y* varies jointly as *x* and *z* and inversely as the square of *w*, and $y = 105$ when $x = 3$, $z = 20$, and $w = 2$.

14. The distance S that an object falls from some point above the ground varies directly as the square of the time t that it falls. If the object falls 4 ft in 0.5 sec, how long will it take the object to fall 64 ft?

Then

$$37,975 = 3.1 \cdot h_2 \cdot 7^2$$
$$250 = h_2$$
$$h_2 = 250 \text{ ft}$$

Method 2. Use a proportion to solve for h_2 without first finding the variation constant.

$$\frac{V_2}{V_1} = \frac{h_2 \cdot g_2^2}{h_1 \cdot g_1^2}$$

$$\frac{37,975}{7750} = \frac{h_2 \cdot 7^2}{100 \cdot 5^2}$$

$$h_2 = 250 \text{ ft}$$

DO EXERCISE 14.

15. a) How much will the 200-lb astronaut weigh when he is 1000 mi from earth?
   b) How far is he from earth when his weight is reduced to 50 lb?

Example 5 (*The weight of an astronaut*). The weight W of an object varies inversely as the square of the distance d from the object to the center of the earth. At sea level (4000 mi from the center of the earth) an astronaut weighs 200 lb. Find his weight when he is 100 mi above the surface of the earth, and the spacecraft is not in motion.

We use the proportion

$$\frac{W_2}{W_1} = \frac{d_1^2}{d_2^2}$$

$$\frac{W_2}{200} = \frac{4000^2}{4100^2}$$

$$16,810,000 \cdot W_2 = 200(16,000,000)$$
$$W_2 = 190 \text{ lb (to the nearest pound)}.$$

DO EXERCISE 15.

EXERCISE SET 2.7

[i] Find an equation of variation where:

1. y varies directly as x, and $y = 0.6$ when $x = 0.4$.

2. y varies inversely as x, and $y = 0.4$ when $x = 0.8$.

3. y varies inversely as the square of x, and $y = 0.15$ when $x = 0.1$.

4. y varies jointly as x and z, and $y = 56$ when $x = 7$ and $z = 10$.

5. y varies jointly as x and z and inversely as w, and $y = \frac{3}{2}$ when $x = 2$, $z = 3$, and $w = 4$.

6. y varies jointly as x and the square of z, and $y = 105$ when $x = 14$ and $z = 5$.

7. y varies jointly as x and z and inversely as the square of w, and $y = \frac{12}{5}$ when $x = 16$, $z = 3$, and $w = 5$.

8. y varies jointly as x and z and inversely as the product of w and p, and $y = \frac{3}{28}$ when $x = 3$, $z = 10$, $w = 7$, and $p = 8$.

9. Suppose y varies directly as x and x is doubled. What is the effect on y?

10. Suppose y varies inversely as x and x is tripled. What is the effect on y?

11. Suppose y varies inversely as the square of x and x is multiplied by n. What is the effect on y?

12. Suppose y varies directly as the square of x and x is multiplied by n. What is the effect on y?

13. The amount of pollution A entering the atmosphere varies directly as the amount of people N living in an area. If 60,000 people cause 42,600 tons of pollutants, how many tons entered the atmosphere in a city with a population of 750,000?

14. The volume V of a given mass of gas varies directly as the temperature T and inversely as the pressure P. If $V = 231 \text{ in}^3$ when $T = 420°$ and $P = 20 \text{ lb/sq in.}$, what is the volume when $T = 300°$ and $P = 15 \text{ lb/sq in.}$?

15. (*The strength of a beam*). The safe load (amount it supports without breaking) L of a beam varies jointly as its width w and the square of its height h and inversely as its length l. If the width and height are doubled at the same time the length is halved, what is the effect on L?

16. For a chord $\overline{PQ}$ through a fixed point A in a circle, the length of $\overline{PA}$ is inversely proportional to the length of $\overline{AQ}$. If the length of $\overline{PA} = 64$ when the length of $\overline{AQ} = 16$, what is the length of $\overline{AQ}$ when the length of $\overline{PA} = 4$?

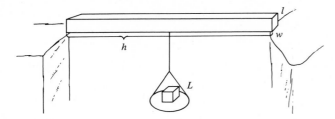

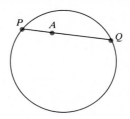

17. ▦ (*A sighting problem*). The distance d that one can see to the horizon varies directly as the square root of the height above sea level. If a person 19.5 m above sea level can see 28.97 km, how high above sea level must one be to see 54.32 km?

18. ▦ (*Electrical resistance*). At a fixed temperature and chemical composition, the resistance of a wire varies directly as the length l and inversely as the square of the diameter d. If the resistance of a certain kind of wire is 0.112 ohm when the diameter is 0.254 cm and the length is 15.24 m, what is the resistance of a wire whose length is 608.7 m and whose diameter is 0.478 cm?

☆ ─────────────────────────────────

19. Show that if p varies directly as q, then q varies directly as p.

20. Show that if u varies inversely as v, then v varies inversely as u; and $1/u$ varies directly as v.

21. The area of a circle varies directly as the square of the length of a diameter. What is the variation constant?

22. P varies directly as the square of t. How does t vary in relationship to P?

CHAPTER 2 REVIEW

Solve.

[2.1, ii] 1. $\dfrac{3x + 2}{7} - \dfrac{5x}{3} = \dfrac{32}{21}$

2. $(p - 3)(3p + 2)(p + 2) = 0$

3. $3x^2 + 2x - 8 = 0$

[2.3, i] 4. $r^2 + 3r - 1 = 0$

5. $3t^2 = 4 + 3t$

[2.1, ii] 6. $\dfrac{5}{2x + 3} + \dfrac{1}{x - 6} = 0$

[2.6, i] 7. $y^4 - 3y^2 + 1 = 0$

8. $x = 2\sqrt{x} - 1$

9. $(x^2 - 1)^2 - (x^2 - 1) - 2 = 0$

10. $t^{2/3} - 10 = 3t^{1/3}$

[2.5, i] 11. $\sqrt{x - 1} - \sqrt{x - 4} = 1$

12. $\sqrt{5x + 1} - 1 = \sqrt{3x}$

[2.3, i] 13. Solve by completing the square: $y^2 - 6y = 16$.

[2.3, ii]   Determine the nature of the solutions of each equation.

14. $4y^2 + 5y + 10 = 0$ 15. $3z^2 + 2z - 1 = 0$

[2.3, iii] 16. Write a quadratic equation whose solutions are -3 and $\frac{1}{2}$.

[2.4, i]   17. Solve $v = \sqrt{2gh}$, for h.

[2.2, i]   18. Solve $\dfrac{1}{a} + \dfrac{1}{b} = \dfrac{1}{t}$, for t.

[2.2, ii]   19. A student scores 73% and 79% on two tests. If the third test counts as if it were two tests, what score must the student make on the third test so the average will be 85%?

20. A can mow a lawn in 4 hr. B can mow it in 2 hr. How long would it take if they worked together?

21. A can mow a lawn in 3 hr. Working together, A and B can mow the lawn in 1 hr. How long would it take B, working alone, to mow the lawn?

22. 18 is 30% of what?   23. What is 9 percent of 50?

[2.4, ii]   24. Two trains leave the same city at right angles. The first train travels 60 km/h. In one hour the trains are 100 km apart. How fast is the second train traveling?

[2.4, ii]   25. In a right triangle the perimeter is 40 and the sum of the squares of the sides is 578. Find the lengths of the sides.

26. A boat travels 2 km upstream and 2 km downstream. The total time for both parts of the trip is 1 hr. The speed of the stream is 2 km/h. What is the speed of the boat in still water? Round to the nearest tenth.

[2.7, i]   27. Find an equation of variation where y varies inversely as the square of x, and $y = 0.005$ when $x = 10$.

[2.7, i]   28. Find an equation of variation where T varies directly as the square of x and inversely as p, and $T = 0.01$ when $x = 6$ and $p = 20$.

[2.7, i]   29. It is theorized that the dividends paid on utilities stocks are inversely proportional to the prime (interest) rate. Recently, the dividends D on the stock of Indianapolis Power and Light were \$2.09 per share and the prime rate was 19%. The prime rate, R, dropped to 17.5%. What dividends were then paid?

[2.7, i]   30. For a body falling freely from rest, the distance (s ft) that the body falls varies directly as the square of the time (t sec). Given that $s = 64$ when $t = 2$, find a formula for s in terms of t. How long will it take the body to fall 900 ft?

31. Suppose a, b, c, and d are nonzero real numbers such that c and d are solutions of

$$x^2 + ax + b = 0$$

and a and b are solutions of

$$x^2 + cx + d = 0.$$

Find $a + b + c + d$.

3
RELATIONS, FUNCTIONS, AND TRANSFORMATIONS

$f\bigl(f(f(x))\bigr)$

$f\Bigl(f\bigl(f(x)\bigr)\Bigr)$

OBJECTIVES

You should be able to:

[i] Find Cartesian products of small sets.
[ii] Indicate or find certain relations.
[iii] Find the domain and range of a relation.

1. Let $A = \{d, e, f\}$ and $B = \{1, 2\}$. List all the ordered pairs in $A \times B$.

3.1 RELATIONS AND ORDERED PAIRS

[i] Cartesian Products

Consider the sets $A = \{1, 2, 3\}$ and $B = \{a, b\}$. From these sets we can pick numbers and form ordered pairs: for example,

$$(1, a), \qquad (2, a), \qquad \text{and} \qquad (3, b).$$

DEFINITION

The Cartesian product of two sets A and B, symbolized $A \times B$, is defined as the set of *all* ordered pairs having the first member from set A and the second member from set B.

Example 1 Find the Cartesian product $A \times B$, where $A = \{1, 2, 3\}$ and $B = \{a, b\}$.

The Cartesian product, $A \times B$, is as follows:

$$(1, a), \quad (2, a), \quad (3, a),$$
$$(1, b), \quad (2, b), \quad (3, b).$$

> **CAUTION!** Be sure in forming Cartesian products that in each pair of $A \times B$ the first member is taken from A and the second member is taken from B.

DO EXERCISE 1.

The sets A and B may be the same.

2. List all of the ordered pairs in the Cartesian square of $\{1, 2, 3, 4\}$. Save the list for later use.

Example 2 Consider the set $\{2, 3, 4, 5\}$. Find the Cartesian product of this set by itself.

The Cartesian product of this set by itself is called a *Cartesian square* and is as follows:

```
5 | (2, 5)  (3, 5)  (4, 5)  (5, 5)
4 | (2, 4)  (3, 4)  (4, 4)  (5, 4)
3 | (2, 3)  (3, 3)  (4, 3)  (5, 3)
2 | (2, 2)  (3, 2)  (4, 2)  (5, 2)
  -----------------------------
      2       3       4       5
```

The headings at the bottom and at the left are for reference only. The Cartesian square consists only of the ordered pairs.

DO EXERCISE 2.

[ii] Relations

In a Cartesian product we can pick out ordered pairs that make up common relations, such as $=$ or $<$ as in the following examples.

Example 3 In the Cartesian product of Example 2 indicate all ordered pairs for which the first member is the same as the second. This set of ordered pairs is the relation =.

(2, 5) (3, 5) (4, 5) (5, 5)
(2, 4) (3, 4) (4, 4) (5, 4)
(2, 3) (3, 3) (4, 3) (5, 3)
(2, 2) (3, 2) (4, 2) (5, 2)

Example 4 In the Cartesian product of Example 2 indicate all ordered pairs for which the first member is less than the second. This set of ordered pairs is the relation <.

(2, 5) (3, 5) (4, 5) (5, 5)
(2, 4) (3, 4) (4, 4) (5, 4)
(2, 3) (3, 3) (4, 3) (5, 3)
(2, 2) (3, 2) (4, 2) (5, 2)

DO EXERCISES 3 AND 4.

There are also many relations that do not have common names and relations with which we are not already familiar. Any time we select a set of ordered pairs from a Cartesian product, we have selected some relation. This is true even if we make the selection at random.

Example 5 The following set of ordered pairs is a relation, but is not a familiar one. It has no common name.

(2, 5) (3, 5) (4, 5) (5, 5)
(2, 4) (3, 4) (4, 4) (5, 4)
(2, 3) (3, 3) (4, 3) (5, 3)
(2, 2) (3, 2) (4, 2) (5, 2)

We shall now make our definition of relation.

DEFINITION

A *relation* from a set *A* to a set *B* is defined to be any set of ordered pairs in *A* × *B*.

[iii] **Domain and Range**

DEFINITION

The set of all first members in a relation is called its *domain.* The set of all second members in a relation is called its *range.*

Example 6 Find the domain and range of the relation = in Example 3.

Domain: {2, 3, 4, 5};   Range: {2, 3, 4, 5}

3. In the Cartesian square of Exercise 2, indicate all ordered pairs for which the first member is the same as the second. This is the relation =.

4. In the Cartesian square of Exercise 2, indicate all ordered pairs for which the first member is greater than the second. This is the relation >.

5. Find the domain and range of the relation in Exercise 3.

6. Find the domain and range of the relation in Exercise 4.

7. Consider the relation whose ordered pairs are (2, 2), (1, 1), (1, 2), and (1, 3). Find the domain and range.

Example 7 Find the domain and range of the relation $<$ in Example 4.

Domain: $\{2, 3, 4\}$;   Range: $\{3, 4, 5\}$

Example 8 Find the domain and range of the relation in Example 5.

Domain: $\{2, 4, 5\}$;   Range: $\{2, 3, 5\}$

DO EXERCISES 5–7.

EXERCISE SET 3.1

[i]

1. List all ordered pairs in the Cartesian product $A \times B$, where $A = \{0, 2\}$ and $B = \{a, b, c\}$. Remember that first coordinates come from A and second coordinates from B.

2. List all ordered pairs in the Cartesian product $A \times B$, where $A = \{1, 3, 5, 9\}$ and $B = \{d, e, f\}$. Remember that first coordinates come from A and second coordinates from B.

[ii] For Exercises 3–8, consider the set $\{-1, 0, 1, 2\}$.

3. Find the set of ordered pairs in the relation $<$ (is less than).

4. Find the set of ordered pairs in the relation $>$ (is greater than).

5. Find the set of ordered pairs in the relation $\leq$ (is less than or equal to).

6. Find the set of ordered pairs in the relation $\geq$ (is greater than or equal to).

7. Find the set of ordered pairs in the relation $=$.

8. Find the set of ordered pairs in the relation $\neq$.

[iii]

9. a) List all the ordered pairs in the Cartesian square $D \times D$, where $D = \{-1, 0, 1, 2\}$.
   b) Graph the relation whose ordered pairs are (0, 0), (1, 1), (0, 1), and (1, 2).
   c) List the domain and the range of this relation.

10. a) List all the ordered pairs in the Cartesian square $E \times E$, where $E = \{-1, 1, 3, 5\}$.
    b) Graph the relation whose ordered pairs are (−1, 1), (1, 1), (−1, 3), and (1, 3).
    c) List the domain and the range of this relation.

OBJECTIVES

You should be able to:

[i] Determine whether an ordered pair of numbers is a solution of an equation with two variables.

[ii] Graph certain equations.

[iii] Given the graph of a relation, describe the domain and the range.

3.2 GRAPHS OF EQUATIONS

Relations in Real Numbers

We are most interested in relations involving $R \times R$, where R is the set of real numbers. The set R is infinite. Thus relations involving R may be infinite, and therefore cannot be indicated by listing the ordered pairs one at a time. We usually indicate such relations with some sort of picture in $R \times R$. This kind of picture is called a *graph*.

Points in the Plane and Ordered Pairs

On a number line each point corresponds to a number. On a plane each point corresponds to a number pair from $R \times R$. To represent $R \times R$ we draw an x-axis and a y-axis perpendicular to each other. Their intersection is called the *origin* and is labeled 0. The arrows show the positive directions.

This is called a Cartesian coordinate system. The first member of an ordered pair is called the *first coordinate*. The second member is

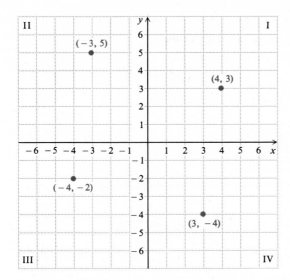

1. Use graph paper.
   a) Draw and label an x-axis and a y-axis.
   b) Label the quadrants.
   c) Plot the points in the relation (3, 2), (−5, −2), (−4, 3).
   d) Find the domain and range of the relation.

called the *second coordinate.** Together these are called the *coordinates of a point.* The axes divide the plane into four *quadrants,* indicated by the Roman numerals.

DO EXERCISE 1.

[i] Solutions of Equations

If an equation has two variables, its solutions are ordered pairs of numbers. A *solution* is an ordered pair which when substituted alphabetically for the variables produces a true equation.

Example 1 Determine whether the following ordered pairs are solutions of the equation $y = 3x - 1$: (−1, −4), (7, 5).

$$
\begin{array}{c|c}
\multicolumn{2}{c}{y \;=\; 3x - 1} \\
\hline
-4 & 3(-1) - 1 \\
-4 & -3 - 1 \\
   & -4
\end{array}
$$

We substitute −1 for x and −4 for y (alphabetical order of variables).

The equation becomes true: (−1, −4) is a solution.

$$
\begin{array}{c|c}
\multicolumn{2}{c}{y \;=\; 3x - 1} \\
\hline
5 & 3 \cdot 7 - 1 \\
5 & 21 - 1 \\
  & 20
\end{array}
$$

We substitute.

The equation becomes false: (7, 5) is not a solution.

DO EXERCISES 2–5.

2. Determine whether (1, 7) is a solution of $y = 2x + 5$.

3. Determine whether (−1, 4) is a solution of $y = 2x + 5$.

4. Determine whether (−2, 5) is a solution of $y = x^2$.

5. Determine whether (4, −5) is a solution of $x^3 - y^2 = 39$.

---

*The first coordinate is sometimes called the *abscissa* and the second coordinate the *ordinate.*

[ii] Graphs of Equations

The solutions of an equation are ordered pairs and thus constitute a relation. To *graph* an equation, or a relation, means to make a drawing of its solutions. Some general suggestions for graphing are as follows.

Graphing suggestions

a) **Use graph paper.**

b) **Label axes with symbols for the variables.**

c) **Use arrows to indicate positive directions.**

d) **Scale the axes; that is, mark numbers on the axes.**

e) **Plot solutions and complete the graph. When finished write down the equation or relation being graphed.**

Example 2 Graph $y = 3x - 1$.

We find some ordered pairs that are solutions. To find an ordered pair, we choose *any* number that is a sensible replacement for x and then determine y. For example, if we choose 2 for x, then $y = 3(2) - 1$, or 5. We have found the solution $(2, 5)$. We continue making choices for x, and finding the corresponding values for y. We make some negative choices for x, as well as positive ones. We keep track of the solutions in a table.

x	y
0	-1
1	2
2	5
-1	-4
-2	-7

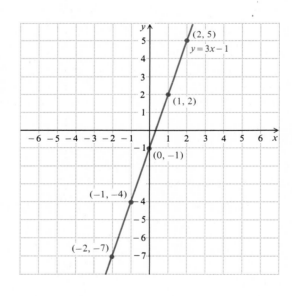

The table gives us the ordered pairs $(0, -1)$, $(1, 2)$, $(2, 5)$, and so on. Next, we plot these points. If we had enough of them, they would make a solid line. We can draw the line with a ruler, and label it $y = 3x - 1$.

Note that the equation $y = 3x - 1$ has an infinite (unending) set of solutions. The *graph of the equation* is a drawing of the relation that is made up of its solutions. Thus the relation consists of all pairs (x, y) such that $y = 3x - 1$ is true. That is, $\{(x, y) | y = 3x - 1\}$.

DO EXERCISE 6.

Example 3 Graph $y = x^2 - 5$.

We select numbers for x and find the corresponding values for y. The table gives us the ordered pairs $(0, -5)$, $(-1, -4)$, and so on.

x	y
0	−5
−1	−4
1	−4
−2	−1
2	−1
−3	4
3	4

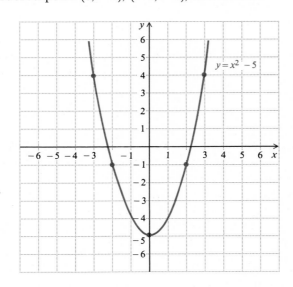

Next we plot these points. We note that as the absolute value of x increases, $x^2 - 5$ also increases. Thus the graph is a curve that rises gradually on either side of the y-axis, as shown above.

This graph shows the relation $\{(x, y) | y = x^2 - 5\}$.

DO EXERCISE 7.

Example 4 Graph the equation $x = y^2 + 1$.

Here we shall select numbers for y and then find the corresponding values for x. This time we arrange them in a horizontal table.

x	2	2	1	5	5
y	−1	1	0	−2	2

We must remember that x is the first coordinate and y is the second coordinate. Thus the table gives us the ordered pairs $(2, -1)$, $(2, 1)$, $(1, 0)$, and so on. We note that as the absolute value of y gradually increases, the value of x also gradually increases. Thus the graph is a curve that stretches farther and farther to the right as it gets farther from the x-axis.

6. Graph $y = -3x + 1$.

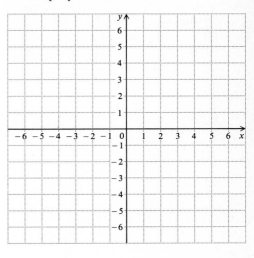

7. Graph $y = 3 - x^2$.

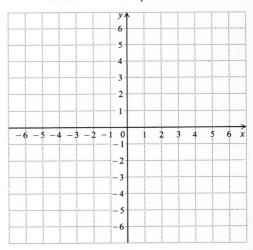

8. Graph $x = y^2 - 5$. [*Hint:* Select values for y and then find the corresponding values of x. When you plot, be sure to find x (horizontally) first.] Compare it with the graph of Example 3.

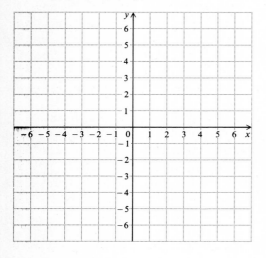

9. Graph $xy = 1$.

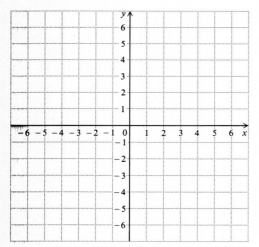

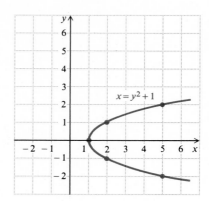

This graph shows the relation $\{(x, y) \mid x = y^2 + 1\}$.

DO EXERCISE 8.

You can always use a calculator to find as many values as desired. This can be especially helpful when you are uncertain about the shape of a graph.

Example 5 Graph the equation $xy = 12$.

We find numbers that satisfy the equation.

x	1	-1	2	-2	3	-3	4	-4	6	-6	12	-12
y	12	-12	6	-6	4	-4	3	-3	2	-2	1	-1

We plot these points and connect them. To see how to do this, note that neither x nor y can be 0. Thus the graph does not cross either axis. As the absolute value of x gets small, the absolute value of y must get large, and vice versa. Thus the graph consists of two curves, as follows.

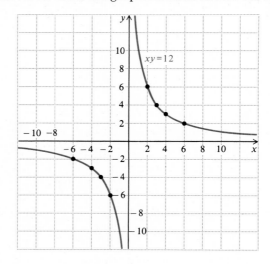

DO EXERCISE 9.

Example 6 Graph the equation $y = |x|$.

We find numbers that satisfy the equation.

x	0	1	−1	2	−2	3	−3	4	−4
y	0	1	1	2	2	3	3	4	4

We plot these points and connect them. To see how to do this, note that as we get farther from the origin, to the left or right, the absolute value of x increases. Thus the graph is a curve that rises to the left and right of the y-axis. It actually consists of parts of two straight lines, as follows.

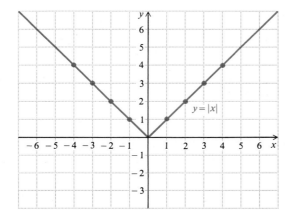

DO EXERCISE 10.

[iii] Domains and Ranges

In each example we see a relation in real numbers, together with its domain and range.

Example 7

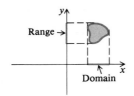

Example 8

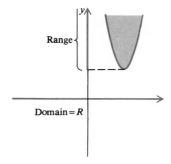

10. Graph the equation $x = |y|$. Compare it with the graph of Example 6.

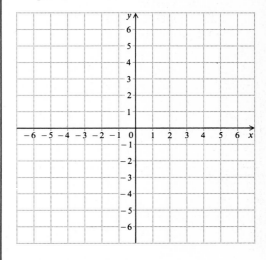

On each diagram in Exercises 11–13:

a) shade (on the x-axis) the domain;
b) shade (on the y-axis) the range.

11.

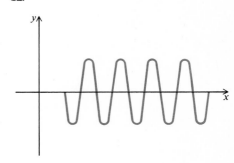

12.

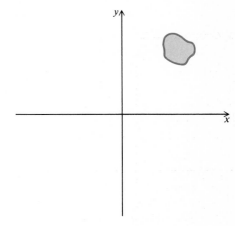

13.

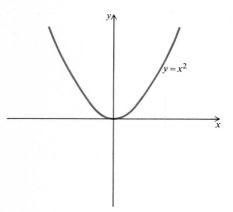

Example 9

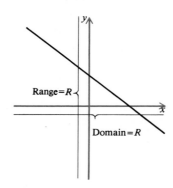

DO EXERCISES 11–13 ON PP. 109 AND 110.

EXERCISE SET 3.2

[i]

1. Determine whether $(-1, -3)$ is a solution of $y = 5x + 2$.

2. Determine whether $(-2, 8)$ is a solution of $y = 3x - 7$.

3. Determine whether $(-2, 7)$ is a solution of $4x + 3y = 12$.

4. Determine whether $(0, -2)$ is a solution of $2x - 3y = 6$.

5. Determine whether $(3, 0)$ is a solution of $x^2 - 2y = 6$.

6. Determine whether $(-2, -1)$ is a solution of $x^2 - 2y = 6$.

[ii] Graph.

7. $y = x + 3$

8. $y = x - 2$

9. $y = 3x - 2$

10. $y = -4x + 1$

11. $y = x^2$

12. $y = -x^2$

13. $y = x^2 + 2$

14. $y = x^2 - 2$

15. $x = y^2 + 2$

16. $x = y^2 - 2$

17. $y = |x + 1|$

18. $y = |x - 1|$

19. $x = |y + 1|$

20. $x = |y - 1|$

21. $xy = 10$

22. $xy = -18$

Graph and compare.

23. $y = x^2 + 1$ and $y = (-x)^2 + 1$

24. $y = x^2 - 2$ and $y = 2 - x^2$

[iii]

25. Graph a relation as follows.

  a) Draw a circle with radius of length 2, centered at (4, 3). Shade the circle and its interior.

  b) Shade (on the x-axis) the domain. Describe the domain.

  c) Shade (on the y-axis) the range. Describe the range.

26. Graph a relation as follows.

  a) Draw a triangle with vertices at (1, 1), (4, 2), and (3, 6). Shade the triangle and its interior.

  b) Shade (on the x-axis) the domain. Describe the domain.

  c) Shade (on the y-axis) the range. Describe the range.

☆

In each of the following, all relations to be considered are in $R \times R$, where R is the set of real numbers.

27. Graph the relation in which the second coordinate is always 2 and the first coordinate may be any real number.

28. Graph the relation in which the first coordinate is always -3, and the second coordinate may be any real number.

29. Graph the relation in which the second coordinate is always 1 more than the first coordinate, and the first coordinate may be any real number.

30. Graph the relation in which the second coordinate is always 1 less than the first coordinate, and the first coordinate may be any real number.

31. Graph the relation in which the second coordinate is always twice the first coordinate, and the first coordinate may be any real number.

32. Graph the relation in which the second coordinate is always half the first coordinate, and the first coordinate may be any real number.

33. Graph the relation in which the second coordinate is always the square of the first coordinate, and the first coordinate may be any real number.

34. Graph the relation in which the first coordinate is always the square of the second coordinate, and the second coordinate may be any real number.

Graph the following equations. The definition of absolute value in Section 1.8 may be helpful for some of these exercises.

35. $y = |x| + x$

36. $y = x|x|$

37. $y = |x^2 - 4|$

38. $y = x^3$

39. $y = \sqrt{x}$

40. $y = |x^3|$

41. $|y| = |x|$

42. $|x| + |y| = 0$

43. $|xy| = 1$

44. Graph the relation $\{(x, y) | 1 < x < 4 \text{ and } -3 < y < -1\}$.

★ ──

45. Graph $|x| + |y| = 1$.

46. Graph the relation $\{(x, y) | |x| \leq 1 \text{ and } |y| \leq 2\}$.

▦ Graph. In each case use your calculator and substitute at least twenty values of x between −5 and 5.

47. $y = \frac{1}{3}x^3 - x + \frac{2}{3}$

48. $y = \frac{1}{3}x^3 - \frac{1}{2}x^2 - 2x + 1$

3.3 FUNCTIONS

[i] Recognizing Graphs of Functions

A *function* is a special kind of relation. It is defined as follows.

DEFINITION

A *function* is a relation in which no two ordered pairs have the same first coordinate and different second coordinates.

In a function, given a member of the domain (a first coordinate) there is one and only one member of the range that goes with it (the second coordinate). Thus each member of the domain *determines* exactly one member of the range. It is easy to recognize the graph of a relation that is a function. If there are two or more points of the graph on the same vertical line, then the relation is not a function. Otherwise it is. Here are some graphs of functions. In graph (c), the solid dot indicates that $(-1, 1)$ belongs to the graph. The open dot indicates that $(-1, -2)$ does not belong to the graph. Thus no vertical line crosses the graph more than once.

OBJECTIVES

You should be able to:

[i] Recognize a graph of a function.

[ii] Use a formula to find function values.

[iii] Find the domain of a function, given by a formula.

[iv] Compose pairs of functions f and g, finding formulas for f ∘ g and g ∘ f.

a)

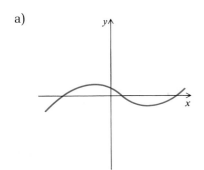

b)

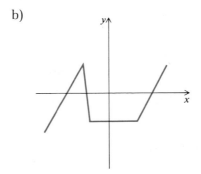

c)

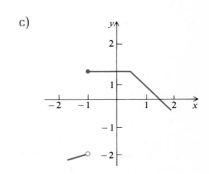

1. Which of the following are graphs of functions?

a)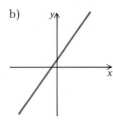

b)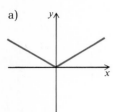

c)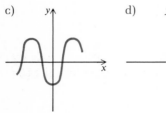

d)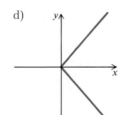

The following are not graphs of functions, because they fail the so-called *vertical line test.* That is, we can find a vertical line that meets the graph in more than one point.

a)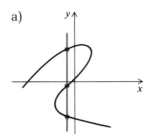

b)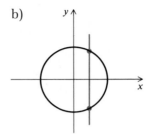

c)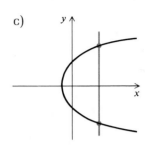

The vertical line test. **If it is possible for a vertical line to meet a graph more than once, the graph is not the graph of a function.**

DO EXERCISE 1.

[ii] Notation for Functions

Functions are often named by letters, such as *f* or *g*. A function *f* is thus a set of ordered pairs. If we represent the first coordinate of a pair by x, then we may represent the second coordinate by $f(x)$. The symbol $f(x)$ is read "*f* of x" or "*f* at x." The number represented by $f(x)$ is called the "value" of the function at x. [*Note:* "$f(x)$" does *not* mean "*f* times x."]

Example 1 Below let us call the function in color g.

(1, 4)	(2, 4)	(3, 4)	(4, 4)
(1, 3)	(2, 3)	(3, 3)	(4, 3)
(1, 2)	(2, 2)	(3, 2)	(4, 2)
(1, 1)	(2, 1)	(3, 1)	(4, 1)

Here $g(1) = 4$, $g(2) = 3$, $g(3) = 2$, and $g(4) = 4$.

Example 2 Let us call the function graphed below *f*. To find function values, we locate x on the x-axis, and then find $f(x)$ on the y-axis.

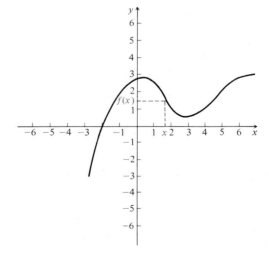

From the graph we can see that $f(-2) = 0$, that $f(0)$ is about 2.7, and that $f(3)$ is about 0.6.

DO EXERCISES 2 AND 3.

Some functions in real numbers can be defined by formulas, or equations. Here are some examples:

$$g(s) = 3, \qquad p(t) = \frac{1}{t}, \qquad f(x) = 3x^2 + 4x - 5, \qquad u(y) = |y| + 3.$$

Function values can be obtained by making substitutions for the variables.

A function such as g above is called a *constant function* because all its function values are the same. The range contains only one number, 3.

Example 3 Let $f(x) = 2x^2 - 3$. We can find function values as follows:

a) $\qquad\qquad f(0) = 2 \cdot 0^2 - 3 = -3;$

b) $\qquad\qquad f(-1) = 2(-1)^2 - 3 = 2 \cdot 1 - 3 = -1;$

c) $\qquad\qquad f(5a) = 2(5a)^2 - 3 = 2 \cdot 25a^2 - 3 = 50a^2 - 3;$

d) $\qquad\quad f(a - 4) = 2(a - 4)^2 - 3 = 2(a^2 - 8a + 16) - 3$
$\qquad\qquad\qquad\qquad = 2a^2 - 16a + 29;$

e) $\dfrac{f(a + h) - f(a)}{h} = \dfrac{[2(a + h)^2 - 3] - [2a^2 - 3]}{h}$

$$= \frac{2a^2 + 4ah + 2h^2 - 3 - 2a^2 + 3}{h}$$

$$= \frac{4ah + 2h^2}{h} = 4a + 2h.$$

If you have trouble finding function values when a formula is given, think of the formula, in the case of Example 3, as

$$f(\quad) = 2(\quad)^2 - 3.$$

Then whatever goes in the blank on the left between parentheses goes in the blank on the right between parentheses.

DO EXERCISE 4.

[iii] Finding the Domain of a Function

When a function in $R \times R$ is given by a formula, the domain is understood to be the set of all real numbers that are sensible replacements.

Example 4 Find the domain of $g(x) = \dfrac{x}{x^2 + 2x - 3}$.

The formula makes sense as long as a replacement for x does not make the denominator 0. To find those replacements that do make the de-

2. In this function, what is $f(1)$? $f(2)$? $f(3)$? $f(4)$?

(1, 4)	(2, 4)	(3, 4)	(4, 4)
(1, 3)	(2, 3)	(3, 3)	(4, 3)
(1, 2)	(2, 2)	(3, 2)	(4, 2)
(1, 1)	(2, 1)	(3, 1)	(4, 1)

What is the domain of this function? What is the range?

3. From the graph, find the following function values approximately: $f(2)$, $f(0)$, $f(-3)$.

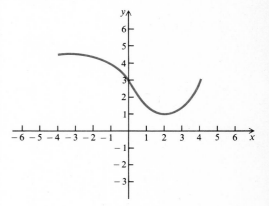

4. $f(x) = 3x^2 + 1$. Find:

a) $f(0)$;

b) $f(1)$;

c) $f(-1)$;

d) $f(2a)$;

e) $f(a + 1)$;

f) $\dfrac{f(a + h) - f(a)}{h}$.

Find the domain of each function.

5. $f(x) = \dfrac{x + 1}{3x^2 + 10x + 8}$

6. $g(x) = \sqrt{10x + 25}$

7. $p(x) = x^3 - 4x^2 + 2x + 8$

nominator 0 we solve $x^2 + 2x - 3 = 0$:

$$x^2 + 2x - 3 = 0$$
$$(x - 1)(x + 3) = 0$$
$$x - 1 = 0 \quad \text{or} \quad x + 3 = 0$$
$$x = 1 \quad \text{or} \quad\quad x = -3.$$

Thus the domain consists of the set of all real numbers except 1 and -3. We can name this set $\{x \mid x \neq 1 \text{ and } x \neq -3\}$.

Example 5 Find the domain of $f(x) = \sqrt{5x - 3}$.

The formula makes sense as long as a replacement makes the radicand nonnegative (no negative number has a square root). Thus to find the domain we solve the inequality $5x - 3 \geq 0$:

$$5x - 3 \geq 0$$
$$5x \geq 3$$
$$x \geq \frac{3}{5}.$$

The domain is $\{x \mid x \geq \frac{3}{5}\}$.

Example 6 Find the domain of $t(x) = x^3 + |x|$.

There are no restrictions on the numbers we can substitute into this formula. We can cube any real number, we can take the absolute value of any real number, and we can add the results. Thus the domain is the entire set of real numbers.

DO EXERCISES 5–7.

Functions, Mappings, and Machines

Functions can be thought of as mappings. A function f *maps* the set of first coordinates (the domain) to the set of second coordinates (the range).

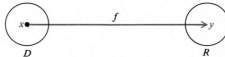

As in the diagram each x in the domain corresponds to (or is mapped onto) just one y in the range. That y is the second coordinate of the ordered pair (x, y).

Example 7 Consider the function f for which $f(x) = 2x + 3$.

Since $f(0)$ is 3, this function maps 0 to 3.

Since $f(3) = 9$, this function maps 3 to 9.

The concept of function is illuminated somewhat by considering a so-called "function machine." In this drawing we see a function machine designed, or programmed, to do the mapping (function) f. The inputs acceptable to the machine are the members of the domain of f. The outputs are, of course, members of the range of f.

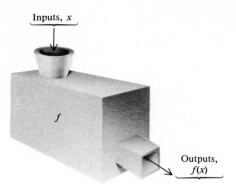

We may sometimes refer to an ordered pair $(x, f(x))$ as an input–output pair. The graph of a function consists of all the input–output pairs.

[iv] Composition of Functions

Functions can be combined in a way called *composition* of functions. Consider, for example,

$$f(x) = x^2 \quad \text{This function squares each input.}$$

and

$$g(x) = x + 1. \quad \text{This function adds 1 to each input.}$$

We define a new function that first does what g does (adds 1) and then does what f does (squares). The new function is called the *composition* of f and g, and is symbolized $f \circ g$. Let us think of hooking two function machines f and g together to get the resultant function machine $f \circ g$.

DEFINITION

The composed function $f \circ g$ is defined as follows: $f \circ g(x) = f\big(g(x)\big)$.

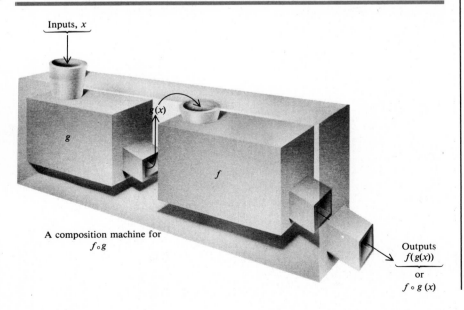

A composition machine for
$f \circ g$

8. Let $f(x) = x - 2$ and $g(x) = 2x^2$.
Find $f \circ g(x)$ and $g \circ f(x)$.

Now let us see how to find a formula for $f \circ g$, and also the reverse composition $g \circ f$.

Example 8 Given that $f(x) = x^2$ and $g(x) = x + 1$, find formulas for $f \circ g(x)$ and $g \circ f(x)$.

By definition of $\circ$,

$$f \circ g(x) = f(g(x)).$$

Now

$$f(g(x)) = f(x + 1) \qquad \text{Substituting } x + 1 \text{ for } g(x)$$
$$= (x + 1)^2 \qquad \text{Substituting } x + 1 \text{ for } x \text{ in the formula for } f(x)$$
$$= x^2 + 2x + 1.$$

Similarly,

$$g \circ f(x) = g(x^2) \qquad \text{Substituting } x^2 \text{ for } f(x)$$
$$= x^2 + 1. \qquad \text{Substituting } x^2 \text{ for } x \text{ in the formula for } g(x)$$

Note that the function $f \circ g$ is not the same as the function $g \circ f$.

In order for functions to be composable, such as f and g above, the outputs of g must be acceptable as inputs for f. In other words, if any number $g(x)$ is not in the domain of f, then x is not in the domain of $f \circ g$. Note also the order of happenings in $f \circ g$. The function $f \circ g$ does *first* what g does, *then* what f does.

DO EXERCISE 8.

9. Let $u(x) = 2x^2$ and $v(x) = 3x + 2$.
Find $u \circ v(x)$ and $v \circ u(x)$.

Example 9 Let $f(x) = 2x$ and $g(x) = x^2 + 1$. Find $f \circ g(x)$ and $g \circ f(x)$.

By definition of $\circ$,

$$f \circ g(x) = f(g(x))$$
$$= f(x^2 + 1) \qquad \text{Substituting } x^2 + 1 \text{ for } g(x)$$
$$= 2(x^2 + 1) \qquad \text{Substituting } x^2 + 1 \text{ for } x \text{ in the formula for } f(x)$$
$$= 2x^2 + 2.$$

Similarly,

$$g \circ f(x) = g(f(x))$$
$$= g(2x) \qquad \text{Substituting } 2x \text{ for } f(x)$$
$$= (2x)^2 + 1 \qquad \text{Substituting } 2x \text{ for } x \text{ in the formula for } g(x)$$
$$= 4x^2 + 1.$$

DO EXERCISE 9.

EXERCISE SET 3.3

[i]

1. Which of the following are graphs of functions?

a)   b)   c)   d)

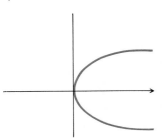   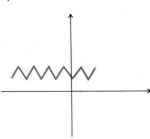   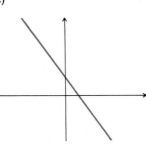   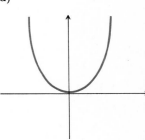

2. Which of the following are graphs of functions? An open circle indicates that the point does not belong to the graph.

a)   b)   c)   d)

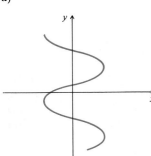   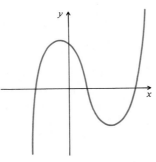   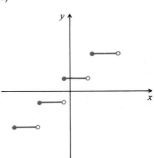   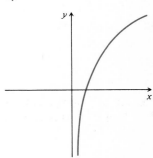

[ii]

3. $f(x) = 5x^2 + 4x$. Find:

   a) $f(0)$;   b) $f(-1)$;

   c) $f(3)$;   d) $f(t)$;

   e) $f(t-1)$;   f) $\dfrac{f(a+h) - f(a)}{h}$.

4. $g(x) = 3x^2 - 2x + 1$. Find:

   a) $g(0)$;   b) $g(-1)$;

   c) $g(3)$;   d) $g(t)$;

   e) $g(a+h)$;   f) $\dfrac{g(a+h) - g(a)}{h}$.

5. $f(x) = 2|x| + 3x$. Find:

   a) $f(1)$;   b) $f(-2)$;

   c) $f(-4)$;   d) $f(2y)$;

   e) $f(a+h)$;   f) $\dfrac{f(a+h) - f(a)}{h}$.

6. $g(x) = x^3 - 2x$. Find:

   a) $g(1)$;   b) $g(-2)$;

   c) $g(-4)$;   d) $g(3y)$;

   e) $g(2+h)$;   f) $\dfrac{g(2+h) - g(2)}{h}$.

7. ▦ $f(x) = 4.3x^2 - 1.4x$. Find:

   a) $f(1.034)$;   b) $f(-3.441)$;

   c) $f(27.35)$;   d) $f(-16.31)$.

8. ▦ $g(x) = \sqrt{2.2|x| + 3.5}$. Find:

   a) $g(17.3)$;   b) $g(-64.2)$;

   c) $g(0.095)$;   d) $g(-6.33)$.

9. $f(x) = \dfrac{x^2 - x - 2}{2x^2 - 5x - 3}$. Find:

   a) $f(0)$;   b) $f(4)$;

   c) $f(-1)$;   d) $f(3)$.

10. $s(x) = \sqrt{\dfrac{3x - 4}{2x + 5}}$. Find:

   a) $s(10)$;   b) $s(2)$;

   c) $s(1)$;   d) $s(-1)$.

[iii] In Exercises 11–20, find the domain of each function.

11. $f(x) = 7x + 4$

12. $f(x) = |3x - 2|$

13. $f(x) = 4 - \dfrac{2}{x}$

14. $f(x) = \sqrt{x - 3}$

15. $f(x) = \sqrt{7x + 4}$

16. $f(x) = \dfrac{1}{9 - x^2}$

17. $f(x) = \dfrac{1}{x^2 - 4}$

18. $f(x) = \dfrac{2x + 6}{x^3 - 4x}$

19. $f(x) = \dfrac{4x^3 + 4}{4x^2 - 5x - 6}$

20. $f(x) = \dfrac{x^3 + 8}{x^2 - 4}$

[iv] In Exercises 21–26, find $f \circ g(x)$ and $g \circ f(x)$.

21. $f(x) = 3x^2 + 2$, $\quad g(x) = 2x - 1$

22. $f(x) = 4x + 3$, $\quad g(x) = 2x^2 - 5$

23. $f(x) = 4x^2 - 1$, $\quad g(x) = \dfrac{2}{x}$

24. $f(x) = \dfrac{3}{x}$, $\quad g(x) = 2x^2 + 3$

25. $f(x) = x^2 + 1$, $\quad g(x) = x^2 - 1$

26. $f(x) = \dfrac{1}{x^2}$, $\quad g(x) = x + 2$

☆ ──────────────────────────────

27. From this graph, find approximately $g(-2)$, $g(-3)$, $g(0)$, and $g(2)$.

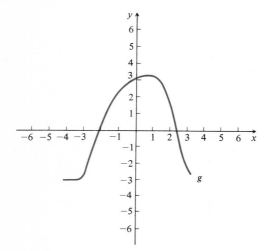

28. From this graph, find approximately $h(-2)$, $h(0)$, $h(3)$, and $h(-3)$.

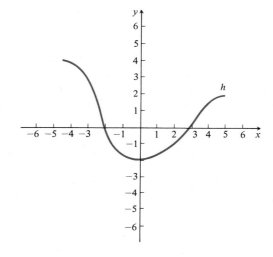

For each function find $\dfrac{f(x + h) - f(x)}{h}$.

29. $f(x) = \dfrac{1}{x}$

30. $f(x) = \dfrac{1}{x^2}$

31. $f(x) = \sqrt{x}$ (Rationalize the numerator.)

Find the domain of each function.

32. $f(x) = \dfrac{\sqrt{x}}{2x^2 - 3x - 5}$

33. $f(x) = \dfrac{\sqrt{x + 3}}{x^2 - x - 2}$

34. $f(x) = \dfrac{\sqrt{x + 1}}{x + |x|}$

35. $f(x) = \sqrt{x^2 + 1}$

36. For $f(x) = \dfrac{1}{1 - x}$, find $f \circ f(x)$ and $f \circ [f \circ f(x)]$.

37. In Exercise 23, find the domain of $f \circ g$ and $g \circ f$.

38. Determine whether the relation $\{(x, y) \mid xy = 0\}$ is a function.

★

39. The *greatest integer function* $f(x) = [x]$ is defined as follows: $[x]$ is the greatest integer that is less than or equal to x. For example, if $x = 3.74$, then $[x] = 3$; and if $x = -0.98$, then $[x] = -1$. Graph the greatest integer function for values of x such that $-5 \leq x \leq 5$.

40. Graph the equation $[y] = [x]$. See Exercise 39. Is this the graph of a function?

3.4 SYMMETRY

Symmetry with Respect to a Line

In the figure points P and P_1 are said to be *symmetric* with respect to line ℓ. They are the same distance from ℓ.

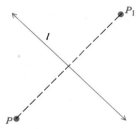

OBJECTIVES

You should be able to:

[i] Given an equation defining a relation, determine whether a graph is symmetric with respect to a coordinate axis.

[ii] Given an equation defining a relation, determine whether the graph is symmetric with respect to the origin.

DEFINITION

> Two points P and P_1 are *symmetric with respect to a line* ℓ if and only if ℓ is the perpendicular bisector of the segment $\overline{PP_1}$. The line ℓ is known as the *line of symmetry.*

We also say that the two points P and P_1 are *reflections* of each other across the line. The line is therefore also known as a *line of reflection*.

Now consider a set of points (geometric figure) as in the colored curve below. This figure is said to be symmetric with respect to the line ℓ, because if you pick any point Q in the set, you can find another point Q_1 in the set such that Q and Q_1 are symmetric with respect to ℓ.

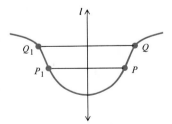

DEFINITION

> A figure, or set of points, is *symmetric with respect to a line* ℓ if and only if for each point Q in the set there exists another point Q_1 in the set for which Q and Q_1 are symmetric with respect to line ℓ.

1. a) Plot the point (3, 2). Let the
   y-axis be a line of symmetry.
   Plot the point symmetric to
   (3, 2). What are its coordinates?

   b) Plot the point (−4, −5). Let the
   y-axis be a line of reflection.
   Plot the image of (−4, −5).
   What are its coordinates?

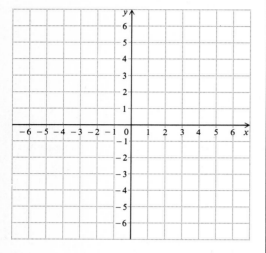

Imagine picking the preceding figure up and flipping it over. Points P and P_1 would be interchanged. Points Q and Q_1 would be interchanged. These are, then, pairs of symmetric points. The entire figure would look exactly like it did before flipping. This means that *each* point of the figure is symmetric with *some* point of the figure. Thus the figure is symmetric with respect to the line. A point and its reflection are known as *images* of each other. Thus P_1 is the image of P, for example. The line ℓ is known as an *axis of symmetry*.

[i] Symmetry with Respect to the Axes

There are special and interesting kinds of symmetry in which a coordinate axis is a line of symmetry. The following example shows figures that are symmetric with respect to an axis and a figure that is not.

Example 1 In graph (a), flipping the figure about the y-axis would not change the figure. In graph (b), flipping the graph about the x-axis would not change the figure. In graph (c), flipping about either axis would change the figure.

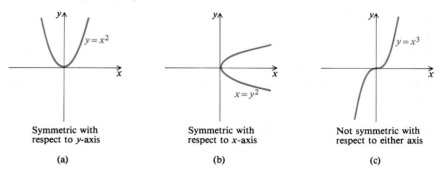

Let us consider a figure like graph (a), symmetric with respect to the y-axis. For every point of the figure there is another point the same distance across the y-axis. The first coordinates of such a pair of points are additive inverses of each other.

Example 2 In the relation $y = x^2$ there are points (2, 4) and (−2, 4). The first coordinates, 2 and −2, are additive inverses of each other, while the second coordinates are the same. For every point of the figure (x, y), there is another point $(−x, y)$.

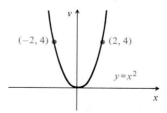

DO EXERCISE 1.

Let us consider a figure symmetric with respect to the x-axis. For every point of such a figure there is another point the same distance across the x-axis. The second coordinates of such a pair of points are additive inverses of each other.

Example 3 In the relation $x = y^2$ there are points $(4, 2)$ and $(4, -2)$. The second coordinates, 2 and -2, are additive inverses of each other, while the first coordinates are the same. For every point of the figure (x, y), there is another point $(x, -y)$.

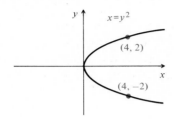

DO EXERCISE 2.

Let us consider a figure symmetric with respect to the y-axis, as in Example 2. Suppose it is defined by an equation. If in this equation we replace x by $-x$, we obtain a new equation, but we get the same figure. This is true because any number x gives us the same y-value as its additive inverse, $-x$.

Let us consider a figure symmetric with respect to the x-axis, as in Example 3. Suppose it is defined by an equation. If we replace y by $-y$ in the equation, we obtain a new equation, but we get the same figure. This is true because any number y gives us the same x-value as $-y$. We thus have a means of testing a relation for symmetry, when it is defined by an equation.

When a relation is defined by an equation:

1. **If replacing x by $-x$ produces an equivalent equation, then the graph is symmetric with respect to the y-axis.**

2. **If replacing y by $-y$ produces an equivalent equation, then the graph is symmetric with respect to the x-axis.**

Example 4 Test $y = x^2 + 2$ for symmetry with respect to the y-axis.

a) Replace x by $-x$.

$$y = x^2 + 2 \qquad (1)$$
$$y = (-x)^2 + 2$$

b) Simplify, if possible.

$$y = (-x)^2 + 2 = x^2 + 2 \qquad (2)$$

c) Is the resulting equation (2) equivalent to the original (1)?

Since the answer is yes, the graph is symmetric with respect to the y-axis.

2. Let the x-axis be a line of symmetry.

a) Plot the point $(4, 3)$. Plot the point symmetric to it. What are its coordinates?

b) Plot the point $(3, -5)$. Plot its image after reflection across the x-axis. What are its coordinates?

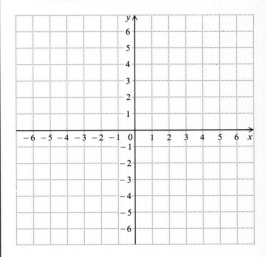

Test the following for symmetry with respect to the coordinate axes.

3. $y = x^2 - 3$

4. $y^2 = x^3$

5. $x^4 = y^2 + 2$

6. $3y^2 + 4x^2 = 12$

7. $a + 3b = 5$

8. $2p^3 + 4q^3 = 1$

Example 5 Test $y = x^2 + 2$ for symmetry with respect to the x-axis.

a) Replace y by $-y$.

$$y = x^2 + 2 \tag{1}$$
$$-y := x^2 + 2$$

b) Simplify, if possible.

The equation is simplified for the most part, although we could multiply on both sides by -1, obtaining

$$y = -(x^2 + 2). \tag{2}$$

c) Is the resulting equation (2) equivalent to the original (1)?

This answer may be obvious to you and is no. To be sure one might use some trial-and-error reasoning as follows. Suppose we substitute 0 for x in equation (1). Then

$$y = 0^2 + 2 = 2,$$

so $(0, 2)$ is a solution of equation (1). For equations (1) and (2) to be equivalent, $(0, 2)$ must also be a solution of equation (2). We substitute to find out:

$$
\begin{array}{c|c}
y = & -(x^2 + 2) \\
\hline
2 & -(0^2 + 2) \\
& -2
\end{array}
$$

Thus $(0, 2)$ is not a solution of equation (2); hence the equations are not equivalent and the graph is not symmetric with respect to the x-axis.

Example 6 Test $x^2 + y^4 + 5 = 0$ for symmetry with respect to the x-axis.

a) Replace y by $-y$.

$$x^2 + y^4 + 5 = 0 \tag{1}$$
$$x^2 + (-y)^4 + 5 = 0$$

b) Simplify, if possible.

$$x^2 + (-y)^4 + 5 = x^2 + y^4 + 5 = 0 \tag{2}$$

c) Is the resulting equation equivalent to the first?

Since the answer is yes, the graph is symmetric with respect to the x-axis.

DO EXERCISES 3–8.

Symmetry with Respect to a Point

Two points are symmetric with respect to a point when they are situated as shown on the following page. That is, the points are the same distance from that point, and all three points are on a line.

DEFINITION

Two points P and P_1 are *symmetric with respect to a point Q if and only* if Q (the point of symmetry) is the midpoint of segment $\overline{PP_1}$.

A *set* of points is symmetric with re-spect to a point when each point in the set is symmetric with some point in the set. This is illustrated here. Imagine sticking a pin in this figure at O and then rotating the figure 180°. Points P and P_1 would be interchanged. Points Q and Q_1 would be interchanged. These are pairs of symmet-ric points. The entire figure would look exactly as it did before rotating. This means that *each* point of the figure is symmetric with *some* point of the figure. Thus the figure is symmetric with respect to the point O.

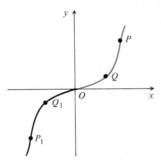

DEFINITION

A set of points is *symmetric with respect to a point B* if and only if for every point P in the set there exists another point P_1 in the set for which P and P_1 are symmetric with respect to B.

[ii] Symmetry with Respect to the Origin

A special kind of symmetry with respect to a point is symmetry with respect to the origin.

Example 7 In graphs (a) and (b), rotating the figure about the origin 180° would not change the figure. In graph (c) such a rotation would change the figure.

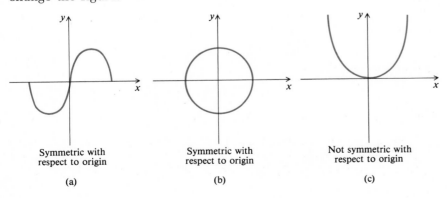

Symmetric with respect to origin

(a)

Symmetric with respect to origin

(b)

Not symmetric with respect to origin

(c)

9. Draw coordinate axes. Let the origin be a point of symmetry.

a) Plot the point (3, 2). Plot the point symmetric to it. What are its coordinates?

b) Plot the point (−4, 3). Plot the point symmetric to it. What are its coordinates?

c) Plot the point (−5, −7). Plot the point symmetric to it. What are its coordinates?

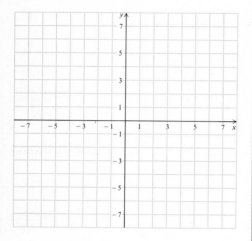

Let us consider figures like (a) and (b) above, symmetric with respect to the origin. For every point of the figure there is another point the same distance across the origin. The first coordinates of such a pair are additive inverses of each other, and the second coordinates are additive inverses of each other.

DO EXERCISE 9.

Example 8 The relation $y = x^3$ is symmetric with respect to the origin. In this relation are the points (2, 8) and (−2, −8). The first coordinates are additive inverses of each other. The second coordinates are additive inverses of each other. For every point of the figure (x, y), there is another point $(-x, -y)$.

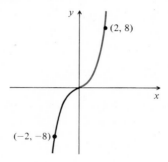

In a relation symmetric with respect to the origin, as in Example 8, if we replace x by −x and y by −y, we obtain a new equation, but we will get the same figure. This is true because whenever a point (x, y) is in the relation, the point $(-x, -y)$ is also in the relation. This gives us a means for testing a relation for symmetry when it is defined by an equation.

When a relation is defined by an equation, if replacing _x_ by −_x_ and replacing _y_ by −_y_ produces an equivalent equation, then the graph is symmetric with respect to the origin.

Example 9 Test $x^2 = y^2 + 2$ for symmetry with respect to the origin.

a) Replace x by −x and y by −y.

$$x^2 = y^2 + 2$$
$$(-x)^2 = (-y)^2 + 2$$

b) Simplify, if possible.

Since $(-x)^2 = x^2$ and $(-y)^2 = y^2$, $(-x)^2 = (-y)^2 + 2$ simplifies to

$$x^2 = y^2 + 2.$$

c) Is the resulting equation equivalent to the original?

Since the answer is yes, the graph is symmetric with respect to the origin.

Example 10 Test $2x + 3y = 8$ for symmetry with respect to the origin.

a) Replace x by $-x$ and y by $-y$.

$$2x + 3y = 8$$
$$2(-x) + 3(-y) = 8$$

b) Simplify, if possible.

$$2(-x) + 3(-y) = -2x - 3y,$$

so

$$-(2x + 3y) = 8 \quad \text{and} \quad 2x + 3y = -8$$

c) Is the resulting equation equivalent to the original?

The equation $2x + 3y = -8$ is *not* equivalent to $2x + 3y = 8$, so the graph is not symmetric with respect to the origin.

DO EXERCISES 10–15.

Test the following for symmetry with respect to the origin.

10. $x^2 + 3y^2 = 4$

11. $x = y$

12. $x = -y$

13. $xy = 5$

14. $ab = -5$

15. $u = |v|$

EXERCISE SET 3.4

[i], [ii] In Exercises 1–12, test for symmetry with respect to the coordinate axes and the origin.

1. $3y = x^2 + 4$

2. $5y = 2x^2 - 3$

3. $y^3 = 2x^2$

4. $3y^3 = 4x^2$

5. $2x^4 + 3 = y^2$

6. $3y^2 = 2x^4 - 5$

7. $2y^2 = 5x^2 + 12$

8. $3x^2 - 2y^2 = 7$

9. $2x - 5 = 3y$

10. $5y = 4x + 5$

11. $3b^3 = 4a^3 + 2$

12. $p^3 - 4q^3 = 12$

[ii] In Exercises 13–24, test for symmetry with respect to the origin.

13. $3x^2 - 2y^2 = 3$

14. $5y^2 = -7x^2 + 4$

15. $5x - 5y = 0$

16. $3x = 3y$

17. $3x + 3y = 0$

18. $7x = -7y$

19. $3x = \dfrac{5}{y}$

20. $3y = \dfrac{7}{x}$

21. $y = |2x|$

22. $3x = |y|$

23. $3a^2 + 4a = 2b$

24. $5v = 7u^2 - 2u$

☆

Consider the following figure for Exercises 25–28.

25. Graph the reflection across the x-axis.

26. Graph the reflection across the y-axis.

27. Graph the reflection across the line $y = x$.

28. Graph the figure formed by reflecting each point through the origin.

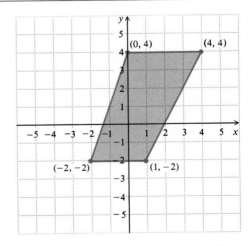

★

29. Consider symmetries with respect to the x-axis, the y-axis, and the origin. Prove that symmetry with respect to any two of these implies symmetry with respect to the other.

OBJECTIVE

You should be able to:

[i] Given the graph of a relation,
graph its transformation under
translations, reflections, stretchings, and shrinkings.

3.5 TRANSFORMATIONS

[i] Given a relation, we can find various ways of altering it to
obtain another relation. Such an alteration is called a *transformation*.
If such an alteration consists merely of moving the graph without
changing its shape or orientation, the transformation is called a *translation*.

Vertical Translations

Consider the relations $y = x^2$ and $y = 1 + x^2$ whose graphs are
shown below. The graph of $y = 1 + x^2$ has the same shape as that of
$y = x^2$, but is moved upward a distance of 1 unit. Consider any equation $y = f(x)$. Adding a constant a to produce $y = a + f(x)$ changes
each function value by the same amount, a, hence produces no change
in the shape of the graph, but merely translates it upward if the constant is positive. If a is negative, the graph will be moved downward.

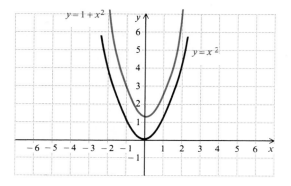

1. a) Graph $y = x^2$. Then graph
$y = 2 + x^2$ and compare.
b) Graph $y = -2 + x^2$ and compare.

DO EXERCISE 1.

Note that $y = 1 + x^2$ is equivalent to $y - 1 = x^2$. Thus the transformation described above amounts to replacing y by $y - 1$ in the original
equation.

THEOREM 1

**In an equation of a relation, replacing y by $y - a$, where a is a con-
stant, translates the graph vertically a distance of $|a|$. If a is positive,
the translation is upward. If a is negative, the translation is downward.***

_____

*In practice, when working with functions, we are more inclined to write $y = a + f(x)$
instead of $y - a = f(x)$, but for relations in general we are more likely to replace y by
$y - a$.

If in an equation we replace y by $y + 3$, this is the same as replacing it by $y - (-3)$. In this case the constant a is -3 and the translation is downward. If we replace y by $y - 5$, the constant a is 5 and the translation is upward.

Example 1 Sketch a graph of $y = |x|$ and then one of $y = -2 + |x|$.

The graph of $y = |x|$ is shown below. Now consider $y = -2 + |x|$. Note that $y = -2 + |x|$ is equivalent to $y + 2 = |x|$ or $y - (-2) = |x|$. This shows that the new equation can be obtained by replacing y in $y = |x|$ by $y - (-2)$, so the translation is downward, 2 units, as shown below.

2. Sketch a graph of $y = -3 + x^2$.

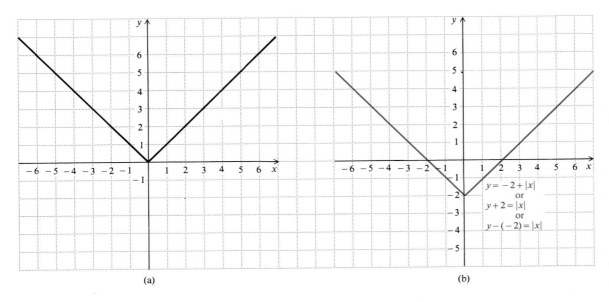

(a)   (b)

DO EXERCISES 2 AND 3.

Horizontal Translations

Replacing y by $y - a$ in an equation translates vertically a distance of $|a|$. The translation is in the positive direction (upward) if a is positive. A similar thing happens in the horizontal direction. If we replace x by $x - b$ everywhere x occurs in the equation, we translate a distance of $|b|$ horizontally. If b is positive, we translate in the positive direction (to the right). If b is negative, we translate in the negative direction (to the left).

THEOREM 2

In an equation of a relation, replacing x by $x - b$, where b is a constant, translates the graph horizontally a distance of $|b|$. If b is positive, the translation is to the right. If b is negative, the translation is to the left.

3. Here is a graph of $y = f(x)$. There is no formula for it. Sketch a graph of $y = 3 + f(x)$.

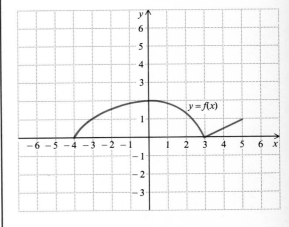

4. Sketch a graph of $y = (x + 3)^2$.

5. Here is a graph of $y = g(x)$. There is no formula for it. Sketch a graph of $y = g(x - 4)$.

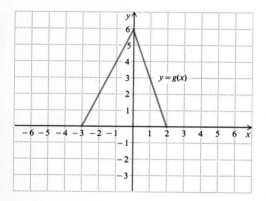

Example 2 Given a graph of $y = |x|$, sketch a graph of $y = |x + 2|$.

Here we note that x is replaced by $x + 2$, or $x - (-2)$. Thus $b = -2$, and the graph will be moved two units in the negative direction (to the left).

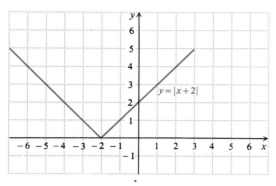

Example 3 A circle centered at the origin with radius of length 1 has an equation $x^2 + y^2 = 1$. If we replace x by $x - 1$ and y by $y + 2$, we translate the circle so that the center is at the point $(1, -2)$.

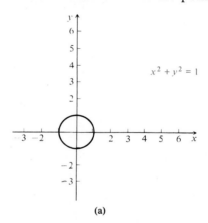

(a)

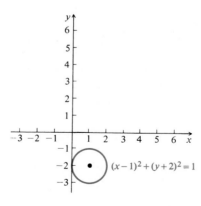

(b)

DO EXERCISES 4 AND 5.

Vertical Stretchings and Shrinkings

Consider the function $y = |x|$. We will compare its graph with that of $y = 2|x|$ and $y = \frac{1}{2}|x|$. The graph of $y = 2|x|$ looks like that of $y = |x|$ but every output is doubled, so the graph is stretched in a vertical direction. The graph of $y = \frac{1}{2}|x|$ is flattened or shrunk in a vertical direction since every output is cut in half.

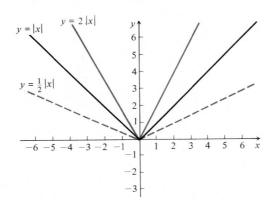

Consider any equation such as $y = f(x)$. Multiplying on the right by the constant 2 will double every function value, thus stretching the graph both ways from the horizontal axis. A similar thing is true for any constant greater than 1. If the constant is between 0 and 1, then the graph will be flattened or shrunk vertically.

DO EXERCISES 6 AND 7.

When we multiply by a negative constant, the graph is reflected across the x-axis as well as being stretched or shrunk.

Example 4 Compare the graphs of $y = |x|$, $y = -2|x|$, and $y = -\frac{1}{2}|x|$.

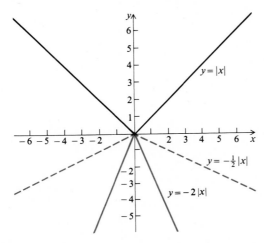

Multiplying by c on the right is, of course, equivalent to dividing by c on the left.

6. Graph $y = x^2$. Graph $y = 2x^2$ and compare.

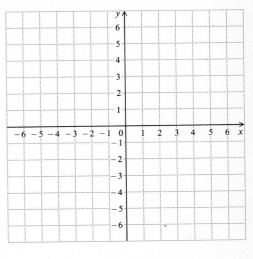

7. Graph $y = \frac{1}{2}x^2$. Compare with $y = x^2$.

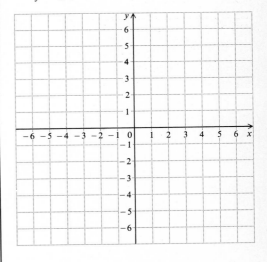

THEOREM 3

In an equation of a relation, dividing y by a constant c does the following to the graph.

1. If $|c| > 1$, the graph is stretched vertically.
2. If $|c| < 1$, the graph is shrunk vertically.
3. If c is negative, the graph is also reflected across the x-axis.*

Note that if $c = -1$, this has the effect of replacing y by $-y$ and we obtain a reflection without stretching or shrinking.

Example 5 Here is a graph of $y = f(x)$. Sketch a graph of $\dfrac{y}{2} = f(x)$ or $y = 2f(x)$.

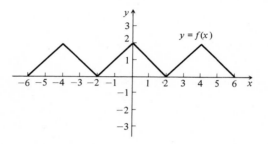

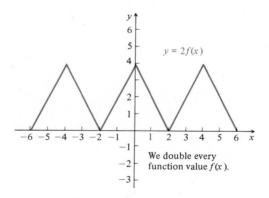

We double every function value $f(x)$.

Example 6 On the following page is a graph of $y = g(x)$. Sketch a graph of $\dfrac{y}{-\dfrac{1}{2}} = g(x)$ or $y = -\dfrac{1}{2}g(x)$.

---

*Again, with functions, we are more inclined to write $y = cf(x)$ instead of $y/c = f(x)$, but for relations in general we are more likely to replace y by y/c.

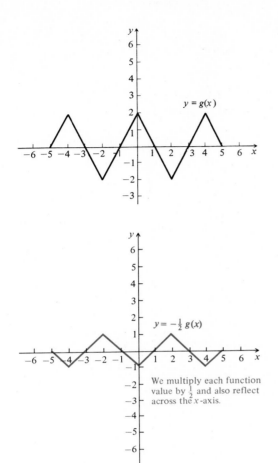

DO EXERCISES 8 AND 9.

Horizontal Stretchings and Shrinkings

For vertical stretchings and shrinkings we divided y by a constant c. Similarly, if we divide x by a constant wherever it occurs, we will get a horizontal stretching or shrinking.

THEOREM 4

In an equation of a relation, dividing x wherever it occurs by d does the following to the graph.

1. If $|d| < 1$, the graph is shrunk horizontally.

2. If $|d| > 1$, the graph is stretched horizontally.

3. If d is negative, the graph is also reflected across the y-axis.*

---

*Again, with functions, we are more inclined to write $y = f(kx)$ instead of $y = f(x/d)$, but for relations in general we usually consider replacing x by x/d.

8. Graph $y = -2x^2$. Compare with $y = x^2$.

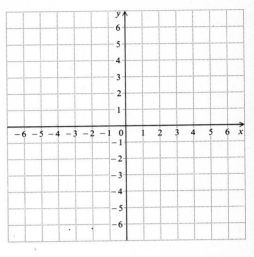

9. Graph $y = -\dfrac{1}{2}x^2$. Compare with $y = x^2$.

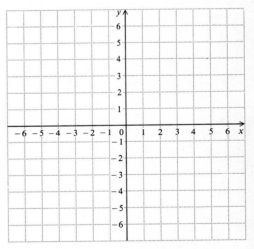

132

Here is a graph of $y = t(x)$. Use graph paper to sketch the following.

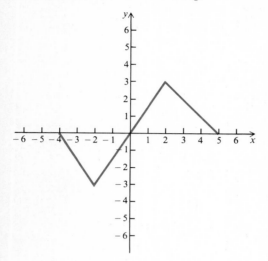

10. Sketch a graph of $y = \frac{1}{2}t(x)$.

11. Sketch a graph of $y = -2t(x)$.

12. Sketch a graph of $y = t(2x)$.

13. Sketch a graph of $y = t\left(-\frac{1}{2}x\right)$.

Note that if $d = -1$, this has the effect of replacing x by $-x$ and we obtain a reflection without stretching or shrinking.

Example 7 Here is a graph of $y = f(x)$. Sketch a graph of $y = f\left(\frac{x}{\frac{1}{2}}\right)$ or $y = f(2x)$, a graph of $y = f\left(\frac{x}{2}\right)$ or $y = f\left(\frac{1}{2}x\right)$, and a graph of $y = f\left(\frac{x}{-2}\right)$ or $y = f\left(-\frac{1}{2}x\right)$.

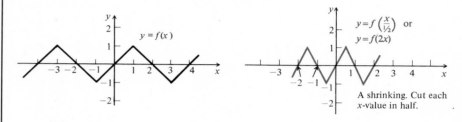

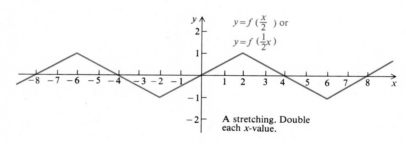

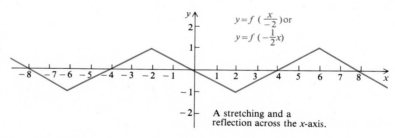

DO EXERCISES 10–13.

EXERCISE SET 3.5

[i] In Exercises 1–14, sketch graphs by transforming the graph of $y = |x|$.

1. $y + 3 = |x|$

2. $y = 2 + |x|$

3. $y = |x - 1|$

4. $y = |x + 2|$

5. $y = -4|x|$

6. $\frac{y}{3} = |x|$

7. $y = \frac{1}{3}|x|$

8. $y = -\frac{1}{4}|x|$

9. $y = |2x|$

10. $y = \left|\frac{x}{3}\right|$

11. $y = |x - 2| + 3$

12. $y = 2|x + 1| - 3$

13. $y = -3|x - 2|$

14. $y = \frac{1}{3}|x + 2| + 1$

Here is a graph of $y = f(x)$. No formula will be given for this function. In Exercises 15-33, sketch graphs by transforming this one.

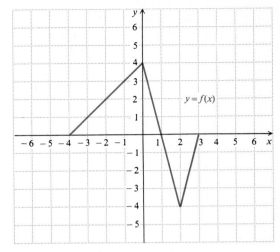

15. $y = 2 + f(x)$

16. $y + 1 = f(x)$

17. $y = f(x - 1)$

18. $y = f(x + 2)$

19. $\dfrac{y}{-2} = f(x)$

20. $y = 3f(x)$

21. $y = \dfrac{1}{3}f(x)$

22. $y = -\dfrac{1}{2}f(x)$

23. $y = f(2x)$

24. $y = f(3x)$

25. $y = f(-2x)$

26. $y = f(-3x)$

27. $y = f\left(\dfrac{x}{-2}\right)$

28. $y = f\left(\dfrac{1}{3}x\right)$

29. $y = f(x - 2) + 3$

30. $y = -3f(x - 2)$

31. $y = 2 \cdot f(x + 1) - 2$

32. $y = \dfrac{1}{2}f(x + 2) - 1$

33. $y = -\dfrac{1}{2}f(x - 3) + 2$

Here is a graph of $y = f(x)$. No formula will be given for this function. In Exercises 34-38, sketch graphs by transforming this one.

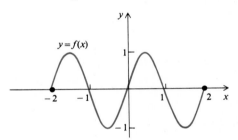

34. $y = -2f(x + 1) - 1$

35. $y = 3f(x + 2) + 1$

36. $y = \dfrac{5}{2}f(x - 3) - 2$

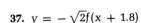

37. $y = -\sqrt{2}f(x + 1.8)$

38. $y = \dfrac{\sqrt{3}}{2} \cdot f(x - 2.5) - 5.3$

★

39. A *linear transformation* of a coordinatized line is one that takes any point x to the point x', where $x' = ax + b$ (a and b constants). For example, the transformation $x' = 3x + 5$ takes the point 2 to the point $3 \cdot 2 + 5$, or 11. Suppose for a particular linear transformation the point 2 goes to 5 and the point 3 goes to 7.

a) Describe the transformation as $x' = ax + b$.

b) This transformation leaves one point fixed. What point is it?

c) The set $\{x \mid -2 \le x \le 1\}$ is mapped to what set under this transformation?

40. How does the linear transformation $x' = x - 3$ transform each of the following sets?

a) $\{x \mid 0 < x < 1\}$

b) $\{x \mid -4 \le x < -1\}$

c) $\{x \mid 8 \le x \le 20\}$

41. Show that any linear transformation $x' = ax + b$, $a \neq 1$, leaves one point fixed.

42. A professor gives a test with 80 possible points. He decides that a score of 55 should be passing. He performs a linear transformation on the scores so that a perfect paper gets 100 and 70 is passing.

a) Describe the linear transformation that the professor used.

b) What grade remains unchanged under the transformation?

43. Suppose the average score on the test of Exercise 42 was 60. What will be the average of the transformed scores?

Given the graph of $y = f(x)$ for Exercises 34–38, graph each of the following.

44. $\dfrac{y}{3} = f\left(2x + \dfrac{1}{2}\right)$

45. $y = -4 \cdot f(5x + 10)$

OBJECTIVES

You should be able to:

[i] Given the graph of a function or a formula, determine whether the function is even, odd, or neither even nor odd.

[ii] Given the graph of a function, determine whether it is periodic, and if it is periodic, determine its period.

[iii] Write interval notation for certain sets.

[iv] Given the graph of a function, determine whether it is continuous over a specified interval, and indicate discontinuities.

[v] Given the graph of a function, determine whether it is increasing, decreasing, or neither increasing nor decreasing.

[vi] Graph functions defined piecewise.

3.6 SOME SPECIAL CLASSES OF FUNCTIONS

[i] Even and Odd Functions

If the graph of a function is symmetric with respect to the y-axis, it is an *even* function. Recall (Section 3.4) that a function will be symmetric to the y-axis if in its equation we can replace x by $-x$ and obtain an equivalent equation. Thus if we have a function given by $y = f(x)$, then $y = f(-x)$ will give the same function if the function is even. In other words, an even function is one for which $f(x) = f(-x)$ for all x in its domain. This is the definition of even function.

DEFINITION

> A function f is an *even* function in case $f(x) = f(-x)$ for all x in the domain of f.

Example 1 Determine whether the function $f(x) = x^2 + 1$ is even.

a) Find $f(-x)$ and simplify.

$$f(-x) = (-x)^2 + 1 = x^2 + 1$$

b) Compare $f(x)$ and $f(-x)$.

Since $f(x) = f(-x)$ for all x in the domain, f is an even function.

Note that the graph is symmetric with respect to the y-axis.

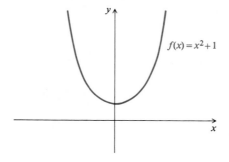

Example 2 Determine whether the function $f(x) = x^2 + 8x^3$ is even.

a) Find $f(-x)$ and simplify.

$$f(-x) = (-x)^2 + 8(-x)^3 = x^2 - 8x^3$$

b) Compare $f(x)$ and $f(-x)$.

Since $f(x)$ and $f(-x)$ are *not* the same for all x in the domain, f is *not* an even function.

DO EXERCISES 1 AND 2.

If the graph of a function is symmetric with respect to the origin, it is an *odd* function. Recall that a function will be symmetric with respect to the origin if in its equation we can replace x by $-x$ and y by $-y$ and obtain an equivalent equation. Thus if we have a function given by $y = f(x)$, then $-y = f(-x)$ will be equivalent if f is an odd function. In other words, an odd function is one for which $f(-x) = -f(x)$ for all x in the domain. Let us make this our definition.

DEFINITION

A function f is an *odd* function when $f(-x) = -f(x)$ for all x in the domain of f.

Example 3 Determine whether $f(x) = x^3$ is an odd function.

a) Find $f(-x)$ and $-f(x)$ and simplify.

$$f(-x) = (-x)^3 = -x^3,$$
$$-f(x) = -x^3$$

b) Compare $f(-x)$ and $-f(x)$.

Since $f(-x) = -f(x)$ for all x in the domain, f is odd.

Note that the graph is symmetric with respect to the origin.

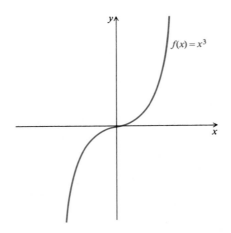

1. Determine whether each function is even.

a)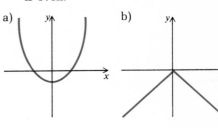

b)

c)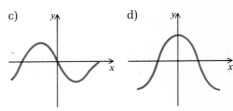

d)

2. Determine whether each function is even.

a) $f(x) = x^2 + 3x$

b) $f(x) = |x|$

c) $f(x) = 3x^2 - x^4$

d) $f(x) = 2x^2 + 1$

3. Determine whether each function is even, odd, or neither even nor odd.

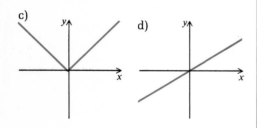

a)   b)

c)   d)

4. Determine whether each function is even, odd, or neither even nor odd.

a) $f(x) = x^3 + 2$
b) $f(x) = x^4 - x^6$
c) $f(x) = x^3 + x$
d) $f(x) = 3x^2 + 3x^5$

e) $f(x) = x^2 - \dfrac{1}{x}$

Example 4 Determine whether $f(x) = x^2 - 4x^3$ is even, odd, or neither even nor odd.

a) Find $f(-x)$ and $-f(x)$ and simplify.

$$f(-x) = (-x)^2 - 4(-x)^3 = x^2 + 4x^3,$$
$$-f(x) = -x^2 + 4x^3$$

b) Compare $f(x)$ and $f(-x)$ to determine whether f is even.

Since $f(x)$ and $f(-x)$ are *not* the same for all x in the domain, f is *not* even.

c) Compare $f(-x)$ and $-f(x)$ to determine whether f is odd.

Since $f(-x)$ and $-f(x)$ are not the same for all x in the domain, f is *not* odd. Thus f is *neither* even nor odd.

DO EXERCISES 3 AND 4.

[ii] Periodic Functions

Certain functions with a repeating pattern are called *periodic*. Here are some examples.

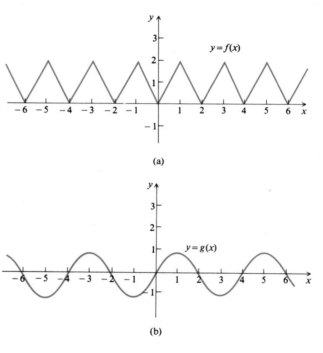

(a)

(b)

The function values of the function f repeat themselves every two units as we move from left to right. In other words, for any x, we have $f(x) = f(x + 2)$. To see this another way, think of the part of the graph between 0 and 2 and note that the rest of the graph consists of copies of it. In terms of translations, if we translate the graph two units to the left or right, the original graph will be obtained.

In the function g, the function values repeat themselves every four units. Hence $g(x) = g(x + 4)$ for any x, and if the graph is translated four units to the left, or right, it will coincide with itself. Or, think of the part of the graph between 0 and 4. The rest of the graph consists of copies of it.

We say that f has a *period* of 2, and that g has a period of 4.

DEFINITION

If a function f has the property that $f(x + p) = f(x)$ whenever x and x + p are in the domain, where p is a positive constant, then f is said to be *periodic*. The smallest positive number p (if there is one) for which $f(x + p) = f(x)$ for all x is called the *period* of the function.

DO EXERCISES 5–7.

[iii] Interval Notation

For a and b real numbers such that $a < b$, we define the *open interval* (a, b) as follows:

(a, b) is the set of all numbers x such that $a < x < b$,

or

$$\{x \mid a < x < b\}.$$

Its graph is as follows:

(a, b) ——○————————○——
          a   b

Note that the endpoints are not included. Be careful not to confuse this notation with that of an ordered pair. The context of the writing should make the meaning clear. If not, we might say "the interval $(-2, 3)$." When we mean an ordered pair, we might say "the pair $(-2, 3)$."

DO EXERCISES 8 AND 9.

The *closed interval* [a, b] is defined as follows:

[a, b] is the set of all x such that $a \leq x \leq b$,

or

$$\{x \mid a \leq x \leq b\}.$$

Its graph is as follows:

[a, b] ——●————————●——
          a   b

Note that endpoints are included. For example, the graph of $[-2, 3]$ is as follows:

[-2, 3] ——●——┼——┼——┼——●——
          -2   -1   0   1   2   3

5. For the function f whose graph was just considered, how does $f(x)$ compare with $f(x + 4)$? with $f(x + 6)$?

6. a) Determine whether this function is periodic.

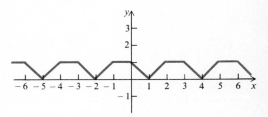

b) If so, what is the period?

7. (*Optional*). For the function t, where $t(x) = 3$, how does $t(x)$ compare with $t(x + 1)$? By the definition, is t periodic? If so, does it have a period?

8. Write interval notation for each set pictured.

a)

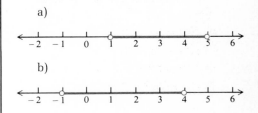

b)

9. Write interval notation for each set.

a) $\{x \mid -2 < x < 3\}$

b) $\{x \mid 0 < x < 1\}$

c) $\left\{x \mid -\frac{1}{4} < x < \sqrt{2}\right\}$

10. Write interval notation for each set pictured.

a)

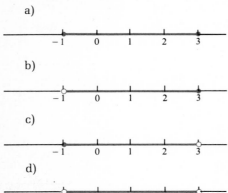

b)

c)

d)

11. Write interval notation for each set.

a) $\left\{x \mid 4 \le x \le 5\frac{1}{2}\right\}$

b) $\{x \mid -3 < x \le 0\}$

c) $\left\{x \mid -\frac{1}{2} \le x < \frac{1}{2}\right\}$

d) $\{x \mid -\pi < x < \pi\}$

12. Is this function continuous

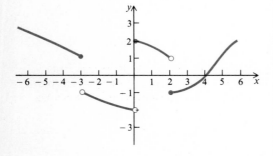

a) in the interval $(-1, 1)$?

b) in the interval $[-2, -1]$?

c) in the interval $(1, 2]$?

d) in the interval $(1, 2)$?

e) in the interval $(-5, 0)$?

13. Where are the discontinuities of the above function?

There are two kinds of *half-open intervals* defined as follows:

a) $\qquad (a, b] = \{x \mid a < x \le b\}.$

This is open on the left. Its graph is as follows:

b) $\qquad [a, b) = \{x \mid a \le x < b\}.$

This is open on the right. Its graph is as follows:

DO EXERCISES 10 AND 11.

[iv] Continuous Functions

Some functions have graphs that are continuous curves, without breaks or holes in them. Such functions are called *continuous functions*.

Example 5 The function f below is continuous because it has no breaks, jumps, or holes in it. The function g has *discontinuities* where $x = -2$ and $x = 5$. The function g is continuous on the interval $[-2, 5)$ or on any interval contained therein. It is also continuous on other intervals, but it is not continuous on an interval such as $[-3, 1]$ or $(3, 7)$.

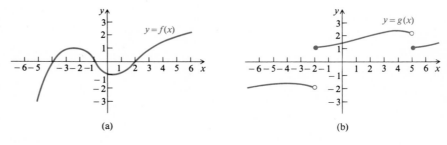

(a)   (b)

DO EXERCISES 12 AND 13.

[v] Increasing and Decreasing Functions

If the graph of a function rises from left to right it is said to be an *increasing* function. If the graph of a function drops from left to right, it is said to be a *decreasing* function. This can be stated more formally.

DEFINITION

1. A function f is an *increasing* function when for all a and b in the domain of f, if $a < b$, then $f(a) < f(b)$.

2. A function f is a *decreasing* function when for all a and b in the domain of f, if $a < b$, then $f(a) > f(b)$.

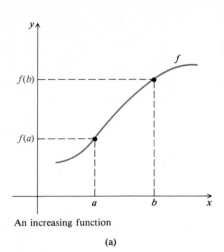

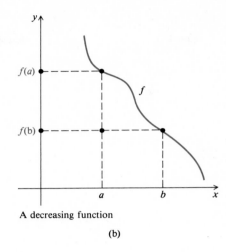

An increasing function
(a)

A decreasing function
(b)

Examples 6–9 Determine whether each function is increasing, decreasing, or neither.

6.

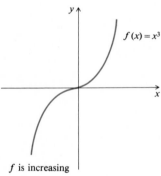

$f(x) = x^3$

f is increasing

7.

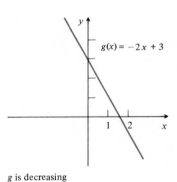

$g(x) = -2x + 3$

g is decreasing

8.

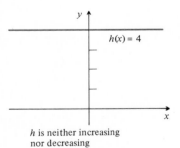

$h(x) = 4$

h is neither increasing nor decreasing

9.

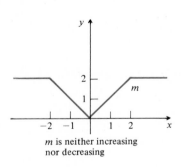

m

m is neither increasing nor decreasing

Note that while the function m is neither increasing nor decreasing, it is increasing on the interval $[0, 2]$ and decreasing on the interval $[-2, 0]$.

DO EXERCISES 14 AND 15.

14. Determine whether each function is increasing, decreasing, or neither increasing nor decreasing.

a)

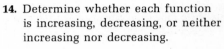

b)

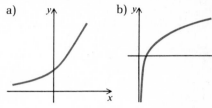

c)

d)

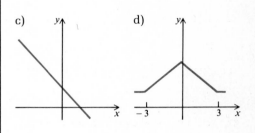

15. For the function in Exercise 14(d), find an interval on which the function is (a) increasing; (b) decreasing.

16. Graph the function defined as follows:

$$f(x) = \begin{cases} x + 3 & \text{for } x \le -2, \\ 1 & \text{for } -2 < x \le 3, \\ x^2 - 10 & \text{for } 3 < x. \end{cases}$$

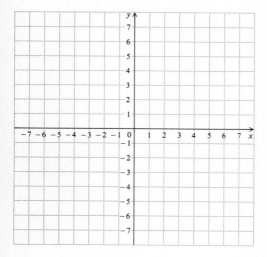

[vi] Functions Defined Piecewise

Sometimes functions are defined piecewise. That is, we have different output formulas for different parts of the domain.

Example 10 Graph the function defined as follows.

$$f(x) = \begin{cases} 4 & \text{for } x \le 0 \\ & \text{(This means that for any input x less than or equal to 0 the output is 4.)} \\ 4 - x^2 & \text{for } 0 < x \le 2 \\ & \text{(This means that for any input x greater than 0 and less than or equal to 2, the output is } 4 - x^2.) \\ 2x - 6 & \text{for } x > 2 \\ & \text{(This means that for any input x greater than 2, the output is } 2x - 6.) \end{cases}$$

See the graph below.

a) We graph $f(x) = 4$ for inputs less than or equal to 0 (that is, $x \le 0$).

b) We graph $f(x) = 4 - x^2$ for inputs greater than 0 and less than or equal to 2 (that is, $0 < x \le 2$).

c) We graph $f(x) = 2x - 6$ for inputs greater than 2 (that is, $x > 2$).

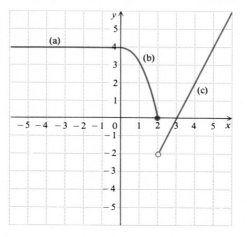

DO EXERCISE 16.

EXERCISE SET 3.6

[i]

1. Determine whether each function is even, odd, or neither even nor odd.

a)

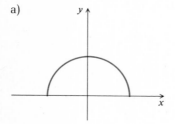

b)

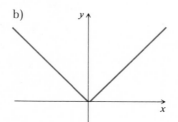

c)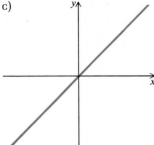

d)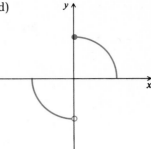

2. Determine whether each function is even, odd, or neither even nor odd.

a)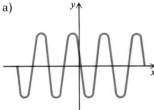

b)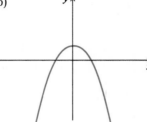

c)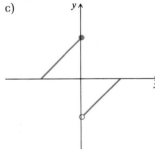

d)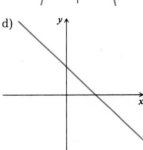

Determine whether each function is even, odd, or neither even nor odd.

3. $f(x) = 2x^2 + 4x$ **4.** $f(x) = -3x^3 + 2x$ **5.** $f(x) = 3x^4 - 4x^2$

6. $f(x) = 5x^2 + 2x^4 - 1$ **7.** $f(x) = 7x^3 + 4x - 2$ **8.** $f(x) = 4x$

9. $f(x) = |3x|$ **10.** $f(x) = x^{24}$ **11.** $f(x) = x^{17}$

12. $f(x) = x + \dfrac{1}{x}$ **13.** $f(x) = x - |x|$ **14.** $f(x) = \sqrt{x}$

15. $f(x) = \sqrt[3]{x}$ **16.** $f(x) = 7$ **17.** $f(x) = 0$

18. $f(x) = \sqrt[3]{x - 2}$ **19.** $f(x) = \sqrt{x^2 + 1}$ **20.** $f(x) = \dfrac{x^2 + 1}{x^3 - x}$

[ii]

21. Determine whether each function is periodic.

a)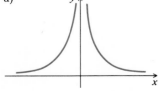

b)

c)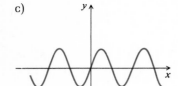

d)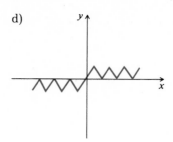

22. Determine whether each function is periodic.

a)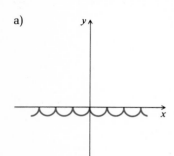

b)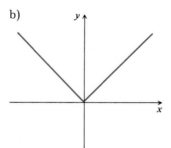

c)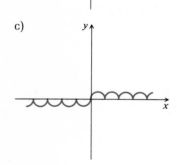

d)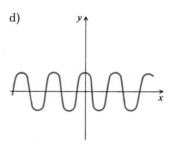

23. What is the period of this function?

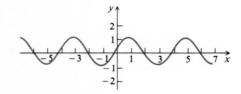

24. What is the period of this function?

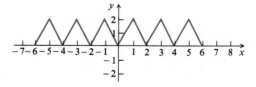

[iii]

25. Write interval notation for each set pictured.

a)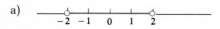

b)

c)

d)

26. Write interval notation for each set pictured.

a)

b)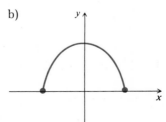

c)

d)

27. Write interval notation for each set.

a) $\{x \mid -2 < x < 4\}$ b) $\left\{x \mid -\frac{1}{4} < x \leq \frac{1}{4}\right\}$

c) $\{x \mid 7 \leq x < 10\pi\}$ d) $\{x \mid -9 \leq x \leq -6\}$

28. Write interval notation for each set.

a) $\{x \mid -5 < x < 0\}$ b) $\{x \mid -\sqrt{2} \leq x < \sqrt{2}\}$

c) $\left\{x \mid -\frac{\pi}{2} < x \leq \frac{\pi}{2}\right\}$ d) $\left\{x \mid -12 \leq x \leq -\frac{1}{2}\right\}$

[iv]

29. Is this function continuous

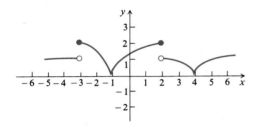

a) in the interval [0, 2]?
b) in the interval (−2, 0)?
c) in the interval [0, 3]?
d) in the interval [−3, −1]?
e) in the interval (−3, −1]?

30. Is this function continuous

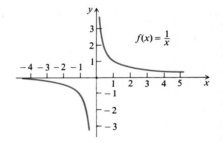

a) in the interval [−3, −1]?
b) in the interval (1, 4)?
c) in the interval [−1, 1]?
d) in the interval [−2, 4]?
e) in the interval (0, 1]?

31. Where are the discontinuities of the function in Exercise 29?

32. Where are the discontinuities of the function in Exercise 30?

[v]

33. Determine whether each function is increasing, decreasing, or neither increasing nor decreasing.

a)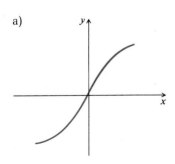

b)

c)

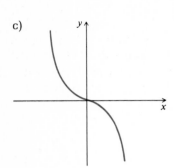

d)

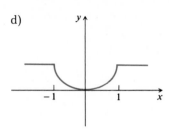

34. Determine whether each function is increasing, decreasing, or neither increasing nor decreasing.

a)

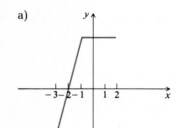

b)

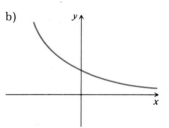

c)

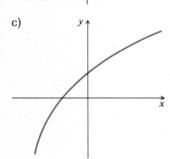

d)

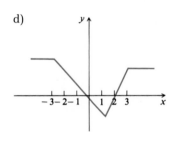

[**vi**] Graph.

35. $f(x) = \begin{cases} 1 & \text{for } x < 0, \\ -1 & \text{for } x \geq 0 \end{cases}$

36. $f(x) = \begin{cases} 2 & \text{for } x \text{ an integer,} \\ -2 & \text{for } x \text{ not an integer} \end{cases}$

37. $f(x) = \begin{cases} 3 & \text{for } x \leq -3, \\ |x| & \text{for } -3 < x \leq 3, \\ -3 & \text{for } x > 3 \end{cases}$

38. $f(x) = \begin{cases} -2x - 6 & \text{for } x \leq -2 \\ 2 - x^2 & \text{for } -2 < x < 2, \\ 2x - 6 & \text{for } x \geq 2 \end{cases}$

39. (*The postage function.*) Postage rates are as follows: 18¢ for the first ounce plus 17¢ for each additional ounce or fraction thereof. Thus if x is the weight of a letter in ounces, then $p(x)$ is the cost of mailing the letter, where

$$p(x) = \begin{cases} 18¢ & \text{if } 0 < x \leq 1, \\ 35¢ & \text{if } 1 < x \leq 2, \\ 52¢ & \text{if } 2 < x \leq 3, \end{cases}$$

and so on, up to 12 ounces, after which postal cost also depends on distance. Graph this function for x such that $0 < x \leq 12$.

40. Graph.

$$f(x) = \begin{cases} 3 + x & \text{for } x \leq 0, \\ \sqrt{x} & \text{for } 0 < x < 4, \\ x^2 - 4x - 1 & \text{for } x \geq 4 \end{cases}$$

☆

41. For the function in Exercise 33(d), find an interval on which the function is (a) increasing; (b) decreasing.

42. For the function in Exercise 34(d), find an interval on which the function is (a) increasing; (b) decreasing.

43. Graph each function. Then determine whether it is increasing, decreasing, or neither increasing nor decreasing.

a) $f(x) = 3x + 4$ b) $f(x) = -3x + 4$

c) $f(x) = x^2 + 1$ d) $f(x) = -2$

e) $f(x) = x^3 + 1$ f) $f(x) = |x|$

44. Graph each function. Then determine whether it is increasing, decreasing, or neither increasing nor decreasing.

a) $f(x) = -2x - 3$ b) $f(x) = 2x - 3$

c) $f(x) = 3x^2$ d) $f(x) = \sqrt{3}$

e) $f(x) = |x| + 2$ f) $f(x) = x^3 - 2$

45. To what interval is each of the following mapped under the linear transformation $x' = x + 2$?

a) $[0, 1]$ b) $(-2, 7]$ c) $(-8, -1)$

46. To what interval is each of the following mapped under the linear transformation $x' = 0.4x - 3$?

a) $[0, 1]$ b) $(-4, -20)$ c) $[6, 11]$

3.7 INVERSES

[i] Inverses of Relations

If in a relation we interchange first and second members in each ordered pair, we obtain a new relation. The new relation is called the *inverse* of the original relation. A relation is shown here in color; its inverse is shaded.

(1, 4)	(2, 4)	(3, 4)	(4, 4)
(1, 3)	(2, 3)	(3, 3)	(4, 3)
(1, 2)	(2, 2)	(3, 2)	(4, 2)
(1, 1)	(2, 1)	(3, 1)	(4, 1)

DO EXERCISE 1.

When a relation is defined by an equation, interchanging x and y produces an equation of the inverse relation.

Example 1 Find an equation of the inverse of $y = x^2 - 5$.

We interchange x and y, and obtain $x = y^2 - 5$. This is an equation of the inverse relation.

DO EXERCISE 2 ON THE FOLLOWING PAGE.

[ii] Graphs of Inverse Relations

Interchanging first and second coordinates in each ordered pair of a relation has the effect of interchanging the x-axis and the y-axis.

DO EXERCISE 3 ON THE FOLLOWING PAGE.

Interchanging the x-axis and the y-axis has the effect of reflecting across the diagonal line whose equation is $y = x$, as shown below. Thus the graphs of a relation and its inverse are always reflections of

OBJECTIVES

You should be able to:

[i] Given an equation defining a relation, write an equation of the inverse relation.

[ii] Given a graph of a relation, sketch a graph of its inverse; or given an equation defining a relation, graph it and then graph its inverse.

[iii] Given an equation defining a relation, determine whether the graph is symmetric with respect to the line $y = x$.

[iv] Given the graph of a relation or an equation for a relation, determine whether its inverse is a function.

[v] Given a function defined by a simple formula, find a formula for its inverse.

[vi] For a function f whose inverse is also a function, quickly find $f^{-1}(f(x))$ and $f(f^{-1}(x))$ for any number x in the domains of the functions.

1. a) Below, graph the relation containing (4, 1), (4, 2), (3, 2), (2, 3), (2, 4), and (1, 4).

b) Interchange the members in each ordered pair to obtain the inverse.

c) Graph the inverse and compare the graphs.

(1, 4)	(2, 4)	(3, 4)	(4, 4)
(1, 3)	(2, 3)	(3, 3)	(4, 3)
(1, 2)	(2, 2)	(3, 2)	(4, 2)
(1, 1)	(2, 1)	(3, 1)	(4, 1)

2. Write an equation of the inverse of each relation.

a) $y = 3x + 2$ b) $y = x$

c) $x^2 + 3y^2 = 4$ d) $y = 5x^2 + 2$

e) $y^2 = 4x - 5$ f) $xy = 5$

3. a) On a piece of thin paper, draw coordinate axes.

b) Draw the line $y = x$.

c) Draw a relation as shown here.

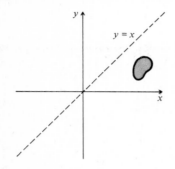

d) Flip the paper over (to interchange the x-axis and y-axis). Look through the paper. How do the graphs of the relation and its inverse compare?

4. Graph the inverse of each relation by reflecting across the line $y = x$.

a)

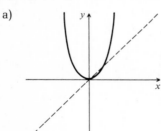

b)

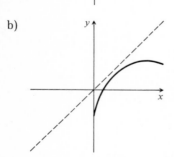

c) Graph $y = 4 - x^2$. Then by reflecting across the line $y = x$, graph its inverse.

each other across the line $y = x$. (This assumes, of course, that the same scale is used on both axes.)

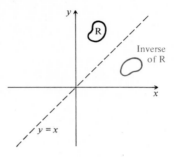

Examples 2–4 In each case a relation is shown in black. The graph of its inverse is shown in color.

2.

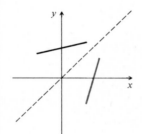

3.

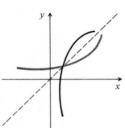

4.

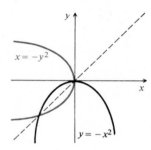

DO EXERCISE 4.

[iii] **Symmetry with Respect to the Line** $y = x$

It can happen that a relation is its own inverse; that is, when x and y are interchanged or the relation is reflected across the line $y = x$, there is no change. Such a relation is symmetric with respect to the line $y = x$. The following are two examples of relations symmetric with respect to the line $y = x$.

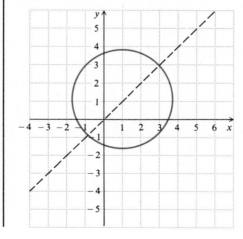

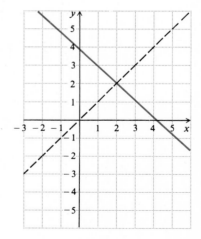

Each of the relations shown in black in Examples 2–4 is *not* symmetric with respect to the line $y = x$.

For a relation defined by an equation: If interchanging *x* and *y* produces an equivalent equation, the relation is its own inverse, and the graph of the equation is symmetric with respect to the line $y = x$.

Example 5 Test $3x + 3y = 5$ for symmetry with respect to the line $y = x$.

a) Interchange x and y. This amounts to replacing each occurrence of x by a y and each y by an x.

$$3x + 3y = 5$$
$$3y + 3x = 5$$

b) Is the resulting equation equivalent to the original?

The commutative law of addition guarantees that the resulting equation is equivalent to the original. Thus the graph is symmetric with respect to the line $y = x$.

Example 6 Test $y = x^2$ for symmetry with respect to the line $y = x$.

a) Interchange x and y.

$$y = x^2$$
$$x = y^2$$

b) Is the resulting equation equivalent to the original?

Note that $(2, 4)$ is a solution of the original equation $y = x^2$, but it is not a solution of the resulting equation $x = y^2$.

y	$= x^2$		x	$= y^2$
4	2^2		2	4^2
	4			16

Thus the equations are not equivalent, so the graph of $y = x^2$ is *not* symmetric with respect to the line $y = x$.

DO EXERCISES 5–12.

[iv] Inverses of Functions

Every function has an inverse, but that inverse may not be a function, as the following examples show.

Example 7 The relation g given by

$$g = \{(5, 3), (3, 1), (-7, 2), (2, 3)\}$$

is a function. Find the inverse of g and determine whether it is a function.

Test the following for symmetry with respect to the line $y = x$.

5. $y = -x$

6. $x + y = 4$

7. $xy = 3$

8. $y = |x|$

9. $3x^2 + 3y^2 = 4$

10. $|x| = |y|$

11. $y = x^3$

12. $x - y = 4$

13. Which of the following have inverses that are functions?

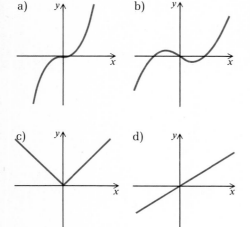

a)   b)

c)   d)

14. Graph the relation $y = x^2 - 1$ and decide whether it is a function. Then graph the inverse and decide whether it is a function.

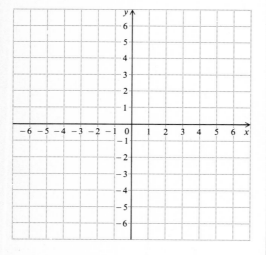

The inverse of g is found by interchanging first and second members in each ordered pair, and is $\{(3, 5), (1, 3), (2, -7), (3, 2)\}$. It is *not* a function because the pairs $(3, 5)$ and $(3, 2)$ have the same first coordinates but different second coordinates.

Example 8 Graph the relation $y = x^2$ and determine whether it is a function. Then graph its inverse and determine whether it is a function.

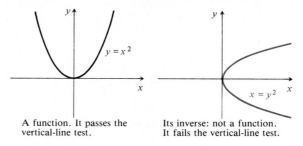

A function. It passes the vertical-line test.

Its inverse: not a function. It fails the vertical-line test.

DO EXERCISES 13 AND 14.

[v] Finding Formulas for Inverses of Functions

All functions have inverses, but in only some cases is the inverse also a function. If the inverse of a function f is also a function, we denote it by f^{-1} (read "f inverse"). [*Caution:* This is *not* exponential notation!] Recall that we obtain the inverse of any relation by reversing the coordinates in each ordered pair. In terms of mapping, let us see what this means. A function f *maps* the set of first coordinates (the domain) D to the set of second coordinates (the range) R. The inverse mapping f^{-1} maps the range of f onto the domain of f. Each y in R is mapped onto just one x in D, provided f^{-1} is a function. Note that the domain of f is the range of f^{-1} and the range of f is the domain of f^{-1}.

Let us consider inverses of functions in terms of function machines. Suppose that the function f programmed into a machine has an inverse that is also a function. Suppose then that the function machine has a reverse switch. When the switch is thrown the machine is then programmed to do the inverse mapping f^{-1}. Inputs then enter at the opposite end and the entire process is reversed.

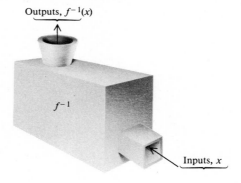

Outputs, $f^{-1}(x)$

f^{-1}

Inputs, x

When a function is defined by a formula, we can sometimes find a formula for its inverse by thinking of interchanging x and y.

Example 9 Given $f(x) = x + 1$, find a formula for $f^{-1}(x)$.

a) Let us think of this as $y = x + 1$.

b) To find the inverse we interchange x and y: $x = y + 1$.

c) Now we solve for y: $y = x - 1$.

d) Thus $f^{-1}(x) = x - 1$.

Note in Example 9 that f maps any x onto $x + 1$ (this function adds 1 to each number of the domain). Its inverse, f^{-1}, maps any number x onto $x - 1$ (this inverse function subtracts 1 from each member of its domain). Thus the function and its inverse do opposite things.

Example 10 Let $g(x) = 2x + 3$. Find a formula for $g^{-1}(x)$.

a) Let us think of this as $y = 2x + 3$.

b) To find the inverse we interchange x and y: $x = 2y + 3$.

c) Now we solve for y: $y = (x - 3)/2$.

d) Thus $g^{-1}(x) = (x - 3)/2$.

Note in Example 10 that g maps any x onto $2x + 3$ (this function doubles each input and then adds 3). Its inverse, g^{-1}, maps any input onto $(x - 3)/2$ (this inverse function subtracts 3 from each input and then divides it by 2). Thus the function and its inverse do opposite things.

Example 11 Consider $f(x) = x^2$. Let us think of this as $y = x^2$.

To find the inverse we interchange x and y: $x = y^2$. Now we solve for y: $y = \pm\sqrt{x}$. Note that for each positive x we get two y's. For example, the pairs $(4, 2)$ and $(4, -2)$ belong to the relation. Look at the following graphs. Note also that the inverse fails the vertical-line test.

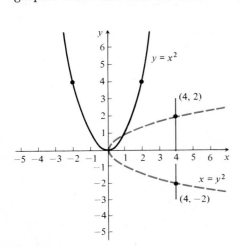
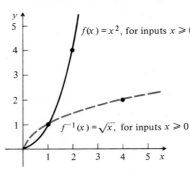

15. $g(x) = 3x - 2$. Find a formula for $g^{-1}(x)$.

If we restrict the domain of $f(x) = x^2$ to nonnegative numbers, then its inverse is a function, $f^{-1}(x) = \sqrt{x}$.

DO EXERCISES 15 AND 16.

[vi] Finding $f^{-1}(f(x))$ and $f(f^{-1}(x))$

Suppose the inverse of a function f is also a function. Let us suppose that we do the mapping f and then do the inverse mapping f^{-1}. We will be back where we started. In other words, if we find $f(x)$ for some x and then find f^{-1} for this number, we will be back at x. In function notation the statement looks like this:

$$f^{-1}(f(x)) = x.$$

This is read "f inverse of f of x equals x." It means, working from the inside out, to take x, then find $f(x)$, and then find the value of f^{-1} for that number. When we do, we will be back where we started, at x. For similar reasons, the following is also true:

$$f(f^{-1}(x)) = x.$$

16. $f(x) = x^2 - 1$ and the domain of f is the set of all nonnegative real numbers. Find a formula for $f^{-1}(x)$.

For the above statements to be true, x must of course be in the domain of the function being considered.

Let us consider function machines. To find $f^{-1}(f(x))$ we use x as an input, when the machine is set on f. We then use the output for an input, running the machine backward. The last output is the same as the first input. The first output must of course be an acceptable input when the machine is set on f^{-1}. Following is a precise statement of the property.

THEOREM 5

> For any function f whose inverse is a function, $f^{-1}(f(a)) = a$ for any a in the domain of f. Also, $f(f^{-1}(a)) = a$ for any a in the domain of f^{-1}.

17. Let $g(x) = 3x - 2$ as in Exercise 15. Find $g^{-1}(g(5))$. Find $g(g^{-1}(5))$ as in Example 12.

Proof. Suppose a is in the domain of f. Then $f(a) = b$, for some b in the range of f. Then the ordered pair (a, b) is in f, and by definition of f^{-1}, (b, a) is in f^{-1}. It follows that $f^{-1}(b) = a$. Then substituting $f(a)$ for b, we get $f^{-1}(f(a)) = a$. A similar proof shows that $f(f^{-1}(a)) = a$, for any a in the domain of f^{-1}.

Example 12 For the function g of Example 10, find $g(4)$. Then find $g^{-1}(g(4))$.

$$g(x) = 2x + 3, \quad \text{so } g(4) = 11.$$

Now
$$g^{-1}(x) = \frac{x - 3}{2},$$

so
$$g^{-1}(11) = \frac{11 - 3}{2} = 4.$$

Thus
$$g^{-1}(g(4)) = 4.$$

DO EXERCISE 17.

Example 13 For the function g of Example 12, find $g^{-1}(g(283))$. Find also $g(g^{-1}(-12{,}045))$.

We note that every real number is in the domain of both g and g^{-1}. Thus we may immediately write the answers, without calculating:

$$g^{-1}(g(283)) = 283,$$
$$g(g^{-1}(-12{,}045)) = -12{,}045.$$

DO EXERCISE 18.

18. Let $g(x) = 3x - 2$ as in Exercise 17. For any number a, find $g^{-1}(g(a))$. Find $g(g^{-1}(a))$.

EXERCISE SET 3.7

[i] Write an equation of the inverse relation.

1. $y = 4x - 5$
2. $y = 3x + 5$
3. $x^2 - 3y^2 = 3$
4. $2x^2 + 5y^2 = 4$
5. $y = 3x^2 + 2$
6. $y = 5x^2 - 4$
7. $xy = 7$
8. $xy = -5$

[ii]

9. Graph $y = x^2 + 1$. Then by reflection across the line $y = x$, graph its inverse.
10. Graph $y = x^2 - 3$. Then by reflection across the line $y = x$, graph its inverse.
11. Graph $y = |x|$. Then by reflection across the line $y = x$, graph its inverse.
12. Graph $x = |y|$. Then by reflection across the line $y = x$, graph its inverse.

[iii] Test for symmetry with respect to the line $y = x$.

13. $3x + 2y = 4$
14. $5x - 2y = 7$
15. $4x + 4y = 3$
16. $5x + 5y = -1$
17. $xy = 10$
18. $xy = 12$
19. $3x = \dfrac{4}{y}$
20. $4y = \dfrac{5}{x}$
21. $y = |2x|$
22. $3x = |2y|$
23. $4x^2 + 4y^2 = 3$
24. $3x^2 + 3y^2 = 5$

[iv]

25. Which of the following have inverses that are functions?

a)   b)   c)   d)

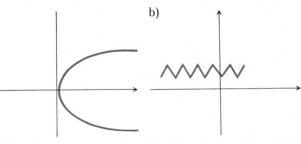   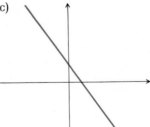   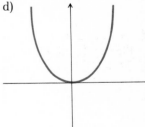

26. Which of the following have inverses that are functions?

a)   b)   c)   d)

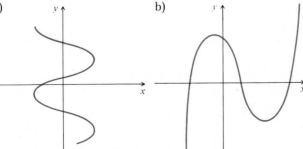   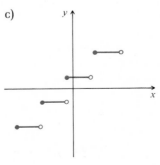   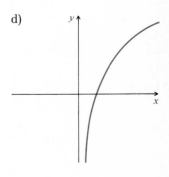

[v]

27. $f(x) = 2x + 5$. Find a formula for $f^{-1}(x)$.
29. $f(x) = \sqrt{x + 1}$. Find a formula for $f^{-1}(x)$.

28. $g(x) = 3x - 1$. Find a formula for $g^{-1}(x)$.
30. $g(x) = \sqrt{x - 1}$. Find a formula for $g^{-1}(x)$.

[vi]

31. $f(x) = 35x - 173$. Find $f^{-1}(f(3))$.
    Find $f(f^{-1}(-125))$.
33. $f(x) = x^3 + 2$. Find $f^{-1}(f(12{,}053))$.
    Find $f(f^{-1}(-17{,}243))$.

32. $g(x) = \dfrac{-173x + 15}{3}$. Find $g^{-1}(g(5))$.
    Find $g(g^{-1}(-12))$.
34. $g(x) = x^3 - 486$. Find $g^{-1}(g(489))$.
    Find $g(g^{-1}(-17{,}422))$.

☆ ──

35. Carefully graph $y = x^2$, using a large scale. Then use the graph to approximate $\sqrt{3.1}$.

36. Carefully graph $y = x^3$, using a large scale. Then use the graph to approximate $\sqrt[3]{-5.2}$.

37. Graph this equation and its inverse. Then test for symmetry with respect to the x-axis, the y-axis, the origin, and the line $y = x$.

$$y = \frac{1}{x^2}$$

★ ──

38. Ice melts at 0° Celsius or 32° Fahrenheit. Water boils at 100° Celsius or 212° Fahrenheit.

   a) What linear transformation converts Celsius temperature to Fahrenheit?

   b) Find the inverse of your answer to (a). Is it a function? What kind of conversion does it accomplish?

   c) At what temperature are the Celsius and Fahrenheit scales the same?

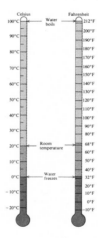

Graph each equation and its inverse. Then test for symmetry with respect to the x-axis, the y-axis, the origin, and the line $y = x$.

39. $|x| - |y| = 1$

40. $y = \dfrac{|x|}{x}$

CHAPTER 3 REVIEW

[3.1, i]   1. List all the ordered pairs in the Cartesian square $G \times G$, where $G = \{1, 3, 5, 7\}$.

[3.1, iii]   2. List the domain and the range of the relation whose ordered pairs are (3, 1), (5, 3), (7, 7), and (3, 5).

[3.2, ii]   Graph.

[3.5, i]   3. $x = |y|$ 4. $y = (x + 1)^2$ 5. $y = |x| - 2$
             6. $f(x) = \sqrt{x}$ 7. $f(x) = \sqrt{x - 2}$ 8. $f(x) = 2\sqrt{x + 3}$

9. $f(x) = \frac{1}{4}\sqrt{x-1} + 2$ $\qquad$ **10.** $|x - y| = 1$

Consider the following relations for Exercises 11–14.

a) $y = 7$ $\qquad$ b) $x^2 + y^2 = 4$ $\qquad$ c) $x^3 = y^3 - y$ $\qquad$ d) $y^2 = x^2 + 3$

e) $x + y = 3$ $\qquad$ f) $x = 3$ $\qquad$ g) $y = x^2$ $\qquad$ h) $y = x^3$

[3.4, i] **11.** Which are symmetric with respect to the x-axis?

[3.4, i] **12.** Which are symmetric with respect to the y-axis?

[3.4, ii] **13.** Which are symmetric with respect to the origin?

[3.7, iii] **14.** Which are symmetric with respect to the line $y = x$?

[3.7, i] Write an equation of the inverse.

15. $y = 3x^2 + 2x - 1$ $\qquad$ **16.** $y = \sqrt{x+2}$

[3.3, i] **17.** Which of the following are graphs of functions?

a)

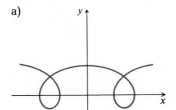

b)

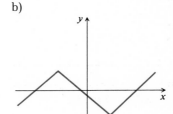

c)

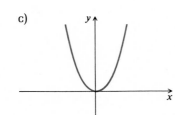

d)

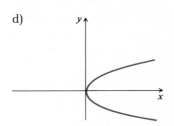

[3.7, iv] **18.** Which of the relations in Exercise 17 have inverses that are functions?

[3.3, ii] Use $f(x) = x^2 - x - 3$ to answer Exercises 19–21. Find:

19. $f(0)$. $\qquad$ **20.** $f(-3)$. $\qquad$ **21.** $f(a + h)$.

[3.3, ii] Use $g(x) = 2\sqrt{x-1}$ to answer Exercises 22–24. Find:

22. $g(1)$. $\qquad$ **23.** $g(5)$. $\qquad$ **24.** $g(a + 2)$.

[3.7, v] **25.** $f(x) = \frac{\sqrt{x}}{2} + 2$. Find a formula for $f^{-1}(x)$.

26. $f(x) = x^2 + 2$. Find a formula for $f^{-1}(x)$.

[3.3, iii] Find the domain of each function.

27. $f(x) = \sqrt{7 - 3x}$ $\qquad$ **28.** $f(x) = \frac{1}{x^2 - 6x + 5}$

29. $f(x) = (x - 9x^{-1})^{-1}$ $\qquad$ **30.** $f(x) = \frac{\sqrt{1-x}}{x - |x|}$

[3.3, iv]   Find $f \circ g(x)$ and $g \circ f(x)$ in Exercises 31 and 32.

31. $f(x) = \dfrac{4}{x^2}$; $g(x) = 3 - 2x$ **32.** $f(x) = 3x^2 + 4x$; $g(x) = 2x - 1$

[3.7, vi]   **33.** $f(x) = x^3 + 2$. Find $f\big(f^{-1}(a)\big)$.

             34. $h(x) = x^{17} + x^{65}$. Find $h^{-1}\big(h(t)\big)$.

[3.5, i]   **35.** Here is a graph of $y = f(x)$.

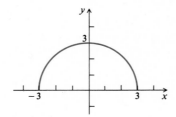

             Sketch graphs of the following.

             a) $y = 1 + f(x)$ b) $y = \dfrac{1}{2}f(x)$ c) $y = f(x + 1)$

[3.6, i]   Use the following to answer Exercises 36–38.

             a)   b)

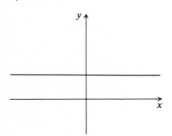   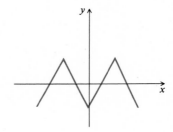

             c) $f(x) = 3x^2 - 2$ d) $f(x) = x + 3$
             e) $f(x) = 3x^3$ f) $f(x) = x^5 - x^3$

             36. Which of the previous are even?   **37.** Which of the previous are odd?

             38. Which of the previous are neither even nor odd?

[3.6, ii]   **39.** Which of the following functions are periodic?

             a)   b)

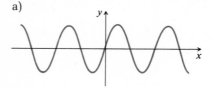   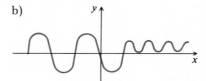

             c)   d)

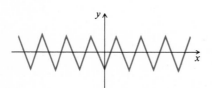   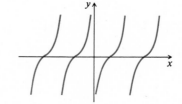

[3.6, ii]   **40.** What is the period of this function?

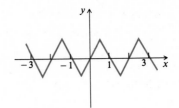

[3.6, iv]   **41.** Is this function continuous
　　　　　　　a) in the interval $[-3, -1]$?
　　　　　　　b) in the interval $[-1, 1]$?

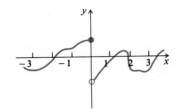

[3.6, iii]   Write interval notation for the following.

42. $\{x \mid -\pi \leq x \leq 2\pi\}$ **43.** $\{x \mid 0 < x \leq 1\}$

[3.6, v]   Use the following for Exercises 44–46.

a)

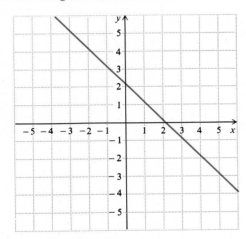

b)

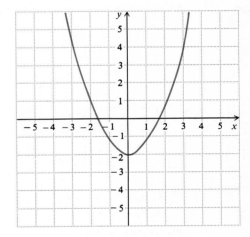

c)

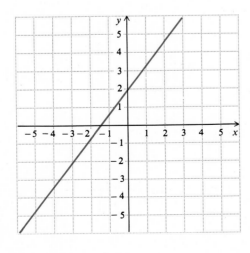

d)

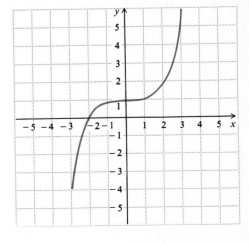

44. Which of the above are increasing?   **45.** Which of the above are decreasing?

46. Which of the above are neither increasing nor decreasing?

[3.6, **vi**]   **47.** Graph

$$f(x) = \begin{cases} x^2 + 2 & \text{for } x < 0, \\ x^3 & \text{for } 0 \le x < 2, \\ -4x + 5 & \text{for } x \ge 2. \end{cases}$$

48. Graph several functions of the type $y = |f(x)|$. Describe a procedure, involving transformations, for graphing such functions.

LINEAR AND QUADRATIC FUNCTIONS AND INEQUALITIES

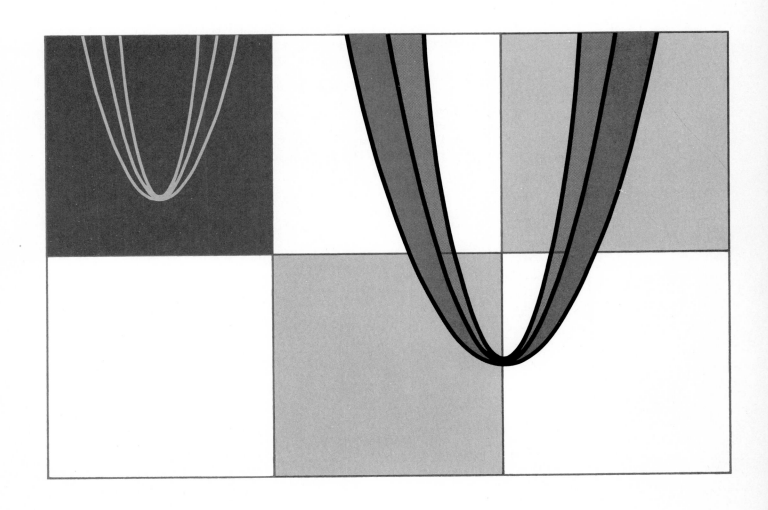

OBJECTIVES

You should be able to:

[i] Determine whether a given equation is linear.
[ii] Graph linear equations.
[iii] Find the slope, if it exists, of the line containing two points.
[iv] Given the slope and coordinates of one point of a line, find an equation of the line.
[v] Given the coordinates of two points, find an equation of the line containing them.
[vi] Given an equation of a line, find its slope, if it exists, and its y-intercept.

1. Determine whether each equation is linear.

a) $3x = 2y + 4$
b) $7y = 11$
c) $5y^2x = 13$
d) $x = \dfrac{4}{y}$

2. Graph $6x - 4y = 12$.

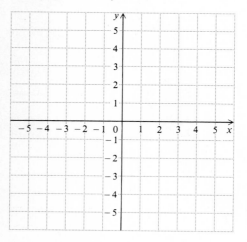

4.1 LINES AND LINEAR FUNCTIONS

[i] Linear Equations

An equation of the type $Ax + By - C = 0$ is called a *linear* equation because its graph is a straight line. In the above, A, B, and C are constants, but A and B cannot both be 0. Any equation that is equivalent to one of this form has a straight-line graph.

Example 1 Determine whether each equation is linear.

a) The equation $5x + 8y = 3$ is linear because it is equivalent to $5x + 8y - 3 = 0$. Here, $A = 5$, $B = 8$, and $C = -3$.

b) The equation $3x^2 - 4y + 5 = 0$ is not linear because x is squared.

c) The equation $4x + 3xy = 25$ is not linear because the product xy occurs.

DO EXERCISE 1.

[ii] Graphs of Linear Equations

Since two points determine a line, we can graph a linear equation by finding two of its points. Then we draw a line through those points.

A third point should always be used as a check. The easiest points to find are often the intercepts (the points where the line crosses the axes).

Example 2 Graph $4x + 5y = 20$.

We set $x = 0$ and find that $y = 4$. Thus $(0, 4)$ is a point of the graph (the y-intercept).

We set $y = 0$ and find that $x = 5$. Thus $(5, 0)$ is a point of the graph (the x-intercept). The graph is shown below. The point $(-2, 5\frac{3}{5})$ was used as a check.

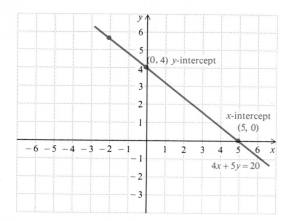

If a graph, such as $y = 7x$, goes through the origin, it has only one intercept, and other points will be needed for graphing.

If an equation has a missing variable ($A = 0$ or $B = 0$), then its graph is parallel to one of the axes.

DO EXERCISE 2.

Example 3

a) Graph $y = 3$.

The graph is shown below.

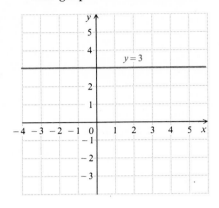

b) Graph $x = -2$.

The graph is shown below.

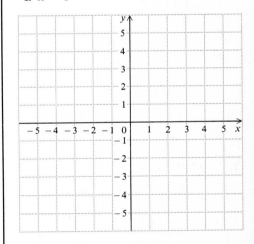

Graph.

3. $3x + 2y = 6$

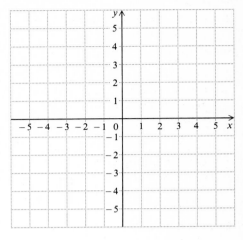

4. $x = 4$

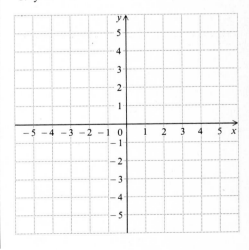

DO EXERCISES 3–5.

[iii] Slope

The graph of a linear equation may slant upward or downward in various ways. Let us see how this relates to equations.

Suppose points P_1 and P_2 with coordinates (x_1, y_1) and (x_2, y_2) are two different points on a line not parallel to an axis. Consider a right triangle as shown with legs parallel to the axes. The point P_3 with coordinates (x_2, y_1) is the third vertex of the triangle. As we move from P_1 to P_2, y changes from y_1 to y_2. The change in y is $y_2 - y_1$. Similarly, the change in x is $x_2 - x_1$. The ratio of these changes is called the *slope*.

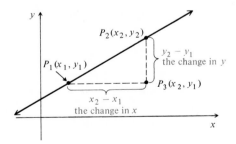

5. $y = -3$

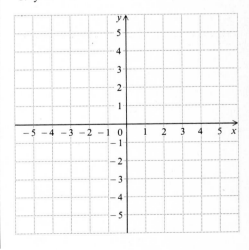

DEFINITION

The *slope m* of a line containing two points (x_1, y_1) and (x_2, y_2) is defined by

$$m = \frac{y_2 - y_1}{x_2 - x_1}.$$

Use graph paper. Graph the lines through these points and find their slopes.

6. (1, 3) and (2, 5)

7. (3, 7) and (5, 3)

8. (1, 1) and (2, 3)

9. (3, 5) and (2, −1)

Use graph paper. Graph the lines through these points and find their slopes.

10. (−1, −1) and (2, −4)

11. (0, 2) and (3, 1)

Find the slopes, if they exist, of the lines containing these points.

12. (4, 6) and (−2, 6)

13. (−3, 5) and (−3, 7)

14. (9, 0) and (6, 0)

Example 4 Graph the line through the points (1, 2) and (3, 6) and find its slope.

Let us call the slope m and let (1, 2) be (x_1, y_1) and (3, 6) be (x_2, y_2). Applying the definition, we obtain

$$m = \frac{y_2 - y_1}{x_2 - x_1} = \frac{6 - 2}{3 - 1} = \frac{4}{2} = 2.$$

Note that we can also use the points in the opposite order, so long as we are consistent. We get the same slope:

$$m = \frac{y_2 - y_1}{x_2 - x_1} = \frac{2 - 6}{1 - 3} = \frac{-4}{-2} = 2.$$

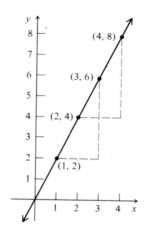

From Example 4 we see that it does not matter in which order we choose the points, so long as we take differences in the same order. From Example 4 we can also see that it does not matter which two points of a line we choose to determine the slope. No matter what points we choose we get the same number for the slope. For example, if we choose (2, 4) and (4, 8) we get for the slope,

$$\frac{8 - 4}{4 - 2} = \frac{4}{2} = 2.$$

DO EXERCISES 6-9.

If a line slants upward from left to right, it has a positive slope, as in Example 4. If a line slants downward from left to right, the change in x and the change in y have opposite signs, so the line has a negative slope.

If a line is horizontal, the change in y for any two points is 0. Thus a horizontal line has zero slope.

If a line is vertical, the change in x for any two points is 0. Thus the slope is not defined, because we cannot divide by zero. A vertical line does not have a slope.

DO EXERCISES 10-14.

[iv] Point-Slope Equations of Lines

Suppose we have a nonvertical line and that the coordinates of one point P_1 are (x_1, y_1). We think of point P_1 as fixed. Suppose, also, that we have a movable point P on the line with coordinates (x, y). Thus the slope would be given by

$$\frac{(y - y_1)}{(x - x_1)} = m. \qquad (1)$$

Note that this is true only when (x, y) is a point different from (x_1, y_1). If we use the multiplication principle, we get*

$$(y - y_1) = m(x - x_1). \qquad \textit{Point-slope equation} \qquad (2)$$

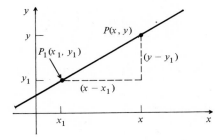

Equation (2) will be true even if $(x, y) = (x_1, y_1)$. Equation (2) is called the *point-slope equation* of a line. Thus if we know the slope of a line and the coordinates of a point on the line, we can find an equation of the line.

Example 5 Find an equation of the line containing the point $(\frac{1}{2}, -1)$ with slope 5.

If we substitute in $(y - y_1) = m(x - x_1)$, we get $y - (-1) = 5(x - \frac{1}{2})$, which simplifies to

$$y + 1 = 5\left(x - \frac{1}{2}\right)$$

or

$$y = 5x - \frac{5}{2} - 1$$

or

$$y = 5x - \frac{7}{2}.$$

DO EXERCISES 15–17.

---

*Hint: It may be easier to remember equation (1), since it relates to the slope formula. We can easily get equation (2) from it.

15. Find an equation of the line containing the point $\left(-2, \frac{1}{4}\right)$ with slope -3.

16. Find an equation of the line with y-intercept $(0, -9)$ and slope $\frac{1}{4}$.

17. Find an equation of the line with x-intercept $(5, 0)$ and slope $-\frac{1}{2}$.

Find an equation of the line containing the given pair of points.

18. $(1, 4)$ and $(3, -2)$

19. $(3, -6)$ and $(0, 4)$

[v] Two-Point Equations of Lines

Suppose a nonvertical line contains the points $P_1(x_1, y_1)$ and $P_2(x_2, y_2)$. The slope of the line is

$$\frac{y_2 - y_1}{x_2 - x_1}.$$

If we substitute $(y_2 - y_1)/(x_2 - x_1)$ for m in the point-slope equation,

$$y - y_1 = m(x - x_1),$$

we have*

$$y - y_1 = \frac{y_2 - y_1}{x_2 - x_1}(x - x_1). \quad \textit{Two-point equation}$$

This is known as the *two-point equation* of a line. Note that either of the two given points can be called P_1 or P_2 and (x, y) is any point on the line.

Example 6 Find an equation of the line containing the points $(2, 3)$ and $(1, -4)$.

If we take $(2, 3)$ as P_1 and use the two-point equation, we get

$$y - 3 = \frac{-4 - 3}{1 - 2}(x - 2),$$

which simplifies to $y = 7x - 11$.

DO EXERCISES 18 AND 19.

[vi] Slope-Intercept Equations of Lines

Suppose a nonvertical line has slope m and y-intercept $(0, b)$. We sometimes, for brevity, refer to the number b as the y-intercept. Let us substitute m and $(0, b)$ in the point-slope equation. We get

$$y - y_1 = m(x - x_1)$$
$$y - b = m(x - 0).$$

This simplifies to

$$y = mx + b. \quad \textit{Slope-intercept equation}$$

This is called the *slope-intercept equation* of a line. The advantage of such an equation is that we can read the slope m and the y-intercept b from the equation.

*Hint: It may be easier to remember

$$\frac{y - y_1}{x - x_1} = \frac{y_2 - y_1}{x_2 - x_1},$$

interpreting each side as slope.

Example 7 Find the slope and the y-intercept of $y = 5x - \frac{1}{4}$.

$$y = 5x - \frac{1}{4}$$

Slope: 5 y-intercept: $-\frac{1}{4}$

Example 8 Find the slope and y-intercept of $y = 8$.

We can rewrite this equation as $y = 0x + 8$. We then see that the slope is 0 and the y-intercept is 8.

DO EXERCISES 20 AND 21.

To find the slope-intercept equation of a line, when given another equation, we solve for y.

Example 9 Find the slope and y-intercept of the line having the equation $3x - 6y - 7 = 0$.

We solve for y, obtaining $y = \frac{1}{2}x - \frac{7}{6}$. Thus the slope is $\frac{1}{2}$ and the y-intercept is $-\frac{7}{6}$.

If a line is vertical it has no slope. Thus it has no slope-intercept equation. Such a line does have a simple equation, however. All vertical lines have equations $x = c$, where c is some constant.

DO EXERCISE 22.

Linear Functions

Any nonvertical straight line is the graph of a function. Such a function is called a *linear function*. Any nonvertical line has an equation $y = mx + b$. Thus a function f is a linear function if and only if it has an equation $f(x) = mx + b$. If the slope m is zero, the equation simplifies to $f(x) = b$. A function like this is called a *constant function*.

20. Find the slope and y-intercept of $y = -7x + 11$.

21. Find the slope and y-intercept of $y = -4$.

22. a) Find the slope-intercept equation of the line whose equation is $-2x + 3y - 6 = 0$.
    b) Find the slope and y-intercept of this line.

EXERCISE SET 4.1

[i]

1. Determine whether each equation is linear.

   a) $3y = 2x - 5$ b) $5x + 3 = 4y$ c) $3y = x^2 + 2$ d) $y = 3$

   e) $xy = 5$ f) $3x^2 + 2y = 4$ g) $3x + \dfrac{1}{y} = 4$ h) $5x - 2 = 4y$

2. Determine whether each equation is linear.

   a) $5y = 3x - 4$ b) $3x + 5 = 7y$ c) $4y = 3x^2 - 4$ d) $4x - \dfrac{2}{y} = 3$

   e) $2xy = 4$ f) $5x^2 + 3y = -4$ g) $x = -4$ h) $6x - 7 = 3y$

[ii] Graph.

3. $8x - 3y = 24$ **4.** $5x - 10y = 50$ **5.** $3x + 12 = 4y$ **6.** $4x - 20 = 5y$
7. $y = -2$ **8.** $2y - 3 = 9$ **9.** $5x + 2 = 17$ **10.** $19 = 5 - 2x$

[iii] Find the slopes of the lines containing these points.

11. $(6, 2)$ and $(-2, 1)$ **12.** $(-2, 1)$ and $(-4, -2)$
13. $(2, -4)$ and $(4, -3)$ **14.** $(5, -3)$ and $(-5, 8)$
15. $(2\pi, 5)$ and $(\pi, 4)$ **16.** $(\sqrt{2}, -4)$ and $(\pi, -4)$

[iv] Find equations of the following lines.

17. Through $(3, 2)$ with $m = 4$ **18.** Through $(4, 7)$ with $m = -2$

19. With y-intercept -5 and $m = 2$ **20.** With y-intercept π and $m = \dfrac{1}{4}$

[v]

21. Containing $(1, 4)$ and $(5, 6)$ **22.** Containing $(-2, 0)$ and $(2, 3)$

[vi] Find the slope and y-intercept of each line.

23. $y = 2x + 3$ **24.** $y = 6 - x$ **25.** $2y = -6x + 10$ **26.** $-3y = -12x + 9$
27. $3x - 4y = 12$ **28.** $5x + 2y = -7$ **29.** $3y + 10 = 0$ **30.** $y = 7$

Find equations of the following lines.

31. ▦ Through $(3.014, -2.563)$ with slope 3.516 **32.** ▦ Through $(-173.4, -17.6)$ with slope -0.00014
33. ▦ Through the points $(1.103, 2.443)$ and $(8.114, 11.012)$ **34.** ▦ Through the points $(473.78, 910.2)$ and $(993.55, 171.43)$

☆ ───

Suppose f is a linear function. Find a formula for $f(x)$ given that:

35. $f(3x) = 3f(x)$.   **36.** $f(kx) = kf(x)$, for some number k.
37. $f(x + 2) = f(x) + 2$.   **38.** $[f(x)]^2$ is linear.

Assume f is a linear function. Are the following true or false?

39. $f(a + b) = f(a) + f(b)$ **40.** $f(ab) = f(a)f(b)$
41. $f(kx) = kf(x)$ **42.** $f(a - b) = f(a) - f(b)$

43. Determine whether these three points are on a line. [*Hint:* Compare the slopes of $\overline{AB}$ and $\overline{BC}$. ($\overline{AB}$ refers to the segment from A to B.)]

$$A(9, 4), \quad B(-1, 2), \quad C(4, 3)$$

44. Determine whether these three points are on a line. (See hint for Exercise 43.)

$$A(-1, -1), \quad B(2, 2), \quad C(-3, -4)$$

45. Use graph paper. Plot the points $A(0, 0)$, $B(8, 2)$, $C(11, 6)$, and $D(3, 4)$. Draw $\overline{AB}$, $\overline{BC}$, $\overline{CD}$, and $\overline{DA}$. Find the slopes of these four segments. Compare the slopes of $\overline{AB}$ and $\overline{CD}$. Compare the slopes of $\overline{BC}$ and $\overline{DA}$. (Figure $ABCD$ is a parallelogram and its opposite sides are parallel.)

46. Use graph paper. Plot the points $E(-2, -5)$, $F(2, -2)$, $G(7, -2)$, and $H(3, -5)$. Draw $\overline{EF}$, $\overline{FG}$, $\overline{GH}$, $\overline{HE}$, $\overline{EG}$, and $\overline{FH}$. Compare the slopes of $\overline{EG}$ and $\overline{FH}$. (Figure $EFGH$ is a rhombus and its diagonals are perpendicular.)

47. (*Fahrenheit temperature as a function of Celsius temperature*). Fahrenheit temperature F is a linear function of Celsius (or Centigrade) temperature C. When C is 0, F is 32. When C is 100, F is 212. Use these data to express F as a linear function of C.

48. (*Celsius temperature as a function of Fahrenheit temperature*). Celsius (Centigrade) temperature C is a linear function of Fahrenheit temperature F. When F is 32, C is 0. When F is 212, C is 100. Use these data to express C as a linear function of F.

49. Suppose P is a nonconstant linear function of Q. Show that Q is a linear function of P.

50. Suppose y is directly proportional to x. Show that y is a linear function of x.

(*Road grade.*) Numbers like 2%, 3%, and 6% are often used to represent the *grade* of a road. Such a number is meant to tell how steep a road up a hill or mountain is. For example, a 3% grade means that for every horizontal distance of 100 ft, the road rises 3 ft. In each case find the road grade and an equation giving the height y of a vehicle in terms of a horizontal distance x.

51.

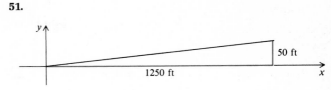

52.

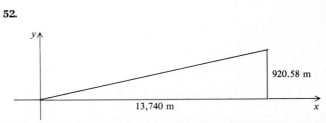

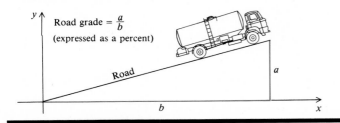

4.2 PARALLEL AND PERPENDICULAR LINES; THE DISTANCE FORMULA

[i] Parallel and Perpendicular Lines

If two lines are vertical, they are parallel. Thus equations such as $x = c_1$ and $x = c_2$ (where c_1 and c_2 are constants) have graphs that are parallel lines. Now we consider nonvertical lines. In order that such lines be parallel, they must have the same slope but different y-intercepts. Thus equations such as $y = mx + b_1$ and $y = mx + b_2$, $b_1 \neq b_2$, have graphs that are parallel lines.

THEOREM 1

Vertical lines are parallel. Nonvertical lines are parallel if and only if they have the same slope and different y-intercepts.

If two equations are equivalent, they represent the same line. Thus if two lines have the same slope and the same y-intercept, they are not really two different lines. They are the same line. In such a case we sometimes speak of "coincident lines."

If one line is vertical and the other is horizontal, such as $x = c_1$ and $y = c_2$, they are perpendicular. Otherwise, how can we tell whether two lines are perpendicular? Consider the figure below.

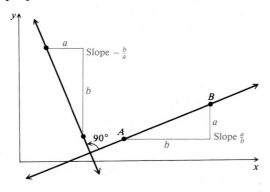

OBJECTIVES

You should be able to:

[i] Given equations of two lines, tell whether they are parallel, perpendicular, or neither.

[ii] Given an equation of a line, find equations of lines parallel or perpendicular to that line and containing a given point.

[iii] Given the coordinates of two points, find the distance between them.

[iv] Determine whether three points with given coordinates are vertices of a right triangle.

[v] Given the coordinates of the endpoints of a segment, find the coordinates of its midpoint.

In each situation slopes of the lines are given. Determine whether they are perpendicular.

1.

$m_1 = 1.25$

$m_2 = -0.79$

2.

$m_1 = 3.2$

$m_2 = -0.3125$

Determine whether each pair of lines is parallel, perpendicular, or neither.

3. $2y - x = 2,$
   $y + 2x = 4.$

4. $3y = 2x + 15,$
   $2y = 3x + 10.$

5. $5y = 3 - 4x,$
   $8x + 10y = 1.$

Consider a line $\overleftrightarrow{AB}$ as shown, with slope a/b. Then think of rotating the figure $90°$ to get a line perpendicular to $\overleftrightarrow{AB}$. For the new line the rise and run are interchanged, but the run is now the additive inverse of what it was. Thus the slope of the new line is $-b/a$. Let us multiply the slopes:

$$\frac{a}{b}\left(-\frac{b}{a}\right) = -1.$$

This is the condition under which lines will be perpendicular.

THEOREM 2

> **Two lines with slopes m_1 and m_2 are perpendicular if and only if $m_1 m_2 = -1$.**

If one line has slope m_1, the slope m_2 of a line perpendicular to it is $-1/m_1$.

DO EXERCISES 1 AND 2.

Example 1 Determine whether the following lines are parallel, perpendicular, or neither.

a) $y + 2 = 5x, \quad 5y + x = -15.$
   We solve for y:

$$y = 5x - 2, \quad y = -\frac{1}{5}x - 3.$$

   The slopes are 5 and $-\frac{1}{5}$. Their product is -1, so the lines are perpendicular.

b) $2y + 4x = 8, \quad 5 + 2x = -y.$
   By solving for y we determine that $m_1 = -2$ and $m_2 = -2$ and the y-intercepts are different, so the lines are parallel.

c) $2x + 1 = y, \quad y + 3x = 4.$
   By solving for y we determine that $m_1 = 2$ and $m_2 = -3$, so the lines are neither parallel nor perpendicular.

DO EXERCISES 3–5.

[ii] Parallel or Perpendicular Lines Through a Point

Example 2 Write equations of the lines perpendicular or parallel to the line $4y - x = 20$ and containing the point $(2, -3)$.

We first solve for y: $y = \frac{1}{4}x + 5$, so the slope is $\frac{1}{4}$. The slope of the perpendicular line is -4.

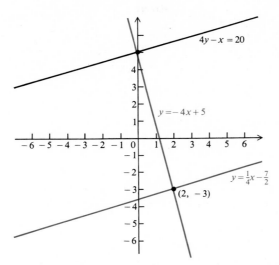

6. Find equations of the lines parallel
and perpendicular to the line
$4 - y = 2x$ and containing the
point $(3, 4)$.

Now we use the point-slope equation to write an equation with slope -4 and containing the point $(2, -3)$:

$$y - y_1 = m(x - x_1)$$
$$y - (-3) = -4(x - 2).$$

This simplifies to $y = -4x + 5$.

The line parallel to the given line will have slope $\frac{1}{4}$. The equation is

$$y - (-3) = \frac{1}{4}(x - 2).$$

This simplifies to $y = \frac{1}{4}x - \frac{7}{2}$.

DO EXERCISES 6 AND 7.

7. Find equations of the lines parallel
and perpendicular to the line
$x = -3$ and containing the point
$(5, -4)$.

Vertical Lines

If a line is vertical, it has no slope. Every vertical line has an equation $x = c$, where c is a constant. Any line parallel to a vertical line must also be vertical, hence must also have an equation $x = k$, where k is a constant.

A line will be perpendicular to a vertical line if and only if that line is horizontal. Every horizontal line has zero slope and has an equation $y = c$, where c is a constant. Thus if a line is vertical, it is simple to determine whether another line is parallel to it or perpendicular to it.

Example 3 Find equations of the lines parallel and perpendicular to the line $x = 4$ and containing the point $(-2, 3)$.

The line $x = 4$ is vertical, so any line parallel to it must be vertical. The line we seek has one x-coordinate, which is -2, so all x-coordinates on the line must be -2. The equation is $x = -2$.

The perpendicular line has a y-coordinate, which is 3, so all y-coordinates must be 3. The equation is $y = 3$.

[iii] The Distance Formula

We develop a formula for finding the distance between two points whose coordinates are known. Suppose the points are on a horizontal line, thus having the same second coordinates. We can find the distance between them by subtracting their first coordinates. This difference may be negative, depending on the order in which we subtract. So to make sure we get a positive number, we take the absolute value of this difference. The distance between two points on a horizontal line (x_1, y) and (x_2, y) is thus $|x_1 - x_2|$. Similarly, the distance between two points on a vertical line (x, y_1) and (x, y_2) is $|y_1 - y_2|$.

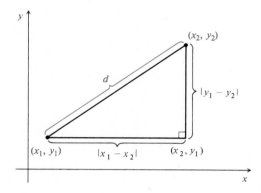

Now consider any two points not on a horizontal or vertical line (x_1, y_1) and (x_2, y_2). These points are vertices of a right triangle, as shown. The other vertex is (x_2, y_1). The legs of this triangle have the lengths $|x_1 - x_2|$ and $|y_1 - y_2|$. Now by the Pythagorean theorem we obtain a relation between the length of the hypotenuse d and the lengths of the legs:

$$d^2 = |x_1 - x_2|^2 + |y_1 - y_2|^2.$$

We may now dispense with the absolute value signs because squares of numbers are never negative. Thus we have

$$d^2 = (x_1 - x_2)^2 + (y_1 - y_2)^2.$$

By taking the square root we obtain the distance between the two points.

THEOREM 3

The distance formula. **The distance between any two points (x_1, y_1) and (x_2, y_2) is given by**

$$d = \sqrt{(x_1 - x_2)^2 + (y_1 - y_2)^2}.$$

Although we derived the distance formula by considering two points not on a horizontal or vertical line, the formula holds for *any* two points.

Example 4 Find the distance between the points $A(-2, 2)$ and $B(4, -3)$ on the islands in the figure.

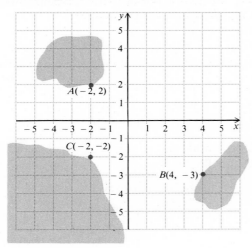

$$d = \sqrt{(-2 - 4)^2 + [2 - (-3)]^2}$$
$$= \sqrt{(-6)^2 + 5^2}$$
$$= \sqrt{36 + 25}$$
$$= \sqrt{61}$$

An approximation for this distance is 7.8.

DO EXERCISES 8–11.

[iv] Vertices of Right Triangles

Example 5 Determine whether the points $A(-2, 2)$, $B(4, -3)$, and $C(-2, -2)$ in the figure of Example 4 are vertices of a right triangle.

First we find the squares of the distances between the points:

$$d_1^2 = (-2 - 4)^2 + [2 - (-3)]^2 = (-6)^2 + 5^2 = 61,$$
$$d_2^2 = [-2 - (-2)]^2 + [2 - (-2)]^2 = 0^2 + 4^2 = 16,$$
$$d_3^2 = [4 - (-2)]^2 + [-3 - (-2)]^2 = 6^2 + (-1)^2 = 37.$$

Since no sum of any two squares is another, it follows by the converse of the Pythagorean theorem that the points are not vertices of a right triangle.

DO EXERCISES 12 AND 13.

[v] Midpoints of Segments

The distance formula can be used to verify or derive a formula for finding the coordinates of the midpoint of a segment when the coordinates of its endpoints are known. We shall not derive this formula but simply state it.

Find the distance between the points of each pair.

8. $(-5, 3)$ and $(2, -7)$

9. $(3, 3)$ and $(-3, -3)$

10. $(9, -5)$ and $(9, 11)$

11. $(0, \pi)$ and $(8, \pi)$

Determine whether the points are vertices of a right triangle.

12. $(11, 1)$, $(6, 6)$, $(2, 2)$

13. $(10, -4)$, $(3, 5)$, $(0, 0)$

Find the midpoints of the segments having endpoints as given.

14. $(-2, 1)$ and $(5, -6)$

15. $(9, -6)$ and $(9, -4)$

THEOREM 4

The midpoint formula. **If the endpoints of a segment are (x_1, y_1) and (x_2, y_2), then the coordinates of the midpoint are**

$$\left(\frac{x_1 + x_2}{2}, \frac{y_1 + y_2}{2}\right).$$

Note that we obtain the coordinates of the midpoint by averaging the coordinates of the endpoints. This is an easy way to remember this formula.

Example 6 Find the midpoint of the segment in Example 4 with endpoints $A(-2, 2)$ and $B(4, -3)$.

Using the midpoint formula, we obtain

$$\left(\frac{-2 + 4}{2}, \frac{2 + (-3)}{2}\right), \text{ or } \left(1, -\frac{1}{2}\right).$$

DO EXERCISES 14 AND 15.

EXERCISE SET 4.2

[i] In Exercises 1–4, determine whether the lines are parallel, perpendicular, or neither.

1. $2x - 5y = -3,$
$2x + 5y = 4$

2. $x + 2y = 5,$
$2x + 4y = 8$

3. $y = 4x - 5,$
$4y = 8 - x$

4. $y = -x + 7,$
$y = x + 3$

[ii] In Exercises 5–10, find an equation of the line containing the given point and parallel to the given line.

5. $(0, 3), 3x - y = 7$

6. $(-4, -5), 2x + y = -4$

7. $(3, 8), x = 2$

8. $(3, -3), x = -1$

9. $(-2, -3), y = 4$

10. $(-3, 2), y = -3$

In Exercises 11–16, find an equation of the line containing the given point and perpendicular to the given line.

11. $(-3, -5), 5x - 2y = 4$

12. $(3, -2), 3x + 4y = 5$

13. $(0, 3), x = 1$

14. $(-2, -2), x = 3$

15. $(-3, -7), y = 2$

16. $(4, -5), y = -1$

17. ▦ Find an equation of the line parallel to the one given, and containing the given point.

$$4.323x - 7.071y = 16.61, \quad (-2.603, 1.818)$$

18. ▦ Find an equation of the line containing the given point and perpendicular to the given line.

$$6.232x + 4.001y = 4.881, \quad (3.149, -2.908)$$

[iii] Find the distance between the points of each pair.

19. $(-3, -2)$ and $(1, 1)$

20. $(5, 9)$ and $(-1, 6)$

21. $(0, -7)$ and $(3, -4)$

22. $(2, 2)$ and $(-2, -2)$

23. $(a, -3)$ and $(2a, 5)$

24. $(5, 2k)$ and $(-3, k)$

25. $(0, 0)$ and (a, b)

26. $(\sqrt{2}, \sqrt{3})$ and $(0, 0)$

27. $(\sqrt{a}, \sqrt{b})$ and $(-\sqrt{a}, \sqrt{b})$

28. $(c - d, c + d)$ and $(c + d, d - c)$

29. ▦ $7.3482, -3.0991)$ and $(18.9431, -17.9054)$

30. ▦ $(-25.414, 175.31)$ and $(275.34, -95.144)$

[iv] Determine whether the points are vertices of a right triangle.

31. $(9, 6), (-1, 2),$ and $(1, -3)$

32. $(-5, -8), (1, 6),$ and $(5, -4)$

[v] Find the midpoints of the segments having the following endpoints.

33. $(-4, 7)$ and $(3, -9)$ **34.** $(4, 5)$ and $(6, -7)$ **35.** (a, b) and $(a, -b)$ **36.** $(-c, d)$ and (c, d)

37. $(-3.895, 8.1212)$ and $(2.998, -8.6677)$ **38.** ▦ $(4.1112, 6.9898)$ and $(5.1928, -6.9143)$

☆ _____

39. Find an equation of the line containing the point $(4, -2)$ and parallel to the line containing $(-1, 4)$ and $(2, -3)$.

40. Find an equation of the line containing the point $(-1, 3)$ and parallel to the line containing $(3, -5)$ and $(-2, -7)$.

41. Find the point on the x-axis that is equidistant from the points $(1, 3)$ and $(8, 4)$.

42. Find the point on the y-axis that is equidistant from the points $(-2, 0)$ and $(4, 6)$.

★ _____

43. Consider any right triangle with base b and height h, situated as shown. Show that the midpoint of the hypotenuse P is equidistant from the three vertices of the triangle.

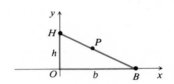

44. Consider any quadrilateral situated as shown. Show that the segments joining the midpoints of the sides, in order as shown, form a parallelogram.

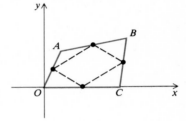

45. Prove that the distance formula holds when two points are on either a vertical line or a horizontal line.

46. The distance formula can be interpreted as a function whose inputs are ordered pairs (P, Q), where P is a pair (x_1, y_1) and Q is a pair (x_2, y_2).

a) What is the domain of the function?

b) What is the range of the function?

c) Under what conditions is an output 0?

4.3 QUADRATIC FUNCTIONS

[i] Properties of Quadratic Functions

If a function can be described by a second-degree polynomial, it is called *quadratic*. The following is a more precise definition.

DEFINITION

A *quadratic* function is a function that can be described as follows:

$$f(x) = ax^2 + bx + c, \quad \text{where} \quad a \neq 0.$$

In this definition we insist that $a \neq 0$; otherwise the polynomial would not be of degree two. One or both of the constants b and c can be 0.

Consider $f(x) = x^2$. This is an even function, so the y-axis is a line of symmetry. The graph opens upward, as shown below. If we multiply by a constant a to get $f(x) = ax^2$, we obtain a vertical stretching or

OBJECTIVES

You should be able to:

[i] Given a quadratic function, find the vertex of its graph, the line of symmetry, and the maximum or minimum value.

[ii] Graph quadratic functions.

[iii] Find the x-intercepts of the graph of a quadratic function.

1. a) Graph $f(x) = 2x^2$.
   b) Does the graph open upward or does it open downward?
   c) What is the line of symmetry?
   d) What is the minimum value?
   e) What is the vertex?

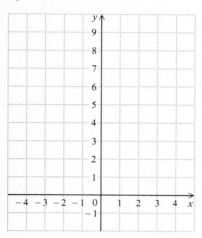

2. a) Graph $f(x) = -0.4x^2$.
   b) Does the graph open upward or does it open downward?
   c) What is the line of symmetry?
   d) What is the maximum value?
   e) What is the vertex?

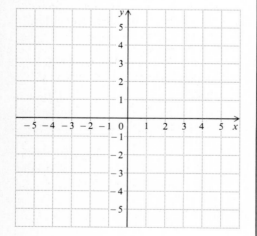

shrinking, and a reflection if a is negative. Examples of this are shown in the figure. Bear in mind that if a is negative the graph opens downward.

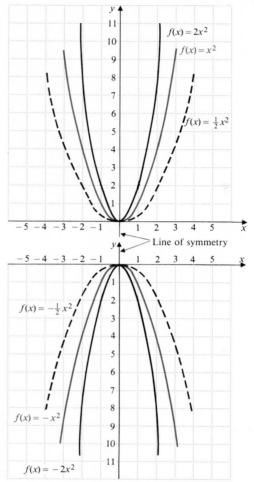

Graphs of quadratic functions. Parabolas.

Graphs of quadratic functions are called *parabolas*. In each parabola shown, the point $(0, 0)$ is the *vertex*.

DO EXERCISES 1 AND 2.

Let us consider $f(x) = a(x - h)^2$. We have replaced x by $x - h$ in ax^2, and will therefore obtain a horizontal translation. The translation will be to the right if h is positive. The examples in the figure illustrate translations, and a summary is given below.

The graph of $f(x) = a(x - h)^2$

a) **opens upward if $a > 0$, downward if $a < 0$;**

b) **has $(h, 0)$ as a vertex;**

c) **has $x = h$ as a line of symmetry;**

d) **has a minimum 0 if $a > 0$, a maximum 0 if $a < 0$.**

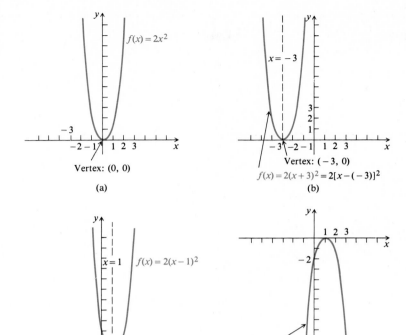

(a)

(b)

(c)

(d)

DO EXERCISES 3 AND 4.

Now consider $f(x) = a(x - h)^2 + k$, or $f(x) - k = a(x - h)^2$. We have replaced $f(x)$ by $f(x) - k$ in the equation $f(x) = a(x - h)^2$. Thus we have a translation. If k is positive, the translation is upward. If k is negative, the translation is downward. Consider these examples. Note that the vertex has been moved off the x-axis.

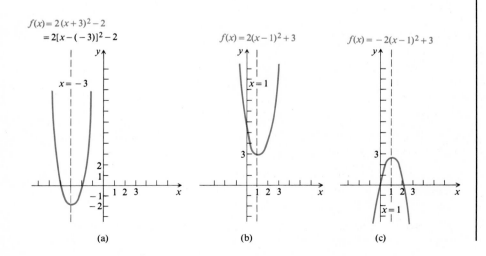

(a)

(b)

(c)

3. a) Graph $f(x) = 3x^2$.
b) Use the graph in (a) to graph $f(x) = 3(x - 2)^2$.
c) What is the vertex?
d) What is the line of symmetry?
e) What is the minimum value?
f) Does the graph open upward or does it open downward?
g) Is the graph of $f(x) = 3(x - 2)^2$ a horizontal translation to the left or to the right?

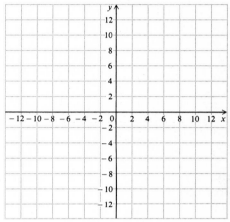

4. a) Graph $f(x) = -3x^2$.
b) Use the graph in (a) to graph
$$f(x) = -3(x + 2)^2$$
$$= -3[x - (-2)]^2.$$
c) What is the vertex?
d) What is the line of symmetry?
e) What is the maximum value?
f) Does the graph open upward or does it open downward?
g) Is the graph of $f(x) = -3(x + 2)^2$ a horizontal translation to the left or to the right?

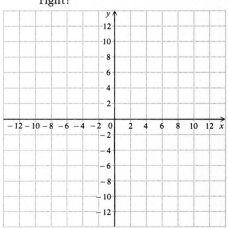

Answer the following questions in Exercises 5–10.

a) What is the vertex?
b) What is the line of symmetry?
c) Is there a maximum? What is it?
d) Is there a minimum? What is it?

5. Graph $f(x) = 3(x - 2)^2 + 4$. Then answer the above questions.

6. Graph $f(x) = -3(x + 2)^2 - 1 = -3[x - (-2)]^2 - 1$. Then answer the above questions.

Without graphing, answer the above questions for each function.

7. $f(x) = (x - 5)^2 + \pi$

8. $f(x) = -3(x - 5)^2$

9. $f(x) = 2\left(x + \frac{1}{4}\right)^2 - 6$

10. $f(x) = -\frac{1}{4}(x + 9)^2 + 3$

The graph of $f(x) = a(x - h)^2 + k$

a) **opens upward if $a > 0$, downward if $a < 0$;**

b) **has (h, k) as a vertex;**

c) **has $x = h$ as the line of symmetry;**

d) **has k as a minimum value (output) if $a > 0$, has k as a maximum value (output) if $a < 0$.**

Thus without graphing, we can determine a lot of information about a function described by $f(x) = a(x - h)^2 + k$. The following table is an example.

FUNCTION	$f(x) = 3(x - \frac{1}{4})^2 - 2$	$g(x) = -3(x + 5)^2 + 7$ $= -3[x - (-5)]^2 + 7$
a) What is the vertex?	$(\frac{1}{4}, -2)$	$(-5, 7)$
b) What is the line of symmetry?	$x = \frac{1}{4}$	$x = -5$
c) Is there a maximum? What is it?	No; graph extends upward; $3 > 0$.	Yes; 7; graph extends downward; $-3 < 0$.
d) Is there a minimum? What is it?	Yes; -2; graph extends upward; $3 > 0$.	No; graph extends downward; $-3 < 0$.

DO EXERCISES 5–10.

Now let us consider a quadratic function $f(x) = ax^2 + bx + c$. Note that it is not in the form $f(x) = a(x - h)^2 + k$. We can put it into that form by *completing the square*.

Example 1 For the function $f(x) = x^2 - 6x + 4$, complete the square.

We consider first the x^2- and x-terms:

$$f(x) = x^2 - 6x \qquad + 4$$

We construct a trinomial square. To do this we take half the coefficient of x and square it. That number is $(-6/2)^2$, or 9. We now add that number to complete the square. We also must subtract it to get an equivalent equation. We can think of this simply as adding $9 - 9$, which is 0.

$$f(x) = x^2 - 6x + 9 - 9 + 4$$
$$= x^2 - 6x + 9 - 5$$

We now factor the trinomial square $x^2 - 6x + 9$ and we are finished:

$$f(x) = (x - 3)^2 - 5.$$

DO EXERCISE 11.

If the coefficient of x^2 is not 1, a preliminary step is needed.

Example 2 For the function $f(x) = 2x^2 + 12x - 1$, complete the square.

Again we consider the x^2- and x-terms, but we begin by factoring out the x^2-coefficient, as follows:

$$f(x) = 2(x^2 + 6x) - 1.$$

Next, we proceed as before, *inside* the parentheses. We take half the coefficient of x and square it. That number is $(6/2)^2$, or 9. We add $9 - 9$, but do it *inside* the parentheses:

$$f(x) = 2(x^2 + 6x + 9 - 9) - 1.$$

We now have an extra, unwanted, term inside the parentheses so we rearrange things a bit:

$$f(x) = 2(x^2 + 6x + 9) - 2 \cdot 9 - 1.$$

We have used the distributive law and multiplication to get the $(-2 \cdot 9)$ outside. We now have

$$f(x) = 2(x^2 + 6x + 9) - 18 - 1$$
$$= 2(x + 3)^2 - 19.$$

To complete the square on a quadratic function $f(x) = ax^2 + bx + c$:

1. **If $a \neq 1$, factor a out of the x^2- and x-terms.**
2. **Take half the resulting coefficient of x and square it to obtain Q. Add $Q - Q$ *inside* the parentheses.**
3. **Multiply to obtain $f(x) = a(x^2 + px + Q) - aQ + c$. Finish by factoring the polynomial and simplifying.**

DO EXERCISE 12.

Example 3 For the function $f(x) = 4x^2 + 12x + 7$, (a) find the vertex and the line of symmetry; and (b) determine whether there is a maximum or minimum value and find that value.

We complete the square:

$$f(x) = 4(x^2 + 3x) + 7$$
$$= 4\left(x^2 + 3x + \frac{9}{4} - \frac{9}{4}\right) + 7 \qquad \left(\frac{3}{2}\right)^2 = \frac{9}{4}$$
$$= 4\left(x^2 + 3x + \frac{9}{4}\right) - 4 \cdot \frac{9}{4} + 7$$
$$= 4\left(x + \frac{3}{2}\right)^2 - 2.$$

11. For the function

$$f(x) = x^2 - 4x + 7,$$

complete the square.

12. Complete the square.

$$f(x) = 3x^2 + 24x + 10$$

13. For the function

$$f(x) = 4x^2 - 12x - 5,$$

a) find the vertex and the line of symmetry;
b) determine whether there is a maximum or minimum value and find that value.

a) The vertex is $(-\frac{3}{2}, -2)$. The line of symmetry goes through this point, hence is the line $x = -\frac{3}{2}$.

b) Since the coefficient of x^2 is positive, the parabola opens upward. Thus we have a minimum value. The value is -2.

DO EXERCISE 13.

Example 4 For the function $f(x) = -2x^2 + 10x - 7$, find the vertex and the maximum or minimum value.

We complete the square:

$$f(x) = -2(x^2 - 5x) - 7$$

$$= -2\left(x^2 - 5x + \frac{25}{4} - \frac{25}{4}\right) - 7 \quad \left(\frac{-5}{2}\right)^2 = \frac{25}{4}$$

$$= -2\left(x^2 - 5x + \frac{25}{4}\right) - (-2)\frac{25}{4} - 7 = -2\left(x - \frac{5}{2}\right)^2 + \frac{11}{2}.$$

The vertex is $(\frac{5}{2}, \frac{11}{2})$. The maximum value is $\frac{11}{2}$, since the coefficient of x^2 is negative.

DO EXERCISE 14.

Graphing Quadratic Functions

We know that the graph of any quadratic function is a parabola. If the coefficient of x^2 is positive, the parabola opens upward. Keeping this in mind, we can calculate and plot a few points and connect them. It also helps to know where the vertex and the line of symmetry are.

14. For the function

$$f(x) = -2x^2 - 10x + 9,$$

find the vertex and the maximum or minimum value.

Example 5 Graph $f(x) = -2x^2 + 10x - 7$.

We note that the coefficient of x^2 is negative. The parabola opens downward. Calculate several input–output pairs. (Note that $f(0)$ is always easy to find.) We then plot the ordered pairs and complete the graph.

x	f(x)
0	−7
−1	−19
1	1
2	5
3	5
4	1

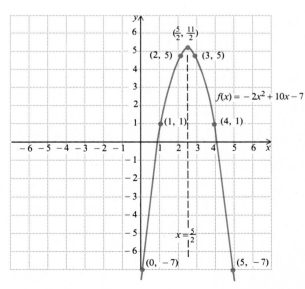

Sometimes, as in this case, the line of symmetry is apparent. (If it is not apparent, we could complete the square as in Example 4.) In this case it is the line $x = \frac{5}{2}$. We can sketch the line, and then:

a) Find the vertex. It is $\left(\frac{5}{2}, f\left(\frac{5}{2}\right)\right)$, or $\left(\frac{5}{2}, \frac{11}{2}\right)$.

b) Find other points to plot by reflecting across the line of symmetry.

For example, the point opposite $(0, -7)$ is $(5, -7)$.

To graph a quadratic function:

1. Note whether the coefficient of x^2 is positive or negative and thus determine whether the curve opens upward or downward.

2. Calculate several input–output pairs. (A calculator can be most useful.) Plot these ordered pairs and complete the graph.

3. Find the vertex and line of symmetry. Use completing the square if they are not apparent. Use reflection across the line of symmetry to find further points to plot.

DO EXERCISES 15 AND 16.

[iii] **x-intercepts**

The points at which a graph crosses the x-axis are called its x-intercepts. These are, of course, the points at which $f(x) = 0$. Thus the x-values at the intercepts* are the solutions of the equation $f(x) = 0$.

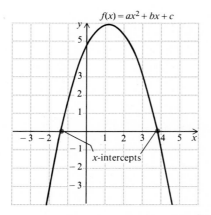

In order to find the x-intercepts of a quadratic function $f(x) = ax^2 + bx + c$ we solve the equation $f(x) = 0$, or

$$ax^2 + bx + c = 0.$$

Example 6 Find the x-intercepts of the graph of $f(x) = x^2 - 2x - 2$.

We solve the equation

$$x^2 - 2x - 2 = 0.$$

---

*Also called *zeros of the function*.

Graph.

15. $f(x) = x^2 - 6x + 4$

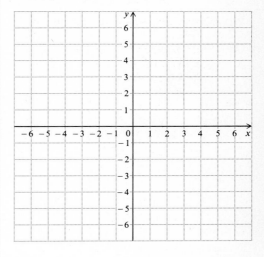

16. $f(x) = -4x^2 + 12x - 5$

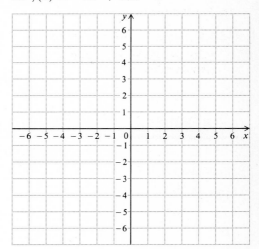

17. Find the x-intercepts.

$$f(x) = x^2 - 2x - 5$$

Find the x-intercepts if they exist.

18. $f(x) = x^2 - 2x - 3$

19. $f(x) = x^2 + 8x + 16$

20. $f(x) = -2x^2 - 4x - 3$

The equation is difficult to factor, so we use the quadratic formula and get $x = 1 \pm \sqrt{3}$. Thus the x-intercepts are $(1 - \sqrt{3}, 0)$ and $(1 + \sqrt{3}, 0)$. For plotting, we approximate, to get $(-0.7, 0)$ and $(2.7, 0)$. We sometimes refer to the x-coordinates as intercepts.

Note that the x-intercepts could also be used when graphing quadratic functions, but they are not essential.

DO EXERCISE 17.

The discriminant, $b^2 - 4ac$, tells us how many real-number solutions the equation $ax^2 + bx + c = 0$ has, so it also indicates how many intercepts there are. Compare.

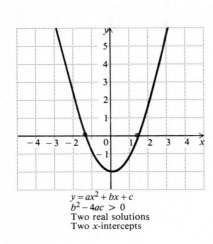

$y = ax^2 + bx + c$
$b^2 - 4ac > 0$
Two real solutions
Two x-intercepts

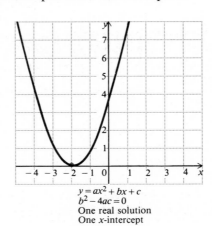

$y = ax^2 + bx + c$
$b^2 - 4ac = 0$
One real solution
One x-intercept

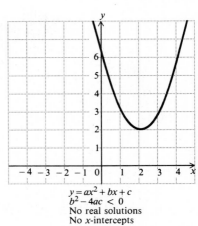

$y = ax^2 + bx + c$
$b^2 - 4ac < 0$
No real solutions
No x-intercepts

DO EXERCISES 18–20.

EXERCISE SET 4.3

[i] For each of the following functions,

a) find the vertex;

b) find the line of symmetry; and

c) determine whether there is a maximum or minimum function value and find that value.

1. $f(x) = x^2$

2. $f(x) = -5x^2$

3. $f(x) = -2(x - 9)^2$

4. $f(x) = 5(x - 7)^2$

5. $f(x) = 2(x - 1)^2 - 4$

6. $f(x) = -(x + 4)^2 - 3$

For each of the following functions,

a) find an equation of the type $f(x) = a(x - h)^2 + k$;

b) find the vertex;

c) determine whether there is a maximum or minimum function value and find that value.

7. $f(x) = -x^2 + 2x + 3$

8. $f(x) = -x^2 + 8x - 7$

9. $f(x) = x^2 + 3x$

10. $f(x) = x^2 - 9x$

11. $f(x) = -\frac{3}{4}x^2 + 6x$

12. $f(x) = \frac{3}{2}x^2 + 3x$

13. $f(x) = 3x^2 + x - 4$

14. $f(x) = -2x^2 + x - 1$

[ii] Graph.

15. $f(x) = -x^2 + 2x + 3$

16. $f(x) = x^2 - 3x - 4$

17. $f(x) = x^2 - 8x + 19$

18. $f(x) = -x^2 - 8x - 17$

19. $f(x) = -\frac{1}{2}x^2 - 3x + \frac{1}{2}$

20. $f(x) = 2x^2 - 4x - 2$

21. $f(x) = 3x^2 - 24x + 50$

22. $f(x) = -2x^2 + 2x + 1$

[iii] Find the x-intercepts.

23. $f(x) = -x^2 + 2x + 3$

24. $f(x) = x^2 - 3x - 4$

25. $f(x) = x^2 - 8x + 5$

26. $f(x) = -x^2 - 3x - 3$

27. $f(x) = -5x^2 + 6x - 5$

28. $f(x) = 2x^2 + x - 5$

☆ _____

Find an equation of the type $f(x) = a(x - h)^2 + k$.

29. $f(x) = ax^2 + bx + c$

30. $f(x) = 3x^2 + mx + m^2$

Graph.

31. $f(x) = |x^2 - 1|$

32. $f(x) = |3 - 2x - x^2|$

Find the maximum or minimum value for each of the following functions.

33. $f(x) = 2.31x^2 - 3.135x - 5.89$

34. $f(x) = -18.8x^2 + 7.92x + 6.18$

35. What is the minimum product of two numbers whose difference is 4.932? What are the numbers?

36. What is the maximum product of two numbers whose sum is 21.355? What are the numbers?

★ _____

37. Find the dimensions and area of the largest rectangle that can be inscribed as shown in a right triangle ABC whose sides have lengths 9 cm, 12 cm, and 15 cm.

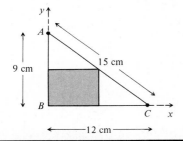

38. A farmer wants to build a rectangular fence near a river, and will use 120 ft of fencing. What is the area of the largest region that can be enclosed? Note that the side next to the river is not fenced.

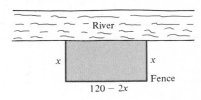

4.4 MATHEMATICAL MODELS

Mathematics is often constructed to fit certain situations. When this is done, we say that we have a *mathematical model*. One of the simplest examples is the natural numbers, i.e., the numbers 1, 2, 3, 4, and so on. These numbers and operations on them were invented so that they would apply to situations in which counting is involved, in some way or other. By working within a mathematical model, we hope to be able to predict what will happen in an actual situation.

OBJECTIVES

You should be able to:

[i] Given a situation that is described by a linear function, find that function from two sets of data and then use that function to make predictions.

[ii] Given a situation described by a quadratic function, use that function to make predictions, including finding a maximum or minimum value of the function.

Example 1 Use the mathematical model of the natural numbers to predict how much money a theater will collect if it charges $2 per admission and 147 people attend.

We translate the problem situation to the language of the mathematical model. In this case, the total number of dollars collected will be the product of the admission price and the number of admissions. We multiply, to get an answer of 294. On this basis we predict that when the owner counts the receipts, there will be $294.

In an example such as the preceding one, we expect our prediction to be exact. In most situations, mathematical models are not that good. Predictions will be only approximate.

Example 2 Use the mathematical model of the natural numbers to predict how much mixture will result from mixing one liter of water and one liter of alcohol.

We solve this problem by adding 1 and 1, to get 2. On this basis, the answer is 2 liters.

In the latter example, our result is not exact. The actual result will be something less than 2 liters, because for some reason there is shrinkage when water and alcohol are mixed.

If a mathematical model does not give precise answers, we ordinarily try revising it so that it does, or we may even look for an entirely new model.

[i] Linear Functions as Mathematical Models

As a result of experiment or experience, we often acquire data that indicate that a function of some sort would be a good mathematical model.

How do we know when a linear function fits a situation?

Example 3 Plot the following data and determine whether a linear equation gives an approximate fit.

Crickets are known to chirp faster when the temperature is higher. Here are some measurements made at different temperatures.

TEMPERATURE, °C	6	8	10	15	20
NUMBER OF CHIRPS per min	11	29	47	75	107

We make a graph with a T- (temperature) axis and an N- (number per minute) axis and plot these data. We see that they do lie approximately on a straight line, so we can use a linear equation in this situation.

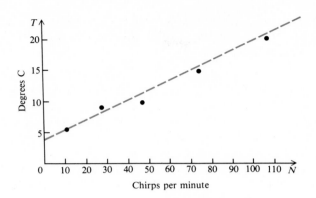

Chirps per minute

Example 4 Wind friction, or *resistance,* increases with speed. Here are some measurements made in a wind tunnel. Plot the data and determine whether a linear equation will give an approximate fit.

VELOCITY, km/h	10	21	34	40	45	52
FORCE OF RESISTANCE, kg	3	4.2	6.2	7.1	15.1	29.0

We make a graph with a V- (velocity) axis and an F- (force) axis and plot the data. They do not lie on a straight line, even approximately. Therefore we cannot use a linear equation in this situation.*

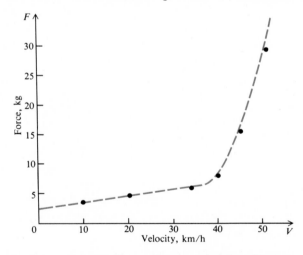

Velocity, km/h

When we have a situation that a linear equation fits, we can find an equation and use it to solve problems and make predictions.

Example 5 When crickets chirp 40 times per minute, the temperature is 10°C. When they chirp 112 times per minute, the temperature is 20°C. We know that a linear equation fits this situation. (a) Find a linear equation that fits the data. (b) From your equation, find the temperature when crickets chirp 76 times per minute; 100 times per minute.

---

*The situation in Example 4 might fit an equation that is not linear. We shall consider such problems later.

1. A class of college students wanted to determine a function from which they can predict a final exam score S from a midterm test score d. To do this they checked with two students who took the class before. One student scored 70 on the midterm and 75 on the final. Another scored 85 on the midterm and 89 on the final.

a) Assuming that S is a linear function of d, express S in terms of d.

a) To find the equation, we use the two known ordered pairs $(40, 10°)$ and $(112, 20°)$. We call these *data points*. We find the two-point equation, using these data points.

$$T - T_1 = \frac{T_2 - T_1}{N_2 - N_1}(N - N_1)$$

$$T - 10 = \frac{20 - 10}{112 - 40}(N - 40) \qquad \text{Substituting}$$

$$T - 10 = \frac{10}{72}(N - 40)$$

$$T - 10 = \frac{5}{36}N - \frac{5}{36} \cdot 40$$

$$T = \frac{5}{36}N + \frac{40}{9}, \quad \text{or} \quad \frac{5N + 160}{36} \qquad \text{Solving for } T$$

b) Using this as a formula, we find T when $N = 76$:

$$T = \frac{5 \cdot 76 + 160}{36} = 15°.$$

When $N = 100$, we get

$$T = \frac{5 \cdot 100 + 160}{36} \approx 18.3°.$$

DO EXERCISE 1.

Equations give decent results only within certain limits. A negative number of chirps per minute would be meaningless. Also, imagine what would happen to a cricket at $-40°$ or at $100°$!

b) Use the equation obtained in (a) to predict a student's final exam score, given that an 81 was scored on the midterm.

Example 6 (*Profit and loss analysis*). Boxowits, Inc., a computer manufacturing firm, is going to produce a new minicalculator. During the first year, the costs for setting up the new production line are $100,000. These are fixed costs such as rent, tools, etc., which must be incurred before any calculators are produced. The additional cost of producing a calculator is $20. This cost is directly related to production, such as material, wages, fuels, and so on. It is a variable cost, according to the number of calculators produced. If x calculators are produced, the variable cost is then $20x$ dollars. The *total* cost of producing x calculators is given by a function C:

$$\begin{aligned} \text{Total cost} &= \text{Fixed cost plus Variable cost} \\ C(x) \quad &= \quad 100{,}000 \quad + \quad 20x. \end{aligned}$$

The firm determines that its revenue (money coming in) from the sales of the calculators is $45 per calculator, or $45x$ dollars, for x calculators. Thus we have a *revenue* function R:

$$R(x) = 45x.$$

The stockholders are most interested in the *profit* function P:

$$\begin{aligned} \text{Profit} &= \text{Revenue} - \text{Total cost} \\ P(x) &= \quad 45x \quad - (100{,}000 + 20x) \\ P(x) &= \quad 25x \quad - 100{,}000. \end{aligned}$$

Find those values of x for which (a) the company will break even; (b) the company will make a profit; and (c) the company will suffer a loss.

In this example, we have a model consisting of three linear functions. The function that will provide the answer to the question is the profit function P. When the profit is 0 the company breaks even. When it is positive the company makes money. When it is negative the company loses money.

Let us first find the value(s) of x that make $P(x) = 0$:

$$0 = P(x) = 25x - 100{,}000$$
$$25x = 100{,}000$$
$$x = 4000.$$

The breakeven point occurs when 4000 calculators are sold. When $x > 4000$ there is a profit. When $x < 4000$ there is a loss.

A graph of the profit function in the preceding example is of interest. In fact, graphs are often of considerable help with mathematical models.

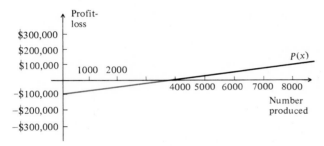

DO EXERCISE 2.

[ii] Quadratic Functions as Mathematical Models

Example 7 (*A projectile problem*). When an object such as a bullet or a ball is shot or thrown upward with an initial velocity v_0, its height is given, approximately, by a quadratic function:

$$s = -4.9t^2 + v_0 t + h.$$

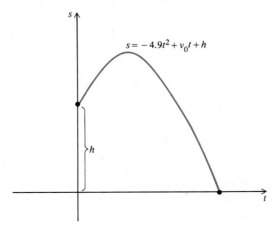

2. Rework Example 6, where $C(x) = 30x + 15{,}000$ and $R(x) = 90x$.

3. A ball is thrown upward from the top of a cliff 12 meters high, at a velocity of 2.8 meters/sec. Find:

a) its maximum height and when it is attained;

b) when it reaches the ground.

In this function, h is the starting height in meters, s is the actual height (also in meters), and t is the time from projection in seconds.

This model is constructed from theoretical principles, rather than experiment. It is based on the assumption that there is no air resistance, and that the force of gravity pulling the object earthward is constant. Neither of these conditions exists precisely, so this model (as is the case with most mathematical models) gives only approximate results.

A model rocket is fired upward. At the end of the burn it has an upward velocity of 49 m/sec and is 155 m high. Find (a) its maximum height and when it is attained; (b) when it reaches the ground.

We will start counting time at the end of the burn. Thus $v_0 = 49$ and $h = 155$. We graph the appropriate function, and begin by completing the square:

$$s = -4.9\left(t^2 - \frac{49}{4.9}t\right) + 155$$

$$= -4.9(t^2 - 10t + 25 - 25) + 155$$
$$= -4.9(t - 5)^2 + 4.9 \times 25 + 155$$
$$= -4.9(t - 5)^2 + 277.5.$$

The vertex of the graph is the point $(5, 277.5)$ and the graph is shown below. The maximum height reached is 277.5 m and it is attained 5 seconds after the end of the burn.

To find when the rocket reaches the ground, we set $s = 0$ in our equation and solve for t:

$$-4.9(t - 5)^2 + 277.5 = 0$$

$$(t - 5)^2 = \frac{277.5}{4.9}$$

$$t - 5 = \sqrt{\frac{277.5}{4.9}} \approx 7.525$$

$$t \approx 12.525.$$

The rocket will reach the ground about 12.525 seconds after the end of the burn.

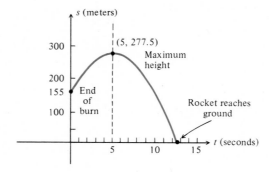

DO EXERCISE 3.

We will fit a quadratic function to a set of data in Chapter 5.

EXERCISE SET 4.4

[i] Solve.

1. (*Life expectancy of females in the United States*). In 1950 the life expectancy of females was 72 years. In 1970 it was 75 years. Let E represent the life expectancy and t the number of years since 1950. ($t = 0$ gives 1950 and $t = 10$ gives 1960.)

a) Assuming E is a linear function of t, express E in terms of t. [*Hint*: Data points are (0, 72) and (20, 75).]

b) Use the equation of (a) to predict the life expectancy of females in 1980; in 1985.

3. (*World record in the 100-meter dash*). In 1920 the world record for the 100-meter dash was 10.43 seconds. In 1970 it was 9.93 seconds. Let R represent the record in the 100-meter dash and t the number of years since 1920.

a) Assuming R is a linear function of t, express R in terms of t.

b) Use the function in (a) to predict the record in 1984; in 1990.

c) According to this model, in what year will the record be 9.0 seconds?

5. (*Profit and loss in the ski business*). A ski manufacturer is planning a new line of skis. For the first year the fixed costs for setting up the new production line are $22,500. Variable costs for producing each pair of skis are estimated to be $40. The sales department projects that 3000 pairs of skis can be sold the first year. The revenue from each pair is to be $85.

a) Formulate a function $C(x)$ for the total cost of producing x pairs of skis.

b) Formulate a function $R(x)$ for the total revenue from the sale of x pairs of skis.

c) Formulate a function $P(x)$ for the profit from the sale of x pairs of skis.

d) What profit or loss will the company realize if 3000 pairs are actually sold?

e) Find the breakeven value of x.

f) Find the values of x that will result in a profit.

g) Find the values of x that will result in a loss.

2. (*Life expectancy of males in the United States*). In 1950 the life expectancy of males was 65 years. In 1970 it was 68 years. Let E represent life expectancy and t the number of years since 1950.

a) Assuming E is a linear function of t, express E in terms of t.

b) Use the equation of (a) to predict the life expectancy of males in 1980; in 1985.

4. (*Natural gas demand*). In 1950 natural gas demand in the United States was 19 quadrillion BTU. In 1960 the demand was 21 quadrillion BTU. Let D represent the demand for natural gas t years after 1950.

a) Assuming D is a linear function of t, express D in terms of t.

b) Use the equation in (a) to predict the natural gas demand in 1988; in 2000.

6. (*Profit and loss.*) A clothing firm has fixed costs of $10,000. To produce x units of a certain kind of suit it costs $20 per unit in addition to the fixed costs. Total revenue from the sale of x suits is $100 per suit.

a) Formulate a function $C(x)$ for the total cost of producing x suits.

b) Formulate a function $R(x)$ for the total revenue from the sale of x suits.

c) Formulate a function $P(x)$ for the profit from the sale of x suits.

d) Find the breakeven value of x.

e) Find the values of x that will result in a loss.

f) Find the values of x that will result in a profit.

[ii] Solve.

7. A rocket is fired upward. At the end of the burn it has an upward velocity of 147 m/sec and is 560 m high. Find (a) its maximum height and when it is attained; (b) when it reaches the ground.

9. The sum of the base and height of a triangle is 20. Find the dimensions for which the area is a maximum.

8. A rocket is fired upward. At the end of the burn it has an upward velocity of 245 m/sec and is 1240 m high. Find (a) its maximum height and when it is attained; (b) when it reaches the ground.

10. The sum of the base and height of a triangle is 14. Find the dimensions for which the area is a maximum.

11. (*Maximizing yield*). An orange grower finds that she gets an average yield of 40 bu per tree when she plants 20 trees on an acre of ground. Each time she adds a tree to an acre the yield per tree decreases by 1 bu, due to congestion. How many trees per acre should she plant for maximum yield?

13. (*Maximizing revenue*). When a theater charges $4 for admission it averages 500 people in attendance. For each 20¢ decrease in admission price the average number of people increases by 30. What should it charge to make the most money?

12. (*Maximizing revenue*). When a theater owner charges $2 for admission he averages 100 people attending. For each 10¢ increase in admission price the average number attending decreases by 1. What should he charge to make the most money?

14. (*Maximizing area*). A farmer wants to enclose two adjacent rectangular regions, as shown below, near a river, one for sheep and one for cattle. No fencing will be used next to the river, but 60 m of fencing will be used. What is the area of the largest region that can be enclosed?

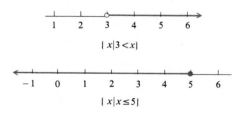

OBJECTIVES

You should be able to:

[i] Find the union and intersection of sets like {1, 2, 3, 4} and {3, 4, 5}; and graph the union and intersection of sets like $\{x \mid x > 1\}$ and $\{x \mid x \leq 3\}$.

[ii] Solve conjunctions and disjunctions of inequalities.

4.5 SETS, SENTENCES, AND INEQUALITIES

[i] Intersections and Unions

The *intersection* of two sets consists of those elements common to the sets. Intersection is illustrated in the figure on the left. Note in the figure on the right that the intersection of two sets may be the empty set. The intersection of sets A and B is indicated as $A \cap B$.

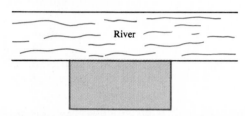

Graphs are set diagrams. They often show pictorially the solution set of an equation or inequality. We can find intersections of solution sets using graphs. In the following example, we find the intersection of the set of all x greater than 3 and the set of all x less than or equal to 5. These sets are indicated, and the symbolism is read, as follows:

$\{x \mid x > 3\}$ "the set of all x such that x is greater than 3,"
$\{x \mid x \leq 5\}$ "the set of all x such that x is less than or equal to 5."

Example 1 Find and graph $\{x \mid x > 3\} \cap \{x \mid x \leq 5\}$.

We graph the two solution sets separately and then find the intersection.

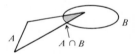

$\{x \mid 3 < x\}$

$\{x \mid x \leq 5\}$

The open circle at 3 indicates that 3 is not in the solution set. The solid circle at 5 indicates that 5 is in the solution set. The intersection is as follows.

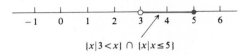

$\{x \mid 3 < x\} \cap \{x \mid x \le 5\}$

Example 2 Graph $\{x \mid -3 \le x\} \cap \{x \mid x \le 0\}$.

Again, we graph the solution sets separately and then find the intersection.

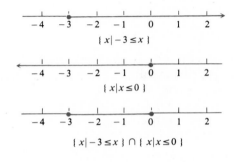

$\{x \mid -3 \le x\}$

$\{x \mid x \le 0\}$

$\{x \mid -3 \le x\} \cap \{x \mid x \le 0\}$

DO EXERCISES 1–6.

The *union* of two sets consists of the members that are in one or both of the sets. Union is illustrated in the following diagram. The union of sets A and B is indicated as $A \cup B$.

$A \cup B$

In the following examples, we find unions of solution sets. Note that we graph the individual sets separately and then combine them.

Example 3 Graph $\{x \mid -1 \le x\} \cup \{x \mid x < 2\}$.

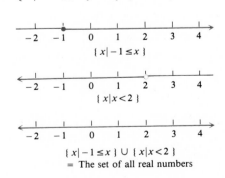

$\{x \mid -1 \le x\}$

$\{x \mid x < 2\}$

$\{x \mid -1 \le x\} \cup \{x \mid x < 2\}$
= The set of all real numbers

Find these intersections.

1. $\{-3, -4, 2, 3, 4\} \cap \{1, 4, -3, 8, 9, 11\}$

2. $\{2, b, c, d, e\} \cap \{1, 2, d, e, f, g\}$

3. Graph $\{x \mid 1 < x\} \cap \{x \mid x \le 3\}$.

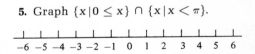

4. Graph $\{x \mid -4 < x\} \cap \{x \mid x < -1\}$.

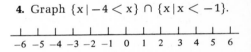

5. Graph $\{x \mid 0 \le x\} \cap \{x \mid x < \pi\}$.

6. Graph $\{x \mid 1 \le x\} \cap \{x \mid x \le -2\}$.

Find these unions.

7. $\{1, 2\} \cup \{2, 3, 4, 5\}$

8. $\{-3, -4, 2, 3, 4\} \cup \{1, 4, -3, 8, 9, 11\}$

9. $\{2, a, b, c\} \cup \{1, 2, c, d, e\}$

10. Graph $\{x \mid x < -3\} \cup \{x \mid x \geq 1\}$.

$$\begin{array}{cccccccccccccc} \llcorner & \mid & \mid & \mid & \mid & \mid & \mid & \mid & \mid & \mid & \mid & \mid & \mid \\ -6 & -5 & -4 & -3 & -2 & -1 & 0 & 1 & 2 & 3 & 4 & 5 & 6 \end{array}$$

11. Graph $\{x \mid x \leq -3\} \cup \{x \mid x \geq 3\}$.

$$\begin{array}{cccccccccccccc} \llcorner & \mid & \mid & \mid & \mid & \mid & \mid & \mid & \mid & \mid & \mid & \mid & \mid \\ -6 & -5 & -4 & -3 & -2 & -1 & 0 & 1 & 2 & 3 & 4 & 5 & 6 \end{array}$$

12. Graph $\{x \mid x > 0\} \cup \{x \mid x < 1\}$.

$$\begin{array}{cccccccccccccc} \llcorner & \mid & \mid & \mid & \mid & \mid & \mid & \mid & \mid & \mid & \mid & \mid & \mid \\ -6 & -5 & -4 & -3 & -2 & -1 & 0 & 1 & 2 & 3 & 4 & 5 & 6 \end{array}$$

13. Graph $\left\{x \mid x > \frac{1}{2}\right\} \cup \left\{x \mid x = \frac{1}{2}\right\}$.

$$\begin{array}{cccccccccccccc} \llcorner & \mid & \mid & \mid & \mid & \mid & \mid & \mid & \mid & \mid & \mid & \mid & \mid \\ -6 & -5 & -4 & -3 & -2 & -1 & 0 & 1 & 2 & 3 & 4 & 5 & 6 \end{array}$$

Abbreviate each conjunction.

14. $4 < x \quad and \quad x < 8$

15. $-3 \leq x \quad and \quad x < 0$

16. $-5 \leq x \quad and \quad x \leq -2$

17. $-1 < x \quad and \quad x \leq -\frac{1}{4}$

Rewrite, using the word *and*.

18. $-\frac{1}{2} < x < 1$

19. $-\frac{17}{3} \leq x < -2$

20. $\frac{19}{4} \leq x \leq \frac{37}{6}$

Example 4 Graph $\{x \mid x \leq -2\} \cup \{x \mid x > 1\}$.

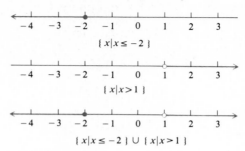

DO EXERCISES 7–13.

[ii] Compound Sentences

When two sentences are joined by the word *and*, a compound sentence is formed. Such a sentence is called a *conjunction*. (The word conjunction used in this way is a logical term; the meaning is not the same as in ordinary grammar.) A conjunction of two sentences is true when both parts are true. Thus the solution set is the intersection of the solution sets of the parts. Consider, for example, the conjunction

$$-2 \leq x \quad and \quad x < 1.$$

Any number that makes this sentence true must make both parts true. We can graph the sentence by graphing the parts and then finding their intersection.

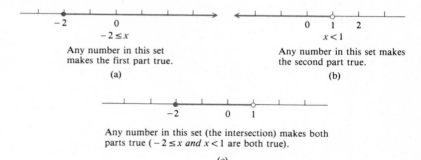

Any number in this set (the intersection) makes both parts true ($-2 \leq x$ *and* $x < 1$ are both true).

(c)

We often abbreviate certain conjunctions of inequalities. In this case,

$$-2 \leq x \quad and \quad x < 1 \qquad \text{is abbreviated} \qquad -2 \leq x < 1.$$

The latter is read "-2 is less than or equal to x and x is less than 1," or "-2 is less than or equal to x is less than 1." Thus

$$\begin{aligned} \{x \mid -2 \leq x < 1\} &= \{x \mid -2 \leq x \quad and \quad x < 1\} \\ &= \{x \mid -2 \leq x\} \cap \{x \mid x < 1\}. \end{aligned}$$

The word *and* corresponds to set *intersection*.

DO EXERCISES 14–20.

Example 5 Solve and graph $-3 < 2x + 5 < 7$.

Method 1. $-3 < 2x + 5$ *and* $2x + 5 < 7$ Rewriting using *and*

$-8 < 2x$ and $2x < 2$ Adding -5

$-4 < x$ and $x < 1$ Multiplying by $\frac{1}{2}$

Method 2. $-3 < 2x + 5 < 7$

$-8 < 2x < 2$

$-4 < x < 1$

The solution set is

$$\{x \mid -4 < x\} \cap \{x \mid x < 1\}, \quad \text{or} \quad \{x \mid -4 < x < 1\}.$$

The graph is as follows.

Example 6 Solve and graph:

$$-4 < \frac{5 - 3x}{2} \le 5.$$

$-8 < 5 - 3x \le 10$ Multiplying by 2

$-13 < -3x \le 5$ Adding -5

$\frac{13}{3} > x \ge -\frac{5}{3}$ Multiplying by $-\frac{1}{3}$

$-\frac{5}{3} \le x < \frac{13}{3}$ $x \ge -\frac{5}{3}$ means $-\frac{5}{3} \le x$, and

$\frac{13}{3} > x$ means $x < \frac{13}{3}$

The solution set is $\{x \mid -\frac{5}{3} \le x < \frac{13}{3}\}$.

The graph is as follows.

DO EXERCISES 21–23.

When two sentences are joined by the word *or*, a compound sentence is formed. Such a sentence is called a *disjunction*. A disjunction of two sentences is true when either part is true. It is also true when both parts are true. The solution set of a disjunction is thus the union of the solution sets of the parts. Consider the disjunction

$$x < -2 \quad or \quad x > \frac{1}{4}.$$

Any number that makes either or both of the parts true makes the disjunction true.

We can graph the sentence by graphing the two parts and then finding their union. The word *or* corresponds to set *union*.

Note that $3x \le 15$ is an abbreviation for $3x < 15$ or $3x = 15$.

Solve. Then graph.

21. $-4 < 3x - 2 \le 10$

22. $-2 < \dfrac{3 - x}{4} < 2$

23. $\dfrac{2}{3} \le 1 - 2x \le \dfrac{5}{3}$

Solve. Then graph.

24. $x + 4 < -3$ or $x + 4 > 3$

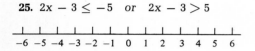

25. $2x - 3 \leq -5$ or $2x - 3 > 5$

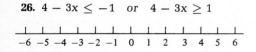

26. $4 - 3x \leq -1$ or $4 - 3x \geq 1$

-6 -5 -4 -3 -2 -1 0 1 2 3 4 5 6

27. $\dfrac{4x + 5}{3} < -2$ or $\dfrac{4x + 5}{3} \geq 2$

-6 -5 -4 -3 -2 -1 0 1 2 3 4 5 6

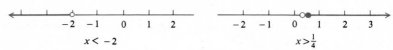

$x < -2$

Any number in this set makes the first part true.

(a)

$x > \frac{1}{4}$

Any number in this set makes the second part true.

(b)

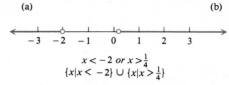

$x < -2 \ or \ x > \frac{1}{4}$

$\{x | x < -2\} \cup \{x | x > \frac{1}{4}\}$

Any number in this set makes one or both of the parts true.

(c)

CAUTION! **There is no compact way to abbreviate disjunctions of inequalities, ordinarily.** *Be careful about this!* **For example, if you try to abbreviate "$-3 < x$ or $x < 4$" as $-3 < x < 4$ you will be** *wrong,* **because $-3 < x < 4$ is an abbreviation for the conjunction $-3 < x$** *and $x < 4$.*

Example 7. Solve. Then graph.

$$2x - 5 < -7 \quad or \quad 2x - 5 > 7$$
$$2x < -2 \quad or \qquad 2x > 12 \qquad \text{Adding 5}$$
$$x < -1 \quad or \qquad x > 6 \qquad \text{Multiplying by } \frac{1}{2}$$

The solution set is $\{x | x < -1 \ or \ x > 6\}$, which is the union $\{x | x < -1\} \cup \{x | x > 6\}$. The graph is as follows.

-3 -2 -1 0 1 2 3 4 5 6 7 8

Example 8 Solve. Then graph.

$$\frac{4 - 3x}{2} < -1 \qquad or \qquad \frac{4 - 3x}{2} \geq 1$$
$$4 - 3x < -2 \qquad or \quad 4 - 3x \geq 2 \qquad \text{Multiplying by 2}$$
$$4 < -2 + 3x \quad or \qquad 4 \geq 2 + 3x \qquad \text{Adding 3x}$$
$$6 < 3x \qquad or \qquad 2 \geq 3x$$
$$2 < x \qquad or \qquad \frac{2}{3} \geq x$$

The solution set is

$$\left\{x | 2 < x \ \ or \ \ \frac{2}{3} \geq x\right\},$$

which is the union

$$\{x | 2 < x\} \cup \left\{x | \frac{2}{3} \geq x\right\}.$$

The graph is as follows.

-1 0 $\frac{2}{3}$ 1 2 3 4

DO EXERCISES 24-27.

EXERCISE SET 4.5

[i] Find these unions or intersections.

1. $\{3, 4, 5, 8, 10\} \cap \{1, 2, 3, 4, 5, 6, 7\}$ **2.** $\{3, 4, 5, 8, 10\} \cup \{1, 2, 3, 4, 5, 6, 7\}$ **3.** $\{0, 2, 4, 6, 8\} \cup \{4, 6, 9\}$

4. $\{0, 2, 4, 6, 8\} \cap \{4, 6, 9\}$ **5.** $\{a, b, c\} \cap \{c, d\}$ **6.** $\{a, b, c\} \cup \{c, d\}$

Graph.

7. $\{x \mid 7 \leq x\} \cup \{x \mid x < 9\}$ **8.** $\left\{x \mid -\frac{1}{2} \leq x\right\} \cup \left\{x \mid x < \frac{1}{2}\right\}$ **9.** $\left\{x \mid -\frac{1}{2} \leq x\right\} \cap \left\{x \mid x < \frac{1}{2}\right\}$

10. $\left\{x \mid x > \frac{1}{4}\right\} \cap \{x \mid 1 \geq x\}$ **11.** $\{x \mid x < -\pi\} \cup \{x \mid x > \pi\}$ **12.** $\{x \mid -\pi \leq x\} \cap \{x \mid x < \pi\}$

13. $\{x \mid x < -7\} \cup \{x \mid x = -7\}$ **14.** $\left\{x \mid x > \frac{1}{2}\right\} \cup \left\{x \mid x = \frac{1}{2}\right\}$

15. $\{x \mid x \geq 5\} \cap \{x \mid x \leq -3\}$ **16.** $\{x \mid x \geq -3\} \cup \{x \mid x \leq 0\}$

[ii] Solve.

17. $-2 \leq x + 1 < 4$ **18.** $-3 < x + 2 \leq 5$ **19.** $5 \leq x - 3 \leq 7$

20. $-1 < x - 4 < 7$ **21.** $-3 \leq x + 4 \leq -3$ **22.** $-5 < x + 2 < -5$

23. $-2 < 2x + 1 < 5$ **24.** $-3 \leq 5x + 1 \leq 3$ **25.** $-4 \leq 6 - 2x < 4$

26. $-3 < 1 - 2x \leq 3$ **27.** $-5 < \frac{1}{2}(3x + 1) \leq 7$ **28.** $\frac{2}{3} \leq -\frac{4}{5}(x - 3) < 1$

29. $3x \leq -6$ *or* $x - 1 > 0$ **30.** $2x < 8$ *or* $x + 3 \geq 1$

31. $2x + 3 \leq -4$ *or* $2x + 3 \geq 4$ **32.** $3x - 1 < -5$ *or* $3x - 1 > 5$

33. $2x - 20 < -0.8$ *or* $2x - 20 > 0.8$ **34.** $5x + 11 \leq -4$ *or* $5x + 11 \geq 4$

35. $x + 14 \leq -\frac{1}{4}$ *or* $x + 14 \geq \frac{1}{4}$ **36.** $x - 9 < -\frac{1}{2}$ *or* $x - 9 > \frac{1}{2}$

☆ ────────────────────────────

Solve.

37. $x \leq 3x - 2 \leq 2 - x$ **38.** $2x \leq 5 - 7x < 7 + x$

39. $(x + 1)^2 > x(x - 3)$ **40.** $(x + 4)(x - 5) > (x + 1)(x - 7)$

41. $(x + 1)^2 \leq (x + 2)^2 \leq (x + 3)^2$ **42.** $(x - 1)(x + 1) < (x + 1)^2 \leq (x - 3)^2$

43. ▦ The length of a rectangle is 15.23 cm. What widths will give a perimeter greater than 40.23 cm and less than 137.8 cm?

44. The height of a triangle is 15 m. What lengths of the base will keep the area less than or equal to 305.4 m² (and, of course, positive)?

45. To get an A in a course, a student's average must be greater than or equal to 90%. It will of course be less than or equal to 100%. On the first three tests a student scored 83%, 87%, and 93%. What scores on the fourth test will produce an A? Is an A possible?

46. In Exercise 45, suppose the scores on the first three tests are 75%, 70%, and 83%. What scores on the fourth test will produce an A? Is an A possible?

Find the domain of each function.

47. $f(x) = \dfrac{\sqrt{x + 2}}{\sqrt{x - 2}}$ **48.** $f(x) = \dfrac{\sqrt{3 - x}}{\sqrt{x + 5}}$

4.6 EQUATIONS AND INEQUALITIES WITH ABSOLUTE VALUE

[i] Absolute value was defined in Chapter 1 (p. 3). An informal way of thinking of absolute value is that it is the distance from 0 of a

OBJECTIVE

You should be able to:

[i] Solve and graph equations and inequalities with absolute value.

Solve and graph.

1. $|x| = 5$

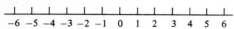

2. $|x| = \frac{1}{4}$

3. $|x| < 5$

4. $|x| \le \frac{1}{4}$

5. $|x| \ge 5$

6. $|x| > \frac{1}{4}$

number on a number line. For example, $|4|$ is 4 because 4 is 4 units from 0; $|-5|$ is 5 because -5 is 5 units from 0. This idea is helpful in solving equations and inequalities with absolute value.

Example 1 Solve $|x| = 3$.

To solve we look for all numbers x whose distance from 0 is 3. There are two of them, so there are two solutions, 3 and -3. The graph is as follows.

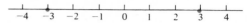

DO EXERCISES 1 AND 2.

Example 2 Solve $|x| < 3$.

This time we look for all numbers x whose distance from 0 is less than 3. These are the numbers between -3 and 3. The solution set and its graph are as follows.

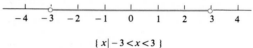

$\{x | -3 < x < 3\}$

DO EXERCISES 3 AND 4.

Example 3 Solve $|x| \ge 3$.

This time we look for all numbers x whose distance from 0 is 3 or greater. The solution set and its graph are as follows.

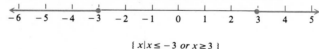

$\{x | x \le -3 \text{ or } x \ge 3\}$

DO EXERCISES 5 AND 6.

The results of the above examples can be generalized as follows.

For any $a > 0$:

i) $|x| = a$ **is equivalent to** $x = -a$ **or** $x = a$.

ii) $|x| < a$ **is equivalent to** $-a < x < a$.

iii) $|x| > a$ **is equivalent to** $x < -a$ **or** $x > a$.

Similar statements hold true for $|x| \le a$ and $|x| \ge a$.

Example 4 Solve $|x - 2| = 3$.

Note that this is a translation of $|x| = 3$, two units to the right. We first graph $|x| = 3$.

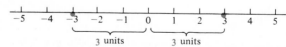

This consists of the numbers that are a distance of 3 from 0. We now translate.

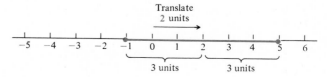

This solution set consists of the numbers that are a distance of 3 from 2 (note that 2 is where 0 went in the translation). The solutions of $|x - 2| = 3$ are -1 and 5.

The results of this example can be generalized as follows.

For any a, $|x - a|$ is the distance from a to x.

Example 5

a) $|x - 5|$ is the distance from 5 to x.

b) $|x + 7|$ is the distance from -7 to x [because $x + 7 = x - (-7)$].

Here are some further examples of solving inequalities.

Example 6 Solve $|x + 2| < 3$.

Method 1. We translate the graph of $|x| < 3$, to the left 2 units.

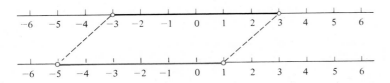

The solution set is $\{x \mid -5 < x < 1\}$.

Method 2. The solutions are those numbers x whose distance from -2 is less than 3. Thus to find the solutions graphically we locate -2. Then we locate those numbers that are less than 3 units to the left and less than 3 units to the right. Thus the solution set is $\{x \mid -5 < x < 1\}$.

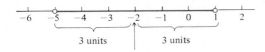

Method 3. We use property (ii), replacing x by x + 2:

$$|x + 2| < 3$$
$$-3 < x + 2 < 3$$
$$-5 < x < 1.$$

The solution set is $\{x \mid -5 < x < 1\}$.

DO EXERCISES 7-9.

The sentences in the next examples are more complicated. Although we could continue using translations and graphs, this actually gets more difficult than if we use properties (i) to (iii).

Example 7 Solve $|2x + 3| = 1$.

$$
\begin{array}{llll}
2x + 3 = -1 & \text{or} & 2x + 3 = 1 & \text{Property (i)} \\
2x = -4 & \text{or} & 2x = -2 & \text{Adding } -3 \\
x = -2 & \text{or} & x = -1 &
\end{array}
$$

The solution set is $\{-2, -1\}$.

Solve. Use all three methods.

7. $|x + 1| = 4$

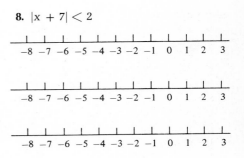

8. $|x + 7| < 2$

9. $|x - 3| > 2$

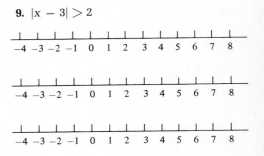

Solve.

10. $|3x - 4| = 7$

11. $\left|5x + \frac{1}{2}\right| < 1$

12. $|3x - 4| > 7$

Example 8 Solve $|2x + 3| \leq 1$.

$$-1 \leq 2x + 3 \leq 1 \qquad \text{Property (ii)}$$
$$-4 \leq 2x \leq -2 \qquad \text{Adding } -3$$
$$-2 \leq x \leq -1$$

The solution set is $\{x \,|\, -2 \leq x \leq -1\}$.

Example 9 Solve $|3 - 4x| > 2$.

$$3 - 4x < -2 \quad \text{or} \quad 3 - 4x > 2 \qquad \text{Property (iii)}$$
$$-4x < -5 \quad \text{or} \quad -4x > -1 \qquad \text{Adding } -3$$
$$x > \frac{5}{4} \quad \text{or} \qquad x < \frac{1}{4}$$

The solution set is $\{x \,|\, x > \frac{5}{4} \text{ or } x < \frac{1}{4}\}$.

DO EXERCISES 10–12.

EXERCISE SET 4.6

[i] Solve and graph.

1. $|x| = 7$ **2.** $|x| = \pi$ **3.** $|x| < 7$ **4.** $|x| \leq \pi$ **5.** $|x| \geq \pi$ **6.** $|x| > 7$

Solve. Use three methods.

7. $|x - 1| = 4$ **8.** $|x - 7| = 5$ **9.** $|x + 8| < 9$ **10.** $|x + 6| \leq 10$

11. $|x + 8| \geq 9$ **12.** $|x + 6| > 10$ **13.** $\left|x - \frac{1}{4}\right| < \frac{1}{2}$ **14.** $|x - 0.5| \leq 0.2$

Solve. Use any method.

15. $|3x| = 1$ **16.** $|5x| = 4$ **17.** $|3x + 2| = 1$ **18.** $|7x - 4| = 8$

19. $|3x| < 1$ **20.** $|5x| \leq 4$ **21.** $|2x + 3| \leq 9$ **22.** $|2x + 3| < 13$

23. $|x - 5| > 0.1$ **24.** $|x - 7| \geq 0.4$ **25.** $\left|x + \frac{2}{3}\right| \leq \frac{5}{3}$ **26.** $\left|x + \frac{3}{4}\right| < \frac{1}{4}$

27. $|6 - 4x| \leq 8$ **28.** $|5 - 2x| > 10$ **29.** $\left|\frac{2x + 1}{3}\right| > 5$ **30.** $\left|\frac{3x + 2}{4}\right| \leq 5$

31. $\left|\frac{13}{4} + 2x\right| > \frac{1}{4}$ **32.** $\left|\frac{5}{6} + 3x\right| < \frac{7}{6}$ **33.** $\left|\frac{3 - 4x}{2}\right| \leq \frac{3}{4}$ **34.** $\left|\frac{2x - 1}{3}\right| \geq \frac{5}{6}$

35. $|x| = -3$ **36.** $|x| < -3$ **37.** $|2x - 4| < -5$ **38.** $|3x + 5| < 0$

39. ▦ $|x - 2.0245| < 0.1011$ **40.** ▦ $|x + 17.217| > 5.0012$

41. ▦ $|3.0147x - 8.9912| \leq 6.0243$ **42.** ▦ $|-2.1437x + 7.8814| \geq 9.1132$

☆ ─────────────────────────────────────

Solve. The definition of absolute value in Section 1.8 may be helpful for some of these exercises.

43. $|4x - 5| = x + 1$ **44.** $|2x + 3| = |x| + 8$ **45.** $||x| - 1| = 3$ **46.** $|x + 2| > x$

47. $|x + 2| \leq |x - 5|$ **48.** $|3x - 1| > 5x - 2$ **49.** $|x| + |x - 1| < 10$ **50.** $|x| - |x - 3| < 7$

51. $|x - 3| + |2x + 5| > 6$

★ ─────────────────────────────────────

52. $|x - 3| + |2x + 5| + |3x - 1| = 12$

Prove the following for any real numbers a and b.

53. $-|a| \leq a \leq |a|$

54. *(The triangle inequality.)* $|a + b| \leq |a| + |b|$

55. Show that if $|a| < \dfrac{e}{2}$ and $|b| < \dfrac{e}{2}$, then $|a + b| < e$.

56. a) Prove that $\left| x - \dfrac{a + b}{2} \right| < \dfrac{b - a}{2}$ is equivalent to $a < x < b$.

Use graphs or the result of (a) to find an inequality with absolute value for each of the following:

b) $-5 < x < 5$ c) $-6 < x < 6$ d) $-1 < x < 7$ e) $-5 < x < 1$

57. Use absolute value to prove that the number halfway between a and b is $\dfrac{a + b}{2}$.

4.7 QUADRATIC AND RATIONAL INEQUALITIES

[i] Quadratic and Certain Other Polynomial Inequalities

Quadratic Inequalities

Inequalities such as the following are called *quadratic* inequalities:

$$x^2 - 2x - 2 > 0, \qquad 3x^2 + 5x + 1 \leq 0.$$

In each case we have a polynomial of degree 2 on the left. One way of solving quadratic inequalities is by graphing.

Example 1 Solve $x^2 + 2x - 3 > 0$.

Consider $f(x) = x^2 + 2x - 3$. Its graph opens upward. Function values will be positive to the left and right of the intercepts as shown. We find the intercepts by solving $f(x) = 0$:

$$x^2 + 2x - 3 = 0$$
$$(x + 3)(x - 1) = 0$$
$$x + 3 = 0 \quad \text{or} \quad x - 1 = 0$$
$$x = -3 \quad \text{or} \quad x = 1.$$

Then the solution set of the inequality is

$$\{x \mid x < -3 \text{ or } x > 1\}.$$

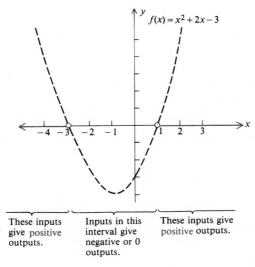

| These inputs give positive outputs. | Inputs in this interval give negative or 0 outputs. | These inputs give positive outputs. |

DO EXERCISE 1.

1. Solve by graphing.

$$x^2 - 3x - 10 < 0$$

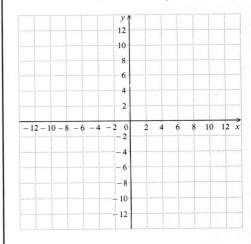

2. Solve by graphing.

$$x^2 + 2x > 4$$

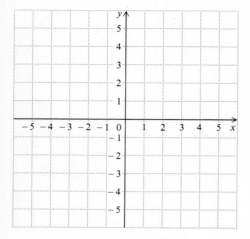

We can solve any quadratic inequality by considering its graph and finding intercepts as in Example 1. In some cases we will need to use the quadratic formula to find the intercepts.

Example 2 Solve $x^2 - 2x \leq 2$.

We first find standard form with 0 on one side:

$$x^2 - 2x - 2 = 0.$$

Consider $f(x) = x^2 - 2x - 2$. Its graph opens upward. Function values will be negative between and including its intercepts as shown. We find the intercepts by solving $f(x) = 0$. This time we will need the quadratic formula. The intercepts are $1 + \sqrt{3}$ and $1 - \sqrt{3}$. The solution set of the inequality is

$$\{x \mid 1 - \sqrt{3} \leq x \leq 1 + \sqrt{3}\}.$$

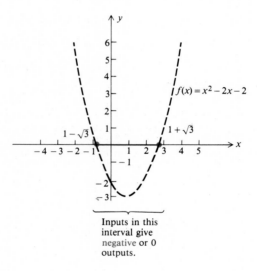

Inputs in this interval give negative or 0 outputs.

DO EXERCISE 2.

It should be pointed out that we need not actually draw graphs as in the preceding examples. Merely visualizing the graph will usually suffice.

Let us now consider another method of solving inequalities that works for polynomials of higher degree.

Example 3 Solve $x^2 + 2x - 3 > 0$.

We factor the inequality, obtaining $(x + 3)(x - 1) > 0$. The solutions of $(x + 3)(x - 1) = 0$ are -3 and 1. They are not solutions of the inequality, but they divide the real-number line in a natural way, pictured as follows. The product $(x + 3)(x - 1)$ is positive or negative, for values other than -3 and 1, depending on the signs of the factors $x + 3$ and $x - 1$. We tabulate signs in these intervals.

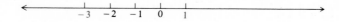

INTERVAL	x + 3	x − 1	PRODUCT: (x + 3)(x − 1)
x < −3	−	−	+
−3 < x < 1	+	−	−
1 < x	+	+	+

For the product $(x + 3)(x − 1)$ to be positive, both of the factors must be positive or both must be negative. We see from the table that the solution set of the inequality is $\{x \mid x > 1\} \cup \{x \mid x < −3\}$, or $\{x \mid x > 1 \text{ or } x < −3\}$.

DO EXERCISES 3 AND 4.

Polynomial Inequalities of Higher Degree

We can extend the preceding method to polynomials with more than two factors.

Example 4 Solve $4x(x + 1)(x − 1) < 0$.

The solutions of $4x(x + 1)(x − 1) = 0$ are $−1, 0$, and 1. They divide the real-number line in a natural way, pictured as follows. The product $4x(x + 1)(x − 1)$ is positive or negative depending on the signs of the factors $4x$, $x + 1$, and $x − 1$. We tabulate signs in these intervals.

INTERVAL	4x	x + 1	x − 1	PRODUCT: 4x(x + 1)(x − 1)
x < −1	−	−	−	−
−1 < x < 0	−	+	−	+
0 < x < 1	+	+	−	−
1 < x	+	+	+	+

The product of three numbers is negative when it has an odd number of negative factors. We see from the table that the solution set is

$$\{x \mid x < −1\} \cup \{x \mid 0 < x < 1\}, \quad \text{or} \quad \{x \mid x < −1 \text{ or } 0 < x < 1\}.$$

DO EXERCISE 5.

Rational Inequalities

We can also use the preceding method when an inequality involves functions that are quotients of polynomials. Such functions are called *rational functions*.

Solve. Use the method of Example 3.

3. $x^2 − 3x > 4$

4. $x^2 − 3x < 4$

5. Solve.

$$7x(x + 3)(x − 2) > 0$$

Solve.

6. $\dfrac{x-3}{x+4} \geq 2$

7. $\dfrac{x}{x-5} \leq 2$

Example 5 Solve $\dfrac{x+1}{x-2} \geq 3$.

We add -3, to get 0 on one side:

$$\frac{x+1}{x-2} - 3 \geq 0.$$

Next, we obtain a single fractional expression:

$$\frac{x+1}{x-2} - 3\frac{x-2}{x-2} = \frac{x+1-3x+6}{x-2} \geq 0$$

$$\frac{-2x+7}{x-2} \geq 0.$$

Let us consider the equality portion:

$$\frac{-2x+7}{x-2} = 0.$$

This has the solution $\frac{7}{2}$. Thus $\frac{7}{2}$ is in the solution set. Next, consider the inequality portion:

$$\frac{-2x+7}{x-2} > 0.$$

The solutions of $-2x+7 = 0$ and $x-2 = 0$ are 2 and $\frac{7}{2}$. They divide the real-number line in a natural way, pictured as follows. The quotient is positive or negative depending on the signs of $-2x+7$ and $x-2$. We tabulate signs in these intervals.

INTERVAL	$-2x+7$	$x-2$	QUOTIENT: $\dfrac{-2x+7}{x-2}$
$x < 2$	+	−	−
$2 < x < \dfrac{7}{2}$	+	+	+
$\dfrac{7}{2} < x$	−	+	−

We see from the table that the solution set of $(-2x+7)/(x-2) > 0$ is $\{x \mid 2 < x < \frac{7}{2}\}$. Thus the solution set of the inequality in question is $\{x \mid 2 < x \leq \frac{7}{2}\}$.

DO EXERCISES 6 AND 7.

EXERCISE SET 4.7

[i] Solve.

1. $x^2 - x - 2 < 0$

2. $x^2 - 4 > 0$

3. $x^2 \geq 1$

4. $x^2 - 2x < 5$

5. $x^2 - 2x + 1 \geq 0$

6. $x^2 + 6x + 9 < 0$

7. $x^2 < 6x - 4$

8. $5x - 2 > x^2$

9. $x^2 - 8x - 20 \leq 0$

10. $x^2 - 8x - 20 \geq 0$

11. $x^2 - 2x > 8$

12. $x^2 - 2x < 8$

13. $4x^2 + 7x < 15$ **14.** $4x^2 + 7x \geq 15$ **15.** $2x^2 + x > 5$ **16.** $2x^2 + x \leq 2$

17. $5x(x + 1)(x - 1) > 0$ **18.** $3x(x + 2)(x - 2) < 0$

19. $(x + 3)(x + 2)(x - 1) < 0$ **20.** $(x - 2)(x - 3)(x + 1) > 0$

[ii] Solve.

21. $\dfrac{1}{4 - x} < 0$ **22.** $\dfrac{-4}{2x + 5} > 0$ **23.** $\dfrac{3x + 2}{x - 3} > 0$ **24.** $\dfrac{5 - 2x}{4x + 3} < 0$

25. $\dfrac{x + 1}{2x - 3} \geq 1$ **26.** $\dfrac{x - 1}{x - 2} \geq 3$ **27.** $\dfrac{x + 1}{x + 2} \leq 3$ **28.** $\dfrac{x + 1}{2x - 3} \leq 1$

29. $(x + 1)(x - 2) > (x + 3)^2$ **30.** $(x - 4)(x + 3) \leq (x - 1)^2$

31. $x^3 - x^2 > 0$ **32.** $x^3 - 4x > 0$ **33.** $x + \dfrac{4}{x} > 4$ **34.** $\dfrac{1}{x^2} \leq \dfrac{1}{x^3}$

35. $\dfrac{1}{x^3} \leq \dfrac{1}{x^2}$ **36.** $x + \dfrac{1}{x} > 2$ **37.** $\dfrac{2 + x - x^2}{x^2 + 5x + 6} < 0$ **38.** $\dfrac{4}{x^2} - 1 > 0$

☆ ———————————————————————————————

Solve.

39. $\left|\dfrac{x + 3}{x - 4}\right| < 2$ **40.** $|x^2 - 5| = 5 - x^2$ **41.** $(7 - x)^{-2} < 0$

42. $(1 - x)^3 > 0$ **43.** $\left|1 + \dfrac{1}{x}\right| < 3$ **44.** $(x + 5)^{-2} > 0$

45. $\left|2 - \dfrac{1}{x}\right| \leq 2 + \left|\dfrac{1}{x}\right|$ **46.** $\dfrac{(x - 2)^2(x - 3)^3(x + 1)}{(x + 2)(4 - x)} \geq 0$

47. $|x^2 + 3x - 1| < 3$ **48.** $|1 + 5x - x^2| \geq 5$

49. The base of a triangle is 4 cm greater than the height. Find the possible heights h such that the area of the triangle will be greater than 10 cm^2.

50. The length of a rectangle is 3 m greater than the width. Find the possible widths w such that the area of the rectangle will be greater than 15 m^2.

51. A company has the following total cost and total revenue functions to use in producing and selling x units of a certain product:

$$R(x) = 50x - x^2, \quad C(x) = 5x + 350.$$

(For reference, see p. 183.)

a) Find the breakeven values.

b) Find the values of x that produce a profit.

c) Find the values of x that result in a loss.

52. A company has the following total cost and total revenue functions to use in producing and selling x units of a certain product:

$$R(x) = 80x - x^2, \quad C(x) = 10x + 600.$$

a) Find the breakeven values.

b) Find the values of x that produce a profit.

c) Find the values of x that result in a loss.

53. Find the numbers k for which the quadratic equation $x^2 + kx + 1 = 0$ has (a) two real-number solutions; (b) no real-number solution.

54. Find the numbers k for which the quadratic equation $2x^2 - kx + 1 = 0$ has (a) two real-number solutions; (b) no real-number solution.

Find the domain of each function.

55. $f(x) = \sqrt{1 - x^2}$ **56.** $f(x) = \dfrac{1}{\sqrt{1 - x^2}}$ **57.** $g(x) = \sqrt{x^2 + 2x - 3}$

CHAPTER 4 REVIEW ————————————————————————

[4.2, iii]   **6.** Find the distance between $(3, 7)$ and $(-2, 4)$.

[4.2, v]   **7.** Find the midpoint of the segment with endpoints $(3, 7)$ and $(-2, 4)$.

[4.2, ii]   Given the point $(1, -1)$ and the line $2x + 3y = 4$:

     8. Find an equation of the line containing the given point and parallel to the given line.

     9. Find an equation of the line containing the given point and perpendicular to the given line.

[4.2, i]   Determine whether the lines are parallel, perpendicular, or neither.

     10. $3x - 2y = 8$,   **11.** $y - 2x = 4$,   **12.** $y = \frac{3}{2}x + 7$,

         $6x - 4y = 2$ $2y - 3x = -7$ $y = -\frac{2}{3}x - 4$

[4.3, i]   For the functions in Questions 13 and 14:

     a) find an equation of the type $f(x) = a(x - h)^2 + k$;

     b) find the vertex;

     c) find the line of symmetry; and

     d) determine whether the second coordinate of the vertex is a maximum or minimum, and find the maximum or minimum.

     13. $f(x) = 3x^2 + 6x + 1$ **14.** $f(x) = -2x^2 - 3x + 6$

[4.3, ii]   **15.** Graph $f(x) = 3x^2 + 6x + 1$.

[4.3, iii]   **16.** Find the x-intercepts: $f(x) = -x^2 - x - 1$.

     Solve.

[4.5, ii]   **17.** $-3x + 2 \le -4$ *and* $x - 1 \le 3$

[4.6, i]   **18.** $|x - 6| < 5$ **19.** $|-3x - 2| > 2$ **20.** $\left|\frac{1}{2}x + 4\right| \le 6$ **21.** $|2x + 5| = 9$

[4.7, i]   **22.** $x^2 - 9 < 0$ **23.** $2x^2 - 3x - 2 > 0$ **24.** $(1 - x)(x + 4)(x - 2) < 0$

[4.7, ii]   **25.** $\dfrac{x - 2}{x + 3} < 4$

[4.4, i]   **26.** (*World records in the mile run*). It has been found experimentally that, within certain restrictions, world records in the mile run can be approximated by a linear function. In 1900 the record for the mile was 4:20 (4 minutes, 20 seconds). In 1975 it was 3:50. Let R represent the record in the mile run, in seconds, and t the number of years since 1900.

     a) Find a linear function that fits the data.

     b) Predict the record in the year 2000.

[4.4, ii]   **27.** The sum of the length and width of a rectangle is 40. Find the dimensions for which the area is a maximum.

     28. A company has the following total cost and total revenue functions to use in producing and selling x units of a product:

$$C(x) = 15x + 200, \qquad R(x) = 100 + x^2.$$

     a) Find the breakeven values.

     b) Find the values of x that produce a profit.

     c) Find the values of x that result in a loss.

     Solve.

     29. $4x \le 5x + 2 < 4 - x$ **30.** $\left|1 - \dfrac{1}{x^2}\right| < 3$ **31.** $(x - 2)^{-3} < 0$

     32. Under what conditions is the square of a linear function a quadratic function?

     Find the domain of each function.

     33. $f(x) = \sqrt{1 - |3x - 2|}$ **34.** $g(x) = \dfrac{1}{\sqrt{5 - |7x + 2|}}$

5

SYSTEMS OF LINEAR EQUATIONS AND INEQUALITIES

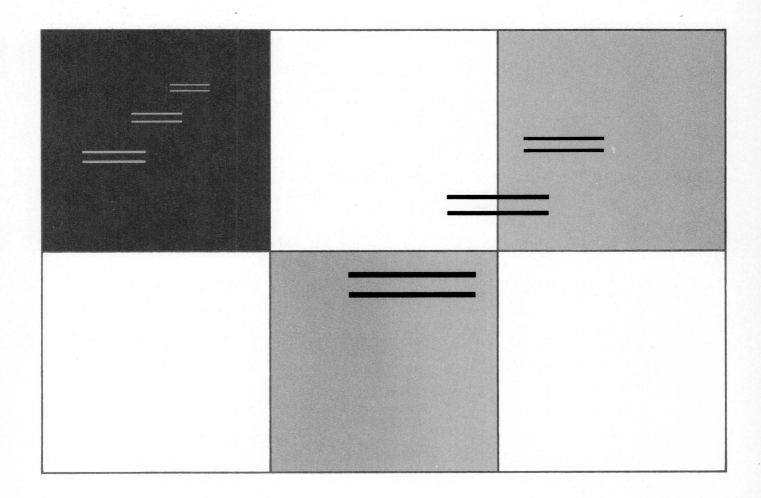

OBJECTIVES

You should be able to:

[i] Determine whether an ordered pair is a solution of a system of equations.

Solve a system of two linear equations in two variables

[ii] by graphing;
[iii] by the substitution method;
[iv] by the addition method.
[v] Solve applied problems by translating them to a system of two linear equations and solving the system.

1. Determine whether $(-3, 2)$ is a solution of the conjunction

$$2x - y = -8 \quad and \quad 3x + 4y = -1.$$

2. Determine whether $(0, \frac{1}{4})$ is a solution of the conjunction

$$5x + 12y = 13 \ and \ \sqrt{2}x + 9y = 10.$$

3. Solve graphically.

$$y - x = 1 \quad and \quad y + x = 3$$

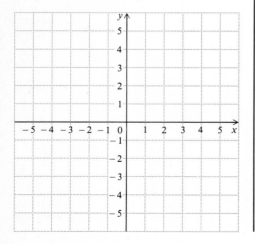

5.1 SYSTEMS OF EQUATIONS IN TWO VARIABLES: UNIQUE SOLUTIONS

[i] Identifying Solutions

Recall that a *conjunction* of sentences is formed by joining sentences with the word *and*. Here is an example:

$$x + y = 11 \quad and \quad 3x - y = 5.$$

A solution of a sentence with two variables such as $x + y = 11$ is an ordered pair. Some pairs in the solution set of $x + y = 11$ are

$$(5, 6), \quad (12, -1), \quad (4, 7), \quad (8, 3).$$

Some pairs in the solution set of $3x - y = 5$ are

$$(0, -5), \quad (4, 7), \quad (-2, -11), \quad (9, 22).$$

The solution set of the sentence

$$x + y = 11 \quad and \quad 3x - y = 5$$

consists of all pairs that make *both* sentences true. That is, it is the *intersection* of the solution sets. Note that $(4, 7)$ is a solution of the conjunction; in fact, it is the only solution.

DO EXERCISES 1 AND 2.

[ii] Solving Systems of Equations Graphically

One way to find solutions is by trial and error. Another way is to graph the equations and look for points of intersection. For example, the graph shows the solution sets of $x + y = 11$ and $3x - y = 5$. Their intersection is the single ordered pair $(4, 7)$.

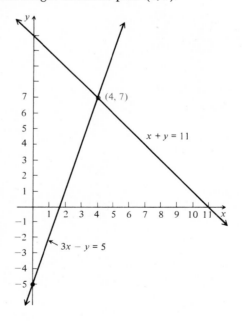

DO EXERCISE 3.

We often refer to a conjunction of equations as a *system* of equations. We usually omit the word *and*, and very often write one equation under the other.

We now consider two algebraic methods for solving systems of linear equations.

[iii] The Substitution Method

To use the substitution method we solve one equation for one of the variables. Then we substitute in the other equation and solve.

Example 1 Solve the system

$$x + y = 11, \tag{1}$$
$$3x - y = 5. \tag{2}$$

First we solve equation (1) for y:*

$$y = 11 - x.$$

Then we substitute $11 - x$ for y in equation (2). This gives an equation in one variable, which we know how to solve from earlier work.

$$3x - (11 - x) = 5$$
$$x = 4.$$

Now substitute 4 for x in either equation (1) or (2) and solve for y. Let us use equation (1):

$$4 + y = 11$$
$$y = 7.$$

The solution is $(4, 7)$. Note the alphabetical listing: 4 for x, 7 for y. *Check:*

$x + y = 11$		$3x - y = 5$	
$4 + 7$	11	$3 \cdot 4 - 7$	5
11		5	

DO EXERCISES 4 AND 5.

[iv] The Addition Method

The *addition method* makes use of the addition and multiplication principles for solving equations. Consider the system

$$3x - 4y = -1,$$
$$-3x + 2y = 0.$$

We add the left-hand sides and obtain $-2y$, and add the right-hand sides† and obtain -1. Thus we obtain a new equation: $-2y = -1$. We

Solve, using the substitution method.

4. $2x + y = 6,$
$3x + 4y = 4$

5. $8x - 3y = -31,$
$2x + 6y = 26$

---

*We could just as well solve for x.

†When we do this we often say that we "added the two equations."

can now solve for y and then substitute into one of the original equations to find x.

$$-2y = -1 \qquad -3x + 2\left(\frac{1}{2}\right) = 0 \qquad \text{Substituting into the}$$
$$y = \frac{1}{2} \qquad -3x + 1 \quad = 0 \qquad \text{second equation}$$
$$-3x = -1$$
$$x = \frac{1}{3}$$

We now know that the solution of the original system is $(\frac{1}{3}, \frac{1}{2})$, because the solutions of this last system are obvious and we know that this system is equivalent to the original. We know that the last system is equivalent to the original, because the only computations we did consisted of using the addition principle, adding only expressions for which all replacements were sensible, and multiplying by a nonzero constant. Each produces results equivalent to the original equations. For this reason we do not need to check results, except to detect errors in computation.

Example 2 Solve:

$$5x + 3y = 7,$$
$$3x - 5y = -23.$$

We first multiply the second equation by 5 to make the x-coefficient a multiple of 5:

$$5x + 3y = 7,$$
$$15x - 25y = -115.$$

Now we multiply the first equation by -3 and add it to the second equation. This gets rid of the x-term:

$$5x + 3y = 7,$$
$$-34y = -136.$$

Next we solve the second equation for y. Then we substitute the result into the first equation to find x.

$$-34y = -136 \qquad 5x + 3(4) = 7$$
$$y = 4 \qquad 5x + 12 = 7$$
$$5x = -5$$
$$x = -1$$

The solution is $(-1, 4)$.

There are some preliminary things that might be done to make the steps of the addition method simpler. One is to first write the equations in the form $Ax + By = C$. Another is to interchange two equations before beginning. For example, if we have the system

$$5x + y = -2,$$
$$x + 7y = 3,$$

we may prefer to write the second equation first. This accomplishes two things. One is that the x-coefficient is 1 in the first equation. Having found y later, this will make solving for x easier. The other is that this makes the x-coefficient in the second equation a multiple of the first.

Something else that might be done is to multiply one or more equations by a power of 10 before beginning in order to eliminate decimal points. For example, if we have the system

$$-0.3x + 0.5y = 0.03,$$
$$0.01x - 0.4y = 1.2,$$

we can multiply both equations by 100 to clear of decimal points, obtaining

$$-30x + 50y = 3,$$
$$x - 40y = 120.$$

Still other procedures are possible. However, it is recommended that no others be used at this point, because we are establishing an *algorithm*, which could be done mechanically by a computer.

Example 3 Solve:

$$5x + y = -2,$$
$$x + 7y = 3.$$

We first interchange the equations so that the x-coefficient of the second equation will be a multiple of the first:

$$x + 7y = 3,$$
$$5x + y = -2.$$

Next, we multiply the first equation by -5 and add the result to the second equation. This gets rid of the x-term in the second equation:

$$x + 7y = 3,$$
$$-34y = -17.$$

Now we solve the second equation for y. Then we substitute the result in the first equation to find x.

$$-34y = -17 \qquad x + 7\left(\frac{1}{2}\right) = 3$$

$$y = \frac{1}{2} \qquad\qquad x + \frac{7}{2} = 3$$

$$x = 3 - \frac{7}{2}, \quad \text{or} \quad -\frac{1}{2}$$

The solution is $(-\frac{1}{2}, \frac{1}{2})$.

DO EXERCISES 6–8.

Solve.

6. $4x + 3y = -6,$
$-4x + 2y = 16$

7. $9x - 2y = -4,$
$3x + 4y = 1$

8. $0.2x + 0.3y = 0.1,$
$0.3x - 0.1y = 0.7$

Solve.

9. The sum of two numbers is 10. The difference is 1. Find the numbers.

10. The sum of two numbers is 1. The difference is 10. Find the numbers.

11. It takes a boat 3 hr to travel 24 km upstream. It takes 2 hr to travel the 24 km downstream. Find the speed of the boat and the speed of the stream.

[v] **Applied Problems**

Translating problems to equations becomes easier in many cases if we translate to more than one equation in more than one variable.

Example 4 An airplane flies the 3000-mi distance from Los Angeles to New York, with a tail wind, in 5 hr. On the return trip, against the wind, it makes the trip in 6 hr. Find the speed of the plane and the speed of the wind.

Translate: Let p = the speed of the plane, and w = the speed of the wind.

Then from $d_1 = r_1 \cdot t_1$, we have, for the trip east,

$$3000 = (p + w)5 = 5p + 5w.$$

From $d_2 = r_2 \cdot t_2$, we have, for the trip west,

$$3000 = (p - w)6 = 6p - 6w.$$

Then we solve the system:

$$5p + 5w = 3000,$$
$$6p - 6w = 3000.$$

The solution is (550, 50); that is, the speed of the plane is 550 mph and the speed of the wind is 50 mph. This checks in the original problem.

DO EXERCISES 9–11.

Example 5 Wine A is 5% alcohol and wine B is 15% alcohol. How many liters of each should be mixed to get a 10-liter mixture that is 12% alcohol?

a) First consider the amount of wine.

Let x = the amount of wine A, and y = the amount of wine B. The amount of the mixture is to be 10 L. We get the equation

$$x + y = 10.$$

b) Second, consider the amount of alcohol.

The amount of alcohol in wine A is 5% $\cdot$ x, and the amount in wine B is 15% $\cdot$ y. We know that the amount in the mixture is 12% $\cdot$ 10. We get the equation

$$5\%x + 15\%y = 12\% \cdot 10,$$

or

$$0.05x + 0.15y = 1.2,$$

or

$$5x + 15y = 120.$$

c) Now we solve the system:

$$x + y = 10,$$
$$5x + 15y = 120.$$

The solution is (3, 7). Thus 3 L of wine A should be mixed with 7 L of wine B.

DO EXERCISE 12.

Solve.

12. Solution A is 25% acid. Solution B is 65% acid. How many liters of each should be mixed to get 8 L of a solution that is 40% acid?

EXERCISE SET 5.1

[i]

1. Determine whether $\left(\frac{1}{2}, 1\right)$ is a solution.

$$3x + y = \frac{5}{2},$$

$$2x - y = \frac{1}{4}$$

2. Determine whether $\left(-2, \frac{1}{4}\right)$ is a solution.

$$x + 4y = -1,$$
$$2x + 8y = -2$$

[ii] Solve graphically.

3. $x + y = 2,$
$3x + y = 0$

4. $x + y = 1,$
$3x + y = 7$

[iii] Solve, using the substitution method.

5. $x - 5y = 4,$
$2x + y = 7$

6. $3x - y = 5,$

$$x + y = \frac{1}{2}$$

[iv] Solve, using the addition method.

7. $x - 3y = 2,$
$6x + 5y = -34$

8. $x + 3y = 0,$
$20x - 15y = 75$

9. $0.3x + 0.2y = -0.9,$
$0.2x - 0.3y = -0.6$

10. $0.2x - 0.3y = 0.3,$
$0.4x + 0.6y = -0.2$

11. $\frac{1}{5}x + \frac{1}{2}y = 6,$

$$\frac{3}{5}x - \frac{1}{2}y = 2$$

12. $\frac{2}{3}x + \frac{3}{5}y = -17,$

$$\frac{1}{2}x - \frac{1}{3}y = -1$$

13. $2a = 5 - 3b,$
$4a = 11 - 7b$

14. $7(a - b) = 14,$
$2a = b + 5$

[v] Solve.

15. Find two numbers whose sum is -10 and whose difference is 1.

16. Find two numbers whose sum is -1 and whose difference is 10.

17. A boat travels 46 km downstream in 2 hr. It travels 51 km upstream in 3 hr. Find the speed of the boat and the speed of the stream.

18. An airplane travels 3000 km with a tail wind in 3 hr. It travels 3000 km with a head wind in 4 hr. Find the speed of the plane and the speed of the wind.

19. Antifreeze A is 18% alcohol. Antifreeze B is 10% alcohol. How many liters of each should be mixed to get 20 L of a mixture that is 15% alcohol?

20. Beer A is 6% alcohol and beer B is 2% alcohol. How many liters of each should be mixed to get 50 L of a mixture that is 3.2% alcohol?

☆ ───

Solve.

21. $\dfrac{x + y}{4} - \dfrac{x - y}{3} = 1,$

$\dfrac{x - y}{2} + \dfrac{x + y}{4} = -9$

22. $\dfrac{x + y}{2} - \dfrac{y - x}{3} = 0,$

$\dfrac{x + y}{3} - \dfrac{x + y}{4} = 0$

Solve. Check by substituting.

23. ▦ $2.35x - 3.18y = 4.82,$
$1.92x + 6.77y = -3.87$

24. ▦ $0.0375x + 0.912y = -1.003,$
$463x - 801y = 946$

Each of the following is a system of equations that is *not* linear. But each is *linear in form,* in that an appropriate substitution, say u for $1/x$ and v for $1/y$, yields a linear system. Solve for the new variable and then solve for the original variable.

25. $\dfrac{1}{x} - \dfrac{3}{y} = 2,$

$\dfrac{6}{x} + \dfrac{5}{y} = -34$

26. $2\sqrt[3]{x} + \sqrt{y} = 0,$
$5\sqrt[3]{x} + 2\sqrt{y} = -5$

27. $3|x| + 5|y| = 30,$
$5|x| + 3|y| = 34$

28. $15x^2 + 2y^3 = 6,$
$25x^2 - 2y^3 = -6$

───

OBJECTIVES

You should be able to:

[i] Determine whether an ordered triple is a solution of a system of equations in three variables.

[ii] Solve systems of linear equations in three or more variables.

[iii] Fit a quadratic function to data when three data points are given.

5.2 SYSTEMS OF EQUATIONS IN THREE OR MORE VARIABLES: UNIQUE SOLUTIONS

[i] Identifying Solutions

A *solution* of a system of equations in three variables is an ordered triple that makes all three equations true.

Example 1 Determine whether $(\tfrac{3}{2}, -4, 3)$ is a solution of the system

$$4x - 2y - 3z = 5,$$
$$-8x - y + z = -5,$$
$$2x + y + 2z = 5.$$

We substitute $(\tfrac{3}{2}, -4, 3)$ into the equations, using alphabetical order.

$4x - 2y - 3z = 5$	
$4 \cdot \dfrac{3}{2} - 2(-4) - 3 \cdot 3$	5
$6 + 8 - 9$	
5	

$-8x - y + z = -5$	
$-8 \cdot \dfrac{3}{2} - (-4) + 3$	-5
$-12 + 4 + 3$	
-5	

$2x + y + 2z = 5$	
$2 \cdot \dfrac{3}{2} + (-4) + 2 \cdot 3$	5
$3 - 4 + 6$	
5	

The triple makes all three equations true, so it is a solution.

DO EXERCISE 1.

[ii] Solving Systems of Equations in Three or More Variables

Graphical methods of solving linear equations in three variables are unsatisfactory, because a three-dimensional coordinate system is required. The substitution method becomes cumbersome for most systems of more than two equations. Therefore, we will use the addition method. It is essentially the same as for systems of two equations.

Example 2 Solve:

$$2x - 4y + 6z = 22, \qquad ①$$
$$4x + 2y - 3z = 4, \qquad ②$$
$$3x + 3y - z = 4. \qquad ③$$

These numbers indicate the equations in the first, second, and third positions, respectively.

We begin by multiplying ③ by 2, to make each x-coefficient a multiple of the first.* Then we have the following:

$$2x - 4y + 6z = 22,$$
$$4x + 2y - 3z = 4,$$
$$6x + 6y - 2z = 8.$$

Next, we multiply ① by −2 and add it to ②. We also multiply ① by −3 and add it to ③. Then we have the following:

$$2x - 4y + 6z = 22,$$
$$10y - 15z = -40,$$
$$18y - 20z = -58.$$

Now we multiply ③ by −5 to make the y-coefficient a multiple of the y-coefficient in ②:

$$2x - 4y + 6z = 22,$$
$$10y - 15z = -40,$$
$$-90y + 100z = 290.$$

Next, we multiply ② by 9 and add it to ③:

$$2x - 4y + 6z = 22,$$
$$10y - 15z = -40,$$
$$-35z = -70.$$

Now we solve ③ for z:

$$2x - 4y + 6z = 22,$$
$$10y - 15z = -40,$$
$$z = 2.$$

---

*By proceeding in this manner, we avoid fractions. The method will work when fractions are allowed, but is more difficult that way.

1. Consider the following system:

$$4x - y + z = 6,$$
$$2x + y + 2z = 3,$$
$$3x - 2y + z = 3.$$

a) Determine whether $(3, 0, \frac{1}{4})$ is a solution.

b) Determine whether $(2, 1, -1)$ is a solution.

2. Solve this system, using exactly the procedure used in the text.

$$x + 2y - z = 5,$$
$$2x - 4y + z = 0,$$
$$3x + 2y + 2z = 3$$

3. Find a quadratic function that fits the data points $(1, 0)$, $(-1, 4)$, and $(2, 1)$.

Next we substitute 2 for z in ②, and solve for y:

$$10y - 15(2) = -40$$
$$10y - 30 = -40$$
$$10y = -10$$
$$y = -1.$$

Finally, we substitute -1 for y and 2 for z in ①:

$$2x - 4(-1) + 6(2) = 22$$
$$2x + 4 + 12 = 22$$
$$2x + 16 = 22$$
$$2x = 6$$
$$x = 3.$$

The solution is $(3, -1, 2)$. To be sure computational errors have not been made, one can check by substituting 3 for x, -1 for y, and 2 for z in all three equations. If all are true, then the triple is a solution.

DO EXERCISE 2.

[iii] Mathematical Models and Applied Problems

In a situation where a quadratic function will serve as a mathematical model, we may wish to find an equation, or formula, for the function. Recall that for a linear model, we can find an equation if we know two data points. For a quadratic function we need three data points.

Example 3 In a certain situation, it is believed that a quadratic function will be a good model. Find an equation of the function, given the data points $(1, -4)$, $(-1, -6)$, and $(2, -9)$.

We wish to find a quadratic function

$$f(x) = ax^2 + bx + c$$

containing the three given data points, i.e., a function for which the equation will be true when we substitute any of the ordered pairs of numbers into it. When we substitute, we get,

$$\text{for } (1, -4), \quad -4 = a \cdot 1^2 + b \cdot 1 + c;$$
$$\text{for } (-1, -6), \quad -6 = a(-1)^2 + b(-1) + c;$$
$$\text{for } (2, -9), \quad -9 = a \cdot 2^2 + b \cdot 2 + c.$$

We now have a system of equations in the three unknowns a, b, and c:

$$a + b + c = -4,$$
$$a - b + c = -6,$$
$$4a + 2b + c = -9.$$

We solve this system of equations, obtaining $(-2, 1, -3)$. Thus the function we are looking for is as follows:

$$f(x) = -2x^2 + x - 3.$$

DO EXERCISE 3.

Example 4 (*The cost of operating an automobile at various speeds*). It is found that the cost of operating an automobile as a function of speed is approximated by a quadratic function. Use the data given to find an equation of the function. Then use the equation to determine the cost of operating the automobile at 60 mph and at 80 mph.

SPEED, in mph	OPERATING COST PER MILE, in cents
10	22
20	20
50	20

We use the three data points to obtain a, b, and c in the equation $f(x) = ax^2 + bx + c$:

$$22 = 100a + 10b + c,$$
$$20 = 400a + 20b + c, \qquad \text{Substituting}$$
$$20 = 2500a + 50b + c.$$

We solve this system of equations, thus obtaining $(0.005, -0.35, 25)$. Thus

$$f(x) = 0.005x^2 - 0.35x + 25.$$

To find the cost of operating at 60 mph, we find $f(60)$:

$$f(60) = 0.005(60)^2 - 0.35(60) + 25 = 22\text{¢}.$$

We also find $f(80)$:

$$f(80) = 0.005(80)^2 - 0.35(80) + 25 = 29\text{¢}.$$

A graph of the cost function is as follows.

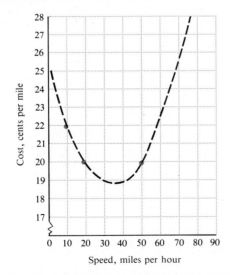

It should be noted that this cost function can give approximate results only within a certain interval. For example, $f(0) = 25$, meaning that it costs 25¢ per mile to stand still, and this, of course, is absurd.

DO EXERCISE 4.

---

4. The following table has values that will fit a quadratic function. Find the number of accidents as a function of age. Then use your model to calculate the number of accidents in which 16-year-olds are involved daily.

AGE OF DRIVER	NUMBER OF ACCIDENTS PER DAY
20	400
40	150
60	400

EXERCISE SET 5.2

[i] Consider the system

$$2x + 3y - 5z = 1,$$
$$6x - 6y + 10z = 3,$$
$$4x - 9y + 5z = 0.$$

1. Determine whether $\left(\frac{1}{2}, \frac{1}{3}, \frac{1}{5}\right)$ is a solution of the system.

2. Determine whether $(-1, 1, 0)$ is a solution of the system.

[ii] Solve.

3. $x + y + z = 2,$
$\quad 6x - 4y + 5z = 31,$
$\quad 5x + 2y + 2z = 13$

4. $x + 6y + 3z = 4,$
$\quad 2x + y + 2z = 3,$
$\quad 3x - 2y + z = 0$

5. $x - y + 2z = -3,$
$\quad x + 2y + 3z = 4,$
$\quad 2x + y + z = -3$

6. $x + y + z = 6,$
$\quad 2x - y - z = -3,$
$\quad x - 2y + 3z = 6$

7. $4a + 9b \quad\quad = 8,$
$\quad 8a \quad\quad + 6c = -1,$
$\quad\quad\quad 6b + 6c = -1$

8. $3p \quad\quad + 2r = 11,$
$\quad\quad q - 7r = 4,$
$\quad p - 6q \quad\quad = 1$

9. $x + y + z + w = 2,$
$\quad 2x + 2y + 4z + w = 1,$
$\quad x - y - z - w = -6,$
$\quad 3x + y - z - w = -2$

10. $x - y + z + w = 0,$
$\quad 2x + 2y + z - w = 5,$
$\quad 3x + y - z - w = -4,$
$\quad x + y - 3z - 2w = -7$

[iii] Solve.

11. (*Curve fitting*). Find numbers a, b, and c such that a quadratic function $ax^2 + bx + c$ fits the data points $(1, 4)$, $(-1, -2)$, and $(2, 13)$.

12. (*Curve fitting*). Find numbers a, b, and c such that a quadratic function $ax^2 + bx + c$ fits the data points $(1, 4)$, $(-1, 6)$, and $(-2, 16)$.

13. (*Predicting earnings*). A business earns $38 in the first week, $66 in the second week, and $86 in the third week. The manager graphs the points $(1, 38)$, $(2, 66)$, and $(3, 86)$ and finds that a quadratic function might fit the data.

a) Find a quadratic function that fits the data.

b) Using your model, predict the earnings for the fourth week.

14. (*Predicting earnings*). A business earns $1000 in its first month, $2000 in the second month, and $8000 in the third month. The manager plots the points $(1, 1000)$, $(2, 2000)$, and $(3, 8000)$ and finds that a quadratic function might fit the data.

a) Find a quadratic function that fits the data.

b) Using your model, predict the earnings for the fourth month.

☆ ─────────────────────────────

Hint for Exercises 15 and 16: Let u represent $1/x$, v represent $1/y$, and w represent $1/z$. First solve for u, v, and w.

15. $\dfrac{2}{x} - \dfrac{1}{y} - \dfrac{3}{z} = -1,$

$\quad \dfrac{2}{x} - \dfrac{1}{y} + \dfrac{1}{z} = -9,$

$\quad \dfrac{1}{x} + \dfrac{2}{y} - \dfrac{4}{z} = 17$

16. $\dfrac{2}{x} + \dfrac{2}{y} - \dfrac{3}{z} = 3,$

$\quad \dfrac{1}{x} - \dfrac{2}{y} - \dfrac{3}{z} = 9,$

$\quad \dfrac{7}{x} - \dfrac{2}{y} + \dfrac{9}{z} = -39$

17. When A, B, and C are working together, they can do a job in 2 hr. When B and C work together, they can do the job in 4 hr. When A and B work together, they can do the job in $\frac{12}{5}$ hr. How long would it take each, working alone, to do the job?

18. Pipes A, B, and C are connected to the same tank. When all three pipes are running, they can fill the tank in 3 hr. When pipes A and C are running, they can fill the tank in 4 hr. When pipes A and B are running, they can fill the tank in 8 hr. How long would it take each, running alone, to fill the tank?

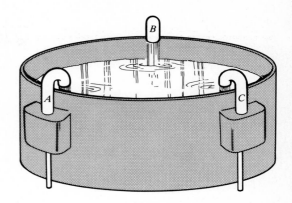

★ ────────────────────────────────

19. ▦ Solve.

$$3.12x + 2.14y - 0.988z = 3.79,$$
$$3.84x - 3.53y + 1.96z = 7.80,$$
$$4.63x - 1.08y + 0.011z = 6.34$$

20. ▦ Solve.

$$1.01x - 0.905y + 2.12z = -2.54,$$
$$1.32x + 2.05y + 2.97z = 3.97,$$
$$2.21x + 1.35y + 0.001z = -3.15$$

5.3 SPECIAL CASES

[i] Inconsistent and Dependent Equations

Inconsistent Equations

DEFINITION

────────────────────────────────

If a system of equations has a solution, we say that it is *consistent*. If a system does not have a solution, we say that it is *inconsistent*.

────────────────────────────────

All of the systems in Sections 5.1 and 5.2 were consistent because they had solutions.

Example 1 Solve:

$$x - 3y = 1,$$
$$-2x + 6y = 5.$$

We multiply the first equation by 2 and add. This gives us

$$x - 3y = 1,$$
$$0 = 7.$$

The second equation says that $0 \cdot x + 0 \cdot y = 7$. There are no numbers x and y for which this is true.

Whenever we obtain a statement such as $0 = 7$, which is obviously false, we will know that the system we are trying to solve has no solutions. It is inconsistent. The solution set is $\emptyset$.

The graphs of the above equations are as follows.

OBJECTIVE

You should be able to:

[i] In solving a system of linear equations in two variables, recognize whether the system is inconsistent or dependent, and describe the solutions, if they exist.

Solve.

1. $3x - 4y = 1$,
  $6x - 8y = 7$

2. Determine whether these systems are consistent or inconsistent.
  a) $3x - y = 2$,
   $6x - 2y = 3$
  b) $x + 4y = 2$,
   $2x - y = 1$

3. Solve.

  $x + 2y - 4z = -3$,
  $2x + y + z = 9$,
  $-x - 2y + 4z = 2$

4. Determine whether these systems are consistent or inconsistent.
  a) $x + 2y + z = 1$,
   $3x + 3y + z = 2$,
   $2x + y = 2$
  b) $x + z = 1$,
    $y + z = 1$,
   $x + y = 1$

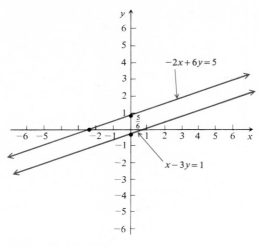

DO EXERCISES 1 AND 2.

Example 2 Solve:

$$x + 3y - 7z = 1, \qquad ①$$
$$2x - 4y + z = 3, \qquad ②$$
$$-2x + 4y - z = 10. \qquad ③$$

Each x-coefficient is a multiple of the first. Thus we can begin by multiplying ① by -2 and adding it to ②. We also multiply ① by 2 and add it to ③:

$$x + 3y - 7z = 1,$$
$$-10y + 15z = 1,$$
$$10y - 15z = 12.$$

Since the y-coefficient of ② is a multiple of the y-coefficient of ③, we proceed by simply adding ② to ③:

$$x + 3y - 7z = 1,$$
$$-10y + 15z = 1,$$
$$0 = 13.$$

Since in ③ we obtain the false equation $0 = 13$, the system is inconsistent; it has no solution. We need work no further.

DO EXERCISES 3 AND 4.

Dependent Equations

DEFINITION

If a system of n linear equations is equivalent to a system of fewer than n of them, we say that the system is *dependent*. If such is not the case, we call the system *independent*.

Consider the following system:

$$2x + 3y = 1,$$
$$4x + 6y = 2.$$

Note that if we multiply the first equation by 2, we get the second equation. Thus the two equations are equivalent. In other words, the conjunction of the *two* equations is equivalent to *one* of the equations. Thus the system is *dependent*.

Consider the following system:

$$x - 4y + 3z = 5,$$
$$x + y + z = 3,$$
$$x + y + z = 3.$$

Since the last two equations are identical, the system of *three* equations is equivalent to the following system of *two* equations:

$$x - 4y + 3z = 5,$$
$$x + y + z = 3.$$

The system is dependent.

In solving a system, how do we know it is dependent? If, at some stage, we find that two of the equations are identical, then we know the system is dependent. If we obtain an obviously true statement, such as $0 = 0$, then we know that the system is dependent. We cannot know whether such a system is consistent or inconsistent without further analysis.

A dependent and consistent system may have an infinite number of solutions. In such a case for systems of two variables we can describe the solutions by expressing one variable in terms of the other.

Example 3 Solve:

$$2x + 3y = 1,$$
$$4x + 6y = 2.$$

We multiply the first equation by -2 and add. This gives us

$$2x + 3y = 1,$$
$$0 + 0 = 0.$$

The last equation contributes nothing, so we consider only the first one. Let us solve it for x. We obtain

$$x = \frac{1}{2} - \frac{3}{2}y, \quad \text{or} \quad \frac{1 - 3y}{2}.$$

We can now describe the ordered pairs in the solution set, in terms of y only, as follows:

$$\left(\frac{1 - 3y}{2}, y \right).$$

Any value we choose for y then gives us a value for x, and thus an ordered pair in the solution set. Some of these solutions are

$$(-4, 3), \quad \left(-\frac{5}{2}, 2 \right), \quad \left(\frac{7}{2}, -2 \right).$$

DO EXERCISES 5 AND 6.

5. Solve. Describe the solutions by expressing one variable in terms of the other. Give three of the ordered pairs in the solution set.

$$-6x + 4y = 10,$$
$$3x - 2y = -5$$

6. Solve. Describe the solutions by expressing one variable in terms of the other. Give three ordered pairs in the solution set.

$$2x + 5y = 1,$$
$$4x + 10y = 2$$

7. Solve, giving the general form of the solutions. Then list three of the solutions.

$$2x + y + z = 9,$$
$$x + 2y - 4z = -3,$$
$$x + y - z = 2$$

When a system of three or more equations is dependent and consistent, we can describe its solutions by expressing one or more of the variables in terms of the others.

Example 4 Solve:

$$x + 2y + 3z = 4, \qquad ①$$
$$2x - y + z = 3, \qquad ②$$
$$3x + y + 4z = 7. \qquad ③$$

Each x-coefficient is a multiple of the first. Thus we can begin solving by multiplying ① by -2 and adding it to ②. We also multiply ① by -3 and add it to ③:

$$x + 2y + 3z = 4,$$
$$-5y - 5z = -5,$$
$$-5y - 5z = -5.$$

Now ② and ③ are identical. We no longer have a system of three equations, but a system of two. Thus we know that the system is dependent. If we were to multiply ② by -1 and add it to ③, we would obtain $0 = 0$. We proceed by multiplying ② by $-\frac{1}{5}$ since the coefficients have the common factor -5:

$$x + 2y + 3z = 4,$$
$$y + z = 1.$$

We will express two of the variables in terms of the other one. Let us choose z. Then we solve ② for y:

$$y = 1 - z.$$

We substitute this value of y in ①, obtaining

$$x + 2(1 - z) + 3z = 4.$$

Solving for x, we get

$$x = 2 - z.$$

The solutions, then, are all of the form

$$(2 - z, 1 - z, z).$$

We obtain the solutions by choosing various values of z. If we let $z = 0$, we obtain the triple $(2, 1, 0)$. If we let $z = 2$, we obtain $(0, -1, 2)$, and so on.

DO EXERCISE 7.

Homogeneous Equations

When all the terms of a polynomial have the same degree, we say that the polynomial is *homogeneous*. Here are some examples:

$$3x^2 + 5y^2, \qquad 4x + 5y - 2z, \qquad 17x^3 - 4y^3 + 57z^3.$$

An equation formed by a homogeneous polynomial set equal to 0 is called a *homogeneous equation*. Let us now consider a system of homogeneous equations.

Example 5 Solve:

$$4x - 3y + z = 0, \qquad ①$$
$$2x \qquad - 3z = 0, \qquad ②$$
$$-8x + 6y - 2z = 0. \qquad ③$$

Any homogeneous system like this always has a solution (can never be inconsistent), because $(0, 0, 0)$ is a solution. This is called the *trivial* solution. There may or may not be other solutions. To find out, we proceed as in the case of nonhomogeneous equations.

First we interchange the first two equations so all the x-coefficients are multiples of the first:

$$2x \qquad - 3z = 0, \qquad ①$$
$$4x - 3y + z = 0, \qquad ②$$
$$-8x + 6y - 2z = 0. \qquad ③$$

Now we multiply ① by -2 and add it to ②. We also multiply ① by 4 and add it to ③:

$$2x \qquad - 3z = 0,$$
$$- 3y + 7z = 0,$$
$$6y - 14z = 0.$$

Next we multiply ② by 2 and add it to ③:

$$2x \qquad - 3z = 0,$$
$$-3y + 7z = 0,$$
$$0 = 0.$$

Now we know that the system is dependent, hence it has an infinite set of solutions. Now we solve ② for y:

$$y = \frac{7}{3}z.$$

Typically, we would now substitute $\frac{7}{3}z$ for y in ①, but since the y-term is missing, we need only solve for x:

$$x = \frac{3}{2}z.$$

We can now describe the members of the solution set as follows:

$$\left(\frac{3}{2}z, \frac{7}{3}z, z\right).$$

Some of the ordered pairs in the solution set are

$$\left(\frac{3}{2}, \frac{7}{3}, 1\right), \qquad \left(3, \frac{14}{3}, 2\right), \qquad \left(-\frac{3}{2}, -\frac{7}{3}, -1\right).$$

DO EXERCISES 8 AND 9.

8. Solve. If there is more than one solution, list three of them.

$$x + y - z = 0,$$
$$x + 2y - 4z = 0,$$
$$2x + y + z = 0$$

9. Solve. If there is more than one solution, list three of them.

$$x + y - z = 0,$$
$$x - y - z = 0,$$
$$x + y + 2z = 0$$

EXERCISE SET 5.3

[i] Solve. If a system has more than one solution, list three of them.

1. $9x - 3y = 15,$
$6x - 2y = 10$

2. $2s - 3t = 9,$
$4s - 6t = 9$

3. $5c + 2d = 24,$
$30c + 12d = 10$

4. $3x + 2y = 18,$
$9x + 6y = 5$

Solve.

5. $3x + 2y = 5,$
$4y = 10 - 6x$

6. $5x + 2 = 7y,$
$-14y + 4 = -10x$

7. $12y - 8x = 6,$
$4x + 3 = 6y$

8. $16x - 12y = 10,$
$6y + 5 = 8x$

Solve. If a system has more than one solution, list three of them.

9. $x + 2y - z = -8,$
$2x - y + z = 4,$
$8x + y + z = 2$

10. $x + 2y - z = 4,$
$4x - 3y + z = 8,$
$5x - y = 12$

11. $2x + y - 3z = 1,$
$x - 4y + z = 6,$
$4x - 16y + 4z = 24$

12. $4x + 12y + 16z = 4,$
$3x + 4y + 5z = 3,$
$x + 8y + 11z = 1$

13. $2x + y - 3z = 0,$
$x - 4y + z = 0,$
$4x - 16y + 4z = 0$

14. $4x + 12y + 16z = 0,$
$3x + 4y + 5z = 0,$
$x + 8y + 11z = 0$

15. $x + y - z = -3,$
$x + 2y + 2z = -1$

16. $x + y + 13z = 0,$
$x - y - 6z = 0$

17. $2x + y + z = 0,$
$x + y - z = 0,$
$x + 2y + 2z = 0$

18. $5x + 4y + z = 0,$
$10x + 8y - z = 0,$
$x - y - z = 0$

19. Classify each of the systems in the odd-numbered exercises 1–17 as consistent or inconsistent, dependent or independent.

20. Classify each of the systems in the even-numbered exercises 2–18 as consistent or inconsistent, dependent or independent.

☆ ──────────────────────────────

Solve.

21. ▦ $4.026x - 1.448y = 18.32,$
$0.724y = -9.16 + 2.013x$

22. ▦ $0.0284y = 1.052 - 8.114x,$
$0.0142y + 4.057x = 0.526$

23. a) Solve:
$x + y + z + w = 4,$
$x + y + z + w = 3,$
$x + y + z + w = 3.$
b) Classify the system as consistent or inconsistent.
c) Classify the system as dependent or independent.

24. a) Solve:
$-8x + 3y + 2z + w = 0,$
$5x + y - z - w = 0,$
$2x + 5y - w = 0,$
$3x - 4y - z = 0.$
b) Classify the system as consistent or inconsistent.
c) Classify the system as dependent or independent.

★ ──────────────────────────────

Determine the constant k such that each system is dependent.

25. $6x - 9y = -3,$
$-4x + 6y = k$

26. $8x - 16y = 20,$
$10x - 20y = k$

5.4 MATRICES AND SYSTEMS OF EQUATIONS

[i] In solving systems of equations, we perform computations with the constants. The variables play no important role in the process. We can simplify writing by omitting the variables. For example, the system

$$3x + 4y = 5,$$
$$x - 2y = 1$$

simplifies to

$$\begin{array}{rrr} 3 & 4 & 5 \\ 1 & -2 & 1 \end{array}$$

if we leave off the variables and omit the operation and equals signs.

In the above example we have written a rectangular array of numbers. Such an array is called a *matrix* (plural, *matrices*). We ordinarily write brackets around matrices, although this is not necessary when calculating with matrices. The following are matrices:

$$\begin{bmatrix} 4 & 1 & 3 & 5 \\ 1 & 0 & 1 & 2 \\ 6 & 3 & -2 & 0 \end{bmatrix}, \quad \begin{bmatrix} 6 & 2 & 1 & 4 & 7 \\ 1 & 2 & 1 & 3 & 1 \\ 4 & 0 & -2 & 0 & -3 \end{bmatrix}, \quad \begin{bmatrix} 1 & 2 \\ 145 & 0 \\ -7 & 9 \\ 8 & 1 \\ 0 & 0 \end{bmatrix}.$$

The *rows* of a matrix are horizontal, and the *columns* are vertical.

$$\begin{bmatrix} 5 & -2 & 2 \\ 1 & 0 & 1 \\ 0 & 1 & 2 \end{bmatrix} \begin{array}{l} \longleftarrow \text{ row 1} \\ \longleftarrow \text{ row 2} \\ \longleftarrow \text{ row 3} \end{array}$$

column 1   column 2   column 3

Let us now use matrices to solve systems of linear equations.

Example 1 Solve:

$$2x - y + 4z = -3,$$
$$x \qquad - 4z = 5,$$
$$6x - y + 2z = 10.$$

We first write a matrix, using only the constants. Note that where there are missing terms we must write 0's:

$$\begin{bmatrix} 2 & -1 & 4 & -3 \\ 1 & 0 & -4 & 5 \\ 6 & -1 & 2 & 10 \end{bmatrix}.$$

We do exactly the same calculations using the matrix that we would do if we wrote the entire equations. The first step, if possible, is to interchange the rows so that each number in the first column below the first number is a multiple of that number. We do this by interchanging rows 1 and 2:

$$\begin{bmatrix} 1 & 0 & -4 & 5 \\ 2 & -1 & 4 & -3 \\ 6 & -1 & 2 & 10 \end{bmatrix}.$$
This corresponds to interchanging equation ① with equation ②.

Next we multiply the first row by -2 and add it to the second row:

$$\begin{bmatrix} 1 & 0 & -4 & 5 \\ 0 & -1 & 12 & -13 \\ 6 & -1 & 2 & 10 \end{bmatrix}.$$
This corresponds to multiplying equation ① by -2 and adding it to equation ②.

Now we multiply the first row by -6 and add it to the third row:

$$\begin{bmatrix} 1 & 0 & -4 & 5 \\ 0 & -1 & 12 & -13 \\ 0 & -1 & 26 & -20 \end{bmatrix}.$$
This corresponds to multiplying equation ① by -6 and adding it to equation ③.

Next we multiply row 2 by -1 and add it to the third row:

$$\begin{bmatrix} 1 & 0 & -4 & 5 \\ 0 & -1 & 12 & -13 \\ 0 & 0 & 14 & -7 \end{bmatrix}.$$
This corresponds to multiplying equation ② by -1 and adding it to equation ③.

If we now put the variables back, we have

$$\begin{aligned} x \quad\quad - \quad 4z &= 5, \\ -y + 12z &= -13, \\ 14z &= -7. \end{aligned}$$

Now we proceed as before. We solve ③ for z and get $z = -\frac{1}{2}$. Next we substitute $-\frac{1}{2}$ for z in ② and solve for y: $-y + 12(-\frac{1}{2}) = -13$, so $y = 7$. Since there is no y-term in ① we need only substitute $-\frac{1}{2}$ for z in ① and solve for x: $x - 4(-\frac{1}{2}) = 5$, so $x = 3$. The solution is $(3, 7, -\frac{1}{2})$.

Note in the preceding that our goal was to get the matrix in the form

$$\begin{bmatrix} a & b & c & d \\ 0 & e & f & g \\ 0 & 0 & h & k \end{bmatrix},$$

where there are just 0's below the *main diagonal,* formed by a, e, and h. Then we put the variables back and complete the solution.

All the operations used in the preceding correspond to operations with the equations and they produce equivalent systems of equations. We call the matrices *row-equivalent,* and the operations that produce them *row-equivalent operations.*

THEOREM 1

Each of the following row-equivalent operations produces equivalent matrices:

a) Interchanging any two rows of a matrix;

b) Multiplying each element of a row by the same nonzero constant;

c) Multiplying each element of a row by a nonzero number and adding the result to another row.

The best overall method for solving systems of equations is by row-equivalent matrices; even computers are programmed to use row-equivalent matrices.

DO EXERCISES 1 AND 2.

1. Solve, using matrices.

$$5x - 2y = -3,$$
$$2x + 5y = -24$$

2. Solve, using matrices.

$$x - 2y + 3z = 4,$$
$$2x - y + z = -1,$$
$$4x + y + z = 1$$

EXERCISE SET 5.4

[i] Solve, using matrices.

1. $4x + 2y = 11,$
$3x - y = 2$

2. $3x - 3y = 11,$
$9x - 2y = 5$

3. $x + 2y - 3z = 9,$
$2x - y + 2z = -8,$
$3x - y - 4z = 3$

4. $x - y + 2z = 0,$
$x - 2y + 3z = -1,$
$2x - 2y + z = -3$

5. $5x - 3y = -2,$
$4x + 2y = 5$

6. $3x + 4y = 7,$
$-5x + 2y = 10$

7. $4x - y - 3z = 1,$
$8x + y - z = 5,$
$2x + y + 2z = 5$

8. $3x + 2y + 2z = 3,$
$x + 2y - z = 5,$
$2x - 4y + z = 0$

9. $p + q + r = 1,$
$p + 2q + 3r = 4,$
$4p + 5q + 6r = 7$

10. $m + n + t = 9,$
$m - n - t = -15,$
$m + n + t = 3$

11. $2x + 2y - 2z - 2w = -10,$
$x + y + z + w = -5,$
$x - y + 4z + 3w = -2,$
$3x - 2y + 2z + w = -6$

12. $2x - 3y + z - w = -8,$
$x + y - z - w = -4,$
$x + y + z + w = 22,$
$x - y - z - w = -14$

☆ ——————

13. A collection of 34 coins consists of dimes and nickels. The total value is $1.90. How many dimes and how many nickels are there?

14. A collection of 43 coins consists of dimes and quarters. The total value is $7.60. How many dimes and how many quarters are there?

15. A collection of 22 coins consists of nickels, dimes, and quarters. The total value is $2.90. There are six more nickels than dimes. How many of each type of coin are there?

16. A collection of 18 coins consists of nickels, dimes, and quarters. The total value is $2.55. There are two more quarters than there are dimes. How many of each type of coin are there?

17. A tobacco dealer has two kinds of tobacco. One is worth $4.05 per lb and the other is worth $2.70 per lb. The dealer wants to blend the two tobaccos to get a 15-lb mixture worth $3.15 per lb. How much of each kind of tobacco should be used?

18. A grocer mixes candy worth $0.80 per lb with nuts worth $0.70 per lb to get a 20-lb mixture worth $0.77 per lb. How many pounds of candy and how many pounds of nuts are used?

Recall the formula $I = prt$, for simple interest.

19. One year $8950 was received in interest from two investments. A certain amount was invested at $12\frac{1}{2}\%$ and $10,000 more than this at 13%. Find the amount of principal invested at each rate.

20. One year a person had some money invested at $13\frac{1}{2}\%$ and another amount at $13\frac{3}{4}\%$. The income from the investments was $3275. The income from the $13\frac{1}{2}\%$ investment was $575 less than that from the $13\frac{3}{4}\%$ investment. How much was invested at each rate?

Solve.

21. ▦ $4.83x + 9.06y = -39.42,$
$-1.35x + 6.67y = -33.99$

22. ▦ $3.11x - 2.04y = -24.39,$
$7.73x + 5.19y = -35.48$

23. ▦ $3.55x - 1.35y + 1.03z = 9.16,$
$-2.14x + 4.12y + 3.61z = -4.50,$
$5.48x - 2.44y - 5.86z = 0.813$

24. ▦ $4.12x - 1.35y - 18.2z = 601.3,$
$-3.41x + 68.9y + 38.7z = 1777,$
$0.955x - 0.813y - 6.53z = 160.2$

★

25. $\sqrt{2}x + \pi y = 3,$
$\pi x - \sqrt{2}y = 1$

26. $ax + by = c,$
$dx + ey = f$

OBJECTIVES

You should be able to:

[i] Evaluate determinants of 2×2 matrices.

[ii] Solve systems of two equations in two variables using Cramer's rule.

[iii] Evaluate determinants of 3×3 matrices.

[iv] Solve systems of three equations in three variables using Cramer's rule.

Evaluate.

1. $\begin{vmatrix} \sqrt{3} & -5 \\ -2 & -\sqrt{3} \end{vmatrix}$

2. $\begin{vmatrix} 1 & 2 \\ 3 & 4 \end{vmatrix}$

3. $\begin{vmatrix} -2 & -3 \\ 4 & x \end{vmatrix}$

5.5 DETERMINANTS AND CRAMER'S RULE

[i] Determinants of Two-by-two Matrices

A matrix of m rows and n columns is called an $m \times n$ matrix (read "m by n"). If a matrix has the same number of rows and columns, it is called a *square* matrix. With every square matrix is associated a number called its determinant, defined as follows for 2×2 matrices.*

DEFINITION

The determinant of the matrix $\begin{bmatrix} a & c \\ b & d \end{bmatrix}$ is denoted $\begin{vmatrix} a & c \\ b & d \end{vmatrix}$ and is defined as follows:

$$\begin{vmatrix} a & c \\ b & d \end{vmatrix} = ad - bc.$$

Example 1 Evaluate $\begin{vmatrix} \sqrt{2} & -3 \\ -4 & -\sqrt{2} \end{vmatrix}$.

$\begin{vmatrix} \sqrt{2} & -3 \\ -4 & -\sqrt{2} \end{vmatrix}$ The arrows indicate the products involved.

$$= \sqrt{2}(-\sqrt{2}) - (-4)(-3) = -2 - 12 = -14$$

DO EXERCISES 1–3.

[ii] Cramer's Rule

Determinants have many uses. One of these is in solving systems of nonhomogeneous linear equations, where the number of variables is

---

*The definition of the determinant of any square matrix is given in Chapter 6.

the same as the number of equations. Let us consider a system of two equations:

$$a_1x + b_1y = c_1,$$
$$a_2x + b_2y = c_2.$$

Using the methods of the preceding sections we can solve. We obtain

$$x = \frac{c_1b_2 - c_2b_1}{a_1b_2 - a_2b_1}, \qquad y = \frac{a_1c_2 - a_2c_1}{a_1b_2 - a_2b_1}.$$

The numerators and denominators of the expressions for x and y are determinants:

$$x = \frac{\begin{vmatrix} c_1 & b_1 \\ c_2 & b_2 \end{vmatrix}}{\begin{vmatrix} a_1 & b_1 \\ a_2 & b_2 \end{vmatrix}}, \qquad y = \frac{\begin{vmatrix} a_1 & c_1 \\ a_2 & c_2 \end{vmatrix}}{\begin{vmatrix} a_1 & b_1 \\ a_2 & b_2 \end{vmatrix}}.$$

The above equations make sense only if the denominator determinant is not 0. If the denominator *is* 0, then one of two things happens.

1. If the denominator is 0 and the other two determinants are also 0, then the system of equations is dependent.

2. If the denominator is 0 and at least one of the other determinants is not 0, then the system is inconsistent.

The equations with determinants above describe *Cramer's rule* for solving systems of equations. To use this rule, we compute the three determinants and compute x and y as shown above. Note that the denominator in both cases contains the coefficients of x and y, in the same position as in the original equations. For x the numerator is obtained by replacing the x-coefficients (the *a*'s) by the *c*'s. For y the numerator is obtained by replacing the y-coefficients (the *b*'s) by the *c*'s.

Example 2 Solve, using Cramer's rule:

$$2x + 5y = 7,$$
$$5x - 2y = -3.$$

$$x = \frac{\begin{vmatrix} 7 & 5 \\ -3 & -2 \end{vmatrix}}{\begin{vmatrix} 2 & 5 \\ 5 & -2 \end{vmatrix}} = \frac{7(-2) - (-3)5}{2(-2) - 5 \cdot 5} = -\frac{1}{29},$$

$$y = \frac{\begin{vmatrix} 2 & 7 \\ 5 & -3 \end{vmatrix}}{\begin{vmatrix} 2 & 5 \\ 5 & -2 \end{vmatrix}} = \frac{2(-3) - 5 \cdot 7}{-29} = \frac{41}{29}$$

The solution is $\left(-\frac{1}{29}, \frac{41}{29}\right)$.

DO EXERCISES 4–6.

Solve, using Cramer's rule.

4. $2x - y = 5,$
$x - 2y = 1$

5. $3x + 4y = -2,$
$5x - 7y = 1$

6. $\sqrt{2}x - \pi y = 3,$
$\pi x + \sqrt{2}y = 4$

Evaluate.

7. $\begin{vmatrix} 3 & 2 & 2 \\ -2 & 1 & 4 \\ 4 & -3 & 3 \end{vmatrix}$

8. $\begin{vmatrix} -5 & 0 & 0 \\ 4 & 2 & 0 \\ -3 & 5 & -6 \end{vmatrix}$

9. $\begin{vmatrix} x & 0 & x \\ 0 & x & 0 \\ 1 & 0 & x \end{vmatrix}$

[iii] Determinants of Three-by-Three Matrices

DEFINITION

The *determinant* of a three-by-three matrix is defined as follows:

$$\begin{vmatrix} a_1 & b_1 & c_1 \\ a_2 & b_2 & c_2 \\ a_3 & b_3 & c_3 \end{vmatrix} = a_1 \cdot \begin{vmatrix} b_2 & c_2 \\ b_3 & c_3 \end{vmatrix} - a_2 \cdot \begin{vmatrix} b_1 & c_1 \\ b_3 & c_3 \end{vmatrix} + a_3 \cdot \begin{vmatrix} b_1 & c_1 \\ b_2 & c_2 \end{vmatrix}.$$

The two-by-two determinants on the right can be obtained from the three-by-three matrix by crossing out the row and column in which the a-coefficient occurs.

Example 3 Evaluate.

$$\begin{vmatrix} -1 & 0 & 1 \\ -5 & 1 & -1 \\ 4 & 8 & 1 \end{vmatrix} = -1 \cdot \begin{vmatrix} 1 & -1 \\ 8 & 1 \end{vmatrix} - (-5) \cdot \begin{vmatrix} 0 & 1 \\ 8 & 1 \end{vmatrix} + 4 \cdot \begin{vmatrix} 0 & 1 \\ 1 & -1 \end{vmatrix}$$

$$= -1(1 + 8) + 5(-8) + 4(-1)$$
$$= -9 - 40 - 4 = -53$$

DO EXERCISES 7–9.

Let us consider three equations in three variables. Consider

$$a_1x + b_1y + c_1z = d_1,$$
$$a_2x + b_2y + c_2z = d_2,$$
$$a_3x + b_3y + c_3z = d_3,$$

and the following determinants:

$$D = \begin{vmatrix} a_1 & b_1 & c_1 \\ a_2 & b_2 & c_2 \\ a_3 & b_3 & c_3 \end{vmatrix}, \qquad D_x = \begin{vmatrix} d_1 & b_1 & c_1 \\ d_2 & b_2 & c_2 \\ d_3 & b_3 & c_3 \end{vmatrix},$$

$$D_y = \begin{vmatrix} a_1 & d_1 & c_1 \\ a_2 & d_2 & c_2 \\ a_3 & d_3 & c_3 \end{vmatrix}, \qquad D_z = \begin{vmatrix} a_1 & b_1 & d_1 \\ a_2 & b_2 & d_2 \\ a_3 & b_3 & d_3 \end{vmatrix}.$$

If we solve the system of equations, we obtain the following:

$$x = \frac{D_x}{D}, \qquad y = \frac{D_y}{D}, \qquad z = \frac{D_z}{D}.$$

Note that we obtain the determinant D_x in the numerator for x from D by replacing the x-coefficients by d_1, d_2, and d_3. A similar thing happens with D_y and D_z. We have thus extended *Cramer's rule* to solve systems of three equations in three variables. As before, when $D = 0$, Cramer's rule cannot be used. If $D = 0$, and D_x, D_y, and D_z are 0, the system is dependent. If $D = 0$ and one of D_x, D_y, or D_z is not zero, then the system is inconsistent.

[iv] **Cramer's Rule**

Example 3 Solve, using Cramer's rule:

$$x - 3y + 7z = 13,$$
$$x + y + z = 1,$$
$$x - 2y + 3z = 4.$$

$$D = \begin{vmatrix} 1 & -3 & 7 \\ 1 & 1 & 1 \\ 1 & -2 & 3 \end{vmatrix} = -10, \qquad D_x = \begin{vmatrix} 13 & -3 & 7 \\ 1 & 1 & 1 \\ 4 & -2 & 3 \end{vmatrix} = 20,$$

$$D_y = \begin{vmatrix} 1 & 13 & 7 \\ 1 & 1 & 1 \\ 1 & 4 & 3 \end{vmatrix} = -6, \qquad D_z = \begin{vmatrix} 1 & -3 & 13 \\ 1 & 1 & 1 \\ 1 & -2 & 4 \end{vmatrix} = -24.$$

Then

$$x = \frac{D_x}{D} = \frac{20}{-10} = -2,$$

$$y = \frac{D_y}{D} = \frac{-6}{-10} = \frac{3}{5},$$

$$z = \frac{D_z}{D} = \frac{-24}{-10} = \frac{12}{5}.$$

The solution is $(-2, \frac{3}{5}, \frac{12}{5})$. In practice, it is not necessary to evaluate D_z. When we have found values for x and y we can substitute them into one of the equations and find z.

DO EXERCISE 10.

10. Solve, using Cramer's rule.

$$x - 3y - 7z = 6,$$
$$2x + 3y + z = 9,$$
$$4x + y = 7$$

EXERCISE SET 5.5

[i] Evaluate.

1. $\begin{vmatrix} -2 & -\sqrt{5} \\ -\sqrt{5} & 3 \end{vmatrix}$

2. $\begin{vmatrix} \sqrt{5} & -3 \\ 4 & 2 \end{vmatrix}$

3. $\begin{vmatrix} x & 4 \\ x & x^2 \end{vmatrix}$

4. $\begin{vmatrix} y^2 & -2 \\ y & 3 \end{vmatrix}$

[iii] Evaluate.

5. $\begin{vmatrix} 3 & 1 & 2 \\ -2 & 3 & 1 \\ 3 & 4 & -6 \end{vmatrix}$

6. $\begin{vmatrix} 3 & -2 & 1 \\ 2 & 4 & 3 \\ -1 & 5 & 1 \end{vmatrix}$

7. $\begin{vmatrix} x & 0 & -1 \\ 2 & x & x^2 \\ -3 & x & 1 \end{vmatrix}$

8. $\begin{vmatrix} x & 1 & -1 \\ x^2 & x & x \\ 0 & x & 1 \end{vmatrix}$

[ii] Solve, using Cramer's rule.

9. $-2x + 4y = 3,$
$3x - 7y = 1$

10. $5x - 4y = -3,$
$7x + 2y = 6$

11. $\sqrt{3}x + \pi y = -5,$
$\pi x - \sqrt{3}y = 4$

12. $\pi x - \sqrt{5}y = 2,$
$\sqrt{5}x + \pi y = -3$

[iv] Solve, using Cramer's rule.

13. $3x + 2y - z = 4,$
$3x - 2y + z = 5,$
$4x - 5y - z = -1$

14. $3x - y + 2z = 1,$
$x - y + 2z = 3,$
$-2x + 3y + z = 1$

15. $6y + 6z = -1,$
$8x + 6z = -1,$
$4x + 9y = 8$

16. $3x + 5y = 2,$
$2x - 3z = 7,$
$4y + 2z = -1$

☆ ───────────────────────────────

Solve.

17. $\begin{vmatrix} x & 5 \\ -4 & x \end{vmatrix} = 24$

18. $\begin{vmatrix} y & 2 \\ 3 & y \end{vmatrix} = y$

19. $\begin{vmatrix} x & -3 \\ -1 & x \end{vmatrix} \geq 0$

20. $\begin{vmatrix} y & -5 \\ -2 & y \end{vmatrix} < 0$

21. $\begin{vmatrix} x + 3 & 4 \\ x - 3 & 5 \end{vmatrix} = -7$

22. $\begin{vmatrix} m + 2 & -3 \\ m + 5 & -4 \end{vmatrix} = 3m - 5$

23. $\begin{vmatrix} 2 & x & 1 \\ 1 & 2 & -1 \\ 3 & 4 & -2 \end{vmatrix} = -6$

24. $\begin{vmatrix} x & 2 & x \\ 3 & -1 & 1 \\ 1 & -2 & 2 \end{vmatrix} = -10$

Rewrite each expression using determinants. Answers may vary.

25. $2L + 2W$

26. $\pi r + \pi h$

27. $a^2 + b^2$

28. $\frac{1}{2} h(a + b)$

29. $2\pi r^2 + 2\pi rh$

30. $x^2 y^2 - Q^2$

───────────────────────────────

OBJECTIVES

You should be able to:

[i] Graph linear inequalities in two variables in the plane.

[ii] Graph systems of linear inequalities and find vertices, if they exist.

1. Determine whether $(1, -4)$ is a solution of the inequality $4x - 5y < 12$.

5.6 SYSTEMS OF INEQUALITIES

[i] Graphs of Inequalities in Two Variables

A *solution* of an inequality in two variables is an ordered pair of numbers that makes the inequality true.

Example 1 Determine whether $(-3, 2)$ is a solution of the inequality $5x - 4y \leq 13$.

We replace x by -3 and y by 2:

$$\begin{array}{c|c} 5x - 4y & \leq \quad 13 \\ \hline 5(-3) - 4 \cdot 2 & 13 \\ -15 - 8 & \\ -23 & \end{array}$$

Since -23 is less than 13, this replacement makes the inequality true.

DO EXERCISE 1.

To graph inequalities, we use our knowledge and skill in graphing equations. In fact, the first step in graphing an inequality is usually the graphing of an equation.

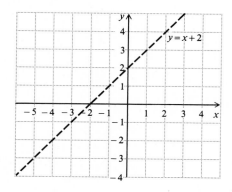

Example 2 Graph $y < x + 2$.

We first graph the equation $y = x + 2$, drawing the line dashed. For any point on the line the value of y is $x + 2$. For any point above the line, y is greater than this; in other words, $y > x + 2$. For any point below the line y is less than $x + 2$, or $y < x + 2$. Thus the graph is the half-plane *below* the line $y = x + 2$. We show this by shading the lower half-plane. The fact that the line is drawn dashed means that points on the line are not in the graph.

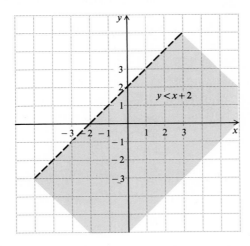

The graph of any linear inequality in two variables is either a half-plane or a half-plane together with the line along the edge.

Example 3 Graph $3y - 2x \geq 1$.

Method 1. Solve for y:

$$y \geq \frac{2x + 1}{3}.$$

Now graph the line $y = (2x + 1)/3$. For any point on the line the value of y is $(2x + 1)/3$. Such values make the inequality true, hence are in the graph. Therefore we draw the line solid this time.

For any point above the line the value of y is greater than $(2x + 1)/3$. Hence any such point is in the graph. The graph of the inequality is the half-plane above the line, together with the line.

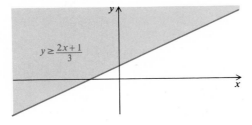

Method 2. We are to graph $3y - 2x \geq 1$.
We graph the line $3y - 2x = 1$, by any method. This time let us find the intercepts. They are $(0, \frac{1}{3})$ and $(-\frac{1}{2}, 0)$. We plot them and then use them to draw the line.

2. Graph $y > -2x$.

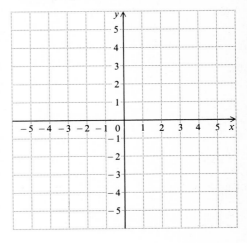

3. Graph $2x + y \geq 2$.

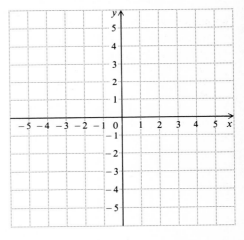

4. Graph $3x - y > -3$.

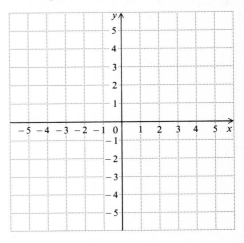

Graph.

5. $x \leq -1$

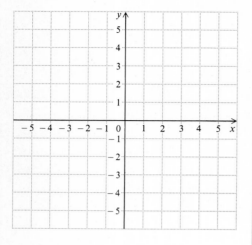

6. $y > -\frac{1}{2}$

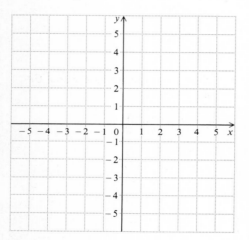

7. $-4 \leq x < 1$

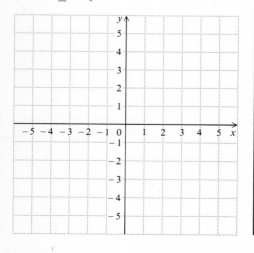

Next, we know that the rest of the graph is a half-plane, but we need to determine which one. All we need to do is choose any point on one of the half-planes, substitute it into the inequality, and see if we get a true sentence. The origin is an easy point to use, if the line doesn't contain the origin. We try $(0, 0)$:

$$3 \cdot 0 - 2 \cdot 0 \geq 1, \quad \text{or} \quad 0 \geq 1.$$

This gives us a false sentence, hence $(0, 0)$ is not a solution. Thus we shade the half-plane not containing $(0, 0)$.

DO EXERCISES 2–4 ON THE PRECEDING PAGE.

Example 4 Graph $x \leq 3$.

We first graph the equation $x = 3$, drawing it solid. Then we shade the appropriate half-plane.

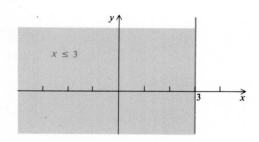

Example 5 Graph $-1 < y \leq 2$.

This is a conjunction of two inequalities:

$$-1 < y \quad \text{and} \quad y \leq 2.$$

It will be true for any y that is both greater than -1 and less than or equal to 2. The graph is as shown at the right.

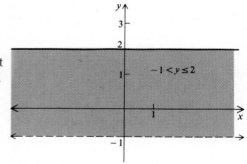

Since our inequality is a conjunction, the graph is the intersection of the graphs of the parts.

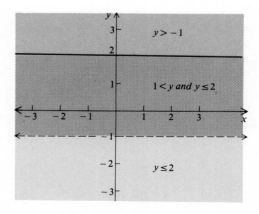

DO EXERCISES 5–8. (NOTE THAT EXERCISE 8 IS ON THE FOLLOWING PAGE.)

[ii] Systems of Inequalities

To get a picture of the solution set of a system or *conjunction* of inequalities, we will graph the inequalities separately and find their intersection. A system of linear inequalities may have a graph that is a polygon and its interior. In certain applications it is important to find the vertices.

Example 6 Graph this system of inequalities. Find the coordinates of any vertices formed.

$$2x + y \geq 2,$$
$$4x + 3y \leq 12,$$
$$\frac{1}{2} \leq x \leq 2,$$
$$y \geq 0$$

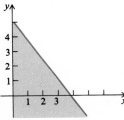

$2x + y \geq 2$

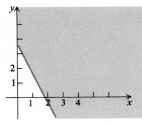

$4x + 3y \leq 12$

$\frac{1}{2} \leq x \leq 2$

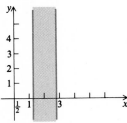

$y \geq 0$

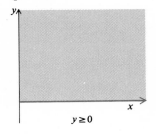

The separate graphs are shown on the left and the graph of the intersection, which is the graph of the system, is shown on the right.

8. Graph $1 \leq y \leq 2\frac{1}{2}$.

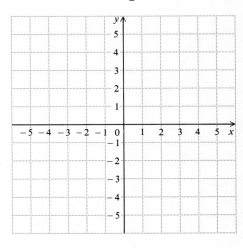

9. Graph. If a polygon is formed, find the vertices.

$$x + y \geq 1,$$
$$y - x \geq 2$$

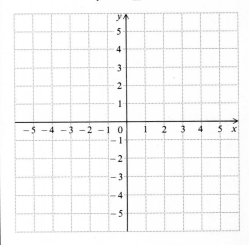

10. Graph. If a polygon is formed, find the vertices.

$$5x + 6y \leq 30,$$
$$0 \leq y \leq 3,$$
$$0 \leq x \leq 4$$

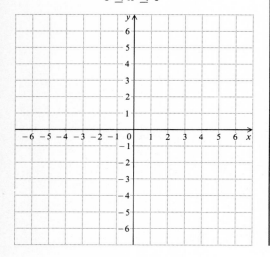

We find the vertex $(\frac{1}{2}, 1)$ by solving the system

$$2x + y = 2,$$
$$x = \frac{1}{2}.$$

We find the vertex $(1, 0)$ by solving the system

$$2x + y = 2,$$
$$y = 0.$$

We find the vertex $(2, 0)$ from the system

$$x = 2,$$
$$y = 0.$$

The vertices $(2, \frac{4}{3})$ and $(\frac{1}{2}, \frac{10}{3})$ were found by solving, respectively, the systems

$$x = 2, \qquad\qquad x = \frac{1}{2},$$
$$\text{and}$$
$$4x + 3y = 12 \qquad\qquad 4x + 3y = 12.$$

DO EXERCISES 9 AND 10. (NOTE THAT EXERCISE 9 IS ON THE PRECEDING PAGE.)

EXERCISE SET 5.6

[i] Graph.

1. $y < x$ **2.** $y \geq x$ **3.** $y + x \geq 0$ **4.** $y + x < 0$

5. $3x - 2y < 6$ **6.** $2x - 5y > 10$ **7.** $2x + 3y \geq 6$ **8.** $x + 2y \leq 4$

9. $3x - 2 \leq 5x + y$ **10.** $2x - 6y \geq 8 + 4y$ **11.** $x < -4$ **12.** $y \geq 5$

13. $0 \leq x < 5\frac{1}{2}$ **14.** $-4 < y < -1$ **15.** $y \geq |x|$ **16.** $y < |x|$

[ii] Graph these systems. Find the coordinates of any vertices formed.

17. $x + y \leq 1,$
$\quad x - y \leq 2$

18. $x + y \leq 3,$
$\quad x - y \leq 4$

19. $y - 2x > 1,$
$\quad y - 2x < 3$

20. $y + x > 0,$
$\quad y + x < 2$

21. $2y - \ x \leq 2,$
$\quad y - 3x \geq -1$

22. $\ x + 3y \geq 9,$
$\quad 3x - 2y \leq 5$

23. $y \leq 2x \ + 1,$
$\quad y \geq -2x + 1,$
$\quad x \leq 2$

24. $x - \ y \leq 2,$
$\quad x + 2y \geq 8,$
$\quad\quad\ y \leq 4$

25. $\ x + 2y \leq 12,$
$\quad 2x + \ y \leq 12,$
$\quad\quad\ x \geq 0,$
$\quad\quad\ y \geq 0$

26. $8x + 5y \leq 40,$
$\quad x + 2y \leq 8,$
$\quad\quad x \geq 0,$
$\quad\quad y \geq 0$

27. $3x + 4y \geq 12,$
$\quad 5x + 6y \leq 30,$
$\quad 1 \leq x \leq 3$

28. $y - 2x \geq 3,$
$\quad y - 2x \leq 5,$
$\quad 6 \leq y \leq 8$

☆
Graph.

29. $y \geq x^2 - 2,$
$\quad y \leq 2 \ - x^2$

30. $x \geq 2y^2 - 1,$
$\quad x < \ y^2$

31. $y < x \ + 1,$
$\quad y \geq x^2$

32. $y \leq -x^2 + 5,$
$\quad y > \frac{1}{2}x^2 - 1$

★
Graph.

33. $|x| + |y| \leq 1$ **34.** $|x + y| \leq 1$ **35.** $|x - y| > 0$ **36.** $|x| > |y|$

5.7 LINEAR PROGRAMMING

[i] Suppose you are taking a test in which different items are worth different numbers of points. Presumably those with higher point values take more time. How many items of each kind should you do, in order to make the best score? A kind of mathematics called *linear programming* (developed during World War II) may provide an answer. Let us consider an example.

Example 1 You are taking a test in which items of type A are worth 10 points and items of type B are worth 15 points. It takes three minutes for each item of type A and six minutes for each item of type B. The total time allowed is 60 minutes and you are not allowed to answer more than 16 questions. Assuming that all of your answers are correct, how many items of each type should you answer to get the best score?

Let x = the number of items of type A, and y = the number of items of type B. The total score T is a function of the two variables x and y:

$$T = 10x + 15y.$$

This function has a domain that is a set of ordered pairs of numbers (x, y). This domain is determined by the following conditions, called *constraints*:

Total number of questions allowed, not more than 16	$x + y \leq 16,$
Time, not more than 60 min	$3x + 6y \leq 60,$
Numbers of items answered will not be negative	$x \geq 0,$
	$y \geq 0.$

We now graph the domain of the function T. This is the graph of the system of inequalities above. We will determine the vertices, if any are formed.

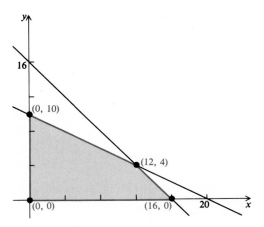

The domain consists of a *convex polygon** and its interior. Under this condition, our linear function does have a maximum value and a minimum value. Moreover, the maximum and minimum values occur

---

*A *convex polygon*, roughly speaking, is one that has no indentations, unlike this one:

1. Find the maximum and minimum values of

$$F = 34x + 6y$$

subject to

$$x + y \leq 6,$$
$$x + y \geq 1,$$
$$1 \leq x \leq 3.$$

at the vertices of the polygon. All we need do to find these values is substitute the coordinates of the vertices in $T = 10x + 15y$.

VERTICES (x, y)	SCORE $T = 10x + 15y$
(0, 0)	0
(16, 0)	160
(12, 4)	180
(0, 10)	150

From the table we see that the minimum value is 0 and the maximum is 180. To get this maximum you would answer 12 items of type A and 4 items of type B.

We now state the main theorem needed to perform linear programming.

THEOREM 2

> If a linear function $F = ax + by + c$ is defined on a domain described by a system of linear inequalities (constraints), then any maximum or minimum value of F will occur at a vertex. If the domain consists of a convex polygon and its interior, then both maximum and minimum values of F actually exist.

2. Find the maximum and minimum values of

$$G = 3x - 5y + 27$$

subject to

$$x + 2y \leq 8,$$
$$0 \leq y \leq 3,$$
$$0 \leq x \leq 6.$$

Example 2 Find the maximum and minimum values of $F = 9x + 40y$ subject to the constraints

$$y - x \geq 1,$$
$$y - x \leq 3,$$
$$2 \leq x \leq 5.$$

We graph the system of inequalities, determine the vertices, and find the function values for those ordered pairs.

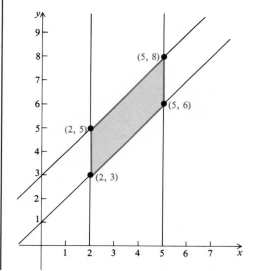

VERTICES (x, y)	SCORE $F = 9x + 40y$
(2, 3)	138
(2, 5)	218
(5, 6)	285
(5, 8)	365

The maximum value of F is 365 when $x = 5$ and $y = 8$. The minimum value of F is 138 when $x = 2$ and $y = 3$.

DO EXERCISES 1 AND 2 ON THE PRECEDING PAGE.

Example 3 A company manufactures motorcycles and bicycles. To stay in business it must produce at least 10 motorcycles each month, but it does not have facilities to produce more than 60 motorcycles. It also does not have facilities to produce more than 120 bicycles. The total production of motorcycles and bicycles cannot exceed 160. The profit on a motorcycle is $134 and on a bicycle is $20. Find the number of each that should be manufactured to maximize profit.

Let x = the number of motorcycles to be produced, and y = the number of bicycles to be produced. The profit P is given by

$$P = \$134x + \$20y,$$

subject to the constraints

$$10 \leq x \leq 60,$$
$$0 \leq y \leq 120,$$
$$x + y \leq 160.$$

We graph the system of inequalities, determine the vertices, and find the function values for those ordered pairs.

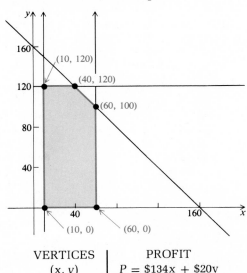

VERTICES (x, y)	PROFIT $P = \$134x + \$20y$
(10, 0)	$ 1,340
(60, 0)	$ 8,040
(60, 100)	$10,040
(40, 120)	$ 7,760
(10, 120)	$ 3,740

Thus the company will make a maximum profit of $10,040 by producing 60 motorcycles and 100 bicycles.

DO EXERCISE 3.

3. A college snack bar cooks and sells hamburgers and hot dogs during the lunch hour. To stay in business it must sell at least 10 hamburgers but cannot cook more than 40. It must also sell at least 30 hot dogs but cannot cook more than 70. It cannot cook more than 90 sandwiches altogether. The profit on a hamburger is $.33 and on a hot dog it is $.21. How many of each kind of sandwich should they sell to make the maximum profit? What is the maximum profit?

EXERCISE SET 5.7

[i] In Exercises 1–4, maximize and also minimize (find maximum and minimum values of the function, and the values of x and y where they occur).

1. $F = 4x + 28y$,
subject to
$5x + 3y \leq 34$,
$3x + 5y \leq 30$,
$x \geq 0$,
$y \geq 0$.

2. $G = 14x + 16y$,
subject to
$3x + 2y \leq 12$,
$7x + 5y \leq 29$,
$x \geq 0$,
$y \geq 0$.

3. $P = 16x - 2y + 40$,
subject to
$6x + 8y \leq 48$,
$0 \leq y \leq 4$,
$0 \leq x \leq 7$.

4. $Q = 24x - 3y + 52$,
subject to
$5x + 4y \geq 20$,
$0 \leq y \leq 4$,
$0 \leq x \leq 3$.

Solve.

5. You are about to take a test that contains questions of type A worth 4 points and questions of type B worth 7 points. You must do at least 5 questions of type A but time restricts doing more than 10. You must do at least 3 questions of type B but time restricts doing more than 10. In total, you can do no more than 18 questions. How many of each type of question must you do to maximize your score? What is this maximum score?

6. You are about to take a test that contains questions of type A worth 10 points and questions of type B worth 25 points. You must do at least 3 questions of type A but time restricts doing more than 12. You must do at least 4 questions of type B but time restricts doing more than 15. In total you can do no more than 20 questions. How many of each type of question must you do to maximize your score? What is this maximum score?

7. A man is planning to invest up to $22,000 in bank X or bank Y or both. He wants to invest at least $2000 but no more than $14,000 in bank X. Bank Y does not insure more than a $15,000 investment so he will invest no more than that in bank Y. The interest in bank X is 6% and in bank Y it is $6\frac{1}{2}$% and this will be simple interest for one year. How much should he invest in each bank to maximize his income? What is the maximum income?

8. A woman is planning to invest up to $40,000 in corporate or municipal bonds or both. The least she is allowed to invest in corporate bonds is $6000 and she does not want to invest more than $22,000 in corporate bonds. She also does not want to invest more than $30,000 in municipal bonds. The interest on corporate bonds is 8% and on municipal bonds it is $7\frac{1}{2}$%. This is simple interest for one year. How much should she invest in each type of bond to maximize her income? What is the maximum income?

9. It takes a tailoring firm 2 hr of cutting and 4 hr of sewing to make a knit suit. To make a worsted suit it takes 4 hr of cutting and 2 hr of sewing. At most, 20 hr per day are available for cutting and, at most, 16 hr per day are available for sewing. The profit on a knit suit is $34 and on a worsted suit is $31. How many of each kind of suit should be made to maximize profit? What is the maximum profit?

10. A pipe tobacco company has 3000 lb of English tobacco, 2000 lb of Virginia tobacco, and 500 lb of Latakia tobacco. To make one batch of SMELLO tobacco it takes 12 lb of English tobacco and 4 lb of Latakia. To make one batch of ROPPO tobacco it takes 8 lb of English and 8 lb of Virginia tobacco. The profit is $10.56 per batch for SMELLO and $6.40 for ROPPO. How many batches of each kind of tobacco should be made to yield maximum profit? What is the maximum profit? *Hint:* Organize the information in a table.

11. An airline with two types of airplanes, P-1 and P-2, has contracted with a tour group to provide accommodations for a minimum of each of 2000 first-class, 1500 tourist, and 2400 economy-class passengers. Airplane P-1 costs $12,000 per mile to operate and can accommodate 40 first-class, 40 tourist, and 120 economy-class passengers, whereas airplane P-2 costs $10,000 per mile to operate and can accommodate 80 first-class, 30 tourist, and 40 economy-class passengers. How many of each type of airplane should be used to minimize the operating cost?

12. A new airplane P-3 becomes available, having an operating cost of $15,000 per mile and accommodating 40 first-class, 80 tourist, and 80 economy-class passengers. If airplane P-1 of Exercise 11 were replaced by airplane P-3, how many of P-2 and P-3 would be needed to minimize the operating cost?

CHAPTER 5 REVIEW

[5.1, iv]   Solve.

[5.3, i]   **1.** $5x - 3y = -4,$
              $\quad 3x - y = -4$

2. $2x + 3y = 2,$
$\quad 5x - y = -29$

3. $x + 5y = 12,$
$\quad 5x + 25y = 12$

4. $\quad 2x - 4y + 3z = -3,$
$\quad -5x + 2y - z = 7,$
$\quad 3x + 2y - 2z = 4$

5. $\quad x + 5y + 3z = 0,$
$\quad 3x - 2y + 4z = 0,$
$\quad 2x + 3y - z = 0$

6. $x - y = 5,$
$\quad y - z = 6,$
$\quad z - w = 7,$
$\quad w + x = 8$

[5.3, i]   **7.** Classify each of the systems in Exercises 1–6 as consistent or inconsistent.

8. Classify each of the systems in Exercises 1–6 as dependent or independent.

[5.1, v]   Solve.

9. The value of 75 coins, consisting of nickels and dimes, is \$5.95. How many of each kind are there?

10. A family invested \$5000, part at 10% and the remainder at 10.5%. The annual income from both investments is \$517. What is the amount invested at each rate?

[5.4, i]   Solve, using matrices. If there is more than one solution, list three of them.

11. $x + 2y = 5,$
$\quad 2x - 5y = -8$

12. $3x + 4y + 2z = 3,$
$\quad 5x - 2y - 13z = 3,$
$\quad 4x + 3y - 3z = 6$

13. $3x + 5y + z = 0,$
$\quad 2x - 4y - 3z = 0,$
$\quad x + 3y + z = 0$

14. $\quad x + y + z + w = -2,$
$\quad -2x + 3y + 2z - 3w = 10,$
$\quad 3x + 2y - z + 2w = -12,$
$\quad 4x - y + z + 2w = 1$

[5.2, iii]   **15.** Find numbers a, b, and c such that the function $y = ax^2 + bx + c$ fits the data points $(0, 3)$, $(1, 0)$, and $(-1, 4)$. Then write the equation.

[5.5, i]   Evaluate.

16. $\begin{vmatrix} 1 & -2 \\ 3 & 4 \end{vmatrix}$

17. $\begin{vmatrix} \sqrt{3} & -5 \\ -3 & -\sqrt{3} \end{vmatrix}$

18. $\begin{vmatrix} -2 & -3 \\ 4 & -x \end{vmatrix}$

[5.5, iii]   Evaluate.

19. $\begin{vmatrix} 2 & -1 & 1 \\ 1 & 2 & -1 \\ 3 & 4 & -3 \end{vmatrix}$

20. $\begin{vmatrix} -5.8 & 7.5 & 4.6 \\ 0 & 2.2 & 8.9 \\ 0 & 0 & 1.3 \end{vmatrix}$

21. $\begin{vmatrix} 3a & 3b & 3c \\ 5a & 5b & 5c \\ d & e & f \end{vmatrix}$

[5.5, ii]   Solve for (x, y) using Cramer's rule.

22. $5x - 2y = 19,$
$\quad 7x + 3y = 15$

23. $ax - by = a^2,$
$\quad bx + ay = ab$

[5.5, iv]   **24.** Solve using Cramer's rule.

$$4x - 5y - z = -1,$$
$$3x + 2y - z = 4,$$
$$3x - 2y + z = 5$$

[5.6, ii]   **25.** Graph this system. Find the coordinates of any vertices formed.

$$2x + y \geq 9,$$
$$4x + 3y \geq 23,$$
$$x + 3y \geq 8,$$
$$x \geq 0,$$
$$y \geq 0$$

[5.7, i]   **26.** Maximize and minimize $T = 6x + 10y$ subject to

$$x + y \leq 10,$$
$$5x + 10y \leq 50,$$
$$x \geq 2,$$
$$y \geq 0.$$

27. You are about to take a test that contains questions of type A worth 7 points and questions of type B worth 12 points. The total number of questions worked must be at least 8. If you know that type A questions take 10 min and type B questions take 8 min and that the maximum time for the test is 80 min, how many of each type of question must you do to maximize your score? What is this maximum score?

28. One year a person invested a total of $40,000, part at 12%, part at 13%, and the rest at $14\frac{1}{2}$%. The total interest received on the investments was $5370. The interest received on the $14\frac{1}{2}$% investment was $1050 more than the interest received on the 13% investment. How much was invested at each rate?

Solve.

29. $\dfrac{2}{3x} + \dfrac{4}{5y} = 8,$

   $\dfrac{5}{4x} - \dfrac{3}{2y} = -6$

30. $\dfrac{3}{x} - \dfrac{4}{y} + \dfrac{1}{z} = -2,$

   $\dfrac{5}{x} + \dfrac{1}{y} - \dfrac{2}{z} = 1,$

   $\dfrac{7}{x} + \dfrac{3}{y} + \dfrac{2}{z} = 19$

Graph.

31. $|x| - |y| \leq 1$

32. $|xy| > 1$

6

MATRICES AND DETERMINANTS

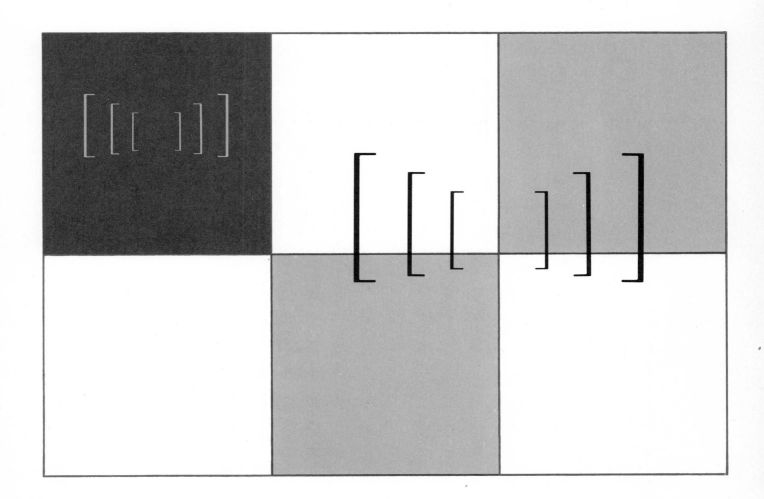

OBJECTIVES

You should be able to:

[i] Add, subtract, and multiply matrices when possible.
[ii] Write a matrix equation equivalent to a system of equations.

Find the dimensions of each matrix.

1. $\begin{bmatrix} -3 & 5 \\ 4 & \frac{1}{4} \\ -\pi & 0 \end{bmatrix}$

2. $\begin{bmatrix} -3 & 0 \\ 0 & 3 \end{bmatrix}$

3. $\begin{bmatrix} 1 & 2 & 3 \\ 0 & 1 & 8 \\ 0 & 0 & 1 \end{bmatrix}$

4. $[\pi \quad \sqrt{2}]$

5. $\begin{bmatrix} -5 \\ \pi \end{bmatrix}$

6. $[-3]$

7. Which of the above are square matrices?

8. Let

$A = \begin{bmatrix} 4 & -1 \\ 6 & -3 \end{bmatrix}$ and $B = \begin{bmatrix} -6 & -5 \\ 7 & 3 \end{bmatrix}$.

a) Find $A + B$.
b) Find $B + A$.

9. Add.

$\begin{bmatrix} -3 & -4 & -5 \\ 0 & 1 & -1 \end{bmatrix} + \begin{bmatrix} 4 & 5 & -5 \\ 2 & 3 & -2 \end{bmatrix}$

6.1 OPERATIONS ON MATRICES*

Dimensions of a Matrix

A matrix of m rows and n columns is called a matrix with *dimensions* $m \times n$ (*read "m by n"*).

Example 1 Find the dimensions of each matrix.

$$\begin{bmatrix} 2 & -3 & 4 \\ -1 & \frac{1}{2} & \pi \end{bmatrix}, \qquad \begin{bmatrix} -3 & 8 & 9 \\ \pi & -2 & 5 \\ -6 & 7 & 8 \end{bmatrix},$$

A 2 × 3 matrix A 3 × 3 matrix

$$[-3 \quad 4], \qquad \begin{bmatrix} 10 \\ -7 \end{bmatrix}.$$

A 1 × 2 matrix A 2 × 1 matrix

DO EXERCISES 1-7.

[i] Operations on Matrices

Matrix Addition

To add matrices, we add the corresponding members. For this to be possible, the matrices must have the same dimensions.

Examples

2. $\begin{bmatrix} -5 & 0 \\ 4 & \frac{1}{2} \end{bmatrix} + \begin{bmatrix} 6 & -3 \\ 2 & 3 \end{bmatrix} = \begin{bmatrix} -5+6 & 0-3 \\ 4+2 & \frac{1}{2}+3 \end{bmatrix}$

$= \begin{bmatrix} 1 & -3 \\ 6 & 3\frac{1}{2} \end{bmatrix}$

3. $\begin{bmatrix} 1 & 3 & 2 \\ -1 & 5 & 4 \\ 6 & 0 & 1 \end{bmatrix} + \begin{bmatrix} -1 & -2 & 1 \\ 1 & -2 & 2 \\ -3 & 1 & 0 \end{bmatrix} = \begin{bmatrix} 0 & 1 & 3 \\ 0 & 3 & 6 \\ 3 & 1 & 1 \end{bmatrix}$

Addition of matrices is both commutative and associative.

DO EXERCISES 8 AND 9.

Zero Matrices

A matrix having zeros for all of its members is called a *zero matrix* and is often denoted by **O**. When a zero matrix is added to another matrix of the same dimensions, that same matrix is obtained. Thus a zero matrix is an *additive identity*.

---

*This chapter is optional.

Example 4

$$\begin{bmatrix} 2 & -1 & 3 \\ 1 & 0 & -1 \end{bmatrix} + \begin{bmatrix} 0 & 0 & 0 \\ 0 & 0 & 0 \end{bmatrix} = \begin{bmatrix} 2 & -1 & 3 \\ 1 & 0 & -1 \end{bmatrix}$$

DO EXERCISE 10.

Inverses and Subtraction

To subtract matrices, we subtract the corresponding members. Of course, the matrices must have the same dimensions for this to be possible.

Example 5

$$\begin{bmatrix} 1 & 2 \\ -2 & 0 \\ -3 & -1 \end{bmatrix} - \begin{bmatrix} 1 & -1 \\ 1 & 3 \\ 2 & 3 \end{bmatrix} = \begin{bmatrix} 0 & 3 \\ -3 & -3 \\ -5 & -4 \end{bmatrix}$$

DO EXERCISES 11 AND 12.

The additive inverse of a matrix can be obtained by replacing each member by its additive inverse. Of course, when two matrices that are inverses of each other are added, a zero matrix is obtained.

Example 6

$$\begin{bmatrix} 1 & 0 & 2 \\ 3 & -1 & 5 \end{bmatrix} + \begin{bmatrix} -1 & 0 & -2 \\ -3 & 1 & -5 \end{bmatrix} = \begin{bmatrix} 0 & 0 & 0 \\ 0 & 0 & 0 \end{bmatrix}$$
$$\mathbf{A} \qquad + \qquad (-\mathbf{A}) \qquad = \qquad \mathbf{O}.$$

DO EXERCISES 13 AND 14.

With numbers, we can subtract by adding an inverse. This is also true of matrices. If we denote matrices by **A** and **B** and an additive inverse by −**B**, this fact can be stated as follows:

$$\mathbf{A} - \mathbf{B} = \mathbf{A} + (-\mathbf{B}).$$

Example 7

$$\begin{bmatrix} 3 & -1 \\ -2 & 4 \end{bmatrix} - \begin{bmatrix} 2 & 1 \\ 3 & -2 \end{bmatrix} = \begin{bmatrix} 1 & -2 \\ -5 & 6 \end{bmatrix}$$
$$\mathbf{A} \qquad - \qquad \mathbf{B}$$

$$\begin{bmatrix} 3 & -1 \\ -2 & 4 \end{bmatrix} + \begin{bmatrix} -2 & -1 \\ -3 & 2 \end{bmatrix} = \begin{bmatrix} 1 & -2 \\ -5 & 6 \end{bmatrix}$$
$$\mathbf{A} \qquad + \qquad (-\mathbf{B})$$

DO EXERCISE 15.

10. Let

$$\mathbf{A} = \begin{bmatrix} 4 & -3 \\ 5 & 8 \end{bmatrix} \quad \text{and} \quad \mathbf{O} = \begin{bmatrix} 0 & 0 \\ 0 & 0 \end{bmatrix}.$$

a) Find **A** + **O**.
b) Find **O** + **A**.

Subtract.

11. $\begin{bmatrix} 1 & 3 & -2 \\ 4 & 0 & 5 \end{bmatrix} - \begin{bmatrix} 2 & -1 & 5 \\ 6 & 4 & -3 \end{bmatrix}$

12. $\begin{bmatrix} 1 & 2 \\ 4 & 1 \\ -5 & 4 \end{bmatrix} - \begin{bmatrix} 7 & -4 \\ 3 & 5 \\ 2 & -1 \end{bmatrix}$

13. Find the additive inverse.

$\begin{bmatrix} 2 & -1 & 5 \\ 6 & 4 & -3 \end{bmatrix}$

14. Add.

$\begin{bmatrix} 2 & -1 & 5 \\ 6 & 4 & -3 \end{bmatrix} + \begin{bmatrix} -2 & 1 & -5 \\ -6 & -4 & 3 \end{bmatrix}$

15. Add. Compare with Exercise 11.

$\begin{bmatrix} 1 & 3 & -2 \\ 4 & 0 & 5 \end{bmatrix} + \begin{bmatrix} -2 & 1 & -5 \\ -6 & -4 & 3 \end{bmatrix}$

Compute these products.

16. $5\begin{bmatrix} 1 & -2 & x \\ 4 & y & 1 \\ 0 & -5 & x^2 \end{bmatrix}$

17. $t\begin{bmatrix} 1 & -1 & 4 & x \\ y & 3 & -2 & y \\ 1 & 4 & -5 & y \end{bmatrix}$

Multiplying Matrices and Numbers

We define a product of a matrix and a number.

DEFINITION

The product of a number k and a matrix A is the matrix, denoted kA, obtained by multiplying each number in A by the number k.

Examples Let

$$A = \begin{bmatrix} -3 & 0 \\ 4 & 5 \end{bmatrix}.$$

Then:

8.
$$3A = 3\begin{bmatrix} -3 & 0 \\ 4 & 5 \end{bmatrix} = \begin{bmatrix} -9 & 0 \\ 12 & 15 \end{bmatrix}$$

9.
$$(-1)A = -1\begin{bmatrix} -3 & 0 \\ 4 & 5 \end{bmatrix} = \begin{bmatrix} 3 & 0 \\ -4 & -5 \end{bmatrix}.$$

DO EXERCISES 16 AND 17.

Products of Matrices

We do not multiply two matrices by multiplying their corresponding members. The motivation for defining matrix products comes from systems of equations.

Let us begin by considering one equation,

$$3x + 2y - 2z = 4.$$

We will write the coefficients on the left side in a 1×3 matrix (a *row* matrix) and the variables in a 3×1 matrix (a *column* matrix). The 4 on the right is written in a 1×1 matrix:

$$\begin{bmatrix} 3 & 2 & -2 \end{bmatrix}\begin{bmatrix} x \\ y \\ z \end{bmatrix} = [4].$$

We can return to our original equation by multiplying the members of the row matrix by those of the column matrix, and adding:

$$\begin{bmatrix} 3 & 2 & -2 \end{bmatrix}\begin{bmatrix} x \\ y \\ z \end{bmatrix} = [3x + 2y - 2z].$$

We define multiplication accordingly. In this special case, we have a row matrix **A** and a column matrix **B**. Their product **AB** is a 1×1 matrix, having the single member 4 (also called $3x + 2y - 2z$).

Example 10 Find the product of these matrices.

$$[3 \quad 2 \quad -1] \begin{bmatrix} 1 \\ -2 \\ 3 \end{bmatrix} = [3 \cdot 1 + 2(-2) + (-1) \cdot 3] = [-4]$$

DO EXERCISE 18.

Let us continue by considering a system of equations:

$$\begin{aligned} 3x + 2y - 2z &= 4, \\ 2x - y + 5z &= 3, \\ -x + y + 4z &= 7. \end{aligned}$$

Consider the following matrices:

$$\begin{bmatrix} 3 & 2 & -2 \\ 2 & -1 & 5 \\ -1 & 1 & 4 \end{bmatrix} \begin{bmatrix} x \\ y \\ z \end{bmatrix} \begin{bmatrix} 4 \\ 3 \\ 7 \end{bmatrix}.$$
$$\quad\quad \mathbf{A} \quad\quad \mathbf{X} \quad \mathbf{B}$$

If we multiply the first row of **A** by the (only) column of **X**, as we did above, we get $3x + 2y - 2z$. If we multiply the second row of **A** by the column in **X**, in the same way, we get the following:

$$[2 \quad -1 \quad 5] \begin{bmatrix} x \\ y \\ z \end{bmatrix} = 2x - y + 5z.$$

Note that the first members are multiplied, the second members are multiplied, the third members are multiplied, and the results are added, to get the single number $2x - y + 5z$. What do we get when we multiply the third row of **A** by the column in **X**?

$$[-1 \quad 1 \quad 4] \begin{bmatrix} x \\ y \\ z \end{bmatrix} = -x + y + 4z$$

We define the product **AX** to be the column matrix

$$\begin{bmatrix} 3x + 2y - 2z \\ 2x - y + 5z \\ -x + y + 4z \end{bmatrix}.$$

Now consider this matrix equation:

$$\begin{bmatrix} 3x + 2y - 2z \\ 2x - y + 5z \\ -x + y + 4z \end{bmatrix} = \begin{bmatrix} 4 \\ 3 \\ 7 \end{bmatrix}.$$

Equality for matrices is the same as for numbers, i.e., a sentence such as $a = b$ says that a and b are two names for the same thing. Thus if the above matrix equation is true, the "two" matrices are really the same one. This means that $3x + 2y - 2z$ is 4, $2x - y + 5z$ is 3, and

18. Multiply.

$$[4 \quad -2 \quad 3] \begin{bmatrix} 2 \\ 3 \\ -5 \end{bmatrix}$$

19. Multiply.

$$\begin{bmatrix} 1 & 4 & 2 \\ -1 & 6 & 3 \\ 3 & 2 & -1 \\ 5 & 0 & 2 \end{bmatrix} \begin{bmatrix} 2 \\ 1 \\ 3 \end{bmatrix}$$

$-x + y + 4z$ is 7, or that

$$3x + 2y - 2z = 4,$$
$$2x - y + 5z = 3,$$

and

$$-x + y + 4z = 7.$$

Thus the matrix equation $\mathbf{AX} = \mathbf{B}$ is equivalent to the original system of equations.

Example 11 Multiply.

$$\begin{bmatrix} 3 & 1 & -1 \\ 1 & 2 & 2 \\ -1 & 0 & 5 \\ 4 & 1 & 2 \end{bmatrix} \begin{bmatrix} 1 \\ 2 \\ 1 \end{bmatrix} = \begin{bmatrix} 3 \cdot 1 + 1 \cdot 2 - 1 \cdot 1 \\ 1 \cdot 1 + 2 \cdot 2 + 2 \cdot 1 \\ -1 \cdot 1 + 0 \cdot 2 + 5 \cdot 1 \\ 4 \cdot 1 + 1 \cdot 2 + 2 \cdot 1 \end{bmatrix} = \begin{bmatrix} 4 \\ 7 \\ 4 \\ 8 \end{bmatrix}$$

DO EXERCISE 19.

In all the examples discussed so far, the second matrix had only one column. If the second matrix has more than one column, we treat it in the same way when multiplying that we treated the single column. The product matrix will have as many columns as the second matrix.

Example 12 Multiply (compare with Example 11).

20. Multiply.

$$\begin{bmatrix} 4 & 1 & 2 \\ -3 & 2 & 3 \\ 2 & 0 & 5 \\ 3 & 1 & 4 \end{bmatrix} \begin{bmatrix} 1 & 4 \\ 2 & 0 \\ -3 & 5 \end{bmatrix}$$

$$\underbrace{\begin{bmatrix} 3 & 1 & -1 \\ 1 & 2 & 2 \\ -1 & 0 & 5 \\ 4 & 1 & 2 \end{bmatrix}}_{\mathbf{A}} \underbrace{\begin{bmatrix} 1 & 0 \\ 2 & 1 \\ 1 & 3 \end{bmatrix}}_{\mathbf{B}} = \begin{bmatrix} 4 & 3 \cdot 0 + 1 \cdot 1 + (-1)3 \\ 7 & 1 \cdot 0 + 2 \cdot 1 + 2 \cdot 3 \\ 4 & -1 \cdot 0 + 0 \cdot 1 + 5 \cdot 3 \\ 8 & 4 \cdot 0 + 1 \cdot 1 + 2 \cdot 3 \end{bmatrix} = \begin{bmatrix} 4 & -2 \\ 7 & 8 \\ 4 & 15 \\ 8 & 7 \end{bmatrix}$$

Same as in Example 11

The rows of **A** multiplied by the second column of **B**

DO EXERCISE 20.

Example 13 Multiply.

$$\begin{bmatrix} 3 & 1 & -1 \\ 2 & 0 & 3 \end{bmatrix} \begin{bmatrix} 1 & 4 & 6 \\ 3 & -1 & 9 \\ 2 & 5 & 1 \end{bmatrix}$$

$$= \begin{bmatrix} 3 \cdot 1 + 1 \cdot 3 - 1 \cdot 2 & 3 \cdot 4 + 1(-1) - 1 \cdot 5 & 3 \cdot 6 + 1 \cdot 9 - 1 \cdot 1 \\ 2 \cdot 1 + 0 \cdot 3 + 3 \cdot 2 & 2 \cdot 4 + 0 \cdot (-1) + 3 \cdot 5 & 2 \cdot 6 + 0 \cdot 9 + 3 \cdot 1 \end{bmatrix}$$

$$= \begin{bmatrix} 4 & 6 & 26 \\ 8 & 23 & 15 \end{bmatrix}$$

If matrix A has n columns and matrix B has n rows, then we can compute the product AB, regardless of the other dimensions. The product will have as many rows as A and as many columns as B.

CAUTION! **Given any two matrices A and B, you may or may not be able to add, subtract, or multiply them. A + B and A − B exist only when the dimensions are the same. AB exists only when the number of columns in A is the same as the number of rows in B.**

For example, consider the matrices

$$A = \begin{bmatrix} 3 & 1 & -1 \\ 2 & 0 & 3 \end{bmatrix} \quad \text{and} \quad B = \begin{bmatrix} 1 & 4 & 6 \\ 3 & -1 & 9 \\ 2 & 5 & 1 \end{bmatrix}.$$

A + B and **A − B** do not exist because the dimensions of **A** and **B** are *not* the same. **AB** does exist because the number of columns in **A**, 3, is the same as the number of rows in **B**, 3. **AB** is given in Example 13. But **BA** does *not* exist because the number of columns in **B**, 3, is not the same as the number of rows in **A**, 2. In this context it can also be pointed out that matrix multiplication is not commutative, **AB** exists, and **BA** does not, so **AB ≠ BA**.

DO EXERCISES 21–24.

[ii] Equivalent Matrix Equations

For later purposes it is important to be able to write a matrix equation equivalent to a system of equations.

Example 14 Write a matrix equation equivalent to this system of equations:

$$\begin{aligned} 4x + 2y - z &= 3, \\ 9x + z &= 5, \\ 4x + 5y - 2z &= 1, \\ x + y + z &= 0. \end{aligned}$$

We write the coefficients on the left in a matrix. We write the product of that matrix by the column matrix containing the variables, and set the result equal to the column matrix containing the constants on the right:

$$\begin{bmatrix} 4 & 2 & -1 \\ 9 & 0 & 1 \\ 4 & 5 & -2 \\ 1 & 1 & 1 \end{bmatrix} \begin{bmatrix} x \\ y \\ z \end{bmatrix} = \begin{bmatrix} 3 \\ 5 \\ 1 \\ 0 \end{bmatrix}$$

DO EXERCISE 25.

A Summary of Properties of Square Matrices

We now list a summary of some of the properties of square matrices of the same dimensions whose elements are real numbers. We make the restriction of square matrices so all additions and multiplications are possible. Some of the proofs will be considered in the exercise set. Note that not all the field properties hold.

---

21. Find **AB** and **BA** if possible.

$$A = \begin{bmatrix} -2 & 4 & 0 \\ -3 & 0 & -8 \end{bmatrix}$$

$$B = \begin{bmatrix} -1 & -2 & -3 \\ 0 & 1 & 0 \\ 4 & 5 & 2 \end{bmatrix}$$

22. Multiply.

$$[4 \quad 1 \quad 0 \quad 2] \begin{bmatrix} 1 & 0 & 1 \\ 2 & -1 & 0 \\ 3 & 5 & 1 \\ 1 & 3 & 0 \end{bmatrix}$$

23. Find **AB** and **BA** and compare.

$$A = \begin{bmatrix} -8 & 3 \\ -4 & 4 \end{bmatrix}$$

$$B = \begin{bmatrix} 1 & -4 \\ 2 & 0 \end{bmatrix}$$

24. Find **AI** and **IA**. Comment.

$$A = \begin{bmatrix} 3 & 2 \\ -1 & 5 \end{bmatrix}$$

$$I = \begin{bmatrix} 1 & 0 \\ 0 & 1 \end{bmatrix}$$

25. Write a matrix equation equivalent to this system of equations.

$$\begin{aligned} 3x + 4y - 2z &= 5, \\ 2x - 2y + 5z &= 3, \\ 6x + 7y - z &= 0 \end{aligned}$$

THEOREM 1

For any square matrices A, B, and C of the same dimensions, the following hold:

COMMUTATIVITY. $A + B = B + A$.

ASSOCIATIVITY. $A + (B + C) = (A + B) + C$, $A(BC) = (AB)C$.

IDENTITY. There exists a unique matrix O, such that

$$A + O = O + A = A.$$

INVERSES. There exists a unique matrix $-A$, such that

$$A + (-A) = -A + A = O.$$

DISTRIBUTIVITY. $A(B + C) = AB + AC$.

For any square matrices A and B, of the same dimensions, and any real numbers k and m,

$$k(A + B) = kA + kB,$$

$$(k + m)A = kA + mA,$$

and

$$(km)A = k(mA).$$

CAUTION! Note that, even with these restrictions, matrix multiplication is still *not* commutative. For example, let

$$A = \begin{bmatrix} 1 & 0 \\ 2 & 0 \end{bmatrix} \quad \text{and} \quad B = \begin{bmatrix} 3 & 4 \\ 0 & 0 \end{bmatrix}.$$

Then

$$AB = \begin{bmatrix} 3 & 4 \\ 6 & 8 \end{bmatrix} \quad \text{and} \quad BA = \begin{bmatrix} 11 & 0 \\ 0 & 0 \end{bmatrix},$$

so $AB \neq BA$.

EXERCISE SET 6.1

[i] For Exercises 1–16, let

$$A = \begin{bmatrix} 1 & 2 \\ 4 & 3 \end{bmatrix}, \quad B = \begin{bmatrix} -3 & 5 \\ 2 & -1 \end{bmatrix}, \quad C = \begin{bmatrix} 1 & -1 \\ -1 & 1 \end{bmatrix}, \quad D = \begin{bmatrix} 1 & 1 \\ 1 & 1 \end{bmatrix},$$

$$E = \begin{bmatrix} 1 & 3 \\ 2 & 6 \end{bmatrix}, \quad F = \begin{bmatrix} 3 & 3 \\ -1 & -1 \end{bmatrix}, \quad O = \begin{bmatrix} 0 & 0 \\ 0 & 0 \end{bmatrix}, \quad \text{and} \quad I = \begin{bmatrix} 1 & 0 \\ 0 & 1 \end{bmatrix}.$$

Find.

1. $A + B$	2. $B + A$	3. $E + O$	4. $2A$
5. $3F$	6. $(-1)D$	7. $3F + 2A$	8. $A - B$
9. $B - A$	10. AB	11. BA	12. OF
13. CD	14. EF	15. AI	16. IA

In Exercises 17–20, let

$$A = \begin{bmatrix} 1 & 0 & -2 \\ 0 & -1 & 3 \\ 3 & 2 & 4 \end{bmatrix}, \quad B = \begin{bmatrix} -1 & -2 & 5 \\ 1 & 0 & -1 \\ 2 & -3 & 1 \end{bmatrix}, \quad C = \begin{bmatrix} -2 & 9 & 6 \\ -3 & 3 & 4 \\ 2 & -2 & 1 \end{bmatrix}, \quad \text{and} \quad I = \begin{bmatrix} 1 & 0 & 0 \\ 0 & 1 & 0 \\ 0 & 0 & 1 \end{bmatrix}.$$

Find.

17. AB   **18. BA**   **19. CI**   **20. IC**

Multiply.

21. $\begin{bmatrix} -3 & 2 \end{bmatrix} \begin{bmatrix} 4 \\ -2 \end{bmatrix}$

22. $\begin{bmatrix} -2 & 0 & 4 \end{bmatrix} \begin{bmatrix} 8 \\ -6 \\ \frac{1}{2} \end{bmatrix}$

23. $\begin{bmatrix} -5 & 1 & 2 \end{bmatrix} \begin{bmatrix} 1 & 3 \\ -1 & 0 \\ 4 & -2 \end{bmatrix}$

24. $\begin{bmatrix} -3 & 2 \\ 0 & 1 \\ -4 & 5 \end{bmatrix} \begin{bmatrix} 4 \\ 2 \end{bmatrix}$

[ii] Write a matrix equation equivalent to each of the following systems of equations.

25. $3x - 2y + 4z = 17,$
$\quad 2x + y - 5z = 13$

26. $3x + 2y + 5z = 9,$
$\quad 4x - 3y + 2z = 10$

27. $\quad x - y + 2z - 4w = 12,$
$\quad 2x - y - z + w = 0,$
$\quad x + 4y - 3z - w = 1,$
$\quad 3x + 5y - 7z + 2w = 9$

28. $\quad 2x + 4y - 5z + 12w = 2,$
$\quad 4x - y + 12z - w = 5,$
$\quad -x + 4y \qquad + 2w = 13,$
$\quad 2x + 10y + z \qquad = 5$

Compute.

29. ▦ $\begin{bmatrix} 3.61 & -2.14 & 16.7 \\ -4.33 & 7.03 & 12.9 \\ 5.82 & -6.95 & 2.34 \end{bmatrix} \begin{bmatrix} 3.05 & 0.402 & -1.34 \\ 1.84 & -1.13 & 0.024 \\ -2.83 & 2.04 & 8.81 \end{bmatrix}$

30. ▦ $\begin{bmatrix} -1.23 & 4.51 & -17.4 \\ 61.2 & -8.81 & 0.123 \\ 14.14 & 6.92 & -14.4 \end{bmatrix} \begin{bmatrix} 4.24 & 16.1 & 41.3 \\ 0.146 & -6.06 & -18.9 \\ -8.43 & 1.12 & 0.0245 \end{bmatrix}$

☆ ───

For Exercises 31 and 32, let

$$\mathbf{A} = \begin{bmatrix} -1 & 0 \\ 2 & 1 \end{bmatrix}, \quad \text{and} \quad \mathbf{B} = \begin{bmatrix} 1 & -1 \\ 0 & 2 \end{bmatrix}.$$

31. Show that $(\mathbf{A} + \mathbf{B})(\mathbf{A} - \mathbf{B}) \neq \mathbf{A}^2 - \mathbf{B}^2$, where

$$\mathbf{A}^2 = \mathbf{AA} \text{ and } \mathbf{B}^2 = \mathbf{BB}.$$

32. Show that $(\mathbf{A} + \mathbf{B})(\mathbf{A} + \mathbf{B}) \neq \mathbf{A}^2 + 2\mathbf{AB} + \mathbf{B}^2$.

★ ───

Let $\mathbf{A} = \begin{bmatrix} a_{11} & a_{12} \\ a_{21} & a_{22} \end{bmatrix}, \quad \mathbf{B} = \begin{bmatrix} b_{11} & b_{12} \\ b_{21} & b_{22} \end{bmatrix}, \quad \mathbf{C} = \begin{bmatrix} c_{11} & c_{12} \\ c_{21} & c_{22} \end{bmatrix}, \quad \mathbf{I} = \begin{bmatrix} 1 & 0 \\ 0 & 1 \end{bmatrix}.$

Prove the following.

33. $\mathbf{A} + \mathbf{B} = \mathbf{B} + \mathbf{A}$

34. $\mathbf{A} + (\mathbf{B} + \mathbf{C}) = (\mathbf{A} + \mathbf{B}) + \mathbf{C}$

35. $k(\mathbf{A} + \mathbf{B}) = k\mathbf{A} + k\mathbf{B}$

36. $(k + m)\mathbf{A} = k\mathbf{A} + m\mathbf{A}$

37. $\mathbf{A}(\mathbf{BC}) = (\mathbf{AB})\mathbf{C}$

38. $\mathbf{AI} = \mathbf{IA} = \mathbf{A}$

6.2 DETERMINANTS OF HIGHER ORDER

Further Notation

We shall define the determinant of any square matrix. To do this we need some new notation. The members, or elements, of a matrix will now be denoted by lower-case letters with two subscripts, as follows:

$$\mathbf{A} = \begin{bmatrix} a_{11} & a_{12} & a_{13} \\ a_{21} & a_{22} & a_{23} \\ a_{31} & a_{32} & a_{33} \end{bmatrix}.$$

OBJECTIVE

You should be able to:

[i] Find a specified element, a_{ij}, of a matrix, its minor, and its cofactor, and be able to expand its determinant about any row or column.

1. For the matrix of Example 1, find a_{11}, a_{13}, a_{22}, a_{31}, and a_{32}.

The element in the ith row and jth column is denoted a_{ij}. We may also name the above matrix **A** as

$$[a_{ij}].$$

Example 1 Consider

$$[a_{ij}] = \begin{bmatrix} -8 & 0 & 6 \\ 4 & -6 & 7 \\ -1 & -3 & 5 \end{bmatrix}.$$

Find a_{12}, a_{23}, and a_{33}.

$$a_{12} = 0 \qquad \text{(This is the intersection of the first row and second column.)}$$

$$a_{23} = 7 \qquad \text{(This is the intersection of the second row and third column.)}$$

$$a_{33} = 5 \qquad \text{(This is the intersection of the third row and third column.)}$$

DO EXERCISE 1.

Minors

We shall restrict our attention to square matrices.

DEFINITION

In a matrix $[a_{ij}]$, the *minor M_{ij}* of an element a_{ij} is the determinant of the matrix found by deleting the ith row and jth column.

Note that a minor is a certain determinant, hence is a number.

Example 2 In the matrix given in Example 1, find M_{11} and M_{23}.

To find M_{11} we delete the first row and the first column:

$$\begin{bmatrix} -8 & 0 & 6 \\ 4 & -6 & 7 \\ -1 & -3 & 5 \end{bmatrix}.$$

We calculate the determinant of the matrix formed by the remaining elements:

$$M_{11} = \begin{vmatrix} -6 & 7 \\ -3 & 5 \end{vmatrix} = (-6) \cdot 5 - (-3) \cdot 7 = -30 - (-21)$$
$$= -30 + 21 = -9.$$

To find M_{23} we delete the second row and the third column:

$$\begin{bmatrix} -8 & 0 & 6 \\ 4 & -6 & 7 \\ -1 & -3 & 5 \end{bmatrix}.$$

We calculate the determinant of the matrix formed by the remaining elements:

$$M_{23} = \begin{vmatrix} -8 & 0 \\ -1 & -3 \end{vmatrix} = -8(-3) - (-1)0 = 24.$$

DO EXERCISE 2.

DEFINITION

In a matrix $[a_{ij}]$, the *cofactor* of an element a_{ij} is denoted A_{ij} and is given by

$$A_{ij} = (-1)^{i+j}M_{ij},$$

where M_{ij} is the minor of a_{ij}. In other words, to find the cofactor of an element, find its minor and multiply it by $(-1)^{i+j}$.

Note that $(-1)^{i+j}$ is 1 if $i + j$ is even and is -1 if $i + j$ is odd. Thus in calculating a cofactor, find the minor. Then add the number of the row and the number of the column. The sum is $i + j$. If this sum is odd, change the sign of the minor. If this sum is even, leave the minor as is.*

Example 3 In the matrix given in Example 1, find A_{11} and A_{23}.

In Example 2 we found that $M_{11} = -9$. In A_{11} the sum of the subscripts is even, so

$$A_{11} = -9.$$

In Example 2 we found that $M_{23} = 24$. In A_{23} the sum of the subscripts is odd, so

$$A_{23} = -24.$$

DO EXERCISE 3.

Evaluating Determinants Using Cofactors

Consider the matrix **A** given by

$$\mathbf{A} = \begin{bmatrix} a_{11} & a_{12} & a_{13} \\ a_{21} & a_{22} & a_{23} \\ a_{31} & a_{32} & a_{33} \end{bmatrix}.$$

The determinant of the matrix, denoted $|\mathbf{A}|$, can be found as follows:

$$|\mathbf{A}| = a_{11}\mathbf{A}_{11} + a_{21}\mathbf{A}_{21} + a_{31}\mathbf{A}_{31}.$$

2. For the matrix of Example 1, find the minors M_{22}, M_{31}, and M_{13}.

3. For the matrix of Example 1, find the cofactors A_{22}, A_{32}, and A_{13}.

---

*$(-1)^{i+j}$ can also be found by counting through the matrix horizontally and/or vertically, starting with a_{11} and $(+)$, saying $+$, $-$, $+$, $-$, and so on, until you come to a_{ij}.

Start here $(+)$

$$\begin{bmatrix} a_{11}^+ \rightarrow {}^- \rightarrow {}^+ \rightarrow {\downarrow}^- \\ {\downarrow}^+ \\ {\downarrow}^- \\ a_{ij}^+ \end{bmatrix}$$ The path does not matter.

4. Consider the matrix of Example 1. Find $|\mathbf{A}|$ by expanding about the second column.

That is, multiply each element of the first column by its cofactor and add:

$$|\mathbf{A}| = a_{11} \cdot \begin{vmatrix} a_{22} & a_{23} \\ a_{32} & a_{33} \end{vmatrix} - a_{21} \cdot \begin{vmatrix} a_{12} & a_{13} \\ a_{32} & a_{33} \end{vmatrix} + a_{31} \cdot \begin{vmatrix} a_{12} & a_{13} \\ a_{22} & a_{23} \end{vmatrix}.$$

The last line is equivalent to the definition on p. 224. It can be shown that $|\mathbf{A}|$ can be found by picking *any* row or column, multiplying each element by its cofactor, and adding. This is called *expanding* the determinant about a row or column. We just expanded $|\mathbf{A}|$ about the first column. We now define the determinant of any square matrix.

DEFINITION

For any $n \times n$ square matrix A ($n > 1$), we define the *determinant* of A, $|\mathbf{A}|$, to be that number found as follows. Pick any row or column. Multiply each element in that row or column by its cofactor and add the results. The determinant of a 1×1 matrix is simply the element of the matrix.

5. Consider the matrix of Example 1. Find $|\mathbf{A}|$ by expanding about the third column.

The value of a determinant will be the same no matter how it is evaluated.

Example 4 Consider the matrix **A** of Example 1. Find $|\mathbf{A}|$ by expanding about the third row.

$$|\mathbf{A}| = (-1)A_{31} + (-3)A_{32} + 5A_{33}$$

$$= (-1)(-1)^{3+1} \cdot \begin{vmatrix} 0 & 6 \\ -6 & 7 \end{vmatrix} + (-3)(-1)^{3+2} \cdot \begin{vmatrix} -8 & 6 \\ 4 & 7 \end{vmatrix}$$

$$+ 5(-1)^{3+3} \cdot \begin{vmatrix} -8 & 0 \\ 4 & -6 \end{vmatrix}$$

$$= (-1) \cdot 1 \cdot [0 \cdot 7 - (-6)6] + (-3)(-1)[-8 \cdot 7 - 4 \cdot 6]$$
$$+ 5 \cdot 1 \cdot [-8(-6) - 4 \cdot 0]$$

$$= -[36] + 3[-80] + 5[48]$$

$$= -36 - 240 + 240$$

$$= -36$$

DO EXERCISES 4 AND 5.

EXERCISE SET 6.2

[i] Use the following matrix for Exercises 1–10.

$$\mathbf{A} = \begin{bmatrix} 7 & -4 & -6 \\ 2 & 0 & -3 \\ 1 & 2 & -5 \end{bmatrix}$$

1. Find a_{11}, a_{32}, and a_{22}.

2. Find a_{13}, a_{31}, and a_{23}.

3. Find M_{11}, M_{32}, and M_{22}.

4. Find M_{13}, M_{31}, and M_{23}.

5. Find A_{11}, A_{32}, and A_{22}.

6. Find A_{13}, A_{31}, and A_{23}.

7. Expand $|\mathbf{A}|$ about the second row.

8. Expand $|\mathbf{A}|$ about the second column.

9. Expand $|\mathbf{A}|$ about the third column.

10. Expand $|\mathbf{A}|$ about the first row.

Use the following matrix for Exercises 11–16.

$$A = \begin{bmatrix} 1 & 0 & 0 & -2 \\ 4 & 1 & 0 & 0 \\ 5 & 6 & 7 & 8 \\ -2 & -3 & -1 & 0 \end{bmatrix}$$

11. Find M_{41} and M_{33}.

12. Find M_{12} and M_{44}.

13. Find A_{24} and A_{43}.

14. Find A_{22} and A_{34}.

15. Expand $|A|$ about the first row.

16. Expand $|A|$ about the third column.

Evaluate.

17. $\begin{vmatrix} 5 & -4 & 2 & -2 \\ 3 & -3 & -4 & 7 \\ -2 & 3 & 2 & 4 \\ -8 & 9 & 5 & -5 \end{vmatrix}$

18. $\begin{vmatrix} x & p & q & r \\ 0 & y & s & t \\ 0 & 0 & z & u \\ 0 & 0 & 0 & w \end{vmatrix}$

☆ —————————————————————

19. If a line contains the points (x_1, y_1) and (x_2, y_2), an equation of the line can be written as follows:

$$\begin{vmatrix} x & y & 1 \\ x_1 & y_1 & 1 \\ x_2 & y_2 & 1 \end{vmatrix} = 0.$$

Prove this.

20. Show that three points (x_1, y_1), (x_2, y_2), and (x_3, y_3) are collinear (on the same straight line) if and only if

$$\begin{vmatrix} x_1 & y_1 & 1 \\ x_2 & y_2 & 1 \\ x_3 & y_3 & 1 \end{vmatrix} = 0.$$

6.3 PROPERTIES OF DETERMINANTS

[i] Evaluation of Certain Determinants

We can simplify the evaluation of certain determinants using the following properties.

THEOREM 2

If a row (or column) of a matrix A has all elements 0, then $|A| = 0$.

Proof. Just expand the determinant about the row (or column) that has all 0's.

Examples Evaluate.

1. $\begin{vmatrix} 0 & 6 \\ 0 & 7 \end{vmatrix} = 0$

2. $\begin{vmatrix} 4 & 5 & -7 \\ 0 & 0 & 0 \\ -3 & 9 & 6 \end{vmatrix} = 0$

DO EXERCISES 1 AND 2.

THEOREM 3

If two rows (or columns) of a matrix A are interchanged to obtain a new matrix B, then $|A| = -|B|$.

OBJECTIVES

You should be able to:

[i] Use properties of determinants to simplify their evaluation.

[ii] Factor certain determinants.

Evaluate.

1. $\begin{vmatrix} 2 & 3 \\ 0 & 0 \end{vmatrix}$

2. $\begin{vmatrix} 2 & 3 & 0 \\ -3 & 7 & 0 \\ 2 & 4 & 0 \end{vmatrix}$

3. $\mathbf{A} = \begin{bmatrix} 5 & -9 \\ -2 & 0 \end{bmatrix}$, $\quad \mathbf{B} = \begin{bmatrix} -2 & 0 \\ 5 & -9 \end{bmatrix}$

a) Find $|\mathbf{A}|$ and $|\mathbf{B}|$.

b) Why does $|\mathbf{A}| = -|\mathbf{B}|$?

4. $\mathbf{C} = \begin{bmatrix} -2 & 3 & 4 \\ 4 & 5 & 6 \\ -9 & -1 & 0 \end{bmatrix}$

$\mathbf{D} = \begin{bmatrix} 4 & 3 & -2 \\ 6 & 5 & 4 \\ 0 & -1 & -9 \end{bmatrix}$

a) Find $|\mathbf{C}|$ and $|\mathbf{D}|$.

b) Why does $|\mathbf{C}| = -|\mathbf{D}|$?

5. Evaluate.

$\begin{vmatrix} -1 & 3 & -7 \\ -1 & 2 & -6 \\ -1 & 3 & -7 \end{vmatrix}$

Solve for x.

6. $\begin{vmatrix} -16 & 32 \\ -5 & -3 \end{vmatrix} = x \cdot \begin{vmatrix} 4 & -8 \\ -5 & -3 \end{vmatrix}$

7. $\begin{vmatrix} -3 & 12 & 2 \\ 5 & -6 & 3 \\ 0 & 18 & 5 \end{vmatrix}$

$= x \cdot \begin{vmatrix} -3 & 2 & 2 \\ 5 & -1 & 3 \\ 0 & 3 & 5 \end{vmatrix}$

Proof. Pick one of the rows (or columns) to be interchanged and expand $|\mathbf{A}|$ about that row. Use that same row to expand $|\mathbf{B}|$. For that row each $(-1)^{i+j}$ has changed signs, so $|\mathbf{A}| = -|\mathbf{B}|$.

Examples

3. $\begin{vmatrix} 6 & 7 & 8 \\ 4 & 1 & 2 \\ 2 & 9 & 0 \end{vmatrix} = -1 \cdot \begin{vmatrix} 6 & 8 & 7 \\ 4 & 2 & 1 \\ 2 & 0 & 9 \end{vmatrix}$ **4.** $\begin{vmatrix} -6 & 8 \\ 4 & -3 \end{vmatrix} = -1 \cdot \begin{vmatrix} 4 & -3 \\ -6 & 8 \end{vmatrix}$

DO EXERCISES 3 AND 4.

THEOREM 4

If two rows (or columns) of a matrix A are the same, then $|\mathbf{A}| = 0$.

Proof. Interchanging the rows (or columns) that are the same does not change $\mathbf{A}$. Thus by Theorem 3, $|\mathbf{A}| = -|\mathbf{A}|$. This is possible only when $|\mathbf{A}| = 0$.

Examples Evaluate.

5. $\begin{vmatrix} 6 & 7 & 8 \\ -2 & 6 & 5 \\ -2 & 6 & 5 \end{vmatrix} = 0$

6. $\begin{vmatrix} -5 & 4 & -5 \\ 3 & 7 & 3 \\ 0 & 12 & 0 \end{vmatrix} = 0$

DO EXERCISE 5.

THEOREM 5

If all the elements of a row (or column) of a matrix A are multiplied by k, $|\mathbf{A}|$ is multiplied by k. Or, if all the elements of a row (or column) of A have a common factor, we can factor it out of the determinant of A.

Examples

7. $\begin{vmatrix} 2 & 4 & 6 \\ -2 & 5 & 9 \\ 4 & -1 & -3 \end{vmatrix} = 3 \cdot \begin{vmatrix} 2 & 4 & 2 \\ -2 & 5 & 3 \\ 4 & -1 & -1 \end{vmatrix}$

8. $\begin{vmatrix} 10 & 25 \\ -4 & -7 \end{vmatrix} = 5 \cdot \begin{vmatrix} 2 & 5 \\ -4 & -7 \end{vmatrix}$

DO EXERCISES 6 AND 7.

Proof of Theorem 5. Expand the determinants about the row (or column) in question. The cofactors are the same and the k can be factored out. Consider the case of a 3×3 matrix and the second column. Let

$$\mathbf{A} = \begin{bmatrix} a_{11} & a_{12} & a_{13} \\ a_{21} & a_{22} & a_{23} \\ a_{31} & a_{32} & a_{33} \end{bmatrix} \quad \text{and} \quad \mathbf{B} = \begin{bmatrix} a_{11} & ka_{12} & a_{13} \\ a_{21} & ka_{22} & a_{23} \\ a_{31} & ka_{32} & a_{33} \end{bmatrix}.$$

Then

$$\begin{aligned} |\mathbf{B}| &= ka_{12}A_{12} + ka_{22}A_{22} + ka_{32}A_{32} \\ &= k(a_{12}A_{12} + a_{22}A_{22} + a_{32}A_{32}) \\ &= k|\mathbf{A}|. \end{aligned}$$

Example 9 Without expanding, find $|\mathbf{A}|$.

$$\mathbf{A} = \begin{bmatrix} -6 & 3 & 8 \\ 15 & -9 & -20 \\ -9 & -1 & 12 \end{bmatrix}$$

$$|\mathbf{A}| = (-3) \cdot \begin{vmatrix} 2 & 3 & 8 \\ -5 & -9 & -20 \\ 3 & -1 & 12 \end{vmatrix} \qquad \text{Factoring } -3 \text{ out of the first column}$$

$$= (-3)(4) \cdot \begin{vmatrix} 2 & 3 & 2 \\ -5 & -9 & -5 \\ 3 & -1 & 3 \end{vmatrix} \qquad \text{Factoring 4 out of the third column}$$

$$= 0 \qquad \text{By Theorem 4}$$

DO EXERCISE 8.

THEOREM 6

If each element in a row (or column) is multiplied by a number k, and the products are added to the corresponding elements of another row (or column), we do not change the value of the determinant.

Example 10 Find a determinant equal to the one on the left by adding three times the second column to the first column.

$$\begin{vmatrix} 0 & 1 & 2 \\ 4 & 5 & 6 \\ 7 & 8 & 9 \end{vmatrix} = \begin{vmatrix} 0 + 3(1) & 1 & 2 \\ 4 + 3(5) & 5 & 6 \\ 7 + 3(8) & 8 & 9 \end{vmatrix} = \begin{vmatrix} 3 & 1 & 2 \\ 19 & 5 & 6 \\ 31 & 8 & 9 \end{vmatrix}$$

Example 11 Find a determinant equal to the one on the left by adding two times the third row to the first row.

$$\begin{vmatrix} 0 & 1 & 2 \\ 4 & 5 & 6 \\ 7 & 8 & 9 \end{vmatrix} = \begin{vmatrix} 0 + 2(7) & 1 + 2(8) & 2 + 2(9) \\ 4 & 5 & 6 \\ 7 & 8 & 9 \end{vmatrix} = \begin{vmatrix} 14 & 17 & 20 \\ 4 & 5 & 6 \\ 7 & 8 & 9 \end{vmatrix}$$

8. Without expanding, find $|\mathbf{A}|$.

$$\mathbf{A} = \begin{bmatrix} 1 & 0 & -7 \\ -8 & 6 & -4 \\ 24 & -18 & 12 \end{bmatrix}$$

9. Find a determinant equal to this one by adding twice the first row to the second row.

$$\begin{vmatrix} -2 & 3 & 4 \\ 1 & 4 & -3 \\ 0 & 9 & 7 \end{vmatrix}$$

Proof of Theorem 6. We prove the theorem for the case of a 3×3 matrix where k times the first column has been added to the third column. Let

$$\mathbf{A} = \begin{bmatrix} a_{11} & a_{12} & a_{13} \\ a_{21} & a_{22} & a_{23} \\ a_{31} & a_{32} & a_{33} \end{bmatrix} \quad \text{and} \quad \mathbf{B} = \begin{bmatrix} a_{11} & a_{12} & ka_{11} + a_{13} \\ a_{21} & a_{22} & ka_{21} + a_{23} \\ a_{31} & a_{32} & ka_{31} + a_{33} \end{bmatrix}.$$

To show that $|\mathbf{A}| = |\mathbf{B}|$, we expand $|\mathbf{B}|$ about the third column:

$$|\mathbf{B}| = (ka_{11} + a_{13})A_{13} + (ka_{21} + a_{23})A_{23} + (ka_{31} + a_{33})A_{33}$$
$$= k(a_{11}A_{13} + a_{21}A_{23} + a_{31}A_{33})$$
$$\quad + (a_{13}A_{13} + a_{23}A_{23} + a_{33}A_{33})$$
$$= k(a_{11}A_{13} + a_{21}A_{23} + a_{31}A_{33}) + |\mathbf{A}|$$
$$= k \cdot \begin{vmatrix} a_{11} & a_{12} & a_{11} \\ a_{21} & a_{22} & a_{21} \\ a_{31} & a_{32} & a_{31} \end{vmatrix} + |\mathbf{A}|$$
$$= k(0) + |\mathbf{A}| \quad \text{By Theorem 4}$$
$$= |\mathbf{A}|.$$

DO EXERCISE 9.

We can use the properties of determinants to simplify their evaluation. We try to find another determinant where in a particular row or column one element is 1 and the rest are 0.

Example 12 Evaluate by first simplifying to a determinant where in one row or column one element is 1 and the rest are 0.

$$\begin{vmatrix} 6 & 2 & 3 \\ 6 & -1 & 5 \\ -2 & 3 & 1 \end{vmatrix}$$

We will try to get two 0's and a 1 in the third row. It already has a 1; that is why we picked the third row. We first factor a 2 out of column 1:

$$2 \cdot \begin{vmatrix} 3 & 2 & 3 \\ 3 & -1 & 5 \\ -1 & 3 & 1 \end{vmatrix}. \quad \text{Theorem 5}$$

Now multiply each element in column 3 by -3 and add the corresponding elements to column 2:

$$2 \cdot \begin{vmatrix} 3 & -7 & 3 \\ 3 & -16 & 5 \\ -1 & 0 & 1 \end{vmatrix}. \quad \text{Theorem 6}$$

Now add the elements in column 3 to the corresponding elements in column 1 (Theorem 6):

$$2 \cdot \begin{vmatrix} 6 & -7 & 3 \\ 8 & -16 & 5 \\ 0 & 0 & 1 \end{vmatrix}.$$

Now evaluate the determinant about the last row:

$$2 \cdot \left(0 \; - \; 0 \; + \; 1 \; \cdot \; \begin{vmatrix} 6 & -7 \\ 8 & -16 \end{vmatrix} \right) = 2 \cdot [6(-16) - 8(-7)] = -80.$$

DO EXERCISES 10 AND 11.

[ii] Factoring Certain Determinants

Example 13 Factor

$$\begin{vmatrix} 1 & x & x^2 \\ 1 & y & y^2 \\ 1 & z & z^2 \end{vmatrix}$$

$$\begin{vmatrix} 1 & x & x^2 \\ 1 & y & y^2 \\ 1 & z & z^2 \end{vmatrix} = \begin{vmatrix} 0 & x-y & x^2-y^2 \\ 1 & y & y^2 \\ 0 & z-y & z^2-y^2 \end{vmatrix}$$

By Theorem 6, adding −1 times the second row to the first row; and −1 times the second row to the third row

$$= (x-y)(z-y) \cdot \begin{vmatrix} 0 & 1 & x+y \\ 1 & y & y^2 \\ 0 & 1 & z+y \end{vmatrix}$$

By Theorem 5, factoring x − y out of the first row and z − y out of the third row

$$= (x-y)(z-y) \cdot \begin{vmatrix} 0 & 0 & x-z \\ 1 & y & y^2 \\ 0 & 1 & z+y \end{vmatrix}$$

By Theorem 6, adding −1 times the third row to the first row

$$= (x-y)(z-y)(x-z) \cdot \begin{vmatrix} 0 & 0 & 1 \\ 1 & y & y^2 \\ 0 & 1 & z+y \end{vmatrix}$$

By Theorem 5, factoring x − z out of the first row

$$= (x-y)(z-y)(x-z)$$

Expanding the determinant about the first row, we get 1.

DO EXERCISE 12.

Evaluate by first simplifying to a determinant where in one row or column one element is 1 and the rest are 0.

10. $\begin{vmatrix} 3 & -1 & 1 \\ 2 & 2 & -4 \\ 2 & 4 & 1 \end{vmatrix}$

11. $\begin{vmatrix} 5 & -4 & 2 & -2 \\ 3 & -3 & -4 & 7 \\ -2 & 3 & 2 & 4 \\ -8 & 9 & 5 & -5 \end{vmatrix}$

12. Factor.

$\begin{vmatrix} a^2 & b^2 & c^2 \\ a & b & c \\ 1 & 1 & 1 \end{vmatrix}$

EXERCISE SET 6.3

[i] Evaluate by first simplifying to a determinant where in one row or column one element is 1 and the rest are 0.

1. $\begin{vmatrix} -4 & 5 \\ 6 & 10 \end{vmatrix}$

2. $\begin{vmatrix} 3 & -9 \\ -2 & 4 \end{vmatrix}$

3. $\begin{vmatrix} 2 & 1 & 1 \\ 2 & -3 & -1 \\ -4 & 5 & 2 \end{vmatrix}$

4. $\begin{vmatrix} 1 & 2 & 4 \\ 2 & 3 & 5 \\ 3 & 1 & 6 \end{vmatrix}$

5. $\begin{vmatrix} 11 & -15 & 20 \\ 16 & 24 & -8 \\ 6 & 9 & 15 \end{vmatrix}$

6. $\begin{vmatrix} 4 & -24 & 15 \\ -3 & 18 & -6 \\ 5 & -4 & 3 \end{vmatrix}$

7. $\begin{vmatrix} -3 & 0 & 2 & 6 \\ 2 & 4 & 0 & -1 \\ -1 & 0 & -5 & 2 \\ 0 & -1 & -2 & -3 \end{vmatrix}$

8. $\begin{vmatrix} -2 & 1 & 0 & 5 \\ 3 & 0 & -4 & -2 \\ 4 & -6 & -8 & -1 \\ 8 & 0 & -2 & -3 \end{vmatrix}$

Find each determinant without expanding.

9. $\begin{vmatrix} x & y & z \\ 0 & 0 & 0 \\ p & q & r \end{vmatrix}$

10. $\begin{vmatrix} 5 & 5 & 5 \\ 3 & 3 & 3 \\ 2 & -7 & 8 \end{vmatrix}$

11. $\begin{vmatrix} 2a & t & -7a \\ 2b & u & -7b \\ 2c & v & -7c \end{vmatrix}$

12. $\begin{vmatrix} a & -1 & 4a \\ b & 2 & 4b \\ x & -3 & 4x \end{vmatrix}$

[ii] Factor.

13. $\begin{vmatrix} x^2 & x & 1 \\ y^2 & y & 1 \\ z^2 & z & 1 \end{vmatrix}$

14. $\begin{vmatrix} 1 & 1 & 1 \\ a & b & c \\ a^2 & b^2 & c^2 \end{vmatrix}$

15. $\begin{vmatrix} x & x^2 & x^3 \\ y & y^2 & y^3 \\ z & z^2 & z^3 \end{vmatrix}$

16. $\begin{vmatrix} 1 & 1 & 1 \\ a & b & c \\ a^3 & b^3 & c^3 \end{vmatrix}$

★

17. Consider a triangle with vertices (x_1, y_1), (x_2, y_2), and (x_3, y_3). The area of this triangle is the absolute value of

$$\frac{1}{2} \cdot \begin{vmatrix} x_1 & y_1 & 1 \\ x_2 & y_2 & 1 \\ x_3 & y_3 & 1 \end{vmatrix}.$$

Prove this. (*Hint:* Look at this drawing. The area of triangle *ABC* is the area of trapezoid *ABDE* plus the area of trapezoid *AEFC* minus the area of trapezoid *BDFC*.)

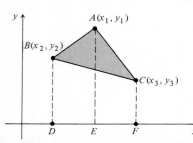

18. Prove that the lines $a_1x + b_1y = c_1$ and $a_2x + b_2y = c_2$ are parallel when

$$\begin{vmatrix} a_1 & b_1 \\ a_2 & b_2 \end{vmatrix} = 0$$

and either

$$\begin{vmatrix} c_1 & b_1 \\ c_2 & b_2 \end{vmatrix} \neq 0 \quad \text{or} \quad \begin{vmatrix} a_1 & c_1 \\ a_2 & c_2 \end{vmatrix} \neq 0.$$

OBJECTIVES

You should be able to:

[i] Calculate the inverse of a square matrix, if it exists, using the cofactor method.

[ii] Calculate the inverse of a square matrix, if it exists, using the Gauss–Jordan reduction method.

[iii] Use matrix inverses to solve systems of *n* equations in *n* variables.

1. Let

$$\mathbf{A} = \begin{bmatrix} a & b \\ c & d \end{bmatrix}$$

and

$$\mathbf{I} = \begin{bmatrix} 1 & 0 \\ 0 & 1 \end{bmatrix}.$$

a) Find **AI**.
b) Find **IA**.
c) Compare **AI** and **IA**.

6.4 INVERSES OF MATRICES

We use the symbol **I** to represent matrices of the type

$$\begin{bmatrix} 1 & 0 \\ 0 & 1 \end{bmatrix},$$

$$\begin{bmatrix} 1 & 0 & 0 \\ 0 & 1 & 0 \\ 0 & 0 & 1 \end{bmatrix}.$$

Note that these are square matrices with 1's extending from the upper left to the lower right along the main diagonal, and 0's elsewhere.

The following property can be shown easily for the 2×2 and 3×3 cases (see Exercise 33 in Exercise Set 6.4).

DO EXERCISE 1.

THEOREM 7

> **For any $n \times n$ matrices A and I, AI = IA = A (I is a multiplicative identity).**

We usually say simply that **I** is an *identity* matrix.

Suppose a matrix **A** has a *multiplicative inverse,* or simply *inverse,* A^{-1}. Then A^{-1} is a matrix for which

$$A \cdot A^{-1} = A^{-1} \cdot A = I.$$

For example, for the matrix

$$A = \begin{bmatrix} 5 & 3 \\ 3 & 2 \end{bmatrix},$$

we have

$$A^{-1} = \begin{bmatrix} 2 & -3 \\ -3 & 5 \end{bmatrix}.$$

We can check that $A \cdot A^{-1} = I$ as follows:

$$A \cdot A^{-1} = \begin{bmatrix} 5 & 3 \\ 3 & 2 \end{bmatrix} \begin{bmatrix} 2 & -3 \\ -3 & 5 \end{bmatrix} = \begin{bmatrix} 1 & 0 \\ 0 & 1 \end{bmatrix} = I.$$

We leave it to the student to verify that $A^{-1} \cdot A = I$.

DO EXERCISE 2.

Calculating Matrix Inverses

In this section we consider two ways of calculating the inverse of a square matrix, if it exists. We shall see later that such inverses exist only when the determinant of the matrix is nonzero.

[i] The Cofactor Method

Before stating this method we need some additional notation.

DEFINITION

The *transpose* of a matrix A, denoted A^t, is found by interchanging the rows and columns of A.

Example 1 Find A^t, B^t, C^t, and D^t.

$$A = \begin{bmatrix} 2 & 4 & 6 \\ 9 & 8 & -2 \\ 0 & -1 & 4 \end{bmatrix}, \quad B = \begin{bmatrix} -3 & 0 & 4 \\ 7 & 1 & 6 \end{bmatrix}, \quad C = \begin{bmatrix} -4 \\ 3 \\ 2 \end{bmatrix}, \quad D = [-1 \quad 2 \quad 3 \quad 0].$$

The transposes are as follows:

$$A^t = \begin{bmatrix} 2 & 9 & 0 \\ 4 & 8 & -1 \\ 6 & -2 & 4 \end{bmatrix},$$

The transpose of a square matrix can be found by "reflecting" across the main diagonal.

$$B^t = \begin{bmatrix} -3 & 7 \\ 0 & 1 \\ 4 & 6 \end{bmatrix}, \quad C^t = [-4 \quad 3 \quad 2], \quad D^t = \begin{bmatrix} -1 \\ 2 \\ 3 \\ 0 \end{bmatrix}.$$

DO EXERCISE 3.

2. Let

$$A = \begin{bmatrix} 3 & 1 & 0 \\ 1 & -1 & 2 \\ 1 & 1 & 1 \end{bmatrix}$$

and

$$A^{-1} = \frac{1}{8} \begin{bmatrix} 3 & 1 & -2 \\ -1 & -3 & 6 \\ -2 & 2 & 4 \end{bmatrix}.$$

a) Find AA^{-1}.
b) Find $A^{-1}A$.
c) Compare AA^{-1} and $A^{-1}A$.

3. Find A^t, B^t, C^t, and D^t.

$$A = \begin{bmatrix} -8 & 1 & -2 \\ -4 & 0 & -1 \\ 6 & 7 & 8 \end{bmatrix},$$

$$B = [-7 \quad 9 \quad 10 \quad 4],$$

$$C = \begin{bmatrix} -20 \\ 11 \end{bmatrix},$$

$$D = \begin{bmatrix} -4 & 5 \\ 1 & 0 \\ 0 & 1 \end{bmatrix}.$$

Find $\mathbf{A}^{-1}$. Use the cofactor method.

4. $\mathbf{A} = \begin{bmatrix} 3 & 5 \\ 2 & 4 \end{bmatrix}$

The following is a procedure for calculating the inverse of a square matrix.

The cofactor method

To find the inverse $\mathbf{A}^{-1}$ of a square matrix $\mathbf{A}$,

a) Find the cofactor of each element;

b) Replace each element by its cofactor;

c) Find the transpose of the matrix found in (b);

d) Multiply the matrix in (c) by $1/|\mathbf{A}|$. The result is $\mathbf{A}^{-1}$.

Example 2 Find $\mathbf{A}^{-1}$.

$$\mathbf{A} = \begin{bmatrix} 3 & 5 \\ 1 & -2 \end{bmatrix}$$

a) Find the cofactor of each element:

$A_{11} = (-1)^{1+1}(-2)$
$\quad = -2,$
$A_{12} = (-1)^{1+2}(1) = -1,$
$A_{21} = (-1)^{2+1}(5) = -5,$
$A_{22} = (-1)^{2+2}(3) = 3.$

The determinant of a 1×1 matrix is just the number. For example, $|[-2]| = -2$. Don't confuse determinant notation with absolute value notation.

b) Replace each element by its cofactor.

$$\mathbf{A} = \begin{bmatrix} 3 & 5 \\ 1 & -2 \end{bmatrix} \qquad \begin{bmatrix} -2 & -1 \\ -5 & 3 \end{bmatrix}$$

5. $\mathbf{A} = \begin{bmatrix} 1 & 0 \\ 2 & -1 \end{bmatrix}$

c) Find the transpose of the matrix found in (b):

The transpose of $\begin{bmatrix} -2 & -1 \\ -5 & 3 \end{bmatrix}$ is $\begin{bmatrix} -2 & -5 \\ -1 & 3 \end{bmatrix}$.

d) Multiply by $\dfrac{1}{|\mathbf{A}|}$:

$$|\mathbf{A}| = 3(-2) - 1(5) = -11,$$

$$\mathbf{A}^{-1} = \frac{1}{-11} \cdot \begin{bmatrix} -2 & -5 \\ -1 & 3 \end{bmatrix} = \begin{bmatrix} \frac{2}{11} & \frac{5}{11} \\ \frac{1}{11} & -\frac{3}{11} \end{bmatrix}.$$

DO EXERCISES 4 AND 5.

Example 3 Find $\mathbf{A}^{-1}$.

$$\mathbf{A} = \begin{bmatrix} 2 & -1 & 1 \\ 1 & -2 & 3 \\ 4 & 1 & 2 \end{bmatrix}$$

a) Find the cofactor of each element.

$$A_{11} = (-1)^{1+1} \cdot \begin{vmatrix} -2 & 3 \\ 1 & 2 \end{vmatrix} = -7, \qquad A_{12} = (-1)^{1+2} \cdot \begin{vmatrix} 1 & 3 \\ 4 & 2 \end{vmatrix} = 10,$$

$$A_{13} = (-1)^{1+3} \cdot \begin{vmatrix} 1 & -2 \\ 4 & 1 \end{vmatrix} = 9, \qquad A_{21} = (-1)^{2+1} \cdot \begin{vmatrix} -1 & 1 \\ 1 & 2 \end{vmatrix} = 3,$$

$$A_{22} = (-1)^{2+2} \cdot \begin{vmatrix} 2 & 1 \\ 4 & 2 \end{vmatrix} = 0, \qquad A_{23} = (-1)^{2+3} \cdot \begin{vmatrix} 2 & -1 \\ 4 & 1 \end{vmatrix} = -6,$$

$$A_{31} = (-1)^{3+1} \cdot \begin{vmatrix} -1 & 1 \\ -2 & 3 \end{vmatrix} = -1, \qquad A_{32} = (-1)^{3+2} \cdot \begin{vmatrix} 2 & 1 \\ 1 & 3 \end{vmatrix} = -5,$$

$$A_{33} = (-1)^{3+3} \cdot \begin{vmatrix} 2 & -1 \\ 1 & -2 \end{vmatrix} = -3.$$

b) Replace each element by its cofactor:

$$\mathbf{A} = \begin{bmatrix} 2 & -1 & 1 \\ 1 & -2 & 3 \\ 4 & 1 & 2 \end{bmatrix} \longrightarrow \begin{bmatrix} -7 & 10 & 9 \\ 3 & 0 & -6 \\ -1 & -5 & -3 \end{bmatrix}.$$

c) Find the transpose of the matrix found in (b).

The transpose of $\begin{bmatrix} -7 & 10 & 9 \\ 3 & 0 & -6 \\ -1 & -5 & -3 \end{bmatrix}$ is $\begin{bmatrix} -7 & 3 & -1 \\ 10 & 0 & -5 \\ 9 & -6 & -3 \end{bmatrix}.$

d) Multiply by $\dfrac{1}{|\mathbf{A}|}$:

$$|\mathbf{A}| = a_{11}A_{11} + a_{21}A_{21} + a_{31}A_{31} \qquad \text{Expanding } |\mathbf{A}| \text{ about the first column}$$

$$= 2(-7) + 1(3) + 4(-1)$$
$$= -15,$$

$$\mathbf{A}^{-1} = \frac{1}{-15} \begin{bmatrix} -7 & 3 & -1 \\ 10 & 0 & -5 \\ 9 & -6 & -3 \end{bmatrix}$$

$$= \begin{bmatrix} \frac{7}{15} & -\frac{1}{5} & \frac{1}{15} \\ -\frac{2}{3} & 0 & \frac{1}{3} \\ -\frac{3}{5} & \frac{2}{5} & \frac{1}{5} \end{bmatrix}.$$

If $|\mathbf{A}|$ is 0, then $1/|\mathbf{A}|$ is not defined and $\mathbf{A}^{-1}$ does not exist.

DO EXERCISE 6.

[ii] The Gauss–Jordan Reduction Method

We now consider a different method. Suppose we want to find the inverse of the matrix of Example 3:

$$\mathbf{A} = \begin{bmatrix} 2 & -1 & 1 \\ 1 & -2 & 3 \\ 4 & 1 & 2 \end{bmatrix}.$$

6. Find $\mathbf{A}^{-1}$. Use the cofactor method.

$$\mathbf{A} = \begin{bmatrix} 3 & 1 & 0 \\ 1 & -1 & 2 \\ 1 & 1 & 1 \end{bmatrix}$$

First we form a new (*augmented*) matrix consisting, on the left, of the matrix **A** and, on the right, of the corresponding identity matrix **I**:

$$\begin{bmatrix} 2 & -1 & 1 \\ 1 & -2 & 3 \\ 4 & 1 & 2 \end{bmatrix} \begin{bmatrix} 1 & 0 & 0 \\ 0 & 1 & 0 \\ 0 & 0 & 1 \end{bmatrix}.$$

The matrix **A** ⟶　　　　　　　　　⟵ The identity matrix **I**

We now proceed in a manner very much like that described in Section 5.4. We attempt to transform **A** to an identity matrix, but whatever operations we perform, we do on the entire augmented matrix. When we finish we will get a matrix like the following:

$$\begin{bmatrix} 1 & 0 & 0 & a & b & c \\ 0 & 1 & 0 & d & e & f \\ 0 & 0 & 1 & g & h & i \end{bmatrix}.$$

The matrix on the right,

$$\begin{bmatrix} a & b & c \\ d & e & f \\ g & h & i \end{bmatrix},$$

will be **A**$^{-1}$.

Example 4 Find **A**$^{-1}$.

$$\mathbf{A} = \begin{bmatrix} 2 & -1 & 1 \\ 1 & -2 & 3 \\ 4 & 1 & 2 \end{bmatrix}$$

a) We find the augmented matrix consisting of **A** and **I**:

$$\begin{bmatrix} 2 & -1 & 1 & 1 & 0 & 0 \\ 1 & -2 & 3 & 0 & 1 & 0 \\ 4 & 1 & 2 & 0 & 0 & 1 \end{bmatrix}.$$

b) We interchange the first and second rows so that the elements of the first column are multiples of the top number on the main diagonal:

$$\begin{bmatrix} 1 & -2 & 3 & 0 & 1 & 0 \\ 2 & -1 & 1 & 1 & 0 & 0 \\ 4 & 1 & 2 & 0 & 0 & 1 \end{bmatrix}.$$

c) Next we obtain 0's in the rest of the first column. We multiply the first row by -2 and add it to the second row. Then we multiply the first row by -4 and add it to the third row:

$$\begin{bmatrix} 1 & -2 & 3 & 0 & 1 & 0 \\ 0 & 3 & -5 & 1 & -2 & 0 \\ 0 & 9 & -10 & 0 & -4 & 1 \end{bmatrix}.$$

d) Next we move down the main diagonal to the number 3. We note that the number below it, 9, is a multiple of 3. We multiply the second row by -3 and add it to the third:

$$\begin{bmatrix} 1 & -2 & 3 & 0 & 1 & 0 \\ 0 & 3 & -5 & 1 & -2 & 0 \\ 0 & 0 & 5 & -3 & 2 & 1 \end{bmatrix}.$$

e) Now we move down the main diagonal to the number 5. We check to see if each number above 5 in the third column is a multiple of 5. Since this is not the case, we multiply the first row by -5:

$$\begin{bmatrix} -5 & 10 & -15 & 0 & -5 & 0 \\ 0 & 3 & -5 & 1 & -2 & 0 \\ 0 & 0 & 5 & -3 & 2 & 1 \end{bmatrix}.$$

f) Now we work back up. We add the third row to the second. We also multiply the third row by 3 and add it to the first:

$$\begin{bmatrix} -5 & 10 & 0 & -9 & 1 & 3 \\ 0 & 3 & 0 & -2 & 0 & 1 \\ 0 & 0 & 5 & -3 & 2 & 1 \end{bmatrix}.$$

g) We move back to the number 3 on the main diagonal. We multiply the first row by -3, so the element on the top of the second column is a multiple of 3:

$$\begin{bmatrix} 15 & -30 & 0 & 27 & -3 & -9 \\ 0 & 3 & 0 & -2 & 0 & 1 \\ 0 & 0 & 5 & -3 & 2 & 1 \end{bmatrix}.$$

h) We multiply the second row by 10 and add it to the first:

$$\begin{bmatrix} 15 & 0 & 0 & 7 & -3 & 1 \\ 0 & 3 & 0 & -2 & 0 & 1 \\ 0 & 0 & 5 & -3 & 2 & 1 \end{bmatrix}.$$

i) Finally, we get all 1's on the main diagonal. We multiply the first row by $\frac{1}{15}$, the second by $\frac{1}{3}$, and the third by $\frac{1}{5}$:

$$\begin{bmatrix} 1 & 0 & 0 & \frac{7}{15} & -\frac{1}{5} & \frac{1}{15} \\ 0 & 1 & 0 & -\frac{2}{3} & 0 & \frac{1}{3} \\ 0 & 0 & 1 & -\frac{3}{5} & \frac{2}{5} & \frac{1}{5} \end{bmatrix}.$$

We now have the matrix **I** on the left. Thus

$$\mathbf{A}^{-1} = \begin{bmatrix} \frac{7}{15} & -\frac{1}{5} & \frac{1}{15} \\ -\frac{2}{3} & 0 & \frac{1}{3} \\ -\frac{3}{5} & \frac{2}{5} & \frac{1}{5} \end{bmatrix}.$$

   With either method the student can check by doing the multiplication $\mathbf{A}^{-1}\mathbf{A}$ or $\mathbf{A}\mathbf{A}^{-1}$. If we cannot obtain the identity matrix on the left using the Gauss–Jordan reduction method, as would be the case when a system has no solution or infinitely many solutions, then $\mathbf{A}^{-1}$ does not exist.

DO EXERCISES 7 AND 8.

Find $\mathbf{A}^{-1}$. Use the Gauss–Jordan reduction method.

7.
$$\mathbf{A} = \begin{bmatrix} 1 & 0 & 1 \\ 2 & 1 & 0 \\ 1 & -1 & 1 \end{bmatrix}$$

8.
$$\mathbf{A} = \begin{bmatrix} 3 & 5 \\ 2 & 4 \end{bmatrix}$$

9. Consider the system

$$4x_1 - 2x_2 = -1,$$
$$x_1 + 5x_2 = 1.$$

a) Write a matrix equation equivalent to the system.
b) Find the coefficient matrix $\mathbf{A}$.
c) Find $\mathbf{A}^{-1}$.
d) Use the inverse of the coefficient matrix to solve the system.

[iii] Solving Systems Using Inverses

We can use matrix inverses to solve certain kinds of systems. Consider the system

$$3x_1 + 5x_2 = -1,$$
$$x_1 - 2x_2 = 4.$$

We write a matrix equation equivalent to this system:

$$\begin{bmatrix} 3 & 5 \\ 1 & -2 \end{bmatrix} \cdot \begin{bmatrix} x_1 \\ x_2 \end{bmatrix} = \begin{bmatrix} -1 \\ 4 \end{bmatrix}.$$

Now we let

$$\begin{bmatrix} 3 & 5 \\ 1 & -2 \end{bmatrix} = \mathbf{A}, \qquad \begin{bmatrix} x_1 \\ x_2 \end{bmatrix} = \mathbf{X}, \quad \text{and} \quad \begin{bmatrix} -1 \\ 4 \end{bmatrix} = \mathbf{B}.$$

Then we have

$$\mathbf{A} \cdot \mathbf{X} = \mathbf{B}.$$

To solve this equation, we first find $\mathbf{A}^{-1}$. We found in Example 2 that

$$\mathbf{A}^{-1} = -\frac{1}{11} \begin{bmatrix} -2 & -5 \\ -1 & 3 \end{bmatrix}.$$

We solve the matrix equation $\mathbf{A} \cdot \mathbf{X} = \mathbf{B}$:

$$\mathbf{A}^{-1} \cdot \mathbf{A} \cdot \mathbf{X} = \mathbf{A}^{-1} \cdot \mathbf{B} \qquad \text{Multiplying by } \mathbf{A}^{-1}$$
$$\mathbf{I} \cdot \mathbf{X} = \mathbf{A}^{-1} \cdot \mathbf{B} \qquad \text{Since } \mathbf{A}^{-1} \cdot \mathbf{A} = \mathbf{I}$$
$$\mathbf{X} = \mathbf{A}^{-1} \cdot \mathbf{B} \qquad \text{Since } \mathbf{I} \text{ is an identity}$$

Now we have, substituting,

$$\mathbf{X} = \mathbf{A}^{-1} \cdot \mathbf{B}$$
$$\begin{bmatrix} x_1 \\ x_2 \end{bmatrix} = -\frac{1}{11} \begin{bmatrix} -2 & -5 \\ -1 & 3 \end{bmatrix} \cdot \begin{bmatrix} -1 \\ 4 \end{bmatrix}$$
$$= -\frac{1}{11} \begin{bmatrix} -18 \\ 13 \end{bmatrix}$$
$$= \begin{bmatrix} \frac{18}{11} \\ -\frac{13}{11} \end{bmatrix}.$$

The solution of the system of equations is therefore $x_1 = \frac{18}{11}$ and $x_2 = -\frac{13}{11}$.

DO EXERCISE 9.

EXERCISE SET 6.4

[i] Find $\mathbf{A}^{-1}$, if it exists. Use the cofactor method. Check your answers by calculating $\mathbf{A}\mathbf{A}^{-1}$ and $\mathbf{A}^{-1}\mathbf{A}$.

1. $\mathbf{A} = \begin{bmatrix} 3 & 2 \\ 5 & 3 \end{bmatrix}$ **2.** $\mathbf{A} = \begin{bmatrix} 3 & 5 \\ 1 & 2 \end{bmatrix}$ **3.** $\mathbf{A} = \begin{bmatrix} 11 & 3 \\ 7 & 2 \end{bmatrix}$ **4.** $\mathbf{A} = \begin{bmatrix} 8 & 5 \\ 5 & 3 \end{bmatrix}$

5. $\mathbf{A} = \begin{bmatrix} 4 & -3 \\ 1 & 2 \end{bmatrix}$ **6.** $\mathbf{A} = \begin{bmatrix} 0 & -1 \\ 1 & 0 \end{bmatrix}$ **7.** $\mathbf{A} = \begin{bmatrix} 3 & 1 & 0 \\ 1 & 1 & 1 \\ 1 & -1 & 2 \end{bmatrix}$ **8.** $\mathbf{A} = \begin{bmatrix} 1 & 0 & 1 \\ 2 & 1 & 0 \\ 1 & -1 & 1 \end{bmatrix}$

9.
$$A = \begin{bmatrix} 1 & -1 & 2 \\ 0 & 1 & 3 \\ 2 & 1 & -2 \end{bmatrix}$$

10.
$$A = \begin{bmatrix} 1 & -1 & 2 \\ 0 & 1 & 2 \\ 1 & -3 & -4 \end{bmatrix}$$

11.
$$A = \begin{bmatrix} 1 & -4 & 8 \\ 1 & -3 & 2 \\ 2 & -7 & 10 \end{bmatrix}$$

12.
$$A = \begin{bmatrix} -2 & 5 & 3 \\ 4 & -1 & 3 \\ 7 & -2 & 5 \end{bmatrix}$$

13.
$$A = \begin{bmatrix} 1 & 2 & 3 & 4 \\ 0 & 1 & 3 & -5 \\ 0 & 0 & 1 & -2 \\ 0 & 0 & 0 & -1 \end{bmatrix}$$

14.
$$A = \begin{bmatrix} -2 & -3 & 4 & 1 \\ 0 & 1 & 1 & 0 \\ 0 & 4 & -6 & 1 \\ -2 & -2 & 5 & 1 \end{bmatrix}$$

[ii] **15–28.** Find A^{-1} of each matrix in Exercises 1–14. Use the Gauss–Jordan reduction method.

[iii] For Exercises 29 and 30, write a matrix equation equivalent to the system. Find the inverse of the coefficient matrix. Use the inverse of the coefficient matrix to solve each system.

29. $7x - 2y = -3,$
$\quad 9x + 3y = 4$

30. $5x_1 + 3x_2 = -2,$
$\quad 4x_1 - x_2 = 1$

31. $x_1 \qquad + x_3 = 1,$
$\quad 2x_1 + x_2 \qquad = 3,$
$\quad x_1 - x_2 + x_3 = 4$

32. $\quad x + 2y + 3z = -1,$
$\quad 2x - 3y + 4z = 2,$
$\quad -3x + 5y - 6z = 4$

☆ ─────────────────────────────

33. Let

$$A = \begin{bmatrix} a & b & c \\ d & e & f \\ g & h & i \end{bmatrix} \quad \text{and} \quad I = \begin{bmatrix} 1 & 0 & 0 \\ 0 & 1 & 0 \\ 0 & 0 & 1 \end{bmatrix}$$

Show that $AI = IA = A$.

In each of the following, state the conditions under which A^{-1} exists. Then find a formula for A^{-1}.

34. $A = [x]$

35. $A = \begin{bmatrix} x & 0 \\ 0 & y \end{bmatrix}$

36.
$$A = \begin{bmatrix} 0 & 0 & x \\ 0 & y & 0 \\ z & 0 & 0 \end{bmatrix}$$

37.
$$A = \begin{bmatrix} x & 1 & 1 & 1 \\ 0 & y & 0 & 0 \\ 0 & 0 & z & 0 \\ 0 & 0 & 0 & w \end{bmatrix}$$

★ ─────────────────────────────

38. Consider

$$a_{11}x + a_{12}y = c_1,$$
$$a_{21}x + a_{22}y = c_2.$$

Use the cofactor method and the equivalent matrix equation $AX = C$ to prove Cramer's rule.

─────────────────────────────

CHAPTER 6 REVIEW

Let

$$A = \begin{bmatrix} 1 & -1 & 0 \\ 2 & 3 & -2 \\ -2 & 0 & 1 \end{bmatrix}, \quad B = \begin{bmatrix} -1 & 0 & 6 \\ 1 & -2 & 0 \\ 0 & 1 & -3 \end{bmatrix}, \quad \text{and} \quad C = \begin{bmatrix} -2 & 0 \\ 1 & 3 \end{bmatrix}.$$

Find each of the following, if possible.

[6.1, i]   **1. A + B**   **2. −3A**   **3. −A**   **4. AB**

[6.4, i]   **5. B + C**   **6. A − B**   **7. A^{-1}**   **8. B^{-1}**

[6.4, **ii**]　　　Find $\mathbf{A}^{-1}$ if it exists.

9. $\mathbf{A} = \begin{bmatrix} -2 & 0 \\ 1 & 3 \end{bmatrix}$

10. $\mathbf{A} = \begin{bmatrix} 0 & 0 & 3 \\ 0 & -2 & 0 \\ 4 & 0 & 0 \end{bmatrix}$

11. $\mathbf{A} = \begin{bmatrix} 1 & 0 & 0 & 0 \\ 0 & 4 & -5 & 0 \\ 0 & 2 & 2 & 0 \\ 0 & 0 & 0 & 1 \end{bmatrix}$

[6.1, **ii**]　　**12.** Write a matrix equation equivalent to this system of equations.

$$3x - 2y + 4z = 13,$$
$$x + 5y - 3z = 7,$$
$$2x - 3y + 7z = -8$$

[6.2, **i**]　　　Evaluate.

13. $\begin{vmatrix} -4 & \sqrt{3} \\ \sqrt{3} & 7 \end{vmatrix}$

14. $\begin{vmatrix} 1 & -1 & 2 \\ -1 & 2 & 0 \\ -1 & 3 & 1 \end{vmatrix}$

15. $\begin{vmatrix} 0 & a & b \\ -a & 0 & c \\ -b & -c & 0 \end{vmatrix}$

16. $\begin{vmatrix} 4 & -7 & 6 & 7 \\ 0 & -3 & 9 & -8 \\ 0 & 0 & -2 & 6 \\ 0 & 0 & 0 & 5 \end{vmatrix}$

[6.3, **i**]　　**17.** Without expanding, show that

$$\begin{vmatrix} 5a & 5b & 5c \\ 3a & 3b & 3c \\ d & e & f \end{vmatrix} = 0.$$

[6.3, **ii**]　　Factor.

18. $\begin{vmatrix} 1 & a & bc \\ 1 & b & ac \\ 1 & c & ab \end{vmatrix}$

19. $\begin{vmatrix} 1 & x^2 & x^3 \\ 1 & y^2 & y^3 \\ 1 & z^2 & z^3 \end{vmatrix}$

20. $\begin{vmatrix} 1 & a & a^2 & a^3 \\ 1 & b & b^2 & b^3 \\ 1 & c & c^2 & c^3 \\ 1 & d & d^2 & d^3 \end{vmatrix}$

21. On the basis of Exercise 16, conjecture and prove a theorem regarding determinants.

EXPONENTIAL AND LOGARITHMIC FUNCTIONS

OBJECTIVES

You should be able to:

[i] Graph exponential and logarithmic equations.
[ii] Given an exponential equation of the type $a^b = c$, write an equivalent logarithmic equation $\log_a c = b$, and vice versa.
[iii] Solve equations like $\log_3 9 = x$, $\log_x 9 = 2$, and $\log_3 x = 2$.
[iv] Simplify expressions like $a^{\log_a x}$ and $\log_a a^x$.

7.1 EXPONENTIAL AND LOGARITHMIC FUNCTIONS

Irrational Exponents

We have defined exponential notation for cases in which the exponent is a rational number. For example $x^{2.34}$, or $x^{234/100}$, means to take the 100th root of x and raise the result to the 234th power. We now consider irrational exponents, such as π or $\sqrt{2}$.

Let us consider 2^π. We know that π has an unending decimal representation,

$$3.1415926535\ldots .$$

Now consider this sequence of numbers:

$$3, \quad 3.1, \quad 3.14, \quad 3.141, \quad 3.1415, \quad 3.14159, \ldots .$$

Each of these numbers is an approximation to π, the more decimal places the better the approximation. Let us use these (rational) numbers to form a sequence as follows:

$$2^3, \quad 2^{3.1}, \quad 2^{3.14}, \quad 2^{3.141}, \quad 2^{3.1415}, \quad 2^{3.14159}, \ldots .$$

Each of the numbers in this sequence is already defined, the exponent being rational. The numbers in this sequence get closer and closer to some real number. We define that number to be 2^π.

We can define exponential notation for any irrational exponent in a similar way. Thus any exponential expression a^x, $a > 0$, now has meaning, whether the exponent is rational or irrational. The usual laws of exponents still hold, in case exponents are irrational. We will not prove that fact here, however.

Exponential Functions

Exponential functions are defined using exponential notation.

DEFINITION

The function $f(x) = a^x$, where a is some positive constant different from 1, is called the *exponential function, base a.*

CAUTION! Here are some exponential functions:

$$f(x) = 2^x, \qquad g(x) = 3^x, \qquad h(x) = (0.178)^x.$$

Note that the variable is the exponent. The following are *not* exponential functions:

$$f(x) = x^2, \qquad g(x) = x^3, \qquad h(x) = x^{0.178}.$$

Note that the variable is not the exponent.

[i] Graphs of Exponential and Logarithmic Functions

Exponential Functions

Example 1 Graph $y = 2^x$. Use the graph to approximate 2^π.

We find some solutions, plot them, and then draw the graph.

x	0	1	2	3	−1	−2	−3
y	1	2	4	8	$\frac{1}{2}$	$\frac{1}{4}$	$\frac{1}{8}$

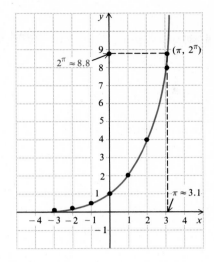

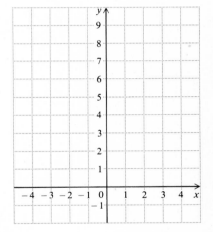

(Note that as x increases, the function values increase. Check this on your calculator. As x decreases, the function values decrease toward 0. Check this on your calculator.)

To approximate 2^{π}, we locate π on the x-axis, at about 3.1. Then we find the corresponding function value. It is about 8.8.

DO EXERCISES 1 AND 2.

Let us now look at some other exponential functions. We will make comparisons, using transformations.

Example 2 Graph $y = 4^x$.

We could plot points and connect them, but let us be more clever. We note that $4^x = (2^2)^x = 2^{2x}$. Thus the function we wish to graph is

$$y = 2^{2x}.$$

Compare this with $y = 2^x$, graphed above. The graph of $y = 2^{2x}$ is a compression, in the x-direction toward the y-axis, of the graph of $y = 2^x$. Knowing this allows us to graph $y = 2^{2x}$ at once. Each point on the graph of 2^x is moved half the distance to the y-axis.

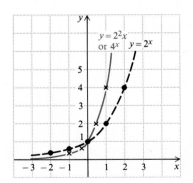

DO EXERCISES 3–5.

1. Use the graph of $y = 2^x$.

   a) Is this function increasing or decreasing?
   b) What is the domain?
   c) What is the range?
   d) What is the y-intercept?
   e) Use the graph to approximate $2^{\sqrt{3}}$ ($\sqrt{3} \approx 1.732$).

2. Graph $y = 4^x$, and compare it with the graph of $y = 2^x$.

   a) Is the function increasing or decreasing?
   b) What is the domain of $y = 4^x$?
   c) What is the range?
   d) What is the y-intercept?
   e) Use the graph to approximate $4^{\sqrt{2}}$ ($\sqrt{2} \approx 1.414$).
   f) Which function increases faster, 2^x or 4^x?

3. Graph $y = 8^x$. Use graph paper. [*Hint:* $8^x = (2^3)^x = 2^{3x}$, hence this is a horizontal compression of the graph of $y = 2^x$.]

4. Graph $y = 1^x$. Use graph paper.

5. Graph $y = \left(\frac{1}{2}\right)^x$. Use graph paper.

6. Graph $y = \left(\frac{1}{3}\right)^x$.

$$\left[\text{Hint: } \left(\frac{1}{3}\right)^x = 3^{-x}.\right]$$

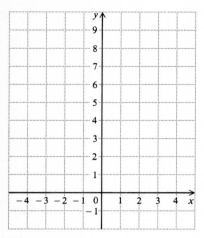

Example 3 Graph $y = \left(\frac{1}{2}\right)^x$.

We could plot points and connect them, but again let us be more clever. We note that $\left(\frac{1}{2}\right)^x = 1/2^x = 2^{-x}$. Thus the function we wish to graph is

$$y = 2^{-x}.$$

Compare this with the graph of $y = 2^x$ in Example 1. The graph of $y = 2^{-x}$ is a reflection, across the y-axis, of the graph of $y = 2^x$. Knowing this allows us to graph $y = 2^{-x}$ at once.

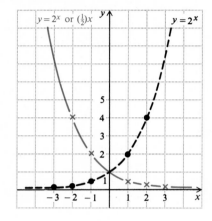

DO EXERCISE 6.

The preceding examples and exercises illustrate exponential functions of various bases. If $a = 1$, then $f(x) = a^x = 1^x = 1$ and the graph is a horizontal line. This is why we exclude 1 as a base for an exponential function. We summarize.

THEOREM 1

1. **When $a > 1$, the function $f(x) = a^x$ is an increasing function. The greater the value of a, the faster the function increases.**
2. **When $0 < a < 1$, the function $f(x) = a^x$ is a decreasing function. The greater the value of a, the more slowly the function decreases.**

It should be noted that for any value of a, the y-intercept is (0, 1).

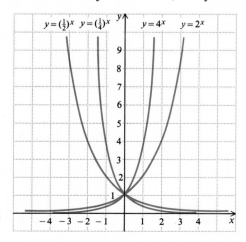

As an application of exponential functions we consider the compound interest formula, $A = P(1 + i)^t$, considered in Section 2.2. Suppose principal P of $1000 is invested at an interest rate of 13%, compounded annually. Then the amount A in the account after time t, in years, is given by the exponential function $A = \$1000(1.13)^t$.

Logarithmic Functions

The inverse of an exponential function, for $a > 0$ and $a \neq 1$, is a function. It is called a *logarithmic* or *logarithm function*. Thus one way to describe a logarithm function is to interchange variables in $y = a^x$:

$$x = a^y.$$

The most useful and interesting logarithmic functions are those for which $a > 1$. The graph of such a function is a reflection of $y = a^x$ across the line $y = x$, as shown below. Note that the domain of a logarithm function is the set of all positive real numbers.

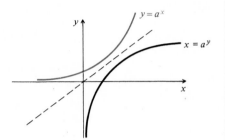

For logarithm functions we use the notation $\log_a (x)$ or $\log_a x$.* That is, we use the symbol $\log_a x$ to denote the second coordinates of a function $x = a^y$. In other words, a logarithmic function can be described as $y = \log_a x$.

DEFINITION

The following are equivalent:

1. $x = a^y$; and
2. $y = \log_a x$ (read "y equals the log, base a, of x").

Thus $\log_a x$ represents the exponent in the equation $x = a^y$, so the logarithm, base a, of a number x is the power to which a is raised to get x.

Example 4 Graph $y = \log_3 x$.

The equation $y = \log_3 x$ is equivalent to $x = 3^y$. The graph is a reflection of $y = 3^x$ across the line $y = x$. We make a table of values for $y = 3^x$ and then interchange x and y.

For $y = 3^x$:

x	0	1	2	-1	-2
y	1	3	9	$\frac{1}{3}$	$\frac{1}{9}$

For $y = \log_3 x$ (or $x = 3^y$):

x	1	3	9	$\frac{1}{3}$	$\frac{1}{9}$
y	0	1	2	-1	-2

---

*The parentheses in $\log_a (x)$ are like those in $f(x)$. In the case of logarithm functions we usually omit the parentheses.

7. Graph $y = \log_2 x$.
What is the domain of this function? What is the range?

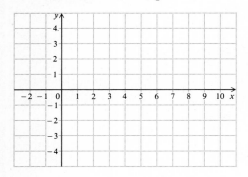

8. Graph $y = \log_4 x$.
What is the domain of this function? What is the range?

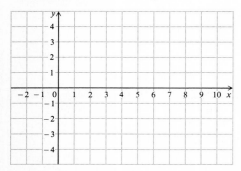

Write equivalent logarithmic equations.

9. $6^0 = 1$

10. $10^{-3} = 0.001$

11. $16^{1/4} = 2$

12. $\left(\dfrac{6}{5}\right)^{-2} = \dfrac{25}{36}$

Write equivalent exponential equations.

13. $\log_2 32 = 5$

14. $\log_{10} 1000 = 3$

15. $\log_{10} 0.01 = -2$

16. $\log_{\sqrt{5}} 5 = 2$

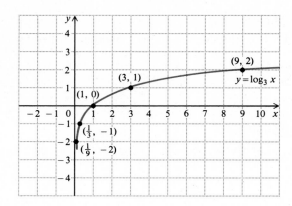

The graph of $y = \log_a x$, for any a, has the x-intercept $(1, 0)$.

DO EXERCISES 7 AND 8.

[ii] Converting Exponential and Logarithmic Equations

It is important to be able to convert from an exponential equation to a logarithmic equation.

Examples Convert to logarithmic equations.

5. $8 = 2^x \rightarrow x = \log_2 8$ It helps, in such conversions, to remember that the *logarithm is the exponent.*

6. $y^{-1} = 4 \rightarrow -1 = \log_y 4$

7. $a^b = c \rightarrow b = \log_a c$

DO EXERCISES 9–12.

It is also important to be able to convert from a logarithmic equation to an exponential equation.

Examples Convert to exponential equations.

8. $y = \log_3 5 \rightarrow 3^y = 5$ Again, it helps to remember that the *logarithm is the exponent.*

9. $-2 = \log_a 7 \rightarrow a^{-2} = 7$

10. $a = \log_b d \rightarrow b^a = d$

DO EXERCISES 13–16.

[iii] Solving Logarithmic Equations

Certain equations containing logarithmic notation can be solved by first converting to exponential notation.

Examples

11. Solve $\log_2 x = -3$.

$\log_2 x = -3$ is equivalent to $2^{-3} = x$. So $x = \frac{1}{8}$.

12. Solve $\log_{27} 3 = x$.

$\log_{27} 3 = x$ is equivalent to $27^x = 3$. Since $27^{1/3} = 3$ we have $x = \frac{1}{3}$.

13. Solve $\log_x 4 = \frac{1}{2}$.

$\log_x 4 = \frac{1}{2}$ is equivalent to $x^{1/2} = 4$. Since $(x^{1/2})^2 = 4^2$, we have $x = 16$.

DO EXERCISES 17–21.

[iv] **Simplifying the Expressions $a^{\log_a x}$ and $\log_a a^x$**

The exponential and logarithm functions are inverses of each other. Let us recall an important fact about functions and their inverses (Section 3.7). If the domains are suitable, then for any x,

$$f\left(f^{-1}(x)\right) = x$$

and

$$f^{-1}\left(f(x)\right) = x.$$

We apply this fact to exponential and logarithm functions. Suppose f is the exponential function, base a:

$$f(x) = a^x.$$

Then f^{-1} is the logarithm function, base a:

$$f^{-1}(x) = \log_a x.$$

Now let us find $f\left(f^{-1}(x)\right)$:

$$f\left(f^{-1}(x)\right) = a^{f^{-1}(x)} = a^{\log_a x} = x.$$

Thus for any suitable base a, $a^{\log_a x} = x$ for any positive number x (negative numbers and 0 do not have logarithms).

Next, let us find $f^{-1}\left(f(x)\right)$:

$$f^{-1}\left(f(x)\right) = \log_a f(x) = \log_a a^x = x.$$

Thus for any suitable base a, $\log_a a^x = x$ for any number x whatever.

These facts are important in simplification and should be learned well.

THEOREM 2

For any number a, suitable as a logarithm base,

1. $a^{\log_a x} = x$, for any positive number x; and
2. $\log_a a^x = x$, for any number x.

Examples Simplify.

14. $2^{\log_2 5} = 5$

15. $10^{\log_{10} t} = t$

16. $\log_e e^{-3} = -3$

17. $\log_{10} 10^{5.6} = 5.6$

DO EXERCISES 22–27.

Solve.

17. $\log_{10} x = 4$

18. $\log_x 81 = 4$

19. $\log_2 16 = x$

20. $\log_3 3 = x$

21. $\log_5 \dfrac{1}{25} = x$

Simplify.

22. $4^{\log_4 3}$

23. $7^{\log_7 \pi}$

24. $b^{\log_b 42}$

25. $\log_5 5^{37}$

26. $\log_e e^M$

27. $\log_{10} 10^{3.2}$

EXERCISE SET 7.1

[i] Graph.

1. a) $y = 4^x$ b) $y = \left(\dfrac{1}{4}\right)^x$ c) $y = \log_4 x$

2. a) $y = 3^x$ b) $y = \left(\dfrac{2}{3}\right)^x$ c) $y = \log_3 x$

3. $y = 5^x$ **4.** $y = 2^x$ **5.** $y = \log_2 x$ **6.** $y = \log_{10} x$

[ii] Write equivalent exponential equations.

7. $\log_2 32 = 5$ **8.** $\log_{10} 1000 = 3$ **9.** $\log_{10} 0.01 = -2$

10. $\log_{\sqrt{5}} 5 = 2$ **11.** $\log_6 6 = 1$ **12.** $\log_b M = N$

Write equivalent logarithmic equations.

13. $6^0 = 1$ **14.** $10^{-3} = 0.001$ **15.** $\left(\dfrac{6}{5}\right)^{-2} = \dfrac{25}{36}$ **16.** $5^4 = 625$

17. $5^{-2} = \dfrac{1}{25}$ **18.** $8^{1/3} = 2$ **19.** $e^{0.08} = 1.0833$ **20.** $10^{0.4771} = 3$

[iii] Solve.

21. $\log_{10} x = 4$ **22.** $\log_3 x = 2$ **23.** $\log_x \dfrac{1}{32} = 5$ **24.** $\log_8 x = \dfrac{1}{3}$

Solve for x.

25. $\log_2 16 = x$ **26.** $\log_3 3 = x$ **27.** $\log_4 2 = x$ **28.** $\log_2 64 = x$

29. $\log_{10} 10^2 = x$ **30.** $\log_3 3^4 = x$ **31.** $\log_\pi \pi = x$ **32.** $\log_a a = x$

33. $\log_{10} 0.001 = x$ **34.** $\log_{10} 1000 = x$

[iv] Simplify.

35. $3^{\log_3 4x}$ **36.** $5^{\log_5 (4x-5)}$ **37.** $\log_Q Q^{\sqrt{5}}$ **38.** $\log_e e^{|x-4|}$

☆ _____

Graph.

39. $y = \log_2 (x + 3)$ **40.** $y = \log_3 (x - 2)$ **41.** $y = 2^x - 1$ **42.** $y = 2^{x-3}$

43. $f(x) = 2^{|x|}$ **44.** $f(x) = \log_3 |x|$ **45.** $f(x) = 2^x + 2^{-x}$ **46.** $f(x) = 2^{-(x-1)}$

What is the domain of each function?

47. $f(x) = 3^x$ **48.** $f(x) = \log_{10} x$ **49.** $f(x) = \log_a x^2$

50. $f(x) = \log_4 x^3$ **51.** $f(x) = \log_{10} (3x - 4)$ **52.** $f(x) = \log_5 |x|$

53. $f(x) = \log_6 (x^2 - 9)$

Solve using graphing.

54. $2^x > 1$ **55.** $3^x \leq 1$ **56.** $\log_2 x < 0$ **57.** $\log_2 x \geq 4$

58. $\log_2 (x - 3) \leq 5$ **59.** $2^{x+3} > 1$

60. ▦ Estimate each of the following to six decimal places.

$$2^3, \quad 2^{3.1}, \quad 2^{3.14}, \quad 2^{3.141}, \quad 2^{3.1415}, \quad 2^{3.14159}$$

61. ▦ Which is larger, 5^π or π^5?   **62.** ▦ Which is larger, $\sqrt{8^3}$ or $8^{\sqrt{3}}$?   **63.** ▦ Graph $y = (0.745)^x$.

★ _____

Graph.

64. $y = 2^{-x^2}$ **65.** $y = 3^{-(x+1)^2}$ **66.** $y = |2^{x^2} - 8|$

67. Solve $3^{3^x} = 1$.

7.2 PROPERTIES OF LOGARITHMIC FUNCTIONS

[i] Let us now establish some basic properties of logarithmic functions.

THEOREM 3

For any positive numbers x and y,

$$\log_a (x \cdot y) = \log_a x + \log_a y,$$

where a is any positive number different from 1.

Theorem 3 says that the logarithm of a *product* is the *sum* of the logarithms of the factors. Note that the base a must remain constant. The logarithm of a sum is *not* the sum of the logarithms of the summands.

Proof of Theorem 3. Since a is positive and different from 1, it can serve as a logarithm base. Since x and y are assumed positive, they are in the domain of the function $y = \log_a x$. Now let $b = \log_a x$ and $c = \log_a y$.

Writing equivalent exponential equations, we have

$$x = a^b \quad \text{and} \quad y = a^c.$$

Next we multiply, to obtain

$$xy = a^b a^c = a^{b+c}.$$

Now writing an equivalent logarithmic equation, we obtain

$$\log_a (xy) = b + c,$$

or

$$\log_a (xy) = \log_a x + \log_a y,$$

which was to be shown.

Example 1 Express as a sum of logarithms and simplify.

$$\log_2 (4 \cdot 16) = \log_2 4 + \log_2 16$$
$$= 2 + 4 = 6$$

DO EXERCISES 1–4.

THEOREM 4

For any positive number x and any number p,

$$\log_a x^p = p \cdot \log_a x,$$

where a is any logarithm base.

Theorem 4 says that the logarithm of a power of a number is the exponent times the logarithm of the number.

Proof of Theorem 4. Let $b = \log_a x$. Then, writing an equivalent exponential equation, we have $x = a^b$. Next we raise both sides of

OBJECTIVE

You should be able to:

[i] Use properties of logarithms to express sums and differences as a single logarithm, and vice versa, and to find certain logarithms given values of others.

Express as a sum of logarithms.

1. $\log_a MN$

2. $\log_5 (25 \cdot 5)$

Express as a single logarithm.

3. $\log_3 7 + \log_3 5$

4. $\log_a C + \log_a A + \log_a B + \log_a I + \log_a N$

Express as a product.

5. $\log_7 4^5$

6. $\log_a \sqrt{5}$

the latter equation to the pth power. This gives us

$$x^p = (a^b)^p, \quad \text{or} \quad a^{bp}.$$

Now we can write an equivalent logarithmic equation,

$$\log_a x^p = \log_a a^{bp} = bp.$$

But $b = \log_a x$, so we have

$$\log_a x^p = p \cdot \log_a x,$$

which was to be shown.

Examples Express as products.

2. $\log_b 9^{-5} = -5 \cdot \log_b 9$

3. $\log_a \sqrt[4]{5} = \log_a 5^{1/4} = \frac{1}{4} \log_a 5$

DO EXERCISES 5 AND 6.

THEOREM 5

For any positive numbers x and y,

$$\log_a \frac{x}{y} = \log_a x - \log_a y,$$

where a is any logarithm base.

Theorem 5 says that the logarithm of a quotient is the difference of the logarithms (i.e., the logarithm of the dividend minus the logarithm of the divisor).

Proof of Theorem 5. $x/y = x \cdot y^{-1}$, so

$$\log_a \frac{x}{y} = \log_a (xy^{-1}).$$

By Theorem 2,

$$\log_a (xy^{-1}) = \log_a x + \log_a y^{-1},$$

and by Theorem 3,

$$\log_a y^{-1} = -1 \cdot \log_a y,$$

so we have

$$\log_a \frac{x}{y} = \log_a x - \log_a y,$$

which was to be shown.

Example 4 Express as sums and differences of logarithms and without exponential notation or radicals.

$$\log_a \frac{\sqrt{17}}{5\pi} = \log_a \sqrt{17} - \log_a 5\pi$$

$$= \log_a 17^{1/2} - (\log_a 5 + \log_a \pi)$$

$$= \frac{1}{2} \log_a 17 - \log_a 5 - \log_a \pi$$

Example 5 Express in terms of logarithms of x, y, and z.

$$\log_a \sqrt[4]{\frac{xy}{z^3}} = \log_a \left(\frac{xy}{z^3}\right)^{1/4}$$

$$= \frac{1}{4} \cdot \log_a \frac{xy}{z^3}$$

$$= \frac{1}{4}[\log_a xy - \log_a z^3]$$

$$= \frac{1}{4}[\log_a x + \log_a y - 3\log_a z]$$

$$= \frac{1}{4}\log_a x + \frac{1}{4}\log_a y - \frac{3}{4}\log_a z$$

Example 6 Express as a single logarithm.

$$\frac{1}{2}\log_a x - 7\log_a y + \log_a z = \log_a \sqrt{x} - \log_a y^7 + \log_a z$$

$$= \log_a \frac{\sqrt{x}}{y^7} + \log_a z$$

$$= \log_a \frac{z\sqrt{x}}{y^7}$$

DO EXERCISES 7–10.

Example 7 Given that $\log_a 2 = 0.301$ and $\log_a 3 = 0.477$, find:

a) $\log_a 6 = \log_a 2 \cdot 3$
$= \log_a 2 + \log_a 3$
$= 0.301 + 0.477 = 0.778;$

b) $\log_a \sqrt{3} = \log_a 3^{1/2} = \frac{1}{2} \cdot \log_a 3 = \frac{1}{2} \cdot 0.477 = 0.2385;$

c) $\log_a \frac{2}{3} = \log_a 2 - \log_a 3 = 0.301 - 0.477 = -0.176;$

d) $\log_a 5$; *No way to find, using Theorems 3–5.*
$(\log_a 5 \neq \log_a 2 + \log_a 3);$

e) $\frac{\log_a 2}{\log_a 3} = \frac{0.301}{0.477} = 0.63.$ Note that we could not use Theorems 3–5; we simply divided.

DO EXERCISE 11.

For any base a, $\log_a a = 1$. This is easily seen by writing an equivalent exponential equation, $a^1 = a$. Similarly, for any base a, $\log_a 1 = 0$. These facts are important and should be remembered well.

THEOREM 6

For any base *a*,

$$\log_a a = 1 \quad \text{and} \quad \log_a 1 = 0.$$

DO EXERCISES 12–15.

---

7. Express as a difference.

a) $\log_a \frac{M}{N}$

b) $\log_c \frac{1}{4}$

8. Express as sums and differences of logarithms and without exponential notation or radicals.

$$\log_{10} \frac{4\pi}{\sqrt{23}}$$

9. Express in terms of logarithms of x, y, and z.

$$\log_a \sqrt{\frac{z^3}{xy}}$$

10. Express as a single logarithm.

$$5\log_a x - \log_a y + \frac{1}{4}\log_a z$$

11. Given that $\log_a 2 = 0.301$ and $\log_a 3 = 0.477$, find:

a) $\log_a 9;$
b) $\log_a \sqrt{2};$
c) $\log_a \sqrt[3]{2};$
d) $\log_a \frac{3}{2};$
e) $\frac{\log_a 3}{\log_a 2}.$

Simplify.

12. $\log_\pi \pi$

13. $\log_9 1$

14. $\log_e 1$

15. $\log_{1/4} \frac{1}{4}$

EXERCISE SET 7.2

[i] Express in terms of logarithms of x, y, and z.

1. $\log_a x^2 y^3 z$

2. $\log_a 5xy^4 z^3$

3. $\log_b \dfrac{xy^2}{z^3}$

4. $\log_c \sqrt[3]{\dfrac{x^4}{y^3 z^2}}$

Express as a single logarithm and simplify if possible.

5. $\dfrac{2}{3}\log_a 64 - \dfrac{1}{2}\log_a 16$

6. $\dfrac{1}{2}\log_a x + 3\log_a y - 2\log_a x$

7. $\log_a 2x + 3(\log_a x - \log_a y)$

8. $\log_a x^2 - 2\log_a \sqrt{x}$

9. $\log_a \dfrac{a}{\sqrt{x}} - \log_a \sqrt{ax}$

10. $\log_a (x^2 - 4) - \log_a (x - 2)$

11. $\log_a (x^3 + y^3) - \log_a (x + y)$

12. $\log_a (x - y) + \log_a (x^2 + xy + y^2)$

Express as a sum and/or difference of logarithms.

13. $\log_a \sqrt{1 - x^2}$

14. $\log_a \dfrac{x + t}{\sqrt{x^2 - t^2}}$

Given $\log_{10} 2 = 0.301$, $\log_{10} 3 = 0.477$, and $\log_{10} 10 = 1$, find:

15. $\log_{10} 4$.

16. $\log_{10} 5$.
$\left(Hint: 5 = \dfrac{10}{2}\right)$

17. $\log_{10} 50$.
$\left(Hint: 50 = \dfrac{100}{2}\right)$

18. $\log_{10} 12$.

19. $\log_{10} 60$.

20. $\log_{10} \dfrac{1}{3}$.

21. $\log_{10} \sqrt{\dfrac{2}{3}}$.

22. $\log_{10} \sqrt[5]{12}$.

23. $\log_{10} 90$.

24. $\log_{10} \dfrac{9}{8}$.

25. $\log_{10} \dfrac{9}{10}$.

26. $\log_{10} \dfrac{1}{4}$.

☆ ──────────────────────────────

Which of the following are false?

27. $\dfrac{\log_a M}{\log_a N} = \log_a M - \log_a N$

28. $\dfrac{\log_a M}{\log_a N} = \log_a \dfrac{M}{N}$

29. $\dfrac{\log_a M}{c} = \log_a M^{1/c}$

30. $\log_N (M \cdot N)^x = x\log_N M + x$

31. $\log_a 2x = 2\log_a x$

32. $\log_a 2x = \log_a 2 + \log_a x$

33. $\log_a (M + N) = \log_a M + \log_a N$

34. $\log_a x^3 = 3\log_a x$

Solve.

35. $\log_\pi \pi^{2x+3} = 4$

36. $3^{\log_3 (8x-4)} = 5$

37. $4^{2\log_4 x} = 7$

38. $8^{2\log_8 x + \log_8 x} = 27$

39. $(x + 3) \cdot \log_a a^x = x$

40. $\log_a x^2 = 2\log_a x$

41. $\log_a 5x = \log_a 5 + \log_a x$

42. $\log_b \dfrac{5}{x + 2} = \log_b 5 - \log_b (x + 2)$

★ ──────────────────────────────

43. If $\log_a x = 2$, what is $\log_a \left(\dfrac{1}{x}\right)$?

44. If $\log_a x = 2$, what is $\log_{1/a} x$?

Prove the following for any base a and any positive number x.

45. $\log_a \left(\dfrac{1}{x}\right) = -\log_a x$

46. $\log_a \left(\dfrac{1}{x}\right) = \log_{1/a} x$

47. Show that $\log_a \left(\dfrac{x + \sqrt{x^2 - 5}}{5}\right) = -\log_a (x - \sqrt{x^2 - 5})$.

48. Graph and compare: $y = \log_2 |x|$ and $y = |\log_2 x|$.

7.3 COMMON LOGARITHMS

Base ten logarithms are know as *common logarithms*. Tables for these logarithms are readily available (Table 2 at the back of this book).

Historical Uses of Logarithms

Before calculators and computers became so readily available, common logarithms were used extensively to do certain kinds of calculations. In fact, that is why logarithms were developed. Today, computations with logarithms are mainly of historical interest; the logarithm *functions* are of modern importance. The study of computations with logarithms can, however, help the student fix the ideas with respect to properties of logarithm functions.

The following is a table of powers of 10, or *logarithms* base 10.

$$
\begin{array}{llll}
1 = 10^{0.0000}, & \text{or} & \log_{10} 1 &= 0.0000 \\
2 = 10^{0.3010}, & \text{or} & \log_{10} 2 &= 0.3010 \\
3 = 10^{0.4771}, & \text{or} & \log_{10} 3 &= 0.4771 \\
4 = 10^{0.6021}, & \text{or} & \log_{10} 4 &= 0.6021 \\
5 = 10^{0.6990}, & \text{or} & \log_{10} 5 &= 0.6990 \\
6 = 10^{0.7782}, & \text{or} & \log_{10} 6 &= 0.7782 \\
7 = 10^{0.8451}, & \text{or} & \log_{10} 7 &= 0.8451 \\
8 = 10^{0.9031}, & \text{or} & \log_{10} 8 &= 0.9031 \\
9 = 10^{0.9542}, & \text{or} & \log_{10} 9 &= 0.9542 \\
10 = 10^{1.0000}, & \text{or} & \log_{10} 10 &= 1.0000 \\
11 = 10^{1.0414}. & \text{or} & \log_{10} 11 &= 1.0414 \\
12 = 10^{1.0792}, & \text{or} & \log_{10} 12 &= 1.0792 \\
13 = 10^{1.1139}, & \text{or} & \log_{10} 13 &= 1.1139 \\
14 = 10^{1.1461}, & \text{or} & \log_{10} 14 &= 1.1461 \\
15 = 10^{1.1761}, & \text{or} & \log_{10} 15 &= 1.1761 \\
16 = 10^{1.2041}, & \text{or} & \log_{10} 16 &= 1.2041
\end{array}
$$

The exponents are approximate, but accurate to four decimal places. To illustrate how logarithms can be used for computation we will use the above table and do some easy calculations.

Example 1 Find 3×4 using the table of exponents.

$$
\begin{aligned}
3 \times 4 &= 10^{0.4771} \times 10^{0.6021} \\
&= 10^{1.0792} \qquad \text{Adding exponents}
\end{aligned}
$$

From the table we see that $10^{1.0792} = 12$, so $3 \times 4 = 12$.

DO EXERCISE 1.

Note from Example 1 that we can find a product by adding the logarithms of the factors and then finding the number having the result as its logarithm. That is, we found the number $10^{1.0792}$. This number is often referred to as the *antilogarithm* of 1.0792. In other words, if

$$ f(x) = \log_{10} x, $$

then

$$ f^{-1}(x) = \text{antilog}_{10} x = 10^x. $$

Using base 10 logarithms, find the following.

2. 4×2

3. $\dfrac{15}{3}$

4. $\sqrt[3]{8}$

5. 3^2

Use Table 2 to find each logarithm.

6. $\log 3.14$

7. $\log 9.99$

8. $\log 4.00$

Use Table 2 to find each antilogarithm.

9. antilog 0.7589

10. antilog 0.0000

11. antilog 0.5587

In other words, an antilogarithm function is simply an exponential function.

Example 2 Find $\frac{14}{2}$, using base 10 logarithms.

$$\log_{10} \frac{14}{2} = \log_{10} 14 - \log_{10} 2$$
$$= 1.1461 - 0.3010$$
$$\log_{10} \frac{14}{2} = 0.8451$$
$$\frac{14}{2} = \text{antilog}_{10}\, 0.8451 = 7$$

Example 3 Find $\sqrt[4]{16}$, using base 10 logarithms.

$$\log_{10} \sqrt[4]{16} = \log_{10} 16^{1/4} = \frac{1}{4} \cdot \log_{10} 16 = \frac{1}{4} \cdot 1.2041$$
$$\log_{10} \sqrt[4]{16} = 0.3010 \qquad \text{Rounded to four decimal places}$$
$$\sqrt[4]{16} = \text{antilog}_{10}\, 0.3010 = 2$$

Example 4 Find 2^3, using base 10 logarithms.

$$\log_{10} 2^3 = 3 \cdot \log_{10} 2 = 3 \cdot 0.3010 = 0.9030$$
$$\log_{10} 2^3 = 0.9030$$
$$2^3 = \text{antilog}_{10}\, 0.9030 \approx 8$$

Note the rounding error.

DO EXERCISES 2–5.

[i] Finding Common Logarithms Using Tables

We often omit the base, 10, when working with common logarithms.

Table 2 contains logarithms of numbers from 1 to 10. Part of that table is shown below. To illustrate the use of the table, let us find log 5.24. We locate the row headed 5.2, then move across to the column headed 4. We find log 5.24 as the colored entry in the table.

x	0	1	2	3	4	5	6	7	8	9
5.0	0.6990	0.6998	0.7007	0.7016	0.7024	0.7033	0.7042	0.7050	0.7059	0.7067
5.1	0.7076	0.7084	0.7093	0.7101	0.7110	0.7118	0.7126	0.7135	0.7143	0.7152
5.2	0.7160	0.7168	0.7177	0.7185	0.7193	0.7202	0.7210	0.7218	0.7226	0.7235
5.3	0.7234	0.7251	0.7259	0.7267	0.7275	0.7284	0.7292	0.7300	0.7308	0.7316
5.4	0.7324	0.7332	0.7340	0.7348	0.7356	0.7364	0.7372	0.7380	0.7388	0.7396

We can find antilogarithms by reversing this process. For example, antilog $0.7193 = 10^{0.7193} = 5.24$. Similarly, antilog $0.7292 = 5.36$.

DO EXERCISES 6–11.

Using Table 2 and scientific notation* we can approximate logarithms of numbers that are not between 1 and 10. First recall that

$$\log_a a^k = k \quad \text{for any number } k. \quad \text{Theorem 2}$$

Thus

$$\log_{10} 10^k = k \quad \text{for any number } k.$$

Examples

5. $\log 52.4 = \log (5.24 \times 10^1)$ Converting to scientific notation
$= \log 5.24 + \log 10^1$ Theorem 3
$= 0.7193 + 1$

6. $\log 0.524 = \log (5.24 \times 10^{-1})$
$= \log 5.24 + \log 10^{-1}$
$= 0.7193 + (-1)$

7. $\log 52{,}400 = \log (5.24 \times 10^4)$
$= \log 5.24 + \log 10^4$
$= 0.7193 + 4$

8. $\log 0.00524 = \log (5.24 \times 10^{-3})$
$= \log 5.24 + \log 10^{-3}$
$= 0.7193 + (-3)$

DO EXERCISES 12 AND 13.

The preceding examples illustrate the importance of using the base 10 for computation. It allows great economy in the printing of tables. If we know the logarithm of a number from 1 to 10 we can multiply that number by any power of ten and easily determine the logarithm of the resulting number. For any base other than 10 this would not be the case. In each of Examples 5–8, the integer part of the logarithm is the exponent in the scientific notation. This integer is called the *characteristic* of the logarithm. The other part of the logarithm, a number between 0 and 1, is called the *mantissa* of the logarithm. Table 2 contains only mantissas.

Example 9 Find log 0.0538, indicating the characteristic and mantissa.

We first write scientific notation for the number:

$$5.38 \times 10^{-2}.$$

Then we find log 5.38. This is the mantissa:

$$\log 5.38 = 0.7308.$$

The characteristic of the logarithm is the exponent -2. Now $\log 0.0538 = 0.7308 + (-2)$, or -1.2692. When negative characteristics occur, it is often best to name the logarithm so that the characteristic and mantissa are preserved. In the preceding example, we have

$$\log 0.0538 = 0.7308 + (-2) = -1.2692,$$

Use scientific notation and Table 2 to find each logarithm.

12. log 289

13. log 0.000289

---

*It may be helpful to review scientific notation in Chapter 1.

Find the following. Use Table 2 and try to write only the answer. Where appropriate, name so that positive mantissas are preserved.

14. log 67,800

15. log 892,000

16. log 45.9

17. log 609,000,000

18. log 0.0782

19. log 0.000111

20. log 0.0079

but the latter notation displays neither the characteristic nor the mantissa. We can rename the characteristic, -2, as $8 - 10$, and then add the mantissa, to obtain

$$8.7308 - 10,$$

thus preserving both mantissa and characteristic.

▦ The characteristic and mantissa are useful when working with logarithm tables, but are not needed on a calculator. For example, on a calculator with a ten-digit readout, we find that

$$\log 0.0538 = -1.269217724,$$

but this shows neither the characteristic nor the mantissa. Check this on your calculator. How can you find the characteristic and mantissa?

Example 10 Find log 0.00687.

We write scientific notation (or at least visualize it):

$$0.00687 = 6.87 \times 10^{-3}.$$

The characteristic is -3, or $7 - 10$. The mantissa, from the table, is 0.8370. Thus $\log 0.00687 = 7.8370 - 10$.

DO EXERCISES 14–20.

Antilogarithms

To find antilogarithms, we reverse the procedure for finding logarithms.

Example 11 Find antilog 2.6085 (or find x such that $\log x = 2.6085$).

$$\text{antilog } 2.6085 = 10^{2.6085}$$
$$= 10^{(2+0.6085)}$$
$$= 10^2 \cdot 10^{0.6085}$$

From the table we can find $10^{0.6085}$, or antilog 0.6085. It is 4.06. Thus we have

$$\text{antilog } 2.6085 = 10^2 \times 4.06, \quad \text{or } 406.$$

Note in this example, we in effect separate the number 2.6085 into an integer and a number between 0 and 1. We use the latter with the table, after which we have scientific notation for our answer.

Example 12 Find antilog 3.7118.

From the table we find that antilog 0.7118 = 5.15. Thus

$$\text{antilog } 3.7118 = 5.15 \times 10^3 \quad \text{Note that 3 is the characteristic.}$$
$$= 5150.$$

Example 13 Find antilog $(7.7143 - 10)$.

The characteristic is -3 and the mantissa is 0.7143.

From the table we find that antilog $0.7143 = 5.18$. Thus

$$\text{antilog } (7.7143 - 10) = 5.18 \times 10^{-3}$$
$$= 0.00518.$$

Example 14 Find antilog -2.2857.

We are to find the antilog of a number, but the number is named so that the mantissa is not apparent. To find the mantissa we add 0, naming it $10 - 10$:

$$-2.2857 = -2.2857 + (10 - 10)$$
$$= (-2.2857 + 10) - 10 = 7.7143 - 10.$$

Then we proceed as in Example 13. The answer is 0.00518.

DO EXERCISES 21-25.

[ii] Calculations with Logarithms (Optional)

The kinds of calculations in which logarithms were helpful historically are multiplication, division, taking powers, and taking roots.

Example 15 Find $\dfrac{0.0578 \times 32.7}{8460}$.

We write a *plan* for the use of logarithms, then look up all the mantissas at one time. Let

$$N = \frac{0.0578 \times 32.7}{8460}.$$

Then

$$\log N = \log 0.0578 + \log 32.7 - \log 8460.$$

This gives us the plan. We use a straight line to indicate addition and a wavy line to indicate subtraction.

Completion of plan:

$$\log 0.0578 = 8.7619 - 10$$
$$\log 32.7 = \underline{1.5145}$$

$$\log \text{numerator} = 10.2764 - 10$$
$$\log 8460 = \underline{3.9274}$$

$$\log \text{fraction} = 6.3490 - 10$$
$$\text{fraction} = 0.000223 \quad \text{Taking antilog}$$

DO EXERCISE 26.

Find the following. Use Table 2. Try to write only the answer.

21. If $\log x = 4.8069$, find x.

22. $10^{4.8069}$

23. $10^{3.9325}$

24. antilog $6.6284 - 10$

25. antilog -1.9788

26. Use logarithms to find

$$\frac{88.4 \times 0.00641}{9.43}.$$

27. $(92.8)^3 \times \sqrt[5]{0.986}$

Example 16 Use logarithms to find $\sqrt[4]{0.325} \times (4.23)^2$.

If

$$N = \sqrt[4]{0.325} \times (4.23)^2,$$

then

$$\log N = \frac{1}{4}\log 0.325 + 2\log 4.23;$$

this yields the plan. Note that powers and roots are involved. In such cases it is easier to first find the following logs:

$$\log 0.325 = 9.5119 - 10$$
$$\log 4.23 = 0.6263.$$

Consider $9.5119 - 10$. Since we will be dividing this number by 4, but -10 is not divisible by 4, we rename this number:

$$\log 0.325 = 39.5119 - 40.$$

Now the negative term is divisible by 4. We divide $\log 0.325$ by 4 and multiply $\log 4.23$ by 2. Then we have

$$\log \sqrt[4]{0.325} = \quad 9.8780 - 10$$
$$\log (4.23)^2 = \quad \underline{1.2526}$$

$$\log \text{product} = 11.1306 - 10$$
$$= \quad 1.1306$$
$$\text{product} = 13.5.$$

DO EXERCISE 27.

EXERCISE SET 7.3

[i] Use Table 2 to find each of the following. Where appropriate, write so that (positive) mantissas are preserved.

1. log 2.46   **2.** log 7.5   **3.** log 347   **4.** log 8720

5. log 52.5   **6.** log 20.8   **7.** log 624,000   **8.** log 13,400

9. log 0.0702   **10.** log 0.640   **11.** log 0.000216   **12.** log 0.173

13. antilog 2.3674   **14.** antilog 1.9222   **15.** antilog $8.2553 - 10$ **16.** antilog $6.6294 - 10$

17. antilog -5.9788 **18.** antilog -2.2628 **19.** $10^{1.4014}$ **20.** $10^{3.6590}$

21. $10^{8.9881-10}$ **22.** $10^{7.5391-10}$

Solve for x.

23. $\log x = 0.6522$ **24.** $\log x = 4.8156$

25. $\log x = 8.1239 - 10$ **26.** $\log x = -1.0218$

[ii] (Optional). Use logarithms to compute. When exact values do not occur in the table, use the nearest values.

27. 3.14×60.4 **28.** 541×0.0152 **29.** $286 \div 1.05$ **30.** $12.8 \div 81.6$

31. $\sqrt{76.9}$ **32.** $\sqrt[3]{56.9}$ **33.** $(1.36)^{4.2}$ **34.** $(0.727)^{3.6}$

35. $\dfrac{70.7 \times (10.6)^2}{18.6 \times \sqrt{276}}$ **36.** $\sqrt[3]{\dfrac{3.24 \times (3.16)^2}{78.4 \times 24.6}}$

☆ ───

Find each of the following. Round to six decimal places.

37. ▦ log 56,789   **38.** ▦ log 0.0111347   **39.** ▦ log (log 3)

Find each antilogarithm. Since $antilog_{10} x = 10^x$, you can find the antilogarithm, base 10, of a number x by raising 10 to the power x.

40. ▦ $10^{0.4356}$ **41.** ▦ $antilog_{10} 7.8943$ **42.** ▦ $antilog_{10} (-7.5689)$

43. ▦ $10^{-3.23445678}$

───

7.4 INTERPOLATION (OPTIONAL*)

Tables are often prepared giving function values for a continuous function. Suppose the table gives four-digit precision. By using a procedure called *interpolation,* we can estimate values between those listed in the table, obtaining four-digit precision.† Interpolation can be done in various ways, the simplest and most common being *linear interpolation.* We describe it now in relation to Table 2 for common logarithms. Remember that what we say applies to a table for *any* continuous function.

[i] Linear Interpolation

Let us consider how a table of values for any function is made. We select members of the domain x_1, x_2, x_3, and so on. Then we compute or somehow determine the corresponding function values $f(x_1)$, $f(x_2)$, $f(x_3)$, and so on. Then we tabulate the results. We might also graph the results.

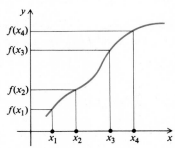

x	x_1	x_2	x_3	x_4	$\cdots$
f(x)	$f(x_1)$	$f(x_2)$	$f(x_3)$	$f(x_4)$	$\cdots$

OBJECTIVE

You should be able to:

[i] Use linear interpolation to find logarithms and antilogarithms from a table.

─────────────────────

*To the instructor: The advent of the hand-held calculator has in one way rendered this section unnecessary, but the skill of interpolating will always be useful because no calculator can contain every table. Thus if interpolating is a goal of your course, then this section should be taught.
†If a calculator were used, there could be some variance in the last decimal place.

1. Find log 4562.

Suppose we want to find the function value $f(x)$ for an x not in the table. If x is halfway between x_1 and x_2, then we can take the number halfway between $f(x_1)$ and $f(x_2)$ as an approximation to $f(x)$. If x is one-fifth of the way between x_2 and x_3, we take the number that is one-fifth of the way between $f(x_2)$ and $f(x_3)$ as an approximation to $f(x)$. What we do is divide the length from x_2 to x_3 in a certain ratio, and then divide the length from $f(x_2)$ to $f(x_3)$ in the same ratio. This is *linear interpolation*.

We can show this geometrically. The length from x_1 to x_2 is divided in a certain ratio by x. The length from $f(x_1)$ to $f(x_2)$ is divided in the same ratio by y. The number y approximates $f(x)$ with the noted error.

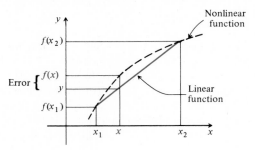

Note the slanted line in the figure. The approximation y comes from this line. This explains the use of the term *linear interpolation*. Let us apply linear interpolation to Table 2 of common logarithms.

Example 1 Find log 34,870.

a) Find the characteristic. Since $34{,}870 = 3.487 \times 10^4$, the characteristic is 4.

b) Find the mantissa. From Table 2 we have:

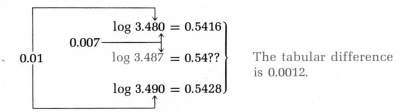

The tabular difference (difference between consecutive values in the table) is 0.0012. Now 3.487 is $\frac{7}{10}$ of the way from 3.480 to 3.490. So we take 0.7 of 0.0012, which is 0.00084, and round it to 0.0008. We add this to 0.5416. The mantissa that results is 0.5424.

c) Add the characteristic and mantissa:

$$\log 34{,}870 = 4.5424.$$

With practice you will take 0.7 of 12, forgetting the zeros but adding in the same way.

DO EXERCISE 1.

Example 2 Find log 0.009543.

a) Find the characteristic. Since $0.009543 = 9.543 \times 10^{-3}$, the characteristic is -3, or $7 - 10$.

b) Find the mantissa. From Table 2 we have:

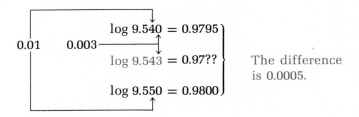

$$\left. \begin{array}{l} \log 9.540 = 0.9795 \\ \log 9.543 = 0.97?? \\ \log 9.550 = 0.9800 \end{array} \right\}$$ The difference is 0.0005.

Now 9.543 is $\frac{3}{10}$ of the way from 9.540 to 9.550, so we take 0.3 of 0.0005, which is 0.00015, and round it to 0.0002. We add this to 0.9795. The mantissa that results is 0.9797.

c) Add the characteristic and the mantissa:

$$\log 0.009543 = 7.9797 - 10.$$

DO EXERCISE 2.

Antilogarithms

We interpolate when finding antilogarithms, using the table in reverse.

Example 3 Find antilog 4.9164.

a) The characteristic is 4. The mantissa is 0.9164.

b) Find the antilog of the mantissa, 0.9164. From Table 2, we have:

$$\left. \begin{array}{l} \text{antilog } 0.9159 = 8.240 \\ \text{antilog } 0.9164 = 8.24? \\ \text{antilog } 0.9165 = 8.250 \end{array} \right\}$$ The difference is 0.010.

The difference between 0.9159 and 0.9165 is 0.0006. Thus 0.9164 is $\frac{0.0005}{0.0006}$, or $\frac{5}{6}$, of the way between 0.9159 and 0.9165. Then antilog 0.9164 is $\frac{5}{6}$ of the way between 8.240 and 8.250, or $\frac{5}{6}(0.010)$, which is $0.00833\ldots$, and round it to 0.008. Thus the antilog of the mantissa is 8.248.

Thus antilog $4.9164 = 8.248 \times 10^4 = 82{,}480$.

DO EXERCISE 3.

2. Find log 0.02387.

3. Find $10^{3.4557}$.

4. Find antilog (6.7749 − 10).

Example 4 Find antilog (7.4122 − 10).

a) The characteristic is −3. The mantissa is 0.4122.

b) Find the antilog of the mantissa, 0.4122. From Table 2 we have:

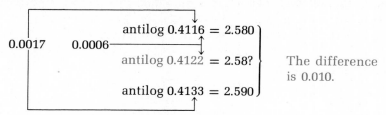

The difference between 0.4116 and 0.4133 is 0.0017. Thus 0.4122 is $\frac{0.0006}{0.0017}$, or $\frac{6}{17}$, of the way between 0.4116 and 0.4133. Then antilog 0.4122 is $\frac{6}{17}$ of the way between 2.580 and 2.590, or $\frac{6}{17}(0.010)$, which is 0.0035, to four places. We round it to 0.004. Thus the antilog of the mantissa is 2.584.

So antilog (7.4122 − 10) = 2.584×10^{-3} = 0.002584.

DO EXERCISE 4.

Calculations

Calculations with logarithms, using Table 2, can be done with four-digit precision when interpolation is used in finding both logarithms and antilogarithms.

EXERCISE SET 7.4

[i] Find each of the following logarithms using interpolation and Table 2.

1. log 41.63   **2.** log 472.1   **3.** log 2.944   **4.** log 21.76

5. log 650.2   **6.** log 37.37   **7.** log 0.1425   **8.** log 0.0904

9. log 0.004257   **10.** log 4518   **11.** log 0.1776   **12.** log 0.08356

13. log 600.6   **14.** log (log 3)   **15.** log (log 5)

Find each of the following antilogarithms using interpolation and Table 2.

16. antilog 1.6350   **17.** antilog 2.3512   **18.** antilog 0.6478   **19.** antilog 1.1624

20. antilog 0.0342   **21.** antilog 4.8453   **22.** antilog 9.8561 − 10 **23.** antilog 8.9659 − 10

24. antilog 7.4128 − 10 **25.** antilog 9.7278 − 10 **26.** antilog 8.2010 − 10 **27.** antilog 7.8630 − 10

☆

Use logarithms and interpolation to do the following calculations. Use four-digit precision. Answers may be checked using a calculator.

28. $\dfrac{35.24 \times (16.77)^3}{12.93 \times \sqrt{276.2}}$

29. $\sqrt[5]{\dfrac{16.79 \times (4.234)^3}{18.81 \times 175.3}}$

7.5 EXPONENTIAL AND LOGARITHMIC EQUATIONS

[i] Exponential Equations

An equation with variables in exponents, such as $3^{2x-1} = 4$, is called an *exponential equation*. We can solve such equations by taking logarithms on both sides and then using Theorem 4.*

Example 1 Solve $3^x = 8$.

$$\log 3^x = \log 8 \qquad \text{Taking log on both sides.}$$
$$\text{Remember } \log m = \log_{10} m.$$
$$x \log 3 = \log 8 \qquad \text{Using Theorem 4}$$
$$x = \frac{\log 8}{\log 3} \qquad \text{Solving for x}$$
$$x \approx \frac{0.9031}{0.4771} \approx 1.8929 \qquad \text{We look up the logs, or find them on a calculator, and divide.}$$

DO EXERCISE 1.

Example 2 Solve $2^{3x-5} = 16$.

$$\log 2^{3x-5} = \log 16 \qquad \text{Taking log on both sides}$$
$$(3x - 5) \log 2 = \log 16 \qquad \text{Using Theorem 4}$$
$$3x - 5 = \frac{\log 16}{\log 2}$$
$$x = \frac{\frac{\log 16}{\log 2} + 5}{3} \approx \frac{\frac{1.2041}{0.3010} + 5}{3} \qquad \begin{array}{l} \text{Solving for x and} \\ \text{evaluating logarithms} \end{array}$$
$$x \approx 3.0001 \qquad \text{Calculating}$$

The answer is approximate because the logarithms are approximate.

DO EXERCISE 2.

Example 3 Solve $\dfrac{e^x + e^{-x}}{2} = t$, for x.

Note that we are to solve for x. However, we have more than one term with x in the exponent. To get a single expression with x in the exponent, we do the following:

$$e^x + e^{-x} = 2t \qquad \text{Multiplying by 2}$$
$$e^x + \frac{1}{e^x} = 2t \qquad \begin{array}{l} \text{Rewriting to eliminate the minus} \\ \text{sign in an exponent} \end{array}$$
$$e^{2x} + 1 = 2t \cdot e^x \qquad \text{Multiplying on both sides by } e^x$$
$$(e^x)^2 - 2t \cdot e^x + 1 = 0$$

*The use of a calculator is recommended for the rest of this chapter.

OBJECTIVES

You should be able to:

[i] Solve exponential and logarithmic equations.

[ii] Solve applied problems involving exponential and logarithmic equations.

1. Solve $2^x = 7$.

2. Solve $4^{2x-3} = 64$, using Table 2.

3. Solve $\dfrac{e^x - e^{-x}}{2} = t$, for x.

This equation is reducible to quadratic, with $u = e^x$. Using the quadratic formula, we obtain

$$e^x = \frac{2t \pm \sqrt{4t^2 - 4}}{2} = t \pm \sqrt{t^2 - 1}$$

$$\log e^x = \log (t \pm \sqrt{t^2 - 1}) \qquad \text{Taking log on both sides}$$
$$x \log e = \log (t \pm \sqrt{t^2 - 1}) \qquad \text{Using Theorem 4}$$
$$x = \frac{\log (t \pm \sqrt{t^2 - 1})}{\log e}.$$

If *e* should be a number suitable as a logarithm base, we might also take base *e* logarithms on both sides. The denominator would then be $\log_e e$, or 1, and we would have

$$x = \log_e (t \pm \sqrt{t^2 - 1}).$$

DO EXERCISE 3.

4. Solve.

$$\log_5 x = 3$$

Logarithmic Equations

Equations that contain logarithmic expressions are called *logarithmic equations*. We solved some such equations in Section 7.1. We did so by converting to an equivalent exponential equation. For example, to solve $\log_2 x = -3$, we convert to $x = 2^{-3}$ and find that $x = \frac{1}{8}$.

To solve logarithmic equations we first try to obtain a single logarithmic expression on one side of the equation and then write an equivalent exponential equation.

Example 4 Solve $\log_3 (5x + 7) = 2$.

We already have a single logarithmic expression, so we write an equivalent exponential equation:

5. Solve.

$$\log_4 (8x - 6) = 3$$

$$5x + 7 = 3^2 \qquad \text{Writing an equivalent exponential equation}$$
$$5x + 7 = 9$$
$$5x = 2$$
$$x = \frac{2}{5}$$

Check:

$$
\begin{array}{c|c}
\log_3 (5x + 7) & = 2 \\
\hline
\log_3 \left(5 \cdot \dfrac{2}{5} + 7\right) & 2 \\
\log_3 (2 + 7) & \\
\log_3 9 & \\
2 &
\end{array}
$$

DO EXERCISES 4 AND 5.

Example 5 Solve $\log x + \log (x - 3) = 1$.

Here we must first obtain a single logarithmic equation:

$$\log x + \log (x - 3) = 1$$
$$\log x (x - 3) = 1 \qquad \text{Using Theorem 3 to obtain a single logarithm}$$
$$\log_{10} x(x - 3) = 1$$
$$x(x - 3) = 10^1 \qquad \text{Converting to an equivalent exponential equation}$$
$$x^2 - 3x = 10$$
$$x^2 - 3x - 10 = 0$$
$$(x + 2)(x - 5) = 0 \qquad \text{Factoring and principle of zero products}$$
$$x = -2 \quad \text{or} \quad x = 5$$

Possible solutions to logarithmic equations must be checked because domains of logarithmic functions consist only of positive numbers.

Check:

$\log x + \log (x - 3) = 1$	$\log x + \log (x - 3) = 1$
$\log (-2) + \log (-2 - 3)$ $\mid$ 1	$\log 5 + \log (5 - 3)$ $\mid$ 1
	$\log 5 + \log 2$
	$\log 10$
	1

The number -2 is not a solution because negative numbers do not have logarithms. The solution is 5.

DO EXERCISE 6.

6. Solve $\log x + \log (x + 3) = 1$.

[ii] Applications

Exponential and logarithmic functions and equations have many applications. We shall consider a few of them.

Example 6 (*Compound interest*). The amount A that principal P will be worth after t years at interest rate i, compounded annually, is given by the formula $A = P(1 + i)^t$. Suppose $4000 principal is invested at 6% interest and yields $5353. For how many years was it invested?

Using the formula $A = P(1 + i)^t$, we have

$$5353 = 4000(1 + 0.06)^t, \quad \text{or} \quad 5353 = 4000(1.06)^t.$$

Solving for t, we have

$$\log 5353 = \log 4000(1.06)^t \qquad \text{Taking log on both sides}$$
$$\log 5353 = \log 4000 + t \log 1.06 \qquad \text{Theorems 3 and 4}$$
$$\frac{\log 5353 - \log 4000}{\log 1.06} = t \qquad \text{Solving for } t$$
$$\frac{3.7286 - 3.6021}{0.0253} = t \qquad \text{Evaluating logarithms}$$
$$5 = t.$$

7. $5000 was invested at 14%, compounded annually, and it yielded $18,540. For how long was it invested?

The money was invested for 5 years.

▦ We can use a calculator for an approximate *check:*

$5353 \;=\; 4000(1.06)^t$
5353 $4000(1.06)^5$
$4000(1.338226)$
5352.904

DO EXERCISE 7.

Example 7 (*Loudness of sound*). The sensation of loudness of sound is not proportional to the energy intensity, but rather is a logarithmic function. *Loudness,* in Bels (after Alexander Graham Bell), of a sound of intensity I is defined to be

$$L = \log \frac{I}{I_0},$$

8. Find the loudness, in decibels, of the sound in a library that is 2510 times as intense as the minimum intensity I_0.

where I_0 is the minimum intensity detectable by the human ear (such as the tick of a watch at 20 ft under quiet conditions). When a sound is 10 times as intense as another, its loudness is 1 Bel greater. If a sound is 100 times as intense as another, it is louder by 2 Bels, and so on. The Bel is a large unit, so a subunit, a *decibel,* is generally used. For L in decibels, the formula is as follows:

$$L = 10 \log \frac{I}{I_0}.$$

a) Find the loudness, in decibels, of the sound in a radio studio, for which the intensity I is 199 times I_0.

We substitute into the formula and calculate, using Table 2:

$$L = 10 \log \frac{199 \cdot I_0}{I_0} = 10 \log 199$$

$$= 10(2.2989)$$
$$= 23 \text{ decibels.}$$

9. Find the loudness, in decibels, of conversational speech, having an intensity that is 10^6 times as intense as the minimum, I_0.

b) Find the loudness of the sound of a heavy truck, for which the intensity is 10^9 times I_0.

$$L = 10 \log \frac{10^9 \cdot I_0}{I_0} = 10 \log 10^9$$

$$= 10 \cdot 9$$
$$= 90 \text{ decibels.}$$

DO EXERCISES 8 AND 9.

Example 8 (*Earthquake magnitude*). The magnitude R (on the Richter scale) of an earthquake of intensity I is defined as follows:

$$R = \log \frac{I}{I_0},$$

where I_0 is a minimum intensity used for comparison. The Mexico City earthquake of 1978 had an intensity $10^{7.85}$ times I_0. What is its magnitude on the Richter scale?

We substitute into the formula:

$$R = \log \frac{10^{7.85} \cdot I_0}{I_0} = \log 10^{7.85} = 7.85.$$

DO EXERCISE 10.

Example 9 (*Forgetting*). Here is a mathematical model from psychology. A group of people take a test and make an average score of S. After a time t they take an equivalent form of the same test. At that time the average score is $S(t)$. According to this model, $S(t)$ is given by the following function:

$$S(t) = A - B \log (t + 1),$$

where t is in months and the constants A and B are determined by experiment in various kinds of learning situations. The model is appropriate only over the interval $[0, 10^{4/B} - 1]$. Students in a zoology class took a final exam, and took equivalent forms of the test at monthly intervals thereafter. The average scores were found to be given by the function

$$S(t) = 78 - 15 \log (t + 1).$$

What was the average score (a) when they took the test originally? (b) after 4 months?

We substitute into the equation defining the function:

a) $S(0) = 78 - 15 \log (0 + 1)$
   $= 78 - 15 \log 1$
   $= 78 - 0 = 78.$ Original score

b) $S(4) = 78 - 15 \log (4 + 1)$
   $= 78 - 15 \log 5$
   $= 78 - 15 \cdot 0.6990$
   $= 78 - 10.49 = 67.51.$ Score after 4 months

DO EXERCISE 11.

Alternate Methods of Solution (Optional)

Following are some other methods of solving exponential equations.

Example 10 Solve $3^x = 8$.

Instead of taking the common logarithm on both sides we take the logarithm, base 3, on both sides. Then we have

$$\log_3 3^x = \log_3 8$$
$$x = \log_3 8.   \text{Theorem 2}$$

10. An earthquake has an intensity that is $10^{7.8}$ times I_0. What is its magnitude on the Richter scale?

11. Students in an accounting class took a final exam and then equivalent forms of the same test at monthly intervals. The average score $S(t)$, after t months, was found to be given by

$$S(t) = 68 - 20 \log (t + 1).$$

a) What was the initial average score?
b) What was the average score after 4 months?
c) What was the average score after 24 months?

12. Solve $2^x = 7$. Use the method of Example 9.

13. Solve $4^{2x-3} = 64$. Use the method of Example 10.

The only problem with this method compared to that of Example 1 is that as yet we do not know how to estimate $\log_3 8$. We shall consider this in Section 7.6.

Example 11 Solve $2^{3x-5} = 16$.

Note that $16 = 2^4$. Then we have

$$2^{3x-5} = 2^4.$$

Since the base is the same, 2, on both sides, the exponents must be the same. Thus

$$3x - 5 = 4$$
$$3x = 9$$
$$x = 3.$$

DO EXERCISES 12 AND 13.

EXERCISE SET 7.5

[i] Solve.

1. $2^x = 32$

2. $3^{x-1} + 3 = 30$

3. $4^{2x} = 8^{3x-4}$

4. $3^{x^2+4x} = \dfrac{1}{27}$

5. $3^{5x} \cdot 9^{x^2} = 27$

6. $4^x = 7$

7. $2^x = 3^{x-1}$

8. $3^{x+2} = 5^{x-1}$

9. $\log x + \log (x - 9) = 1$

10. $\log x - \log (x + 3) = -1$

11. $\log (x + 9) - \log x = 1$

12. $\log (2x + 1) - \log (x - 2) = 1$

13. $\log_4 (x + 3) + \log_4 (x - 3) = 2$

14. $\log_8 (x + 1) - \log_8 x = \log_8 4$

15. $\log x^2 = (\log x)^2$

16. $(\log_3 x)^2 - \log_3 x^2 = 3$

17. $\log_3 (\log_4 x) = 0$

18. $\log (\log x) = 2$

19. Solve for x. (*Hint:* Use $\log_e$ in the last step.)

$$\frac{e^x - e^{-x}}{2} = t$$

20. Solve for x.

$$3^x + 3^{-x} = t$$

WATCH-OUT FOR nnrln 2 ANSWERS!

21. Solve for x.

$$\frac{5^x - 5^{-x}}{5^x + 5^{-x}} = t$$

22. Solve for x.

$$\frac{e^x + e^{-x}}{e^x - e^{-x}} = t$$

[ii] Solve.

23. (*Doubling time*). How many years will it take an investment of $1000 to double itself when interest is compounded annually at 6%?

24. (*Tripling time*). How many years will it take an investment of $1000 to triple itself when interest is compounded annually at 5%?

25. Find the loudness of the sound of an automobile, having an intensity 3,100,000 times I_0.

26. Find the loudness of the sound of a dishwasher, having an intensity 2,500,000 times I_0.

27. Find the loudness of the threshold of sound pain, for which the intensity is 10^{14} times I_0.

28. Find the loudness of a jet aircraft, having an intensity 10^{12} times I_0.

29. The Los Angeles earthquake of 1971 had an intensity $10^{6.7}$ times I_0. What was its magnitude on the Richter scale?

30. The San Francisco earthquake of 1906 had an intensity $10^{8.25}$ times I_0. What was its magnitude on the Richter scale?

31. An earthquake has a magnitude of 5 on the Richter scale. What is its intensity?

32. An earthquake has a magnitude of 7 on the Richter scale. What is its intensity?

33. Students in an industrial mathematics course take a final exam and are then retested at monthly intervals. The forgetting function is given by the equation $S(t) = 82 - 18 \log (t + 1)$.
   a) What was the average score originally on the final exam?
   b) What was the average score after 5 months had elapsed?

34. Students graduating from a cosmetology curriculum take a final exam and are then retested at monthly intervals. The forgetting function is given by the equation $S(t) = 75 - 20 \log (t + 1)$.
   a) What was the average score originally on the final exam?
   b) What was the average score after 6 months had elapsed?

35. Refer to Exercise 33. How much time will elapse before the average score has decreased to 64?

36. Refer to Exercise 34. How much time will elapse before the average score has decreased to 61?

37. In chemistry, pH is defined as follows:

$$\text{pH} = -\log [H^+],$$

where $[H^+]$ is the hydrogen ion concentration in moles per liter. For example, the hydrogen ion concentration in milk is 4×10^{-7} moles per liter, so

$$\text{pH} = -\log (4 \times 10^{-7}) = -[\log 4 + (-7)] \approx 6.4.$$

For tomatoes, $[H^+]$ is about 6.3×10^{-5}. Find the pH.

38. For eggs, $[H^+]$ is about 1.6×10^{-8}. Find the pH.

☆

Solve.

39. $\log \sqrt{x} = \sqrt{\log x}$

40. $\log_5 \sqrt{x^2 + 1} = 1$

41. $(\log_a x)^{-1} = \log_a x^{-1}$

42. $|\log_5 x| = 2$

43. $\log_3 |x| = 2$

44. $\dfrac{(e^{3x+1})^2}{e^4} = e^{10x}$

45. $\dfrac{\sqrt{(e^{2x} \cdot e^{-5x})^{-4}}}{e^x \div e^{-x}} = e^7$

46. $\log x^{\log x} = 4$

47. Solve $y = ax^n$, for n. Use $\log_x$.

48. Solve $y = ke^{at}$, for t. Use $\log_e$.

49. Solve for t. Use $\log_e$.

$$P = P_0 e^{rt/100}$$

50. Solve for t. Use $\log_e$.

$$I = \frac{E}{R}(1 - e^{-(Rt/L)})$$

51. Solve for t.

$$T = T_0 + (T_1 - T_0)10^{-kt}$$

52. Solve for n. Use $\log_V$.

$$PV^n = c$$

53. Solve for Q.

$$\log_a Q = \frac{1}{3} \log_a y + b$$

54. Solve for y.

$$\log_a y = 2x + \log_a x$$

Solve for x.

55. $x^{\log x} = \dfrac{x^3}{100}$

56. $x^{\log x} = 100x$

★

Solve.

57. $|\log_5 x| + 3 \log_5 |x| = 4$

58. $|\log_a x| = \log_a |x|$

59. $(0.5)^x < \dfrac{4}{5}$

60. $8x^{0.3} - 8x^{-0.3} = 63$

61. Solve the system of equations.
$$5^{x+y} = 100,$$
$$3^{2x-y} = 1000$$

62. Given that
$$\log_2 [\log_3 (\log_4 x)] =$$
$$\log_3 [\log_2 (\log_4 y)] =$$
$$\log_4 [\log_3 (\log_2 z)] = 0,$$
find $x + y + z$.

63. If $2 \log_3 (x - 2y) = \log_3 x + \log_3 y$, find $\dfrac{x}{y}$.

64. Find the ordered pair (x, y) for which $4^{\log_{16} 27} = 2^x 3^y$.

1. ▦ Find $\left(1 + \dfrac{1}{n}\right)^n$ when $n = 5$.

2. ▦ Find $\left(1 + \dfrac{1}{n}\right)^n$ when $n = 10$.

7.6 THE NUMBER e; NATURAL LOGARITHMS; CHANGE OF BASE

The Number e

One of the most important numbers is a certain irrational number, with a nonrepeating decimal representation. This number is usually named e, and it arises in a number of different ways. One of these is the following.

Consider the compound interest formula

$$A = p(1 + r)^t,$$

where A is the amount that the investment is worth, p is the principal, r is the rate of interest, and t is the time in years. Interest is usually compounded more often than once a year, say n times per year. In this event we alter the formula. The rate of interest for an *interest period* is r/n and the number of interest periods is $n \cdot t$, for t years. Thus we have

$$A = p\left(1 + \frac{r}{n}\right)^{nt}.$$

Suppose you invest one dollar for one year at 100%. Since $r = 100\%$, or 1, this makes the formula simple:

$$A = \left(1 + \frac{1}{n}\right)^n.$$

The amount A is now a function of n. The more often interest is compounded, the greater A becomes. Let us look at some values of this function.

COMPOUNDING	n	A
Annually	1	$(1 + \frac{1}{1})^1 = 2$
Semiannually	2	$(1 + \frac{1}{2})^2 = 2.25$
Quarterly	4	$(1 + \frac{1}{4})^4 = 2.44\ldots$

According to this table, for semiannual compounding the dollar grows to $2.25 in one year. For quarterly compounding it grows to $2.44. What would you expect to happen if interest were compounded daily, or once each minute?

DO EXERCISES 1 AND 2.

Let us look at some more function values.

▦ COMPOUNDING	n	A
Daily	365	$2.71456\ldots$
Once per hour	8,760	$2.71812\ldots$
Once per minute	525,600	$2.71828\ldots$

This result may be surprising. The function values approach a limit as n increases. No matter how often interest is compounded your investment will not amount to more than $2.72 for the year. The limit that these function values approach is the number called e:

$$e = 2.718281828459\ldots.$$

Natural Logarithms and Exponential Functions

[i] Exponential Functions, Base e

The exponential function, base e, is important in many applications, and its inverse, the logarithm function base e, is also important in mathematical theory and in applications as well. Logarithms to the base e are called *natural logarithms*. Most scientific calculators have an $\boxed{e^x}$ key. Table 4 at the back of the book gives function values for e^x and e^{-x}. Using a calculator or these tables we can construct graphs of $y = e^x$ and $y = \log_e x$.

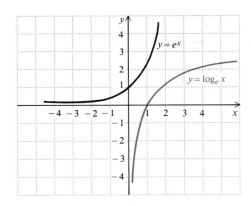

DO EXERCISES 3 AND 4.

[ii] Natural Logarithms

The number $\log_e x$ is abbreviated ln x; that is, ln x $= \log_e x$. The following is a restatement of the basic theorems of logarithms in terms of natural logarithms.

$$\ln xy = \ln x + \ln y \qquad \textbf{(Theorem 3)}$$

$$\ln x^p = p \ln x \qquad \textbf{(Theorem 4)}$$

$$\ln \frac{x}{y} = \ln x - \ln y \qquad \textbf{(Theorem 5)}$$

$$\ln e^k = k \qquad \textbf{(Theorem 2)}$$

If you have an $\boxed{\ln}$ key on your calculator, you can find such logarithms directly; for example, to six decimal places,

$$\ln 5.24 = 1.656321, \quad \ln 52.4 = 3.958907, \quad \text{and} \quad \ln 0.001277 = -6.663242.$$

From ln 5.24 and ln 52.4 we note that natural logarithms do not have characteristics and mantissas. Common logarithms have characteristics and mantissas because our numeration system is based on 10. For any base other than 10, logarithms have neither characteristics nor mantissas.

DO EXERCISES 5–8.

3. Using the same axes, graph
   a) $y = 2^x$,
   b) $y = 4^x$,
   c) $y = e^x$.
   Use Table 4 to obtain values of e^x.

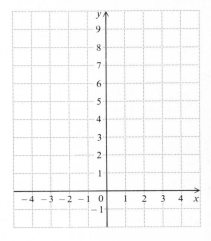

4. Graph $y = e^{-x}$.

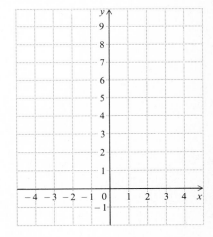

▦ Find each logarithm. Round to six decimal places.

5. ln 2

6. ln 100

7. ln 0.07432

8. ln 0.9999

Find each logarithm. Use Table 7.

9. ln 8.13

10. ln 81,300

11. ln 0.0813

12. ln 2000

13. ln 0.0001

Solve for t.

14. $e^t = 80$

15. $e^{-0.06t} = 0.07$

If you do not have a calculator with a natural logarithm key, you can use Table 7, part of which is shown below, for some values of ln x.

	ln x									
x	0.00	0.01	0.02	0.03	0.04	0.05	0.06	0.07	0.08	0.09
5.0	1.6094	1.6114	1.6134	1.6154	1.6174	1.6194	1.6214	1.6233	1.6253	1.6273
5.1	1.6292	1.6312	1.6332	1.6351	1.6371	1.6390	1.6409	1.6429	1.6448	1.6467
5.2	1.6487	1.6506	1.6525	1.6544	1.6563	1.6582	1.6601	1.6620	1.6639	1.6658
5.3	1.6677	1.6696	1.6715	1.6734	1.6752	1.6771	1.6790	1.6808	1.6827	1.6845
5.4	1.6864	1.6882	1.6901	1.6919	1.6938	1.6956	1.6974	1.6993	1.7011	1.7029

Example 1 Find ln 5.24. Use Table 7.

To find ln 5.24, locate the row headed 5.2, then move across to the column headed 0.04. Note the colored number in the table.

$$\ln 5.24 = 1.6563$$

We can find natural logarithms of numbers not in the table as follows. We first write scientific notation for the number.

Example 2 Find ln 5240. Use Table 7.

$$
\begin{aligned}
\ln 5240 &= \ln (5.24 \times 10^3) \\
&= \ln 5.24 + \ln 10^3 \quad &\text{Theorem 3} \\
&= \ln 5.24 + 3 \ln 10 \quad &\text{Theorem 4} \\
&= 1.6563 + 6.9078 \quad &\text{Find ln 5.24 in the body of Table 7} \\
& &\text{and 3 ln 10 at the bottom.} \\
&= 8.5641
\end{aligned}
$$

Example 3 Find ln 0.000524. Use Table 7.

$$
\begin{aligned}
\ln 0.000524 &= \ln (5.24 \times 10^{-4}) \\
&= \ln 5.24 + \ln 10^{-4} \quad &\text{Theorem 3} \\
&= \ln 5.24 - 4 \ln 10 \quad &\text{Theorem 4} \\
&= 1.6563 - 9.2103 \quad &\text{Find ln 5.24 in the body of Table} \\
& &\text{7 and 4 ln 10 at the bottom.} \\
&= -7.5540
\end{aligned}
$$

DO EXERCISES 9–13.

[iii] Exponential Equations

Natural logarithms can be used to solve exponential equations involving base e.

Example 4 Solve $e^t = 40$.

We could take the common logarithm on both sides as we did in Section 7.5, but taking the natural logarithm makes the simplification on the left easier. We also know how to find ln 40.

$$
\begin{aligned}
e^t &= 40 \\
\ln e^t &= \ln 40 \quad &\text{Taking the natural log on both sides} \\
t &= \ln 40 \quad &\text{Theorem 2} \\
t &\approx 3.7 \quad &\text{Using a calculator or Table 7}
\end{aligned}
$$

DO EXERCISES 14 AND 15.

[iv] Applications

There are many applications of exponential functions, base e. We consider a few.

Example 5 (*Population growth*). One mathematical model for describing population growth is the formula

$$P = P_0 e^{kt},$$

where P_0 is the number of people at time 0, P is the number of people at time t, and k is a positive constant depending on the situation. The population of the United States in 1970 was 208 million. In 1980 it was 225 million. Use these data to find the value of k and then use the model to predict the population in 2000.

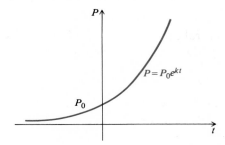

Time t begins with 1970. That is, $t = 0$ in 1970, $t = 10$ in 1980, and so on. Substituting the data into the formula, we get

$$225 = 208e^{k \cdot 10}.$$

We solve for k:

$\ln 225 = \ln 208 e^{10k}$ Taking the natural logarithm on both sides
$\ln 225 = \ln 208 + \ln e^{10k}$ Theorem 3
$\ln 225 = \ln 208 + 10k$ Theorem 2

$$k = \frac{\ln 225 - \ln 208}{10}$$ Solving for k

$$k = \frac{5.4161 - 5.3375}{10}$$ Using a calculator or Table 7

$$k = 0.008$$ Calculating

To find the population in 2000, we will use $P_0 = 208$ (population in the year 1970). We then have

$$P = 208e^{0.008t}.$$

In 2000, t will be 30, so we have

$$P = 208e^{0.008(30)}$$

$$= 208e^{0.24}.$$

Using a calculator or Table 4, we find that $e^{0.24} = 1.2712$. Multiplying by 208 gives us about 264 million. This is our prediction for the population of the United States in the year 2000.

DO EXERCISE 16.

16. The population of Tempe, Arizona, was 25,000 in 1960. In 1969 it was 52,000.

   a) Use these data to determine k in the growth model.
   b) Use these data to predict the population of Tempe in 1984.

17. Radioactive bismuth has a half-life of 5 days. A scientist buys 224 grams of it. How much of it will remain radioactive in 30 days?

Example 6 (*Radioactive decay*). In a radioactive substance such as radium, some of the atoms are always ceasing to become radioactive. Thus the amount of a radioactive substance decreases. This is called radioactive *decay*. A model for radioactive decay is as follows:

$$N = N_0 e^{-kt},$$

where N_0 is the amount of a radioactive substance at time 0, N is the amount at time t, and k is a positive constant depending on the situation. Strontium 90 has a *half-life* of 25 years. This means that half of a sample of the substance will cease to become radioactive in 25 yr. Find k in the formula and then use the formula to find how much of a 36-gram sample will remain after 100 years.

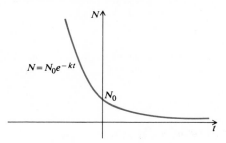

When $t = 25$ (half-life), N will be half of N_0, so we have

$$\frac{1}{2}N_0 = N_0 e^{-25k} \quad \text{or} \quad \frac{1}{2} = e^{-25k}.$$

We take the natural log on both sides:

$$\ln \frac{1}{2} = \ln e^{-25k} = -25k.$$

Thus

$$k = -\frac{\ln 0.5}{25} \approx 0.0277.$$

Now to find the amount remaining after 100 years, we use the formula

$$N = 36e^{-0.0277 \cdot 100}$$
$$= 36e^{-2.77}.$$

From a calculator or Table 4 we find that $e^{-2.8} = 0.0608$, and thus

$$N \approx 2.2 \text{ grams.}$$

DO EXERCISE 17.

Example 7 (*Atmospheric pressure*). Under standard conditions of temperature, the atmospheric pressure at height h is given by

$$P = P_0 e^{-kh},$$

where P is the pressure, P_0 is the pressure where $h = 0$, and k is a positive constant. Standard sea level pressure is 1013 millibars. Suppose that the pressure at 18,000 ft is half that at sea level. Then find k and find the pressure at 1000 ft.

At **18,000** ft P is half of P_0, so we have

$$\frac{P_0}{2} = P_0 e^{-18,000k}.$$

$506.5 = 1013\, e^{-18000\,ft}$

Taking natural logarithms on both sides, we obtain

$$\ln \frac{1}{2} = -18,000k, \quad \text{or} \quad -\ln 2 = -18,000k.$$

Then

$$k = \frac{\ln 2}{18,000} = 3.85 \times 10^{-5}.$$

To find the pressure at 1000 ft we use the fact that $P_0 = 1013$ and we let $h = 1000$:

$$P = 1013 e^{-3.85 \times 10^{-5} \times 1000} = 1013 e^{-0.0385}.$$

Rounding the exponent to -0.04 and using a calculator or Table 4, we calculate the pressure to be **973 millibars** at **1000 ft**.

DO EXERCISE 18.

[v] Logarithm Tables and Change of Base

The following theorem shows how we can change logarithm bases. The theorem can be applied to find the logarithm of a number to any base, using a table of common or natural logarithms.

THEOREM 7

For any bases a and b, and any positive number M,

$$\log_b M = \frac{\log_a M}{\log_a b}.$$

Proof. Let $x = \log_b M$. Then $b^x = M$, so that

$$\log_a M = \log_a b^x, \quad \text{or} \quad x \log_a b.$$

We now solve for x:

$$x = \log_b M = \frac{\log_a M}{\log_a b}.$$

This is the desired formula.

Example 8 Find $\log_3 81$.

$$\log_3 81 = \frac{\log_{10} 81}{\log_{10} 3} = \frac{1.9085}{0.4771} \approx 4.0$$

Example 9 Find $\log_5 346$.

$$\log_5 346 = \frac{\log_{10} 346}{\log_{10} 5} = \frac{2.5391}{0.6990} \approx 3.6325$$

DO EXERCISES 19–21.

18. Calculate the atmospheric pressure at 10,000 ft.

19. Find $\log_5 125$.

20. Find $\log_6 4870$.

21. Use 2.718 for e. Find $\log_{10} e$.

22. Find $\log_e 1030$.

If in Theorem 7 we let $a = 10$ and $b = e$, we obtain a formula for changing from common to natural logarithms:

$$\log_e M = \frac{\log_{10} M}{\log_{10} e}.$$

Now $\log_{10} e \approx 0.4343$, so we have the following.

THEOREM 8

$$\log_e M = \frac{\log_{10} M}{0.4343}$$

23. Find $\log_e 0.457$.

Example 10 Find $\log_e 257$.

$$\log_e 257 = \frac{\log_{10} 257}{0.4343} = \frac{2.4099}{0.4343} = 5.5489$$

DO EXERCISES 22 AND 23.

We can make use of this change-of-base procedure to solve certain exponential equations.

24. Solve $2^x = 7$.

Example 11 Solve $3^x = 8$.

We take logarithms, base 3, on both sides. Then we have

$$\log_3 3^x = \log_3 8$$
$$x = \log_3 8 \qquad \text{Theorem 2}$$

To get an approximation we change to base 10 and divide:

$$x = \log_3 8 = \frac{\log_{10} 8}{\log_{10} 3} = \frac{0.9031}{0.4771} = 1.8929.$$

Compare this with Examples 1 and 10 in Section 7.5.

DO EXERCISE 24.

EXERCISE SET 7.6

[i] Graph.

1. $y = e^{2x}$ **2.** $y = e^{0.5x}$ **3.** $y = e^{-2x}$

4. $y = e^{-0.5x}$ **5.** $y = 1 - e^{-x}$, for $x \geq 0$ **6.** $y = 2(1 - e^{-x})$, for $x \geq 0$

[ii] Find each natural logarithm to four decimal places. Use a calculator or Table 7. Answers in the back of the book have been found using Table 7. If you use a calculator, there may be some variance in the last decimal place.

7. ln 1.88	**8.** ln 18.8	**9.** ln 0.0188	**10.** ln 0.188
11. ln 2.13	**12.** ln 213	**13.** ln 0.213	**14.** ln 0.00213
15. ln 4500	**16.** ln 81,000	**17.** ln 0.00056	**18.** ln 0.999
19. ln 0.08	**20.** ln 0.0471	**21.** ln 980,000	**22.** ln 765,000,000

[iii] Solve.

23. $e^t = 100$

24. $e^t = 1000$

25. $e^x = 60$

26. $e^k = 90$

27. $e^{-t} = 0.1$

28. $e^{-t} = 0.01$

29. $e^{-0.02k} = 0.06$

30. $e^{0.07t} = 2$

[iv] Solve.

31. (*Population growth*). The population of Dallas was 680,000 in 1960. In 1969 it was 815,000. Find k in the growth formula and estimate the population in 1990.

32. (*Population growth*). The population of Kansas City was 475,000 in 1960. In 1970 it was 507,000. Find k in the growth formula and estimate the population in 2000.

33. (*Radioactive decay*). The half-life of polonium is 3 minutes. After 30 minutes, how much of a 410-gram sample will remain radioactive?

34. (*Radioactive decay*). The half-life of a lead isotope is 22 yr. After 66 yr, how much of a 1000-gram sample will remain radioactive?

35. (*Radioactive decay*). A certain radioactive substance decays from 66,560 grams to 6.5 grams in 16 days. What is its half-life?

36. Ten grams of uranium will decay to 2.5 grams in 496,000 years. What is its half-life?

37. (*Radiocarbon dating*). Carbon-14, an isotope of carbon, has a half-life of 5750 years. Organic objects contain carbon-14 as well as nonradioactive carbon, in known proportions. When a living organism dies, it takes in no more carbon. The carbon-14 decays, thus changing the proportions of the kinds of carbon in the organism. By determining the amount of carbon-14 it is possible to determine how long the organism has been dead, hence how old it is.

a) How old is an animal bone that has lost 30% of its carbon-14?

b) A mummy discovered in the pyramid Khufu in Egypt had lost 46% of its carbon-14. Determine its age.

38. (*Radiocarbon dating*).

a) How old is an animal bone that has lost 20% of its carbon-14?

b) The Statue of Zeus at Olympia in Greece is one of the seven wonders of the world. It is made of gold and ivory. The ivory was found to have lost 35% of its carbon-14. Determine the age of the statue.

39. (*Atmospheric pressure*). What is the pressure at the top of Mt. Shasta in California, 14,162 ft high?

40. (*Atmospheric pressure*). Blood will boil when atmospheric pressure goes below about 62 millibars. At what altitude, in an unpressurized vehicle, will a pilot's blood boil?

41. (*Consumer price index*). The *consumer price index* compares the costs of goods and services over various years. The base year is 1967. The same goods and services that cost $100 ($P_0$) in 1967 cost $184.50 in 1977. Assuming the exponential model:

a) Find the value k and write the equation.

b) Estimate what the same goods and services will cost in 1987.

c) When did the same goods and services cost double that of 1967?

42. (*Cost of a double-dip ice cream cone*). In 1970 the cost of a double-dip ice cream cone was 52¢. In 1978 it was 66¢. Assuming the exponential model:

a) Find the value k and write the equation.

b) Estimate the cost of a cone in 1986.

c) When will the cost of a cone be twice that of 1978?

[v] Find the following.

43. $\log_4 20$

44. $\log_8 0.99$

45. $\log_5 0.78$

46. $\log_{12} 15,000$

47. $\log_e 12$ (Do not use Table 7.)

48. $\log_e 0.77$ (Do not use Table 7.)

☆

Solve for t.

49. $P = P_0 e^{kt}$

50. $P = P_0 e^{-kt}$

Verify each of the following.

51. $\ln x = \dfrac{\log x}{\log e} \approx 2.3026 \log x$

52. $\log x = \dfrac{\ln x}{\ln 10} \approx 0.4343 \ln x$

53. Which is larger, e^{π} or π^{e}?

54. Which is larger, $e^{\sqrt{\pi}}$ or $\sqrt{e^{\pi}}$?

55. Given $f(x) = (1 + x)^{1/x}$, find $f(1)$, $f(0.5)$, $f(0.2)$, $f(0.1)$, $f(0.01)$, and $f(0.001)$ to six decimal places. This sequence of numbers approaches the number e.

56. Given $f(t) = t^{1/(t-1)}$, find $f(0.5)$, $f(0.9)$, $f(0.99)$, $f(0.999)$, and $f(0.9999)$ to six decimal places. This sequence of numbers approaches the number e.

★

57. Find a (simple) formula for radioactive decay involving H, the half-life.

58. The time required for a population to double is called the *doubling time*.

a) Find an expression relating T, the doubling time, to k, the constant in the growth equation.

b) Find a (simple) formula for population growth involving T.

Prove the following for any logarithm bases a and b.

59. $\log_a b = \dfrac{1}{\log_b a}$

60. $a^{(\log_b M) \div (\log_b a)} = M$

61. $a^{(\log_b M)(\log_b a)} = M^{(\log_b a)^2}$

62. $\log_a (\log_a x) = \log_a (\log_b x) - \log_a (\log_b a)$

CHAPTER 7 REVIEW

[7.1, i]   Graph.

[7.6, i]   **1.** $y = \log_2 (x - 1)$ **2.** $y = e^{0.4x}$ **3.** $y = e^{-x} - 3$

[7.1, ii]   **4.** Write an exponential equation equivalent to $\log_8 \dfrac{1}{4} = -\dfrac{2}{3}$.

5. Write a logarithmic equation equivalent to $7^{2.3} = x$.

[7.1, iii]   Solve.

6. $\log_x 64 = 3$ **7.** $\log_{16} 4 = x$ **8.** $\log_5 125 = x$

[7.2, i]   **9.** Write an equivalent expression containing a single logarithm.

$$\frac{1}{2} \log_b a + \frac{3}{2} \log_b c - 4 \log_b d$$

[7.2, i]   Given that $\log_a 2 = 0.301$, $\log_a 3 = 0.477$, and $\log_a 7 = 0.845$, find:

10. $\log_a 18$.   **11.** $\log_a \dfrac{7}{2}$.   **12.** $\log_a \dfrac{1}{4}$.   **13.** $\log_a \sqrt{3}$.

14. Express in terms of logarithms of M and N: $\log \sqrt[3]{M^2/N}$.

[7.3, i]   Using Table 2, find:

15. $\log 26.3$.   **16.** $\log 0.00806$.   **17.** $10^{1.8686}$.   **18.** antilog $(8.4409 - 10)$.

[7.6, v]   **19.** $\log_5 290$ (round to the nearest tenth).

[7.3, ii]   **20.** Use logarithms to compute $(0.0524)^2 \cdot \sqrt{0.0638}$.

[7.6, ii]   Using Table 7, find:

21. $\ln 8.9$ **22.** $\ln 560$.   **23.** $\ln 0.0462$.

[7.1, iv]   **24.** Simplify $\log_{12} 12^{x^2+1}$.

[7.5, i]　　　Solve.

25. $3^{1-x} = 9^{2x}$　　　　　　　　　　　　　**26.** $\log (x^2 - 1) - \log (x - 1) = 1$

[7.5, ii]　　**27.** How many years will it take an investment of $1000 to double if interest is compounded annually at 13%?

[7.5, ii]　　**28.** What is the loudness, in decibels, of a sound whose intensity is 1000 times I_0?

[7.6, iv]　　**29.** The half-life of a radioactive substance is 15 days. How much of a 25-gram sample will remain radioactive after 30 days?

[7.5, i]　　　Solve.

30. $\log x^2 = \log x$　　　　　　　　　　**31.** $\log_2 (x - 1) + \log_2 (x + 1) = 3$

32. $\log 2 + 2 \log x = \log (5x + 3)$　　　　**33.** $\log x = \ln x$

34. $|\log_4 x| = 3$

35. Graph $y = |\log_3 x|$.　　　　　　　　　**36.** Graph $y = |e^x - 4|$.

Find the domain.

37. $f(x) = \dfrac{17}{\sqrt{5 \ln x - 6}}$　　　　　　　**38.** $f(x) = \dfrac{8}{e^{4x} - 10}$

8

THE
CIRCULAR
FUNCTIONS

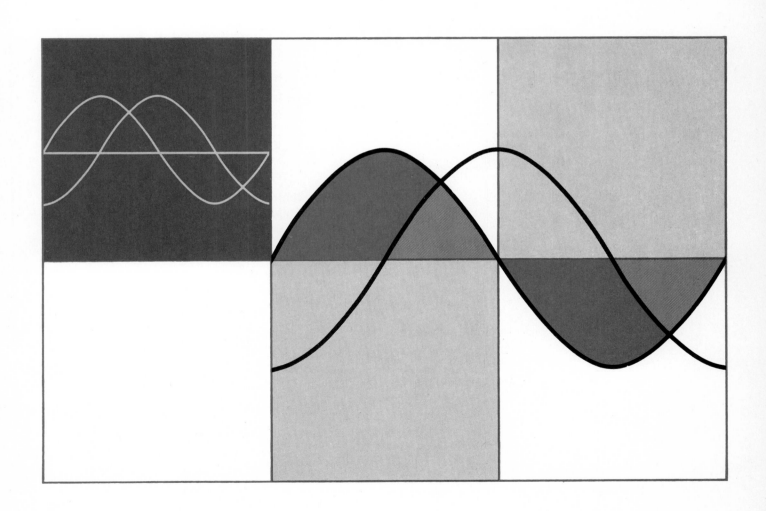

You should be able to:

[i] Define and determine the six trig-onometric ratios, or functions, and use them to find measures of parts of triangles.

[ii] Use these functions to solve sim-ple problems involving right tri-angles.

1. Find $m \angle A$.

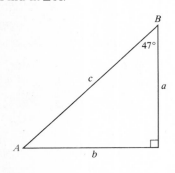

8.1 TRIANGLES*

In this chapter we consider an important class of functions called *trigonometric* functions, or *circular* functions. They are based, histori-cally, on certain properties of triangles. In this section we consider some of those properties. We focus our attention on right triangles.

The angles and sides of a right triangle are often labeled as shown, with $\angle C$ a right angle and c the length of the hypotenuse. Side a is opposite $\angle A$ and side b is opposite $\angle B$.

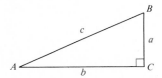

In any triangle the sum of the angle measures is 180°. Thus in a right triangle the measures of the two acute angles add up to 90°.

Example 1 In the triangle, find $m \angle B$ (the measure of $\angle B$).

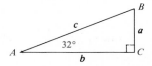

We know that $m \angle A + m \angle B = 90°$ and that $m \angle A = 32°$, so

$$m \angle B = 90° - 32° = 58°.$$

DO EXERCISE 1.

Similar Triangles

Triangles are similar if their corresponding angles have the same measure. In the following triangles, if we know that $\angle A$ and $\angle A'$ are the same size, then we know that *all* corresponding angles have the same measure; hence the triangles are similar.

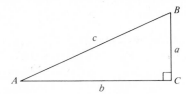 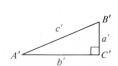

---

*The reader who wishes to approach trigonometry from the functional point of view first, and then consider triangles, can postpone this section. The reader who wishes a brief introduction to triangles before studying the functions can read this section and then proceed.

There are several paths through the trigonometry. For example, it is possible to treat triangles thoroughly before delving deeply into analytic trigonometry. For a thorough discussion of this, see the preface.

In similar triangles corresponding sides are in the same ratio (are proportional). For the triangles above, this means the following:

$$\frac{a}{a'} = \frac{b}{b'} = \frac{c}{c'}.$$

Similar triangles can be used to determine distances without measuring directly.

Example 2 Find the height of the flagpole using a rod and a measuring device as shown.

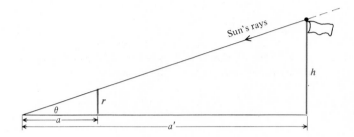

Place the rod of length r so that the top of its shadow coincides with the shadow of the top of the flagpole. We know that we have similar triangles, because the acute angle θ at the ground is the same in both triangles. We measure and find that $a = 3$ ft, that $a' = 30$ ft, and that $r = 2$ ft.

Now $h/r = a'/a$ by similar triangles, so

$$h = r\frac{a'}{a} = 2\left(\frac{30}{3}\right)$$
$$= 20 \text{ ft.}$$

DO EXERCISES 2 AND 3.

[i] Trigonometric Ratios

In a right triangle as shown below, the ratio a/c of the side *opposite* θ to the hypotenuse depends on the size of θ. In other words, this ratio is a *function* of θ.

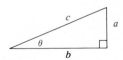

This function is called the *sine* function. Another such function, called the *cosine* function, is the ratio b/c, the ratio of the side *adjacent* to θ to the hypotenuse. There are other such ratios, or trigonometric functions, defined as follows.

2. Find b.

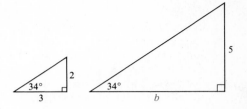

3. Find the height of this building.

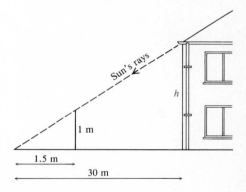

4. Find the following.

a) sin θ, cos θ, tan θ, cot θ, sec θ, csc θ

b) sin ϕ, cos ϕ, tan ϕ, cot ϕ, sec ϕ, csc ϕ

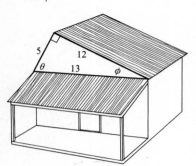

FUNCTION	ABBREVIATION FOR FUNCTION VALUE	DEFINING RATIO
sine	sin (θ)* or sin θ	$\dfrac{\text{side opposite } \theta}{\text{hypotenuse}}$
cosine	cos (θ) or cos θ	$\dfrac{\text{side adjacent } \theta}{\text{hypotenuse}}$
tangent	tan (θ) or tan θ	$\dfrac{\text{side opposite } \theta}{\text{side adjacent } \theta}$
cotangent	cot (θ) or cot θ	$\dfrac{\text{side adjacent } \theta}{\text{side opposite } \theta}$
secant	sec (θ) or sec θ	$\dfrac{\text{hypotenuse}}{\text{side adjacent } \theta}$
cosecant	csc (θ) or csc θ	$\dfrac{\text{hypotenuse}}{\text{side opposite } \theta}$

Note that the function values do not depend on the size of a triangle, but only on the size of the angle θ.

Example 3 In this triangle find:

a) the trigonometric function values for θ.

b) the trigonometric function values for ϕ.

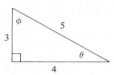

$$\sin \theta = \frac{\text{side opposite } \theta}{\text{hypotenuse}} = \frac{3}{5}$$

$$\cos \theta = \frac{\text{side adjacent } \theta}{\text{hypotenuse}} = \frac{4}{5}$$

$$\tan \theta = \frac{\text{side opposite } \theta}{\text{side adjacent } \theta} = \frac{3}{4}$$

$$\cot \theta = \frac{\text{side adjacent } \theta}{\text{side opposite } \theta} = \frac{4}{3}$$

$$\sec \theta = \frac{\text{hypotenuse}}{\text{side adjacent } \theta} = \frac{5}{4}$$

$$\csc \theta = \frac{\text{hypotenuse}}{\text{side opposite } \theta} = \frac{5}{3}$$

$$\sin \phi = \frac{\text{side opposite } \phi}{\text{hypotenuse}} = \frac{4}{5}$$

$$\cos \phi = \frac{\text{side adjacent } \phi}{\text{hypotenuse}} = \frac{3}{5}$$

$$\tan \phi = \frac{\text{side opposite } \phi}{\text{side adjacent } \phi} = \frac{4}{3}$$

$$\cot \phi = \frac{\text{side adjacent } \phi}{\text{side opposite } \phi} = \frac{3}{4}$$

$$\sec \phi = \frac{\text{hypotenuse}}{\text{side adjacent } \phi} = \frac{5}{3}$$

$$\csc \phi = \frac{\text{hypotenuse}}{\text{side opposite } \phi} = \frac{5}{4}$$

DO EXERCISE 4.

Values for the trigonometric functions have been worked out in great detail and are given in tables and by some calculators. When

---

*For function values we generally write $f(x)$. For these functions we often omit parentheses, as in sin θ.

these values are available we can use them to help determine distances or angles without actually measuring them directly.

[ii] **Using the Trigonometric Functions**

Example 4 An observer stands on level ground, 200 meters from the base of a TV tower, and looks up at an angle of 26.5° to see the top of the tower. How high is the tower above the observer's eye level?

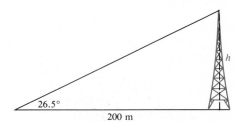

26.5°

200 m

We draw a diagram and see that a right triangle is formed. We use one of the trigonometric functions. This time the tangent function is most convenient. From the definition of the tangent function, we have

$$\frac{h}{200} = \tan 26.5°.$$

Then $h = 200 \tan 26.5°$. We find, from a table, that $\tan 26.5° = 0.499$, approximately. Thus $h = 200 \times 0.499 = 99.8$ m.

DO EXERCISE 5.

Calculators

Scientific calculators contain tables of trigonometric functions. It is recommended that students not use them until after the trigonometric tables have been studied (Section 8.5). In the meantime, calculators will be most useful for arithmetic calculations.

Example 5 A kite flies at a height of 60 ft when 130 ft of string is out. Assuming that the string is in a straight line, what is the angle that it makes with the ground?

130 ft

60 ft

θ

We draw a diagram and then use the most convenient trigonometric function. A portion of a sine table is given below.

sin 26.5° = 0.4462	sin 28.5° = 0.4772
sin 27° = 0.4540	sin 29° = 0.4848
sin 27.5° = 0.4617	sin 29.5° = 0.4924
sin 28° = 0.4695	sin 30° = 0.5000

5. An observer stands 120 meters from a tree, and finds that the line of sight to the top of the tree is 32.3° above the horizontal. The tangent of 32.3° is 0.632. Find the height of the tree above eye level.

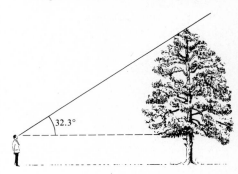

32.3°

6. A guy wire is 13.6 meters long, and is fastened from the ground to a pole 6.5 meters above the ground. What angle does the wire make with the ground?

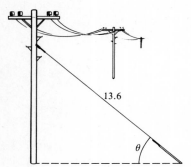

From the definition of the sine function, we have

$$\frac{60}{130} = \sin \theta$$

$$= 0.4615.$$

From the table, we find that θ is about 27.5°.

DO EXERCISE 6.

EXERCISE SET 8.1

[i]

1. Find $m \angle B$.

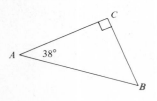

2. Find $m \angle A$.

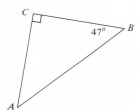

3. Find b.

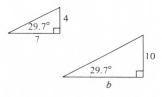

4. Find c.

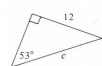

5. Find the trigonometric function values for θ.

6. Find the trigonometric function values for ϕ.

7. ▦ Find the trigonometric function values for θ. Round to four decimal places.

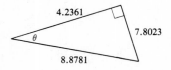

8. ▦ Find the trigonometric function values for ϕ. Round to four decimal places.

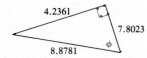

[ii] For Exercises 9–14, use the following values.

sin 36° = 0.5878	cos 36° = 0.8090	tan 36° = 0.7265	cot 36° = 1.376
sin 36.5° = 0.5948	cos 36.5° = 0.8039	tan 36.5° = 0.7400	cot 36.5° = 1.351
sin 37° = 0.6018	cos 37° = 0.7986	tan 37° = 0.7536	cot 37° = 1.327
sin 37.5° = 0.6088	cos 37.5° = 0.7934	tan 37.5° = 0.7673	cot 37.5° = 1.303
sin 38° = 0.6157	cos 38° = 0.7880	tan 38° = 0.7813	cot 38° = 1.280

9. a) Find ∠B.*
   b) Find *a*.
   c) Find *b*.

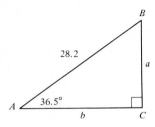

10. a) Find ∠B.
   b) Find *a*.
   c) Find *b*.

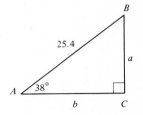

11. A guy wire is attached to a pole, and makes an angle of 37.5° with the ground. Find:

   a) the angle *B* that the wire makes with the pole;

   b) the distance *b* from *A* to the pole;

   c) the length of the wire.

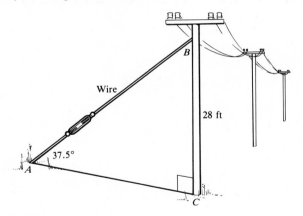

12. A cable across a river is attached at points *B* and *C*. A picnic grill at *A*, together with *B* and *C*, form the vertices of a right triangle. Find:

   a) the angle at *A*;

   b) the width of the river, *a*;

   c) the distance *c*.

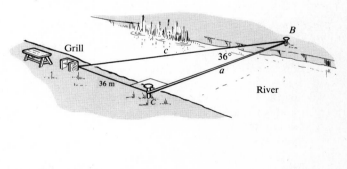

☆

13. Find *h*.

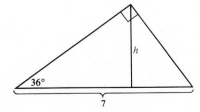

14. Find *a*.

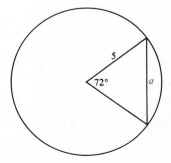

*Instead of saying "find *m* ∠ *B*," we often say "find ∠ *B*."

15. A 45° right triangle is half of a square. Use the Pythagorean theorem to find the length of a diagonal and then find:

    a) sin 45°

    b) cos 45°;

    c) tan 45°.

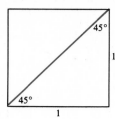

16. A 30°–60° right triangle is half of an equilateral triangle. Using this fact, find:

    a) sin 30°;

    b) cos 30°;

    c) sin 60°;

    d) cos 60°.

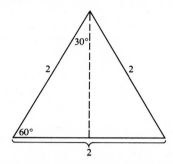

OBJECTIVES

You should be able to:

[i] Given the coordinates of a point on a circle centered at the origin, find its reflection across the u-axis, the v-axis, and the origin.

[ii] For the unit circle, state the coordinates of the intercepts and the midpoints of the arcs in each quadrant.

[iii] Given a real number that is an integral number of 4ths or 6ths of π, find on the unit circle the point determined by it. Given a point on the unit circle that is an integral number of 8ths or 12ths of the distance around the circle, find the real numbers between -2π and 2π that determine that point.

[iv] Given a real number n and the coordinates of the point it determines on the unit circle, state the coordinates of the point determined by $-n$.

8.2 THE UNIT CIRCLE

To develop the important circular functions we will use a circle with a radius of length 1. Such a circle is called a *unit circle*. When its center is at the origin of a coordinate system, it has an equation

$$u^2 + v^2 = 1.$$

(We use u and v for the variables here because we wish to reserve x and y for other purposes later.)

[i] Reflections

Any circle centered at the origin is symmetric with respect to the u-axis, the v-axis, and the origin.

Example 1 Suppose the point (1, 2) is on a circle centered at the origin. Find its reflection across the u-axis, the v-axis, and the origin.

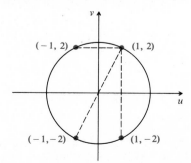

The reflection across the u-axis is $(1, -2)$.
The reflection across the v-axis is $(-1, 2)$.
The reflection across the origin is $(-1, -2)$.
By symmetry, all these points are on the circle.

DO EXERCISES 1 AND 2.

[ii] Special Points

The intercepts (points of intersection with the axes) of the unit circle A, B, C, and D have coordinates as shown. Let us find the coordinates of point E, which is the intersection of the circle with the line $u = v$. Note that this point divides the arc in the first quadrant in half.

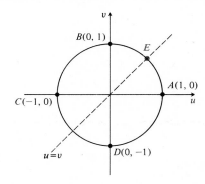

Since for any point (u, v) on the circle $u^2 + v^2 = 1$, and for point E we have $u = v$, we know that the coordinates of E satisfy $u^2 + u^2 = 1$ (substituting u for v in the equation of the circle). Let us solve this equation:

$$u^2 + u^2 = 1, \quad \text{or } 2u^2 = 1$$

$$u^2 = \frac{1}{2}$$

$$u = \sqrt{\frac{1}{2}}, \quad \text{or } \frac{\sqrt{1}}{\sqrt{2}}, \quad \text{or } \frac{1}{\sqrt{2}}.$$

Now

$$\frac{1}{\sqrt{2}} = \frac{1}{\sqrt{2}} \cdot \frac{\sqrt{2}}{\sqrt{2}} = \frac{\sqrt{2}}{2}.$$

Thus the coordinates of point E are $\left(\dfrac{\sqrt{2}}{2}, \dfrac{\sqrt{2}}{2}\right)$.

DO EXERCISE 3.

[iii] Distances on the Unit Circle

The circumference of a circle of radius r is $2\pi r$. Thus for the unit circle the circumference is 2π. If a point starts at A and travels counter-

1. The point $(3, -4)$ is on a circle centered at the origin. Find the coordinates of its reflection across

   a) the u-axis,
   b) the v-axis, and
   c) the origin.
   d) Are these points on the circle? Why?

2. The point $(-5, -2)$ is on a circle centered at the origin. Find the coordinates of its reflection across

   a) the u-axis,
   b) the v-axis, and
   c) the origin.
   d) Are these points on the circle? Why?

3. Find the coordinates of the reflection of point E across

   a) the u-axis,
   b) the v-axis, and
   c) the origin.
   d) Are these points on the circle? Why?

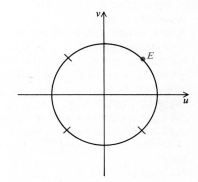

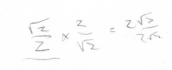

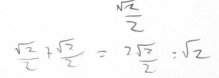

4. How far will a point travel if it goes

a) $\frac{1}{4}$ of the way around the unit circle?

b) $\frac{3}{8}$ of the way around the unit circle?

c) $\frac{3}{4}$ of the way around the unit circle?

5. How far will a point move on the unit circle, counterclockwise, going from point A to

a) M?   b) N?
c) P?   d) Q?

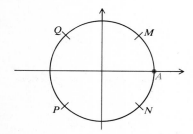

6. How far will a point move on the unit circle, counterclockwise, in going from A to

a) F?   b) G?
c) H?   d) J?

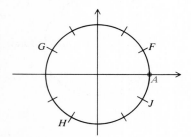

clockwise around the circle, it will travel a distance of 2π. If it travels halfway around the circle, it will travel a distance of π.

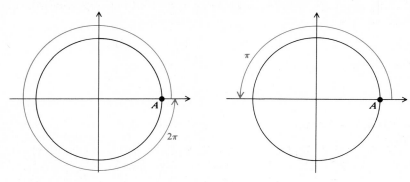

DO EXERCISE 4.

If a point travels $\frac{1}{8}$ of the way around the circle, it will travel a distance of $\left(\frac{1}{8}\right) \cdot 2\pi$, or $\pi/4$. Note that the stopping point is $\frac{1}{4}$ of the way from A to C and $\frac{1}{2}$ of the way from A to B.

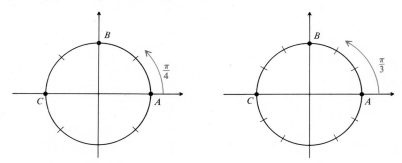

If a point travels $\frac{1}{6}$ of the way around the circle ($\frac{2}{12}$ of the way, as shown), it will travel a distance of $\left(\frac{1}{6}\right) \cdot 2\pi$, or $\pi/3$. Note that the stopping point is $\frac{1}{3}$ of the way from A to C and $\frac{2}{3}$ of the way from A to B.

DO EXERCISES 5 AND 6.

A point may travel completely around the circle and then continue. For example, if it goes around once and then continues $\frac{1}{4}$ of the way around, it will have traveled a distance of $2\pi + (\pi/2)$, or $\left(\frac{5}{2}\right)\pi$.

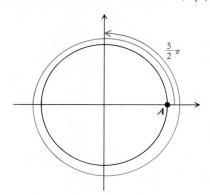

Any positive number thus determines a point on the unit circle. For the number 35, for example, we start at A and travel counterclockwise a distance of 35. The point at which we stop is the point "determined" by the number 35.

DO EXERCISES 7 AND 8.

[iv] Negative Inputs

Negative numbers also determine points on the unit circle. For a negative number, we move clockwise around the circle. Points for $-\pi/4$ and $-3\pi/2$ are shown. Any negative number thus determines a point on the unit circle. The same is true for any positive number. The number 0 determines the point A. Hence there is a point on the unit circle for every real number.

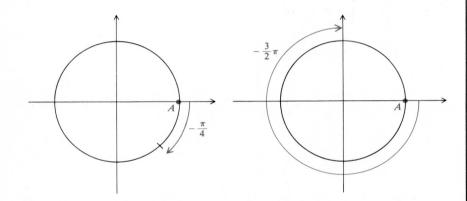

Points on the unit circle determined by two numbers that are additive inverses are situated as shown. The points are reflections of each other across the u-axis. Thus their first coordinates are the same and their second coordinates are additive inverses of each other.

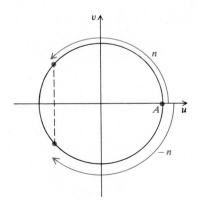

DO EXERCISE 9.

7. On this unit circle, mark the points determined by the following numbers:

a) $\frac{\pi}{4}$; b) $\frac{7}{4}\pi$; c) $\frac{9}{4}\pi$; d) $\frac{13}{4}\pi$.

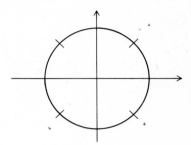

8. On this unit circle, mark the points determined by the following numbers:

a) $\frac{\pi}{6}$; b) $\frac{7}{6}\pi$; c) $\frac{13}{6}\pi$; d) $\frac{25}{6}\pi$.

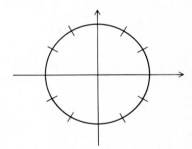

9. On this unit circle, mark the points determined by the following numbers:

a) $-\frac{\pi}{2}$; b) $-\frac{3}{4}\pi$; c) $-\frac{7}{4}\pi$;

d) -2π.

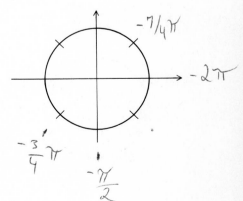

EXERCISE SET 8.2

[i]

1. The point $(-5, 2)$ is on a circle centered at the origin. Find the coordinates of its reflection across (a) the u-axis, (b) the v-axis, and (c) the origin. (d) Are these points on the circle? Why?

2. The point $(3, -5)$ is on a circle centered at the origin. Find the coordinates of its reflection across (a) the u-axis, (b) the v-axis, and (c) the origin. (d) Are these points on the circle? Why?

[ii]

3. Find the coordinates of these points on a unit circle.

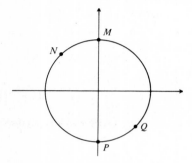

4. Find the coordinates of these points on a unit circle.

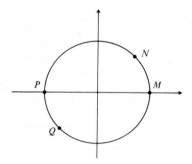

[iii] For each of Exercises 5–10, sketch a unit circle such as the one shown below.

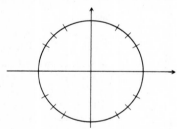

5. Mark the points determined by (a) $\frac{\pi}{4}$, (b) $\frac{3}{2}\pi$, (c) $\frac{3}{4}\pi$, (d) π, (e) $\frac{11}{4}\pi$, and (f) $\frac{17}{4}\pi$.

6. Mark the points determined by (a) $\frac{\pi}{2}$, (b) $\frac{5}{4}\pi$, (c) 2π, (d) $\frac{9}{4}\pi$, (e) $\frac{13}{4}\pi$, and (f) $\frac{23}{4}\pi$.

7. Mark the points determined by (a) $\frac{\pi}{6}$, (b) $\frac{2}{3}\pi$, (c) $\frac{7}{6}\pi$, (d) $\frac{10}{6}\pi$, (e) $\frac{14}{6}\pi$, and (f) $\frac{23}{4}\pi$.

8. Mark the points determined by (a) $\frac{\pi}{3}$, (b) $\frac{5}{6}\pi$, (c) $\frac{11}{6}\pi$,

9. Mark the points determined by (a) $-\frac{\pi}{2}$, (b) $-\frac{3}{4}\pi$, (c) $-\frac{5}{6}\pi$, (d) $-\frac{5}{2}\pi$, (e) $-\frac{17}{6}\pi$, and (f) $-\frac{9}{4}\pi$.

10. Mark the points determined by (a) $-\frac{3}{2}\pi$, (b) $-\frac{\pi}{3}$, (c) $-\frac{7}{6}\pi$, (d) $-\frac{13}{6}\pi$, (e) $-\frac{19}{6}\pi$, and (f) $-\frac{11}{4}\pi$.

(d) $\frac{13}{6}\pi$, (e) $\frac{23}{6}\pi$, and (f) $\frac{33}{6}\pi$.

11. Find two real numbers between -2π and 2π that determine each of these points.

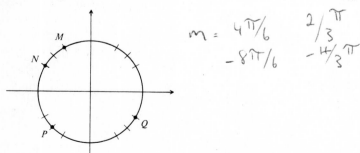

12. Find two real numbers between -2π and 2π that determine each of these points.

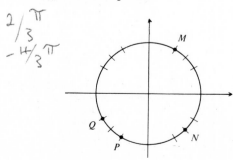

[iv]

13. The number $\frac{\pi}{6}$ determines a point on the unit circle with coordinates

$$\left(\frac{\sqrt{3}}{2}, \frac{1}{2}\right).$$

What are the coordinates of the point determined by $-\frac{\pi}{6}$?

14. The number $\frac{\pi}{3}$ determines a point on the unit circle with coordinates

$$\left(\frac{1}{2}, \frac{\sqrt{3}}{2}\right).$$

What are the coordinates of the point determined by $-\frac{\pi}{3}$?

☆ 15. For the points given in Exercise 11, find a real number between 2π and 4π that determines each point.

16. For the points given in Exercise 12, find a real number between -2π and -4π that determines each point.

17. A point on the unit circle has u-coordinate $-\frac{1}{2}$. What is its v-coordinate?

18. A point on the unit circle has u-coordinate $-\frac{\sqrt{3}}{2}$. What is its v-coordinate?

19. ▦ A point on the unit circle has u-coordinate 0.25671. What is its v-coordinate?

20. ▦ A point on the unit circle has v-coordinate -0.88041. What is its u-coordinate?

8.3 THE SINE AND COSINE FUNCTIONS

[i] The Sine Function

Every real number s determines a point T on the unit circle, as shown. To the real number s we assign the second coordinate of point T as a function value. The function thus defined is called the *sine function*. Function values are denoted by "sine s" or, more briefly, "sin s".* By considering the unit circle, we see that

$$\sin 0 = 0, \qquad \sin \frac{\pi}{2} = 1, \qquad \sin \frac{\pi}{4} = \sin \frac{3}{4}\pi = \frac{\sqrt{2}}{2}.$$

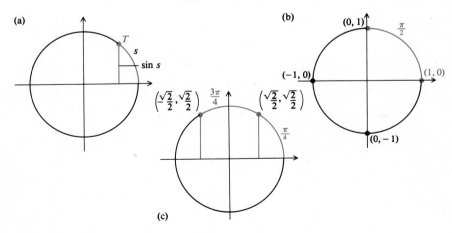

*The sine function is a generalization of what was done in Section 8.1. To help visualize this, we could draw a triangle as shown here. Its hypotenuse has length 1, so $\sin \theta = v/1$, or v. Note also that the arc length s determines the angle θ, so we think of $\sin s$ rather than $\sin \theta$.

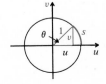

1. Using a unit circle, find:

    a) $\sin \pi$;   b) $\sin \frac{3}{2}\pi$;

    c) $\sin \frac{\pi}{4}$;   d) $\sin\left(-\frac{\pi}{4}\right)$;

    e) $\sin \frac{2\pi}{3}$;   f) $\sin\left(-\frac{\pi}{6}\right)$.

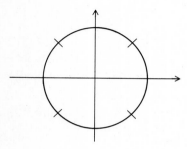

2. Is the sine function periodic? If so, what is its period?

3. Is the sine function even? Is it odd?

4. Does the sine function appear to be continuous?

5. What is the domain of the sine function?

6. What is the range of the sine function?

DO EXERCISE 1.

The sine function also has the following values, as we shall later verify:

$$\sin \frac{\pi}{6} = \frac{1}{2},$$

$$\sin \frac{\pi}{3} = \frac{\sqrt{3}}{2}.$$

[ii] Properties of the Sine Function

Function values increase to a maximum of 1 at $\pi/2$, then decrease to 0 at π, decrease further to -1 at $(\frac{3}{2})\pi$, then increase to 0 at 2π, and so on. The graph of the sine function is shown below. Note that on the s-axis, π appears at about 3.14.

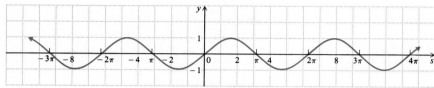

The sine function

The following is a device that could be constructed to draw a graph of the sine function. A light shines horizontally, casting a shadow of the ball fastened to the wheel at P. The wheel rotates at a constant speed and the paper on which the shadow falls moves at a constant speed. The shadow of the ball traces a sine graph.

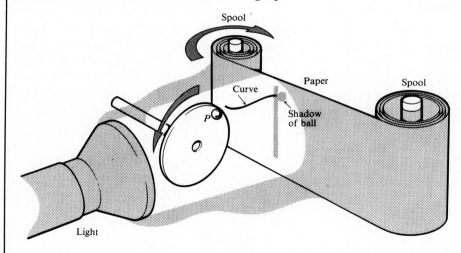

DO EXERCISES 2–6.

From the graph of the sine function, certain properties are apparent.

1. **The sine function is periodic, with period 2π.**

2. **It is an odd function, because it is symmetric with respect to the origin. Thus we know that sin (−s) = −sin s for all real numbers s.**

3. **It is continuous everywhere.**

4. **The domain is the set of all real numbers. The range is the set of all real numbers from −1 to 1, inclusive.**

The unit circle can be used to verify some of the above, as well as to determine other important properties of the sine function.

Let us verify that the period is 2π. Consider any real number s and its point T on the unit circle. Now we increase s by 2π. The point for s + 2π is again the point T. Hence for any real number s, sin s = sin (s + 2π). There is no number smaller than 2π for which this is true. Hence the period is 2π. Actually, we know that sin s = sin (s + 2kπ), where k is any integer.*

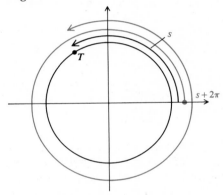

Some Identities

Consider any real number s and its additive inverse −s. These numbers determine points on the unit circle that are symmetric with respect to the u-axis. Hence their second coordinates are additive inverses of each other. Thus we know for any number s that sin (−s) = −sin s, so the sine function is odd.

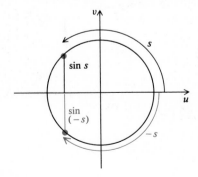

DO EXERCISE 7.

7. A real number s determines the point T as shown.

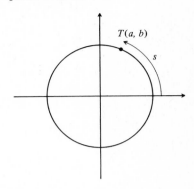

a) Find the point for s + π. What are its coordinates?

b) Find the point for s − π. What are its coordinates?

---

*If k = 1, we have sin s = sin (s + 2π). If k = 3, we have sin s = sin (s + 6π). If k = −5, we have sin s = sin (s − 10π), and so on.

Consider any real number s and its point T on the unit circle. Then $s + \pi$ and $s - \pi$ both determine a point T_1 symmetric with T with respect to the origin. Thus the sines of $s + \pi$ and $s - \pi$ are the same, and are the additive inverse of $\sin s$. This gives us another important property.

For any real number s, $\sin (s + \pi) = -\sin s$, and $\sin (s - \pi) = -\sin s$.

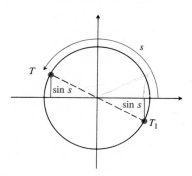

Let us add to our list a property concerning $\sin (\pi - s)$. Since $\pi - s$ is the additive inverse of $s - \pi$, $\sin (\pi - s) = \sin [-(s - \pi)]$. Since the sine function is odd, the latter is equal to $-\sin (s - \pi)$. By the second property in the preceding paragraph, it is equal to $-(-\sin s)$ or $\sin s$. Thus our last property is

$$\sin (\pi - s) = \sin s.$$

Equations that hold for all sensible replacements for the variables are known as *identities*. We shall use the sign $\equiv$ instead of $=$ to indicate that an equation is an identity. We now have the following identities concerning the sine function.

$$\sin s \equiv \sin (s + 2k\pi), \ k \text{ any integer}$$
$$\sin (-s) \equiv -\sin s$$
$$\sin (s \pm \pi) \equiv -\sin s$$
$$\sin (\pi - s) \equiv \sin s$$

The *amplitude* of a periodic function is half the difference between its maximum and minimum function values. It is always positive. The maximum value of the sine function can be seen to be 1, either from the graph or the unit circle, whereas the minimum value is -1. Thus the amplitude of the sine function is 1.

[iii] The Cosine Function

Every real number s determines a point T on the unit circle. To the real number s we assign the first coordinate of point T as a function value. This defines the function known as the *cosine function*. Func-

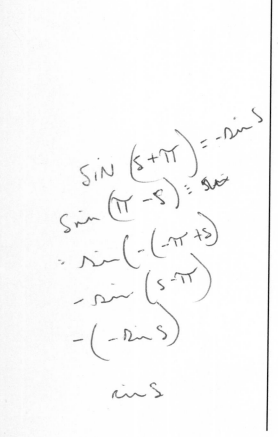

tion values are denoted by "cosine s" or, more briefly, "cos s." By considering the unit circle, we see that*

$$\cos 0 = 1,$$

$$\cos \frac{\pi}{2} = 0,$$

$$\cos \frac{\pi}{4} = \frac{\sqrt{2}}{2},$$

$$\cos \frac{3}{4}\pi = -\frac{\sqrt{2}}{2}.$$

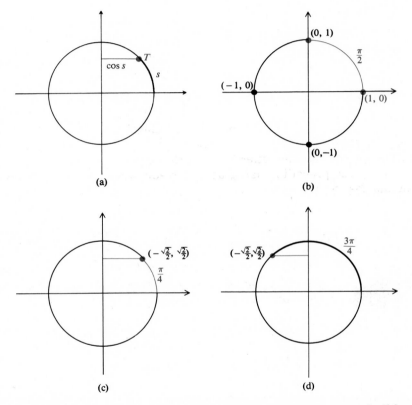

(a)   (b)

(c)   (d)

The cosine function also has the following values, as we shall later verify:

$$\cos \frac{\pi}{6} = \frac{\sqrt{3}}{2}, \cos \frac{\pi}{3} = \frac{1}{2}.$$

DO EXERCISE 8.

---

*The cosine function is also a generalization of what was done in Section 8.1. The triangle here has a hypotenuse of length 1, so $\cos \theta = u/1$, or u. Since the arc length s determines the angle θ, we think of cos s rather than cos θ.

8. Using a unit circle, find:

a) $\cos \pi$;

b) $\cos \frac{3}{2}\pi$;

c) $\cos \frac{\pi}{4}$;

d) $\cos\left(-\frac{\pi}{4}\right)$;

e) $\cos \frac{5\pi}{6}$;

f) $\cos\left(-\frac{\pi}{3}\right)$.

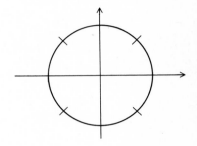

9. Is the cosine function periodic? If so, what is its period?

The graph of the cosine function is as follows. Note that on the s-axis, π appears at about 3.14.

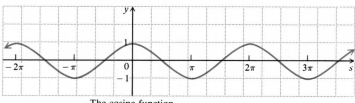

The cosine function

DO EXERCISES 9–13.

It should be noted that for any point on the unit circle, the coordinates are (cos s, sin s).

10. Is the cosine function even? Is it odd?

[iv] Properties of the Cosine Function

From the graph of the cosine function, certain properties are apparent.

1. The cosine function is periodic, with period 2π.

2. It is an even function, because it is symmetric with respect to the y-axis. Thus we know that cos $(-s) = \cos s$ for all real numbers s.

11. Does the cosine function appear to be continuous?

3. It is continuous everywhere.

4. The domain is the set of all real numbers. The range is the set of all real numbers from -1 to 1 inclusive.

5. The amplitude is 1.

The unit circle can be used to verify that the period is 2π and, in fact, that cos $s = \cos (s + 2k\pi)$ for any integer k.

12. What is the domain of the cosine function?

Consider any real number s and its additive inverse. They determine points symmetric with respect to the u-axis, hence they have the same first coordinates. Thus we know that for any number s, cos $(-s) = \cos s$, and that the cosine function is even.

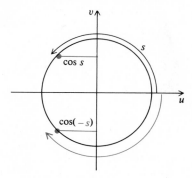

13. What is the range of the cosine function?

Let us consider cos s and cos $(s \pm \pi)$. Recall that s and $s \pm \pi$ determine points that are symmetric with respect to the origin. Thus their coordinates are additive inverses of each other. This verifies that cos $(s \pm \pi) = -\cos s$ for all real numbers s.

Let us consider $\cos(\pi - s)$. This is equal to $\cos[-(s - \pi)]$. Since the cosine function is even, this is equal to $\cos(s - \pi)$, and by the above definition this is in turn equal to $-\cos s$. Hence we have $\cos(\pi - s) = -\cos s$. We now list the identities that we have developed for the cosine function.

$$\cos s \equiv \cos(s + 2k\pi), \; k \text{ any integer}$$
$$\cos(-s) \equiv \cos s$$
$$\cos(s \pm \pi) \equiv -\cos s$$
$$\cos(\pi - s) \equiv -\cos s$$

Instead of the letter s, we can use any letter. We usually use x, as is the custom with most functions.

EXERCISE SET 8.3

[i], [iii]

1. Construct a graph of the sine function. The lower set of axes is shown here at half size, so on other paper, copy the axes at twice the size shown below. Then from the unit circle (shown full size here) transfer vertical distances with a compass.

2. Construct a graph of the cosine function. Follow the instructions for Exercise 1, except that horizontal distances on the unit circle will be transferred with a compass.

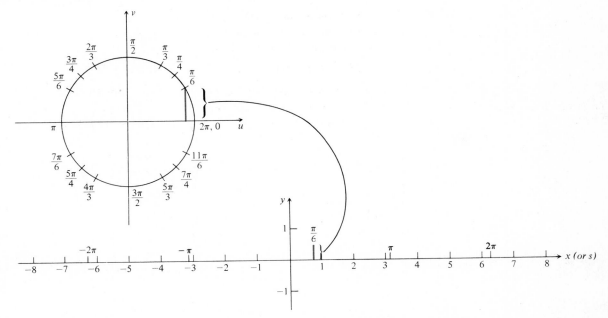

3. a) Sketch a graph of $y = \sin x$.

b) By reflecting the graph in (a), sketch a graph of $y = \sin(-x)$.

c) By reflecting the graph in (a), sketch a graph of $y = -\sin x$.

d) How do the graphs in (b) and (c) compare?

4. a) Sketch a graph of $y = \cos x$.

b) By reflecting the graph in (a), sketch a graph of $y = \cos(-x)$.

c) By reflecting the graph in (a), sketch a graph of $y = -\cos x$.

d) How do the graphs in (a) and (b) compare?

5. a) Sketch a graph of $y = \sin x$.

   b) By translating, sketch a graph of $y = \sin (x + \pi)$.

   c) By reflecting the graph of (a), sketch a graph of $y = -\sin x$.

   d) How do the graphs of (b) and (c) compare?

7. a) Sketch a graph of $y = \cos x$.

   b) By translating, sketch a graph of $y = \cos (x + \pi)$.

   c) By reflecting the graph of (a), sketch a graph of $y = -\cos x$.

   d) How do the graphs in (b) and (c) compare?

6. a) Sketch a graph of $y = \sin x$.

   b) By translating, sketch a graph of $y = \sin (x - \pi)$.

   c) By reflecting the graph of (a), sketch a graph of $y = -\sin x$.

   d) How do the graphs of (b) and (c) compare?

8. a) Sketch a graph of $y = \cos x$.

   b) By translating, sketch a graph of $y = \cos (x - \pi)$.

   c) By reflecting the graph of (a), sketch a graph of $y = -\cos x$.

   d) How do the graphs in (b) and (c) compare?

[ii], [iv] Complete.

9. $\cos (-x) \equiv$ _____

11. $\sin (x + \pi) \equiv$ _____

13. $\cos (\pi - x) \equiv$ _____

15. $\cos (x + 2k\pi) \equiv$ _____

17. $\cos (x - \pi) \equiv$ _____

10. $\sin (-x) \equiv$ _____

12. $\sin (x - \pi) \equiv$ _____

14. $\sin (\pi - x) \equiv$ _____

16. $\sin (x + 2k\pi) \equiv$ _____

18. $\cos (x + \pi) \equiv$ _____

[i]

19. Find:

   a) $\sin \dfrac{\pi}{4}$;   b) $\sin \pi$;   c) $\sin \dfrac{\pi}{6}$;

   d) $\sin \dfrac{5}{4}\pi$;   e) $\sin \dfrac{3}{2}\pi$;

   f) $\sin \dfrac{5}{6}\pi$;   g) $\sin \left(-\dfrac{\pi}{4}\right)$;

   h) $\sin (-\pi)$;   j) $\sin \left(-\dfrac{5}{4}\pi\right)$;

   k) $\sin \left(-\dfrac{3}{2}\pi\right)$;   m) $\sin \left(-\dfrac{\pi}{3}\right)$.

20. Find:

   a) $\sin \dfrac{\pi}{2}$;   b) $\sin \dfrac{\pi}{3}$;   c) $\sin \dfrac{3}{4}\pi$;

   d) $\sin \dfrac{7}{4}\pi$;   e) $\sin 2\pi$;   f) $\sin \left(-\dfrac{\pi}{2}\right)$;

   g) $\sin \left(-\dfrac{3}{4}\pi\right)$;   h) $\sin \left(-\dfrac{7}{4}\pi\right)$;

   j) $\sin (-2\pi)$;   k) $\sin \left(-\dfrac{2}{3}\pi\right)$;

   m) $\sin \left(-\dfrac{\pi}{6}\right)$.

[iii]

21. Find:

   a) $\cos \dfrac{\pi}{4}$;   b) $\cos \pi$;   c) $\cos \dfrac{5}{4}\pi$;

   d) $\cos \dfrac{\pi}{6}$;   e) $\cos \dfrac{3}{2}\pi$;   f) $\cos \dfrac{5}{6}\pi$;

   g) $\cos \left(-\dfrac{\pi}{4}\right)$;   h) $\cos (-\pi)$;

   j) $\cos \left(-\dfrac{5}{4}\pi\right)$;   k) $\cos \left(-\dfrac{3}{2}\pi\right)$;

   m) $\cos \left(-\dfrac{\pi}{3}\right)$.

22. Find:

   a) $\cos \dfrac{\pi}{2}$;   b) $\cos \dfrac{\pi}{3}$;   c) $\cos \dfrac{3}{4}\pi$;

   d) $\cos \dfrac{7}{4}\pi$;   e) $\cos 2\pi$;   f) $\cos \left(-\dfrac{\pi}{2}\right)$;

   g) $\cos \left(-\dfrac{3}{4}\pi\right)$;   h) $\cos \left(-\dfrac{\pi}{6}\right)$;

   j) $\cos \left(-\dfrac{7}{4}\pi\right)$;   k) $\cos \left(-\dfrac{2}{3}\pi\right)$.

☆

23. For which numbers is

a) $\sin x = 1$?

b) $\sin x = -1$?

25. Solve for x.

$$\sin x = 0$$

27. Find $f \circ g$ and $g \circ f$, where $f(x) = x^2 + 2x$ and $g(x) = \cos x$.

29. ▦ Calculate the following. Do not use the trigonometric function keys.

a) $\cos \dfrac{\pi}{6}$ b) $\cos \dfrac{\pi}{4}$

24. For which numbers is

a) $\cos x = 1$?

b) $\cos x = -1$?

26. Solve for x.

$$\cos x = 0$$

28. ▦ Calculate the following. Do not use the trigonometric function keys.

a) $\sin \dfrac{\pi}{4}$ b) $\sin \dfrac{\pi}{3}$

30. Show that the sine function does not have the *linearity* property, i.e., that it is not true that $\sin(x + y) = \sin x + \sin y$ for all numbers x and y.

8.4 THE OTHER CIRCULAR FUNCTIONS

[i] Definitions

There are four other circular functions, all defined in terms of the sine and cosine functions. Their definitions are as follows.

DEFINITIONS

The tangent function:	$\tan s \equiv \dfrac{\sin s}{\cos s}$
The cotangent function:	$\cot s \equiv \dfrac{\cos s}{\sin s}$
The secant function:	$\sec s \equiv \dfrac{1}{\cos s}$
The cosecant function:	$\csc s \equiv \dfrac{1}{\sin s}$

The Tangent Function

The tangent of a number is the quotient of its sine by its cosine.

Example 1 Find $\tan 0$.

$$\tan 0 = \frac{\sin 0}{\cos 0} = \frac{0}{1} = 0$$

Example 2 Find $\tan \dfrac{\pi}{4}$.

$$\tan \frac{\pi}{4} = \frac{\sin \dfrac{\pi}{4}}{\cos \dfrac{\pi}{4}} = \frac{\dfrac{\sqrt{2}}{2}}{\dfrac{\sqrt{2}}{2}} = 1$$

OBJECTIVES

You should be able to:

[i] Define the tangent, cotangent, secant, and cosecant functions; state their properties; sketch their graphs; and find function values (when they exist) for multiples of $\pi/6$, $\pi/4$, and $\pi/3$.

[ii] Given values of the sine and cosine, calculate the values of the other functions.

[iii] Verify simple identities.

1. Find $\tan \pi$.

2. Find $\tan \frac{3}{4}\pi$.

3. Find $\tan \frac{\pi}{2}$.

4. Find three other real numbers at which the tangent function is undefined.

Example 3 Find $\tan \frac{\pi}{3}$.

$$\tan \frac{\pi}{3} = \frac{\sin \frac{\pi}{3}}{\cos \frac{\pi}{3}} = \frac{\frac{\sqrt{3}}{2}}{\frac{1}{2}} = \sqrt{3}$$

Not every real number has a tangent.

Example 4 Show that $\tan \frac{\pi}{2}$ does not exist.

$$\tan \frac{\pi}{2} = \frac{\sin \frac{\pi}{2}}{\cos \frac{\pi}{2}} = \frac{1}{0},$$

but division by 0 is undefined. Thus $\pi/2$ is not in the domain of the tangent function, and $\tan(\pi/2)$ does not exist.

DO EXERCISES 1–4.

The tangent function is undefined for any number whose cosine is 0. Thus it is undefined for $(\pi/2) + k\pi$, where k is any integer. In the first quadrant the function values are positive. In the second quadrant, however, the sine and cosine values have opposite signs, so the tangent values are negative. Tangent values are also negative in the fourth quadrant, and they are positive in quadrant three.

A graph of the tangent function is shown below. Note that the function value is 0 when x = 0, and the values increase as x increases toward $\pi/2$. As we approach $\pi/2$, the denominator becomes very small, so the tangent values become very large. In fact, they increase without bound. The dashed vertical line at $\pi/2$ is not a part of the graph. The graph approaches this vertical line closely, however. The vertical dashed lines are known as *asymptotes* to the curve.

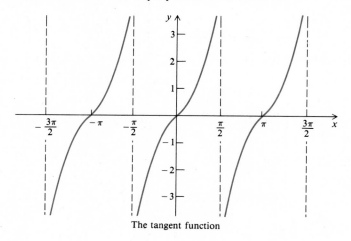

The tangent function

The tangent function is periodic, with a period of π. Its domain is the set of all real numbers except $(\pi/2) + k\pi$, k any integer. Its range is the set of all real numbers. It is continuous except where it is not defined.

DO EXERCISES 5–9.

The Cotangent Function

The cotangent of a number is the quotient of its cosine by its sine. Thus the cotangent function is undefined for any number whose sine is 0. The graph of this function is shown here.

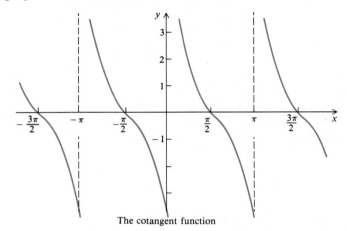

The cotangent function

DO EXERCISES 10–14.

The cotangent function is periodic, with a period of π. Its domain is the set of all real numbers except $k\pi$, k any integer. Its range is the set of all real numbers. It is continuous except where it is not defined.

Like the tangent function, the cotangent has positive function values in quadrants I and III and negative function values in quadrants II and IV. The tangent and cotangent are reciprocals, hence the following identities hold.

$$\tan x \equiv \frac{1}{\cot x}, \qquad \cot x \equiv \frac{1}{\tan x}$$

The Secant Function

The secant and cosine functions are reciprocals.

Example 5 Find $\sec \frac{\pi}{4}$.

$$\sec \frac{\pi}{4} = \frac{1}{\cos \frac{\pi}{4}} = \frac{1}{\frac{\sqrt{2}}{2}} = \sqrt{2}$$

5. What is the period of the tangent function?

6. What is the domain of the tangent function?

7. What is the range of the tangent function?

8. Does the tangent function appear to be even? odd?

9. Find three numbers at which the cotangent function is undefined.

10. What is the period of the cotangent function?

11. What is the domain of the cotangent function?

12. What is the range of the cotangent function?

13. In which quadrants are the function values of the cotangent function positive? negative?

14. Does the cotangent function appear to be odd? even?

15. Find sec $\frac{\pi}{4}$.

16. Find sec π.

17. Find three numbers at which the secant function is undefined.

18. What is the period of the cosecant function?

19. What is the domain of the cosecant function?

20. What is the range of the cosecant function?

21. In which quadrants are the function values of the cosecant function positive? negative?

22. Does the cosecant function appear to be even? odd?

This function is undefined for those numbers x for which $\cos x = 0$. A graph of the secant function is shown below. The cosine graph is also shown for reference, since these functions are reciprocals.

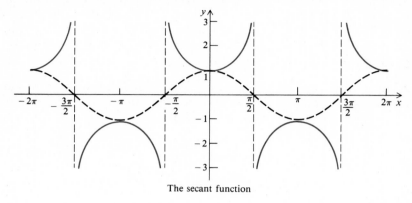

The secant function

DO EXERCISES 15–17.

The secant function is periodic, with period 2π. Its domain is the set of all real numbers except $(\pi/2) + k\pi$, k any integer. Its range consists of all real numbers 1 and greater, in addition to all real numbers -1 and less. It is continuous wherever it is defined.

The Cosecant Function

Since the cosecant and sine functions are reciprocals, the cosecant will be undefined wherever the sine function is 0. The functions will be positive together and negative together. A graph of the cosecant function is shown below. The sine graph is shown for reference, since these functions are reciprocals.

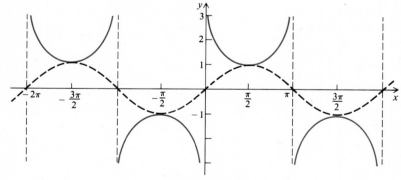

The cosecant function

DO EXERCISES 18–22.

The cosecant function is periodic, with period 2π. Its domain is the set of all real numbers except $k\pi$, k any integer. Its range is the same as for the secant function. It is continuous wherever it is defined.

Signs of the Functions

Suppose a real number s determines a point on the unit circle in the first quadrant. Since both coordinates are positive there, the function values for all the circular functions will be positive. In the second quadrant the first coordinate is negative and the second coordinate is positive. Thus the sine function has positive values and the cosine has negative values. The secant, being the reciprocal of the cosine, also has negative values, and the cosecant, being the reciprocal of the sine, has positive values. The tangent and cotangent both have negative values in the second quadrant.

DO EXERCISE 23.

The following diagram summarizes, showing the quadrants in which the circular function values are positive. This diagram need not be memorized, because the signs can readily be determined by reference to a unit circle.

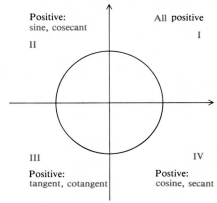

Positive:
sine, cosecant

II

All positive

I

III

Positive:
tangent, cotangent

IV

Positive:
cosine, secant

[ii] Calculating Function Values

Given values for sine and cosine, we can calculate the other function values.

▦ **Example 6** Given that $\sin (\pi/9) \approx 0.34202$ and $\cos (\pi/9) \approx 0.93969$, calculate the other function values for $\pi/9$.

$$\tan \frac{\pi}{9} = \frac{\sin \dfrac{\pi}{9}}{\cos \dfrac{\pi}{9}} \approx \frac{0.34202}{0.93969} \approx 0.36397$$

$$\cot \frac{\pi}{9} = \frac{\cos \dfrac{\pi}{9}}{\sin \dfrac{\pi}{9}} \approx \frac{0.93969}{0.34202} \approx 2.74747$$

$$\sec \frac{\pi}{9} = \frac{1}{\cos \dfrac{\pi}{9}} \approx \frac{1}{0.93969} \approx 1.06418$$

$$\csc \frac{\pi}{9} = \frac{1}{\sin \dfrac{\pi}{9}} \approx \frac{1}{0.34202} \approx 2.92381$$

23. In the third quadrant, what are the signs of the six circular functions?

$\cos(\alpha + \beta) = \cos\alpha \cos\beta$

$\sin\left(x + \frac{\pi}{2}\right) = \cos x$

▦ **24.** Given that $\sin (\pi/11) = 0.28173$ and $\cos (\pi/11) = 0.95949$, calculate the other function values for $\pi/11$.

▦ **Example 7** Given that $\sin (2/3)\pi \approx 0.86603$ and $\cos (2/3)\pi \approx -0.50000$, calculate the other function values for $(2/3)\pi$.

$$\tan \frac{2}{3}\pi = \frac{\sin \frac{2}{3}\pi}{\cos \frac{2}{3}\pi} \approx \frac{0.86603}{-0.50000} \approx -1.73206$$

$$\cot \frac{2}{3}\pi = \frac{\cos \frac{2}{3}\pi}{\sin \frac{2}{3}\pi} \approx \frac{-0.50000}{0.86603} \approx -0.57735$$

$$\sec \frac{2}{3}\pi = \frac{1}{\cos \frac{2}{3}\pi} \approx \frac{1}{-0.50000} \approx -2.0000$$

$$\csc \frac{2}{3}\pi = \frac{1}{\sin \frac{2}{3}\pi} \approx \frac{1}{0.86603} \approx 1.15469$$

DO EXERCISE 24.

[iii] Some Identities

The tangent function appears to be odd. Let us investigate by finding $\tan (-x)$. By definition of the tangent function,

$$\tan (-x) \equiv \frac{\sin (-x)}{\cos (-x)}.$$

But we know that $\sin (-x) \equiv -\sin x$ and $\cos (-x) \equiv \cos x$. Hence

$$\tan (-x) \equiv \frac{\sin (-x)}{\cos (-x)} \equiv \frac{-\sin x}{\cos x} \equiv -\tan x.$$

Thus the tangent function is indeed odd. Similarly, it can be shown that the cotangent function is odd, the secant function is even, and the cosecant function is odd. Each such demonstration gives us an identity.

By use of the previously developed identities involving the sine and cosine functions we can develop a number of other identities.

Example 8 Show that $\tan (x + \pi) \equiv \tan x$.

$$\tan (x + \pi) \equiv \frac{\sin (x + \pi)}{\cos (x + \pi)} \qquad \text{By definition of the tangent function}$$

$$\equiv \frac{-\sin x}{-\cos x} \qquad \text{Using } \sin (x + \pi) \equiv -\sin x \text{ and } \cos (x + \pi) \equiv -\cos x$$

$$\equiv \frac{\sin x}{\cos x}$$

$$\equiv \tan x$$

Example 9 Show that $\sec(\pi - x) \equiv -\sec x$.

$$\sec(\pi - x) \equiv \frac{1}{\cos(\pi - x)} \qquad \text{By definition of the secant function}$$

$$\equiv \frac{1}{-\cos x} \qquad \text{Using } \cos(\pi - x) \equiv -\cos x$$

$$\equiv -\frac{1}{\cos x}$$

$$\equiv -\sec x$$

Example 10 Show that $\csc(x - \pi) \equiv -\csc x$.

$$\csc(x - \pi) \equiv \frac{1}{\sin(x - \pi)} \qquad \text{By definition of the cosecant function}$$

$$\equiv \frac{1}{-\sin x} \qquad \text{Using } \sin(x - \pi) \equiv -\sin x$$

$$\equiv -\csc x$$

DO EXERCISES 25–27.

25. Show that $\cot(-x) \equiv -\cot x$.

26. Show that $\tan(x - \pi) \equiv \tan x$.

27. Show that $\csc(\pi - x) \equiv \csc x$.

EXERCISE SET 8.4

[i] Find the following function values.

1. $\cot \dfrac{\pi}{4}$

2. $\tan\left(-\dfrac{\pi}{4}\right)$

3. $\tan \dfrac{\pi}{6}$

4. $\cot\left(\dfrac{5\pi}{6}\right)$

5. $\sec \dfrac{\pi}{4}$

6. $\csc \dfrac{3\pi}{4}$

7. $\tan \dfrac{3\pi}{2}$

8. $\cot \pi$

9. $\tan \dfrac{2\pi}{3}$

10. $\cot\left(-\dfrac{2\pi}{3}\right)$

[ii]

11. ▦ Complete this table of approximate function values. Do not use trigonometric function keys. Round to five decimal places.

	$\pi/16$	$\pi/8$	$\pi/6$	$\pi/4$	$3\pi/8$	$7\pi/16$
sin	0.19509	0.38268			0.92388	0.98079
cos	0.98079	0.92388			0.38268	0.19509
tan						
cot						
sec						
csc						

12. Complete this table of approximate function values. Do not use trigonometric function keys. Round to five decimal places.

	$-\pi/16$	$-\pi/8$	$-\pi/6$	$-\pi/4$	$-\pi/3$
sin	-0.19509	-0.38268	-0.50000	-0.70711	-0.86603
cos	0.98079	0.92388	0.86603	0.70711	0.50000
tan					
cot					
sec					
csc					

In Exercises 13–16, use graph paper. Use the tables above, plus your other knowledge of the functions, to make graphs.

13. Graph the tangent function between -2π and 2π.

14. Graph the cotangent function between -2π and 2π.

15. Graph the secant function between -2π and 2π.

16. Graph the cosecant function between -2π and 2π.

17. Of the six circular functions, which are even?

18. Of the six circular functions, which are odd?

19. Which of the six circular functions have period 2π?

20. Which of the six circular functions have period π?

21. In which quadrants is the tangent function positive? negative?

22. In which quadrants is the cotangent function positive? negative?

23. In which quadrants is the secant function positive? negative?

24. In which quadrants is the cosecant function positive? negative?

[iii] Verify the following identities.

25. $\sec(-x) \equiv \sec x$

26. $\csc(-x) \equiv -\csc x$

27. $\cot(x + \pi) \equiv \cot x$

28. $\cot(x - \pi) \equiv \cot x$

29. $\sec(x + \pi) \equiv -\sec x$

30. $\tan(\pi - x) \equiv -\tan x$

☆

31. Verify the identity $\sec(x - \pi) \equiv -\sec x$ graphically.

32. Verify the identity $\tan(x + \pi) \equiv \tan x$ graphically.

33. Describe how the graphs of the tangent and cotangent functions are related.

34. Describe how the graphs of the secant and cosecant functions are related.

35. Which pairs of circular functions have the same zeros? (A "zero" of a function is an input that produces an output of 0.)

36. Describe how the asymptotes of the tangent, cotangent, secant, and cosecant functions are related to the inputs that produce outputs of 0.

37. Graph $f(x) = |\tan x|$.

★

38. Solve $\cos x \le \sec x$.

39. Solve $\sin x > \csc x$.

OBJECTIVES

You should be able to:

[i] State the Pythagorean identities and derive variations of them.

[ii] State the cofunction identities for the sine and cosine, and derive cofunction identities for the other functions.

8.5 SOME RELATIONS AMONG THE CIRCULAR FUNCTIONS

We have already seen that there are some relations among the six circular functions. In fact, the sine and cosine functions are used to define the others. There are certain other important relations among these functions. These are given by identities.

[i] The Pythagorean Identities

Recall that the equation of the unit circle in a u-v plane is

$$u^2 + v^2 = 1.$$

For any point on the unit circle, the coordinates u and v satisfy this equation. Suppose a real number s determines a point T on the unit circle, with coordinates (u, v). Then $u = \cos s$ and $v = \sin s$. Substituting these in the equation of the unit circle gives us the identity

$$\sin^2 s + \cos^2 s \equiv 1.$$

This is true because the coordinates of a point on the unit circle are $(\cos s, \sin s)$. When exponents are used with the circular functions, we write $\sin^2 s$ instead of $(\sin s)^2$. This identity relates the sine and cosine of any real number s. It is an important identity, known as one of the *Pythagorean* identities. We now develop another. We will divide the above identity by $\sin^2 s$:

$$\frac{\sin^2 s}{\sin^2 s} + \frac{\cos^2 s}{\sin^2 s} \equiv \frac{1}{\sin^2 s}.$$

Simplifying, we get

$$1 + \cot^2 s \equiv \csc^2 s.$$

This relation is valid for any number s for which $\sin^2 s \neq 0$, since we divided by $\sin^2 s$. But the numbers for which $\sin^2 s = 0$ (or $\sin s = 0$) are exactly the ones for which the cotangent and cosecant functions are undefined. Hence our new equation holds for all numbers s for which $\cot s$ and $\csc s$ are defined, and is thus an identity.

DO EXERCISE 1.

The third Pythagorean identity, obtained by dividing the first by $\cos^2 s$, is

$$1 + \tan^2 s \equiv \sec^2 s.$$

The Pythagorean identities should be memorized. Certain variations of them, although most useful, need not be memorized, because they are so easily obtained from these three.

Example 1 Derive identities that give $\cos^2 x$ and $\cos x$ in terms of $\sin x$.

We begin with $\sin^2 x + \cos^2 x \equiv 1$, and solve for $\cos^2 x$. We have $\cos^2 x \equiv 1 - \sin^2 x$. This gives $\cos^2 x$ in terms of $\sin x$. If we now take the principal square root, we obtain

$$|\cos x| \equiv \sqrt{1 - \sin^2 x}.$$

1. Consider $\sin^2 s + \cos^2 s \equiv 1$.
   a) Divide by $\cos^2 s$ and simplify.
   b) For what values of $\cos s$ is the new equation meaningless?
   c) Does the new equation hold for all values of s for which the functions are defined? (Is it an identity?)

2. Derive an identity for $\sin^2 x$ in terms of $\cos x$.

We can also express this as

$$\cos x \equiv \pm \sqrt{1 - \sin^2 x},$$

with the understanding that the sign must be determined by the quadrant in which the point for x lies.

DO EXERCISES 2 AND 3.

[ii] The Cofunction Identities

3. Derive an identity for $\sin x$ in terms of $\cos x$.

The sine and cosine are called *cofunctions* of each other. Similarly, the tangent and cotangent are cofunctions and the secant and cosecant are cofunctions. Another class of identities gives functions in terms of their cofunctions at numbers differing by $\pi/2$. We consider first the sine and cosine. Consider this graph.

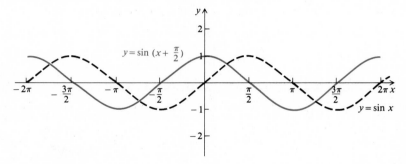

4. a) Graph $y = \cos x$.
  b) Translate, to obtain a graph of

$$y = \cos\left(x - \frac{\pi}{2}\right).$$

  c) Graph $y = \sin x$.
  d) How do the graphs of (b) and (c) compare?
  e) Write the identity thus established.

The graph of $y = \sin x$ has been translated to the left a distance of $\pi/2$, to obtain the graph of $y = \sin(x + \pi/2)$. The latter is also a graph of the cosine function. Thus we obtain the identity

$$\sin\left(x + \frac{\pi}{2}\right) \equiv \cos x.$$

DO EXERCISE 4.

By means similar to that above, we obtain the identity

$$\cos\left(x - \frac{\pi}{2}\right) \equiv \sin x.$$

Consider the following graph. The graph of $y = \sin x$ has been translated to the right a distance of $\pi/2$, to obtain the graph of $y = \sin(x - \pi/2)$. The latter is a reflection of the cosine function across the x-axis. In other words, it is a graph of $y = -\cos x$. We thus obtain the following identity:

$$\sin\left(x - \frac{\pi}{2}\right) \equiv -\cos x.$$

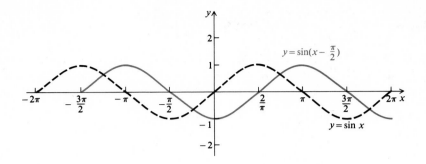

DO EXERCISE 5.

By means similar to that above, we obtain the identity

$$\cos\left(x + \frac{\pi}{2}\right) \equiv -\sin x.$$

We have now established the following cofunction identities.*

$$\sin\left(x \pm \frac{\pi}{2}\right) \equiv \pm\cos x$$

$$\cos\left(x \pm \frac{\pi}{2}\right) \equiv \mp\sin x$$

These should be learned. The other cofunction identities can be obtained easily, using these and the definitions of the other circular functions.

Example 2 Derive an identity relating $\sin(\pi/2 - x)$ to $\cos x$.

Since the sine function is odd, we know that

$$\sin\left(\frac{\pi}{2} - x\right) \equiv \sin\left[-\left(x - \frac{\pi}{2}\right)\right] \equiv -\sin\left(x - \frac{\pi}{2}\right).$$

Now consider the identity already established,

$$\sin\left(x - \frac{\pi}{2}\right) \equiv -\cos x.$$

This is equivalent to

$$-\sin\left(x - \frac{\pi}{2}\right) \equiv \cos x,$$

and we now have

$$\sin\left(\frac{\pi}{2} - x\right) \equiv \cos x.$$

DO EXERCISE 6.

---

*There are four identities in this list. Two of them are obtained by taking the top signs, the other two by taking the bottom signs.

5. a) Graph $y = \cos x$.
   b) Translate, to obtain a graph of
      $$y = \cos\left(x + \frac{\pi}{2}\right).$$
   c) Graph $y = \sin x$.
   d) Reflect the graph of c, to obtain a graph of $y = -\sin x$.
   e) How do the graphs of (b) and (d) compare?
   f) Write the identity thus established.

6. Prove that $\cos\left(\frac{\pi}{2} - x\right) \equiv \sin x$.

7. Find an identity for $\cot\left(x - \dfrac{\pi}{2}\right)$.

Example 3 Find an identity for $\tan\left(x + \dfrac{\pi}{2}\right)$.

By definition of the tangent function,

$$\tan\left(x + \frac{\pi}{2}\right) \equiv \frac{\sin\left(x + \dfrac{\pi}{2}\right)}{\cos\left(x + \dfrac{\pi}{2}\right)}.$$

Using the cofunction identities above, we obtain

$$\frac{\sin\left(x + \dfrac{\pi}{2}\right)}{\cos\left(x + \dfrac{\pi}{2}\right)} \equiv \frac{\cos x}{-\sin x} \equiv -\frac{\cos x}{\sin x} \equiv -\cot x.$$

Thus the identity we seek is

$$\tan\left(x + \frac{\pi}{2}\right) \equiv -\cot x.$$

Example 4 Find an identity for $\sec\left(\dfrac{\pi}{2} - x\right)$.

By definition of the secant function,

$$\sec\left(\frac{\pi}{2} - x\right) \equiv \frac{1}{\cos\left(\dfrac{\pi}{2} - x\right)}.$$

If we proceed as in Example 2, the following can be established: $\cos(\pi/2 - x) \equiv \cos(x - \pi/2)$. Then we obtain

$$\frac{1}{\cos\left(\dfrac{\pi}{2} - x\right)} \equiv \frac{1}{\sin x} \equiv \csc x.$$

Thus the identity we seek is

$$\sec\left(\frac{\pi}{2} - x\right) \equiv \csc x.$$

DO EXERCISE 7.

EXERCISE SET 8.5

[i]

1. From the identity $1 + \cot^2 x \equiv \csc^2 x$, obtain two other identities.

2. From the identity $1 + \tan^2 x \equiv \sec^2 x$, obtain two other identities.

3. a) Derive an identity for $\csc x$ in terms of $\cot x$.
b) Derive an identity for $\cot x$ in terms of $\csc x$.

4. a) Derive an identity for $\tan x$ in terms of $\sec x$.
b) Derive an identity for $\sec x$ in terms of $\tan x$.

[ii]

5. Find an identity for $\tan\left(x - \dfrac{\pi}{2}\right)$.

6. Find an identity for $\cot\left(x + \dfrac{\pi}{2}\right)$.

7. Find an identity for $\sec\left(\frac{\pi}{2} - x\right)$.

8. Find an identity for $\csc\left(\frac{\pi}{2} - x\right)$.

9. For the six circular functions, list the cofunction identities involving $x \pm \frac{\pi}{2}$.

10. In which quadrant is $\sin x = \sqrt{1 - \cos^2 x}$?
11. ▦ Given that $\sin x = 0.1425$ in quadrant II, find the other circular function values.
12. ▦ Given that $\tan x = 0.8012$ in quadrant III, find the other circular function values.
13. ▦ Given that $\cot x = 0.7534$ and $\sin x > 0$, find the other circular function values.
14. ▦ Given that $\sin\frac{\pi}{8} = 0.38268$, use identities to find:

   a) $\cos\frac{\pi}{8}$; b) $\cos\frac{5\pi}{8}$; c) $\sin\frac{5\pi}{8}$; d) $\sin\frac{-3\pi}{8}$; e) $\cos\frac{-3\pi}{8}$.

15. Using sines, cosines, and a determinant, write an equation of the unit circle.
16. The coordinates of a point on the unit circle are $(\cos s, \sin s)$, for some suitable s. Show that for a circle with center at the origin of radius r, the coordinates of any point on the circle are $(r\cos s, r\sin s)$.

8.6 GRAPHS

[i], [ii] Variations of Basic Graphs

We will consider graphs of some variations of the sine and cosine functions. In particular, we are interested in the following:

$$y = A + B\sin(Cx - D),$$

and

$$y = A + B\cos(Cx - D),$$

where A, B, C, and D are constants, some of which may be 0. These constants have the effect of translating, stretching, reflecting, or shrinking the basic graphs. Let us examine an equation to see what each of these constants does.* We first change the form a bit.

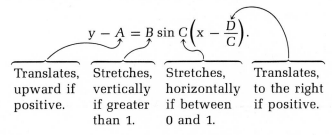

|Translates, upward if positive. | Stretches, vertically if greater than 1. | Stretches, horizontally if between 0 and 1. | Translates, to the right if positive.|

The amplitude of the sine or cosine function is multiplied by B, when that constant is present. The period will be affected by the constant C. If $C = 2$, for example, the period will be divided by 2. If $C = \frac{1}{2}$, the period will be multiplied by 2. In general, the new period will be $2\pi/C$.

*You may wish to review pp. 126–134.

OBJECTIVES

You should be able to:
[i] Sketch graphs of $y = A + B\sin(Cx + D)$ and $y = A + B\cos(Cx + D)$, for various values of the constants A, B, C, and D.
[ii] For functions like these, determine the amplitude, period, and phase shift.
[iii] By addition of ordinates, graph sums of functions.

1. Sketch a graph of $y = -2 + \cos x$.

Example 1 The graph of $y = 3 + \sin x$ is a translation of $y = \sin x$ upward 3 units ($A = 3$, $B = C = 1$, and $D = 0$).

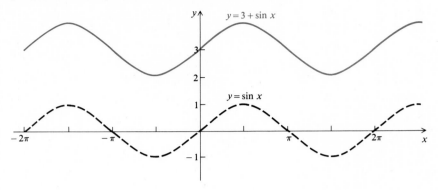

DO EXERCISE 1.

Example 2 The graph of $y = 2 \sin x$ is a vertical stretching of the graph of $y = \sin x$. The amplitude of this function is 2 ($A = 0$, $B = 2$, $C = 1$, and $D = 0$).

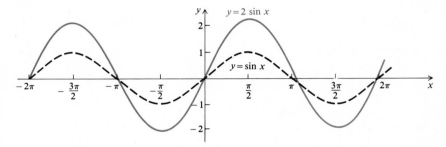

2. Sketch a graph of $y = -2 \cos x$. What is the amplitude?

    If the constant B in $y = B \sin x$ is negative, there will also be a reflection across the x-axis. If the absolute value of B is less than 1, then there will be a vertical shrinking. $|B|$ will be the amplitude.

Example 3 The graph of $y = -(\frac{1}{2}) \sin x$ is a vertical shrinking and a reflection of the graph of $y = \sin x$. The amplitude is $\frac{1}{2}$.

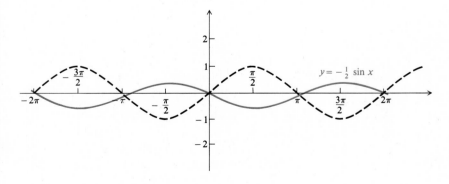

DO EXERCISE 2.

Example 4 The graph of $y = \sin 2x$ is a horizontal shrinking of the graph of $y = \sin x$. The period of this function is π (half the period of $y = \sin x$).

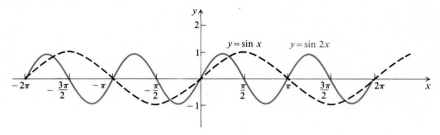

DO EXERCISES 3 AND 4.

Example 5 The graph of $y = \sin (2x - \pi)$ is the same as that of $y = \sin 2(x - \pi/2)$. We merely factored the 2 out of the parentheses. The $\pi/2$ translates the graph of $y = \sin 2x$ a distance of $\pi/2$ to the right. The 2 in front of the parentheses shrinks by half, making the period π.

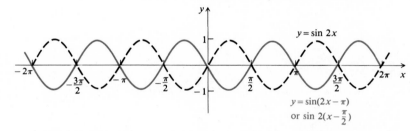

DO EXERCISE 5.

The quantity D/C translates to the right or left, and is called the *phase shift*. In Example 5, the phase shift is $\pi/2$. This means that the graph of $\sin 2x$ has been shifted forward a distance of $\pi/2$.

Example 6 The graph of $y = 3 \sin [2x + (\pi/2)]$ has an amplitude of 3.

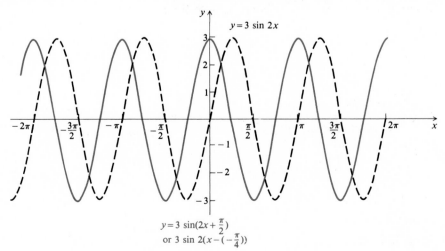

3. Sketch a graph of $y = \cos \frac{1}{2}x$. What is the period?

4. Sketch a graph of $y = \sin (-2x)$. What is the period?

5. Sketch a graph of $y = \cos (2x + \pi)$. What is the period? What is the phase shift?

6. Sketch a graph of
$$y = 3 \cos\left(2x - \frac{\pi}{2}\right).$$

What is the amplitude? What is the period? What is the phase shift?

The period is π and the phase shift is $-\pi/4$. This means that the graph of $3 \sin 2x$ has been shifted back a distance of $\pi/4$.

DO EXERCISE 6.

The oscilloscope is an electronic device that draws graphs on a cathode-ray tube. By manipulating the controls, one can change such things as the amplitude, period, and phase. The oscilloscope has many applications, and the trigonometric functions play a major role in many of them.

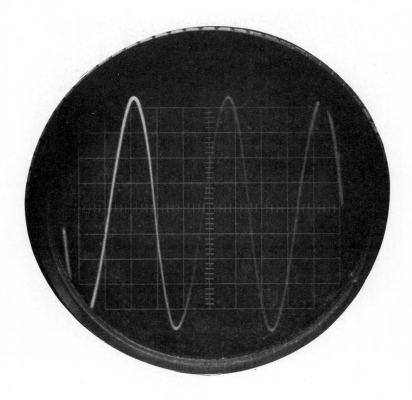

[iii] Graphs of Sums

A function that is a sum of two functions can often be graphed by a method called *addition of ordinates*. We graph the two functions separately and then add the second coordinates (called *ordinates*) graphically. A compass may be helpful.

Example 7 Graph $y = 2 \sin x + \sin 2x$.

We graph $y = 2 \sin x$ and $y = \sin 2x$ using the same axes.

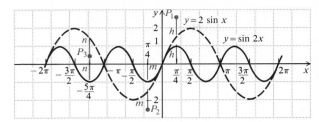

7. By addition of ordinates, sketch a graph of $y = \cos x + \sin 2x$.

Now we graphically add some ordinates to obtain points on the graph we seek. At $x = \pi/4$ we transfer the distance h, which is the value of $\sin 2x$, upward to add it to the value of $2 \sin x$. Point P_1 is on the graph we seek. At $x = -\pi/4$ we do a similar thing, but the distance m is negative. Point P_2 is on our graph. At $x = -(\frac{5}{4})\pi$ the distance n is negative, so we in effect subtract it from the value of $2 \sin x$. Point P_3 is on the graph we seek. We continue to plot points in this fashion and then connect them to get the desired graph, shown here.

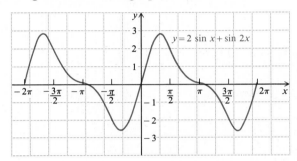

DO EXERCISE 7.

A "sawtooth" function, such as the one shown on this oscilloscope, has numerous applications—for example, in television circuits. This sawtooth function can be approximated extremely well by adding, electronically, several sine and cosine functions.

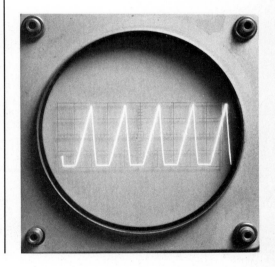

EXERCISE SET 8.6

[i] In Exercises 1–28, use graph paper. Sketch graphs of the functions.

1. $y = 2 + \sin x$

2. $y = -3 + \cos x$

3. $y = \frac{1}{2} \sin x$

4. $y = \frac{1}{3} \cos x$

5. $y = 2 \cos x$

6. $y = 3 \cos x$

7. $y = -\frac{1}{2} \cos x$

8. $y = -2 \sin x$

9. $y = \cos 2x$

10. $y = \cos 3x$

11. $y = \cos (-2x)$

12. $y = \cos (-3x)$

13. $y = \sin \frac{1}{2} x$

14. $y = \cos \frac{1}{3} x$

15. $y = \sin \left(-\frac{1}{2} x \right)$

16. $y = \cos \left(-\frac{1}{3} x \right)$

17. $y = \cos(2x - \pi)$

18. $y = \sin(2x + \pi)$

19. $y = 2\cos\left(\frac{1}{2}x - \frac{\pi}{2}\right)$

20. $y = 4\sin\left(\frac{1}{4}x + \frac{\pi}{8}\right)$

21. $y = -3\cos(4x - \pi)$

22. $y = -3\sin\left(2x + \frac{\pi}{2}\right)$

[iii]

23. $y = 2\cos x + \cos 2x$

24. $y = 3\cos x + \cos 3x$

25. $y = \sin x + \cos 2x$

26. $y = 2\sin x + \cos 2x$

27. $y = 3\cos x - \sin 2x$

28. $y = 3\sin x - \cos 2x$

[ii] In Exercises 29–34, determine the amplitude, period, and phase shift.

29. $y = 3\cos\left(3x - \frac{\pi}{2}\right)$

30. $y = 4\sin\left(4x - \frac{\pi}{3}\right)$

31. $y = -5\cos\left(4x + \frac{\pi}{3}\right)$

32. $y = -4\sin\left(5x + \frac{\pi}{2}\right)$

33. $y = \frac{1}{2}\sin(2\pi x + \pi)$

34. $y = -\frac{1}{4}\cos(\pi x - 4)$

☆ ─────────────────────────────────

35. Graph $y = \sec^2 x$.

36. Graph $y = \tan x \csc x$.

37. Graph $y = 3\cos 2(x + \pi/4)$ by first converting to a sine.

★ ─────────────────────────────────

38. Graph $y = x + \sin x$. (*Hint:* Use addition of ordinates. You might also wish to use a calculator to find some function values to plot.)

39. Graph $y = 2^{-x}\sin x$, for $0 \le x \le 4\pi$. Functions like this have applications in the theory of damped oscillations. (*Hint:* First graph $y = 2^{-x}$ and also $y = -(2^{-x})$. Then graph $y = \sin x$ and consider multiplying the ordinates. You may wish to use a calculator to find some function values to plot.)

─────────────────────────────────

OBJECTIVES

You should be able to:

[i] Multiply and factor expressions containing trigonometric expressions.

[ii] Compute with and simplify expressions containing trigonometric expressions.

[iii] Compute with radical expressions containing trigonometric expressions, including rationalizing numerators or denominators.

[iv] Solve equations containing trigonometric expressions for the values of those expressions.

8.7 ALGEBRAIC AND TRIGONOMETRIC MANIPULATIONS

The circular functions and variations of them are called *trigonometric functions*. Trigonometric expressions such as $\sin 2x$ or $\tan(x - \pi)$ represent numbers, just as algebraic expressions such as $3x$ or $3x^2 - 2$ represent numbers. Thus we can manipulate expressions containing trigonometric expressions in much the same way as we manipulate purely algebraic expressions.

[i] **Multiplying and Factoring**

Example 1 Multiply and simplify $\cos y(\tan y - \sec y)$.

$$\cos y(\tan y - \sec y) = \cos y \tan y - \cos y \sec y$$

The multiplication has now been accomplished, but we can simplify:

$$\cos y \tan y - \cos y \sec y = \cos y \frac{\sin y}{\cos y} - \cos y \frac{1}{\cos y}$$

$$= \sin y - 1.$$

In Example 1 it has been illustrated how we can use certain identities to accomplish simplification. There is no rule for doing such simplifications, but it is often helpful to put everything in terms of sines and cosines, as we did in Example 1.

Example 2 Factor and simplify $\sin^2 x \cos^2 x + \cos^4 x$.

$$\sin^2 x \cos^2 x + \cos^4 x = \cos^2 x \, (\sin^2 x + \cos^2 x)$$

The factoring has been accomplished, but we can now simplify, using the Pythagorean identity $\sin^2 x + \cos^2 x \equiv 1$:

$$\cos^2 x \, (\sin^2 x + \cos^2 x) = \cos^2 x.$$

Example 3 Factor and simplify $\tan x + \cos \left(x - \dfrac{\pi}{2} \right)$.

$$\tan x + \cos \left(x - \frac{\pi}{2} \right) = \sin x \, \frac{1}{\cos x} + \sin x.$$

Here we have used the identity $\cos (x - \pi/2) \equiv \sin x$, as well as the definition of the tangent function. Now we factor as follows:

$$\sin x \left(\frac{1}{\cos x} + 1 \right) \quad \text{or} \quad \sin x \, (\sec x + 1).$$

DO EXERCISES 1–3.

[ii] **Simplifying**

Example 4 Simplify $\dfrac{\sin x - \sin x \cos x}{\sin x + \sin x \tan x}$.

$$\frac{\sin x - \sin x \cos x}{\sin x + \sin x \tan x} = \frac{\sin x \, (1 - \cos x)}{\sin x \, (1 + \tan x)} \qquad \text{Factoring}$$

$$= \frac{1 - \cos x}{1 + \tan x} \qquad \text{Simplifying}$$

Example 5 Subtract and simplify $\dfrac{2}{\sin x - \cos x} - \dfrac{3}{\sin x + \cos x}$.

$$\frac{2}{\sin x - \cos x} - \frac{3}{\sin x + \cos x} = \frac{2}{\sin x - \cos x} \cdot \frac{\sin x + \cos x}{\sin x + \cos x}$$

$$- \frac{3}{\sin x + \cos x} \cdot \frac{\sin x - \cos x}{\sin x - \cos x}$$

$$= \frac{2(\sin x + \cos x) - 3(\sin x - \cos x)}{\sin^2 x - \cos^2 x}$$

$$= \frac{-\sin x + 5 \cos x}{\sin^2 x - \cos^2 x}$$

1. Multiply and simplify.

$$\sin x (\cot x + \csc x)$$

2. Factor and simplify.

$$\sin^3 x + \sin x \cos^2 x$$

3. Factor and simplify.

$$\cot x - \sin \left(\frac{\pi}{2} - x \right)$$

4. Simplify.

$$\frac{\cos x + \sin x \cos x}{\cos x - \cos x \cot x}$$

5. Add and simplify.

$$\frac{2}{\sin x - \cos x} + \frac{3}{\sin x + \cos x}$$

6. Add and simplify.

$$\frac{2}{\cot^3 x - 2 \cot^2 x} + \frac{2}{\cot x - 2}$$

7. Multiply and simplify.

$$\sqrt{\tan x \sin^2 x} \cdot \sqrt{\tan x \sin x}$$

8. Rationalize the numerator.

$$\sqrt{\frac{\cos x}{3}}$$

Example 6 Add and simplify $\dfrac{2}{\tan^3 x - 2 \tan^2 x} + \dfrac{2}{\tan x - 2}$.

$$\frac{2}{\tan^3 x - 2 \tan^2 x} + \frac{2}{\tan x - 2} = \frac{2}{\tan^2 x (\tan x - 2)} + \frac{2}{\tan x - 2} \cdot \frac{\tan^2 x}{\tan^2 x}$$

$$= \frac{2 + 2 \tan^2 x}{\tan^2 x (\tan x - 2)} = \frac{2(1 + \tan^2 x)}{\tan^2 x (\tan x - 2)}$$

Now we use the Pythagorean identity $1 + \tan^2 x \equiv \sec^2 x$ in the numerator, and obtain for that numerator,

$$2 \sec^2 x, \quad \text{or} \quad 2 \frac{1}{\cos^2 x}.$$

Thus we have

$$\frac{2}{\cos^2 x \dfrac{\sin^2 x}{\cos^2 x} (\tan x - 2)},$$

and finally

$$\frac{2}{\sin^2 x (\tan x - 2)}.$$

DO EXERCISES 4–6.

[iii] **Radical Expressions**

When radicals occur, the use of absolute value is sometimes necessary, but its consideration can be complex. In the examples and exercises that follow, we shall assume that all quantities are nonnegative.

Example 7 Multiply and simplify $\sqrt{\sin^3 x \cos x} \cdot \sqrt{\cos x}$.

$$\sqrt{\sin^3 x \cos x} \cdot \sqrt{\cos x} = \sqrt{\sin^3 x \cos^2 x}$$
$$= \sin x \cos x \sqrt{\sin x}$$

Example 8 Rationalize the denominator: $\sqrt{\dfrac{2}{\tan x}}$.

$$\sqrt{\frac{2}{\tan x}} = \sqrt{\frac{2}{\tan x} \cdot \frac{\tan x}{\tan x}} = \sqrt{\frac{2 \tan x}{\tan^2 x}} = \frac{\sqrt{2 \tan x}}{\tan x}$$

DO EXERCISES 7 AND 8.

[iv] **Equations**

Example 9 Solve for $\tan x$: $\tan^2 x + \tan x = 56$.

$\tan^2 x + \tan x - 56 = 0$	Reducible to quadratic; let $u = \tan x$
$(\tan x + 8)(\tan x - 7) = 0$	Factoring
$\tan x + 8 = 0 \quad$ or $\quad \tan x - 7 = 0$	
$\tan x = -8 \quad$ or $\quad \tan x = 7$	

Example 10 Solve for sec x: $\sec^2 x - \frac{3}{4}\sec x = \frac{1}{2}$.

$$\sec^2 x - \frac{3}{4}\sec x - \frac{1}{2} = 0$$

We now use the quadratic formula, and obtain

$$\sec x = \frac{\frac{3}{4} \pm \sqrt{\frac{9}{16} - 4\cdot 1 \cdot\left(-\frac{1}{2}\right)}}{2} = \frac{\frac{3}{4} \pm \sqrt{\frac{41}{16}}}{2}$$

$$\sec x = \frac{\frac{3}{4} \pm \frac{\sqrt{41}}{4}}{2} = \frac{3 \pm \sqrt{41}}{8}.$$

Example 11 Solve for sin x:

$$8\sin^2 x + 2\cos\left(x + \frac{\pi}{2}\right) - 3 = 0.$$

We first use the identity $\cos(x + \pi/2) \equiv -\sin x$, to get everything in terms of sin x:

$$8\sin^2 x - 2\sin x - 3 = 0.$$

Now we factor:

$$(4\sin x - 3)(2\sin x + 1) = 0$$
$$4\sin x - 3 = 0 \quad\text{or}\quad 2\sin x + 1 = 0$$
$$\sin x = \frac{3}{4} \quad\text{or}\quad \sin x = -\frac{1}{2}.$$

Note that for the manipulations in Examples 9–11 no knowledge of trigonometry is necessary. A symbol such as sin x or tan x plays the same role as a single letter such as *u* or *v*. Later, when trigonometric equations are considered, we shall solve for x rather than sin x or tan x.

DO EXERCISES 9–11.

9. Solve for cot x.
$$\cot^2 x + \cot x = 12$$

10. Solve for cos x.
$$\cos^2 x - \frac{1}{4}\cos x = \frac{3}{2}$$

11. Solve for cos x.
$$6\cos^2 x + \sin\left(\frac{\pi}{2} - x\right) - 2 = 0$$

EXERCISE SET 8.7A

[i] In Exercises 1–20, multiply and simplify.

1. $(\sin x - \cos x)(\sin x + \cos x)$
2. $(\tan y - \cot y)(\tan y + \cot y)$
3. $\tan x\,(\cos x - \csc x)$
4. $\cot x\,(\sin x + \sec x)$
5. $\cos y \sin y\,(\sec y + \csc y)$
6. $\tan y \sin y\,(\cot y - \csc y)$
7. $(\sin x + \cos x)(\csc x - \sec x)$
8. $(\sin x + \cos x)(\sec x + \csc x)$
9. $(\sin y - \cos y)^2$
10. $(\sin y + \cos y)^2$
11. $(1 + \tan x)^2$
12. $(1 + \cot x)^2$
13. $(\sin y - \csc y)^2$
14. $(\cos y + \sec y)^2$
15. $(\cos x - \sec x)(\cos^2 x + \sec^2 x + 1)$
16. $(\sin x + \csc x)(\sin^2 x + \csc^2 x - 1)$
17. $(\cot x - \tan x)(\cot^2 x + 1 + \tan^2 x)$
18. $(\cot y + \tan y)(\cot^2 y - 1 + \tan^2 y)$
19. $\sin\left(\frac{\pi}{2} - x\right)[\sec x - \cos x]$
20. $\cos\left(\frac{\pi}{2} - x\right)[\csc x - \sin x]$

In Exercises 21–40, factor and simplify.

21. $\sin x \cos x + \cos^2 x$

22. $\sec x \csc x - \csc^2 x$

23. $\sin^2 y - \cos^2 y$

24. $\tan^2 y - \cot^2 y$

25. $\tan x + \sin(\pi - x)$

26. $\cot x - \cos(\pi - x)$

27. $\sin^4 x - \cos^4 x$

28. $\tan^4 x - \sec^4 x$

29. $3 \cot^2 y + 6 \cot y + 3$

30. $4 \sin^2 y + 8 \sin y + 4$

31. $\csc^4 x + 4 \csc^2 x - 5$

32. $-8 + \tan^4 x - 2 \tan^2 x$

33. $\sin^3 y + 27$

34. $1 - 125 \tan^3 y$

35. $\sin^3 y - \csc^3 y$

36. $\cos^3 x - \sec^3 x$

37. $\sin x \cos y - \cos\left(x + \dfrac{\pi}{2}\right) \tan y$

38. $\cos\left(x - \dfrac{\pi}{2}\right) \tan y + \sin x \cot y$

39. $\cos(\pi - x) + \cot x \sin\left(x - \dfrac{\pi}{2}\right)$

40. $\sin(\pi - x) - \tan x \cos\left(\dfrac{\pi}{2} - x\right)$

EXERCISE SET 8.7B

[ii] Simplify.

1. $\dfrac{\sin^2 x \cos x}{\cos^2 x \sin x}$

2. $\dfrac{\cos^2 x \sin x}{\sin^2 x \cos x}$

3. $\dfrac{4 \sin x \cos^3 x}{18 \sin^2 x \cos x}$

4. $\dfrac{30 \sin^3 x \cos x}{6 \cos^2 x \sin x}$

5. $\dfrac{\cos^2 x - 2 \cos x + 1}{\cos x - 1}$

6. $\dfrac{\sin^2 x + 2 \sin x + 1}{\sin x + 1}$

7. $\dfrac{\cos^2 x - 1}{\cos x - 1}$

8. $\dfrac{\sin^2 x - 1}{\sin x + 1}$

9. $\dfrac{4 \tan x \sec x + 2 \sec x}{6 \sin x \sec x + 2 \sec x}$

10. $\dfrac{6 \tan x \sin x - 3 \sin x}{9 \sin^2 x + 3 \sin x}$

11. $\dfrac{\cos^2 x - 1}{\sin\left(\dfrac{\pi}{2} - x\right) - 1}$

12. $\dfrac{\sin^2 x - 1}{\cos\left(\dfrac{\pi}{2} - x\right) + 1}$

13. $\dfrac{\sin x - \cos\left(x - \dfrac{\pi}{2}\right) \cos x}{-\sin x - \cos\left(x - \dfrac{\pi}{2}\right) \tan x}$

14. $\dfrac{\cos x - \sin\left(\dfrac{\pi}{2} - x\right) \sin x}{\cos x - \cos(\pi - x) \tan x}$

15. $\dfrac{\sin^4 x - \cos^4 x}{\sin^2 x - \cos^2 x}$

16. $\dfrac{\sec^4 x - \tan^4 x}{\sec^2 x + \tan^2 x}$

17. $\dfrac{\cos^2 x + 2 \sin\left(x - \dfrac{\pi}{2}\right) + 1}{\sin\left(\dfrac{\pi}{2} - x\right) - 1}$

18. $\dfrac{\sin^2 x - 2 \cos\left(x - \dfrac{\pi}{2}\right) + 1}{\cos\left(\dfrac{\pi}{2} - x\right) - 1}$

19. $\dfrac{\sin^2 y \cos\left(y + \dfrac{\pi}{2}\right)}{\cos^2 y \cos\left(\dfrac{\pi}{2} - y\right)}$

20. $\dfrac{\cos^2 y \sin\left(y + \dfrac{\pi}{2}\right)}{\sin^2 y \sin\left(\dfrac{\pi}{2} - y\right)}$

21. $\dfrac{2 \sin^2 x}{\cos^3 x} \cdot \left(\dfrac{\cos x}{2 \sin x}\right)^2$

22. $\dfrac{4 \cos^3 x}{\sin^2 x} \cdot \left(\dfrac{\sin x}{4 \cos x}\right)^2$

23. $\dfrac{3 \sin x}{\cos^2 x} \cdot \dfrac{\cos^2 x + \cos x \sin x}{\cos^2 x - \sin^2 x}$

24. $\dfrac{5 \cos x}{\sin^2 x} \cdot \dfrac{\sin^2 x - \sin x \cos x}{\sin^2 x - \cos^2 x}$

25. $\dfrac{\tan^2 y}{\sec y} \div \dfrac{3 \tan^3 y}{\sec y}$

26. $\dfrac{\cot^3 y}{\csc y} \div \dfrac{4 \cot^2 y}{\csc y}$

27. $\dfrac{1}{\sin^2 y - \cos^2 y} - \dfrac{2}{\cos y + \sin y}$

28. $\dfrac{3}{\cos y - \sin y} - \dfrac{2}{\sin^2 y - \cos^2 y}$

29. $\left(\dfrac{\sin x}{\cos x}\right)^2 - \dfrac{1}{\cos^2 x}$

30. $\left(\dfrac{\cot x}{\csc x}\right)^2 + \dfrac{1}{\csc^2 x}$

31. $\dfrac{\sin^2 x - 9}{2 \cos x + 1} \cdot \dfrac{10 \cos x + 5}{3 \sin x + 9}$

32. $\dfrac{9 \cos^2 x - 25}{2 \cos x - 2} \cdot \dfrac{\cos^2 x - 1}{6 \cos x - 10}$

EXERCISE SET 8.7C

[iii] In Exercises 1–10, simplify.

1. $\sqrt{\sin^2 x \cos x} \cdot \sqrt{\cos x}$

2. $\sqrt{\cos^2 x \sin x} \cdot \sqrt{\sin x}$

3. $\sqrt{\sin^3 y} + \sqrt{\sin y \cos^2 y}$

4. $\sqrt{\cos y \sin^2 y} - \sqrt{\cos^3 y}$

5. $\sqrt{\sin^2 x + 2 \cos x \sin x + \cos^2 x}$

6. $\sqrt{\tan^2 x - 2 \tan x \sin x + \sin^2 x}$

7. $(1 - \sqrt{\sin y})(\sqrt{\sin y} + 1)$

8. $(2 - \sqrt{\tan y})(\sqrt{\tan y} + 2)$

9. $\sqrt{\sin x}\,(\sqrt{2 \sin x} + \sqrt{\sin x \cos x})$

10. $\sqrt{\cos x}\,(\sqrt{3 \cos x} - \sqrt{\sin x \cos x})$

In Exercises 11–18, rationalize the denominator.

11. $\sqrt{\dfrac{\sin x}{\cos x}}$

12. $\sqrt{\dfrac{\cos x}{\sin x}}$

13. $\sqrt{\dfrac{\sin x}{\cot x}}$

14. $\sqrt{\dfrac{\cos x}{\tan x}}$

15. $\sqrt{\dfrac{\cos^2 x}{2 \sin^2 x}}$

16. $\sqrt{\dfrac{\sin^2 x}{3 \cos^2 x}}$

17. $\sqrt{\dfrac{1 + \sin x}{1 - \sin x}}$

18. $\sqrt{\dfrac{1 - \cos x}{1 + \cos x}}$

[iv]

19. Solve for tan x.
$$\tan^2 x + 4 \tan x = 21$$

20. Solve for sec x.
$$\sec^2 x - 7 \sec x = -10$$

21. Solve for sin x.
$$8 \sin^2 x - 2 \sin \dot{x} = 3$$

22. Solve for cos x.
$$6 \cos^2 x + 17 \cos x = -5$$

23. Solve for cot x.
$$\cot^2 x + 9 \cot x - 10 = 0$$

24. Solve for csc x.
$$\csc^2 x + 3 \csc x - 10 = 0$$

25. Solve for sin x.
$$\sin^2 x - \cos\left(x + \frac{\pi}{2}\right) = 6$$

26. Solve for sin x.
$$2 \cos^2\left(x - \frac{\pi}{2}\right) - 3 \cos\left(x + \frac{\pi}{2}\right) - 2 = 0$$

27. Solve for tan x.
$$\tan^2 x - 6 \tan x = 4$$

28. Solve for cot x.
$$\cot^2 x + 8 \cot x = 5$$

29. Solve for sec x.
$$6 \sec^2 x - 5 \sec x - 2 = 0$$

30. Solve for csc x.
$$2 \csc^2 x - 3 \csc x - 4 = 0$$

CHAPTER 8 REVIEW

The point $(3, -2)$ is on a circle centered at the origin. Find the coordinates of its reflections across:

[8.2, i] **1.** The y-axis. **2.** The origin. **3.** The x-axis.

On a unit circle, mark the points determined by:

[8.2, iii] **4.** $\dfrac{7\pi}{6}$. **5.** $\dfrac{3\pi}{4}$. **6.** $\dfrac{\pi}{6}$. **7.** $\dfrac{9\pi}{4}$.

[8.3, i] **8.** Sketch a graph of $y = \sin x$.

[8.3, ii] **9.** What is the domain of the sine function?

[8.3, iii] **10.** Sketch a graph of $y = \cos x$.

[8.3, iv] **11.** What is the period of the cosine function?

[8.3, i] **12.** Sketch a graph of $y = \sin\left(x + \dfrac{\pi}{2}\right)$. Use the axes of Exercise 8.

[8.3, iii] **13.** Complete the following table.

	$\pi/6$	$\pi/4$	$\pi/3$	$\pi/2$	$3\pi/4$	$5\pi/4$
sin x						
cos x						

[8.4, i] **14.** Sketch a graph of $y = \csc x$.

15. What is the period of the cosecant function?

16. What is the range of the cosecant function?

17. In which quadrants are the signs of the sine and the tangent the same?

[8.4, iii] **18.** Verify the following identity: $\cot(x - \pi) \equiv \cot x$.

Complete these Pythagorean identities.

[8.5, i] **19.** $\sin^2 x + \cos^2 x \equiv$ _____

20. $1 + \cot^2 x \equiv$ _____

[8.5, ii] Complete these cofunction identities.

21. $\cos\left(x + \dfrac{\pi}{2}\right) \equiv$ _____

22. $\cos\left(\dfrac{\pi}{2} - x\right) \equiv$ _____

23. $\sin\left(x - \dfrac{\pi}{2}\right) \equiv$ _____

[8.5, i] **24.** Express tan x in terms of sec x.

[8.6, i] **25.** Sketch a graph of $y = 3 + \cos\left(x - \dfrac{\pi}{4}\right)$.

[8.6, ii] **26.** What is the phase shift of the function in Exercise 25?

27. What is the period of the function in Exercise 25?

[8.6, iii] **28.** Sketch a graph of $y = 3\cos x + \sin x$ for values of x between 0 and 2π.

Simplify.

[8.7, i] **29.** $\cos x\,(\tan x + \cot x)$

[8.7, ii]   **30.** $\dfrac{\csc x \,(\sin^2 x + \cos^2 x \tan x)}{\sin x + \cos x}$

[8.7, iii]   **31.** Rationalize the denominator.

$$\sqrt{\dfrac{\tan x}{\sec x}}$$

[8.7, iv]   **32.** Solve for $\tan x$: $3 \tan^2 x - 2 \tan x - 2 = 0$.

33. Does $5 \sin x = 7$ have a solution for x? Why or why not?

34. For what values of x in $\left(0, \dfrac{\pi}{2}\right]$ is the following true? $\sin x < x$

9

TRIGONOMETRIC FUNCTIONS, ROTATIONS, AND ANGLES

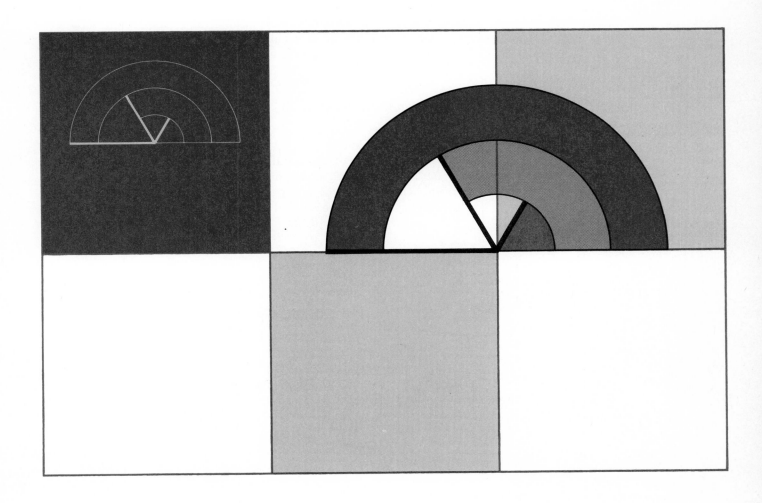

OBJECTIVES

You should be able to:

[i] Given the measure of an angle or rotation in degrees or radians, tell in which quadrant the terminal side lies.

[ii] Convert from degree measure to radian measure, and convert from radian measure to degree measure.

[iii] Find the length of an arc of a circle, given the measure of its central angle and the length of a radius; also find the measure of a central angle of a circle, given the length of its arc and the length of a radius.

1. In which quadrant does the terminal side of each angle lie?

  a) 47°
  b) 212°
  c) −43°
  d) −135°
  e) 365°
  f) −365°
  g) 740°

2. How many degrees are there in

  a) one revolution?
  b) half of a revolution?
  c) one-fourth of a revolution?
  d) one-eighth of a revolution?
  e) one-sixth of a revolution?
  f) one-twelfth of a revolution?

9.1 ROTATIONS AND ANGLES

We shall consider rotations abstractly. That is, instead of considering a specific rotating object, such as a wheel or part of a machine, we will consider a rotating ray, with its endpoint at the origin of an xy-plane. The ray starts in position along the positive half of the x-axis. A counterclockwise rotation will be called positive. Clockwise rotations will be called negative.

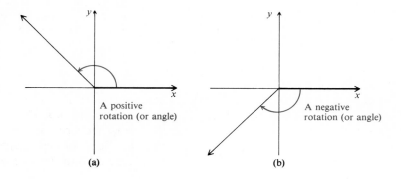

(a) A positive rotation (or angle)   (b) A negative rotation (or angle)

Note that the rotating ray and the positive half of the x-axis form an angle. Thus we often speak of "rotations" and "angles" interchangeably. The rotating ray is often called the *terminal side* of the angle, and the positive half of the x-axis is called the *initial side*.

[i] Measures of Rotations or Angles

The measure of an angle, or rotation, may be given in degrees. For example, a complete revolution has a measure of 360°, half a revolution has a measure of 180°, and so on.* We also speak of an *angle* of 90°, or 720°, or −240°, and so on. An angle between 0° and 90° has its terminal side in the first quadrant. An angle between 90° and 180° has its terminal side in the second quadrant. An angle between 0° and −90° has its terminal side in the fourth quadrant, and so on.

DO EXERCISES 1 AND 2.

[ii] Radian Measure

Consider a circle with its center at the origin and radius of length 1 (a *unit* circle). The distance around this circle from the initial side of an angle to the terminal side can be used as a measure of the rotation. This kind of angle measure is very useful. The unit is called a *radian*. One radian is about 57°.

------

*Subunits are *minutes* (60 minutes = 1°) and *seconds* (60 seconds = 1 minute).

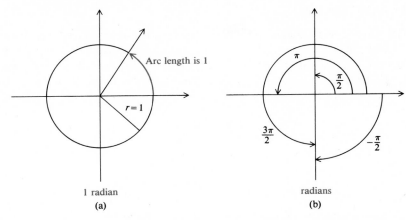

1 radian
(a)

radians
(b)

Since the circumference of a circle is $2\pi r$, and in this case $r = 1$, a rotation of $360°$ (1 revolution) has a measure of 2π radians. Half of a revolution is a rotation of $180°$ or π radians. A quarter revolution is a rotation of $90°$, or $\pi/2$ radians, and so on. To convert between degrees and radians we can use the notion of "multiplying by one" (see Section 1.11). Note the following:

$$\frac{1 \text{ revolution}}{1 \text{ revolution}} = 1 = \frac{2\pi \text{ radians}}{360 \text{ degrees}} = \frac{\pi \text{ radians}}{180 \text{ degrees}};$$

also

$$\frac{180 \text{ degrees}}{\pi \text{ radians}} = 1.$$

When a rotation is given in radians, the word "radians" is optional and is most often omitted. Thus if no unit is given for a rotation, it is understood to be in radians.

Example 1 Convert $60°$ to radians.

$$60° = 60° \cdot \frac{\pi \text{ radians}}{180°}$$

$$= \frac{60°}{180°} \pi \text{ radians}$$

$$= \frac{\pi}{3} \text{ radians, or } \frac{\pi}{3}$$

Using 3.14 for π, we find $\pi/3$ radians is about 1.047 radians.

Example 2 Convert $3\pi/4$ radians to degrees.

$$\frac{3\pi}{4} \text{ radians} = \frac{3\pi}{4} \text{ radians} \cdot \frac{180°}{\pi \text{ radians}}$$

$$= \frac{3\pi}{4\pi} \cdot 180° = 135°$$

DO EXERCISES 3–6.

3. Convert to radian measure. Leave answers in terms of π.

a) $225°$
b) $315°$
c) $-720°$

4. Convert to radian measure. Do not leave answers in terms of π (use 3.14 for π).

a) $72\frac{1}{2}°$ (a safe working angle for a certain ladder)
b) $300°$
c) $-315°$

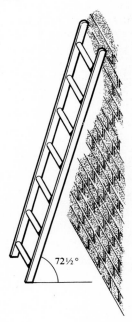

$72\frac{1}{2}°$

5. Convert to degree measure.

a) $4\pi/3$
b) $5\pi/2$
c) $-4\pi/5$

6. In which quadrant does the terminal side of each angle lie?

a) $5\pi/4$
b) $17\pi/8$
c) $-\pi/15$
d) 37.3π

7. Find the length of an arc of a circle with 10-cm radius, associated with a central angle of measure $11\pi/6$. (Use 3.14 for π.)

8. Find the radian measure of a rotation where a point 2.5 cm from the center of rotation travels 15 cm.

[iii] Arc Length and Central Angles

Radian measure can be determined using a circle other than a unit circle. In the following drawing a unit circle is shown along with another circle. The angle shown is a central angle of both circles; hence the arcs that it intercepts have their lengths in the same ratio as the radii of the circles. The radii of the circles are r and 1.

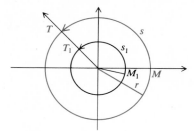

The corresponding arc lengths are MT and M_1T_1, or more simply, s and s_1. We therefore have the proportion

$$\frac{s}{r} = \frac{s_1}{1}.$$

Now s_1 is the radian measure of the rotation in question. It is more common to use a Greek letter, such as θ, for the measure of an angle or rotation. We commonly use the letter s for arc length. Adopting this convention, the above proportion becomes $\theta = s/r$. In any circle, arc length, central angle, and length of radius are related in this fashion. Or, in general, the following is true.

The radian measure θ of a rotation is the ratio of the distance s traveled by a point at a radius r from the center of rotation, to the length of the radius r:

$$\theta = \frac{s}{r}.$$

Example 3 Find the length of an arc of a circle of 5-cm radius associated with a central angle of $\pi/3$ radians.

$$\theta = \frac{s}{r}, \quad \text{or} \quad s = r\theta$$

Therefore, $s = 5 \cdot \dfrac{\pi}{3}$ cm or, using 3.14 for π, about 5.23 cm.

Example 4 Find the measure of a rotation in radians where a point 2 m from the center of rotation travels 4 m.

$$\theta = \frac{s}{r} = \frac{4\,\text{m}}{2\,\text{m}} = 2 \qquad \text{The unit is understood to be radians.}$$

A look at Examples 3 and 4 will show why the word "radian" is most often omitted. In Example 4, we have the division 4 m/2 m, which simplifies to the number 2, since m/m = 1. From this point of view it would seem preferable to omit the word "radians." In Example 3, had we used the word "radians" all the way through, our answer would have come out to be 5.23 cm-radians. It is a distance we seek; hence we know the unit should be centimeters. Thus we must omit the word "radians." A measure in radians is simply a number, so it is usually preferable to omit the word "radians."

CAUTION! In using the formula $\theta = s/r$ you must make sure that θ is in radians and that s and r are expressed in the same unit.

DO EXERCISES 7–9. (NOTE THAT EXERCISES 7 AND 8 ARE ON THE PRECEDING PAGE.)

9. Find the radian measure of a rotation where a point 24 in. from the center of rotation travels 3 ft.

EXERCISE SET 9.1

[i] For angles of the following measures, state in which quadrant the terminal side lies.

1. 34°   **2.** 320°   **3.** −120°   **4.** 175°

5. $\dfrac{\pi}{3}$ **6.** $-\dfrac{3}{4}\pi$ **7.** $\dfrac{11}{4}\pi$ **8.** $\dfrac{19}{4}\pi$

[ii] Convert to radian measure. Leave answers in terms of π.

9. 30°   **10.** 15°   **11.** 60°

12. 200°   **13.** 75°   **14.** 300°

15. 37.71°   **16.** 12.73°   **17.** 214.6°   **18.** 73.87°

Convert to radian measure. Do not leave answers in terms of π. Use 3.14 for π.

19. 120°   **20.** 240°   **21.** 320°

22. 75°   **23.** 200°   **24.** 300°

25. ▦ 117.8°   **26.** ▦ 231.2°   **27.** ▦ 1.354°   **28.** ▦ 327.9°

Convert these radian measures to degree measure.

29. 1 radian   **30.** 2 radians   **31.** 8π

32. -12π **33.** $\dfrac{3}{4}\pi$ **34.** $\dfrac{5}{4}\pi$

35. ▦ 1.303   **36.** ▦ 2.347   **37.** ▦ 0.7532π **38.** ▦ −1.205π

[iii]

39. In a circle with 120-cm radius, an arc 132 cm long subtends a central angle of how many radians? how many degrees, to the nearest degree?

40. In a circle with 200-cm radius, an arc 65 cm long subtends a central angle of how many radians? how many degrees, to the nearest degree?

41. Through how many radians does the minute hand of a clock rotate in 50 min?

42. A wheel on a car has a 14-in. radius. Through what angle (in radians) does the wheel turn while the car travels 1 mi?

43. In a circle with 10-m radius, how long is an arc associated with a central angle of 1.6 radians?

44. In a circle with 5-m radius, how long is an arc associated with a central angle of 2.1 radians?

[ii]

45. Certain positive angles are marked here in degrees. Find the corresponding radian measures.

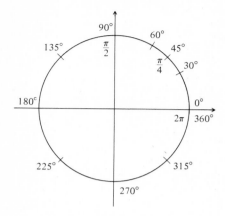

46. Certain negative angles are marked here in degrees. Find the corresponding radian measures.

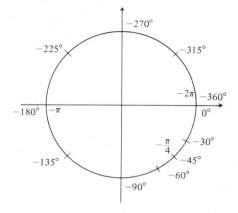

☆ ────────────────────────────────

47. The *grad* is a unit of angle measure similar to a degree. A right angle has a measure of 100 grads. Convert the following to grads.

a) 48° b) 153° c) $\frac{\pi}{8}$ radians d) $\frac{5\pi}{7}$ radians

48. The *mil* is a unit of angle measure. A right angle has a measure of 1600 mils. Convert the following to degrees, minutes, and seconds.

a) 100 mils   b) 350 mils

49. On the earth, one degree of latitude is how many kilometers? how many miles? (Assume that the radius of the earth is 6400 km, or 4000 miles, approximately.)

50. One minute of latitude on the earth is equivalent to one *nautical mile*. Find the circumference and radius of the earth in nautical miles.

51. An astronaut on the moon observes the earth, about 240,000 miles away. The diameter of the earth is about 8000 miles. Find the angle α.

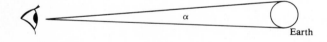

52. The circumference of the earth was computed by Eratosthenes (276–195 B.C.). He knew the distance from Aswan to Alexandria to be about 500 miles. From each town he observed the sun at noon, finding the angular difference to be 7°12′ (7 degrees, 12 minutes). Do Eratosthenes' calculation.

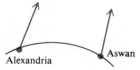

★ ────────────────────────────────

53. What is the angle between the hands of a clock at 7:45?

54. At what time between noon and 1:00 P.M. are the hands of a clock perpendicular?

────────────────────────────────

OBJECTIVES

You should be able to:

[i] Convert between linear and angular speed.
[ii] Find total distance or total angle, when speed, radius, and time are given.

9.2 ANGULAR SPEED

Speed is defined to be distance traveled per unit of time. Similarly, *angular speed* is defined to be amount of rotation per unit of time. For example, we might speak of the angular speed of a wheel as 150 revolutions per minute or the angular speed of the earth as 2π radians per day. The Greek letter ω (omega) is usually used for angular speed. Thus angular speed is defined as

$$\omega = \frac{\theta}{t}.$$

[i] Relating Linear and Angular Speed

For many applications it is important to know a relationship between angular speed and linear speed. For example, we might wish to find the linear speed of a point on the earth, knowing its angular speed. Or, we might wish to know the linear speed of an earth satellite, knowing its angular speed. To develop the relationship we seek, we recall the relation between angle and distance from the preceding section, $\theta = s/r$. This is equivalent to

$$s = r\theta.$$

We divide by the time, t, to obtain

$$\frac{s}{t} = r\frac{\theta}{t}.$$

Now s/t is linear speed v, and θ/t is angular speed ω. We thus have the relation we seek.

The linear speed v of a point a distance of r from the center of rotation is given by

$$v = r\omega,$$

where ω is the angular speed in radians per unit of time.

In deriving this formula we used the equation $s = r\theta$, in which the units for s and r must be the same and θ must be in radians. So, for our new formula $v = r\omega$, the units of distance for v and r must be the same, ω must be in radians per unit of time, and the units of time must be the same for v and ω.

Example 1 An earth satellite in a circular orbit 1200 km high makes one complete revolution every 90 min. What is its linear speed? Use 6400 km for the length of a radius of the earth.

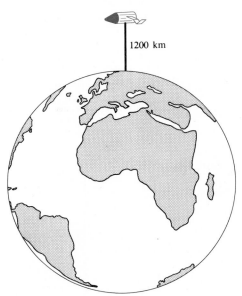

1200 km

1. A wheel with 12-cm diameter is rotating at 10 revolutions per second. What is the velocity of a point on the rim?

2. In using $v = r\omega$, if v is in cm/sec, what must be the units for r and ω?

3. In using $v = r\omega$, if ω is in radians/yr and r is in km, what must be the units for v?

4. The old oaken bucket is being raised at 3 ft/sec. The radius of the drum is 10 in. What is the angular speed of the handle?

We will use the formula $v = r\omega$; thus we shall need r and ω.

$r = 6400 \text{ km} + 1200 \text{ km}$ Radius of earth plus height of satellite

$= 7600 \text{ km}$,

$\omega = \dfrac{2\pi \text{ radians}}{90 \text{ min}} = \dfrac{\pi}{45 \text{ min}}$ We have, as usual, omitted the word "radians."

Now, using $v = r\omega$, we have

$$v = 7600 \text{ km} \cdot \frac{\pi}{45 \text{ min}} = \frac{7600\pi}{45} \cdot \frac{\text{km}}{\text{min}}.$$

Using 3.14 for π, we obtain $v = 530 \dfrac{\text{km}}{\text{min}}.$*

DO EXERCISE 1.

Example 2 An anchor is being hoisted at 2 ft/sec, the chain being wound around a capstan with a 1.8-yd diameter. What is the angular speed of the capstan?

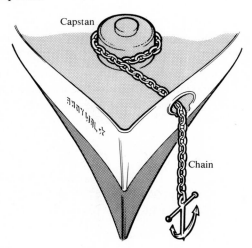

We will use the formula $\omega = v/r$, taking care to use the proper units. Since v is in ft/sec, we need r in ft. Then ω will be in radians/sec:

$$r = \frac{1.8}{2} \text{ yd} \cdot \frac{3 \text{ ft}}{\text{yd}} = 2.7 \text{ ft},$$

$$\omega = \frac{v}{r} = \frac{2}{2.7} = 1.11 \text{ radians/sec.}$$

CAUTION! In applying the formula $v = r\omega$, we must be sure that the units for v and r are the same and that ω is in radians per unit of time. The units of time must be the same for v and ω.

DO EXERCISES 2–4.

---
*It may be appropriate at this point to review Section 1.11 on handling units.

[ii] Total Distance and Total Angle

The formulas $\theta = s/r$ and $v = r\omega$ can be used in combination to find distances and angles in various situations involving rotational motion.

Example 3 A car is traveling at 45 mph. Its tires have a 20-in. radius. Find the angle through which a wheel turns in 5 sec.

Recall that $\omega = \theta/t$ or $\theta = \omega t$. Thus we can find θ if we know ω and t. To find ω we use $v = r\omega$. For convenience we will first convert 45 mph to ft/sec:

$$v = 45 \, \frac{\text{mi}}{\text{hr}} \cdot \frac{1 \, \text{hr}}{60 \, \text{min}} \cdot \frac{1 \, \text{min}}{60 \, \text{sec}} \cdot \frac{5280 \, \text{ft}}{1 \, \text{mi}}$$

$$= 66 \, \frac{\text{ft}}{\text{sec}}.$$

Now $r = 20$ in. We shall convert to ft, since v is in ft/sec:

$$r = 20 \, \text{in.} \cdot \frac{1 \, \text{ft}}{12 \, \text{in.}}$$

$$= \frac{20}{12} \, \text{ft} = \frac{5}{3} \, \text{ft}.$$

Using $v = r\omega$, we have

$$66 \, \frac{\text{ft}}{\text{sec}} = \frac{5}{3} \, \text{ft} \cdot \omega, \quad \text{so} \quad \omega = 39.6 \, \frac{\text{radians}}{\text{sec}}.$$

Then

$$\theta = \omega t = 39.6 \, \frac{\text{radians}}{\text{sec}} \cdot 5 \, \text{sec} = 198 \, \text{radians}.$$

DO EXERCISE 5.

5. The diameter of a wheel of a car is 30 in. When the car is traveling 60 mph (88 ft/sec), how many revolutions does the wheel make in 1 sec?

EXERCISE SET 9.2

[i]

1. A flywheel is rotating 7 radians/sec. It has a 15-cm diameter. What is the speed of a point on its rim, in cm/min?

2. A wheel is rotating at 3 radians/sec. The wheel has a 30-cm radius. What is the speed of a point on its rim, in m/min?

3. A $33\frac{1}{3}$-rpm record has a radius of 15 cm. What is the linear velocity of a point on the rim, in cm/sec?

4. A 45-rpm record has a radius of 8.7 cm. What is the linear velocity of a point on the rim, in cm/sec?

5. The earth has a 4000-mile radius and rotates one revolution every 24 hours. What is the linear speed of a point on the equator, in mph?

6. The earth is 93,000,000 miles from the sun and traverses its orbit, which is nearly circular, every 365.25 days. What is the linear velocity of the earth in its orbit, in mph?

7. A wheel has a 32-cm diameter. The speed of a point on its rim is 11 m/s. What is its angular speed?

8. A horse on a merry-go-round is 7 m from the center and travels at 10 km/h. What is its angular speed?

9. (*Determining the speed of a river*). A water wheel has a 10-ft radius. To get a good approximation to the speed of the river, you count the revolutions of the wheel and find that it makes 14 revolutions per minute. What is the speed of the river?

10. (*Determining the speed of a river*). A water wheel has a 10-ft radius. To get a good approximation to the speed of the river, you count the revolutions of the wheel and find that it makes 16 rpm. What is the speed of the river?

[ii]

11. The wheels of a bicycle have a 24-in. diameter. When the bike is being ridden so that the wheels make 12 rpm, how far will the bike travel in 1 min?

12. The wheels of a car have a 15-in. radius. When the car is being driven so that the wheels make 10 revolutions/sec, how far will the car travel in one minute?

13. A car is traveling 30 mph. Its wheels have a 14-in. radius. Find the angle through which a wheel rotates in 10 sec.

14. A car is traveling 40 mph. Its wheels have a 15-in. radius. Find the angle through which a wheel rotates in 12 sec.

☆ ───

15. Two pulleys, 50 cm and 30 cm in diameter, respectively, are connected by a belt. The larger pulley makes 12 revolutions per minute. Find the angular speed of the smaller pulley in radians/sec.

16. One gear wheel turns another, the teeth being on the rims. The wheels have 40-cm and 50-cm radii and the smaller wheel rotates at 20 rpm. Find the angular speed of the larger wheel in radians/sec.

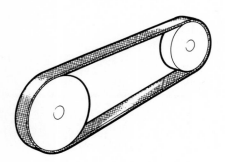

17. An airplane engine is idling at 800 rpm. When the throttle is opened, it takes 4.3 seconds for the speed to come up to 2500 rpm. What was the angular acceleration (a) in rpm per second? (b) in radians per second per second?

18. The linear speed of an airplane, flying at low level over the sea, is 175 knots (nautical miles per hour). It accelerates to 325 knots in 12 seconds.

   a) What was its linear acceleration in knots per second?

   b) What was its angular acceleration in radians per second per second? (*Hint:* One nautical mile is equivalent to one minute of latitude.)

9.3 TRIGONOMETRIC FUNCTIONS OF ANGLES OR ROTATIONS

[i] Function Values Defined

We now extend the domains of the functions defined on p. 306 to angles or rotations of any size. To do this, we consider a triangle with one vertex at the origin of a coordinate system and one side along the x-axis. The point R can be in any of the four quadrants, or on one of the axes. The sides of the triangle have lengths x, y, and r, as shown, and the angle θ is always measured from the positive half of the x-axis. The numbers x and y can be positive, negative, or zero, but the number r will always be considered positive.

OBJECTIVES

You should be able to:

[i] Given a triangle in standard position with known sides, in any quadrant, determine the six function values.
[ii] Find the function values, without tables, for any angle whose terminal side makes an angle of 30°, 45°, or 60° with the x-axis.
[iii] Find the function values for any angle whose terminal side lies on a coordinate axis.

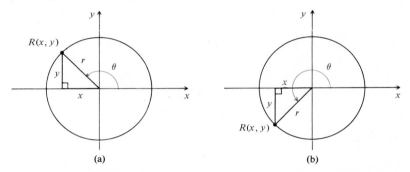

(a)   (b)

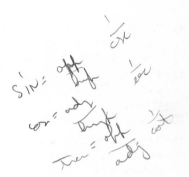

The values of the six trigonometric functions are as follows.

$$\sin \theta = \frac{y}{r} = \frac{\text{second coordinate}}{\text{radius}}$$

$$\cos \theta = \frac{x}{r} = \frac{\text{first coordinate}}{\text{radius}}$$

$$\tan \theta = \frac{y}{x} = \frac{\text{second coordinate}}{\text{first coordinate}}$$

$$\cot \theta = \frac{x}{y} = \frac{\text{first coordinate}}{\text{second coordinate}}$$

$$\sec \theta = \frac{r}{x} = \frac{\text{radius}}{\text{first coordinate}}$$

$$\csc \theta = \frac{r}{y} = \frac{\text{radius}}{\text{second coordinate}}$$

Example 1 Find the six trigonometric function values for the angle shown.

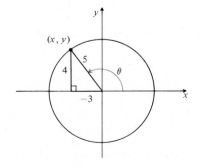

1. Find the six trigonometric function values for the angle shown here.

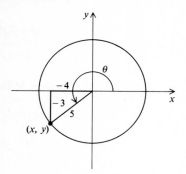

$$\sin \theta = \frac{y}{r} = \frac{4}{5} = 0.8, \qquad \cos \theta = \frac{x}{r} = \frac{-3}{5} = -0.6,$$

$$\tan \theta = \frac{y}{x} = \frac{4}{-3} \approx -1.33, \qquad \cot \theta = \frac{x}{y} = \frac{-3}{4} = -0.75,$$

$$\sec \theta = \frac{r}{x} = \frac{5}{-3} \approx -1.67, \qquad \csc \theta = \frac{r}{y} = \frac{5}{4} = 1.25.$$

DO EXERCISE 1.

[ii] Function Values for Some Special Angles

Our knowledge of triangles enables us to determine function values for certain angles. Let us first recall the Pythagorean theorem about right triangles. It says that in any right triangle, as shown, $a^2 + b^2 = c^2$, where c is the length of the hypotenuse.

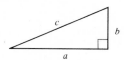

A 45° right triangle is half of a square, so its legs are the same length. Let us consider such a triangle whose legs have length 1. Then its hypotenuse has length c given by

$$1^2 + 1^2 = c^2, \quad \text{or} \quad c^2 = 2, \quad \text{or} \quad c = \sqrt{2}.$$

Such a triangle is shown below. From this diagram we can easily determine the function values for 45° or $\pi/4$:

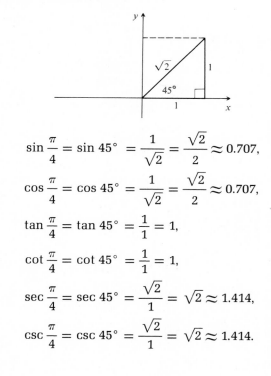

$$\sin \frac{\pi}{4} = \sin 45° = \frac{1}{\sqrt{2}} = \frac{\sqrt{2}}{2} \approx 0.707,$$

$$\cos \frac{\pi}{4} = \cos 45° = \frac{1}{\sqrt{2}} = \frac{\sqrt{2}}{2} \approx 0.707,$$

$$\tan \frac{\pi}{4} = \tan 45° = \frac{1}{1} = 1,$$

$$\cot \frac{\pi}{4} = \cot 45° = \frac{1}{1} = 1,$$

$$\sec \frac{\pi}{4} = \sec 45° = \frac{\sqrt{2}}{1} = \sqrt{2} \approx 1.414,$$

$$\csc \frac{\pi}{4} = \csc 45° = \frac{\sqrt{2}}{1} = \sqrt{2} \approx 1.414.$$

These function values should be memorized. The decimal values need not be, and it is sufficient to memorize the first three, since the others are their reciprocals.

$$\sin 45° = \frac{\sqrt{2}}{2}, \qquad \cos 45° = \frac{\sqrt{2}}{2}, \qquad \tan 45° = 1.$$

Next we consider an equilateral triangle, each of whose sides has length 2. If we take half of it as shown, we obtain a right triangle having a hypotenuse of length 2 and a leg of length 1. The other leg has length a, given by the Pythagorean theorem as follows:

$$a^2 + 1^2 = 2^2, \quad \text{or} \quad a^2 = 3, \quad \text{or} \quad a = \sqrt{3}.$$

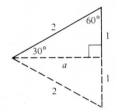

The acute angles of this triangle have measures of 30° and 60°. We can now determine function values for these angles. The function values for 30° or $\pi/6$ are as follows:

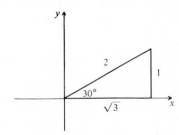

$$\sin \frac{\pi}{6} = \sin 30° = \frac{1}{2},$$

$$\cos \frac{\pi}{6} = \cos 30° = \frac{\sqrt{3}}{2},$$

$$\tan \frac{\pi}{6} = \tan 30° = \frac{1}{\sqrt{3}} = \frac{\sqrt{3}}{3},$$

$$\cot \frac{\pi}{6} = \cot 30° = \sqrt{3},$$

$$\sec \frac{\pi}{6} = \sec 30° = \frac{2}{\sqrt{3}} = \frac{2\sqrt{3}}{3},$$

$$\csc \frac{\pi}{6} = \csc 30° = 2.$$

We obtain the function values for 60° or $\pi/3$ by repositioning the triangle:

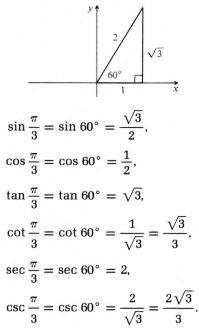

$$\sin \frac{\pi}{3} = \sin 60° = \frac{\sqrt{3}}{2},$$

$$\cos \frac{\pi}{3} = \cos 60° = \frac{1}{2},$$

$$\tan \frac{\pi}{3} = \tan 60° = \sqrt{3},$$

$$\cot \frac{\pi}{3} = \cot 60° = \frac{1}{\sqrt{3}} = \frac{\sqrt{3}}{3},$$

$$\sec \frac{\pi}{3} = \sec 60° = 2,$$

$$\csc \frac{\pi}{3} = \csc 60° = \frac{2}{\sqrt{3}} = \frac{2\sqrt{3}}{3}.$$

The function values for 30° and 60° should be memorized, but again it is sufficient to learn only those for the sine, cosine, and tangent, since the others are their reciprocals.

$$\sin 30° = \frac{1}{2}, \qquad \cos 30° = \frac{\sqrt{3}}{2}, \qquad \tan 30° = \frac{\sqrt{3}}{3},$$

$$\sin 60° = \frac{\sqrt{3}}{2}, \qquad \cos 60° = \frac{1}{2}, \qquad \tan 60° = \sqrt{3}.$$

Signs of the Functions

Function values of the generalized trigonometric functions can be positive, negative, or zero, depending on where the terminal side of the angle lies. In the first quadrant, all function values are positive. In the second quadrant, first coordinates are negative and second coordinates are positive.

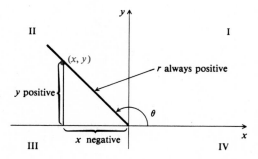

Thus if θ is in quadrant II,

$\sin \theta = \dfrac{y}{r}$ is positive, because y and r are positive;

$\cos \theta = \dfrac{x}{r}$ is negative, because x is negative and r is positive;

$\tan \theta = \dfrac{y}{x}$ is negative, because y is positive and x is negative;

and so on.

The following diagram summarizes, showing the quadrants in which the function values are positive. The diagram need not be memorized, because the signs can readily be determined as in the diagram above.

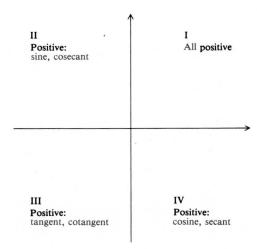

II
Positive:
sine, cosecant

I
All positive

III
Positive:
tangent, cotangent

IV
Positive:
cosine, secant

Example 2 Find the trigonometric function values for 210°.

We draw a diagram showing the terminal side of a 210° angle. We now have a right triangle with a 30° angle as shown. We can read off function values.

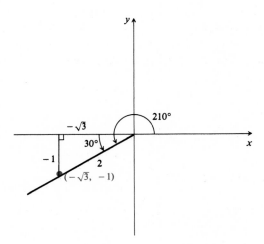

2. Find the trigonometric function values for 120°.

$$\sin 210° = \frac{-1}{2} = -\frac{1}{2}$$

$$\cos 210° = \frac{-\sqrt{3}}{2} = -\frac{\sqrt{3}}{2}$$

$$\tan 210° = \frac{-1}{-\sqrt{3}} = \frac{1}{\sqrt{3}} \text{ or } \frac{\sqrt{3}}{3}$$

$$\cot 210° = \frac{-\sqrt{3}}{-1} = \sqrt{3}$$

$$\sec 210° = \frac{2}{-\sqrt{3}} = -\frac{2}{\sqrt{3}} \text{ or } -\frac{2\sqrt{3}}{3}$$

$$\csc 210° = \frac{2}{-1} = -2$$

DO EXERCISES 2 AND 3.

[iii] Terminal Side on an Axis

Now let us suppose that the terminal side of an angle falls on one of the axes. In that case, one of the coordinates is zero. The definitions of the functions still apply, but in some cases functions will not be defined because a denominator will be 0.

3. Find the trigonometric function values for −135°.

Example 3 Find the trigonometric function values for 180°, or π radians.

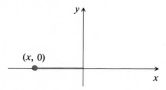

We note that the first coordinate is negative, that the second coordinate is 0, and that x and r have the same absolute value (r always being positive). Thus we have:

$$\sin 180° = \frac{0}{r} = 0;$$

$$\cos 180° = \frac{x}{r} = -1 \qquad \text{Since } |x| = |r|, \text{ but x and r have opposite signs}$$

$$\tan 180° = \frac{y}{x} = \frac{0}{x} = 0;$$

$$\cot 180° = \frac{x}{y} = \frac{x}{0}; \qquad \text{Thus cot 180° is undefined.}$$

$$\sec 180° = \frac{r}{x} = -1; \qquad \text{The reciprocal of cos 180°}$$

$$\csc 180° = \frac{r}{0}. \qquad \text{Thus csc 180° is undefined.}$$

Example 4 The valve cap on a bicycle wheel is 24.5 in. from the center of the wheel. From the position shown, the wheel starts rolling. After the wheel has turned 390°, how far above the ground is the valve cap? Assume that the outer radius of the tire is 26 in.

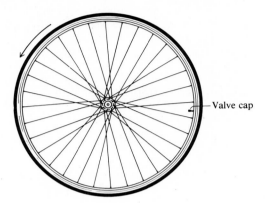

Valve cap

We draw the terminal side of an angle of 390°. Since $\sin 390° = \frac{1}{2}$, the valve cap is $\frac{1}{2} \times 24.5$ in. above the center of the wheel. The distance above the ground is then $12.25 + 26$ in. or 38.25 in.

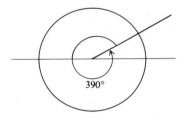

390°

DO EXERCISES 4 AND 5.

[ii], [iii] Function Values for Any Angle

We are now in a position to be able to determine the trigonometric function values for many other angles. If the terminal side of an angle falls on one of the axes, the function values are 0 or 1 or −1 or else are undefined. Thus we can determine function values for any multiple of 90°, or $\pi/2$. We can also determine the function values for any angle whose terminal side makes a 30°, 45°, or 60° angle with the x-axis. Consider, for example, an angle of 150°, or $5\pi/6$. The terminal side makes a 30° angle with the x-axis, since $180° - 150° = 30°$. As the diagram shows, triangle ONR is congruent to triangle $ON'R'$; hence the

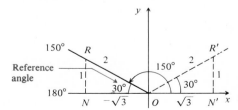

4. Find the trigonometric function values for 270°, or $3\pi/2$ radians.

5. The seats of a ferris wheel are 35 ft from the center of the wheel. When you board the wheel you are 5 ft above the ground. After you have rotated through an angle of 765°, how far above the ground are you?

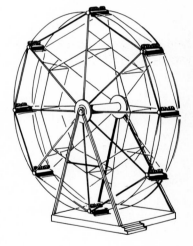

6. Given that

$$\sin 36° = 0.6283,$$
$$\cos 36° = 0.8090,$$
$$\tan 36° = 0.7265,$$

find the trigonometric function values for 324°.

ratios of the sides of the two triangles are the same except perhaps for sign. We could determine the function values directly from triangle *ONR*, but this is not necessary. If we remember that the sine is positive in quadrant II and that the cosine and tangent are negative, we can simply use the values for 30°, prefixing the appropriate sign. The triangle *ONR* is called a *reference* triangle and its acute angle at the origin is called a *reference* angle.

In general, to find the function values of an angle, we find them for the reference angle and prefix the appropriate sign.

Example 5 Find the sine, cosine, and tangent of 600°, or 10π/3.

We find the multiple of 180° nearest 600°:

$$180° \times 2 = 360°,$$
$$180° \times 3 = 540°,$$

and

$$180° \times 4 = 720°.$$

The nearest multiple is 540°. The difference between 600° and 540° is 60°. This gives us the reference angle.

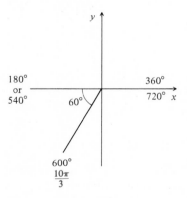

We recall that sin 60° = $\sqrt{3}/2$, cos 60° = $\frac{1}{2}$, and tan 60° = $\sqrt{3}$. We also note that the sine and cosine are negative in the third quadrant, and that the tangent is positive. Hence we have

$$\sin \frac{10\pi}{3} = \sin 600° = -\frac{\sqrt{3}}{2},$$

$$\cos \frac{10\pi}{3} = \cos 600° = -\frac{1}{2},$$

$$\tan \frac{10\pi}{3} = \tan 600° = \sqrt{3}.$$

DO EXERCISE 6.

EXERCISE SET 9.3

Before beginning, make a list of the function values of 30°, 60°, and 45°. Then memorize your list. In Exercises 1–4, find the six trigonometric function values for the angle θ as shown.

[i]

1.
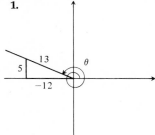

2.
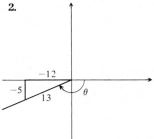

3.
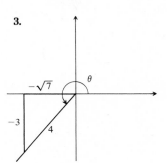

4.
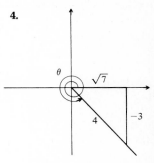

[ii] In Exercises 5 and 6, use the fact that $\sqrt{2} \approx 1.414$ and $\sqrt{3} \approx 1.732$.

5. Find decimal values for the six trigonometric functions of 30°, or $\frac{\pi}{6}$.

6. Find decimal values for the six trigonometric functions of 60°, or $\frac{\pi}{3}$.

[ii], [iii] Find the following if they exist.

7. cos 180°

8. sin 360°

9. $\tan \frac{\pi}{2}$

10. cot π

11. sec 720°

12. csc 720°

13. $\sin \left(-\frac{3\pi}{4} \right)$

14. $\cos \frac{3\pi}{4}$

15. $\sin \frac{5\pi}{6}$

16. $\cos \frac{5\pi}{6}$

17. tan 240°

18. cot 240°

19. $\sec \frac{7\pi}{4}$

20. $\csc \frac{7\pi}{4}$

21. tan (−315°)

22. cot (−315°)

23. sin (−210°)

24. cos (−210°)

25. $\tan \frac{7\pi}{6}$

26. $\cot \frac{7\pi}{6}$

27. $\sin \frac{11\pi}{4}$

28. $\cos \frac{11\pi}{4}$

29. $\tan \frac{11\pi}{6}$

30. $\cot \frac{11\pi}{6}$

☆ ───

31. Compile a list of function values for the sine function, from −2π to 2π. Then graph the function.

32. Is the sine function an even function? an odd function? periodic?

33. Compile a list of function values for the cosine function, from −2π to 2π. Then graph the function.

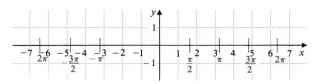

34. Is the cosine function an even function? an odd function? periodic?

★
_____

35. This diagram shows a piston of a steam engine, driving a drive wheel of a locomotive. The radius of the drive wheel (from the center of the wheel to the pin P) is R, and the length of the rod is L. Suppose that the drive wheel is rotating at a speed of ω radians per second (so that $\theta = \omega t$). Show that the distance of pin Q from 0, the center of the wheel, is a function of time, given by

$$x = \sqrt{L^2 - R^2 \sin^2 \omega t} + R \cos \omega t.$$

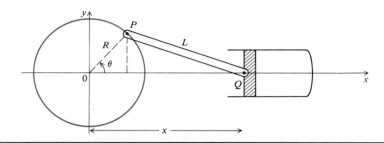

OBJECTIVES

You should be able to:

[i] Given a function value for an angle and the quadrant in which the terminal side lies, find the other five function values.

[ii] Given the function values for an acute angle, find the function values for its complement.

9.4 RELATIONSHIPS AMONG FUNCTION VALUES

[i] The Six Functions Related

When we know one of the trigonometric function values for an angle, we can find the other five if we know the quadrant in which the terminal side lies. The idea is to sketch a triangle in the appropriate quadrant, use the Pythagorean theorem to find the lengths of its sides as needed, and then read off the ratios of its sides.

Example 1 Given that $\tan \theta = -\frac{2}{3}$ and that θ is in the second quadrant, find the other function values.

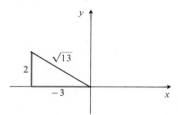

We first sketch a second quadrant triangle. Since $\tan \theta = -\frac{2}{3}$, we make the legs of lengths 2 and 3. The hypotenuse must then have length $\sqrt{13}$. Now we can read off the appropriate ratios:

$$\sin \theta = \frac{2}{\sqrt{13}}, \qquad \cos \theta = -\frac{3}{\sqrt{13}},$$

$$\tan \theta = -\frac{2}{3}, \qquad \cot \theta = -\frac{3}{2},$$

$$\sec \theta = -\frac{\sqrt{13}}{3}, \qquad \csc \theta = \frac{\sqrt{13}}{2}.$$

Example 2 Given that $\cot \theta = 2$ and θ is in the third quadrant, find $\sin \theta$, $\cos \theta$, and $\tan \theta$.

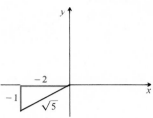

Again, we sketch a triangle and label the sides, then read off the appropriate ratios:

$$\sin \theta = -\frac{1}{\sqrt{5}},$$

$$\cos \theta = -\frac{2}{\sqrt{5}},$$

$$\tan \theta = \frac{1}{2}.$$

DO EXERCISES 1 AND 2.

[ii] Cofunctions and Complements

In a right triangle the acute angles are complementary, since the sum of all three angle measures is 180° and the right angle accounts for 90° of this total. Thus if one acute angle of a right triangle is θ, the other is $90° - \theta$, or $\pi/2 - \theta$. Note that the sine of $\angle A$ is also the cosine of $\angle B$, its complement:

$$\sin \theta = \frac{a}{c}, \qquad \text{cosine } (90° - \theta) = \frac{a}{c}.$$

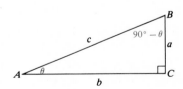

Similarly, the tangent of $\angle A$ is the cotangent of its complement and the secant of $\angle A$ is the cosecant of its complement.

These pairs of functions are called *cofunctions*. The name *cosine* originally meant the sine of the complement. The name *cotangent* meant the tangent of the complement and *cosecant* meant the secant of the complement. A complete list of the cofunction relations is as follows. Equations that hold for all sensible replacements for the variables are known as *identities*. The sign $\equiv$ is used instead of $=$ to indicate that an equation is an identity.

$$\sin \theta \equiv \cos (90° - \theta), \qquad \cos \theta \equiv \sin (90° - \theta),$$
$$\tan \theta \equiv \cot (90° - \theta), \qquad \cot \theta \equiv \tan (90° - \theta),$$
$$\sec \theta \equiv \csc (90° - \theta), \qquad \csc \theta \equiv \sec (90° - \theta).$$

1. Given $\cos \theta = \frac{3}{4}$ and that the terminal side is in quadrant IV, find the other function values.

2. Given $\cot \theta = -3$ and that the terminal side is in quadrant II, find the other function values.

3. Given that

$$\sin 75° = 0.9659, \quad \cos 75° = 0.2588,$$
$$\tan 75° = 3.732, \quad \cot 75° = 0.2679,$$
$$\sec 75° = 3.864, \quad \csc 75° = 1.035,$$

find the function values for the complement of 75°.

Example 3 Given that

$$\sin 18° = 0.3090, \quad \cos 18° = 0.9511,$$
$$\tan 18° = 0.3249, \quad \cot 18° = 3.078,$$
$$\sec 18° = 1.051, \quad \csc 18° = 3.236,$$

find the six function values for 72°.

Since 72° and 18° are complements, we have sin 72° = cos 18°, etc., and the function values are

$$\sin 72° = 0.9511, \quad \cos 72° = 0.3090,$$
$$\tan 72° = 3.078, \quad \cot 72° = 0.3249,$$
$$\sec 72° = 3.236, \quad \csc 72° = 1.051.$$

DO EXERCISE 3.

EXERCISE SET 9.4

[i] In Exercises 1–6, a function value and a quadrant are given. Find the other five function values.

1. $\sin \theta = -\dfrac{1}{3}$, III

2. $\sin \theta = -\dfrac{1}{5}$, IV

3. $\cos \theta = \dfrac{3}{5}$, IV

4. $\cos \theta = -\dfrac{4}{5}$, II

5. $\cot \theta = -2$, IV

6. $\tan \theta = 5$, III

[ii]

7. Given that
$$\sin 65° = 0.9063, \quad \cos 65° = 0.4226,$$
$$\tan 65° = 2.145, \quad \cot 65° = 0.4663,$$
$$\sec 65° = 2.366, \quad \csc 65° = 1.103,$$
find the six function values for 25°.

8. Given that
$$\sin 32° = 0.5299, \quad \cos 32° = 0.8480,$$
$$\tan 32° = 0.6249, \quad \cot 32° = 1.600,$$
$$\sec 32° = 1.179, \quad \csc 32° = 1.887,$$
find the six function values for 58°.

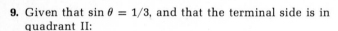

9. Given that sin θ = 1/3, and that the terminal side is in quadrant II:

a) find the other function values for θ;

b) find the six function values for $\pi + \theta$;

c) find the six function values for $\pi - \theta$;

d) find the six function values for $2\pi - \theta$.

10. Given that cos θ = 4/5, and that the terminal side is in quadrant IV:

a) find the other function values for θ;

b) find the six function values for $\pi + \theta$;

c) find the six function values for $\pi - \theta$;

d) find the six function values for $2\pi - \theta$.

11. ▦ Given that sin 27° = 0.45399:

a) find the other function values for 27°;

b) find the six function values for 63°.

12. ▦ Given that cot 54° = 0.72654:

a) find the other function values for 54°;

b) find the six function values for 36°.

13. ▦ Given that cos 38° = 0.78801, find the six function values for 128°.

14. ▦ Given that tan 73° = 3.27085, find the six function values for 343°.

15. For any acute angle θ, we know that cos (90° − θ) = sin θ. Consider angles other than acute angles. Does this relation still hold, and if so to what extent?

16. Consider the equation sin (x + π/2) = cos x, where x is an angle in radians. Does this relation hold for acute angles? Does it hold for other than acute angles?

9.5 TABLES OF TRIGONOMETRIC FUNCTIONS

[i] Finding Function Values

By use of certain formulas, theoretically determined, tables of values for the trigonometric functions have been constructed. Table 3 at the back of the book is such a table. A portion of it is shown below. This table gives the function values for angles from 0° to 90° only. Values for other angles can always be determined from these, since any angle has an acute reference angle. Note that angle measures are given in degrees and minutes (60 minutes = 1°) as well as radians. Values in the table are given for intervals of 10′ (10 minutes) and are correct to four digits.

Degrees	Radians	Sin	Cos	Tan	Cot	Sec	Csc		
56°00′	.6283	.5878	.8090	.7265	1.376	1.236	1.701	.9425	54°00′
10	312	901	073	310	368	239	695	396	50
20	341	925	056	355	360	241	688	367	40
30	.6370	.5948	.8039	.7400	1.351	1.244	1.681	.9338	30
40	400	972	021	445	343	247	675	308	20
50	429	995	004	490	335	249	668	279	10
37°00′	.6458	.6018	.7986	.7536	1.327	1.252	1.662	.9250	53°00′
10	487	041	969	581	319	255	655	221	50
20	516	065	951	627	311	258	649	192	40
30	.6545	.6088	.7934	.7673	1.303	1.260	1.643	.9163	30
40	574	111	916	720	295	263	636	134	20
50	603	134	898	766	288	266	630	105	10
44°00′	.7679	.6947	.7193	.9657	1.036	1.390	1.440	.8029	46°00′
10	709	967	173	713	030	394	435	999	50
20	738	988	153	770	024	398	431	970	40
30	.7767	.7009	.7133	.9827	1.018	1.402	1.427	.7941	30
40	796	030	112	884	012	406	423	912	20
50	825	050	092	942	006	410	418	883	10
45°00′	.7854	.7071	.7071	1.000	1.000	1.414	1.414	.7854	45°00′
		Cos	Sin	Cot	Tan	Csc	Sec	Radians	Degrees

The headings on the left of Table 3 range from 0° to 45° only. Thus the table may seem to be only half of what it should be. Function values from 45° to 90° are the cofunction values of their complements, however. Hence these can be found without further entries. For angles from 45° to 90° the headings on the right are used, together with the headings at the bottom. For example, sin 37° is found to be 0.6018 using the top and left headings. The cosine of 53° (the complement of 37°) is found also to be 0.6018, using the bottom and right headings.

Example 1 Find cos 37°20′.

We find 37°20′ in the left column and then *cos* at the top. At the intersection of this row and column we find the entry we seek:

$$\cos 37°20' = 0.7951.$$

Example 2 Find cot 0.9192.

Since no degree symbol is given, we know that the angle is in radians. We do not find 0.9192 on the left, under radians. This is because the angle is greater than 45° or $\pi/4$ radians. Therefore, we will use the right and bottom headings. At the right, under radians, we find 0.9192. Next

OBJECTIVES

You should be able to:
[i] Use Table 3 to find function values.
[ii] Convert between degrees and minutes and degrees and tenths.
[iii] Use Table 3 with interpolation.
[iv] Use Table 3 in reverse.
[v] Use Table 3 to find angles of any size, and in reverse.

Use Table 3 to find the following.

1. sin 15°20′

2. cot 64°50′

3. cos 0.4451

4. tan 0.8319

Use Table 3 to answer the following.

5. How big must an angle be in order that it first differs from its sine in the fourth decimal place? the third decimal place?

we find *cot* at the bottom. At the intersection of this row and column we find the entry we seek:

$$cot\ 0.9192 = 0.7627.$$

DO EXERCISES 1–4.

Function Values for Small Angles

Consider this diagram showing a unit circle and a small angle.

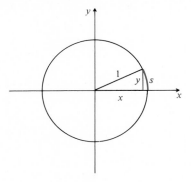

Since we are assuming *s* to be small, the length of arc *s* is very nearly the same as *y*. Also *x* is very nearly 1. Since $y = \sin s$, we see that $s \approx \sin s$. Since $\tan s = y/x$ and $x \approx 1$, we see that $\tan s \approx s$. These relations also hold if *s* is negative with small absolute value, and are useful in certain applications.

If $|s|$ is a small number, $s \approx \sin s \approx \tan s$ (a small angle—in radians— is approximately equal to its sine and its tangent).

DO EXERCISE 5.

[ii] Minutes Versus Tenths and Hundredths

It is usual to express the measure of an angle in degrees and minutes, but it is often useful to have the measure in degrees, tenths, hundredths, etc. Thus we will need to know how to convert between these.

Examples

3. Convert 16.35° to degrees and minutes.

$$16.35° = 16° + 0.35 \times 1°$$

Now

$$0.35 \times 1° = 0.35 \times 60' = 21'.$$

So

$$16.35° = 16°21'.$$

4. Convert $34°37'$ to degrees and decimal parts of degrees.

$$34°37' = 34° + \frac{37°}{60},$$

$$\frac{37}{60} = 0.617$$

so

$$34°37' = 34.617°.$$

DO EXERCISES 6 AND 7.

[iii] **Interpolation**

The process of linear interpolation, described in Section 7.4, can be applied to tables of any function. In particular, we can use it with tables of the trigonometric functions.

Example 5 Find $\tan 27°43'$.

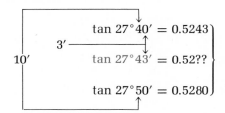

The tabular difference is 0.0037.

The tabular difference is 0.0037, and $43'$ is $\frac{3}{10}$ of the way from $40'$ to $50'$, so we take $\frac{3}{10}$ of 0.0037. This is 0.0011. Thus

$$\tan 27°43' = 0.5243 + 0.0011, \quad \text{or} \quad 0.5254.$$

Interpolation for decreasing functions is about the same as for increasing functions, but there is a slight difference as the next example shows.

Example 6 Find $\cot 29°44'$.

$$
\begin{array}{l}
\cot 29°40' = 1.756 \\
\cot 29°44' = 1.7?? \\
\cot 29°50' = 1.744
\end{array}
$$

4' ⎤ 10'

The tabular difference is 0.012.

We take 0.4 of the tabular difference, 0.012, and obtain 0.0048, or 0.005. Because the cotangent function is *decreasing* in the interval $(0°, 90°]$, we *subtract* 0.005 from 1.756. Thus $\cot 29°44' = 1.751$.

Once the process of interpolation is understood, it will not be necessary to write as much as we did in the above examples. After a bit of

6. Convert $37.45°$ to degrees and minutes.

7. Convert $43°55'$ to degrees and decimal parts of degrees.

Use Table 3 for the following.

8. Find sin 38°47′.

9. Find cot 27°45′.

10. Find the acute angle for which tan θ = 2.394, in both degrees and radians.

11. Find the acute angle for which cot θ = 1.819, in both degrees and radians.

practice, you will find that interpolation is rather easy, and you can accomplish some of the steps without writing.

DO EXERCISES 8 AND 9.

[iv] Using the Table in Reverse

Let us look at an example of using the tables in reverse, that is, given a function value, to find the measure of an angle.

Example 7 Given tan $B = 0.3727$, find B (between 0° and 90°).

$$
10' \begin{cases} \begin{aligned} \tan 20°20' &= 0.3706 \\ \tan B \quad\;\; &= 0.3727 \\ \tan 20°30' &= 0.3739 \end{aligned} \end{cases} \Bigg\} \quad\; 0.0021
$$

The tabular difference is 0.0033.

We find that B is $\frac{21}{33}$ or $\frac{7}{11}$ of the way from 20°20′ to 20°30′. The tabular difference is 10′, so we take $\frac{7}{11}$ of 10′ and obtain 6′. Thus $B = 20°26′$, or 0.3566 radian.

DO EXERCISES 10 AND 11.

[v] Function Values for Angles of Any Size

Table 3 gives function values for angles from 0° to 90°. For any other angle we can still use the table to find the function values. To do this, we first find the reference angle, which is the angle that the terminal side makes with the x-axis. We then look up the function values for the reference angle in the table and use the appropriate sign, depending on the quadrant in which the terminal side lies.

Example 8 Find sin 285°40′.

We first find the reference angle. To do this we find the multiple of 180° that is nearest 285°40′. The multiples of 180° are 180°, 360°, and so on. Now 285°40′ is nearest 360°. We subtract to find the reference angle.

$$
\begin{aligned}
359°60' & \\
-285°40' & \\
\hline
74°20' &
\end{aligned}
$$

Now we find that sin 74°20′ = 0.9628. The terminal side is in quadrant IV, where the sine is negative. Hence

$$\sin 285°41' = -0.9628.$$

It is helpful to make a diagram, like the following, showing the reference angle.

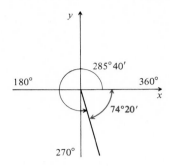

DO EXERCISES 12 AND 13.

Example 9 Given that $\tan A = -0.3727$, find A between 270° and 360°.

We first find the reference angle, exactly as in Example 7, ignoring the fact that $\tan A$ is negative. The reference angle is thus 20°26′.

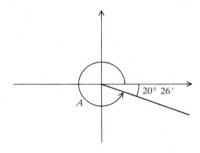

The angle A itself is found by subtracting 20°26′ from 360°.

$$\begin{array}{r} 359°60' \\ -20°26' \\ \hline 339°34' \end{array} \quad \text{Subtracting}$$

DO EXERCISE 14.

▦ The Use of Calculators

Scientific calculators contain tables of trigonometric functions—usually the sine, cosine, and tangent only. You can enter an angle and obtain any of these function values, and by taking their reciprocals, find the other three trigonometric function values.

Important: Some calculators require angles to be entered in degrees. On others either degrees or radians can be used. Be sure to use the unit appropriate to the calculator you are using.

To use the tables in reverse, e.g., to find an angle when its sine is given, first enter the sine value. Then press the key marked $\boxed{\sin^{-1}}$ (on some calculators). On other calculators the key is marked $\boxed{\text{arcsin}}$. On still others, two keys must be pressed in sequence: $\boxed{\text{arc}}$ and then $\boxed{\sin}$. Be sure to read the instructions for the calculator you are using.

Use Table 3 for the following.

12. Find cos 563°20′.

13. Find cos (−729°50′).

14. Given that $\sin A = -0.7304$, find A between 180° and 270°.

EXERCISE SET 9.5

[i] Use Table 3 to find the following.

1. sin 13°20′   **2.** sin 41°40′   **3.** cos 56°30′   **4.** cos 71°50′

5. tan 0.3956   **6.** tan 0.7010   **7.** cot 0.9134   **8.** cot 1.0443

[ii] Convert to degrees and minutes.

9. 46.38°   **10.** 85.21°   **11.** 67.84°   **12.** 38.48°

Convert to degrees and decimal parts of degrees.

13. 45°45′   **14.** 36°17′   **15.** 76°53′   **16.** 12°23′

[iii] Use Table 3 to find the following.

17. sin 28°13′   **18.** sin 36°42′   **19.** cos 53°53′   **20.** cos 80°33′

21. cos 75.15°   **22.** cos 81.91°   **23.** sin 17.43°   **24.** sin 38.72°

25. $\sin \dfrac{5\pi}{7}$ **26.** $\cos \dfrac{4\pi}{5}$

[iv] For each of the following, find θ in degrees and minutes between 0° and 90°.

27. sin θ = 0.2363   **28.** sin θ = 0.3854   **29.** cos θ = 0.3719

30. cos θ = 0.6361   **31.** cos θ = 0.3538   **32.** cos θ = 0.9678

For each of the following, find θ in radians between 0 and $\pi/2$.

33. tan θ = 0.4699   **34.** cot θ = 1.621

35. sec θ = 1.457   **36.** csc θ = 1.173

[v] Use Table 3 to find the following.

37. sin 292°40′   **38.** cos 472°40′   **39.** cos 514°30′   **40.** tan 349°28′

Find θ in degrees and minutes in the interval indicated.

41. sin θ = 0.2363,   (90°, 180°)   **42.** sin θ = −0.3854,   (270°, 360°)

43. cos θ = −0.3719,   (180°, 270°)   **44.** cos θ = 0.6361,   (360°, 450°)

Find θ in radians in the interval indicated.

45. cos θ = −0.3538, $\left(\dfrac{\pi}{2}, \pi\right)$ **46.** cos θ = 0.9678, $\left(-\dfrac{\pi}{2}, 0\right)$

47. sec θ = 1.457, $\left(\dfrac{3\pi}{2}, 2\pi\right)$ **48.** csc θ = −1.173, $\left(\pi, \dfrac{3\pi}{2}\right)$

☆ ——————————————————————————————————

49. How big must an angle be in order that it first differs from its tangent in the fourth decimal place?

50. How big must an angle be in order that it first differs in the third decimal place?

51. ▦ Angles are measured in degrees, minutes (60′ = 1 degree), and seconds (60 seconds = 1′). Seconds are denoted ″. Convert radian measure of $\pi/7$ to degrees, minutes, and seconds.

52. ▦ Convert 61°38′22″ to degrees and decimal parts of degrees.

53. ▦ The formula

$$\sin x = x - \frac{x^3}{6} + \frac{x^5}{120}$$

gives an approximation for sine values when x is in radians. Calculate sin 0.5 and compare your answer with Table 3.

54. Using values from the table, graph the tangent function between −90° and 90°. What is the domain of the tangent function?

55. Use the fact that $\sin \theta \approx \theta$ when θ is small to calculate the diameter of the sun. The sun is 93 million miles from the earth and the angle it subtends at the earth's surface is about 31′59″.

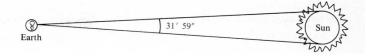

56. When light passes from one substance to another, rays are bent, depending on the speed of light in those substances. For example, light traveling in air is bent as it enters water. Angle i is called the angle of *incidence* and angle r is called the angle of *refraction*. Snell's law states that $\sin i/\sin r$ is constant. The constant is called the *index of refraction*. The index of refraction of a certain crystal is 1.52. Light strikes it, making an angle of incidence of 27°. What is the angle of refraction?

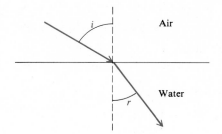

57. ▦ The function $\sin x/x$ (with x in radians) approaches a limit as x approaches 0. What is it?

58. a) Find $\log \sin 57°$ (the logarithm of the sine of 57°).
b) What is the domain of the log sin function?

9.6 FURTHER RELATIONS AMONG THE FUNCTIONS (OPTIONAL)*

[i] Some Important Identities

Consider the unit circle, whose equation is $x^2 + y^2 = 1$. For any point (x, y) on the circle, the second coordinate is $\sin \theta$ and the first coordinate is $\cos \theta$, since the length of the hypotenuse is 1. Thus we have the following.

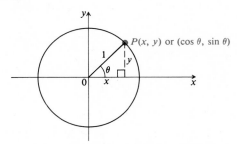

For any point *P* on the unit circle, the coordinates are (cos θ, sin θ), where θ is any angle having $\overrightarrow{OP}$ for its terminal side.

If we substitute $\cos \theta$ and $\sin \theta$ for x and y, respectively, in the equation of the unit circle, we obtain the following important identity.

$$\sin^2 \theta + \cos^2 \theta \equiv 1$$

---

*This section should be omitted, except in the "minimal course" (see the preface).

OBJECTIVES

You should be able to:

[i] Given a function value and a quadrant, find the other trigonometric function values.
[ii] Determine whether trigonometric functions are even or odd.
[iii] Construct graphs of the trigonometric functions and determine the domain, range, and period.
[iv] Derive cofunction identities.

▥ **1.** Given that $\sin\theta = -0.47$ and that θ is in quadrant IV, use identities to find the other function values.

For any point P on the unit circle, $\tan\theta = y/x$ and $\cot\theta = x/y$, provided denominators do not become 0. Since $x = \cos\theta$ and $y = \sin\theta$, we have the following:

$$\tan\theta = \frac{y}{x} = \frac{\sin\theta}{\cos\theta} \qquad \text{and} \qquad \cot\theta = \frac{x}{y} = \frac{\cos\theta}{\sin\theta}.$$

Since the secant and cosine values are reciprocals and the cosecant and sine values are reciprocals, we have the following:

$$\sec\theta = \frac{1}{\cos\theta} \qquad \text{and} \qquad \csc\theta = \frac{1}{\sin\theta}.$$

We now list four useful identities.

$$\tan\theta \equiv \frac{\sin\theta}{\cos\theta} \qquad \cot\theta \equiv \frac{\cos\theta}{\sin\theta}$$

$$\sec\theta \equiv \frac{1}{\cos\theta} \qquad \csc\theta \equiv \frac{1}{\sin\theta}$$

▥ **Example 1** Given that $\cos\theta = -0.61$ and that θ is in the second quadrant, use identities to find the other function values.

We begin with $\sin^2\theta + \cos^2\theta \equiv 1$, solving for $\sin\theta$:

$\sin\theta \equiv \pm\sqrt{1 - \cos^2\theta}$ The sign is to be chosen according to the quadrant—in this case, +.

$\sin\theta = \sqrt{1 - (-0.61)^2} = 0.7924.$

We proceed to calculate, using the appropriate identities:

$$\tan\theta = \frac{\sin\theta}{\cos\theta} = \frac{0.7924}{-0.61} = -1.2990,$$

$$\cot\theta = \frac{\cos\theta}{\sin\theta} = \frac{-0.61}{0.7924} = -0.7698,$$

$$\sec\theta = \frac{1}{\cos\theta} = \frac{1}{-0.61} = -1.6393,$$

$$\csc\theta = \frac{1}{\sin\theta} = \frac{1}{0.7924} = 1.2620.$$

DO EXERCISE 1.

Some More Identities

The identity $\sin^2\theta + \cos^2\theta \equiv 1$ is one of the so-called *Pythagorean* identities, because it can be derived from the Pythagorean property of right triangles. There are two other Pythagorean identities, obtained from this one by dividing. In one case we divide by $\sin^2\theta$ and in the other we divide by $\cos^2\theta$.

$$\frac{\sin^2\theta}{\sin^2\theta} + \frac{\cos^2\theta}{\sin^2\theta} \equiv \frac{1}{\sin^2\theta} \qquad \text{Dividing by } \sin^2\theta$$

Simplifying, we obtain $1 + \cot^2 \theta \equiv \csc^2 \theta$. We list the Pythagorean identities thus derived:

$$1 + \tan^2 \theta \equiv \sec^2 \theta, \qquad 1 + \cot^2 \theta \equiv \csc^2 \theta.$$

DO EXERCISE 2.

[ii] Even and Odd Functions

Some of the trigonometric functions are even and some of them are odd. This fact gives us some other useful identities. We determine evenness or oddness for the sine and cosine functions by considering an angle θ and its inverse $-\theta$.

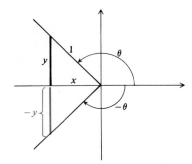 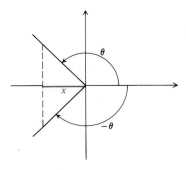

The diagrams show only special cases, but it is true in general that $\sin(-\theta) = -\sin \theta$ and $\cos(-\theta) = \cos \theta$. Thus the cosine function is even and the sine function is odd.

$$\sin(-\theta) \equiv -\sin \theta, \qquad \cos(-\theta) \equiv \cos \theta$$

DO EXERCISE 3.

Example 2 Determine whether the tangent function is even or odd.

We know that $\tan \theta \equiv \sin \theta / \cos \theta$. We substitute from the identities just derived:

$$\begin{aligned}
\tan(-\theta) &\equiv \frac{\sin(-\theta)}{\cos(-\theta)} \\
&\equiv \frac{-\sin \theta}{\cos \theta} \\
&\equiv -\frac{\sin \theta}{\cos \theta} \\
&\equiv -\tan \theta.
\end{aligned}$$

The tangent function is therefore odd.

2. Given that $\tan \theta = 1.43$ and that θ is in quadrant III, find the other function values.

3. Make a diagram to show that $\sin(-\theta) = -\sin \theta$ when θ is between $0°$ and $90°$.

A similar derivation shows that the cotangent function is also odd. We thus have the following:

$$\tan(-\theta) \equiv -\tan\theta,$$
$$\cot(-\theta) \equiv -\cot\theta.$$

[iii] Graphs of the Functions

Graphs of the trigonometric functions can be constructed by plotting and connecting points, using values obtained from the tables and keeping in mind the signs of the functions in the various quadrants. Following are graphs of four of the functions. The angles are in radians, and we have chosen to use x instead of θ, as is customary when graphing functions.

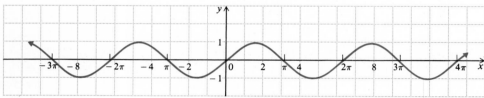

The sine function

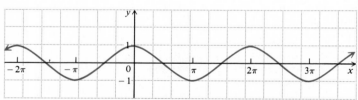

The cosine function

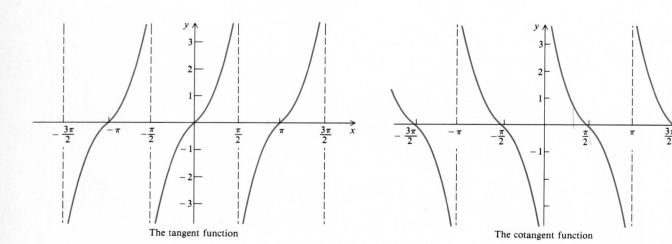

The tangent function   The cotangent function

DO EXERCISE 4.

[iv] The Cofunction Identities

Another class of identities gives functions in terms of their cofunctions at numbers differing by $\pi/2$. We consider first the sine and cosine. Consider this graph.

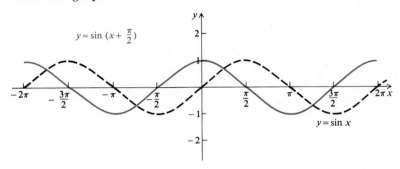

The graph of $y = \sin x$ has been translated to the left a distance of $\pi/2$, to obtain the graph of $y = \sin(x + \pi/2)$. The latter is also a graph of the cosine function. Thus we obtain the identity

$$\sin\left(x + \frac{\pi}{2}\right) \equiv \cos x.$$

By means similar to that above, we obtain the identity

$$\cos\left(x - \frac{\pi}{2}\right) \equiv \sin x.$$

Consider the following graph. The graph of $y = \sin x$ has been translated to the right a distance of $\pi/2$, to obtain the graph of $y = \sin(x - \pi/2)$. The latter is a reflection of the cosine function across the x-axis. In other words, it is a graph of $y = -\cos x$. We thus obtain the following identity:

$$\sin\left(x - \frac{\pi}{2}\right) \equiv -\cos x.$$

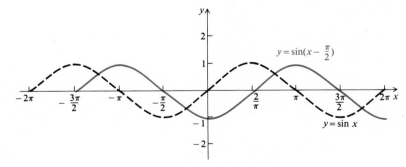

By a means similar to that above, we obtain the identity

$$\cos\left(x + \frac{\pi}{2}\right) \equiv -\sin x.$$

4. Using values from Table 3, plot points and draw a graph of $y = \tan \theta$ between $-\pi/4$ and $\pi/4$, θ in radians.

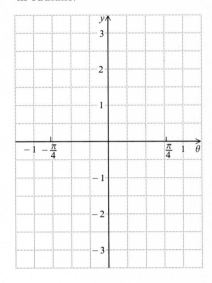

5. Find an identity for cot $(\pi/2 - x)$.

We have now established the following cofunction identities:*

To Be Memorized

$$\sin\left(x \pm \frac{\pi}{2}\right) \equiv \pm\cos x,$$

$$\cos\left(x \pm \frac{\pi}{2}\right) \equiv \mp\sin x.$$

These should be learned. The other cofunction identities can be obtained easily from them.

Example 3 Derive an identity relating $\sin(\pi/2 - x)$ to $\cos x$.

Since the sine function is odd, we know that

$$\sin\left(\frac{\pi}{2} - x\right) \equiv \sin - \left(x - \frac{\pi}{2}\right) \equiv -\sin\left(x - \frac{\pi}{2}\right).$$

Now consider the identity already established,

$$\sin\left(x - \frac{\pi}{2}\right) \equiv -\cos x.$$

This is equivalent to

$$-\sin\left(x - \frac{\pi}{2}\right) \equiv \cos x,$$

and we now have

$$\sin\left(\frac{\pi}{2} - x\right) \equiv \cos x.$$

DO EXERCISE 5.

EXERCISE SET 9.6

[i]

1. ▦ Given that $\sin x = 0.1425$ in quadrant II, find the other circular function values, using identities.

2. ▦ Given that $\tan x = 0.8012$ in quadrant III, find the other circular function values, using identities.

3. ▦ Given that $\cot x = 0.7534$ and $\sin x > 0$, find the other circular function values, using identities.

4. ▦ Given that $\sin\dfrac{\pi}{8} = 0.38268$, use identities to find:

a) $\cos\dfrac{\pi}{8}$;   b) $\cos\dfrac{5\pi}{8}$;   c) $\sin\dfrac{5\pi}{8}$;

d) $\sin\dfrac{-3\pi}{8}$;   e) $\cos\dfrac{-3\pi}{8}$.

[ii]

5. Show that the cotangent function is odd.

6. Determine whether the secant and cosecant functions are even or odd.

7. Make a diagram to show that $\sin(-\theta) = -\sin\theta$ when θ is between $180°$ and $270°$.

8. Make a diagram to show that $\cos(-\theta) = \cos\theta$ when θ is between $-360°$ and $-540°$.

---

*There are four identities in this list. Two of them are obtained by taking the top signs, the other two by taking the bottom signs.

[iii]

9. Using function values from a table, construct a graph of the sine function between -2π and 2π.

10. Using function values from a table, construct a graph of the cosine function between -2π and 2π.

11. For the sine function, determine the domain, range, and period.

12. For the cosine function, determine the domain, range, and period.

13. For the tangent function, determine the domain, range, and period.

14. For the cotangent function, determine the domain, range, and period.

[iv]

15. Find an identity for $\cos\left(\dfrac{\pi}{2} - x\right)$.

16. Find an identity for $\tan\left(\dfrac{\pi}{2} - x\right)$.

17. Find an identity for $\tan\left(x - \dfrac{\pi}{2}\right)$.

18. Find an identity for $\tan\left(x + \dfrac{\pi}{2}\right)$.

19. Find an identity for $\sec\left(x - \dfrac{\pi}{2}\right)$.

20. Find an identity for $\csc\left(\dfrac{\pi}{2} - x\right)$.

☆ ——————————————————————————————

21. Graph the secant function.

22. Graph the cosecant function.

23. Graph $y = \tan\dfrac{x}{2}$.

24. Graph $y = \cot 2x$.

25. The coordinates of any point on the unit circle are $(\cos s, \sin s)$, for some suitable number s. Show that for a circle with center at the origin of radius r, the coordinates of any point on the circle are $(r\cos s, r\sin s)$.

CHAPTER 9 REVIEW

[9.1, i] For angles of the following measures, state in which quadrant the terminal side lies, convert to radian measure in terms of π, and convert to radian measure not in terms of π. Use 3.14 for π.

[9.1, ii] **1.** 87° **2.** 145° **3.** 30° **4.** −30°

Convert to degree measure.

[9.1, ii] **5.** $\dfrac{3\pi}{2}$ **6.** 4π

[9.1, iii] **7.** Find the length of an arc of a circle, given a central angle of $\pi/4$ and a radius of 7 cm.

[9.1, iii] **8.** An arc 18 m long on a circle of radius 8 m subtends an angle of how many radians? how many degrees, to the nearest degree?

[9.2, i] **9.** A phonograph record revolves at 45 rpm. What is the linear velocity in cm/min of a point 4 cm from the center?

[9.2, i] **10.** An automobile wheel has a diameter of 26 in. If the car travels 30 mph, what is the angular velocity in radians/hr of a point on the edge of the wheel?

[9.3, i] **11.** Find the six trigonometric function values for the angle θ as shown.

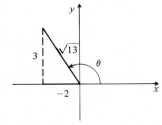

[9.3, ii]   **12.** Complete the following table.

θ	0°	30°	45°	60°	90°	270
sin θ						
cos θ						
tan θ						
cot θ						
sec θ						
csc θ						

[9.3, ii]   Find the following exactly. Do not use the table.

     13. sin 495°   **14.** tan −315°   **15.** cot 210°

     16. Graph the sine function and state its domain, range, and period.

[9.4, i]   Given that tan $\theta = 2/\sqrt{5}$ and that the terminal side is in quadrant III, find the following.

     17. sin θ **18.** cos θ **19.** cot θ **20.** sec θ **21.** csc θ

[9.4, ii]   Given that sin $\theta = 0.6820$, cos $\theta = 0.7314$, tan $\theta = 0.9325$, cot $\theta = 1.0724$, sec $\theta = 1.3673$, and csc $\theta = 1.4663$, find the following.

     22. sin (90° − θ)   **23.** cos (90° − θ)   **24.** tan (90° − θ)

     25. cot (90° − θ)   **26.** sec (90° − θ)   **27.** csc (90° − θ)

[9.5, ii]   **28.** Convert to degrees and minutes: 22.20°.

     29. Convert to degrees and decimal parts of degrees: 47°33′.

     Use Table 3 to find the following.

[9.5, i]   **30.** cos 8°20′   **31.** tan 27°14′   **32.** sec 44°30′

[9.5, iii]   **33.** cot $\theta = 3.450$ and 0° $< \theta <$ 90°. Find θ.

     34. sin $\theta = 0.6293$ and 0° $< \theta <$ 90°. Find θ.

[9.6, i]   **35.** Given that cos x = −0.1425 in quadrant II, find the other circular function values, using identities.

[9.6, ii]   **36.** a) List the circular functions that are even.

          b) List the circular functions that are odd.

[9.6, iii]   **37.** Graph the sine and cosine functions using the same axes.

[9.6, iv]   **38.** Find an identity for csc $\left(x - \dfrac{\pi}{2}\right)$.

     39. Graph y $= 3 \sin \dfrac{x}{2}$ and determine the domain, range, and period.

TRIGONOMETRIC IDENTITIES, INVERSE FUNCTIONS, AND EQUATIONS

2θ 2θ 2θ

2θ 2θ 2θ

OBJECTIVES

You should be able to:

[i] Use the sum and difference identities to find function values.
[ii] Simplify expressions using the sum and difference identities.
[iii] Find angles between lines, knowing their slopes.

10.1 SUM AND DIFFERENCE FORMULAS

[i] Difference Identities

We now develop some important identities involving sums or differences of two numbers (or angles), first an identity for the cosine of the difference of two numbers. We shall use the Greek letters α (alpha) and β (beta) for these numbers. Let us consider a real number α in the interval $[\pi/2, \pi]$ and a real number β in the interval $[0, \pi/2]$. These determine points A and B on the unit circle as shown. The arc length s is $\alpha - \beta$, and we know that $0 \leq s \leq \pi$. Recall that the coordinates of A are $(\cos \alpha, \sin \alpha)$, and the coordinates of B are $(\cos \beta, \sin \beta)$.

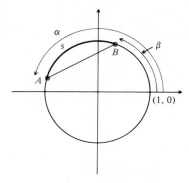

Using the distance formula, we can write an expression for the square of the distance AB:

$$AB^2 = (\cos \alpha - \cos \beta)^2 + (\sin \alpha - \sin \beta)^2.$$

This can be simplified as follows:

$$
\begin{aligned}
AB^2 &= \cos^2 \alpha - 2 \cos \alpha \cos \beta + \cos^2 \beta \\
&\quad + \sin^2 \alpha - 2 \sin \alpha \sin \beta + \sin^2 \beta \\
&= (\sin^2 \alpha + \cos^2 \alpha) + (\sin^2 \beta + \cos^2 \beta) \\
&\quad - 2(\cos \alpha \cos \beta + \sin \alpha \sin \beta) \\
&= 2 - 2(\cos \alpha \cos \beta + \sin \alpha \sin \beta).
\end{aligned}
$$

Now let us imagine rotating the circle above so that point B is at $(1, 0)$. The coordinates of point A are now $(\cos s, \sin s)$. The distance AB has not changed.

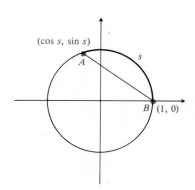

Again we use the distance formula to write an expression for the square of AB:

$$AB^2 = (\cos s - 1)^2 + (\sin s - 0)^2.$$

This simplifies as follows:

$$\begin{aligned}AB^2 &= \cos^2 s - 2\cos s + 1 + \sin^2 s \\ &= (\sin^2 s + \cos^2 s) + 1 - 2\cos s \\ &= 2 - 2\cos s.\end{aligned}$$

Equating our two expressions for AB^2 and simplifying, we obtain

$$\cos s = \cos \alpha \cos \beta + \sin \alpha \sin \beta. \qquad (1)$$

But $s = \alpha - \beta$, so we have the equation

$$\cos (\alpha - \beta) = \cos \alpha \cos \beta + \sin \alpha \sin \beta. \qquad (2)$$

This formula holds for any numbers α and β for which $\alpha - \beta$ is the length of the shortest arc from A to B; in other words, when $0 \leq \alpha - \beta \leq \pi$.

Example 1 Find $\cos \left(\dfrac{3\pi}{4} - \dfrac{\pi}{3} \right)$.

$$\begin{aligned}\cos \left(\frac{3\pi}{4} - \frac{\pi}{3} \right) &= \cos \frac{3\pi}{4} \cos \frac{\pi}{3} + \sin \frac{3\pi}{4} \sin \frac{\pi}{3} \\ &= -\frac{\sqrt{2}}{2} \cdot \frac{1}{2} + \frac{\sqrt{2}}{2} \cdot \frac{\sqrt{3}}{2} \\ &= \frac{\sqrt{2}}{4}(\sqrt{3} - 1) \quad \text{or} \quad \frac{\sqrt{6} - \sqrt{2}}{4}\end{aligned}$$

Example 2 Find $\cos 15°$.

$$\begin{aligned}\cos 15° &= \cos (45° - 30°) \\ &= \cos 45° \cos 30° + \sin 45° \sin 30° \\ &= \frac{\sqrt{2}}{2} \cdot \frac{\sqrt{3}}{2} + \frac{\sqrt{2}}{2} \cdot \frac{1}{2} \\ &= \frac{\sqrt{2}}{4}(\sqrt{3} + 1) \quad \text{or} \quad \frac{\sqrt{6} + \sqrt{2}}{4}\end{aligned}$$

DO EXERCISES 1 AND 2.

Formula (1) above holds when s is the length of the shortest arc from A to B. Given any real numbers α and β, the length of the shortest arc from A to B is not always $\alpha - \beta$. However, $\cos s$ is always equal to $\cos (\alpha - \beta)$. We will not prove this in detail, but the following diagrams help to show how a proof would be done. The length of the shortest arc from A to B always turns out to be $\alpha - \beta$ or $\beta - \alpha$, plus some multiple of 2π.

1. Find $\cos \left(\dfrac{\pi}{2} - \dfrac{\pi}{6} \right)$.

2. Find $\cos 105°$ as $\cos (150° - 45°)$.

Simplify.

3. $\sin \dfrac{\pi}{3} \sin\left(-\dfrac{\pi}{4}\right) + \cos\left(-\dfrac{\pi}{4}\right) \cos \dfrac{\pi}{3}$

4. $\cos 37° \cos 12° + \sin 12° \sin 37°$

5. $\cos \alpha \cos(-\beta) + \sin \alpha \sin(-\beta)$

(a)

(b)

(c)

Since $s = |\alpha - \beta| + 2k\pi$, where k is an integer, we have

$$\cos s = \cos(|\alpha - \beta| + 2k\pi)$$
$$= \cos|\alpha - \beta|,$$

since the cosine function has a period of 2π. Now let us recall the identity $\cos(-s) \equiv \cos s$. From this identity we see that we can drop the absolute value signs and then we know that $\cos s = \cos(\alpha - \beta)$ for any real numbers α and β. Thus formula (2) holds for all real numbers α and β. That formula is thus the identity we sought.

$$\cos(\alpha - \beta) \equiv \cos \alpha \cos \beta + \sin \alpha \sin \beta$$

[ii] Simplification

Example 3 Simplify

$$\sin\left(-\dfrac{5\pi}{2}\right) \sin \dfrac{\pi}{2} + \cos \dfrac{\pi}{2} \cos\left(-\dfrac{5\pi}{2}\right).$$

This is equal to

$$\cos \dfrac{\pi}{2} \cos\left(-\dfrac{5\pi}{2}\right) + \sin \dfrac{\pi}{2} \sin\left(-\dfrac{5\pi}{2}\right).$$

By the identity above we can simplify to

$$\cos\left[\dfrac{\pi}{2} - \left(-\dfrac{5\pi}{2}\right)\right], \quad \text{or} \quad \cos \dfrac{6\pi}{2}, \quad \text{or} \quad \cos 3\pi,$$

which is -1.

DO EXERCISES 3–5.

[i], [ii] Other Formulas

The other sum and difference formulas we seek follow easily from the one we have just derived. Let us consider $\cos(\alpha + \beta)$. This is equal to $\cos[\alpha - (-\beta)]$, and by the identity on the preceding page, we have

$$\cos(\alpha + \beta) \equiv \cos \alpha \cos(-\beta) + \sin \alpha \sin(-\beta).$$

But $\cos(-\beta) \equiv \cos \beta$ and $\sin(-\beta) \equiv -\sin \beta$, so the identity we seek is the following.

$$\cos(\alpha + \beta) \equiv \cos \alpha \cos \beta - \sin \alpha \sin \beta.$$

To develop an identity for the sine of a sum, we recall the cofunction identity $\sin \theta \equiv \cos(\pi/2 - \theta)$. In this identity we shall substitute $\alpha + \beta$ for θ, obtaining $\sin(\alpha + \beta) \equiv \cos[\pi/2 - (\alpha + \beta)]$. We can now use the identity for the cosine of a difference. We get

$$\sin(\alpha + \beta) \equiv \cos\left[\frac{\pi}{2} - (\alpha + \beta)\right]$$

$$\equiv \cos\left[\left(\frac{\pi}{2} - \alpha\right) - \beta\right]$$

$$\equiv \cos\left(\frac{\pi}{2} - \alpha\right)\cos \beta + \sin\left(\frac{\pi}{2} - \alpha\right)\sin \beta$$

$$\equiv \sin \alpha \cos \beta + \cos \alpha \sin \beta.$$

Thus the identity we seek is

$$\sin(\alpha + \beta) \equiv \sin \alpha \cos \beta + \cos \alpha \sin \beta.$$

DO EXERCISES 6 AND 7.

To find a formula for the sine of a difference, we can use the identity just derived, substituting $-\beta$ for β. We obtain

$$\sin(\alpha - \beta) \equiv \sin \alpha \cos \beta - \cos \alpha \sin \beta.$$

A formula for the tangent of a sum can be derived as follows, using identities already established:

$$\tan(\alpha + \beta) \equiv \frac{\sin(\alpha + \beta)}{\cos(\alpha + \beta)}$$

$$\equiv \frac{\sin \alpha \cos \beta + \cos \alpha \sin \beta}{\cos \alpha \cos \beta - \sin \alpha \sin \beta} \cdot \frac{\dfrac{1}{\cos \alpha \cos \beta}}{\dfrac{1}{\cos \alpha \cos \beta}}$$

$$\equiv \frac{\dfrac{\sin \alpha \cos \beta}{\cos \alpha \cos \beta} + \dfrac{\cos \alpha \sin \beta}{\cos \alpha \cos \beta}}{\dfrac{\cos \alpha \cos \beta}{\cos \alpha \cos \beta} - \dfrac{\sin \alpha \sin \beta}{\cos \alpha \cos \beta}}$$

$$\equiv \frac{\dfrac{\sin \alpha}{\cos \alpha} + \dfrac{\sin \beta}{\cos \beta}}{1 - \dfrac{\sin \alpha \sin \beta}{\cos \alpha \cos \beta}} \equiv \frac{\tan \alpha + \tan \beta}{1 - \tan \alpha \tan \beta}.$$

6. Find $\sin\left(\dfrac{\pi}{4} + \dfrac{\pi}{3}\right)$.

7. Simplify.

$$\sin \alpha \cos(-\beta) + \cos \alpha \sin(-\beta)$$

8. Derive the formula for tan $(\alpha - \beta)$.

Similarly, a formula for the tangent of a difference can be established. Following is a list of the sum and difference formulas. These should be memorized.

$$\cos (\alpha \mp \beta) \equiv \cos \alpha \cos \beta \pm \sin \alpha \sin \beta^*$$

$$\sin (\alpha \mp \beta) \equiv \sin \alpha \cos \beta \mp \cos \alpha \sin \beta$$

$$\tan (\alpha \mp \beta) \equiv \frac{\tan \alpha \mp \tan \beta}{1 \pm \tan \alpha \tan \beta}$$

DO EXERCISE 8.

The identities involving sines and tangents can be used in the same way as those involving cosines in the earlier examples.

9. Find tan 75° as tan (45° + 30°).

Example 4 Find tan 15°.

$$\tan 15° = \tan (45° - 30°) = \frac{\tan 45° - \tan 30°}{1 + \tan 45° \tan 30°}$$

$$= \frac{1 - \sqrt{3}/3}{1 + \sqrt{3}/3} = \frac{3 - \sqrt{3}}{3 + \sqrt{3}}$$

Example 5 Simplify $\sin \frac{\pi}{3} \cos \pi + \sin \pi \cos \frac{\pi}{3}$.

This is equal to

$$\sin \frac{\pi}{3} \cos \pi + \cos \frac{\pi}{3} \sin \pi.$$

Thus by the fourth identity in the list above, we can simplify to

$$\sin \left(\frac{\pi}{3} + \pi \right), \quad \text{or} \quad \sin \frac{4\pi}{3}, \quad \text{or} \quad -\frac{\sqrt{3}}{2}.$$

10. Simplify $\sin \frac{\pi}{2} \cos \frac{\pi}{3} - \sin \frac{\pi}{3} \cos \frac{\pi}{2}$.

DO EXERCISES 9 AND 10.

[iii] Angles Between Lines

Recall that a nonvertical line has an equation $y = mx + b$, where m is the slope. Note that such a line makes an angle with the positive half of the x-axis whose tangent is the slope of the line, as in tan $\theta = m$.

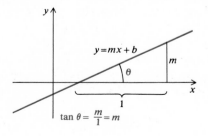

*There are six identities here, half of them obtained by using the colored signs.

One of the identities just developed gives us an easy way to find an angle formed by two lines. We shall consider the case in which neither line is vertical. Thus we shall consider two lines with equations:

$$l_1 : y = m_1 x + b_1 \quad \text{and} \quad l_2 : y = m_2 x + b_2.$$

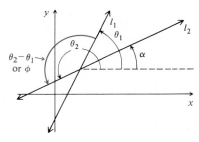

The slopes m_1 and m_2 are the tangents of the angles θ_1 and θ_2 that the lines form with the positive half of the x-axis. Thus we have $m_1 = \tan \theta_1$ and $m_2 = \tan \theta_2$. We wish to find the measure of the smallest angle through which l_1 can be rotated in the positive direction to get to l_2. In this case that angle is $\theta_2 - \theta_1$, or ϕ. We proceed as follows:

$$\tan \phi = \tan (\theta_2 - \theta_1) = \frac{\tan \theta_2 - \tan \theta_1}{1 + \tan \theta_2 \tan \theta_1}$$

$$= \frac{m_2 - m_1}{1 + m_2 m_1}.$$

Suppose we had taken the lines in the reverse order. Then the smallest *positive* angle from l_1 to l_2 would be ϕ, as shown here. Note that $\tan \theta_1 = m_1$.

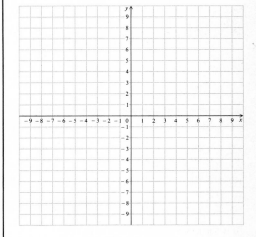

Since $\theta_2 = \alpha + 180°$, $\tan \theta_2 = \tan \alpha$ and $\tan \theta_2 = m_2$. Thus the same formula for $\tan \phi$ derived above holds. In the first case ϕ is an acute angle, so $\tan \phi$ will be positive. In the second case ϕ is obtuse, so $\tan \phi$ will be negative. Thus we have a general formula for finding the angle from one line to another.

THEOREM 1

The smallest positive angle ϕ from a line l_1 to a line l_2 (where neither line is vertical) can be obtained from

$$\tan \phi = \frac{m_2 - m_1}{1 + m_2 m_1},$$

where m_1 and m_2 are the slopes of l_1 and l_2, respectively. If $\tan \phi$ is positive, ϕ is acute. If $\tan \phi$ is negative, then ϕ is obtuse.

11. Find the smallest positive angle from l_1 to l_2. Sketch a graph and show this angle.

$$l_1 : 3y = \sqrt{3}x + 6,$$
$$l_2 : y + 5 = \sqrt{3}x.$$

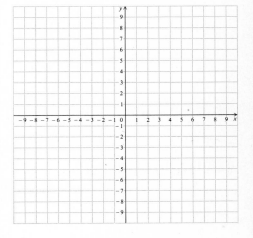

12. Find the smallest positive angle from l_1 to l_2. Sketch a graph and show this angle.

$$l_1 : y = \sqrt{3}x - 4,$$
$$l_2 : 3y = \sqrt{3}x + 3.$$

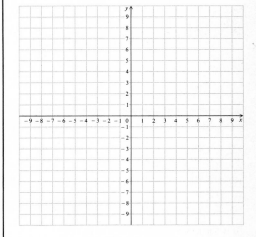

13. A loading ramp rises 4 ft with a run of 10 ft. A wire from the top of the building is attached to the ground 8 ft from the building. Find the angle θ that the wire makes with the ramp.

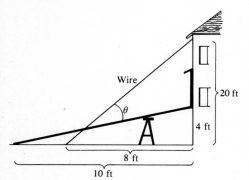

Example 6 Two lines have slopes as follows: l_1 has slope $\sqrt{3}/3$, l_2 has slope $\sqrt{3}$. How large is the smallest positive angle from l_1 to l_2?

We have $m_1 = \sqrt{3}/3$ and $m_2 = \sqrt{3}$. Then

$$\tan \phi = \frac{\sqrt{3} - \dfrac{\sqrt{3}}{3}}{1 + \sqrt{3} \cdot \dfrac{\sqrt{3}}{3}} = \frac{\dfrac{3\sqrt{3} - \sqrt{3}}{3}}{\dfrac{3+3}{3}} = \frac{2\sqrt{3}}{6} = \frac{\sqrt{3}}{3}.$$

Since $\tan \phi$ is positive, ϕ is acute. Hence

$$\phi = \frac{\pi}{6} \quad \text{or} \quad 30°.$$

Example 7 Find the smallest positive angle from l_1 to l_2:

$$l_1 : 2y = x + 3, \qquad l_2 : 3y + 4x + 6 = 0.$$

We first solve for y to obtain the slopes:

$$l_1 : y = \frac{1}{2}x + \frac{3}{2}, \quad \text{so} \quad m_1 = \frac{1}{2};$$

$$l_2 : y = -\frac{4}{3}x - 2, \quad \text{so} \quad m_2 = -\frac{4}{3};$$

$$\tan \phi = \frac{-\dfrac{4}{3} - \dfrac{1}{2}}{1 + \left(-\dfrac{4}{3}\right)\left(\dfrac{1}{2}\right)} = \frac{\dfrac{-8-3}{6}}{\dfrac{6-4}{6}} = -\frac{11}{2} = -5.5.$$

Since $\tan \phi$ is negative, we know that ϕ is obtuse. From the table we find an angle whose tangent is 5.5 to be about 79°40'. This is a reference angle. Subtracting from 180°, we obtain

$$\phi = 100°20', \qquad \text{approximately.}$$

DO EXERCISES 11–13. (NOTE THAT EXERCISES 11 AND 12 ARE ON THE PRECEDING PAGE.)

EXERCISE SET 10.1

[i] In Exercises 1–8, use the sum and difference identities to evaluate.

1. $\sin 75°$
(*Hint:* $75° = 45° + 30°$.)

2. $\cos 75°$

3. $\sin 15°$
(*Hint:* $15° = 45° - 30°$.)

4. $\cos 15°$

5. $\sin 105°$
(*Hint:* Use the results of Exercises 1 and 2.)

6. $\cos 105°$

7. $\tan 75°$

8. $\cot 15°$

In Exercises 9–14, assume that $\sin u = \frac{3}{5}$ and $\sin v = \frac{4}{5}$ and that u and v are between 0 and $\pi/2$. Then evaluate.

9. $\sin (u + v)$

10. $\sin (u - v)$

11. $\cos (u + v)$

12. $\cos (u - v)$

13. $\tan (u + v)$

14. $\tan (u - v)$

In Exercises 15 and 16, assume that $\sin \theta = 0.6249$ and $\cos \phi = 0.1102$, and that θ and ϕ are both first-quadrant angles. Evaluate.

15. ▦ $\sin (\theta + \phi)$

16. ▦ $\cos (\theta + \phi)$

[ii] In Exercises 17–26, simplify or evaluate.

17. $\sin 37° \cos 22° + \cos 37° \sin 22°$

18. $\cos 37° \cos 22° - \sin 22° \sin 37°$

19. $\dfrac{\tan 20° + \tan 32°}{1 - \tan 20° \tan 32°}$

20. $\dfrac{\tan 35° - \tan 12°}{1 + \tan 35° \tan 12°}$

21. $\sin(\alpha + \beta) + \sin(\alpha - \beta)$

22. $\sin(\alpha + \beta) - \sin(\alpha - \beta)$

23. $\cos(\alpha + \beta) + \cos(\alpha - \beta)$

24. $\cos(\alpha + \beta) - \cos(\alpha - \beta)$

25. $\cos(u + v) \cos v + \sin(u + v) \sin v$

26. $\sin(u - v) \cos v + \cos(u - v) \sin v$

[iii] In Exercises 27–34, find the angle from l_1 to l_2.

27. $l_1 : 3y = \sqrt{3}x + 2$,
$l_2 : 3y + \sqrt{3}x = -3$

28. $l_1 : 3y = \sqrt{3}x + 3$,
$l_2 : y = \sqrt{3}x + 2$

29. $l_1 : 2x = 3 - 2y$,
$l_2 : x + y = 5$

30. $l_1 : 2x = 3 + 2y$,
$l_2 : x - y = 5$

31. $l_1 : 2x - 5y + 1 = 0$,
$l_2 : 3x + y - 7 = 0$

32. $l_1 : 2x + y - 4 = 0$,
$l_2 : y - 2x + 5 = 0$

33. $l_1 : y = 3$,
$l_2 : x + y = 5$

34. $l_1 : y = 5$,
$l_2 : x - y = 2$

☆ ───────────────────────────

35. Find an identity for $\sin 2\theta$.
(Hint: $2\theta = \theta + \theta$.)

36. Find an identity for $\cos 2\theta$.
(Hint: $2\theta = \theta + \theta$.)

37. Derive an identity for $\cot(\alpha + \beta)$ in terms of $\cot \alpha$ and $\cot \beta$.

38. Derive an identity for $\cot(\alpha - \beta)$ in terms of $\cot \alpha$ and $\cot \beta$.

The cofunction identities can be derived from the sum and difference formulas. Derive identities for the following.

39. $\sin\left(\dfrac{\pi}{2} - x\right)$

40. $\cos\left(\dfrac{\pi}{2} - x\right)$

Find the slope of line l_1, where m_2 is the slope of line l_2 and ϕ is the smallest positive angle from l_1 to l_2.

41. $m_2 = \dfrac{4}{3}$, $\phi = 45°$

42. $m_2 = \dfrac{2}{3}$, ϕ is the smallest angle having slope $\dfrac{5}{2}$

43. Line l_1 contains the points $(-2, 4)$ and $(5, -1)$. Find the slope of line l_2 such that the angle from l_1 to l_2 is 45°.

44. Line l_1 contains $(3, -1)$ and $(-4, 2)$. Find the slope of line l_2 such that the angle from l_1 to l_2 is $-45°$.

45. ▦ Line l_1 contains the points $(-2.123, 3.899)$ and $(-4.892, -0.9012)$ and line l_2 contains $(0, -3.814)$ and $(5.925, 4.013)$. Find the smallest positive angle from l_1 to l_2.

46. Line l_1 contains the points $(-3, 7)$ and $(-3, -2)$. Line l_2 contains $(0, -4)$ and $(2, 6)$. Find the smallest positive angle from l_1 to l_2.

47. In a circus a guy wire A is attached to the top of a 30-ft pole. Wire B is used for performers to walk up to the tight wire, 10 ft above the ground. Find the angle ϕ between the wires.

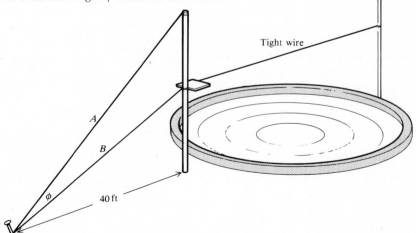

★ ───────────────────────────

48. Find an identity for $\cos(\alpha + \beta)$ involving only cosines.

49. Find an identity for $\sin(\alpha + \beta + \gamma)$.

OBJECTIVES

You should be able to:

[i] Use the double-angle identities to find function values of twice an angle when one function value is known for that angle.

[ii] Use the half-angle identities to find the function values of half an angle when one function value is known for that angle.

[iii] Simplify certain trigonometric expressions using the half-angle and double-angle formulas.

1. Given $\sin \theta = \frac{3}{5}$ and that θ is in the first quadrant, what is $\sin 2\theta$?

10.2 SOME IMPORTANT IDENTITIES

Two important classes of trigonometric identities are known as the half-angle identities and the double-angle identities.

[i] Double-Angle Identities

To develop these identities we shall use the sum formulas from the preceding section. We first develop a formula for $\sin 2\theta$. Recall that

$$\sin (\alpha + \beta) \equiv \sin \alpha \cos \beta + \cos \alpha \sin \beta.$$

We shall consider a number θ and substitute it for both α and β in this identity. We obtain

$$\sin (\theta + \theta) \equiv \sin 2\theta \equiv \sin \theta \cos \theta + \cos \theta \sin \theta$$
$$\equiv 2 \sin \theta \cos \theta.$$

The identity we seek is

$$\sin 2\theta \equiv 2 \sin \theta \cos \theta.$$

Example 1 If $\sin \theta = \frac{3}{8}$ and θ is in the first quadrant, what is $\sin 2\theta$?

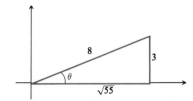

From the figure we see that $\cos \theta = \sqrt{55}/8$. Thus

$$\sin 2\theta = 2 \sin \theta \cos \theta = 2 \cdot \frac{3}{8} \cdot \frac{\sqrt{55}}{8} = \frac{3\sqrt{55}}{32}.$$

DO EXERCISE 1.

Double-angle identities for the cosine and tangent functions can be derived in much the same way as the above identity:

$$\cos (\alpha + \beta) \equiv \cos \alpha \cos \beta - \sin \alpha \sin \beta$$
$$\cos 2\theta \equiv \cos (\theta + \theta) \equiv \cos \theta \cos \theta - \sin \theta \sin \theta$$
$$\equiv \cos^2 \theta - \sin^2 \theta$$
$$\cos 2\theta \equiv \cos^2 \theta - \sin^2 \theta.$$

Now we derive an identity for $\tan 2\theta$:

$$\tan (\alpha + \beta) \equiv \frac{\tan \alpha + \tan \beta}{1 - \tan \alpha \tan \beta}$$

$$\tan 2\theta \equiv \tan (\theta + \theta) \equiv \frac{\tan \theta + \tan \theta}{1 - \tan \theta \tan \theta} \equiv \frac{2 \tan \theta}{1 - \tan^2 \theta}$$

$$\tan 2\theta \equiv \frac{2 \tan \theta}{1 - \tan^2 \theta}.$$

Example 2 Given that $\tan \theta = -\frac{3}{4}$ and θ is in the second quadrant, find $\sin 2\theta$, $\cos 2\theta$, $\tan 2\theta$, and the quadrant in which 2θ lies.

By drawing a diagram as shown, we find that $\sin \theta = \frac{3}{5}$ and $\cos \theta = -\frac{4}{5}$.

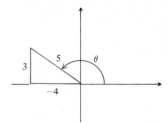

Now,

$$\sin 2\theta = 2 \sin \theta \cos \theta$$

$$= 2 \cdot \frac{3}{5} \cdot \left(-\frac{4}{5}\right) = -\frac{24}{25};$$

$$\cos 2\theta = \cos^2 \theta - \sin^2 \theta = \left(-\frac{4}{5}\right)^2 - \left(\frac{3}{5}\right)^2$$

$$= \frac{16}{25} - \frac{9}{25} = \frac{7}{25};$$

$$\tan 2\theta = \frac{2 \tan \theta}{1 - \tan^2 \theta} = \frac{2 \cdot \left(-\frac{3}{4}\right)}{1 - \left(-\frac{3}{4}\right)^2}$$

$$= -\frac{\frac{3}{2}}{1 - \frac{9}{16}} = -\frac{24}{7}.$$

Since $\sin 2\theta$ is negative and $\cos 2\theta$ is positive, we know that 2θ is in quadrant IV. Note that $\tan 2\theta$ could have been found more easily in this case by dividing the values of $\sin 2\theta$ and $\cos 2\theta$.

DO EXERCISE 2.

Two other useful identities for $\cos 2\theta$ can be derived easily, as follows:

$$\cos 2\theta \equiv \cos^2 \theta - \sin^2 \theta$$
$$\equiv (1 - \sin^2 \theta) - \sin^2 \theta \qquad \text{Using } \sin^2 \theta + \cos^2 \theta \equiv 1$$
$$\equiv 1 - 2 \sin^2 \theta.$$

Similarly,

$$\cos 2\theta \equiv \cos^2 \theta - \sin^2 \theta$$
$$\equiv \cos^2 \theta - (1 - \cos^2 \theta)$$
$$\equiv 2 \cos^2 \theta - 1.$$

2. Given $\cos \theta = -\frac{5}{13}$ and that θ is in the third quadrant, find $\sin 2\theta$, $\cos 2\theta$, and $\tan 2\theta$. Also determine the quadrant in which 2θ lies.

3. Find a formula for $\cos 3\theta$ in terms of function values of θ.

Solving these two identities for $\sin^2 \theta$ and $\cos^2 \theta$, respectively, we obtain two more identities that are often useful. Following is a list of double-angle identities. It should be memorized.

$$\sin 2\theta \equiv 2 \sin \theta \cos \theta$$
$$\cos 2\theta \equiv \cos^2 \theta - \sin^2 \theta$$
$$\equiv 1 - 2 \sin^2 \theta$$
$$\equiv 2 \cos^2 \theta - 1$$
$$\tan 2\theta \equiv \frac{2 \tan \theta}{1 - \tan^2 \theta}$$
$$\sin^2 \theta \equiv \frac{1 - \cos 2\theta}{2}$$
$$\cos^2 \theta \equiv \frac{1 + \cos 2\theta}{2}$$

By division and the last two identities, it is always easy to deduce the following identity, which is also often useful:

$$\tan^2 \theta \equiv \frac{1 - \cos 2\theta}{1 + \cos 2\theta}.$$

From the basic identities (in the lists to be memorized), others can be obtained.

Example 3 Find a formula for $\sin 3\theta$ in terms of function values of θ.

$$\sin 3\theta \equiv \sin (2\theta + \theta)$$
$$\equiv \sin 2\theta \cos \theta + \cos 2\theta \sin \theta$$
$$\equiv (2 \sin \theta \cos \theta) \cos \theta + (2 \cos^2 \theta - 1) \sin \theta$$
$$\equiv 2 \sin \theta \cos^2 \theta + 2 \sin \theta \cos^2 \theta - \sin \theta$$
$$\sin 3\theta \equiv 4 \sin \theta \cos^2 \theta - \sin \theta$$

4. Find a formula for $\sin^3 x$ in terms of function values of x or $2x$, raised only to the first power.

Example 4 Find a formula for $\cos^3 x$ in terms of function values of x or $2x$, raised only to the first power.

$$\cos^3 x \equiv \cos^2 x \cos x \equiv \frac{1 + \cos 2x}{2} \cos x$$

DO EXERCISES 3 AND 4.

[ii] Half-Angle Identities

To develop these identities, we use three of the previously developed ones. As shown below, we take square roots and replace θ by $\phi/2$.

$$\sin^2 \theta \equiv \frac{1 - \cos 2\theta}{2} \longrightarrow \left| \sin \frac{\phi}{2} \right| \equiv \sqrt{\frac{1 - \cos \phi}{2}}$$

$$\cos^2 \theta \equiv \frac{1 + \cos 2\theta}{2} \longrightarrow \left| \cos \frac{\phi}{2} \right| \equiv \sqrt{\frac{1 + \cos \phi}{2}}$$

$$\tan^2 \theta \equiv \frac{1 - \cos 2\theta}{1 + \cos 2\theta} \longrightarrow \left| \tan \frac{\phi}{2} \right| \equiv \sqrt{\frac{1 - \cos \phi}{1 + \cos \phi}}$$

The half-angle formulas are those on the right above. We can eliminate the absolute value signs by introducing $\pm$ signs, with the understanding that we use $+$ or $-$ depending on the quadrant in which the angle lies. We thus obtain these formulas in the following form.

$$\sin \frac{\phi}{2} \equiv \pm \sqrt{\frac{1 - \cos \phi}{2}}$$

$$\cos \frac{\phi}{2} \equiv \pm \sqrt{\frac{1 + \cos \phi}{2}}$$

$$\tan \frac{\phi}{2} \equiv \pm \sqrt{\frac{1 - \cos \phi}{1 + \cos \phi}}$$

These formulas should be memorized.

There are two other formulas for $\tan (\phi/2)$ that are often useful. They can be obtained as follows:

$$\left| \tan \frac{\phi}{2} \right| \equiv \sqrt{\frac{1 - \cos \phi}{1 + \cos \phi}} \equiv \sqrt{\frac{1 - \cos \phi}{1 + \cos \phi} \cdot \frac{1 + \cos \phi}{1 + \cos \phi}}$$

$$\equiv \sqrt{\frac{1 - \cos^2 \phi}{(1 + \cos \phi)^2}}$$

$$\equiv \sqrt{\frac{\sin^2 \phi}{(1 + \cos \phi)^2}} \equiv \frac{|\sin \phi|}{|1 + \cos \phi|}.$$

Now $1 + \cos \phi$ cannot be negative because $\cos \phi$ is never less than -1. Hence the absolute value signs are not necessary in the denominator. As the following graph shows, $\tan (\phi/2)$ and $\sin \phi$ have the same sign for all ϕ for which $\tan (\phi/2)$ is defined.

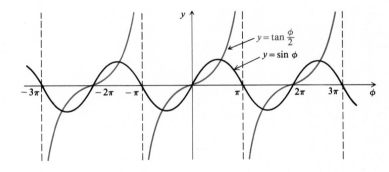

Thus we can dispense with the other absolute value signs, and obtain the formula we seek. A second formula can be obtained similarly.

GET THIS UNDERSTOOD

$$\tan \frac{\phi}{2} \equiv \frac{\sin \phi}{1 + \cos \phi}, \qquad \tan \frac{\phi}{2} \equiv \frac{1 - \cos \phi}{\sin \phi}$$

These formulas have the advantage that they give the sign of $\tan (\phi/2)$ directly.

5. Find $\cos 15°$.

Example 5 Find $\sin 15°$.

$$\sin 15° = \sin \frac{30°}{2} = \pm\sqrt{\frac{1 - \cos 30°}{2}}$$

$$= \pm\sqrt{\frac{1 - (\sqrt{3}/2)}{2}}$$

$$= \pm\sqrt{\frac{2 - \sqrt{3}}{4}} = \frac{\sqrt{2 - \sqrt{3}}}{2}$$

(The expression is positive, because $15°$ is in the first quadrant.)

Example 6 Find $\tan \frac{\pi}{8}$.

$$\tan \frac{\pi}{8} = \tan \frac{\frac{\pi}{4}}{2} = \frac{\sin \frac{\pi}{4}}{1 + \cos \frac{\pi}{4}} = \frac{\frac{\sqrt{2}}{2}}{1 + \frac{\sqrt{2}}{2}}$$

$$= \frac{\sqrt{2}}{2 + \sqrt{2}}$$

$$= \sqrt{2} - 1 \qquad \text{Rationalizing the denominator}$$

DO EXERCISES 5 AND 6.

6. Find $\tan \frac{\pi}{12}$.

[iii] Simplification

Many simplifications of trigonometric expressions are possible through the use of the identities we have developed.

Example 7 Simplify

$$\frac{1 - \cos 2x}{4 \sin x \cos x}.$$

We search the list of identities, attempting to find some substitution that will simplify the expression. In this case, one might note that the denominator is $2(2 \sin x \cos x)$ and thus find a simplification, since $2 \sin x \cos x \equiv \sin 2x$. We now have

$$\frac{1 - \cos 2x}{2 \sin 2x}.$$

Now

$$\frac{1 - \cos \phi}{\sin \phi} \equiv \tan \frac{\phi}{2}.$$

Using this we obtain

$$\frac{1}{2} \cdot \frac{1 - \cos 2x}{\sin 2x} \equiv \frac{1}{2} \tan x.$$

Sum–Product Identities

On occasion, it is convenient to convert a product of trigonometric expressions to a sum, or the reverse. The following identities are useful in this connection. Proofs are left as exercises.

KNOW

$$\sin u \cdot \cos v \equiv \frac{1}{2}[\sin (u + v) + \sin (u - v)]$$

$$\cos u \cdot \sin v \equiv \frac{1}{2}[\sin (u + v) - \sin (u - v)]$$

$$\cos u \cdot \cos v \equiv \frac{1}{2}[\cos (u - v) + \cos (u + v)]$$

$$\sin u \cdot \sin v \equiv \frac{1}{2}[\cos (u - v) - \cos (u + v)]$$

$$\sin x + \sin y \equiv 2 \sin \frac{x + y}{2} \cos \frac{x - y}{2}$$

$$\sin x - \sin y \equiv 2 \cos \frac{x + y}{2} \sin \frac{x - y}{2}$$

$$\cos y + \cos x \equiv 2 \cos \frac{x + y}{2} \cos \frac{x - y}{2}$$

$$\cos y - \cos x \equiv 2 \sin \frac{x + y}{2} \sin \frac{x - y}{2}$$

EXERCISE SET 10.2

[i] In Exercises 1–6, find sin 2θ, cos 2θ, tan 2θ, and the quadrant in which 2θ lies.

1. $\sin \theta = \frac{4}{5}$ (θ in quadrant I)

2. $\sin \theta = \frac{5}{13}$ (θ in quadrant I)

3. $\cos \theta = -\frac{4}{5}$ (θ in quadrant III)

4. $\cos \theta = -\frac{3}{5}$ (θ in quadrant III)

5. $\tan \theta = \frac{4}{3}$ (θ in quadrant III)

6. $\tan \theta = \frac{3}{4}$ (θ in quadrant III)

7. Find a formula for sin 4θ in terms of function values of θ.

8. Find a formula for cos 4θ in terms of function values of θ.

9. Find a formula for $\sin^4 \theta$ in terms of function values of θ or 2θ or 4θ, raised only to the first power.

10. Find a formula for $\cos^4 \theta$ in terms of function values of θ or 2θ or 4θ, raised only to the first power.

[ii]

11. Find sin 75° without using tables. (*Hint:* 75 = 150/2.)

12. Find cos 75° without using tables.

13. Find tan 75° without using tables.

14. Find tan 67.5° without using tables. (*Hint:* 67.5 = 135/2.)

15. Find $\sin \frac{5\pi}{8}$ without using tables.

16. Find $\cos \frac{5\pi}{8}$ without using tables.

[i], [ii] In Exercises 17–22, given that sin $\theta = 0.3416$ and that θ is in the first quadrant, find:

17. ▦ sin 2θ.

18. ▦ cos 2θ.

19. ▦ sin 4θ.

20. ▦ cos 4θ.

21. ▦ $\sin \frac{\theta}{2}$.

22. ▦ $\cos \frac{\theta}{2}$.

[iii] Simplify.

23. $\dfrac{\sin 2x}{2 \sin x}$

24. $\dfrac{\sin 2x}{2 \cos x}$

25. $1 - 2 \sin^2 \dfrac{x}{2}$

26. $2 \cos^2 \dfrac{x}{2} - 1$

27. $2 \sin \dfrac{x}{2} \cos \dfrac{x}{2}$

28. $2 \sin 2x \cos 2x$

29. $\cos^2 \dfrac{x}{2} - \sin^2 \dfrac{x}{2}$

30. $\cos^4 x - \sin^4 x$

31. $(\sin x + \cos x)^2 - \sin 2x$

32. $(\sin x - \cos x)^2 + \sin 2x$

33. $2 \sin^2 \dfrac{x}{2} + \cos x$

34. $2 \cos^2 \dfrac{x}{2} - \cos x$

35. $(-4 \cos x \sin x + 2 \cos 2x)^2 + (2 \cos 2x + 4 \sin x \cos x)^2$

36. $(-4 \cos 2x + 8 \cos x \sin x)^2 + (8 \sin x \cos x + 4 \cos 2x)^2$

37. $2 \sin x \cos^3 x + 2 \sin^3 x \cos x$

38. $2 \sin x \cos^3 x - 2 \sin^3 x \cos x$

☆ _____

39. Prove the first four of the sum–product identities. (*Hint:* They follow from the sum and difference formulas given in Section 10.1.)

40. Prove the last four of the sum–product identities. (*Hint:* Use the results of Exercise 39.)

★ _____

41. Graph $f(x) = \cos^2 x - \sin^2 x$.

42. Graph $y = |\sin x \cos x|$.

OBJECTIVES

You should be able to:

[i] Use the basic identities to prove other identities.

[ii] Given an expression $c \sin ax + d \cos ax$, find an equivalent expression $A \sin (ax + b)$.

[iii] Graph an equation

$$y = c \sin ax + d \cos ax$$

by first expressing the right-hand side in terms of the sine function only.

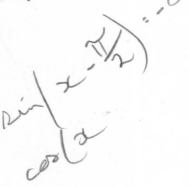

10.3 PROVING IDENTITIES

Basic Identities

One should remember certain trigonometric identities. Then as the occasion arises, other identities can be proved using those memorized. Following is a minimal list of identities that should be learned. Note that most formulas involving cotangents, secants, and cosecants are not in the list. That is because these functions are reciprocals of the sine, cosine, and tangent, and thus formulas for the former can easily be derived from those for the latter.

$$\sin (-x) \equiv -\sin x,$$
$$\cos (-x) \equiv \cos x,$$
$$\tan (-x) \equiv -\tan x$$

Pythagorean identities

$$\sin^2 x + \cos^2 x \equiv 1,$$
$$1 + \tan^2 x \equiv \sec^2 x,$$
$$1 + \cot^2 x \equiv \csc^2 x$$

Cofunction identities

$$\sin \left(x \pm \frac{\pi}{2} \right) \equiv \pm\cos x,$$
$$\cos \left(x \pm \frac{\pi}{2} \right) \equiv \mp\sin x$$

Sum and difference identities

$$\sin(\alpha \pm \beta) \equiv \sin \alpha \cos \beta \pm \cos \alpha \sin \beta,$$
$$\cos(\alpha \pm \beta) \equiv \cos \alpha \cos \beta \mp \sin \alpha \sin \beta,$$
$$\tan(\alpha \pm \beta) \equiv \frac{\tan \alpha \pm \tan \beta}{1 \mp \tan \alpha \tan \beta}$$

Double-angle identities

$$\sin 2x \equiv 2 \sin x \cos x,$$
$$\cos 2x \equiv \cos^2 x - \sin^2 x \equiv 1 - 2 \sin^2 x$$
$$\equiv 2 \cos^2 x - 1,$$
$$\tan 2x \equiv \frac{2 \tan x}{1 - \tan^2 x},$$
$$\sin^2 x \equiv \frac{1 - \cos 2x}{2},$$
$$\cos^2 x \equiv \frac{1 + \cos 2x}{2}$$

Half-angle identities

$$\sin \frac{x}{2} \equiv \pm \sqrt{\frac{1 - \cos x}{2}},$$
$$\cos \frac{x}{2} \equiv \pm \sqrt{\frac{1 + \cos x}{2}},$$
$$\tan \frac{x}{2} \equiv \pm \sqrt{\frac{1 - \cos x}{1 + \cos x}} \equiv \frac{\sin x}{1 + \cos x} \equiv \frac{1 - \cos x}{\sin x}$$

[i] Proving Identities

In proving other identities, using these, it is wise to consult the list to gain an idea of how to proceed. It is usually helpful to express all function values in terms of sines and cosines. *It is most important* in proving an identity to work with one side at a time. The idea is to simplify each side separately until the same expression is obtained in the two cases.

Example 1 Prove the following identity.

$$\tan^2 x - \sin^2 x \equiv \sin^2 x \tan^2 x$$

$\dfrac{\sin^2 x}{\cos^2 x} - \sin^2 x$	$\sin^2 x \dfrac{\sin^2 x}{\cos^2 x}$
$\dfrac{\sin^2 x - \sin^2 x \cos^2 x}{\cos^2 x}$	$\dfrac{\sin^4 x}{\cos^2 x}$
$\dfrac{\sin^2 x (1 - \cos^2 x)}{\cos^2 x}$	
$\dfrac{\sin^2 x \sin^2 x}{\cos^2 x}$	
$\dfrac{\sin^4 x}{\cos^2 x}$	

Prove the following identities.

1. $\cot^2 x - \cos^2 x \equiv \cos^2 x \cot^2 x$

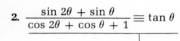

In the above proof, a vertical line has been drawn as a reminder to work separately with the two sides. Everything has been expressed in terms of sines and cosines and then the two sides are simplified. We work with each side separately because, as we saw in Chapter 2, "doing the same thing to both sides of an equation" does not always yield equivalent equations.

Example 2 Prove the following identity.

$$\frac{\sin 2\theta}{\sin \theta} - \frac{\cos 2\theta}{\cos \theta} \equiv \sec \theta$$

$$\frac{2 \sin \theta \cos \theta}{\sin \theta} - \frac{\cos^2 \theta - \sin^2 \theta}{\cos \theta} \quad \bigg| \quad \frac{1}{\cos \theta}$$

$$\frac{2 \cos^2 \theta - \cos^2 \theta + \sin^2 \theta}{\cos \theta}$$

$$\frac{\cos^2 \theta + \sin^2 \theta}{\cos \theta}$$

$$\frac{1}{\cos \theta}$$

DO EXERCISES 1 AND 2.

HINTS FOR PROVING IDENTITIES

1. Work with one side at a time.

2. Work with the more complex side first.

3. Do the algebraic manipulations, such as adding and subtracting.

4. Converting all expressions to sines and cosines is often helpful.

5. Try something! Identities are never proved by staring.

---

2. $\dfrac{\sin 2\theta + \sin \theta}{\cos 2\theta + \cos \theta + 1} \equiv \tan \theta$

[ii] An Identity for $c \sin ax + d \cos ax$

An expression $c \sin ax + d \cos ax$, where a, c, and d are constants, is equivalent to an expression $A \sin (ax + b)$, where A and b are also constants. To prove this we shall look for the numbers A and b. First, let us consider the numbers $c/\sqrt{c^2 + d^2}$ and $d/\sqrt{c^2 + d^2}$. These numbers are the coordinates of a point on the unit circle because

$$\left(\frac{c}{\sqrt{c^2 + d^2}}\right)^2 + \left(\frac{d}{\sqrt{c^2 + d^2}}\right)^2 = 1.$$

Thus there is a number b for which these numbers are the cosine and sine, respectively. In other words,

$$\frac{c}{\sqrt{c^2 + d^2}} = \cos b,$$

and

$$\frac{d}{\sqrt{c^2 + d^2}} = \sin b,$$

for some number b. Now we will consider the expression $c \sin ax + d \cos ax$, and multiply it by 1, as follows:

$$\frac{\sqrt{c^2 + d^2}}{\sqrt{c^2 + d^2}}[c \sin ax + d \cos ax]$$

$$\equiv \sqrt{c^2 + d^2}\left[(\sin ax) \cdot \frac{c}{\sqrt{c^2 + d^2}} + (\cos ax)\frac{d}{\sqrt{c^2 + d^2}}\right]$$

$$\equiv \sqrt{c^2 + d^2}\,[\sin ax \cos b + \cos ax \sin b].$$

Now, by the identity for the sine of a sum, we obtain

$$\sqrt{c^2 + d^2}\,\sin (ax + b).$$

Thus we have an identity as follows.

THEOREM 2

> **For any numbers a, c, and d,**
>
> $$c \sin ax + d \cos ax \equiv A \sin (ax + b),$$
>
> **where $A = \sqrt{c^2 + d^2}$ and b is a number whose cosine is c/A and whose sine is d/A.**

Example 3 Find an expression equivalent to $\sqrt{3} \sin x + \cos x$ that involves only the sine function.

In this expression, referring to the theorem above, we see that $c = \sqrt{3}$, $d = 1$, and $a = 1$. From this we find A and b:

$$A = \sqrt{c^2 + d^2} = \sqrt{3 + 1} = 2, \quad \cos b = \frac{\sqrt{3}}{2}, \quad \text{and} \quad \sin b = \frac{1}{2}.$$

Thus we can use $30°$ or $\pi/6$ for b. Thus

$$\sqrt{3} \sin x + \cos x \equiv 2 \sin \left(x + \frac{\pi}{6}\right).$$

Example 4 Find an expression equivalent to $\sqrt{3} \sin (\pi/4)t + \cos (\pi/4)t$ involving only the sine function.

In this case $c = \sqrt{3}$, $d = 1$, and $a = \pi/4$.

$$A = \sqrt{3 + 1} = 2, \quad \cos b = \frac{\sqrt{3}}{2}, \quad \text{and} \quad \sin b = \frac{1}{2}.$$

Thus we have

$$\sqrt{3} \sin \frac{\pi}{4}t + \cos \frac{\pi}{4}t \equiv -2 \sin \left(\frac{\pi}{4}t + \frac{\pi}{6}\right).$$

Find an expression involving only the sine function, equivalent to each of the following.

3. $\sin 2x + \cos 2x$

4. $12 \sin x - 5 \cos x$

5. Graph $y = \sqrt{3} \sin x + \cos x$.

Example 5 Find an expression equivalent to $3 \sin 2x - 4 \cos 2x$ involving only the sine function.

In this case $c = 3$, $d = -4$, and $a = 2$.

$$A = \sqrt{3^2 + (-4)^2} = 5, \qquad \cos b = \frac{3}{5}, \quad \text{and} \quad \sin b = -\frac{4}{5}.$$

Thus we have $3 \sin 2x - 4 \cos 2x \equiv 5 \sin (2x + b)$, where b is a number whose cosine is $\frac{3}{5}$ and whose sine is $-\frac{4}{5}$.

DO EXERCISES 3 AND 4.

[iii] Graphing

Example 6 Graph $y = \sin x - \cos x$.

We first transform the right-hand side to an expression involving only the sine function. Since $c = 1$ and $d = -1$, we have

$$A = \sqrt{1^2 + (-1)^2} = \sqrt{2}, \qquad \cos b = \frac{1}{\sqrt{2}}, \quad \text{and} \quad \sin b = -\frac{1}{\sqrt{2}}.$$

So $b = -\pi/4$ and we have $y = \sqrt{2} \sin (x - \pi/4)$. This can be graphed by translating the sine graph $\pi/4$ units to the right and stretching it vertically.

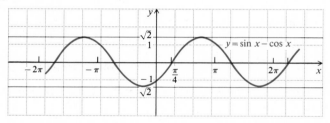

Skill in making graphs like this is important in such fields as engineering, electricity, and physics.

DO EXERCISE 5.

1 To 32.

EXERCISE SET 10.3

[i] In Exercises 1–32, prove the identities.

1. $\csc x - \cos x \cot x \equiv \sin x$

2. $\sec x - \sin x \tan x \equiv \cos x$

3. $\dfrac{1 + \cos \theta}{\sin \theta} + \dfrac{\sin \theta}{\cos \theta} \equiv \dfrac{\cos \theta + 1}{\sin \theta \cos \theta}$

4. $\dfrac{1}{\sin \theta \cos \theta} - \dfrac{\cos \theta}{\sin \theta} \equiv \dfrac{\sin \theta \cos \theta}{1 - \sin^2 \theta}$

5. $\dfrac{1 - \sin x}{\cos x} \equiv \dfrac{\cos x}{1 + \sin x}$

6. $\dfrac{1 - \cos x}{\sin x} \equiv \dfrac{\sin x}{1 + \cos x}$

7. $\dfrac{1 + \tan \theta}{1 + \cot \theta} \equiv \dfrac{\sec \theta}{\csc \theta}$

8. $\dfrac{\cot \theta - 1}{1 - \tan \theta} \equiv \dfrac{\csc \theta}{\sec \theta}$

9. $\dfrac{\sin x + \cos x}{\sec x + \csc x} \equiv \dfrac{\sin x}{\sec x}$

10. $\dfrac{\sin x - \cos x}{\sec x - \csc x} \equiv \dfrac{\cos x}{\csc x}$

11. $\dfrac{1 + \tan \theta}{1 - \tan \theta} + \dfrac{1 + \cot \theta}{1 - \cot \theta} \equiv 0$

12. $\dfrac{\cos^2 \theta + \cot \theta}{\cos^2 \theta - \cot \theta} \equiv \dfrac{\cos^2 \theta \tan \theta + 1}{\cos^2 \theta \tan \theta - 1}$

13. $\dfrac{1 + \cos 2\theta}{\sin 2\theta} \equiv \cot \theta$

14. $\dfrac{2 \tan \theta}{1 + \tan^2 \theta} \equiv \sin 2\theta$

15. $\sec 2\theta \equiv \dfrac{\sec^2 \theta}{2 - \sec^2 \theta}$

16. $\cot 2\theta \equiv \dfrac{\cot^2 \theta - 1}{2 \cot \theta}$

17. $\dfrac{\sin (\alpha + \beta)}{\cos \alpha \cos \beta} \equiv \tan \alpha + \tan \beta$

18. $\dfrac{\cos (\alpha - \beta)}{\cos \alpha \sin \beta} \equiv \tan \alpha + \cot \beta$

19. $1 - \cos 5\theta \cos 3\theta - \sin 5\theta \sin 3\theta \equiv 2 \sin^2 \theta$

20. $2 \sin \theta \cos^3 \theta + 2 \sin^3 \theta \cos \theta \equiv \sin 2\theta$

21. $\dfrac{\tan \theta + \sin \theta}{2 \tan \theta} \equiv \cos^2 \dfrac{\theta}{2}$

22. $\dfrac{\tan \theta - \sin \theta}{2 \tan \theta} \equiv \sin^2 \dfrac{\theta}{2}$

23. $\cos^4 x - \sin^4 x \equiv \cos 2x$

24. $\dfrac{\cos^4 x - \sin^4 x}{1 - \tan^4 x} \equiv \cos^4 x$

25. $\dfrac{\tan 3\theta - \tan \theta}{1 + \tan 3\theta \tan \theta} \equiv \dfrac{2 \tan \theta}{1 - \tan^2 \theta}$

26. $\left(\dfrac{1 + \tan \theta}{1 - \tan \theta} \right)^2 \equiv \dfrac{1 + \sin 2\theta}{1 - \sin 2\theta}$

27. $\dfrac{\cos^3 x - \sin^3 x}{\cos x - \sin x} \equiv \dfrac{2 + \sin 2x}{2}$

28. $\dfrac{\sin^3 t + \cos^3 t}{\sin t + \cos t} \equiv \dfrac{2 - \sin 2t}{2}$

29. $\sin (\alpha + \beta) \sin (\alpha - \beta) \equiv \sin^2 \alpha - \sin^2 \beta$

30. $\cos (\alpha + \beta) \cos (\alpha - \beta) \equiv \cos^2 \alpha - \sin^2 \beta$

31. $\cos (\alpha + \beta) + \cos (\alpha - \beta) \equiv 2 \cos \alpha \cos \beta$

32. $\sin (\alpha + \beta) + \sin (\alpha - \beta) \equiv 2 \sin \alpha \cos \beta$

[ii] In Exercises 33–38, find an equivalent expression involving the sine function only.

33. $\sin 2x + \sqrt{3} \cos 2x$

34. $\sqrt{3} \sin 3x - \cos 3x$

35. $4 \sin x + 3 \cos x$

36. $4 \sin 2x - 3 \cos 2x$

37. $6.75 \sin 0.374x + 4.08 \cos 0.374x$

38. $97.81 \sin 0.8081x - 4.89 \cos 0.8081x$

[iii]

39. Graph $y = \sin 2x - \cos 2x$.

40. Graph $y = \sin x + \sqrt{3} \cos x$.

☆ ──

41. Show that $\log (\cos x - \sin x) + \log (\cos x + \sin x) \equiv \log \cos 2x$.

42. The following equation occurs in the study of mechanics:

$$\sin \theta = \frac{I_1 \cos \phi}{\sqrt{(I_1 \cos \phi)^2 + (I_2 \sin \phi)^2}}.$$

It can happen that $I_1 = I_2$. Assuming that this happens, simplify the equation.

43. In the theory of alternating current, the following equation occurs:

$$R = \frac{1}{\omega C(\tan \theta + \tan \phi)}.$$

Show that this equation is equivalent to $R = \dfrac{\cos \theta \cos \phi}{\omega C \sin (\theta + \phi)}$.

44. Show that

$$\begin{vmatrix} \cos x & \sin x \\ -\sin x & \cos x \end{vmatrix} \equiv \begin{vmatrix} \cos x & -\sin x \\ \sin x & \cos x \end{vmatrix} \equiv 1.$$

45. Show that

$$\begin{bmatrix} \cos x & \sin x \\ -\sin x & \cos x \end{bmatrix} \begin{bmatrix} \cos y & \sin y \\ -\sin y & \cos y \end{bmatrix} \equiv \begin{bmatrix} \cos (x + y) & \sin (x + y) \\ -\sin (x + y) & \cos (x + y) \end{bmatrix}.$$

★ ──

46. In electrical theory the following equations occur:

$$E_1 = \sqrt{2}\, E_t \cos \left(\theta + \frac{\pi}{P} \right), \qquad E_2 = \sqrt{2}\, E_t \cos \left(\theta - \frac{\pi}{P} \right).$$

Assuming that these equations hold, show that

$$\frac{E_1 + E_2}{2} = \sqrt{2}\, E_t \cos \theta \cos \frac{\pi}{P} \quad \text{and} \quad \frac{E_1 - E_2}{2} = -\sqrt{2}\, E_t \sin \theta \sin \frac{\pi}{P}.$$

OBJECTIVES

You should be able to:

[i] Given a number a, find all values of arcsin a, arccos a, and arctan a, in degrees; and in radians, if a is a number for which use of the table is not required.

[ii] Find principal values of the inverses of the trigonometric functions.

1. Sketch a graph of $y = \cos^{-1} x$. Is this relation a function?

2. Sketch a graph of $y = \cot^{-1} x$. Is this relation a function?

10.4 INVERSES OF THE TRIGONOMETRIC FUNCTIONS

Recall that to obtain the inverse of any relation we interchange the first and second members of each ordered pair in the relation. If a relation is defined by an equation, say in x and y, interchanging x and y produces an equation of the inverse relation. The graphs of a relation and its inverse are reflections of each other across the line $y = x$. The x- and y-axes are interchanged in such a reflection. Let us consider the inverse of the sine function, $y = \sin x$. The inverse may be denoted several ways, as follows:

$$x = \sin y, \qquad y = \sin^{-1} x, \qquad y = \arcsin x.$$

Thus $\sin^{-1} x$ is a number whose sine is x.

The notation arcsin x arises because it is the length of an arc on the unit circle for which the sine is x. The notation $\sin^{-1} x$ is not exponential notation. It does *not* mean $1/\sin x$! Either of the latter two kinds of notation above can be read "the inverse sine of x" or "the arc sine of x" or "the number (or angle) whose sine is x." Notation is chosen similarly for the inverses of the other trigonometric functions: $\cos^{-1} x$ or arccos x, $\tan^{-1} x$ or arctan x, and so on.

Graphs of $y = \sin^{-1} x$ and $y = \tan^{-1} x$ are shown below. Note that these relations are not functions.

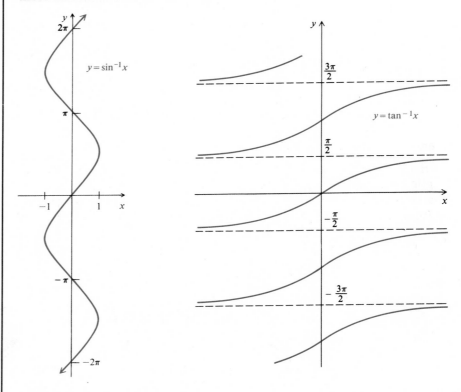

DO EXERCISES 1 AND 2.

[i] Finding Inverse Values

We illustrate finding inverse values using both a graph and the unit circle. In practice the unit circle is easier to use.

Example 1 Find all values of arcsin $\frac{1}{2}$.

On the graph of $y = \text{arcsin } x$, we draw a vertical line at $x = \frac{1}{2}$ as shown. It intersects the graph at points whose y-value is arcsin $\frac{1}{2}$. Some of the numbers whose sine is $\frac{1}{2}$ are seen to be $\pi/6$, $5\pi/6$, $-7\pi/6$, and so on. From the graph we can see that $\pi/6$ plus any multiple of 2π is such a number. Also $5\pi/6$ plus any multiple of 2π is such a number. The complete set of values is given by $\pi/6 + 2k\pi$, k an integer, and $5\pi/6 + 2k\pi$, k an integer.

Use a unit circle to find all values of the following.

3. arccos $\dfrac{\sqrt{2}}{2}$

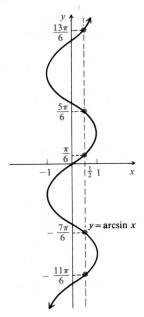

4. $\sin^{-1} \dfrac{\sqrt{3}}{2}$

Example 2 Find all values of arcsin $\frac{1}{2}$.

On the unit circle there are two points at which the sine is $\frac{1}{2}$. The arc length for the point in the first quadrant is $\pi/6$ plus any multiple of 2π. The arc length for the point in the second quadrant is $5\pi/6$ plus any multiple of 2π. Hence we obtain all values of arcsin $\frac{1}{2}$ as follows:

$$\frac{\pi}{6} + 2k\pi \quad \text{or} \quad \frac{5\pi}{6} + 2k\pi, \quad k \text{ an integer.}$$

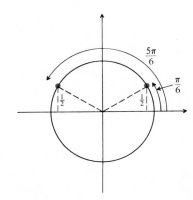

DO EXERCISES 3 AND 4.

5. Find, in degrees, all values of $\sin^{-1}(0.4226)$.

Example 3 Find all values of $\cos^{-1}(-0.9397)$ in degrees.

From Table 3 we find that the angle whose cosine is 0.9397 is 20°. This is the reference angle. We sketch this on a unit circle to find the two points where the cosine is −0.9397. The angles are 160° and 200°, plus any multiple of 360°. Thus the values of $\cos^{-1}(-0.9397)$ are

$$160° + k \cdot 360° \quad \text{or} \quad 200° + k \cdot 360°,$$

where k is any integer.

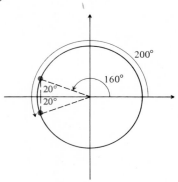

Example 4 Find all values of arctan 1 (see the figure below).

We find the two points on the unit circle at which the tangent is 1. These points are opposite ends of a diameter. Hence the arc lengths differ by π. Thus we have for all values of arctan 1,

$$\frac{\pi}{4} + k\pi, \qquad k \text{ an integer.}$$

6. Find all values of arctan (-1).

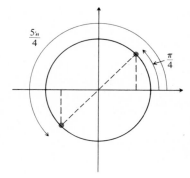

DO EXERCISES 5 AND 6.

[ii] Principal Values

The inverses of the trigonometric functions are not themselves functions. However, if we restrict the ranges of these relations, we can obtain functions. These graphs show how this restriction is made.

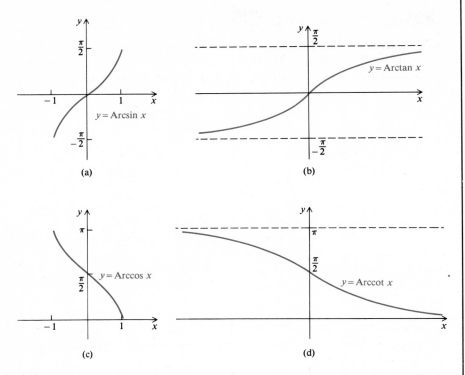

These relations with their ranges so restricted are functions, and the values in these restricted ranges are called *principal values*. To denote principal values we capitalize, as follows:

$$\text{Arcsin } x, \qquad \text{Sin}^{-1} x, \qquad \text{Arccos } x, \qquad \text{Cos}^{-1} x,$$

and so on. Thus whereas arcsin $(\frac{1}{2})$ represents an infinite set of numbers, Arcsin $(\frac{1}{2})$ represents the single number $\pi/6$.

Note that for the function $y = \text{Arcsin } x$ the range is the interval $[-\pi/2, \pi/2]$. For the function $y = \text{Arctan } x$ the range is $(-\pi/2, \pi/2)$. For the function $y = \text{Arccos } x$ the range is $[0, \pi]$, and for $y = \text{Arccot } x$ the range is $(0, \pi)$. These diagrams show where principal values are found on the unit circle. The restricted ranges should be memorized.

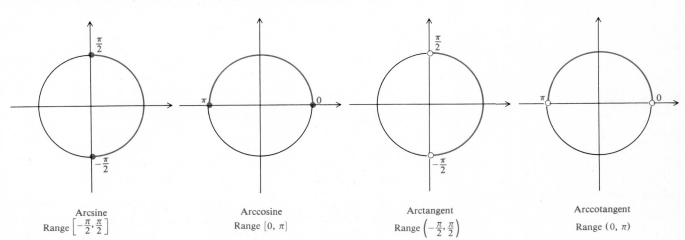

Find the following.

7. $\text{Arcsin } \dfrac{\sqrt{3}}{2}$

8. $\text{Cos}^{-1}\left(-\dfrac{\sqrt{2}}{2}\right)$

9. $\text{Arccot } (-1)$

10. $\text{Tan}^{-1} (-1)$

Example 5 Find $\text{Arcsin } \sqrt{2}/2$ and $\text{Cos}^{-1}\left(-\frac{1}{2}\right)$.

In the restricted range as shown in the figure, the only number whose sine is $\sqrt{2}/2$ is $\pi/4$. Hence $\text{Arcsin } \sqrt{2}/2 = \pi/4$. The only number whose cosine is $-\frac{1}{2}$ in the restricted range is $2\pi/3$. Hence $\text{Cos}^{-1}\left(-\frac{1}{2}\right) = 2\pi/3$.

DO EXERCISES 7–10.

▦ The Use of Calculators

Scientific calculators can be used to find inverses of trigonometric function values. The values obtained from a calculator are *principal* values, and the typical calculator provides values for only the Arcsin, Arccos, and Arctan functions.

Some calculators give inverse function values in either radians or degrees. Some give values only in degrees. The key strokes involved in finding inverse function values vary with the calculator. Be sure to read the instructions for the calculator you are using.

EXERCISE SET 10.4

[i] Find all values of the following without using tables.

1. $\arcsin \dfrac{\sqrt{2}}{2}$
2. $\arcsin \dfrac{\sqrt{3}}{2}$
3. $\cos^{-1} \dfrac{\sqrt{2}}{2}$
4. $\cos^{-1} \dfrac{\sqrt{3}}{2}$

5. $\sin^{-1}\left(-\dfrac{\sqrt{2}}{2}\right)$
6. $\sin^{-1}\left(-\dfrac{\sqrt{3}}{2}\right)$
7. $\arccos\left(-\dfrac{\sqrt{2}}{2}\right)$
8. $\arccos\left(-\dfrac{\sqrt{3}}{2}\right)$

9. $\arctan \sqrt{3}$
10. $\arctan \dfrac{\sqrt{3}}{3}$
11. $\cot^{-1} 1$
12. $\cot^{-1} \sqrt{3}$

13. $\arctan\left(-\dfrac{\sqrt{3}}{3}\right)$
14. $\arctan (-\sqrt{3})$
15. $\text{arccot } (-1)$
16. $\text{arccot } (-\sqrt{3})$

17. $\text{arcsec } 1$
18. $\text{arcsec } 2$
19. $\csc^{-1} 1$
20. $\csc^{-1} 2$

Use tables to find, in degrees, all values of the following.

21. $\arcsin 0.3907$
22. $\arcsin 0.9613$
23. $\sin^{-1} 0.6293$
24. $\sin^{-1} 0.8746$

25. $\arccos 0.7990$
26. $\arccos 0.9265$
27. $\cos^{-1} 0.9310$
28. $\cos^{-1} 0.2735$

29. $\tan^{-1} 0.3673$
30. $\tan^{-1} 1.091$
31. $\cot^{-1} 1.265$
32. $\cot^{-1} 0.4770$

33. $\sec^{-1} 1.167$
34. $\sec^{-1} 1.440$
35. $\text{arccsc } 6.277$
36. $\text{arccsc } 1.111$

[ii] Find the following without using tables.

37. $\text{Arcsin } \dfrac{\sqrt{2}}{2}$
38. $\text{Arcsin } \dfrac{1}{2}$
39. $\text{Cos}^{-1} \dfrac{1}{2}$
40. $\text{Cos}^{-1} \dfrac{\sqrt{2}}{2}$

41. $\text{Sin}^{-1}\left(-\dfrac{\sqrt{3}}{2}\right)$
42. $\text{Sin}^{-1}\left(-\dfrac{1}{2}\right)$
43. $\text{Arccos}\left(-\dfrac{\sqrt{2}}{2}\right)$
44. $\text{Arccos}\left(-\dfrac{\sqrt{3}}{2}\right)$

45. $\text{Tan}^{-1}\left(-\dfrac{\sqrt{3}}{3}\right)$
46. $\text{Tan}^{-1} (-\sqrt{3})$
47. $\text{Arccot}\left(-\dfrac{\sqrt{3}}{3}\right)$
48. $\text{Arccot} (-\sqrt{3})$

Find the following, in degrees, using tables.

49. $\text{Arcsin } 0.2334$
50. $\text{Arcsin } 0.4514$
51. $\text{Sin}^{-1} (-0.6361)$
52. $\text{Sin}^{-1} (-0.8192)$

53. $\text{Arccos } (-0.8897)$
54. $\text{Arccos } (-0.2924)$
55. $\text{Tan}^{-1} (-0.4074)$
56. $\text{Tan}^{-1} (-0.2401)$

57. $\text{Cot}^{-1} (-5.396)$
58. $\text{Cot}^{-1} (-1.319)$

10.5 COMBINATIONS OF TRIGONOMETRIC FUNCTIONS AND THEIR INVERSES

[i] Immediate Simplification

In practice, various combinations of trigonometric functions and their inverses arise. We shall discuss evaluating and simplifying such expressions. Let us consider, for example,

sin Arcsin x   (the sine of a number whose sine is x).

Recall, from Chapter 3, that if a function f has an inverse that is also a function, then $f(f^{-1}(x)) = x$ and also $f^{-1}(f(x)) = x$ for all x in the domains of the functions. Since the sine and arcsine relations are inverses of each other, we might suspect that

$$\text{sin Arcsin } x = x$$

for all x in the domain of the Arcsin relation. However, we cannot apply the result of Chapter 3 here because the inverse of the sine function is not a function. Let us study sin Arcsin x using a unit circle. Consider a number a in the domain of the Arcsine function (a number from −1 to 1). Then Arcsin a is the length of an arc s as shown. Its sine is a. In other words, s = Arcsin a.

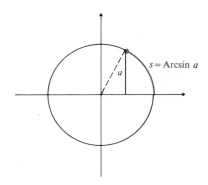

Recalling that s = Arcsin a, and taking the sine on both sides, we have

$$\text{sin } s = \text{sin Arcsin } a.$$

From the drawing we see that sin s = a. Substituting, we have

$$\text{sin Arcsin } a = a.$$

Thus it is true that sin Arcsin x $\equiv$ x. The sensible replacements are those numbers in the domain of the Arcsine function, that is, numbers from −1 to 1.

Similar identities hold for the other trigonometric functions:

---

sin Arcsin $x \equiv x$,	**cot Arccot** $x \equiv x$,
cos Arccos $x \equiv x$,	**sec Arcsec** $x \equiv x$,
tan Arctan $x \equiv x$,	**csc Arccsc** $x \equiv x$.

Find the following.

1. Arcsin sin $\dfrac{3\pi}{2}$

2. Sin^{-1} sin $\dfrac{\pi}{6}$

3. Arctan tan $\dfrac{3\pi}{4}$

4. Cos^{-1} cos $\dfrac{5\pi}{4}$

5. Arccos cos $\dfrac{5\pi}{6}$

6. sin Arctan 1

7. cos Arcsin $\dfrac{1}{2}$

8. Cos^{-1} sin $\dfrac{\pi}{6}$

9. Sin^{-1} tan $\dfrac{\pi}{4}$

Now let us consider Arcsin sin x. We might also suspect that this is equal to x for any x, but this is not true unless x is in the range of the Arcsine function.

Example 1 Find Arcsin sin $3\pi/4$.

We first find sin $3\pi/4$. It is $\sqrt{2}/2$. Next we find Arcsin $\sqrt{2}/2$. It is $\pi/4$. So Arcsin sin $3\pi/4 = \pi/4$.

DO EXERCISES 1–5.

For any x in the range of the Arcsine function, we do have Arcsin sin x = x. Similar conditions hold for the other functions.

THEOREM 3

> The following are true if x is in the range of the inverse function. Otherwise they are not.
>
> $$\text{Arcsin sin } x = x, \qquad \text{Arccot cot } x = x,$$
> $$\text{Arccos cos } x = x, \qquad \text{Arcsec sec } x = x,$$
> $$\text{Arctan tan } x = x, \qquad \text{Arccsc csc } x = x.$$

[ii] Simplifying Combinations

Now we consider some other kinds of combinations.

Example 2 Find sin Arctan (-1).

We first find Arctan (-1). It is $-\pi/4$. Now we find the sine of this, sin $(-\pi/4) = -\sqrt{2}/2$, so sin Arctan $(-1) = -\sqrt{2}/2$.

Example 3 Find Cos^{-1} sin $\pi/2$.

We first find sin $\pi/2$. It is 1. Now we find Cos^{-1} 1. It is 0, so Cos^{-1} sin $\pi/2 = 0$.

DO EXERCISES 6–9.

Now let us consider

$$\cos \text{Arcsin } \frac{3}{5}.$$

Without using tables, we cannot find Arcsin $\frac{3}{5}$. However, we can still evaluate the entire expression without using tables. In a case like this we sketch a triangle as shown. The angle θ in this triangle is an angle whose sine is $\frac{3}{5}$ (it is Arcsin $\frac{3}{5}$). We wish to find the cosine of this angle. Since the triangle is a right triangle, we can find the length of the base, b. It is 4. Thus we know that $\cos \theta = b/5$ or $\frac{4}{5}$. Therefore

$$\cos \text{Arcsin } \frac{3}{5} = \frac{4}{5}.$$

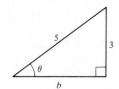

Example 4 Find sin Arccot $x/2$.

We draw a right triangle whose sides have lengths x and 2, so that $\cot \theta = x/2$. We find the length of the hypotenuse and then read off the sine ratio.

$$\sin \text{Arccot } \frac{x}{2} = \frac{2}{\sqrt{x^2 + 2^2}}$$

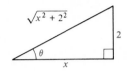

Example 5 Find cos Arctan p.

We draw a right triangle, two of whose sides have lengths p and 1, so that $\tan \theta = p/1$. We find the length of the other side and then read off the cosine ratio.

$$\cos \text{Arctan } p = \frac{1}{\sqrt{1 + p^2}}$$

The idea in Examples 4 and 5 is to sketch a triangle, two of whose sides have the appropriate ratio. Then use the Pythagorean theorem to find the length of the third side and read off the desired ratio.

DO EXERCISES 10–12.

In some cases the use of certain identities is needed to evaluate expressions.

Example 6 Evaluate $\sin (\text{Sin}^{-1} \frac{1}{2} + \text{Cos}^{-1} \frac{4}{5})$.

To simplify the use of an identity we shall make substitutions. Let

$$u = \text{Sin}^{-1} \frac{1}{2} \quad \text{and} \quad v = \text{Cos}^{-1} \frac{4}{5}.$$

Then we have $\sin (u + v)$. By a sum formula, this is equivalent to

$$\sin u \cos v + \cos u \sin v.$$

Now we make the reverse substitutions and obtain

$$\sin \text{Sin}^{-1} \frac{1}{2} \cdot \cos \text{Cos}^{-1} \frac{4}{5} + \cos \text{Sin}^{-1} \frac{1}{2} \cdot \sin \text{Cos}^{-1} \frac{4}{5}.$$

This immediately simplifies to

$$\frac{1}{2} \cdot \frac{4}{5} + \cos \text{Sin}^{-1} \frac{1}{2} \cdot \sin \text{Cos}^{-1} \frac{4}{5}.$$

Find the following.

10. cos Arctan $\dfrac{b}{3}$

11. sin Arctan 2

$$\left(Hint: 2 = \frac{2}{1}. \right)$$

12. tan Arcsin t

Evaluate.

13. $\cos \left(\text{Sin}^{-1} \dfrac{\sqrt{3}}{2} - \text{Cos}^{-1} \dfrac{1}{2} \right)$

14. $\tan \left(\dfrac{1}{2} \text{Arcsin} \dfrac{3}{5} \right)$

(*Hint:* Let Arcsin $\frac{3}{5} = u$ and use a half-angle formula.)

Now $\cos \text{Sin}^{-1} \frac{1}{2}$ readily simplifies to $\sqrt{3}/2$. To find $\sin \text{Cos}^{-1} \frac{4}{5}$, we shall need a triangle.

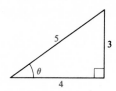

We find that $\sin \text{Cos}^{-1} \frac{4}{5} = \frac{3}{5}$. Our expression is now simplified to

$$\frac{1}{2} \cdot \frac{4}{5} + \frac{\sqrt{3}}{2} \cdot \frac{3}{5},$$

which simplifies to

$$\frac{4 + 3\sqrt{3}}{10}.$$

DO EXERCISES 13 AND 14.

EXERCISE SET 10.5

[i] Evaluate or simplify.

1. $\sin \text{Arcsin } 0.3$

2. $\cos \text{Arccos } 0.2$

3. $\tan \text{Tan}^{-1} (-4.2)$

4. $\cot \text{Cot}^{-1} (-1.5)$

5. $\text{Arcsin} \sin \dfrac{2\pi}{3}$

6. $\text{Arccos} \cos \dfrac{3\pi}{2}$

7. $\text{Sin}^{-1} \sin \left(-\dfrac{3\pi}{4} \right)$

8. $\text{Cos}^{-1} \cos \left(-\dfrac{\pi}{4} \right)$

9. $\text{Sin}^{-1} \sin \dfrac{\pi}{5}$

10. $\text{Cos}^{-1} \cos \dfrac{\pi}{7}$

11. $\text{Tan}^{-1} \tan \dfrac{2\pi}{3}$

12. $\text{Cot}^{-1} \cot \dfrac{2\pi}{3}$

[ii]

13. $\sin \text{Arctan } \sqrt{3}$

14. $\sin \text{Arctan } \dfrac{\sqrt{3}}{3}$

15. $\cos \text{Arcsin } \dfrac{\sqrt{3}}{2}$

16. $\cos \text{Arcsin } \dfrac{\sqrt{2}}{2}$

17. $\tan \text{Cos}^{-1} \dfrac{\sqrt{2}}{2}$

18. $\tan \text{Cos}^{-1} \dfrac{\sqrt{3}}{2}$

19. $\text{Cos}^{-1} \sin \dfrac{\pi}{3}$

20. $\text{Cos}^{-1} \sin \pi$

21. $\text{Arcsin} \cos \dfrac{\pi}{6}$

22. $\text{Arcsin} \cos \dfrac{\pi}{4}$

23. $\text{Sin}^{-1} \tan \dfrac{\pi}{4}$

24. $\text{Sin}^{-1} \tan \left(-\dfrac{\pi}{4} \right)$

25. $\sin \text{Arctan } \dfrac{x}{2}$

26. $\sin \text{Arctan } \dfrac{a}{3}$

27. $\tan \text{Cos}^{-1} \dfrac{3}{x}$

28. $\cot \text{Sin}^{-1} \dfrac{5}{y}$

29. $\cot \text{Sin}^{-1} \dfrac{a}{b}$

30. $\tan \text{Cos}^{-1} \dfrac{p}{q}$

31. $\cos \text{Tan}^{-1} \dfrac{\sqrt{2}}{3}$

32. $\cos \text{Tan}^{-1} \dfrac{\sqrt{3}}{4}$

33. $\tan \text{Arcsin } 0.1$

34. $\tan \text{Arcsin } 0.2$

35. $\cot \text{Cos}^{-1} (-0.2)$

36. $\cot \text{Cos}^{-1} (-0.3)$

37. $\sin \text{Arccot } y$

38. $\sin \text{Arccot } x$

39. $\cos \text{Arctan } t$

40. $\sin \text{Arctan } t$

41. $\cot \text{Sin}^{-1} y$

42. $\tan \text{Cos}^{-1} y$

43. $\sin \text{Cos}^{-1} x$

44. $\cos \text{Sin}^{-1} x$

45. $\tan \left(\dfrac{1}{2} \text{Arcsin} \dfrac{4}{5} \right)$

46. $\tan \left(\dfrac{1}{2} \text{Arcsin} \dfrac{1}{2} \right)$

47. $\cos \left(\dfrac{1}{2} \text{Arcsin} \dfrac{1}{2} \right)$

48. $\cos \left(\dfrac{1}{2} \text{Arcsin} \dfrac{\sqrt{3}}{2} \right)$

49. $\sin \left(2 \text{Cos}^{-1} \dfrac{3}{5} \right)$

50. $\sin \left(2 \text{Cos}^{-1} \dfrac{1}{2} \right)$

51. $\cos \left(2 \text{Sin}^{-1} \dfrac{5}{13} \right)$

52. $\cos \left(2 \text{Cos}^{-1} \dfrac{4}{5} \right)$

53. $\sin \left(\text{Sin}^{-1} \dfrac{1}{2} + \text{Cos}^{-1} \dfrac{3}{5} \right)$

54. $\sin \left(\text{Sin}^{-1} \dfrac{1}{2} - \text{Cos}^{-1} \dfrac{4}{5} \right)$

55. $\cos \left(\text{Sin}^{-1} \dfrac{\sqrt{2}}{2} + \text{Cos}^{-1} \dfrac{3}{5} \right)$

56. $\cos \left(\text{Sin}^{-1} \dfrac{4}{5} - \text{Cos}^{-1} \dfrac{1}{2} \right)$

57. $\sin (\text{Sin}^{-1} x + \text{Cos}^{-1} y)$

58. $\sin (\text{Sin}^{-1} x - \text{Cos}^{-1} y)$

59. $\cos (\text{Sin}^{-1} x + \text{Cos}^{-1} y)$

60. $\cos (\text{Sin}^{-1} x - \text{Cos}^{-1} y)$

61. ▦ $\sin (\text{Sin}^{-1} 0.6032 + \text{Cos}^{-1} 0.4621)$

62. ▦ $\cos (\text{Sin}^{-1} 0.7325 - \text{Cos}^{-1} 0.4838)$

☆

63. An observer's eye is at point A, looking at a mural of height h, with the bottom of the mural y feet above the eye. The eye is x feet from the wall. Write an expression for θ in terms of x, y, and h.

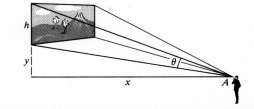

64. ▦ Evaluate the expression given in Exercise 63 when $x = 20$ ft, $y = 7$ ft, and $h = 25$ ft.

10.6 TRIGONOMETRIC EQUATIONS

[i] Simple Equations

When an equation contains a trigonometric expression with a variable such as sin x, it is called a *trigonometric equation*. To solve such an equation, we find all replacements for the variable that make the equation true.

Example 1 Solve $2 \sin x = 1$.

We first solve for sin x:

$$\sin x = \frac{1}{2}.$$

Now we note that the solutions are those numbers having a sine of $\frac{1}{2}$. We look for them. The unit circle is helpful. There are just two points on it for which the sine is $\frac{1}{2}$, as shown.

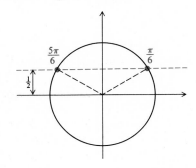

1. Solve. Give answers in both degrees and radians.

$$2 \cos x = 1$$

They are the points for $\pi/6$ and $5\pi/6$. These numbers, plus any multiple of 2π, are the solutions

$$\frac{\pi}{6} + 2k\pi \quad \text{or} \quad \frac{5\pi}{6} + 2k\pi,$$

where k is any integer. In degrees, the solutions are

$$30° + k \cdot 360° \quad \text{or} \quad 150° + k \cdot 360°,$$

where k is any integer.

DO EXERCISE 1.

Example 2 Solve $4 \cos^2 x = 1$.

We first solve for $\cos x$:

$$\cos^2 x = \frac{1}{4}$$

$$|\cos x| = \frac{1}{2} \qquad \text{Taking the principal square root}$$

$$\cos x = \pm \frac{1}{2}.$$

Now we use the unit circle to find those numbers having a cosine of $\frac{1}{2}$ or $-\frac{1}{2}$. The solutions are $\pi/3$, $2\pi/3$, $4\pi/3$, $5\pi/3$, plus any multiple of 2π.

2. Find all solutions in $[0, 2\pi)$. Give answers in both degrees and radians.

$$4 \sin^2 x = 1$$

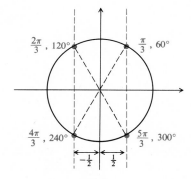

DO EXERCISE 2.

In solving trigonometric equations, it is usually sufficient to find just the solutions from 0 to 2π. We then remember that any multiple of 2π may be added to obtain the rest of the solutions.

The following example illustrates that when we look for solutions in the interval $[0, 2\pi)$ we must be cautious.

Example 3 Solve $2 \sin 2x = 1$ in the interval $[0, 2\pi)$.

We first solve for $\sin 2x$: $\sin 2x = \frac{1}{2}$.

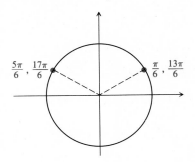

Using the unit circle we find points for which $\sin 2x = \frac{1}{2}$. However, since x is to run from 0 to 2π, we must let 2x run from 0 to 4π. These values of 2x are $\pi/6$, $5\pi/6$, $13\pi/6$, and $17\pi/6$. Thus the desired values of x in $[0, 2\pi)$ are half of these. Therefore,

$$x = \frac{\pi}{12}, \quad \frac{5\pi}{12}, \quad \frac{13\pi}{12}, \quad \frac{17\pi}{12}.$$

DO EXERCISE 3.

[ii] Using Tables

In solving some trigonometric equations, it is necessary to use tables. In such cases, we usually find answers in degrees.

Example 4 Solve $2 + \sin x = 2.5299$ in $[0, 360°)$.

We first solve for $\sin x$:

$$\sin x = 0.5299.$$

From the table we find the reference angle, $x = 32°$. Since $\sin x$ is positive, the solutions are to be found in the first and second quadrants. The solutions are

$$32° \quad \text{and} \quad 148°.$$

Example 5 Solve $\sin x - 1 = -1.5299$ in $[0, 360°)$.

We first solve for $\sin x$:

$$\sin x = -0.5299.$$

From the table we find the reference angle, $32°$. Since $\sin x$ is negative, the solutions are to be found in the third and fourth quadrants. The solutions are

$$212° \quad \text{and} \quad 328°.$$

DO EXERCISES 4 AND 5.

3. Find all solutions in $[0, 2\pi)$. Leave answers in terms of π.

$$2 \cos 2x = 1$$

Solve. Give answers in degrees.

4. $2 + \cos x = 2.7660$

5. $\cos x - 1 = -1.7660$

Solve.

6. $8\cos^2\theta + 2\cos\theta = 1$

7. $2\cos^2\phi + \cos\phi = 0$

[iii] **Using Algebraic Techniques**

In solving trigonometric equations, we can expect to apply some algebra before concerning ourselves with the trigonometric part. In the next examples, we begin by factoring. The equations are reducible to quadratic.

Example 6 Solve $8\cos^2\theta - 2\cos\theta = 1$.

Since we will be using the principle of zero products, we first obtain a 0 on one side of the equation.

$$8\cos^2\theta - 2\cos\theta - 1 = 0$$
$$(4\cos\theta + 1)(2\cos\theta - 1) = 0 \qquad \text{Factoring}$$
$$4\cos\theta + 1 = 0 \quad \text{or} \quad 2\cos\theta - 1 = 0 \qquad \text{Principle of zero products}$$

$$\cos\theta = -\frac{1}{4} = -0.25 \quad \text{or} \quad \cos\theta = \frac{1}{2}$$

$$\theta = 104°30', \ 255°30' \quad \text{or} \quad \theta = 60°, \ 300° \quad \left[\text{or } \frac{\pi}{3}, \frac{5\pi}{3}\right]$$

The solutions in $[0, 360°)$ are $104°30'$, $255°30'$, $60°$, and $300°$.

It may be helpful in accomplishing the algebraic part of solving trigonometric equations to make substitutions as in Chapter 2. If we use such an approach, the algebraic part of Example 6 would look like this:

$$8\cos^2\theta - 2\cos\theta = 1.$$

Let $u = \cos\theta$.

$$8u^2 - 2u = 1$$
$$(4u + 1)(2u - 1) = 0$$
$$4u + 1 = 0 \quad \text{or} \quad 2u - 1 = 0$$
$$u = -\frac{1}{4} \quad \text{or} \quad u = \frac{1}{2}$$

Example 7 Solve $2\sin^2\phi + \sin\phi = 0$.

$$\sin\phi(2\sin\phi + 1) = 0 \qquad \text{Factoring}$$
$$\sin\phi = 0 \quad \text{or} \quad 2\sin\phi + 1 = 0 \qquad \text{Principle of zero products}$$

$$\sin\phi = 0 \quad \text{or} \quad \sin\phi = -\frac{1}{2}$$

$$\phi = 0, \pi \quad \text{or} \quad \phi = \frac{7\pi}{6}, \frac{11\pi}{6}$$

The solutions in $[0, 2\pi)$ are 0, π, $7\pi/6$, and $11\pi/6$.

DO EXERCISES 6 AND 7.

In case a trigonometric equation is quadratic but difficult or impossible to factor, we use the quadratic formula.

Example 8 Solve $10 \sin^2 x - 12 \sin x - 7 = 0$ in $[0, 360°)$.

It may help to make the substitution $u = \sin x$, in order to obtain $10u^2 - 12u - 7 = 0$. Then use the quadratic formula to find u, or $\sin x$.

$$\sin x = \frac{12 \pm \sqrt{144 + 280}}{20} \qquad \text{Using the quadratic formula}$$

$$= \frac{12 \pm \sqrt{424}}{20} = \frac{12 \pm 2\sqrt{106}}{20} = \frac{6 \pm \sqrt{106}}{10}$$

$$= \frac{6 \pm 10.296}{10}$$

$$\sin x = 1.6296 \quad \text{or} \quad \sin x = -0.4296$$

Since sines are never greater than 1, the first of the equations on the preceding line has no solution. We look up the other number and find the reference angle to be $25°30'$ to the nearest $10'$. Thus the solutions are $205°30'$ and $334°30'$.

DO EXERCISE 8.

8. Solve using the quadratic formula.

$$10 \cos^2 x - 10 \cos x - 7 = 0$$

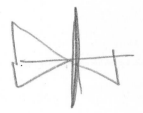

EXERCISE SET 10.6

[i], [ii] Solve, finding all solutions.

1. $\sin x = \dfrac{\sqrt{3}}{2}$ **2.** $\cos x = \dfrac{\sqrt{3}}{2}$ **3.** $\cos x = \dfrac{1}{\sqrt{2}}$

4. $\tan x = \sqrt{3}$ **5.** $\sin x = 0.3448$ **6.** $\cos x = 0.6406$

Solve, finding all solutions in $[0, 2\pi)$ or $[0°, 360°)$.

7. $\cos x = -0.5495$ **8.** $\sin x = -0.4279$

9. $2 \sin x + \sqrt{3} = 0$ **10.** $\sqrt{3} \tan x + 1 = 0$

11. $2 \tan x + 3 = 0$ **12.** $4 \sin x - 1 = 0$

[iii]

13. $4 \sin^2 x - 1 = 0$ **14.** $2 \cos^2 x = 1$

15. $\cot^2 x - 3 = 0$ **16.** $\csc^2 x - 4 = 0$

17. $2 \sin^2 x + \sin x = 1$ **18.** $2 \cos^2 x + 3 \cos x = -1$

19. $\cos^2 x + 2 \cos x = 3$ **20.** $2 \sin^2 x - \sin x = 3$

21. $4 \sin^3 x - \sin x = 0$ **22.** $2 \cos^2 x - \sqrt{3} \cos x = 0$

23. $2 \sin^2 \theta + 7 \sin \theta = 4$ **24.** $2 \sin^2 \theta - 5 \sin \theta + 2 = 0$

25. $6 \cos^2 \phi + 5 \cos \phi + 1 = 0$ **26.** $2 \sin^2 \phi + \sin \phi - 1 = 0$

27. $2 \sin t \cos t + 2 \sin t - \cos t - 1 = 0$ **28.** $2 \sin t \tan t + \tan t - 2 \sin t - 1 = 0$

29. $\cos 2x \sin x + \sin x = 0$ **30.** $\sin 2x \cos x - \cos x = 0$

☆ ───

Solve, restricting solutions to $[0, 2\pi)$ or $[0°, 360°)$ where sensible to do so.

31. $|\sin x| = \dfrac{\sqrt{3}}{2}$

32. $|\cos x| = \dfrac{1}{2}$

33. $\sqrt{\tan x} = \sqrt[4]{3}$

34. $12 \sin x - 7\sqrt{\sin x} + 1 = 0$

35. $16 \cos^4 x - 16 \cos^2 x + 3 = 0$

36. $\ln \cos x = 0$

37. $e^{\sin x} = 1$

38. $\sin \ln x = -1$

39. $e^{\ln \sin x} = 0$

40. Solve this system of equations for x and y:

$$x \cos \theta - y \sin \theta = 0,$$
$$x \sin \theta + y \cos \theta = 1.$$

41. ▦ Solve this system graphically:

$$y = 2 \sin \frac{x}{2},$$
$$y = \text{Tan}^{-1} x.$$

42. Solve graphically: $\sin x = \tan \dfrac{x}{2}$.

──

OBJECTIVE

You should be able to:

[i] Solve trigonometric equations
requiring the use of identities.

10.7 IDENTITIES IN SOLVING EQUATIONS

[i] When a trigonometric equation involves more than one function, we may use identities to put it in terms of a single function. This can usually be done in several ways. In the following three examples, we illustrate by solving the same equation three different ways.

Example 1 Solve the equation $\sin x + \cos x = 1$.

To express $\cos x$ in terms of $\sin x$, we use the identity

$$\sin^2 x + \cos^2 x \equiv 1.$$

From this we obtain $\cos x \equiv \pm \sqrt{1 - \sin^2 x}$. Now we substitute in the original equation:

$$\sin x \pm \sqrt{1 - \sin^2 x} = 1.$$

This is a radical equation. We get the radical alone on one side and then square both sides:

$$1 - \sin^2 x = 1 - 2 \sin x + \sin^2 x.$$

This simplifies to

$$-2 \sin^2 x + 2 \sin x = 0.$$

Now we factor:

$$-2 \sin x (\sin x - 1) = 0.$$

Using the principle of zero products, we have

$$-2 \sin x = 0 \quad \text{or} \quad \sin x - 1 = 0,$$
$$\sin x = 0 \quad \text{or} \quad \sin x = 1.$$

The values of x in $[0, 2\pi)$ satisfying these are

$$x = 0, \quad x = \pi, \quad \text{or} \quad x = \frac{\pi}{2}.$$

Now we check these in the original equation. We find that π does not check but the other values do. Thus the solutions are 0 and $\pi/2$.

> **CAUTION!** **It is important to check when solving. Values are often obtained that are not solutions of the original equation.**

DO EXERCISE 1.

Example 2 Solve $\sin x + \cos x = 1$.

This time we will square both sides, obtaining

$$\sin^2 x + 2 \sin x \cos x + \cos^2 x = 1.$$

Since $\sin^2 x + \cos^2 x \equiv 1$, we can simplify to

$$2 \sin x \cos x = 0.$$

Now we use the identity $2 \sin x \cos x \equiv \sin 2x$, and obtain

$$\sin 2x = 0.$$

The values in $[0, 2\pi)$ satisfying this equation are 0, $\pi/2$, π, and $3\pi/2$. Only 0 and $\pi/2$ check.

DO EXERCISE 2.

Example 3 Solve $\sin x + \cos x = 1$.

Let us recall the identity $c \sin ax + d \cos ax \equiv A \sin (ax + b)$, where $A = \sqrt{c^2 + d^2}$ and b is a number whose cosine is c/A and whose sine is d/A. In this case $c = d = a = 1$, so $A = \sqrt{2}$ and $b = \pi/4$. Thus our equation is equivalent to

$$\sqrt{2} \sin \left(x + \frac{\pi}{4} \right) = 1.$$

Then

$$\sin \left(x + \frac{\pi}{4} \right) = \frac{1}{\sqrt{2}} = \frac{\sqrt{2}}{2}$$

$$x + \frac{\pi}{4} = \frac{\pi}{4} \quad \text{or} \quad x + \frac{\pi}{4} = \frac{3\pi}{4}$$

$$x = 0 \quad \text{or} \quad x = \frac{\pi}{2}.$$

Both values check.

DO EXERCISE 3.

1. Solve $\sin x - \cos x = 1$ as in Example 1.

2. Solve $\sin x - \cos x = 1$, as in Example 2.

3. Solve $\sin x - \cos x = 1$, as in Example 3.

4. Solve $\tan^2 x \cos x - \cos x = 0$.

5. Solve $\sin 2x + \cos x = 0$.

Example 4 Solve $2\cos^2 x \tan x - \tan x = 0$.

First, we factor:

$$\tan x (2\cos^2 x - 1) = 0.$$

Now we use the identity $2\cos^2 x - 1 \equiv \cos 2x$:

$$\tan x \cos 2x = 0$$

$\tan x = 0$ or $\cos 2x = 0$ Principle of zero products

$x = 0, \pi$ or $2x = \dfrac{\pi}{2}, \dfrac{3\pi}{2}, \dfrac{5\pi}{2}, \dfrac{7\pi}{2}$

$x = 0, \pi$ or $x = \dfrac{\pi}{4}, \dfrac{3\pi}{4}, \dfrac{5\pi}{4}, \dfrac{7\pi}{4}.$

All values check. The solutions in $[0, 2\pi)$ are 0, π, $\pi/4$, $3\pi/4$, $5\pi/4$, and $7\pi/4$.

DO EXERCISE 4.

Example 5 Solve $\cos 2x + \sin x = 1$.

We first use the identity $\cos 2x \equiv 1 - 2\sin^2 x$, to get

$$1 - 2\sin^2 x + \sin x = 1$$
$$-2\sin^2 x + \sin x = 0$$
$$\sin x(1 - 2\sin x) = 0 \quad \text{Factoring}$$

$\sin x = 0$ or $1 - 2\sin x = 0$ Principle of zero products

$\sin x = 0$ or $\sin x = \dfrac{1}{2}$

$x = 0, \pi$ or $x = \dfrac{\pi}{6}, \dfrac{5\pi}{6}.$

All values check. The solutions in $[0, 2\pi)$ are 0, π, $\pi/6$, and $5\pi/6$.

DO EXERCISE 5.

Example 6 Solve $\sin 5\theta \cos 2\theta - \cos 5\theta \sin 2\theta = \sqrt{2}/2$.

We use the identity

$$\sin(\alpha - \beta) \equiv \sin\alpha\cos\beta - \cos\alpha\sin\beta$$

to get $\sin 3\theta = \sqrt{2}/2$. Then

$$3\theta = \dfrac{\pi}{4}, \dfrac{3\pi}{4}, \dfrac{9\pi}{4}, \dfrac{11\pi}{4}, \dfrac{17\pi}{4}, \dfrac{19\pi}{4}$$

and

$$\theta = \dfrac{\pi}{12}, \dfrac{\pi}{4}, \dfrac{3\pi}{4}, \dfrac{11\pi}{12}, \dfrac{17\pi}{12}, \dfrac{19\pi}{12}.$$

Let us check $\pi/12$:

$$\sin\dfrac{5\pi}{12}\cos\dfrac{\pi}{6} - \cos\dfrac{5\pi}{12}\sin\dfrac{\pi}{6}.$$

We again use the difference identity to obtain $\sin(5\pi/12 - \pi/6)$. This is equal to sin 3π/12 or sin π/4, and this is equal to $\sqrt{2}/2$. Hence π/12 checks. All of the above answers check.

DO EXERCISE 6.

Example 7 Solve $\tan^2 x + \sec x - 1 = 0$.

We use the identity $1 + \tan^2 x \equiv \sec^2 x$. Substituting, we get

$$\sec^2 x - 1 + \sec x - 1 = 0,$$

or

$$\sec^2 x + \sec x - 2 = 0$$
$$(\sec x + 2)(\sec x - 1) = 0 \quad \text{Factoring}$$
$$\sec x = -2 \quad \text{or} \quad \sec x = 1 \quad \text{Principle of zero products}$$
$$x = \frac{2}{3}\pi, \frac{4}{3}\pi \quad \text{or} \quad x = 0.$$

All of these values check. The solutions in $[0, 2\pi)$ are 0, 2π/3, and 4π/3.

Example 8 Solve $\sin x = \sin (2x - \pi)$.

We can use the identity $\sin (y - \pi) \equiv -\sin y$ or the identity $\sin (\alpha - \beta) \equiv \sin \alpha \cos \beta - \cos \alpha \sin \beta$ with the right-hand side. We get

$$\sin x = -\sin 2x.$$

Next we use the identity $\sin 2x \equiv 2 \sin x \cos x$, and obtain

$$\sin x = -2 \sin x \cos x, \quad \text{or} \quad 2 \sin x \cos x + \sin x = 0$$
$$\sin x (2 \cos x + 1) = 0 \quad \text{Factoring}$$

$$\sin x = 0 \quad \text{or} \quad \cos x = -\frac{1}{2} \quad \text{Principle of zero products}$$

$$x = 0, \pi \quad \text{or} \quad x = \frac{2\pi}{3}, \frac{4\pi}{3}.$$

All of these values check, so the solutions in $[0, 2\pi)$ are 0, π, 2π/3, and 4π/3.

DO EXERCISE 7.

6. Solve
$$\sin 3\theta \cos \theta - \cos 3\theta \sin \theta = \frac{1}{2}.$$

7. Solve $\cos x = \cos (\pi - 2x)$.

EXERCISE SET 10.7

[i] Find all solutions of the following equations in $[0, 2\pi)$.

1. $\tan x \sin x - \tan x = 0$

2. $2 \sin x \cos x + \sin x = 0$

3. $2 \sec x \tan x + 2 \sec x + \tan x + 1 = 0$

4. $2 \csc x \cos x - 4 \cos x - \csc x + 2 = 0$

5. $\sin 2x - \cos x = 0$

6. $\cos 2x - \sin x = 1$

7. $\sin 2x \sin x - \cos x = 0$

8. $\sin 2x \cos x - \sin x = 0$

9. $\sin 2x + 2 \sin x \cos x = 0$

10. $\cos 2x \sin x + \sin x = 0$

11. $\cos 2x \cos x + \sin 2x \sin x = 1$

12. $\sin 2x \sin x - \cos 2x \cos x = -\cos x$

13. $\sin 4x - 2 \sin 2x = 0$

14. $\sin 4x + 2 \sin 2x = 0$

15. $\sin 2x + 2\sin x - \cos x - 1 = 0$

16. $\sin 2x + \sin x + 2\cos x + 1 = 0$

17. $\sec^2 x = 4\tan^2 x$

18. $\sec^2 x - 2\tan^2 x = 0$

19. $\sec^2 x + 3\tan x - 11 = 0$

20. $\tan^2 x + 4 = 2\sec^2 x + \tan x$

21. $\cot x = \tan(2x - 3\pi)$

22. $\tan x = \cot(2x + \pi)$

23. $\cos(\pi - x) + \sin\left(x - \dfrac{\pi}{2}\right) = 1$

24. $\sin(\pi - x) + \cos\left(\dfrac{\pi}{2} - x\right) = 1$

25. $\dfrac{\cos^2 x - 1}{\sin\left(\dfrac{\pi}{2} - x\right) - 1} = \dfrac{\sqrt{2}}{2} + 1$

26. $\dfrac{\sin^2 x - 1}{\cos\left(\dfrac{\pi}{2} - x\right) + 1} = \dfrac{\sqrt{2}}{2} - 1$

27. $2\cos x + 2\sin x = \sqrt{6}$

28. $2\cos x + 2\sin x = \sqrt{2}$

29. $\sqrt{3}\cos x - \sin x = 1$

30. $\sqrt{2}\cos x - \sqrt{2}\sin x = 2$

☆ ──

Solve.

31. $\text{Arccos } x = \text{Arccos } \dfrac{3}{5} - \text{Arcsin } \dfrac{4}{5}$

32. $\text{Sin}^{-1} x = \text{Tan}^{-1}\dfrac{1}{3} + \text{Tan}^{-1}\dfrac{1}{2}$

33. Solve graphically: $\sin x - \cos x = \cot x$.

★ ──

34. Solve this system:

$$\sin x + \cos y = 1,$$
$$2\sin x = \sin 2y.$$

CHAPTER 10 REVIEW

[10.1, i]   Use the sum and difference formulas to write equivalent expressions. You need not simplify.

1. $\cos\left(x + \dfrac{3\pi}{2}\right)$

2. $\tan(45° - 30°)$

[10.1, ii]   **3.** Simplify $\cos 27° \cos 16° + \sin 27° \sin 16°$.

4. Given $\tan \alpha = \sqrt{3}$, $\sin \beta = \sqrt{2}/2$, and α and β are between 0 and $\pi/2$, evaluate $\tan(\alpha - \beta)$ exactly.

[10.2, ii]   **5.** Find $\sin \dfrac{\pi}{8}$.

[10.2, iii]   **6.** Simplify $\dfrac{\sin 2\theta}{\sin^2 \theta}$.

[10.3, i]   **7.** Prove the identity: $\tan 2\theta \equiv \dfrac{2\tan\theta}{1 - \tan^2\theta}$.

[10.4, i]   **8.** Find, in radians, all values of $\sin^{-1}\dfrac{1}{2}$.

9. Use a table to find, in degrees, all values of $\cot^{-1} 0.1584$.

[10.4, ii]   **10.** Find $\text{Sin}^{-1}\left(-\dfrac{\sqrt{2}}{2}\right)$.

[10.5, ii]   **11.** Evaluate or simplify $\tan\left(\text{Arctan }\dfrac{7}{8}\right)$.

[10.1, iii]   **12.** Find the angle, to the nearest 10 minutes, from l_1 to l_2, given the equations

$$l_1: y = 2x - 4 \quad \text{and} \quad l_2: x - y = 2.$$

[10.6, iii]   **13.** Solve: $\sin^2 x - 7\sin x = 0$, solutions in $[0, 2\pi)$.

[10.3, iii]   **14.** Find an equivalent expression involving the sine function only: $6\sin 3x + 2\cos 3x$.

15. Graph $y + 1 = 2\cos^2 x$.

16. Graph $f(x) = 2\text{Sin}^{-1}\left(x + \dfrac{\pi}{2}\right)$.

11

TRIANGLES, VECTORS, AND APPLICATIONS

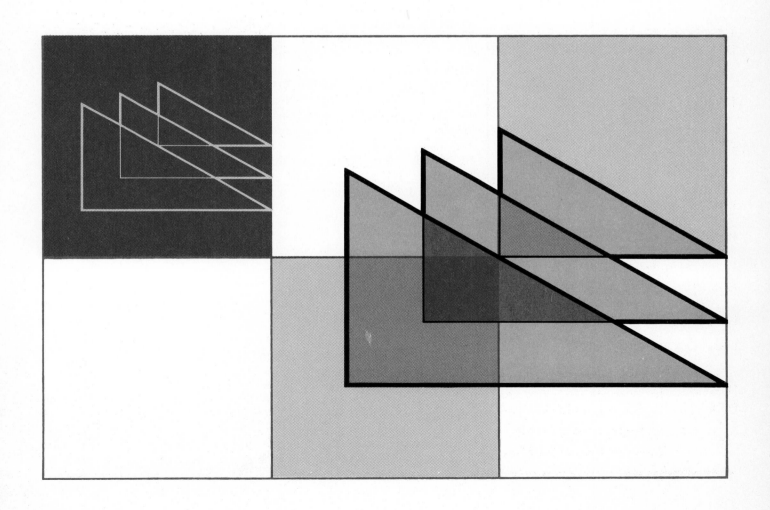

OBJECTIVES

You should be able to:
[i] Solve right triangles.
[ii] Solve applied problems involving the solution of right triangles.

1. Find the lengths a and b in this triangle. Assume that angle A is given to the nearest minute, and length AB to four-digit precision.

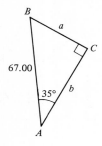

2. Solve this triangle. Use four-digit precision.

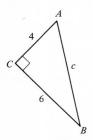

11.1 SOLVING RIGHT TRIANGLES AND APPLICATIONS

[i] Solving Triangles

In Section 8.1 the trigonometric functions were defined and the solving of right triangles was introduced. Tables of the trigonometric functions were considered in Chapter 9. We continue consideration of solving right triangles, a topic important in many applications of trigonometry. The word *trigonometry* actually originally meant "triangle measurement."

The precision to which we can find an angle, using a ratio of sides, such as the sine ratio, depends upon how precisely we know the lengths of the sides. The following table shows the relationship.

NUMBER OF DIGITS IN RATIO	PRECISION OF ANGLE MEASURE
4	To nearest minute
3	To nearest ten minutes
2	To nearest degree

The table can also be read in reverse. For example, if we know an angle to the nearest ten minutes, then the number of significant digits in a value obtained from the table is 3, and hence the number of significant digits obtained for the length of a side of a triangle is also 3.

When we have four-digit precision, but a number itself has fewer than four digits, we often indicate the precision by writing 0's after the decimal point, as in 18.00 or 7.000.

In some of the exercises of this section, the precision is not realistic. For example, one would scarcely consider the distance from Los Angeles to Chicago to the nearest thousandth of a mile. The exercises may use unwarranted precision, from the standpoint of reality, in order to provide valid practice in calculating.

Example 1 Find the length b in this triangle. Use four-digit precision.

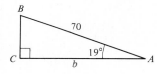

The known side is the hypotenuse. The side we seek is adjacent to the known angle. Thus we shall use the cosine: $\cos A = b/70$. We solve for b:

$$b = 70 \times \cos 19° = 70 \times 0.9455$$
$$b = 66.19.$$

DO EXERCISES 1 AND 2.

When we solve a triangle, we find the *measures* of its sides and angles not already known. We sometimes shorten this to saying that we "find the angles" or "find the sides."

[ii] Applications of Solving Triangles

In many applied problems, unknown parts of right triangles are to be found.

Example 2 *Finding cloud height.* A device for measuring cloud height at night consists of a vertical beam of light, which makes a spot on the clouds. The spot is viewed from a point 135 ft away. The angle of elevation is 67°40′. (The angle between the horizontal and a line of sight is called an *angle of elevation* or an *angle of depression,* the latter if the line of sight is below the horizontal.) Find the height of the clouds.

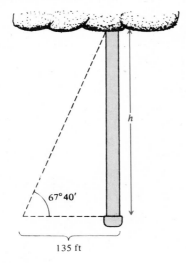

135 ft

From the drawing we have

$$\frac{h}{135} = \tan 67°40′$$

$$h = 135 \times \tan 67°40′$$

$$= 135 \times 2.434$$

$$= 329 \text{ ft.}$$

Note that distances are precise to three digits and the angle to the nearest ten minutes.

Following is a general procedure for solving problems involving triangles.

To solve a triangle problem:
1. **Draw a sketch of the problem situation.**
2. **Look for triangles and sketch them in.**
3. **Mark the known and unknown sides and angles.**
4. **Express the desired side or angle in terms of known trigonometric ratios. Then solve.**

DO EXERCISE 3.

3. The length of a guy wire to a pole is 37.7 ft. It makes an angle of 71°20′ with the ground, which is horizontal. How high above the ground is it attached to the pole?

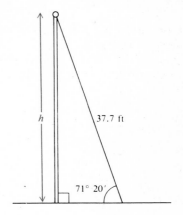

4. A downslope distance is measured to be 241.3 ft and the angle of depression α is measured to be 5°15′. Find the horizontal distance.

Calculators

A calculator would be most convenient for use with Example 2. The same is true for most of the examples and exercises of this chapter, and the use of a calculator is recommended, although not essential, except for certain exercises marked ▦.

If function values are obtained from a calculator, rather than Table 3, it should be kept in mind that the number of decimal places will be different; hence answers may vary because of rounding-error differences. The answers in the back of the book are computed using Table 3.

It is vital to note whether the calculator requires parts of degrees to be entered in tenths and hundredths, rather than minutes and seconds.

Example 3 (*Surveying.*) Horizontal distances must often be measured, even though terrain is not level. One way of doing it is as follows. Distance down a slope is measured with a surveyor's tape, and the distance d is measured by making a level sighting from A to a pole held vertically at B, or the angle α is measured by an instrument placed at A. Suppose that a slope distance L is measured to be 121.3 ft and the angle α is measured to be 3°25′. Find the horizontal distance from A to B.

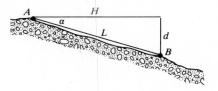

From the drawing we see that $H/L = \cos \alpha$. Thus $H = L \cos \alpha$, and in this case,

$$H = 121.3 \times 0.9982$$
$$= 121.1 \text{ ft.}$$

DO EXERCISE 4.

Example 4 In aerial navigation, directions are given in degrees, clockwise from north. Thus east is 90°, south is 180°, and so on. An airplane leaves an airport and travels for 100 mi in a direction 300°. How far north of the airport is the plane then? how far west?

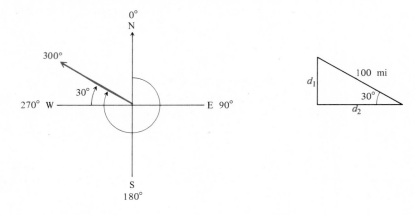

The direction of flight is as shown. In the triangle, d_1 is the northerly distance and d_2 is the westerly distance. Then

$$\frac{d_1}{100} = \sin 30° \qquad \text{and} \qquad \frac{d_2}{100} = \cos 30°,$$

$$d_1 = 100 \sin 30° = 100 \times 0.5$$
$$= 50 \text{ mi (to the nearest mile)},$$
$$d_2 = 100 \cos 30° = 100 \times 0.866$$
$$= 87 \text{ mi} \quad \text{(to the nearest mile)}.$$

Example 5 In surveying and some other applications, directions, or *bearings,* are given by reference to north or south using an acute angle. For example, N 40°W means 40° west of north and S 30°E means 30° east of south. A forest ranger at point A sights a fire directly south of him. A second ranger at a point B, 7 miles east, sights the fire at a bearing of S 27°20′W. How far from A is the fire?

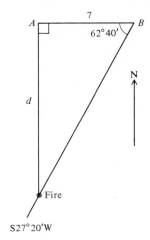

From the drawing we see that the desired distance d is part of a right triangle, as shown. We have

$$\frac{d}{7} = \tan 62°40′$$
$$d = 7 \tan 62°40′ = 7 \times 1.935$$
$$= 13.5 \text{ mi.}$$

DO EXERCISES 5 AND 6.

▦ **Example 6** From an observation tower two markers are viewed on the ground. The markers and the base of the tower are on a line and the observer's eye is 65.3 ft above the ground. The angles of depression to the markers are 53°10′ and 27°50′. How far is it from one marker to the other?

We first make a sketch. We look for right triangles and sketch them in. Then we mark the known information on the sketch. The distance we seek is d, which is $d_1 - d_2$.

5. An airplane flies 150 km from an airport in a direction of 115°. It is then how far east of the airport? how far south?

6. Directly east of a lookout station there is a small forest fire. The bearing of this fire from a station 12 km south of the first is N 57°10′ E. How far is the fire from the southerly lookout station?

7. From an airplane flying 7500 ft above level ground, one can see two towns directly to the east. The angles of depression to the towns are 5°10′ and 77°30′. How far apart are the towns, to the nearest mile?

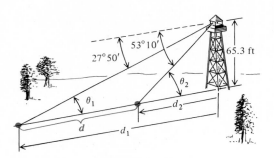

From the right triangles in the drawing, we have

$$\frac{d_1}{65.3} = \cot \theta_1 \quad \text{and} \quad \frac{d_2}{65.3} = \cot \theta_2.$$

Then $d_1 = 65.3 \cot 27°50′$ and $d_2 = 65.3 \cot 53°10′$. We could calculate these and subtract, but the use of a calculator is more efficient if we leave all calculations until last.

$$
\begin{aligned}
d = d_1 - d_2 &= 65.3 \cot 27°50′ - 65.3 \cot 53°10′ \\
&= 65.3(\cot 27°50′ - \cot 53°10′) \\
&= 65.3(\cot 27.83° - \cot 53.17°) \quad \text{Converting to decimals} \\
&= 65.3(1.894 - 0.7490) \\
&= 74.8 \text{ ft}
\end{aligned}
$$

DO EXERCISE 7.

EXERCISE SET 11.1

[i] In Exercises 1–18, standard lettering for a right triangle will be used: A, B, and C are the angles, C being the right angle. The sides opposite A, B, and C are a, b, and c, respectively. Solve the triangles, using three-digit precision.

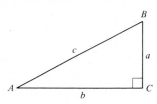

1. $A = 36°10′$, $a = 27.2$

2. $A = 87°40′$, $a = 9.73$

3. $B = 12°40′$, $b = 98.1$

4. $B = 69°50′$, $b = 127$

5. $A = 17°20′$, $b = 13.6$

6. $A = 78°40′$, $b = 1340$

7. $B = 23°10′$, $a = 0.0345$

8. $B = 69°20′$, $a = 0.00488$

9. $A = 47°30′$, $c = 48.3$

10. $A = 88°50′$, $c = 3950$

11. $B = 82°20′$, $c = 0.982$

12. $B = 56°30′$, $c = 0.0447$

13. $a = 12.0$, $b = 18.0$

14. $a = 10.0$, $b = 20.0$

15. $a = 16.0$, $c = 20.0$

16. $a = 15.0$, $c = 45.0$

17. $b = 1.80$, $c = 4.00$

18. $b = 100$, $c = 450$

[ii]

19. A guy wire to a pole makes an angle of 73°10′ with the level ground, and is 14.5 ft from the pole at the ground. How far above the ground is the wire attached to the pole?

20. A kite string makes an angle of 31°40′ with the (level) ground and 455 ft of string is out. How high is the kite?

21. A road rises 3 m per 100 horizontal m. What angle does it make with the horizontal?

22. A kite is 120 ft high when 670 ft of string is out. What angle does the kite make with the ground?

23. What is the angle of elevation of the sun when a 6-ft man casts a 10.3-ft shadow?

24. What is the angle of elevation of the sun when a 35-ft mast casts a 20-ft shadow?

25. From a balloon 2500 ft high, a command post is seen with an angle of depression of 7°40′. How far is it from a point on the ground below the balloon to the command post?

26. From a lighthouse 55 ft above sea level, the angle of depression to a small boat is 11°20′. How far from the foot of the lighthouse is the boat?

27. An airplane travels at 120 km/h for 2 hr in a direction of 243° from Chicago. At the end of this time, how far south of Chicago is the plane?

28. An airplane travels at 150 km/h for 2 hr in a direction of 138° from Omaha. At the end of this time, how far east of Omaha is the plane?

29. Ship A is due west of a lighthouse. Ship B is 12 km south of ship A. From ship B the bearing to the lighthouse is N 63°20′E. How far is ship A from the lighthouse?

30. Lookout station A is 15 km west of station B. The bearing from A to a fire directly south of B is S 37°50′E. How far is the fire from B?

31. From a balloon two km high, the angles of depression to two towns in line with the balloon are 81°20′ and 13°40′. How far apart are the towns?

32. From a balloon 1000 m high, the angles of depression to two artillery posts, in line with the balloon, are 11°50′ and 84°10′. How far apart are the artillery posts?

33. A weather balloon is directly west of two observing stations 10 km apart. The angles of elevation of the balloon from the two stations are 17°50° and 78°10′. How high is the balloon?

34. From two points south of a hill on level ground and 1000 ft apart, the angles of elevation of the hill are 12°20′ and 82°40′. How high is the hill?

35. Show that the area of a right triangle is $\frac{1}{4}c^2 \sin 2A$.

36. Show that the area of a right triangle is $bc \sin A/2$.

★ ───────────────────────────────────

37. Find a formula for the distance to the horizon, as a function of the height of the observer above the earth. Calculate the distance to the horizon from an airplane at an altitude of 1000 ft.

38. In finding horizontal distance from slope distance (see Example 3) $H = L - C$, where C is a correction. Show that a good approximation to C is $d^2/2L$.

11.2 THE LAW OF SINES

[i] The trigonometric functions can be used to solve triangles that are not right triangles (oblique triangles). In order to solve oblique triangles we need to derive some properties, one of which is called the *law of sines*.

We shall consider any oblique triangle. It may or may not have an obtuse angle. We consider both cases, but the derivations are essentially the same.

OBJECTIVES

You should be able to:

[i] Use the law of sines to solve any triangle, given a side and two angles.

[ii] Use the law of sines to solve triangles, given two sides and an angle opposite one of them, finding two solutions when they exist, and recognizing when a solution does not exist.

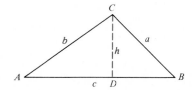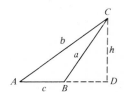

The triangles are lettered in the standard way, with angles A, B, and C and the sides opposite them a, b, and c, respectively. We have drawn an altitude from vertex C. It has length h. In either triangle we now have, from triangle ADC,

$$\frac{h}{b} = \sin A \quad \text{or} \quad h = b \sin A.$$

From triangle DBC on the left we have $h/a = \sin B$, or $h = a \sin B$. On the right we have $h/a = \sin (180° - B) = \sin B$. So in either kind of triangle we now have

$$h = b \sin A \quad \text{and} \quad h = a \sin B.$$

Thus it follows that

$$b \sin A = a \sin B.$$

We divide by $\sin A \sin B$ to obtain

$$\frac{a}{\sin A} = \frac{b}{\sin B}.$$

There is no danger of dividing by 0 here because we are dealing with triangles whose angles are never 0° or 180°.

If we were to consider an altitude from vertex A in the triangles shown above, the same argument would give us

$$\frac{b}{\sin B} = \frac{c}{\sin C}.$$

We combine these results to obtain the law of sines, which holds for right triangles as well as oblique triangles.

THEOREM 1

The law of sines. **In any triangle ABC,**

$$\frac{a}{\sin A} = \frac{b}{\sin B} = \frac{c}{\sin C}$$

(the sides are proportional to the sines of the opposite angles).

Solving Triangles (AAS)

When two angles and a side of any triangle are known, the law of sines can be used to solve the triangle.

▦ **Example 1** In triangle ABC, $a = 4.56$, $A = 43°$, and $C = 57°$. Solve the triangle.

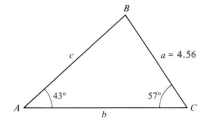

We first draw a sketch. We find B, as follows:

$$B = 180° - (43° + 57°) = 80°.$$

We can now find the other two sides, using the law of sines:

$$\frac{c}{\sin C} = \frac{a}{\sin A}$$

$$c = \frac{a \sin C}{\sin A} = \frac{4.56 \sin 57°}{\sin 43°}$$

$$= \frac{4.56 \times 0.8387}{0.6820} = 5.61,$$

$$\frac{b}{\sin B} = \frac{a}{\sin A}$$

$$b = \frac{a \sin B}{\sin A} = \frac{4.56 \sin 80°}{0.6820}$$

$$= \frac{4.56 \times 0.9848}{0.6820} = 6.58.$$

We have now found the unknown parts of the triangle: $B = 80°$, $c = 5.61$, and $b = 6.58$.

DO EXERCISE 1.

[ii] The Ambiguous Case (SSA)

When two sides of a triangle and an angle opposite one of them are known, the law of sines can be used to solve the triangle. However, there may be more than one solution. Thus this is known as the ambiguous case. Suppose a, b, and A are given. Then the various possibilities are as shown in the four cases below.

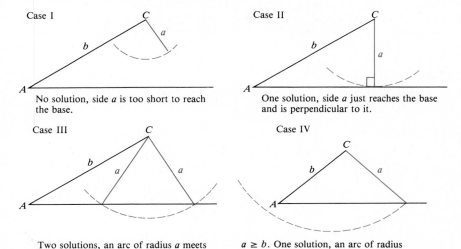

Case I
No solution, side a is too short to reach the base.

Case II
One solution, side a just reaches the base and is perpendicular to it.

Case III
Two solutions, an arc of radius a meets the base at two points.

Case IV
$a \geq b$. One solution, an arc of radius a meets the base at just one point, other than A.

The following examples correspond to the four possibilities just described.

1. Solve this triangle.

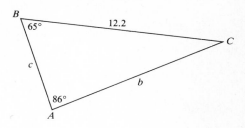

2. Solve this triangle.

$a = 40$, $b = 12$, $B = 57°$

Example 2 (*Case I.*) In triangle ABC, $a = 15$, $b = 25$, and $A = 47°$. Solve the triangle.

We look for B:

$$\frac{a}{\sin A} = \frac{b}{\sin B}.$$

Then

$$\sin B = \frac{b \sin A}{a} = \frac{25 \sin 47°}{15} = \frac{25 \times 0.7314}{15} = 1.219.$$

Since there is no angle having a sine greater than 1, there is no solution.

DO EXERCISE 2.

Example 3 (*Case II.*) In triangle ABC, $a = 12$, $b = 5$, and $B = 24°38'$. Solve the triangle.

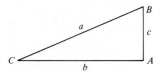

We look for A:

$$\frac{a}{\sin A} = \frac{b}{\sin B}$$

$$\sin A = \frac{a \sin B}{b} = \frac{12 \sin 24°38'}{5}$$

$$= \frac{12 \times 0.4168}{5} = 1.000$$

$$A = 90°.$$

Then $C = 90° - 24°38' = 65°22'$. Since $c/a = \cos B$,

$$c = a \cos B = 12 \times 0.9090 = 10.9.$$

DO EXERCISE 3.

3. Solve this triangle.

$a = 3$, $b = 4$, $A = 48°35'$

Example 4 (*Case III.*) In triangle ABC, $a = 20$, $b = 15$, and $B = 30°$. Solve the triangle.

We look for A:

$$\frac{a}{\sin A} = \frac{b}{\sin B}$$

$$\sin A = \frac{a \sin B}{b} = \frac{20 \sin 30°}{15}$$

$$\sin A = \frac{20 \times 0.5}{15} = 0.6667.$$

There are two angles less than 180° having a sine of 0.6667. They are 42° and 138°, to the nearest degree. This gives us two possible solutions.

Possible solution 1. $A = 42°$. Then $C = 180° - (30° + 42°) = 108°$.

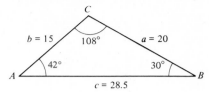

We now find c:

$$\frac{c}{\sin C} = \frac{b}{\sin B}$$

$$c = \frac{b \sin C}{\sin B} = \frac{15 \sin 108°}{\sin 30°}$$

$$= \frac{15 \times 0.9511}{0.5} = 28.5.$$

These parts make a triangle, as shown; hence we have a solution.

Possible solution 2. $A = 138°$. Then $C = 12°$.

We now find c:

$$c = \frac{b \sin C}{\sin B} = \frac{15 \sin 12°}{\sin 30°} = \frac{15 \times 0.2079}{0.5} = 6.2.$$

These parts make a triangle; hence we have a second solution.

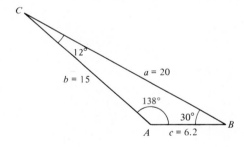

DO EXERCISE 4.

Example 5 (*Case IV.*) In triangle ABC, $a = 25$, $b = 10$, and $A = 42°$. Solve the triangle.

We look for B:

$$\frac{b}{\sin B} = \frac{a}{\sin A}$$

$$\sin B = \frac{b \sin A}{a} = \frac{10 \sin 42°}{25}$$

$$= \frac{10 \times 0.6691}{25} = 0.2676.$$

Then $B = 15°30'$ or $B = 164°30'$. Since $a > b$, we know there is only one solution. If we had not noticed this, we could tell it now. An angle

5. In triangle ABC, $b = 20$, $c = 10$, and $B = 38°$. Solve the triangle.

of $164°30'$ cannot be an angle of this triangle because it already has an angle of $42°$, and these two would total more than $180°$.

$$C = 180° - (42° + 15°30') = 122°30',$$

$$\frac{c}{\sin C} = \frac{a}{\sin A},$$

$$c = \frac{a \sin C}{\sin A} = \frac{25 \sin 122°30'}{\sin 42°} \qquad \text{Solving for } c$$

$$c = \frac{25 \times 0.8434}{0.6691} = 31.5$$

DO EXERCISE 5.

EXERCISE SET 11.2

[i] In Exercises 1–12, solve the triangle ABC.

1. $A = 133°$, $B = 30°$, $b = 18$ **2.** $B = 120°$, $C = 30°$, $a = 16$ **3.** $B = 38°$, $C = 21°$, $b = 24$

4. $A = 131°$, $C = 23°$, $b = 10$ **5.** $A = 68°30'$, $C = 42°40'$, $c = 23.5$ **6.** $B = 118°20'$, $C = 45°40'$, $b = 42.1$

[ii]

7. $A = 36°$, $a = 24$, $b = 34$ **8.** $C = 43°$, $c = 28$, $b = 27$ **9.** $A = 116°20'$, $a = 17.2$, $c = 13.5$

10. $A = 47°50'$, $a = 28.3$, $b = 18.2$ **11.** $C = 61°10'$, $c = 30.3$, $b = 24.2$ **12.** $B = 58°40'$, $a = 25.1$, $b = 32.6$

[i], [ii]

13. Points A and B are on opposite sides of a lunar crater. Point C is 50 meters from A. The measure of $\angle BAC$ is determined to be $112°$ and the measure of $\angle ACB$ is determined to be $42°$. What is the width of the crater?

14. A guy wire to a pole makes a $71°$ angle with level ground. At a point 25 ft farther from the pole than the guy wire, the angle of elevation of the top of the pole is $37°$. How long is the guy wire?

15. A pole leans away from the sun at an angle of $7°$ to the vertical. When the angle of elevation of the sun is $51°$, the pole casts a shadow 47 ft long on level ground. How long is the pole?

16. A vertical pole stands by a road that is inclined $10°$ to the horizontal. When the angle of elevation of the sun is $23°$, the pole casts a shadow 38 ft long directly downhill along the road. How long is the pole?

17. A reconnaissance airplane leaves its airport on the east coast of the United States and flies in a direction of $085°$. Because of bad weather it returns to another airport 230 km to the north of its home base. For the return it flies in a direction of $283°$. What was the total distance it flew?

18. Lookout station B is 10.2 km east of station A. The bearing of a fire from A is S $10°40'$ W. The bearing of the fire from B is S $31°20'$ W. How far is the fire from A? from B?

19. A boat leaves a lighthouse A and sails 5.1 km. At this time it is sighted from lighthouse B, 7.2 km west of A. The bearing of the boat from B is N $65°10'$ E. How far is the boat from B?

20. An airplane leaves airport A and flies 200 km. At this time its bearing from airport B, 250 km to the west, is $120°$. How far is the airplane from B?

☆

21. Prove that the area of a parallelogram is the product of two sides and the sine of the included angle.

22. Prove that the area of a quadrilateral is half the product of the lengths of its diagonals and the sine of an angle between the diagonals.

★

23. When two objects, such as ships, airplanes, or runners, move in straight-line paths, if the distance between them is decreasing and if the bearing from one of them to the other is constant, they will collide. ("Constant bearing means collision," as mariners put it.) Prove that this statement is true.

11.3 THE LAW OF COSINES

[i] A second property of triangles important in solving oblique triangles is called the *law of cosines*. To derive this property we consider any triangle ABC placed on a coordinate system. We place the origin at one of the vertices, say C, and the positive half of the x-axis along one of the sides, say CB. Then the coordinates of B are $(a, 0)$, and the coordinates of A are $(b \cos C, b \sin C)$.

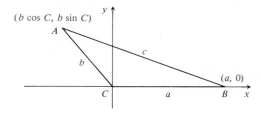

We will next use the distance formula to determine c^2:

$$c^2 = (b \cos C - a)^2 + (b \sin C - 0)^2.$$

Now we multiply and simplify:

$$c^2 = b^2 \cos^2 C - 2ab \cos C + a^2 + b^2 \sin^2 C$$
$$c^2 = a^2 + b^2(\sin^2 C + \cos^2 C) - 2ab \cos C$$
$$c^2 = a^2 + b^2 - 2ab \cos C.$$

Had we placed the origin at one of the other vertices, we would have obtained

$$a^2 = b^2 + c^2 - 2bc \cos A$$

or

$$b^2 = a^2 + c^2 - 2ac \cos B.$$

This result can be summarized as follows.

THEOREM 2

The law of cosines. **In any triangle ABC,**

$$a^2 = b^2 + c^2 - 2bc \cos A,$$
$$b^2 = a^2 + c^2 - 2ac \cos B, \quad \text{and}$$
$$c^2 = a^2 + b^2 - 2ab \cos C.$$

(In any triangle, the square of a side is the sum of the squares of the other two sides, minus twice the product of those sides and the cosine of the included angle.)

Only one of the above formulas need be memorized. The other two can be obtained by a change of letters.

Solving Triangles (SAS)

When two sides of a triangle and the included angle are known, we can use the law of cosines to find the third side. The law of cosines or the law of sines can then be used to finish solving the triangle.

OBJECTIVES

You should be able to:

[i] Use the law of cosines, with the law of sines, to solve any triangle, given two sides and the included angle.

[ii] Use the law of cosines to solve any triangle, given three sides.

1. In triangle ABC, $b = 18$, $c = 28$, and $A = 122°$. Solve the triangle.

Example 1 In triangle ABC, $a = 24$, $c = 32$, and $B = 115°$. Solve the triangle.

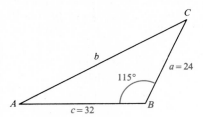

We first find the third side. From the law of cosines,

$$b^2 = a^2 + c^2 - 2ac \cos B$$
$$= 24^2 + 32^2 - 2 \cdot 24 \cdot 32(-0.4226)$$
$$= 2249.$$

Then $b = \sqrt{2249} = 47.4$. Next, we use the law of sines to find a second angle:

$$\frac{a}{\sin A} = \frac{b}{\sin B} \quad \text{and} \quad \sin A = \frac{a \sin B}{b}$$

$$\sin A = \frac{24 \sin 115°}{47.4} = \frac{24 \times 0.9063}{47.4} = 0.4589.$$

Thus

$$A = 27°20'.$$

The third angle is now easy to find:

$$C = 180° - (115° + 27°20') = 37°40'.$$

DO EXERCISE 1.

[ii] Solving Triangles (SSS)

When all three sides of a triangle are known, the law of cosines can be used to solve the triangle.

Example 2 In triangle ABC, $a = 18$, $b = 25$, and $c = 12$. Solve the triangle.

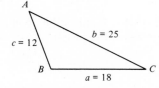

Let us find angle B. We select the formula from the law of cosines that contains $\cos B$; in other words, $b^2 = a^2 + c^2 - 2ac \cos B$. We solve this for $\cos B$ and substitute:

$$\cos B = \frac{a^2 + c^2 - b^2}{2ac} = \frac{18^2 + 12^2 - 25^2}{2 \cdot 18 \cdot 12} = -0.3634.$$

Then $B = 111°20'$. Similarly, we shall find angle A:*

$$a^2 = b^2 + c^2 - 2bc \cos A$$

$$\cos A = \frac{b^2 + c^2 - a^2}{2bc} = \frac{25^2 + 12^2 - 18^2}{2 \cdot 25 \cdot 12} = 0.7417.$$

Thus $A = 42°10'$. Then $C = 180° - (111°20' + 42°10') = 26°30'$.

DO EXERCISE 2.

2. In triangle ABC, $a = 25$, $b = 10$, and $c = 20$. Solve the triangle.

EXERCISE SET 11.3

In Exercises 1–12, solve the triangles.
[i]

1. $A = 30°$, $b = 12$, $c = 24$ **2.** $C = 60°$, $a = 15$, $b = 12$ **3.** $A = 133°$, $b = 12$, $c = 15$

4. $B = 116°$, $a = 31$, $c = 25$ **5.** $B = 72°40'$, $c = 16$, $a = 78$ **6.** $A = 24°30'$, $b = 68$, $c = 14$

[ii]

7. $a = 12$, $b = 14$, $c = 20$ **8.** $a = 22$, $b = 22$, $c = 35$ **9.** $a = 3.3$, $b = 2.7$, $c = 2.8$

10. $a = 16$, $b = 20$, $c = 32$ **11.** $a = 2.2$, $b = 4.1$, $c = 2.4$ **12.** $a = 3.6$, $b = 6.2$, $c = 4.1$

[i], [ii]

13. A ship leaves a harbor and sails 15 nautical mi east. It then sails 18 nautical mi in a direction of S 27°E. How far is it, then, from the harbor and in what direction?

14. An airplane leaves an airport and flies west 147 km. It then flies 200 km in a direction of 220°. How far is it, then, from the airport and in what direction?

15. Two ships leave harbor at the same time. The first sails N 15°W at 25 knots (a knot is one nautical mile per hour). The second sails N 32°E at 20 knots. After 2 hr, how far apart are the ships?

16. Two airplanes leave an airport at the same time. The first flies 150 km/h in a direction of 320°. The second flies 200 km/h in a direction of 200°. After 3 hr, how far apart are the planes?

17. A hill is inclined 5° to the horizontal. A 45-ft pole stands at the top of the hill. How long a rope will it take to reach from the top of the pole to a point 35 ft downhill from the base of the pole?

18. A hill is inclined 15° to the horizontal. A 40-ft pole stands at the top of the hill. How long a rope will it take to reach from the top of the pole to a point 68 ft downhill from the base of the pole?

19. A piece of wire 5.5 m long is bent into a triangular shape. One side is 1.5 m long and another is 2 m long. Find the angles of the triangle.

20. A triangular lot has sides 120 ft long, 150 ft long, and 100 ft long. Find the angles of the lot.

21. A slow-pitch softball diamond is a square 60 ft on a side. The pitcher's mound is 46 ft from home. How far is it from the pitcher's mound to first base?

22. A baseball diamond is a square 90 ft on a side. The pitcher's mound is 60.5 ft from home. How far does the pitcher have to run to cover first?

23. The longer base of an isosceles trapezoid measures 14 ft. The nonparallel sides measure 10 ft, and the base angles measure 80°.

a) Find the length of a diagonal.

b) Find the area.

24. An isosceles triangle has a vertex angle of 38° and this angle is included by two sides, each measuring 20 ft. Find the area of the triangle.

25. A field in the shape of a parallelogram has sides that measure 50 yd and 70 yd. One angle of the field measures 78°. Find the area of the field.

26. An aircraft takes off to fly a 180-mile trip. After flying 75 miles it is 10 miles off course. How much should the heading be corrected to then fly straight to the destination, assuming no wind correction?

---
*The law of sines could be used at this point.

☆ ───

27. From the top of a hill 20 ft above the surface of a lake, a tree across the lake is sighted. The angle of elevation is 11°. From the same position, the top of the tree is seen by reflection in the lake. The angle of depression is 14°. What is the height of the tree?

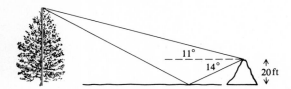

28. A bridge is being built across a canyon. The length of the bridge is 5045 ft. From the deepest point in the canyon, the angles of elevation of the ends of the bridge are 78° and 72°. How deep is the canyon?

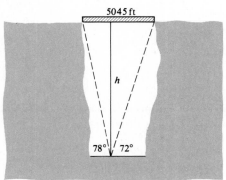

29. Find a formula for the area of an isosceles triangle in terms of the congruent sides and their included angle. Under what conditions will the area of a triangle with fixed congruent sides be a maximum?

30. (*Surveying.*) In surveying, a series of bearings and distances is called a *traverse*. In the following, measurements are taken as shown.

a) Compute the bearing BC.

b) Compute the distance and bearing AC.

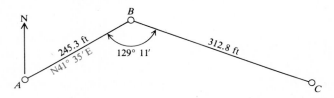

★ ───

31. Show that in any triangle *ABC*,

$$a^2 + b^2 + c^2 = 2(bc \cos A + ac \cos B + ab \cos C).$$

32. Show that in any triangle *ABC*,

$$\frac{\cos A}{a} + \frac{\cos B}{b} + \frac{\cos C}{c} = \frac{a^2 + b^2 + c^2}{2abc}.$$

───

OBJECTIVES

You should be able to:

[i] Given two vectors, find their sum, or resultant.

[ii] Solve applied problems involving finding sums of vectors.

11.4 VECTORS

In many applications there arise certain quantities in which a direction is specified. Any such quantity having a *direction* and a *magnitude* is called a *vector quantity,* or *vector*. Here are some examples of vector quantities.

Displacement. An object moves a certain distance in a certain direction.

A train travels 100 mi to the northeast.

A person takes 5 steps to the west.

A batter hits a ball 100 m along the left-field foul line.

Velocity.   An object travels at a certain speed in a certain direction.

The wind is blowing 15 mph from the northwest.

An airplane is traveling 450 km/h in a direction of 243°.

Force.   A push or pull is exerted on an object in a certain direction.

A 15-kg force is exerted downward on the handle of a jack.

A 25-lb upward force is required to lift a box.

A wagon is being pulled up a 30° incline, requiring an effort of 200 kg.

We shall represent vectors abstractly by directed line segments, or arrows. The length is chosen, according to some scale, to represent the magnitude of the vector, and the direction of the arrow represents the direction of the vector. For example, if we let 1 cm represent 5 km/h, then a 15-km/h wind from the northwest would be represented by an arrow 3 cm long, as shown.

[i] Vector Addition

To "add" vectors, we find a single vector that would have the same effect as the vectors combined. We shall illustrate vector addition for displacements, but it is done the same way for any vector quantities. Suppose a person takes 4 steps east and then 3 steps north. He will then be 5 steps from the starting point in the direction shown. The *sum* of the two vectors is the vector 5 steps in magnitude and in the direction shown. The sum is also called the *resultant* of the two vectors.

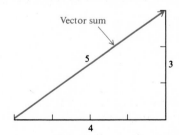

In general, if we have two vectors **a** and **b**,* we can add them as in the above example. That is, we place the tail of one arrow at the head of the other and then find the vector that forms the third side of a

---

*It is usual to denote vectors using boldface type, and sometimes in handwriting to write a bar over the letters.

1. Two forces of 5.0 kg and 14.0 kg act at right angles to each other. Find the resultant, specifying the angle that it makes with the larger force.

triangle. Or, we can place the tails of the arrows together, complete a parallelogram, and find the diagonal of the parallelogram.

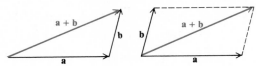

Example 1 Forces of 15 kg and 25 kg act on an object at right angles to each other. Find their *sum,* or *resultant,* giving the angle that it makes with the larger force.

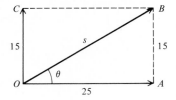

We make a drawing, this time a rectangle, using s for the length of vector **OB**. Since *OAB* is a right triangle, we have

$$\tan \theta = \frac{15}{25} = 0.6.$$

Thus θ, the angle the resultant makes with the larger force, is 31°. Now $s/15 = \csc \theta$, or $s = 15 \csc \theta = 15 \times 1.942 = 29.1$. Thus the resultant **OB** has a magnitude of 29.1 kg and makes an angle of 31° with the larger force.

DO EXERCISE 1.

[ii] **Applications**

Example 2 An airplane heads in a direction of 100° at 180-km/h airspeed while a wind is blowing 40 km/h from 220°. Find the speed of the airplane over the ground and the direction of its track over the ground.

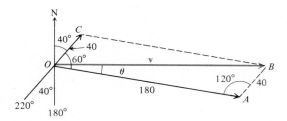

We first make a drawing. The wind is represented by **OC** and the velocity vector of the airplane by **OA**. The resultant velocity is **v**, the sum of the two vectors. We denote the length of **v** by $|\mathbf{v}|$. By the law of cosines in $\triangle OAB$, we have

$$|\mathbf{v}|^2 = 40^2 + 180^2 - 2 \cdot 40 \cdot 180 \cos 120°$$
$$= 41{,}200.$$

Thus $|\mathbf{v}|$ is 203 km/h. By the law of sines in the same triangle,

$$\frac{40}{\sin \theta} = \frac{203}{\sin 120°},$$

or

$$\sin \theta = \frac{40 \sin 120°}{203} = 0.1706.$$

Thus $\theta = 10°$ to the nearest degree. Therefore the ground speed of the airplane is 203 km/h and its track is in a direction of 90°.

DO EXERCISE 2.

2. An airplane heads in a direction of 90° at 140-km/h airspeed. The wind is 50 km/h from 210°. Find the direction and speed of the airplane over the ground.

EXERCISE SET 11.4

[i] In Exercises 1–8, magnitudes of vectors **a** and **b** are given and the angle between the vectors θ. Find the resultant, giving the direction by specifying to the nearest degree the angle it makes with vector **a**.

1. $|\mathbf{a}| = 45$, $|\mathbf{b}| = 35$, $\theta = 90°$

2. $|\mathbf{a}| = 54$, $|\mathbf{b}| = 43$, $\theta = 90°$

3. $|\mathbf{a}| = 10$, $|\mathbf{b}| = 12$, $\theta = 67°$

4. $|\mathbf{a}| = 25$, $|\mathbf{b}| = 30$, $\theta = 75°$

5. $|\mathbf{a}| = 20$, $|\mathbf{b}| = 20$, $\theta = 117°$

6. $|\mathbf{a}| = 30$, $|\mathbf{b}| = 30$, $\theta = 123°$

7. $|\mathbf{a}| = 23$, $|\mathbf{b}| = 47$, $\theta = 27°$

8. $|\mathbf{a}| = 32$, $|\mathbf{b}| = 74$, $\theta = 72°$

[ii]

9. Two forces of 5 kg and 12 kg act on an object at right angles. Find the magnitude of the resultant and the angle it makes with the smaller force.

10. Two forces of 30 kg and 40 kg act on an object at right angles. Find the magnitude of the resultant and the angle it makes with the smaller force.

11. Forces of 420 kg and 300 kg act on an object. The angle between the forces is 50°. Find the resultant, giving the angle it makes with the larger force.

12. Forces of 410 kg and 600 kg act on an object. The angle between the forces is 47°. Find the resultant, giving the angle it makes with the smaller force.

13. A balloon is rising 12 ft/sec while a wind is blowing 18 ft/sec. Find the speed of the balloon and the angle it makes with the horizontal.

14. A balloon is rising 10 ft/sec while a wind is blowing 5 ft/sec. Find the speed of the balloon and the angle it makes with the horizontal.

15. A boat heads 35°, propelled by a force of 750 lb. A wind from 320° exerts a force of 150 lb on the boat. How large is the resultant force and in what direction is the boat moving?

16. A boat heads 220°, propelled by a 650-lb force. A wind from 080° exerts a force of 100 lb on the boat. How large is the resultant force, and in what direction is the boat moving?

17. A ship sails N 80°E for 120 nautical mi, then S 20°W for 200 nautical mi. How far is it then from the starting point, and in what direction?

18. An airplane flies 032° for 210 km, then 280° for 170 km. How far is it, then, from the starting point, and in what direction?

19. A motorboat has a speed of 15 km/h. It crosses a river whose current has a speed of 3 km/h. In order to cross the river at right angles, the boat should be pointed in what direction?

20. An airplane has an airspeed of 150 km/h. It is to make a flight in a direction of 080° while there is a 25-km/h wind from 350°. What should be the airplane's heading?

OBJECTIVES

You should be able to:

[i] Resolve vectors into components.
[ii] Add vectors using components.
[iii] Change from rectangular to polar notation for vectors.
[iv] Change from polar to rectangular notation for vectors.
[v] Do simple manipulations in vector algebra.

11.5 VECTORS AND COORDINATES

[i] Components of Vectors

Given a vector, it is often convenient to reverse the addition procedure, that is, to find two vectors whose sum is the given vector. Usually the two vectors we seek will be perpendicular. The two vectors we find are called *components* of the given vector.

Example 1 A certain vector **a** has a magnitude of 130 and is inclined 40° with the horizontal. Resolve the vector into horizontal and vertical components.

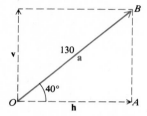

We first make a drawing showing horizontal and vertical vectors whose sum is the given vector **a**. From △OAB, we see

$$|\mathbf{h}| = 130 \cos 40° = 99.6$$

and

$$|\mathbf{v}| = 130 \sin 40° = 83.6.$$

These are the components we seek.

Example 2 An airplane is flying at 200 km/h in a direction of 305°. Find the westerly and northerly components of its velocity.

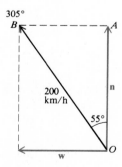

We first make a drawing showing westerly and northerly vectors whose sum is the given velocity. From △OAB, we see that

$$|\mathbf{n}| = 200 \cos 55° = 115 \text{ km/h},$$
$$|\mathbf{w}| = 200 \sin 55° = 164 \text{ km/h}.$$

Example 3 A ten-pound block is sitting on a 30° incline. Find the components of the weight parallel and perpendicular to the incline.

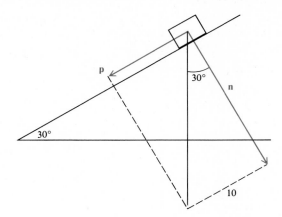

We make a drawing showing the components. From the drawing we see that

$$|\mathbf{p}| = 10 \sin 30° = 5 \,\text{lb},$$
$$|\mathbf{n}| = 10 \cos 30° = 8.67 \,\text{lb}.$$

DO EXERCISES 1 AND 2.

[ii] **Adding Vectors Using Components**

Given a vector, we can resolve it into components. Given components of a vector, we can find the vector. We simply find the vector sum of the components.

Example 4 A vector **v** has a westerly component of 12 and a southerly component of 16. Find the vector.

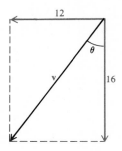

From the drawing we see that $\tan \theta = \frac{12}{16} = 0.75$. Thus $\theta = 36°50'$. Then $|\mathbf{v}| = 12 \csc 36°50' = 20$. Thus the vector has a magnitude of 20 and a direction of S 36°50′W.

DO EXERCISE 3.

1. A vector of magnitude 100 points southeast. Resolve the vector into easterly and southerly components.

2. Two boys are pulling on a rope attached to a tree with a force of 45 lb in a direction S 35°W. Find the westerly and southerly components of the force.

3. A wind has an easterly component (*from* the east) of 10 km/h and a southerly component (*from* the south) of 16 km/h. Find the magnitude and direction of the wind.

4. One vector has components of 7 up and 9 to the left. A second vector has components of 3 down and 12 to the left. Find the sum, expressing it

a) in terms of components;
b) giving magnitude and direction.

When we have two vectors resolved into components, we can add the two vectors by adding the components.

Example 5 A vector **v** has a westerly component of 3 and a northerly component of 5. A second vector **w** has an easterly component of 8 and a northerly component of 4. Find the components of the sum. Find the sum.

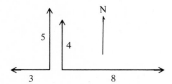

Adding the east–west components, we obtain 5 east. Adding the north–south components, we obtain 9 north. The components of **v** + **w** are 5 east and 9 north. We now add these components:

$$\tan \theta = \frac{9}{5} = 1.8,$$

$$\theta = 61°,$$

$$|\mathbf{v} + \mathbf{w}| = 9 \csc 61° = 10.3.$$

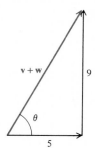

The sum **v** + **w** has a magnitude of 10.3 and a direction of N 29°E.

DO EXERCISE 4.

Add the vectors, giving answers as ordered pairs.

5. (7, 2) and (5, −10)

Analytic Representation of Vectors

If we place a coordinate system so that the origin is at the tail of an arrow representing a vector, we say that the vector is in *standard position*. Then if we know the coordinates of the other end of the vector, we know the vector. The coordinates will be the x-component and the y-component of the vector. Thus we can consider an ordered pair (a, b) to be a vector. When vectors are given in this form, it is easy to add them. We simply add the respective components.

6. (−2, 7) and (14, −3)

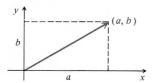

Example 6 Find **u** + **v**, where **u** = (3, −7) and **v** = (4, 2).

The sum is found by adding the x-components and then adding the y-components: **u** + **v** = (3, −7) + (4, 2) = (7, −5).

DO EXERCISES 5 AND 6.

[iii] **Polar Notation**

Vectors can also be specified by giving their length and direction. When this is done, we say that we have polar notation for the vector. An example of polar notation is $(15, 260°)$. The angle is measured from the positive half of the x-axis counterclockwise. In Section 11.6, we consider a coordinate system (called a *polar coordinate system*) in which polar notation is very natural.

Example 7 Find polar notation for the vector **v**, where **v** $= (7, -2)$.

From the diagram, we see the reference angle ϕ.

$$\tan \phi = \frac{-2}{7} = -0.2857$$

$$\phi = -16°,$$
$$|\mathbf{v}| = \sqrt{(-2)^2 + 7^2} = 7.28.$$

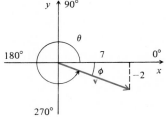

Thus polar notation for **v** is

$$(7.28, 344°) \quad \text{or} \quad (7.28, -16°).$$

DO EXERCISE 7.

[iv] Given polar notation for a vector we can also find the ordered pair notation (rectangular notation).

Example 8 Find rectangular notation for the vector **w**, where **w** $= (14, 155°)$.

From the drawing we have

$$x = 14 \cos 155° = -12.7,$$
$$y = 14 \sin 155° = 5.92.$$

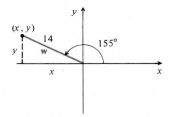

Thus rectangular notation for **w** is $(-12.7, 5.92)$.

DO EXERCISE 8.

7. Find polar notation for the vector $(-8, 3)$.

8. Find rectangular notation for the vector $(15, 341°)$.

9. For $\mathbf{u} = (-3, 5)$ and $\mathbf{v} = (4, 2)$, find the following.

a) $\mathbf{u} + \mathbf{v}$
b) $\mathbf{v} - \mathbf{u}$
c) $5\mathbf{u} - 2\mathbf{v}$
d) $|3\mathbf{u} + 2\mathbf{v}|$

[v] Properties of Vectors and Vector Algebra

With vectors represented in rectangular notation (a, b), it is easy to determine many of their properties. We consider the set of all ordered pairs of real numbers (a, b). This is the set of all vectors on a plane. Consider the sum of any two vectors.

$$\mathbf{u} = (a, b), \qquad \mathbf{v} = (c, d),$$
$$\mathbf{u} + \mathbf{v} = (a + c, b + d), \qquad \mathbf{v} + \mathbf{u} = (c + a, d + b).$$

We have just shown that addition of vectors is commutative. Other properties are likewise not difficult to show. The system of vectors is not a field, but it does have some of the familiar field properties.

PROPERTIES OF VECTORS

COMMUTATIVITY. Addition of vectors is commutative.

ASSOCIATIVITY. Addition of vectors is associative.

IDENTITY. There is an additive identity, the vector $(0, 0)$, or $\mathbf{0}$.

INVERSES. Every vector (a, b) has an additive inverse $(-a, -b)$.

SUBTRACTION. We define $\mathbf{u} - \mathbf{v}$ to be the vector which when added to $\mathbf{v}$ gives $\mathbf{u}$. It follows that if $\mathbf{u} = (a, b)$ and $\mathbf{v} = (c, d)$, then $\mathbf{u} - \mathbf{v} = (a - c, b - d)$. Thus subtraction is always possible. Also, $\mathbf{u} - \mathbf{v} = \mathbf{u} + (-\mathbf{v})$.

LENGTH, OR ABSOLUTE VALUE. If $\mathbf{u} = (a, b)$, then it follows that $|\mathbf{u}| = \sqrt{a^2 + b^2}$.

A vector can be multiplied by a real number. A real number in this context is called a *scalar*. For example, $2\mathbf{v}$ is a vector in the same direction as $\mathbf{v}$ but twice as long. Analytically, we define scalar multiplication as follows.

DEFINITION

Scalar multiplication. If r is any real number and $\mathbf{v} = (a, b)$, then $r\mathbf{v} = (ar, br)$.

Example 9 Do the following calculations, where $\mathbf{u} = (4, 3)$ and $\mathbf{v} = (-5, 8)$: (a) $\mathbf{u} + \mathbf{v}$; (b) $\mathbf{u} - \mathbf{v}$; (c) $3\mathbf{u} - 4\mathbf{v}$; (d) $|3\mathbf{u} - 4\mathbf{v}|$.

a) $\mathbf{u} + \mathbf{v} = (4, 3) + (-5, 8) = (-1, 11)$

b) $\mathbf{u} - \mathbf{v} = \mathbf{u} + (-\mathbf{v}) = (4, 3) + (5, -8) = (9, -5)$

c) $3\mathbf{u} - 4\mathbf{v} = 3(4, 3) - 4(-5, 8) = (12, 9) - (-20, 32)$
$= (32, -23)$

d) $|3\mathbf{u} - 4\mathbf{v}| = \sqrt{32^2 + (-23)^2} = \sqrt{1553} \approx 39.41$

DO EXERCISE 9.

EXERCISE SET 11.5

[ii] In Exercises 1–22, rectangular notation for vectors is given. In Exercises 1–6, add, giving rectangular notation for the answers.

1. $(3, 7) + (2, 9)$

2. $(5, -7) + (-3, 2)$

3. $(17, 7.6) + (-12.2, 6.1)$

4. $(-15.2, 37.1) + (7.9, -17.8)$

5. $(-650, -750) + (-12, 324)$

6. $(-354, -973) + (-75, 256)$

[iii] In Exercises 7–14, find polar notation.

7. $(3, 4)$

8. $(4, 3)$

9. $(10, -15)$

10. $(17, -10)$

11. $(-3, -4)$

12. $(-4, -3)$

13. $(-10, 15)$

14. $(-17, 10)$

[iv] In Exercises 15–22, find rectangular notation.

15. $(4, 30°)$

16. $(8, 60°)$

17. $(10, 235°)$

18. $(15, 210°)$

19. $(20, 330°)$

20. $(20, 200°)$

21. $(100, -45°)$

22. $(150, -60°)$

[i]

23. A vector **u** with magnitude 150 is inclined upward 52° from the horizontal. Find the horizontal and vertical components of **u**.

24. A vector **u** with magnitude 170 is inclined downward 63° from the horizontal. Find the horizontal and vertical components of **u**.

25. An airplane is flying 220° at 250 km/h. Find the southerly and westerly components of its velocity **v**.

26. A wind is blowing from 310° at 25 mph. Find the southerly and easterly components of the wind velocity **v**.

[ii]

27. A force **f** has a westerly component of 25 kg and a southerly component of 35 kg. Find the magnitude and direction of **f**.

28. A force **f** has an upward component of 65 kg and a component to the left of 90 kg. Find the magnitude and direction of **f**.

29. Vector **u** has a westerly component of 15 and a northerly component of 22. Vector **v** has a northerly component of 6 and an easterly component of 8. Find (a) the components of **u** + **v**; (b) the magnitude and direction of **u** + **v**.

30. Vector **u** has a component of 18 upward and a component of 12 to the left. Vector **v** has a component of 5 downward and a component of 35 to the right. Find (a) the components of **u** + **v**; (b) the magnitude and direction of **u** + **v**.

[v] In Exercises 31–40, do the calculations for the following vectors: $\mathbf{u} = (3, 4)$, $\mathbf{v} = (5, 12)$, and $\mathbf{w} = (-6, 8)$.

31. $3\mathbf{u} + 2\mathbf{v}$

32. $3\mathbf{v} - 2\mathbf{w}$

33. $(\mathbf{u} + \mathbf{v}) - \mathbf{w}$

34. $\mathbf{u} - (\mathbf{v} + \mathbf{w})$

35. $|\mathbf{u}| + |\mathbf{v}|$

36. $|\mathbf{u}| - |\mathbf{v}|$

37. $|\mathbf{u} + \mathbf{v}|$

38. $|\mathbf{u} - \mathbf{v}|$

39. $2|\mathbf{u} + \mathbf{v}|$

40. $2|\mathbf{u}| + 2|\mathbf{v}|$

☆

41. If **PQ** is any vector, what is **PQ** + **QP**?

42. The *inner product* of vectors **u** • **v** is a scalar, defined as follows: $\mathbf{u} \cdot \mathbf{v} = |\mathbf{u}||\mathbf{v}| \cos \theta$, where θ is the angle between the vectors. Show that vectors **u** and **v** are perpendicular if and only if $\mathbf{u} \cdot \mathbf{v} = 0$.

★

43. Show that scalar multiplication is distributive over vector addition.

44. Let $\mathbf{u} = (3, 4)$. Find a vector that has the same direction as **u** but length 1.

45. Prove that for any vectors **u** and **v**, $|\mathbf{u} + \mathbf{v}| \leq |\mathbf{u}| + |\mathbf{v}|$.

OBJECTIVES

You should be able to:

[i] Plot points, given their polar co-ordinates, and determine polar coordinates of points on a graph.

[ii] Convert from rectangular to polar coordinates and from polar to rectangular coordinates.

[iii] Graph simple polar equations.

11.6 POLAR COORDINATES

[i] In graphing we locate a point with an ordered pair of numbers (a, b). We may consider any such ordered pair to be a vector. From our work with vectors we know that we could also locate a point with a vector, given a length and a direction. When we use rectangular notation to locate points, we describe the coordinate system as *rectangular*.* When we use polar notation, we describe the coordinate system as *polar*. As this diagram shows, any point has rectangular coordinates (x, y) and has polar coordinates (r, θ). On a polar graph the origin is called the *pole* and the positive half of the x-axis is called the *polar axis*.

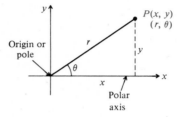

To plot points on a polar graph, we usually locate θ first, then move a distance r from the pole in that direction. If r is negative, we move in the opposite direction. Polar-coordinate graph paper, shown below, facilitates plotting. Point B illustrates that θ may be in radians. Point E illustrates that the polar coordinates of a point are not unique.

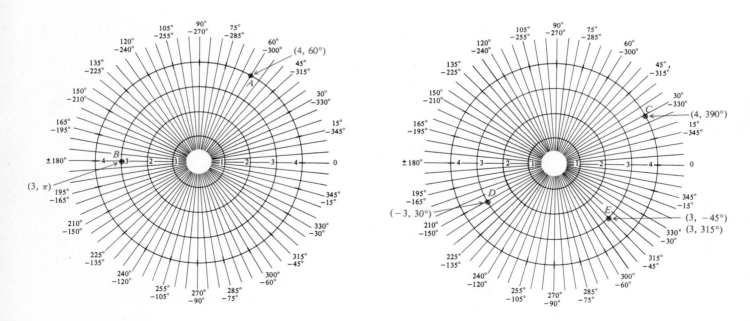

---

*Also called *Cartesian*, after the French mathematician René Descartes.

DO EXERCISES 1–4.

[ii] Polar and Rectangular Equations

Some curves have simpler equations in polar coordinates than in rectangular coordinates. For others the reverse is true. The following relationships between the two kinds of coordinates allow us to convert an equation from rectangular to polar coordinates:

$$x = r \cos \theta, \qquad y = r \sin \theta.$$

Example 1 Convert to a polar equation: $x^2 + y^2 = 25$.

$$\begin{aligned}
(r \cos \theta)^2 + (r \sin \theta)^2 &= 25 \qquad \text{Substituting for x and y} \\
r^2 \cos^2 \theta + r^2 \sin^2 \theta &= 25 \\
r^2(\cos^2 \theta + \sin^2 \theta) &= 25 \\
r^2 &= 25 \\
r &= 5
\end{aligned}$$

Example 1 illustrates that the polar equation of a circle centered at the origin is much simpler than the rectangular equation.

Example 2 Convert to a polar equation: $2x - y = 5$.

$$\begin{aligned}
2(r \cos \theta) - (r \sin \theta) &= 5 \\
r(2 \cos \theta - \sin \theta) &= 5
\end{aligned}$$

DO EXERCISES 5 AND 6.

The relationships that we need to convert from polar equations to rectangular equations can easily be determined from the triangle in the diagram on p. 450. They are as follows:

$$r = \pm\sqrt{x^2 + y^2},$$

$$\sin \theta = \frac{\pm y}{\sqrt{x^2 + y^2}},$$

$$\cos \theta = \frac{\pm x}{\sqrt{x^2 + y^2}},$$

$$\tan \theta = \frac{y}{x}$$

To convert to rectangular notation, we substitute as needed from either of the two lists, choosing the + or − signs as appropriate in each case.

Example 3 Convert to a rectangular (Cartesian) equation: $r = 4$.

$$\begin{aligned}
+\sqrt{x^2 + y^2} &= 4 \qquad \text{Substituting for r} \\
x^2 + y^2 &= 16 \qquad \text{Squaring}
\end{aligned}$$

In squaring like this, we must be careful not to introduce solutions of the equation that are not already present. In this example we did not, because the graph is a circle of radius 4 centered at the origin, in both cases.

1. Plot the following points.

   a) $(3, 60°)$ d) $(5, -60°)$

   b) $(0, 10°)$ e) $(2, 3\pi/2)$

   c) $(-5, 120°)$

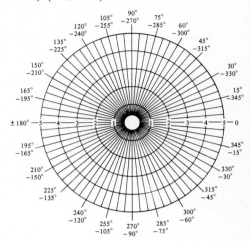

2. Find polar coordinates of each of these points. Give two answers for each.

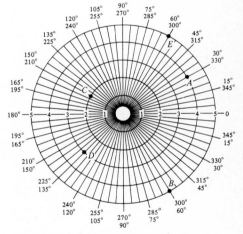

3. Find polar coordinates of these points.

   a) $(3, 3)$ c) $(-3, 3\sqrt{3})$

   b) $(0, -4)$ d) $(2\sqrt{3}, -2)$

4. Find rectangular coordinates of these points.

   a) $(5, 30°)$ c) $(-5, 45°)$

   b) $(10, \pi/3)$ d) $(-8, -5\pi/6)$

Convert to polar equations.

5. $2x + 5y = 9$

6. $x^2 + y^2 + 8x = 0$

Convert to Cartesian equations.

7. $r = 7$

8. $r \sin \theta = 5$

9. $r - 3 \cos \theta = 5 \sin \theta$

10. Graph $r = 1 - \sin \theta$.

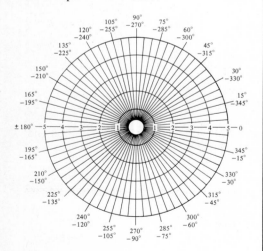

Example 4 Convert to a rectangular equation: $r \cos \theta = 6$.

From our first list we know that $x = r \cos \theta$, so we have $x = 6$.

Example 5 Convert to a rectangular equation: $r = 2 \cos \theta + 3 \sin \theta$.

Let us first multiply on both sides by r:

$$r^2 = 2r \cos \theta + 3r \sin \theta$$
$$x^2 + y^2 = 2x + 3y \qquad \text{Substituting}$$

DO EXERCISES 7–9.

[iii] Graphing Polar Equations

To graph a polar equation, we usually make a table of values, choosing values of θ and calculating corresponding values of r. We plot the points and then complete the graph, as in the rectangular case. A difference arises in the polar case, because as θ increases sufficiently, points will begin to repeat and the curve will be traced again and again. When such a point is reached, the curve is complete.

Example 6 Graph $r = \cos \theta$.

We first make a table of values.

θ	0°	30°	45°	60°	90°	120°	135°	150°	180°
r	1	0.866	0.707	0.5	0	−0.5	−0.707	−0.866	−1

We plot these points and note that the last point is the same as the first. (Points having negative r are plotted in color.)

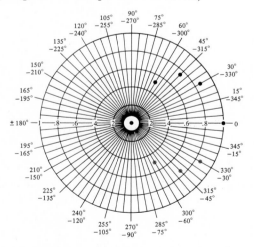

We try another point, $(-0.866, 210°)$, and find that it has already been plotted. Since the repeating has started, we plot no more points, but draw the graph, which is a circle as shown.

DO EXERCISE 10.

EXERCISE SET 11.6

[i] Use polar coordinate paper. Graph the following points.

1. $(4, 30°)$ **2.** $(5, 45°)$ **3.** $(0, 37°)$ **4.** $(0, 48°)$

5. $(-6, 150°)$ **6.** $(-5, 135°)$ **7.** $(-8, 210°)$ **8.** $(-5, 270°)$

9. $(3, -30°)$ **10.** $(6, -45°)$ **11.** $(7, -315°)$ **12.** $(4, -270°)$

13. $(-3, -30°)$ **14.** $(-6, -45°)$ **15.** $(-3.2, 27°)$ **16.** $(-6.8, 34°)$

17. $\left(6, \dfrac{\pi}{4}\right)$ **18.** $\left(5, \dfrac{\pi}{6}\right)$ **19.** $\left(4, \dfrac{3\pi}{2}\right)$ **20.** $\left(3, \dfrac{3\pi}{4}\right)$

21. $\left(-6, \dfrac{\pi}{4}\right)$ **22.** $\left(-5, \dfrac{\pi}{6}\right)$ **23.** $\left(-4, -\dfrac{3\pi}{2}\right)$ **24.** $\left(-3, -\dfrac{3\pi}{4}\right)$

[ii] Find polar coordinates of the following points.

25. $(4, 4)$ **26.** $(5, 5)$ **27.** $(0, 5)$ **28.** $(0, -3)$

29. $(4, 0)$ **30.** $(-5, 0)$ **31.** $(3, 3\sqrt{3})$ **32.** $(-3, -3\sqrt{3})$

33. $(\sqrt{3}, 1)$ **34.** $(-\sqrt{3}, 1)$ **35.** $(3\sqrt{3}, 3)$ **36.** $(4\sqrt{3}, -4)$

Find Cartesian coordinates of the following points.

37. $(4, 45°)$ **38.** $(5, 60°)$ **39.** $(0, 23°)$ **40.** $(0, -34°)$

41. $(-3, 45°)$ **42.** $(-5, 30°)$ **43.** $(6, -60°)$ **44.** $(3, -120°)$

45. $\left(10, \dfrac{\pi}{6}\right)$ **46.** $\left(12, \dfrac{3\pi}{4}\right)$ **47.** $\left(-5, \dfrac{5\pi}{6}\right)$ **48.** $\left(-6, \dfrac{3\pi}{4}\right)$

Convert to polar equations.

49. $3x + 4y = 5$ **50.** $5x + 3y = 4$ **51.** $x = 5$ **52.** $y = 4$

53. $x^2 + y^2 = 36$ **54.** $x^2 + y^2 = 16$ **55.** $x^2 - 4y^2 = 4$ **56.** $x^2 - 5y^2 = 5$

Convert to rectangular equations.

57. $r = 5$ **58.** $r = 8$ **59.** $\theta = \dfrac{\pi}{4}$ **60.** $\theta = \dfrac{3\pi}{4}$

61. $r \sin \theta = 2$ **62.** $r \cos \theta = 5$ **63.** $r = 4 \cos \theta$ **64.** $r = -3 \sin \theta$

65. $r - r \sin \theta = 2$ **66.** $r + r \cos \theta = 3$ **67.** $r - 2 \cos \theta = 3 \sin \theta$ **68.** $r + 5 \sin \theta = 7 \cos \theta$

[iii] Graph.

69. $r = 4 \cos \theta$ **70.** $r = 4 \sin \theta$ **71.** $r = 1 - \cos \theta$

72. $r = 1 - \sin \theta$ **73.** $r = \sin 2\theta$ **74.** $r = \sin 3\theta$

★ ———

75. Convert to a rectangular equation: $r = \sec^2 \dfrac{\theta}{2}$.

76. The center of a regular hexagon is at the origin and one vertex is the point $(4, 0°)$. Find the coordinates of the other vertices.

77. Graph $r = 2 \cos 3\theta$.

78. Graph $r = 3 \sin 2\theta$.

11.7 FORCES IN EQUILIBRIUM

[i] When several forces act through the same point on an object, their vector sum must be **0** in order for a balance to occur. When a balance occurs, then the object is either stationary or moving in a straight line without acceleration. The fact that the vector sum must be **0** for a balance and vice versa allows us to solve many applied problems involving forces. In these problems we usually resolve vectors

OBJECTIVE

You should be able to:

[i] Solve applied problems in which several forces acting at a point are in equilibrium.

1. A weight of 300 lb is supported by two ropes at a point A. One rope is horizontal and the other makes an angle of 30° with the horizontal. Find the forces (of tension) in the two ropes.

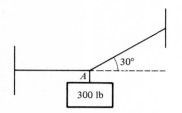

into components. For a balance, the sum of the components must be **0** separately.*

Example 1 A 200-lb sign is hanging from the end of a horizontal hinged boom, supported by a cable from the end of the boom and inclined 35° with the horizontal. Find the force (tension) in the cable and the force (compression) in the boom.

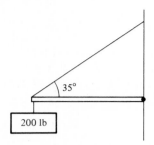

At the end of the boom we have three forces acting at a point: the weight of the sign acting down, the cable pulling up and to the right, and the boom pushing to the left. We draw a force diagram showing these.

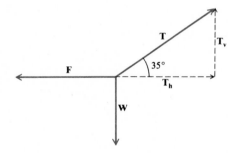

We have called the tension in the cable **T**, and the compression in the boom **F** and the weight **W**. We resolve the tension in the cable into horizontal and vertical components. We have a balance, so **F** + **T** + **W** = **0**. In fact, the sum of the horizontal components must be **0**, and the sum of the vertical components must also be **0**. Thus we know that $|\mathbf{T_v}| = 200$ lb and that $|\mathbf{F}| = |\mathbf{T_h}|$. Then we have

$$|\mathbf{T_v}| = 200 = |\mathbf{T}| \sin 35°$$

$$|\mathbf{T}| = \frac{|\mathbf{T_v}|}{\sin 35°} = \frac{200}{0.5736} = 349 \text{ lb.}$$

Then $|\mathbf{F}| = |\mathbf{T_h}| = 349 \cos 35° = 286$ lb. The answer is that the tension in the cable is 349 lb and the compression in the boom is 286 lb.

DO EXERCISE 1.

---

*For a balance, the rotational tendency of the forces must also be zero. This will always be the case if the forces are concurrent (act through the same point). In this section we consider only applications in which forces are concurrent.

Example 2 A block weighing 100 kg rests on a 25° incline. Find the components of its weight perpendicular and parallel to the incline.

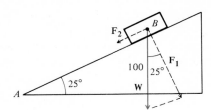

The weight **W** is 100 kg acting downward as shown. We draw $\mathbf{F_1}$ and $\mathbf{F_2}$, the components. The angle at B has the same measure as the angle at A because their sides are respectively perpendicular. Thus we have

$$|\mathbf{F_1}| = 100 \cos 25° = 90.6 \text{ kg} \qquad \text{and} \qquad |\mathbf{F_2}| = 100 \sin 25° = 42.3 \text{ kg}.$$

DO EXERCISE 2.

In a situation like that of Example 2, the force necessary to hold the block on the incline is a force the size of $\mathbf{F_2}$ but opposite in direction. The force with which the block pushes on the incline is $\mathbf{F_1}$.

Example 3 A 500-kg block is suspended by two ropes as shown. Find the tension in each rope.

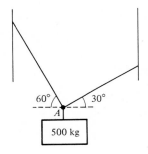

At point A there are three forces acting: the block is pulling down, and the two ropes are pulling upward and outward. We draw a force diagram showing these.

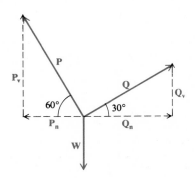

2. A block weighing 100 kg rests on a 30° incline. Find the components of its weight parallel to and perpendicular to the incline.

3. Two ropes support a 1000-kg weight, as shown. Find the tension in each rope.

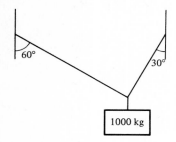

1000 kg

We have called the forces exerted by the ropes **P** and **Q**. Their horizontal components have magnitudes $|\mathbf{P}|\cos 60°$ and $|\mathbf{Q}|\cos 30°$. Since there is a balance, these must be the same:

$$|\mathbf{P}|\cos 60° = |\mathbf{Q}|\cos 30°.$$

The vertical components in the ropes have magnitudes $|\mathbf{P}|\sin 60°$ and $|\mathbf{Q}|\sin 30°$. Since there is a balance, these must total 500 lb, the weight of the block:

$$|\mathbf{P}|\sin 60° + |\mathbf{Q}|\sin 30° = 500.$$

We now have two equations with unknowns $|\mathbf{P}|$ and $|\mathbf{Q}|$:

$$|\mathbf{P}|\cos 60° - |\mathbf{Q}|\cos 30° = 0, \tag{1}$$
$$|\mathbf{P}|\sin 60° + |\mathbf{Q}|\sin 30° = 500. \tag{2}$$

To solve, we begin by solving equation (1) for P:

$$|\mathbf{P}| = |\mathbf{Q}|\frac{\cos 30°}{\cos 60°}. \tag{3}$$

Next, we substitute in equation (2):

$$|\mathbf{Q}|\frac{\cos 30° \sin 60°}{\cos 60°} + |\mathbf{Q}|\sin 30° = 500$$

$$|\mathbf{Q}|\frac{\cos 30° \sin 60°}{\cos 60°} + |\mathbf{Q}|\frac{\sin 30° \cos 60°}{\cos 60°} = 500$$

$$|\mathbf{Q}|(\cos 30° \sin 60° + \sin 30° \cos 60°) = 500 \cos 60°$$

$$|\mathbf{Q}|(\sin 90°) = 500 \cos 60° \quad \text{Using an identity}$$

$$|\mathbf{Q}| = 500 \cdot \frac{1}{2} = 250 \text{ kg}.$$

Then substituting in equation (3), we obtain

$$|\mathbf{P}| = 250\frac{\cos 30°}{\cos 60°} = 250\frac{\sqrt{3}/2}{1/2} = 250\sqrt{3} = 433 \text{ kg}.$$

DO EXERCISE 3.

EXERCISE SET 11.7

[i]

1. A 150-lb sign is hanging from the end of a hinged boom, supported by a cable inclined 42° with the horizontal. Find the tension in the cable and the compression in the boom.

2. A 300-lb sign is hanging from the end of a hinged boom, supported by a cable inclined 51° from the horizontal. Find the tension in the cable and the compression in the boom (see the figure for Exercise 1).

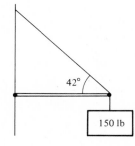

42°

150 lb

3. A weight of 200 kg is supported by a frame made of two rods and hinged at *A*, *B*, and *C* (see the figure below). Find the forces exerted by the two rods.

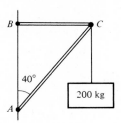

4. A weight of 300 kg is supported by the hinged frame of Exercise 3, but where the angle shown is 50°. Find the forces exerted by the two rods.

5. The force due to air movement on an airplane wing in flight is 2800 lb, acting at an angle of 28° with the vertical. What is the lift (vertical component) and what is the drag (horizontal component)?

6. The force due to air movement on an airplane wing in flight is 3500 lb, acting at an angle of 32° with the vertical. What is the lift (vertical component) and what is the drag (horizontal component)?

7. A 100-kg block of ice rests on a 37° incline. What force parallel to the incline is necessary to keep it from sliding down?

8. What force is necessary to pull a 6500-kg truck up a 7° incline?

9. The moon's gravity is $\frac{1}{6}$ that of the earth. To simulate the moon's gravity an incline is used, as shown in the figure to the right. The astronaut is suspended in a sling and is free to move horizontally. The component of his weight perpendicular to the incline should be $\frac{1}{6}$ of his weight. What must the angle θ be?

10. To simulate gravity on a planet having $\frac{3}{8}$ of the earth's gravity, what should the angle θ be? (See the figure and explanation for Exercise 9.)

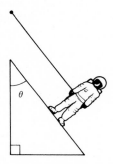

11. A 400-kg block is suspended by two ropes as shown. Find the tension in each rope.

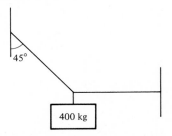

12. A 1000-lb block is suspended by two ropes as shown. Find the tension in each rope.

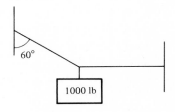

13. A 2000-kg block is suspended by two ropes as shown. Find the tension in each rope.

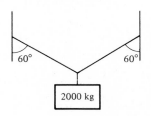

14. A 1500-lb block is suspended by two ropes as shown. Find the tension in each rope.

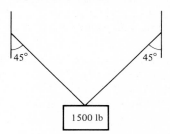

15. A 2500-kg block is suspended by two ropes as shown. Find the tension in each rope.

16. A 2000-kg block is suspended by two ropes as shown. Find the tension in each rope.

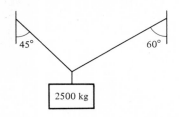

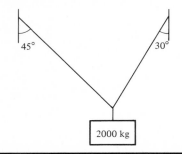

CHAPTER 11 REVIEW

In Exercises 1 and 2 solve the triangles. Angle C is a right angle. Use three-digit precision.

[11.1, i]   **1.** $a = 7.3$, $c = 8.6$ **2.** $a = 30.5$, $B = 51°10'$.

3. One leg of a right triangle bears east. The hypotenuse is 734 cm long and bears N 57°20′E. Find the perimeter.

[11.2, i]   **4.** Solve triangle ABC, where $B = 118°20'$, $C = 27°30'$, and $b = 0.974$.

5. In an isosceles triangle, the base angles each measure 52°20′ and the base is 513 cm long. Find the lengths of the other two sides.

[11.3, i]   **6.** In triangle ABC, $a = 3.7$, $c = 4.9$, and $B = 135°$. Find b.

[11.1, ii]   **7.** An observer's eye is 6 ft above the floor. A mural is being viewed. The bottom of the mural is at floor level. The observer looks downward 13° to see the bottom and upward 17° to see the top. How tall is the mural?

[11.2, ii]   **8.** A parallelogram has sides of lengths 3.21 cm and 7.85 cm. One of its angles measures 147°. Find the area of the parallelogram.

[11.5, ii]   **9.** A car is moving north at 45 mph. A ball is thrown from the car at 37 mph in a direction of N 45°E. Find the speed and direction of the ball over the ground.

[11.7, i]   **10.** A block weighing 150 lb rests on an inclined plane. The plane makes an angle of 45° with the ground. Find the components of the weight parallel and perpendicular to the plane.

[11.5, iv]   **11.** Find rectangular notation for the vector $(20, -200°)$.

[11.5, iii]   **12.** Find polar notation for the vector $(-2, 3)$.

[11.6, ii]   **13.** Convert to a polar equation: $x^2 + y^2 + 2x - 3y = 0$.

[11.5, ii]   **14.** Vector **u** has a component of 12 lb upward and a component of 18 lb to the right. Vector **v** has components of 35 lb downward and 35 lb to the left. Find the components of **u** + **v** and the magnitude and direction of **u** + **v**.

[11.5, v]   **15.** Do the following calculations for **u** = $(4, 3)$ and **v** = $(-3, 4)$.
   a) $4\mathbf{u} - 3\mathbf{v}$
   b) $|\mathbf{u} + \mathbf{v}|$

[11.7, i]   **16.** A crane is supporting a 500-kg steel beam. The boom of the crane is inclined 30° from the vertical. Find the compression in the boom of the crane.

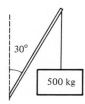

17. A parallelogram has sides of lengths 3.42 and 6.97. Its area is 18.4. Find the sizes of its angles.

12
IMAGINARY AND COMPLEX NUMBERS

i *i* *i*

i *i* *i*

OBJECTIVES

You should be able to:

[i] Express imaginary numbers (square roots of negative numbers) in terms of i, and simplify.

[ii] Add, subtract, and multiply complex numbers, expressing the answer as $a + bi$. Also factor sums of squares.

[iii] Determine whether a complex number is a solution of an equation.

[iv] Use the fact that for equality of complex numbers, the real parts must be the same and the imaginary parts must be the same.

Express in terms of i.

1. $\sqrt{-6}$

2. $-\sqrt{-10}$

3. $\sqrt{-4}$

4. $-\sqrt{-25}$

5. Simplify $\sqrt{-5}\sqrt{-2}$.

Simplify.

6. $\dfrac{\sqrt{-22}}{\sqrt{-2}}$

7. $\dfrac{\sqrt{-21}}{\sqrt{3}}$

8. $\sqrt{-16} + \sqrt{-9}$

9. $\sqrt{-25} - \sqrt{-4}$

10. $\sqrt{-17} + \sqrt{-9}$

12.1 IMAGINARY AND COMPLEX NUMBERS

[i] Imaginary Numbers

Negative numbers do not have square roots in the system of real numbers. Certain equations have no solutions. A new kind of number, called *imaginary*, was invented so that negative numbers would have square roots and certain equations would have solutions. These numbers were devised, starting with an imaginary unit, named i, with the agreement that $i^2 = -1$ or $i = \sqrt{-1}$. All other imaginary numbers can be expressed as a product of i and a real number.

Example 1 Express $\sqrt{-5}$ and $-\sqrt{-7}$ in terms of i.
$$\sqrt{-5} = \sqrt{-1 \cdot 5} = \sqrt{-1}\sqrt{5} = i\sqrt{5},$$
$$-\sqrt{-7} = -\sqrt{-1 \cdot 7} = -\sqrt{-1}\sqrt{7} = -i\sqrt{7}$$

It is also to be understood that the imaginary unit obeys the familiar laws of real numbers, such as the commutative and associative laws.

Example 2 Simplify $\sqrt{-3}\sqrt{-7}$.

Important. We first express the two imaginary numbers in terms of i:
$$\sqrt{-3}\sqrt{-7} = i\sqrt{3} \cdot i\sqrt{7}.$$

Now, rearranging and combining, we have
$$i^2\sqrt{3}\sqrt{7} = -1 \cdot \sqrt{21}$$
$$= -\sqrt{21}.$$

Had we not expressed imaginary numbers in terms of i at the outset, we would have obtained $\sqrt{21}$ instead of $-\sqrt{21}$.

> *Important:* **All imaginary numbers must be expressed in terms of i before simplifying.**

DO EXERCISES 1–5.

Example 3 Simplify $-\sqrt{20}/\sqrt{-5}$.
$$\frac{-\sqrt{20}}{\sqrt{-5}} = \frac{-\sqrt{20}}{i\sqrt{5}} \cdot \frac{i}{i} = \frac{-i\sqrt{20}}{i^2\sqrt{5}}$$
$$= \frac{-i}{-1}\sqrt{\frac{20}{5}} = 2i$$

Example 4 Simplify $\sqrt{-9} + \sqrt{-25}$.
$$\sqrt{-9} + \sqrt{-25} = i\sqrt{9} + i\sqrt{25}$$
$$= 3i + 5i = (3 + 5)i$$
$$= 8i$$

DO EXERCISES 6–10.

Powers of i

Let us look at the powers of i:

$$i^2 = -1, \qquad\qquad i^5 = i^4 \cdot i = 1 \cdot i = i,$$
$$i^3 = i^2 \cdot i = -1 \cdot i = -i, \qquad i^6 = i^5 \cdot i = i \cdot i = -1,$$
$$i^4 = i^2 \cdot i^2 = -1(-1) = 1, \qquad i^7 = i^6 \cdot i = -1 \cdot i = -i.$$

The first four powers of i are all different, but thereafter there is a repeating pattern, in cycles of four. Note that $i^4 = 1$ and that all powers of i^4, such as i^8, i^{16}, and so on, are 1. To find a higher power we express it in terms of the nearest power of 4 less than the given one.

Example 5 Simplify i^{17} and i^{23}.

$$i^{17} = i^{16} \cdot i$$

Since i^{16} is a power of i^4 and $i^4 = 1$, we know $i^{16} = 1$ so

$$i^{17} = 1 \cdot i = i,$$
$$i^{23} = i^{20} \cdot i^3.$$

Since i^{20} is a power of i^4, $i^{20} = 1$, so $i^{23} = 1 \cdot i^3 = 1 \cdot (-i) = -i$.

DO EXERCISES 11–13.

[ii] Complex Numbers

The equation $x^2 + 1 = 0$ has no solution in real numbers, but it has the imaginary solutions i and $-i$. There are still rather simple-looking equations that do not have either real or imaginary solutions. For example, $x^2 - 2x + 2 = 0$ does not. If we allow sums of real and imaginary numbers, however, this equation and many others have solutions, as we shall show.

In order that more equations will have solutions, we invent a new system of numbers called the *system of complex numbers*.* A complex number is a sum of a real number and an imaginary number.

DEFINITION

The set of complex numbers consists of all numbers $a + bi$, where a and b are real numbers.

For the complex number $a + bi$ we say that the *real part* is a and the *imaginary part* is bi.

We also agree that the familiar properties of real numbers (the *field* properties) hold also for complex numbers†. We list them.

Simplify.

11. i^{25}

12. i^{18}

13. i^{31}

---

*One may wonder why we do not invent a system of imaginary numbers. That would not make sense because products of imaginary numbers are not necessarily imaginary. Consider, for example, $i^2 = -1$.

†In a more rigorous treatment, addition and multiplication are defined for complex numbers. It is then proved that the field properties hold.

Simplify.

14. $(8 - i) + (4 + 2i)$

15. $(9 + 2i) - (4 + 3i)$

16. $(2 + 4i)(3 + i)$

17. $(4 + 5i) + (4 - 5i)$

18. $2i(4 + 3i)$

19. $(5 + 6i) - (5 + 3i)$

Factor.

20. $x^2 + 4$

21. $9 + y^2$

> **COMMUTATIVITY.** Addition and multiplication are commutative.
>
> **ASSOCIATIVITY.** Addition and multiplication are associative.
>
> **DISTRIBUTIVITY.** Multiplication is distributive over addition and also over subtraction.
>
> **IDENTITIES.** The additive identity is $0 + 0i$, or 0. The multiplicative identity is $1 + 0i$, or 1.
>
> **ADDITIVE INVERSES.** Every complex number $a + bi$ has the additive inverse $-a - bi$.
>
> **MULTIPLICATIVE INVERSES.** Every nonzero complex number has a multiplicative inverse, or reciprocal.

The complex-numbers system is an extension of the real-number system. Any number $a + bi$, where a and b are real numbers, is a complex number. The number b can be 0, in which case we have $a + 0i$, which simplifies to the real number a. Thus the complex numbers include all of the real numbers. They also include all of the imaginary numbers, because any imaginary number bi is equal to $0 + bi$.

Calculations

Since in the system of complex numbers the field properties hold, calculations are much the same as for real numbers. The primary difference is that one must remember that $i^2 = -1$.

Example 6 Simplify $(8 + 6i) + (3 + 2i)$.

$$(8 + 6i) + (3 + 2i) = 8 + 3 + 6i + 2i$$
$$= 11 + 8i$$

Example 7 Simplify $(1 + 2i)(1 + 3i)$.

$$(1 + 2i)(1 + 3i) = 1 + 6i^2 + 2i + 3i$$
$$= 1 - 6 + 2i + 3i \qquad i^2 = -1$$
$$= -5 + 5i$$

Example 8 Simplify $(3 + 2i) - (5 - 2i)$.

$$(3 + 2i) - (5 - 2i) = (3 + 2i) - 5 + 2i$$
$$= 3 - 5 + 2i + 2i$$
$$= -2 + 4i$$

DO EXERCISES 14–19.

In the field of real numbers, a sum of two squares cannot be factored. In the system of complex numbers a sum of squares is always factorable.

Example 9 Factor $x^2 + y^2$.

$$(x + yi)(x - yi)$$

A check by multiplying will show that this is correct.

DO EXERCISES 20 AND 21.

[iii] Solutions of Equations

In the system of complex numbers, a great many equations have solutions. In fact, any equation $P(x) = 0$, where $P(x)$ is a nonconstant polynomial, has a solution.*

Example 10 Determine whether $1 + i$ is a solution of $x^2 - 2x + 2 = 0$.

$$x^2 - 2x + 2 = 0$$

$(1 + i)^2 - 2(1 + i) + 2$	0	Substituting
$1 + 2i + i^2 - 2 - 2i + 2$		
$1 + 2i - 1 - 2 - 2i + 2$		
0		

The number $1 + i$ is a solution.

DO EXERCISE 22.

[iv] Equality for Complex Numbers

An equation $a + bi = c + di$ will be true if and only if the real parts are the same and the imaginary parts are the same. In other words, we have the following.

$$a + bi = c + di \quad \text{if and only if} \quad a = c \text{ and } b = d.$$

Example 11 Suppose that $3x + yi = 5x + 1 + 2i$. Find x and y.

We equate the real parts: $3x = 5x + 1$. Solving this equation, we obtain $x = -\frac{1}{2}$. We equate the imaginary parts: $yi = 2i$. Thus $y = 2$.

DO EXERCISE 23.

22. Determine whether $1 - i$ is a solution of $x^2 - 2x + 2 = 0$.

23. Given that $3x + 1 + (y + 2)i = 2x + 2yi$, find x and y.

EXERCISE SET 12.1

[i] In Exercises 1–6, express in terms of i.

1. $\sqrt{-15}$　　　　　　　**2.** $\sqrt{-17}$　　　　　　　**3.** $\sqrt{-16}$

4. $\sqrt{-25}$　　　　　　　**5.** $-\sqrt{-12}$　　　　　　　**6.** $-\sqrt{-20}$

In Exercises 7–20, simplify. Leave answers in terms of i in Exercises 7–10.

7. $\sqrt{-16} + \sqrt{-25}$　　　　**8.** $\sqrt{-36} - \sqrt{-4}$　　　　**9.** $\sqrt{-7} - \sqrt{-10}$

10. $\sqrt{-5} + \sqrt{-7}$　　　　**11.** $\sqrt{-5}\sqrt{-11}$　　　　**12.** $\sqrt{-7}\sqrt{-8}$

13. $-\sqrt{-4}\sqrt{-5}$　　**14.** $-\sqrt{-9}\sqrt{-7}$　　**15.** $\dfrac{-\sqrt{5}}{\sqrt{-2}}$　　**16.** $\dfrac{\sqrt{-7}}{-\sqrt{5}}$

17. $\dfrac{\sqrt{-9}}{-\sqrt{4}}$　　**18.** $\dfrac{-\sqrt{25}}{\sqrt{-16}}$　　**19.** $\dfrac{-\sqrt{-36}}{\sqrt{-9}}$　　**20.** $\dfrac{\sqrt{-25}}{-\sqrt{-16}}$

[ii] In Exercises 21–38, simplify.

21. $(2 + 3i) + (4 + 2i)$　　**22.** $(5 - 2i) + (6 + 3i)$　　**23.** $(4 + 3i) + (4 - 3i)$　　**24.** $(2 + 3i) + (-2 - 3i)$

25. $(8 + 11i) - (6 + 7i)$　　**26.** $(9 - 5i) - (4 + 2i)$　　**27.** $2i - (4 + 3i)$　　**28.** $3i - (5 + 2i)$

*See Chapter 13 for more about this.

29. $(1 + 2i)(1 + 3i)$ **30.** $(1 + 4i)(1 - 3i)$ **31.** $(1 + 2i)(1 - 3i)$ **32.** $(2 + 3i)(2 - 3i)$

33. $3i(4 + 2i)$ **34.** $5i(3 - 4i)$ **35.** $(2 + 3i)^2$ **36.** $(3 - 2i)^2$

37. i^{13} **38.** i^{72}

Factor.

39. $4x^2 + 25y^2$ **40.** $16a^2 + 49b^2$

[iii]

41. Determine whether $1 + 2i$ is a solution of $x^2 - 2x + 5 = 0$.

42. Determine whether $1 - 2i$ is a solution of $x^2 - 2x + 5 = 0$.

[iv] In Exercises 43 and 44, solve for x and y.

43. $4x + 7i = -6 + yi$ **44.** $-4 + (x + y)i = 2x - 5y + 5i$

☆

45. A function f is defined as follows: $f(z) = z^2 - 4z + i$. Find $f(3 + i)$.

46. A function g is defined as follows: $g(z) = 2z^2 + z - 2i$. Find $g(2 - i)$.

47. Show that the general rule for radicals, in real numbers, $\sqrt{a \cdot b} = \sqrt{a} \cdot \sqrt{b}$, does not hold for complex numbers.

48. Show that the general rule for radicals, in real numbers, $\sqrt{\dfrac{a}{b}} = \dfrac{\sqrt{a}}{\sqrt{b}}$ does not hold for complex numbers.

49. Multiply.

$$\begin{bmatrix} i & 0 \\ 0 & 3i \end{bmatrix} \begin{bmatrix} i & 2i \\ 1 - i & 4 - i \end{bmatrix}$$

★

50. Solve $z^2 = -2i$. (*Hint*: Let $z = a + bi$.)

51. Solve $\dfrac{z^2}{4} = i$. (*Hint*: Let $z = a + bi$.)

OBJECTIVES

You should be able to:

[i] Find the conjugate of a complex number and divide complex numbers.

[ii] Find the reciprocal of a complex number and express it in the form $a + bi$.

[iii] Given a polynomial in z, find a polynomial in $\bar{z}$ that is its conjugate.

Find the conjugate of each number.

1. $7 + 2i$ **2.** $6 - 4i$

3. $-5i$ **4.** $3i$

5. -3 **6.** 8

12.2 CONJUGATES AND DIVISION

[i] Conjugates

We shall define the conjugate of a complex number as follows.

DEFINITION

The *conjugate* of a complex number $a + bi$ is $a - bi$, and the conjugate of $a - bi$ is $a + bi$.

We illustrate:

The conjugate of $3 + 4i$ is $3 - 4i$.
The conjugate of $5 - 7i$ is $5 + 7i$.
The conjugate of $5i$ is $-5i$.
The conjugate of 6 is 6.

DO EXERCISES 1–6.

Division

Fractional notation is useful for division of complex numbers. We also use the notion of conjugates.

Example 1 Divide $4 + 5i$ by $1 + 4i$.

We shall write fractional notation and then multiply by 1.

$$\frac{4 + 5i}{1 + 4i} = \frac{4 + 5i}{1 + 4i} \cdot \frac{1 - 4i}{1 - 4i} \qquad \text{Note that } 1 - 4i \text{ is the conjugate of the divisor.}$$

$$= \frac{(4 + 5i)(1 - 4i)}{1^2 - 4^2 i^2}$$

$$= \frac{24 - 11i}{1 + 16} = \frac{24 - 11i}{17} \qquad i^2 = -1$$

$$= \frac{24}{17} - \frac{11}{17} i$$

The procedure of the last example allows us always to find a quotient of two numbers and express it in the form $a + bi$. This is true because the product of a number and its conjugate is always a real number (we will prove this later), giving us a real-number denominator.

DO EXERCISES 7 AND 8.

[ii] Reciprocals

We can find the reciprocal, or multiplicative inverse, of a complex number by division. The reciprocal of a complex number z is $1/z$.

Example 2 Find the reciprocal of $2 - 3i$ and express it in the form $a + bi$.

a) The reciprocal of $2 - 3i$ is $\dfrac{1}{2 - 3i}$.

b) We can express it in the form $a + bi$ as follows:

$$\frac{1}{2 - 3i} = \frac{1}{2 - 3i} \cdot \frac{2 + 3i}{2 + 3i} = \frac{2 + 3i}{2^2 - 3^2 i^2}$$

$$= \frac{2 + 3i}{4 + 9} = \frac{2}{13} + \frac{3}{13} i.$$

DO EXERCISE 9.

[iii] Properties of Conjugates

We may use a single letter for a complex number. For example, we could shorten $a + bi$ to z. To denote the conjugate of a number, we use a bar. The conjugate of z is $\bar{z}$. Or, the conjugate of $a + bi$ is $\overline{a + bi}$. Of course, by definition of conjugates, $\overline{a + bi} = a - bi$ and $\overline{a - bi} = a + bi$. We have already noted that the product of a number and its conjugate is always a real number. Let us state this formally and prove it.

THEOREM 1

For any complex number z, $z \cdot \bar{z}$ is a real number.

Divide.

7. $\dfrac{1 + 3i}{3 + 2i}$

8. $\dfrac{2 + i}{3 - 2i}$

9. Find the reciprocal of $3 + 4i$ and express it in the form $a + bi$.

10. Compare $\overline{(3 + 2i) + (4 - 5i)}$ and $\overline{(3 + 2i)} + \overline{(4 - 5i)}$.

Proof. Let $z = a + bi$. Then

$$z \cdot \bar{z} = (a + bi)(a - bi)$$
$$= a^2 - b^2 i^2$$
$$= a^2 + b^2.$$

Since a and b are real numbers, so is $a^2 + b^2$. Thus $z \cdot \bar{z}$ is real.

The sum of a number and its conjugate is also always real. We state this as the second property.

THEOREM 2

For any complex number z, $z + \bar{z}$ is a real number.

Proof. Let $z = a + bi$. Then

$$z + \bar{z} = (a + bi) + (a - bi)$$
$$= 2a.$$

Since a is a real number, $2a$ is real. Thus $z + \bar{z}$ is real.

Now we consider the conjugate of a sum and compare it with the sum of the conjugates.

Example 3 Compare $\overline{(2 + 4i) + (5 + i)}$ and $\overline{(2 + 4i)} + \overline{(5 + i)}$.

a) $\overline{(2 + 4i) + (5 + i)} = \overline{7 + 5i}$ Adding the complex numbers
                        $= 7 - 5i$ Taking the conjugate

b) $\overline{(2 + 4i)} + \overline{(5 + i)} = (2 - 4i) + (5 - i)$ Taking conjugates
                       $= 7 - 5i$ Adding

DO EXERCISE 10.

Taking the conjugate of a sum gives the same result as adding the conjugates. Let us state and prove this.

THEOREM 3

For any complex numbers z and w, $\overline{z + w} = \bar{z} + \bar{w}$.

Proof. Let $z = a + bi$ and $w = c + di$. Then

$$\overline{z + w} = \overline{(a + bi) + (c + di)}$$
$$= \overline{(a + c) + (b + d)i},$$

by adding. We now take the conjugate and obtain $(a + c) - (b + d)i$. Now

$$\bar{z} + \bar{w} = \overline{(a + bi)} + \overline{(c + di)}$$
$$= (a - bi) + (c - di),$$

taking the conjugates. We will now add to obtain $(a + c) - (b + d)i$, the same result as before. Thus $\overline{z + w} = \bar{z} + \bar{w}$.

Let us next consider the conjugate of a product.

Example 4 Compare $\overline{(3 + 2i)(4 - 5i)}$ and $\overline{(3 + 2i)} \cdot \overline{(4 - 5i)}$.

$$\overline{(3 + 2i)(4 - 5i)} = \overline{22 - 7i} \qquad \text{Multiplying}$$
$$= 22 + 7i, \qquad \text{Taking the conjugate}$$
$$\overline{(3 + 2i)} \cdot \overline{(4 - 5i)} = (3 - 2i)(4 + 5i) \qquad \text{Taking conjugates}$$
$$= 22 + 7i \qquad \text{Multiplying}$$

DO EXERCISE 11.

The conjugate of a product is the product of the conjugates. This is our next result.

THEOREM 4

For any complex numbers z and w, $\overline{z \cdot w} = \overline{z} \cdot \overline{w}$.

Proof. Let $z = a + bi$ and $w = c + di$. Then

$$\overline{z \cdot w} = \overline{(a + bi)(c + di)}$$
$$= \overline{(ac - bd) + (bc + ad)i}. \qquad \text{Multiplying}$$

Taking the conjugate, we obtain $(ac - bd) - (bc + ad)i$. Now

$$\overline{z} \cdot \overline{w} = \overline{(a + bi)} \cdot \overline{(c + di)} = (a - bi)(c - di),$$

taking conjugates. By multiplication we obtain $(ac - bd) - (bc + ad)i$, the same result as before. Thus $\overline{z \cdot w} = \overline{z} \cdot \overline{w}$.

Let us now consider conjugates of powers, using the preceding result.

Example 5 Show that for any complex number z, $\overline{z^2} = \overline{z}^2$.

$$\overline{z^2} = \overline{z \cdot z} \qquad \text{By the definition of exponents}$$
$$= \overline{z} \cdot \overline{z} \qquad \text{By Theorem 4}$$
$$= \overline{z}^2 \qquad \text{By the definition of exponents}$$

DO EXERCISE 12.

We now state our next result.

THEOREM 5

For any complex number z, $\overline{z^n} = \overline{z}^n$. In other words, the conjugate of a power is the power of the conjugate. We understand that the exponent is a natural number.

The conjugate of a real number $a + 0i$ is $a - 0i$, and both are equal to a. Thus a real number is its own conjugate. We state this as our next result.

THEOREM 6

If z is a real number, then $\overline{z} = z$.

11. Compare $\overline{(2 + 5i)(1 + 3i)}$ and $\overline{(2 + 5i)} \cdot \overline{(1 + 3i)}$.

12. Show that for any complex number z, $\overline{z^3} = \overline{z}^3$.

Find a polynomial in $\bar{z}$ that is the conjugate of each of the following.

13. $5z^3 + 4z^2 - 2z + 1$

14. $7z^5 - 3z^3 + 8z^2 + z$

Conjugates of Polynomials

Given a polynomial in z, where z is a variable for a complex number, we can find its conjugate in terms of $\bar{z}$.

Example 6 Find a polynomial in $\bar{z}$ that is the conjugate of $3z^2 + 2z - 1$.

We write an expression for the conjugate and then use the properties of conjugates.

$$
\begin{aligned}
\overline{3z^2 + 2z - 1} &= \overline{3z^2} + \overline{2z} - \overline{1} && \text{By Theorem 3} \\
&= \overline{3z^2} + \overline{2} \cdot \overline{z} - \overline{1} && \text{By Theorem 4} \\
&= 3\overline{z^2} + 2\overline{z} - 1 && \text{The conjugate of a real number} \\
& && \text{is the number itself.} \\
&= 3\bar{z}^2 + 2\bar{z} - 1 && \text{By Theorem 5}
\end{aligned}
$$

DO EXERCISES 13 AND 14.

EXERCISE SET 12.2

[i] In Exercises 1–16, simplify.

1. $\dfrac{4 + 3i}{1 - i}$ **2.** $\dfrac{2 - 3i}{5 - 4i}$ **3.** $\dfrac{\sqrt{2} + i}{\sqrt{2} - i}$ **4.** $\dfrac{\sqrt{3} + i}{\sqrt{3} - i}$

5. $\dfrac{3 + 2i}{i}$ **6.** $\dfrac{2 + 3i}{i}$ **7.** $\dfrac{i}{2 + i}$ **8.** $\dfrac{3}{5 - 11i}$

9. $\dfrac{1 - i}{(1 + i)^2}$ **10.** $\dfrac{1 + i}{(1 - i)^2}$ **11.** $\dfrac{3 - 4i}{(2 + i)(3 - 2i)}$ **12.** $\dfrac{(4 - i)(5 + i)}{(6 - 5i)(7 - 2i)}$

13. $\dfrac{1 + i}{1 - i} \cdot \dfrac{2 - i}{1 - i}$ **14.** $\dfrac{1 - i}{1 + i} \cdot \dfrac{2 + i}{1 + i}$ **15.** $\dfrac{3 + 2i}{1 - i} + \dfrac{6 + 2i}{1 - i}$ **16.** $\dfrac{4 - 2i}{1 + i} + \dfrac{2 - 5i}{1 + i}$

[ii] In Exercises 17–24, find the reciprocal and express it in the form $a + bi$.

17. $4 + 3i$ **18.** $4 - 3i$ **19.** $5 - 2i$ **20.** $2 + 5i$

21. i **22.** $-i$ **23.** $-4i$ **24.** $5i$

[iii] In Exercises 25–28, find a polynomial in $\bar{z}$ that is the conjugate.

25. $3z^5 - 4z^2 + 3z - 5$ **26.** $7z^4 + 5z^3 - 12z$

27. $4z^7 - 3z^5 + 4z$ **28.** $5z^{10} - 7z^8 + 13z^2 - 4$

☆ _____

29. Solve $z + 6\bar{z} = 7$. **30.** Solve $5z - 4\bar{z} = 7 + 8i$.

31. Let $z = a + bi$. Find $\frac{1}{2}(z + \bar{z})$. **32.** Let $z = a + bi$. Find $\frac{1}{2}i(\bar{z} - z)$.

33. Find $f(3 + i)$, where $f(z) = \dfrac{\bar{z}}{z - 1}$. **34.** Show that for any complex number z, $z + \bar{z}$ is real.

★ _____

35. Let $z = a + bi$.

Find a general expression for $\dfrac{1}{z}$.

36. Let $z = a + bi$ and $w = c + di$.

Find a general expression for $\dfrac{w}{z}$.

37. State and prove a theorem about the conjugate of a polynomial.

12.3 EQUATIONS AND COMPLEX NUMBERS

[i] Linear Equations

Linear equations with complex coefficients are solved in the same way as equations with real-number coefficients. The steps used in solving depend on the field properties in each case.

Example 1 Solve $3ix + 4 - 5i = (1 + i)x + 2i$.

$$3ix - (1 + i)x = 2i - (4 - 5i) \quad \text{Adding } -(1 + i)x \text{ and } -(4 - 5i)$$
$$(-1 + 2i)x = -4 + 7i \quad \text{Simplifying}$$
$$x = \frac{-4 + 7i}{-1 + 2i} \quad \text{Dividing}$$
$$x = \frac{-4 + 7i}{-1 + 2i} \cdot \frac{-1 - 2i}{-1 - 2i} \quad \text{Simplifying}$$
$$x = \frac{18 + i}{5} = \frac{18}{5} + \frac{1}{5}i$$

DO EXERCISE 1.

[ii] Quadratic Equations

In Chapter 2 we derived the quadratic formula for equations with real-number coefficients. A glance at that derivation shows that all the steps can also be done with complex-number coefficients with one possible exception. We need to know that every nonzero complex number has two square roots that are additive inverses of each other. Such is the case, as will be shown later. Thus the quadratic formula holds for equations with complex coefficients.

THEOREM 7

Any equation $ax^2 + bx + c = 0$, where $a \neq 0$ and a, b, and c are complex numbers, has solutions $\dfrac{-b \pm \sqrt{b^2 - 4ac}}{2a}$.

Example 2 Solve $x^2 + (1 + i)x - 2i = 0$.

We note that $a = 1$, $b = 1 + i$, and $c = -2i$. Thus

$$x = \frac{-(1 + i) \pm \sqrt{(1 + i)^2 - 4 \cdot 1 \cdot (-2i)}}{2 \cdot 1} \quad \text{Substituting in the formula}$$
$$= \frac{-1 - i \pm \sqrt{10i}}{2}. \quad \text{Simplifying}$$

DO EXERCISE 2.

In Example 1 we have not attempted to evaluate $\sqrt{10i}$. That will be done later (Example 6). Let us say, however, that $10i$ has two square roots that are additive inverses of each other. The same is true of every

3. Solve.
$$5x^2 + 6x + 5 = 0$$

nonzero complex number. There is no problem of negative numbers not having square roots, because in the system of complex numbers there are no positive or negative numbers. Whenever we speak of positive or negative numbers, we will be referring to the system of real numbers.

[iii] Quadratic Equations with Real Coefficients

Since any real number a is $a + 0i$, we can consider real numbers to be special kinds of complex numbers. If a quadratic equation has real coefficients, the quadratic formula always gives solutions in the system of complex numbers. Recall that in using the formula, we take the square root of $b^2 - 4ac$. If this *discriminant* is negative, the solutions will have nonzero imaginary parts.

Example 3 Solve $x^2 + 2x + 5 = 0$.

We note that $a = 1$, $b = 2$, and $c = 5$. Thus

$$x = \frac{-2 \pm \sqrt{2^2 - 4 \cdot 1 \cdot 5}}{2 \cdot 1} = \frac{-2 \pm \sqrt{-16}}{2} = \frac{-2 \pm 4i}{2}.$$

The solutions are $-1 + 2i$ and $-1 - 2i$.

4. Find an equation having i and $1 + i$ as solutions.

It is easy to see from Example 3 that when there are nonzero imaginary parts, an equation has two solutions that are conjugates of each other.

DO EXERCISE 3.

[iv] Writing Equations with Specified Solutions

The principle of zero products for real numbers states that a product is 0 if and only if at least one of the factors is 0. This principle also holds for complex numbers, as we will show in Section 12.4. Since the principle holds for complex numbers we can write equations having specified solutions.

Example 4 Find an equation having the numbers 1, i, and $-i$ as solutions.

The factors we use will be $x - 1$, $x - i$, and $x + i$. Next we set the product of these factors equal to 0:

$$(x - 1)(x - i)(x + i) = 0.$$

5. Find an equation having 2, i, and $-i$ as solutions.

Now we multiply and simplify:

$$(x - 1)(x^2 - i^2) = 0$$
$$(x - 1)(x^2 + 1) = 0$$
$$x^3 - x^2 + x - 1 = 0.$$

Example 5 Find an equation having -1, i, and $1 + i$ as solutions.

$$(x + 1)(x - i)[x - (1 + i)] = 0$$
$$(x^2 + x - ix - i)(x - 1 - i) = 0$$
$$x^3 - 2ix^2 - ix - 2x - 1 + i = 0$$

DO EXERCISES 4 AND 5.

[v] **Finding Square Roots**

In Example 2 we used the quadratic formula, but did not evaluate $\sqrt{10i}$. We wish to find complex numbers z for which $z^2 = 10i$. We do so in the following example.

6. Solve.

$$z^2 = 3 - 4i$$

Example 6 Solve $z^2 = 10i$.

Let $z = x + yi$. Then we have $(x + yi)^2 = 10i$, or $x^2 + 2xyi + y^2i^2 = 10i$, or $x^2 - y^2 + 2xyi = 0 + 10i$. Since the real parts must be the same and the imaginary parts the same, $x^2 - y^2 = 0$ and $2xy = 10$. Here we have two equations in real numbers. We can find a solution by solving the second equation for y and substituting in the first. Since

$$y = \frac{5}{x},$$

then

$$x^2 - \left(\frac{5}{x}\right)^2 = 0$$

$$x^2 - \frac{25}{x^2} = 0 \quad \text{or} \quad x^4 - 25 = 0.$$

We now have an equation reducible to quadratic:

$$(x^2 + 5)(x^2 - 5) = 0 \qquad \text{Factoring}$$
$$x^2 + 5 = 0 \quad \text{or} \quad x^2 - 5 = 0. \qquad \text{Principle of zero products}$$

The real solutions are $\sqrt{5}$ and $-\sqrt{5}$. Now, going back to $2xy = 10$, we see that if $x = \sqrt{5}$, then $y = \sqrt{5}$, and if $x = -\sqrt{5}$, then $y = -\sqrt{5}$. Thus the solutions of our equation (and the square roots of $10i$) are $\sqrt{5} + \sqrt{5}i$ and $-\sqrt{5} - \sqrt{5}i$.

DO EXERCISE 6.

EXERCISE SET 12.3

[i] Solve.

1. $(3 + i)x + i = 5i$

2. $(2 + i)x - i = 5 + i$

3. $2ix + 5 - 4i = (2 + 3i)x - 2i$

4. $5ix + 3 + 2i = (3 - 2i)x + 3i$

5. $(1 + 2i)x + 3 - 2i = 4 - 5i + 3ix$

6. $(1 - 2i)x + 2 - 3i = 5 - 4i + 2x$

7. $(5 + i)x + 1 - 3i = (2 - 3i)x + 2 - i$

8. $(5 - i)x + 2 - 3i = (3 - 2i)x + 3 - i$

[ii] In Exercises 9–14, solve using the quadratic formula. You need not evaluate square roots.

9. $x^2 + (1 - i)x + i = 0$

10. $x^2 + (1 + i)x - i = 0$

11. $2x^2 + ix + 1 = 0$

12. $2x^2 - ix + 1 = 0$

13. $3x^2 + (1 + 2i)x + 1 - i = 0$

14. $3x^2 + (1 - 2i)x + 1 + i = 0$

[iii] Solve.

15. $x^2 - 2x + 5 = 0$

16. $x^2 - 4x + 5 = 0$

17. $x^2 - 4x + 13 = 0$

18. $x^2 - 6x + 13 = 0$

19. $x^2 + 3x + 4 = 0$

20. $3x^2 + x + 2 = 0$

[iv] Find an equation having the specified solutions.

21. $2i, -2i$ **22.** $3i, -3i$ **23.** $1 + i, 1 - i$ **24.** $2 + i, 2 - i$

25. $2 + 3i, 2 - 3i$ **26.** $4 + 3i, 4 - 3i$ **27.** $3, i$ **28.** $5, i$

29. $1, 3i, -3i$ **30.** $1, 2i, -2i$ **31.** $2, 1 + i, i$ **32.** $3, 1 - i, i$

33. $i, 2i, -i$ **34.** $i, -2i, -i$

[v] Solve.

35. $z^2 = 4i$ **36.** $z^2 = -4i$ **37.** $z^2 = 3 + 4i$ **38.** $z^2 = 5 - 12i$

Solve. (*Hint:* Factor first.)

39. $x^3 - 8 = 0$ **40.** $x^3 + 8 = 0$

Solve.

41. $2x + 3y = 7 - 7i,$
$\quad 3x - 2y = 4 + 9i$

42. $3x + 4y = 11 + 9i,$
$\quad 2x - y = -5i$

★

Solve for x and y.

43. $\log x + 2^y i = 2 + 16i$ **44.** $\sin^2 x + i \cos^2 y = 1 + \dfrac{i}{2}$ **45.** $e^{2x} + ie^y = 1 + i$

OBJECTIVES

You should be able to:

[i] Graph a complex number $a + bi$ and graph the sum of two numbers.

[ii] Given rectangular, or binomial, notation for a complex number, find polar notation.

[iii] Given polar notation for a complex number, find binomial notation.

[iv] Use polar notation to multiply and divide complex numbers.

12.4 GRAPHICAL REPRESENTATION AND POLAR NOTATION

[i] Graphs

The real numbers can be graphed on a line. Complex numbers are graphed on a plane. We graph a complex number $a + bi$ in the same way that we graph an ordered pair of real numbers (a, b). In place of an x-axis we have a *real axis,* and in place of a y-axis we have an *imaginary axis.*

Examples Graph the following.

1. $3 + 2i$
2. $-4 + 5i$
3. $-5 - 4i$

See Fig. 12.1.

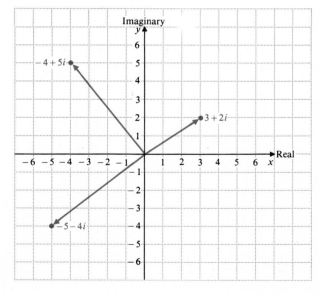

Figure 12.1

Horizontal distances correspond to the real part of a number. Vertical distances correspond to the imaginary part. Graphs of the numbers are shown above as vectors. The horizontal component of the vector for $a + bi$ is a and the vertical component is b. Complex numbers are sometimes used in the study of vectors.

DO EXERCISE 1.

Adding complex numbers is like adding vectors using components. For example, to add $3 + 2i$ and $5 + 4i$ we add the real parts and the imaginary parts to obtain $8 + 6i$. Graphically, then, the sum of two complex numbers looks like a vector sum. It is the diagonal of a parallelogram.

Example 4 Show graphically $2 + 2i$ and $3 - i$. Show also their sum.

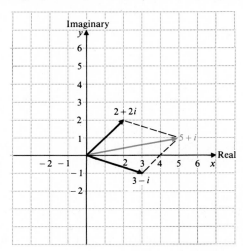

DO EXERCISE 2.

[ii] Polar Notation for Complex Numbers

In Chapter 11 we studied polar notation for vectors, in which we use the length of a vector and the angle that it makes with the x-axis. In a similar way we develop polar notation for complex numbers. From the diagram we can see that the length of the vector for $a + bi$ is $\sqrt{a^2 + b^2}$. Note that this quantity is a nonnegative real number. It is called the *absolute value* of $a + bi$, and is denoted $|a + bi|$.

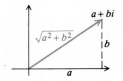

DEFINITION

The *absolute value* of a complex number $a + bi$ is denoted $|a + bi|$ and is defined to be $\sqrt{a^2 + b^2}$.

1. Graph these complex numbers.

a) $5 - 3i$
b) $-3 + 4i$
c) $-5 - 2i$
d) $5 + 5i$

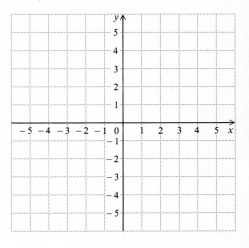

2. Graph each pair of complex numbers and graph their sum.

a) $2 + 3i$, $1 - 5i$
b) $-5 + 2i$, $-1 - 4i$

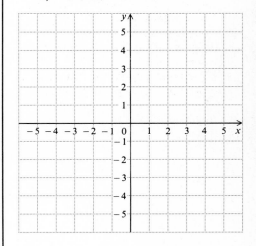

3. Find the following absolute values.

a) $|4 - 3i|$

b) $|-12 - 5i|$

4. Write rectangular notation for

$$\sqrt{2}\,(\cos 315° + i \sin 315°).$$

5. Write binomial notation for

$$2 \operatorname{cis}\left(-\frac{\pi}{6}\right).$$

Example 5 Find $|3 + 4i|$.

$$|3 + 4i| = \sqrt{3^2 + 4^2} = \sqrt{9 + 16} = 5$$

DO EXERCISE 3.

Now let us consider any complex number $a + bi$. Suppose that its absolute value is r. Let us also suppose that the angle that the vector makes with the real axis is θ. As this diagram shows, we have

$$a = r \cos\theta \quad \text{and} \quad b = r \sin\theta.$$

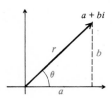

Thus

$$a + bi = r \cos\theta + ir \sin\theta$$
$$= r(\cos\theta + i \sin\theta).$$

This is polar notation for $a + bi$. The angle θ is called the *argument*.

Polar notation for complex numbers is also called *trigonometric notation*, and is often abbreviated $r \operatorname{cis}\theta$.

[ii], [iii] **Change of Notation**

To change from polar notation to *binomial*, or *rectangular*, notation $a + bi$, we proceed as in Section 11.6, recalling that $a = r \cos\theta$ and $b = r \sin\theta$.

Example 6 Write binomial notation for $2(\cos 120° + i \sin 120°)$.

$$a = 2 \cos 120° = -1,$$
$$b = 2 \sin 120° = \sqrt{3}$$

Thus $2(\cos 120° + i \sin 120°) = -1 + i\sqrt{3}$.

Example 7 Write binomial notation for $\sqrt{8} \operatorname{cis} \dfrac{7\pi}{4}$.

$$a = \sqrt{8} \cos \frac{7\pi}{4} = \sqrt{8} \cdot \frac{1}{\sqrt{2}} = 2,$$

$$b = \sqrt{8} \sin \frac{7\pi}{4} = \sqrt{8} \cdot \frac{-1}{\sqrt{2}} = -2$$

Thus $\sqrt{8} \operatorname{cis} \dfrac{7\pi}{4} = 2 - 2i$.

DO EXERCISES 4 AND 5.

To change from binomial notation to polar notation, we remember that $r = \sqrt{a^2 + b^2}$ and θ is an angle for which $\sin \theta = b/r$ and $\cos \theta = a/r$.

Example 8 Find polar notation for $1 + i$.

We note that $a = 1$ and $b = 1$. Then
$$r = \sqrt{1^2 + 1^2} = \sqrt{2},$$
$$\sin \theta = \frac{1}{\sqrt{2}} \quad \text{and} \quad \cos \theta = \frac{1}{\sqrt{2}}.$$

Thus $\theta = \pi/4$, or $45°$, and we have
$$1 + i = \sqrt{2} \operatorname{cis} \frac{\pi}{4} \quad \text{or} \quad 1 + i = \sqrt{2} \operatorname{cis} 45°.$$

Example 9 Find polar notation for $\sqrt{3} - i$.
$$r = \sqrt{(\sqrt{3})^2 + (-1)^2} = 2,$$
$$\sin \theta = -\frac{1}{2} \quad \text{and} \quad \cos \theta = \frac{\sqrt{3}}{2}.$$

Thus $\theta = 11\pi/6$, or $330°$, and we have
$$\sqrt{3} - i = 2 \operatorname{cis} \frac{11\pi}{6} = 2 \operatorname{cis} 330°.$$

In changing to polar notation, note that there are many angles satisfying the given conditions. We ordinarily choose the smallest positive angle.

DO EXERCISES 6 AND 7.

[iv] Multiplication and Polar Notation

Multiplication of complex numbers is somewhat easier to do with polar notation than with rectangular notation. We simply multiply the absolute values and add the arguments. Let us state this more formally and prove it.

THEOREM 8

For any complex numbers $r_1 \operatorname{cis} \theta_1$ and $r_2 \operatorname{cis} \theta_2$,
$$(r_1 \operatorname{cis} \theta_1)(r_2 \operatorname{cis} \theta_2) = r_1 \cdot r_2 \operatorname{cis} (\theta_1 + \theta_2).$$

Proof. Let us first multiply $a_1 + b_1 i$ by $a_2 + b_2 i$:
$$(a_1 + b_1 i)(a_2 + b_2 i) = (a_1 a_2 - b_1 b_2) + (a_2 b_1 + a_1 b_2)i.$$

Recall that
$$a_1 = r_1 \cos \theta_1, \qquad b_1 = r_1 \sin \theta_1,$$
and
$$a_2 = r_2 \cos \theta_2, \qquad b_2 = r_2 \sin \theta_2.$$

6. Write polar notation for $1 - i$.

7. Write polar notation for
$$-3\sqrt{2} - 3\sqrt{2}i.$$

We substitute these in the product above, to obtain

$$r_1(\cos\theta_1 + i\sin\theta_1) \cdot r_2(\cos\theta_2 + i\sin\theta_2)$$
$$= (r_1 r_2 \cos\theta_1 \cos\theta_2 - r_1 r_2 \sin\theta_1 \sin\theta_2)$$
$$+ (r_1 r_2 \sin\theta_1 \cos\theta_2 + r_1 r_2 \cos\theta_1 \sin\theta_2)i.$$

This simplifies to

$$r_1 r_2(\cos\theta_1 \cos\theta_2 - \sin\theta_1 \sin\theta_2) + r_1 r_2(\sin\theta_1 \cos\theta_2 + \cos\theta_1 \sin\theta_2)i.$$

Now, using identities for sums of angles, we simplify, obtaining

$$r_1 r_2 \cos(\theta_1 + \theta_2) + r_1 r_2 \sin(\theta_1 + \theta_2)i,$$

or

$$r_1 r_2 \operatorname{cis}(\theta_1 + \theta_2),$$

which was to be shown.

To divide complex numbers we do the reverse of the above. We state that fact, but will omit the proof.

THEOREM 9

For any complex numbers $r_1 \operatorname{cis}\theta_1$ and $r_2 \operatorname{cis}\theta_2$, $(r_2 \neq 0)$,

$$\frac{r_1 \operatorname{cis}\theta_1}{r_2 \operatorname{cis}\theta_2} = \frac{r_1}{r_2}\operatorname{cis}(\theta_1 - \theta_2).$$

Example 10 Find the product of $3\operatorname{cis}40°$ and $7\operatorname{cis}20°$.

$$3\operatorname{cis}40° \cdot 7\operatorname{cis}20° = 3 \cdot 7\operatorname{cis}(40° + 20°) = 21\operatorname{cis}60°$$

Example 11 Find the product of $2\operatorname{cis}\pi$ and $3\operatorname{cis}\left(-\frac{\pi}{2}\right)$.

$$2\operatorname{cis}\pi \cdot 3\operatorname{cis}\left(-\frac{\pi}{2}\right) = 2 \cdot 3\operatorname{cis}\left(\pi - \frac{\pi}{2}\right) = 6\operatorname{cis}\frac{\pi}{2}.$$

Example 12 Convert to polar notation and multiply: $(1 + i)(\sqrt{3} - i)$.

We first find polar notation (see Examples 8 and 9):

$$1 + i = \sqrt{2}\operatorname{cis}45°, \qquad \sqrt{3} - i = 2\operatorname{cis}330°.$$

We now multiply, using Theorem 8:

$$(\sqrt{2}\operatorname{cis}45°)(2\operatorname{cis}330°) = 2 \cdot \sqrt{2}\operatorname{cis}375° \quad \text{or} \quad 2\sqrt{2}\operatorname{cis}15°.$$

Example 13 Divide $2\operatorname{cis}\pi$ by $4\operatorname{cis}\frac{\pi}{2}$.

$$\frac{2\operatorname{cis}\pi}{4\operatorname{cis}\frac{\pi}{2}} = \frac{2}{4}\operatorname{cis}\left(\pi - \frac{\pi}{2}\right)$$

$$= \frac{1}{2}\operatorname{cis}\frac{\pi}{2}$$

Example 14 Convert to polar notation and divide: $(1 + i)/(1 - i)$.

We first convert to polar notation:

$$1 + i = \sqrt{2} \text{ cis } 45°, \qquad \text{See Example 8.}$$
$$1 - i = \sqrt{2} \text{ cis } 315°.$$

We now divide, using Theorem 9:

$$\frac{1 + i}{1 - i} = \frac{\sqrt{2} \text{ cis } 45°}{\sqrt{2} \text{ cis } 315°} = 1 \cdot \text{cis } (45° - 315°)$$

$$= 1 \cdot \text{cis } (-270°) \quad \text{or} \quad 1 \cdot \text{cis } 90°.$$

DO EXERCISES 8–12.

The Principle of Zero Products

In the system of complex numbers, when we multiply by 0 the result is 0. This is half of the principle of zero products. We have yet to show that the converse is true: that if a product is zero, at least one of the factors must be zero. With polar notation that is easy to do. The number 0 has polar notation $0 \text{ cis } \theta$, where θ can be any angle. Now consider a product $r_1 \text{ cis } \theta_1 \cdot r_2 \text{ cis } \theta_2$, where neither factor is 0. This means that $r_1 \neq 0$ and $r_2 \neq 0$. Since these are nonzero real numbers, their product is not zero, and the product of the complex numbers $r_1 r_2 \text{ cis } (\theta_1 + \theta_2)$ is not zero. Thus the principle of zero products holds in complex numbers.

Multiply.

8. $5 \text{ cis } 25° \cdot 4 \text{ cis } 30°$

9. $8 \text{ cis } \pi \cdot \dfrac{1}{2} \text{ cis } \dfrac{\pi}{4}$

10. Convert to polar notation and multiply.

$$(1 + i)(2 + 2i)$$

11. Divide.

$$10 \text{ cis } \frac{\pi}{2} \div 5 \text{ cis } \frac{\pi}{4}$$

12. Convert to polar notation and divide.

$$\frac{\sqrt{3} - i}{1 + i}$$

EXERCISE SET 12.4

[i] In Exercises 1–8, graph each pair of complex numbers and their sum.

1. $3 + 2i, \ 2 - 5i$

2. $4 + 3i, \ 3 - 4i$

3. $-5 + 3i, \ -2 - 3i$

4. $-4 + 2i, \ -3 - 4i$

5. $2 - 3i, \ -5 + 4i$

6. $3 - 2i, \ -5 + 5i$

7. $-2 - 5i, \ 5 + 3i$

8. $-3 - 4i, \ 6 + 3i$

[iii] In Exercises 9–16, find rectangular notation.

9. $3(\cos 30° + i \sin 30°)$

10. $6(\cos 150° + i \sin 150°)$

11. $10 \text{ cis } 270°$

12. $5 \text{ cis } (-60°)$

13. $\sqrt{8}\left(\cos \dfrac{\pi}{4} + i \sin \dfrac{\pi}{4}\right)$

14. $5\left(\cos \dfrac{\pi}{3} + i \sin \dfrac{\pi}{3}\right)$

15. $\sqrt{8} \text{ cis } \dfrac{5\pi}{4}$

16. $\sqrt{8} \text{ cis } \left(-\dfrac{\pi}{4}\right)$

[ii] In Exercises 17–22, find polar notation.

17. $1 - i$

18. $\sqrt{3} + i$

19. $10\sqrt{3} - 10i$

20. $-10\sqrt{3} + 10i$

21. -5

22. $-5i$

[iv] In Exercises 23–30, convert to polar notation and then multiply or divide.

23. $(1 - i)(2 + 2i)$

24. $(1 + i\sqrt{3})(1 + i)$

25. $(2\sqrt{3} + 2i)(2i)$

26. $(3\sqrt{3} - 3i)(2i)$

27. $\dfrac{1 - i}{1 + i}$

28. $\dfrac{1 - i}{\sqrt{3} - i}$

29. $\dfrac{2\sqrt{3} - 2i}{1 + \sqrt{3}i}$

30. $\dfrac{3 - 3\sqrt{3}i}{\sqrt{3} - i}$

☆

31. Show that for any complex number z, $|z| = |-z|$. (*Hint:* Let $z = a + bi$.)

32. Show that for any complex number z, $|z| = |\bar{z}|$. (*Hint:* Let $z = a + bi$.)

33. Show that for any complex number z, $|z\bar{z}| = |z|^2$.

34. Show that for any complex number z, $|z^2| = |z|^2$.

35. Show that for any complex numbers z and w, $|z \cdot w| = |z| \cdot |w|$. (*Hint:* Let $z = r_1 \operatorname{cis} \theta_1$ and $w = r_2 \operatorname{cis} \theta_2$.)

36. Show that for any complex number z and any nonzero complex number w, $\left|\dfrac{z}{w}\right| = \dfrac{|z|}{|w|}$. (Use the hint for Exercise 35.)

37. On a complex plane, graph $|z| = 1$.

38. On a complex plane, graph $z + \bar{z} = 3$.

★

39. Find polar notation for $(\cos \theta + i \sin \theta)^{-1}$.

40. Compute $\begin{bmatrix} i & 0 \\ 0 & -i \end{bmatrix}^3$.

OBJECTIVES

You should be able to:

[i] Use DeMoivre's theorem to raise complex numbers to powers.

[ii] Find the nth roots of a complex number.

1. Find $(1 - i)^{10}$.

2. Find $(\sqrt{3} + i)^4$.

12.5 DeMOIVRE'S THEOREM

[i] Powers of Complex Numbers

An important theorem about powers and roots of complex numbers is named for the French mathematician DeMoivre (1667–1754). Let us consider a number $r \operatorname{cis} \theta$ and its square:

$$(r \operatorname{cis} \theta)^2 = (r \operatorname{cis} \theta)(r \operatorname{cis} \theta) = r \cdot r \operatorname{cis} (\theta + \theta) = r^2 \operatorname{cis} 2\theta.$$

Similarly, we see that

$$(r \operatorname{cis} \theta)^3 = r \cdot r \cdot r \operatorname{cis} (\theta + \theta + \theta) = r^3 \operatorname{cis} 3\theta.$$

The generalization of this is DeMoivre's theorem.

THEOREM 10

DeMoivre's theorem. **For any complex number $r \operatorname{cis} \theta$ and any natural number n, $(r \operatorname{cis} \theta)^n = r^n \operatorname{cis} n\theta$.**

Example 1 Find $(1 + i)^9$.

We first find polar notation: $1 + i = \sqrt{2} \operatorname{cis} 45°$. Then

$$\begin{aligned}(1 + i)^9 &= (\sqrt{2} \operatorname{cis} 45°)^9 \\ &= \sqrt{2}^9 \operatorname{cis} 9 \cdot 45° \\ &= 2^{9/2} \operatorname{cis} 405° \\ &= 16\sqrt{2} \operatorname{cis} 45°.\end{aligned}$$

405° has the same terminal side as 45°.

DO EXERCISES 1 AND 2.

[ii] **Roots of Complex Numbers**

As we shall see, every nonzero complex number has two square roots. A number has three cube roots, four fourth roots, and so on. In general a nonzero complex number has n different nth roots. They can be found by the formula that we now state and prove.

THEOREM 11

The nth roots of a complex number r cis θ are given by

$$r^{1/n} \text{ cis} \left(\frac{\theta}{n} + k \cdot \frac{360°}{n} \right),$$

where $k = 0, 1, 2, \ldots, n - 1$.

We show that this formula gives us n different roots, using DeMoivre's theorem. We take the expression for the nth roots and raise it to the nth power, to show that we get r cis θ:

$$\left[r^{1/n} \text{ cis} \left(\frac{\theta}{n} + k \cdot \frac{360°}{n} \right) \right]^n = (r^{1/n})^n \text{ cis} \left(\frac{\theta}{n} \cdot n + k \cdot n \cdot \frac{360°}{n} \right)$$

$$= r \text{ cis} (\theta + k \cdot 360°) = r \text{ cis } \theta.$$

Thus we know that the formula gives us nth roots for any natural number k. Next we show that there are at least n different roots. To see this, consider substituting 0, 1, 2, and so on, for k. When $k = n$ the cycle begins to repeat, but from 0 to $n - 1$ the angles obtained and their sines and cosines are all different. There cannot be more than n different nth roots. That fact follows from the *fundamental theorem of algebra,* considered in the next chapter.

Example 2 Find the square roots of $2 + 2\sqrt{3}i$.

We first find polar notation: $2 + 2\sqrt{3}i = 4 \text{ cis } 60°$. Then

$$(4 \text{ cis } 60°)^{1/2} = 4^{1/2} \text{ cis} \left(\frac{60°}{2} + k \cdot \frac{360°}{2} \right), \quad k = 0, 1,$$

$$= 2 \text{ cis} \left(30° + k \cdot \frac{360°}{2} \right), \quad k = 0, 1.$$

Thus the roots are $2 \text{ cis } 30°$ and $2 \text{ cis } 210°$, or

$$\sqrt{3} + i \quad \text{and} \quad -\sqrt{3} - i.$$

DO EXERCISE 3.

In Example 2 it should be noted that the two square roots of a number were additive inverses of each other. The same is true of the square roots of any complex number. To see this, let us find the square roots of any complex number r cis θ.

$$(r \text{ cis } \theta)^{1/2} = r^{1/2} \text{ cis} \left(\frac{\theta}{2} + k \cdot \frac{360°}{2} \right), \quad k = 0, 1,$$

$$= r^{1/2} \text{ cis } \frac{\theta}{2} \quad \text{or} \quad r^{1/2} \text{ cis} \left(\frac{\theta}{2} + 180° \right)$$

3. Find the square roots.

a) $2i$

b) $10i$

4. Find and graph the cube roots of −1.

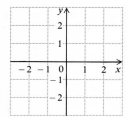

Now let us look at the two numbers on a graph. They lie on a line, so if one number has binomial notation $a + bi$, the other has binomial notation $−a − bi$. Hence their sum is 0 and they are additive inverses of each other.

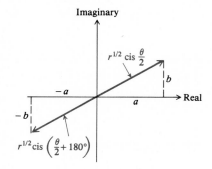

Example 3 Find the cube roots of 1. Locate them on a graph.

$$1 = 1 \operatorname{cis} 0°$$

$$(1 \operatorname{cis} 0°)^{1/3} = 1^{1/3} \operatorname{cis}\left(\frac{0°}{3} + k \cdot \frac{360°}{3}\right), \quad k = 0, 1, 2.$$

The roots are $1 \operatorname{cis} 0°$, $1 \operatorname{cis} 120°$, and $1 \operatorname{cis} 240°$, or

$$1, \quad -\frac{1}{2} + \frac{\sqrt{3}}{2}i, \quad \text{and} \quad -\frac{1}{2} - \frac{\sqrt{3}}{2}i.$$

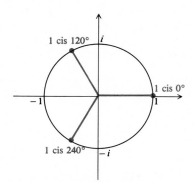

Note in the example that the graphs of the cube roots lie equally spaced about a circle. The same is true of the nth roots of any complex number.

DO EXERCISE 4.

EXERCISE SET 12.5

[i] In Exercises 1–6, raise the number to the power and write polar notation for the answer.

1. $\left(2 \operatorname{cis} \frac{\pi}{3}\right)^3$

2. $\left(3 \operatorname{cis} \frac{\pi}{2}\right)^4$

3. $\left(2 \operatorname{cis} \frac{\pi}{6}\right)^6$

4. $\left(2 \operatorname{cis} \frac{\pi}{5}\right)^5$

5. $(1 + i)^6$

6. $(1 - i)^6$

In Exercises 7–14, raise the number to the power and write rectangular notation for the answer.

7. $(2 \text{ cis } 240°)^4$

8. $(2 \text{ cis } 120°)^4$

9. $(1 + \sqrt{3}i)^4$

10. $(-\sqrt{3} + i)^6$

11. $\left(\dfrac{1}{\sqrt{2}} + \dfrac{1}{\sqrt{2}}i\right)^{10}$

12. $\left(\dfrac{1}{\sqrt{2}} - \dfrac{1}{\sqrt{2}}i\right)^{12}$

13. $\left(\dfrac{\sqrt{3}}{2} + \dfrac{1}{2}i\right)^{12}$

14. $\left(\dfrac{\sqrt{3}}{2} - \dfrac{1}{2}i\right)^{14}$

[ii]

15. Solve $x^2 = -1 + \sqrt{3}i$. That is, find the square roots of $-1 + \sqrt{3}i$.

16. Solve $x^2 = -\sqrt{3} - i$. That is, find the square roots of $-\sqrt{3} - i$.

17. Solve $x^3 = i$.

18. ▦ Solve $x^3 = 68.4321$.

19. Find and graph the fourth roots of 16.

20. Find and graph the fourth roots of i.

☆ ──

In Exercises 21 and 22, solve. Evaluate square roots.

21. $x^2 + (1 - i)x + i = 0$

22. $3x^2 + (1 + 2i)x + 1 - i = 0$

CHAPTER 12 REVIEW

Simplify.

[12.1, ii] **1.** $(2 - 2i)(3 + 4i)$

2. $(3 - 5i) - (2 - i)$

3. $(6 + 2i) + (-4 - 3i)$

4. $\dfrac{2 - 3i}{1 - 3i}$

[12.1, iv] **5.** Solve for x and y: $4x + 2i = 8 - (2 + y)i$.

[12.2, iii] **6.** Find a polynomial in $\bar{z}$ that is the conjugate of $3z^3 + z - 7$.

[12.3, iv] **7.** Find an equation having the solutions $1 - 2i$, $1 + 2i$.

Solve.

[12.3, iii] **8.** $5x^2 - 4x + 1 = 0$

[12.3, ii] **9.** $x^2 + 3ix - 1 = 0$

[12.3, v] **10.** Find the square roots of $4i$.

[12.4, i] **11.** Graph the pair of complex numbers $-3 - 2i$, $4 + 7i$, and their sum.

[12.4, iii] **12.** Find rectangular notation for $2(\cos 135° + i \sin 135°)$.

[12.4, ii] **13.** Find polar notation for $1 + i$.

[12.4, iv] **14.** Find the product of $7 \text{ cis } 18°$ and $10 \text{ cis } 32°$.

[12.5, ii] **15.** Find the cube roots of $1 + i$.

16. Let $f(z) = 3z^2 - 2\bar{z} + 5i$. Find $f(2 - i)$.

17. Solve $3z + 2\bar{z} = 5 + 2i$.

18. Solve
$2x - 3y = 7 + 7i$,
$3x + 2y = 4 - 9i$.

[12.5, ii] **19.** Solve $x^3 - 27 = 0$.

20. Graph $|z - (1 + i)| = 1$.

POLYNOMIALS AND RATIONAL FUNCTIONS

$P(P(P(x)))$

$P(P(P(x)))$

OBJECTIVES

You should be able to:

[i] Determine the degree of a polynomial.

[ii] Determine whether a given number is a root of a polynomial.

[iii] By division, determine whether one polynomial is a factor of another.

[iv] Given a polynomial $P(x)$ and a divisor $d(x)$, determine the quotient and remainder and express $P(x)$ as

$$P(x) = d(x) \cdot Q(x) + R(x).$$

Determine the degree of each polynomial.

1. $x^5 - x^3 + x^2 + 2i$

2. $3x - i$

3. $3x^2 - \sqrt{2}$

4. 0.5

5. 0

13.1 POLYNOMIALS AND POLYNOMIAL FUNCTIONS

[i] We have already considered polynomials. We now make a formal definition.

DEFINITION

A *polynomial* in one variable is any expression of the type

$$a_n x^n + a_{n-1} x^{n-1} + \cdots + a_2 x^2 + a_1 x + a_0,$$

where n is a nonnegative integer and $a_0, \ldots, a_n$, are numbers.

The numbers mentioned in this definition are complex numbers, but in certain cases may be restricted to real numbers, rational numbers, or integers. Some or all of them may be 0.

Terminology

In a polynomial, the numbers $a_0, \ldots, a_n$ are called *coefficients*. The first nonzero coefficient, a_n, is called the *leading coefficient*. The number n is called the *degree* of the polynomial, except that the polynomial consisting only of the number 0 has no degree.

Example 1 We list some polynomials and their degrees.

POLYNOMIAL	DEGREE
$5x^3 - 3x^2 + i$	3
$-4x + \sqrt{2}$	1
25	0
0	No degree

DO EXERCISES 1–5.

[ii] **Zeros or Roots of Polynomials**

When a number is substituted for the variable in a polynomial, the result is some unique number. Thus every polynomial defines a function. We often refer to polynomials, therefore, using function notation, such as $P(x)$. If a number a makes a polynomial 0, i.e., $P(a) = 0$, then a is said to be a *zero* of $P(x)$. Note that a is also a solution of the equation $P(x) = 0$. We often say that a is a *root* of the equation $P(x) = 0$ and also a *root* of the polynomial $P(x)$.*

Example 2 $P(x) = x^3 + 2x^2 - 5x - 6$. Is 3 a *zero* of $P(x)$?

We substitute 3 into the polynomial:

$$P(3) = 3^3 + 2(3)^2 - 5 \cdot 3 - 6 = 24.$$

Since $P(3) \neq 0$, 3 is not a zero of the polynomial.

---

*Terminology varies from author to author. Traditionally, one referred to the *zeros* of a function and the *roots* of an equation. Subsequently it was found to be convenient in exposition to speak also of the "roots" of a function, especially a polynomial function. In some books the traditional distinction is still made today. We have found that the freer use of terminology facilitates simplification of the exposition.

Example 3 Is -1 a root of $P(x)$ in Example 1?

$$P(-1) = (-1)^3 + 2(-1)^2 - 5(-1) - 6 = 0.$$

Since $P(-1) = 0$, -1 is a root of the polynomial.

DO EXERCISES 6–8.

[iii] Division of Polynomials and Factors

When we divide one polynomial by another we obtain a quotient and a remainder. If the remainder is 0, then the divisor is a *factor* of the dividend.

Example 4 Divide, to find whether $x^2 + 9$ is a factor of $x^4 - 81$.

$$
\begin{array}{r}
x^2 - 9 \\
\hline
x^2 + 9 \,)\overline{x^4 \qquad\qquad - 81} \\
\underline{x^4 \quad + 9x^2} \\
-9x^2 \quad - 81 \\
\underline{-9x^2 \quad - 81} \\
0
\end{array}
$$

Note how spaces have been left for missing terms in the dividend.

Since the remainder is 0, we know that $x^2 + 9$ is a factor.

Example 5 Divide, to find whether $x^2 + 3x - 1$ is a factor of $x^4 - 81$.

$$
\begin{array}{r}
x^2 - 3x + 10 \\
\hline
x^2 + 3x - 1 \,)\overline{x^4 \qquad\qquad\qquad - 81} \\
\underline{x^4 + 3x^3 - \quad x^2} \\
-3x^3 + \quad x^2 \\
\underline{-3x^3 - \quad 9x^2 + \quad 3x} \\
10x^2 - \quad 3x - 81 \\
\underline{10x^2 + 30x - 10} \\
-33x - 71
\end{array}
$$

Since the remainder is not 0, we know that $x^2 + 3x - 1$ is not a factor of $x^4 - 81$.

DO EXERCISES 9 AND 10.

[iv] In general, when we divide a polynomial $P(x)$ by a divisor $d(x)$ we obtain some polynomial $Q(x)$ for a quotient and some polynomial $R(x)$ for a remainder. The remainder must either be 0 or have degree less than that of $d(x)$. To check a division we multiply the quotient by the divisor and add the remainder, to see if we get the dividend. Thus these polynomials are related as follows:

$$P(x) = d(x) \cdot Q(x) + R(x).$$

Example 6 If $P(x) = x^4 - 81$, $d(x) = x^2 + 9$, then $Q(x) = x^2 - 9$ and $R(x) = 0$, and

$$\underbrace{x^4 - 81}_{P(x)} = \underbrace{(x^2 + 9)}_{d(x)} \cdot \underbrace{(x^2 - 9)}_{Q(x)} + \underbrace{0}_{R(x)}.$$

6. Determine whether the following numbers are roots of the polynomial $x^2 - 4x - 21$.

a) 7

b) 3

c) -7

7. Determine whether the following numbers are roots of the polynomial $x^4 - 16$.

a) 2

b) -2

c) -1

d) 0

8. Determine whether the following numbers are roots of the polynomial $x^2 + 1$.

a) 1

b) -1

c) i

d) $-i$

9. By division, determine whether the following polynomials are factors of the polynomial $x^4 - 16$.

a) $x - 2$

b) $x^2 + 3x - 1$

10. By division, determine whether the following polynomials are factors of the polynomial $x^3 + 2x^2 - 5x - 6$.

a) $x + 1$

b) $x - 3$

c) $x^2 + 3x - 1$

11. Divide $x^3 + 2x^2 - 5x - 6$ by $x - 3$. Then express the dividend as

$$P(x) = d(x) \cdot Q(x) + R(x).$$

Example 7 If $P(x) = x^4 - 81$, $d(x) = x^2 + 3x - 1$, then $Q(x) = x^2 - 3x + 10$ and $R(x) = -33x - 71$, and

$$\underbrace{x^4 - 81}_{P(x)} = \underbrace{(x^2 + 3x - 1)}_{d(x)} \cdot \underbrace{(x^2 - 3x + 10)}_{Q(x)} + \underbrace{(-33x - 71)}_{R(x)}$$

DO EXERCISE 11.

EXERCISE SET 13.1

[i] Determine the degree of each polynomial.

1. $x^4 - 3x^2 + 1$

2. $2x^5 - x^4 + \frac{1}{4}x - 7$

3. $-2x + 5$

4. $3x - \sqrt{\pi}$

5. $2x^2 - 3x + 4$

6. $\frac{1}{4}x^2 - 7$

7. 3

8. 0

[ii]

9. Determine whether 2, 3, and -1 are roots, or zeros, of

$$P(x) = x^3 + 6x^2 - x - 30.$$

10. Determine whether 2, 3, and -1 are roots, or zeros, of

$$P(x) = 2x^3 - 3x^2 + x - 1.$$

[iii]

11. For $P(x)$ in Exercise 9, which of the following are factors of $P(x)$?

a) $x - 2$　　b) $x - 3$　　c) $x + 1$

12. For $P(x)$ in Exercise 10, which of the following are factors of $P(x)$?

a) $x - 2$　　b) $x - 3$　　c) $x + 1$

[iv] In each of the following, a polynomial $P(x)$ and a divisor $d(x)$ are given. Find the quotient $Q(x)$ and the remainder $R(x)$ when $P(x)$ is divided by $d(x)$ and express $P(x)$ in the form $d(x) \cdot Q(x) + R(x)$.

13. $P(x) = x^3 + 6x^2 - x - 30$,
$d(x) = x - 2$

14. $P(x) = 2x^3 - 3x^2 + x - 1$,
$d(x) = x - 2$

15. $P(x)$ as in Exercise 13,
$d(x) = x - 3$

16. $P(x)$ as in Exercise 14,
$d(x) = x - 3$

17. $P(x) = x^3 - 8$,
$d(x) = x + 2$

18. $P(x) = x^3 + 27$,
$d(x) = x + 1$

19. $P(x) = x^4 + 9x^2 + 20$,
$d(x) = x^2 + 4$

20. $P(x) = x^4 + x^2 + 2$,
$d(x) = x^2 + x + 1$

21. $P(x) = 5x^7 - 3x^4 + 2x^2 - 3$,
$d(x) = 2x^2 - x + 1$

22. $P(x) = 6x^5 + 4x^4 - 3x^2 + x - 2$,
$d(x) = 3x^2 + 2x - 1$

☆ ────────────────────────────────

23. For $P(x) = x^5 - 64$,

a) find $P(2)$;

b) find the remainder when $P(x)$ is divided by $x - 2$, and compare your answer to (a);

c) find $P(-1)$;

d) find the remainder when $P(x)$ is divided by $x + 1$, and compare your answer to (c).

24. For $P(x) = x^3 + x^2$,

a) find $P(-1)$;

b) find the remainder when $P(x)$ is divided by $x + 1$, and compare your answer to (a);

c) find $P(2)$;

d) find the remainder when $P(x)$ is divided by $x - 2$, and compare your answer to (c).

★ ────────────────────────────────

25. For $P(x) = 2x^2 - ix + 1$,

a) find $P(-i)$;

b) find the remainder when $P(x)$ is divided by $x + i$.

26. For $P(x) = 2x^2 + ix - 1$,

a) find $P(i)$;

b) find the remainder when $P(x)$ is divided by $x - i$.

27. Under what conditions is a polynomial function of real variables an even function, i.e., $P(x) = P(-x)$ for all x? Does your answer still hold for complex variables? Explain.

28. Under what conditions is a polynomial function of real variables an odd function, i.e., $-P(x) = P(-x)$ for all x? Does your answer still hold for complex variables?

29. Find a rule for determining the degree of the product of two polynomials with real coefficients.

30. Find a rule for determining the degree of the sum of two polynomials with real coefficients.

13.2 THE REMAINDER AND FACTOR THEOREMS

Some of the exercises in the preceding exercise set illustrate the following theorem.

THEOREM 1

The remainder theorem. **If a number r is substituted for x in the polynomial $P(x)$, then the result $P(r)$ is the remainder that would be obtained by dividing $P(x)$ by $x - r$.**

Proof. The equation $P(x) = d(x) \cdot Q(x) + R(x)$ is the basis of this proof. If we divide $P(x)$ by $x - r$, we obtain a quotient $Q(x)$ and a remainder $R(x)$ related as follows:

$$P(x) = (x - r) \cdot Q(x) + R(x).$$

The remainder $R(x)$ must either be 0 or have degree less than $x - r$. Thus $R(x)$ must be a constant. Let us call this constant R. In the above expression we get a true sentence whenever we replace x by any number. Let us replace x by r. We get

$$P(r) = (r - r) \cdot Q(r) + R$$
$$P(r) = \quad 0 \cdot Q(r) + R$$
$$P(r) = R.$$

This tells us that the function value $P(r)$ is the remainder obtained when we divide $P(x)$ by $x - r$.

Theorem 1 motivates us to find a rapid way of dividing by $x - r$, in order to find function values.

[i] Synthetic Division

To streamline division, we can arrange the work so that duplicate and unnecessary writing are avoided. Consider the following.

A.
$$
\begin{array}{r}
4x^2 + 5x + 11 \\
x - 2\overline{)4x^3 - 3x^2 + x + 7} \\
\underline{4x^3 - 8x^2} \\
5x^2 + x \\
\underline{5x^2 - 10x} \\
11x + 7 \\
\underline{11x - 22} \\
29
\end{array}
$$

B.
$$
\begin{array}{r}
4 \quad 5 \quad 11 \\
1 - 2\overline{)4 - 3 + 1 + 7} \\
\underline{4 - 8} \\
5 + 1 \\
\underline{5 - 10} \\
11 + 7 \\
\underline{11 - 22} \\
29
\end{array}
$$

OBJECTIVES

You should be able to:

[i] Use synthetic division to find the quotient and remainder when a polynomial is divided by $x - r$.

[ii] Use the remainder theorem to find a function value $P(r)$ when a polynomial $P(x)$ is divided by $x - r$.

[iii] Determine whether $x - r$ is a factor of $P(x)$ by determining whether $P(r) = 0$.

Use synthetic division to find the quotient and remainder.

1. $(x^3 + 6x^2 - x - 30) \div (x - 2)$

2. $(x^3 - 2x^2 + 5x - 4) \div (x + 2)$

3. $(y^3 + 1) \div (y + 1)$

4. Let $P(x) = x^5 - 2x^4 - 7x^3 + x^2 + 20$. Use synthetic division to find

  a) $P(10)$;
  b) $P(-8)$.

The division in B is the same as that in A, but we wrote only the coefficients. The color numerals are duplicated, so we look for an arrangement in which they are not duplicated. We can also simplify things by using the additive inverse of -2 and then adding instead of subtracting. When things are thus "collapsed," we have the algorithm known as *synthetic division*.

C. (Synthetic Division)

$$
\begin{array}{r|rrrr}
2 & 4 & -3 & 1 & 7 \\
  &   & 8 & 10 & 22 \\
\hline
  & 4 & 5 & 11 & 29
\end{array}
$$

We "bring down" the 4. Then we multiply it by the 2, to get 8, and add to get 5. We then multiply 5 by 2 to get 10, add, and so on. The last number, 29, is the remainder. The others are the coefficients of the quotient.

Example 1 Use synthetic division to find the quotient and remainder:

$$(2x^3 + 7x^2 - 5) \div (x + 3).$$

First note that $x + 3 = x - (-3)$.

$$
\begin{array}{r|rrrr}
-3 & 2 & 7 & 0 & -5 \\
   &   & -6 & -3 & 9 \\
\hline
   & 2 & 1 & -3 & 4
\end{array}
$$

Note: We must write 0's for missing terms.

The quotient is $2x^2 + x - 3$. The remainder is 4.

DO EXERCISES 1–3.

[ii] **Function Values for Polynomials**

We can now apply synthetic division to the finding of function values for polynomials.

Example 2 $P(x) = 2x^5 - 3x^4 + x^3 - 2x^2 + x - 8$. Find $P(10)$.

By Theorem 1, $P(10)$ is the remainder when $P(x)$ is divided by $x - 10$. We use synthetic division to find that remainder.*

$$
\begin{array}{r|rrrrrr}
10 & 2 & -3 & 1 & -2 & 1 & -8 \\
   &   & 20 & 170 & 1710 & 17{,}080 & 170{,}810 \\
\hline
   & 2 & 17 & 171 & 1708 & 17{,}081 & 170{,}802
\end{array}
$$

Thus $P(10) = 170{,}802$.

DO EXERCISE 4.

---

*Compare this with the work involved in a direct calculation! ▦ A calculator is most useful when finding polynomial function values by synthetic division. In Example 2, one would begin by entering 2. Then multiply by 10, add the result to -3, multiply that result by 10, and so on. The number 10 can be stored and recalled at each step where needed if the calculator has a memory. Since only the last result is needed, no intermediate values need be recorded.

Example 3 Determine whether -4 is a zero or root of $P(x)$, where $P(x) = x^3 + 8x^2 + 8x - 32$.

We use synthetic division and Theorem 1 to find $P(-4)$.

$$
\begin{array}{r|rrrr}
-4 & 1 & 8 & 8 & -32 \\
 & & -4 & -16 & 32 \\
\hline
 & 1 & 4 & -8 & 0
\end{array}
$$

Since $P(-4) = 0$, the number -4 is a root of $P(x)$.

DO EXERCISE 5.

[iii] Finding Factors of Polynomials

We now consider the following useful corollary of the remainder theorem.

THEOREM 2

The factor theorem. **For a polynomial $P(x)$, if $P(r) = 0$, then the polynomial $x - r$ is a factor of $P(x)$.**

Proof. If we divide $P(x)$ by $x - r$, we obtain a quotient and remainder, related as follows:

$$P(x) = (x - r) \cdot Q(x) + P(r).$$

Then if $P(r) = 0$, we have

$$P(x) = (x - r) \cdot Q(x),$$

so $x - r$ is a factor of $P(x)$.

This theorem is very useful in factoring polynomials, and hence in the solving of equations.

Example 4 Let $P(x) = x^3 + 2x^2 - 5x - 6$. Factor $P(x)$ and thus solve the equation $P(x) = 0$.

We look for linear factors of the form $x - r$. Let us try $x - 1$. We use synthetic division to see whether $P(1) = 0$.

$$
\begin{array}{r|rrrr}
1 & 1 & 2 & -5 & -6 \\
 & & 1 & 3 & -2 \\
\hline
 & 1 & 3 & -2 & -8
\end{array}
$$

We know that $x - 1$ is not a factor of $P(x)$. We try $x + 1$.

$$
\begin{array}{r|rrrr}
-1 & 1 & +2 & -5 & -6 \\
 & & -1 & -1 & 6 \\
\hline
 & 1 & 1 & -6 & 0
\end{array}
$$

We know that $x + 1$ is one factor and the quotient, $x^2 + x - 6$, is another. Thus

$$P(x) = (x + 1)(x^2 + x - 6).$$

5. Let $P(x) = x^3 + 6x^2 - x - 30$. Using synthetic division, determine whether the given numbers are roots of $P(x)$.

a) 2

b) 5

c) -3

6. Determine whether $x - \frac{1}{2}$ is a factor of $4x^4 + 2x^3 + 8x - 1$.

7. Determine whether $x + 5$ is a factor of $x^4 + 625$.

8. a) Let $P(x) = x^3 + 6x^2 - x - 30$. Determine whether $x - 2$ is a factor of $P(x)$.
b) Find another factor of $P(x)$.
c) Find a complete factorization of $P(x)$.

The trinomial is easily factored, so we have

$$P(x) = (x + 1)(x + 3)(x - 2).$$

To solve the equation $P(x) = 0$, we use the principle of zero products. The solutions are -1, -3, and 2.

DO EXERCISES 6–8.

EXERCISE SET 13.2

[i] Use synthetic division to find the quotient and remainder.

1. $(2x^4 + 7x^3 + x - 12) \div (x + 3)$

2. $(x^3 - 7x^2 + 13x + 3) \div (x - 2)$

3. $(x^3 - 2x^2 - 8) \div (x + 2)$

4. $(x^3 - 3x + 10) \div (x - 2)$

5. $(x^4 - 1) \div (x - 1)$

6. $(x^5 + 32) \div (x + 2)$

7. $(2x^4 + 3x^2 - 1) \div \left(x - \frac{1}{2}\right)$

8. $(3x^4 - 2x^2 + 2) \div \left(x - \frac{1}{4}\right)$

9. $(x^4 - y^4) \div (x - y)$

10. $(x^3 + 3ix^2 - 4ix - 2) \div (x + i)$

[ii] Use synthetic division to find the function values.

11. $P(x) = x^3 - 6x^2 + 11x - 6$
Find $P(1)$, $P(-2)$, $P(3)$.

12. $P(x) = x^3 + 7x^2 - 12x - 3$
Find $P(-3)$, $P(-2)$, $P(1)$.

13. $P(x) = 2x^5 - 3x^4 + 2x^3 - x + 8$
Find $P(20)$ and $P(-3)$.

14. $P(x) = x^5 - 10x^4 + 20x^3 - 5x - 100$
Find $P(-10)$ and $P(5)$.

[iii] Using synthetic division, determine whether the numbers are roots of the polynomials.

15. $-3, 2$; $P(x) = 3x^3 + 5x^2 - 6x + 18$

16. $-4, 2$; $P(x) = 3x^3 + 11x^2 - 2x + 8$

17. $-3, \frac{1}{2}$; $P(x) = x^3 - \frac{7}{2}x^2 + x - \frac{3}{2}$

18. $i, -i, -2$; $P(x) = x^3 + 2x^2 + x + 2$

Factor the polynomial $P(x)$. Then solve the equation $P(x) = 0$.

19. $P(x) = x^3 + 4x^2 + x - 6$

20. $P(x) = x^3 + 5x^2 - 2x - 24$

21. $P(x) = x^3 - 6x^2 + 3x + 10$

22. $P(x) = x^3 + 2x^2 - 13x + 10$

23. $P(x) = x^3 - x^2 - 14x + 24$

24. $P(x) = x^3 - 3x^2 - 10x + 24$

25. $P(x) = x^4 - 22x^2 + 51x - 30$

26. $P(x) = x^4 + 11x^3 + 41x^2 + 61x + 30$

☆ _____

Solve.

27. $x^3 + 2x^2 - 13x + 10 > 0$

28. $x^4 - x^3 - 19x^2 + 49x - 30 < 0$

29. Find k so that $x + 2$ is a factor of $x^3 - kx^2 + 3x + 7k$.

30. ▓ Given that $f(x) = 2.13x^5 - 42.1x^3 + 17.5x^2 + 0.953x - 1.98$, find $f(3.21)$
a) by synthetic division;   b) by substitution.

31. For what values of k will the remainder be the same when $x^2 + kx + 4$ is divided by $x - 1$ or $x + 1$?

32. When $x^2 - 3x + 2k$ is divided by $x + 2$ the remainder is 7. Find the value of k.

★ _____

33. Devise a way to use the method of synthetic division when the divisor is a polynomial such as $bx - r$.

34. Prove that $x - a$ is a factor of $x^n - a^n$, for any natural number n.

13.3 THEOREMS ABOUT ROOTS

The Fundamental Theorem of Algebra

A linear, or first-degree, polynomial $ax + b$ (where $a \neq 0$, of course) has just one root, $-b/a$. From Chapter 12 we know that any quadratic polynomial with complex numbers for coefficients has at least one, and at most two, roots. The following theorem is a generalization. A proof is beyond this text.

THEOREM 3

The fundamental theorem of algebra. **Every polynomial of degree greater than 0, with complex coefficients, has at least one root in the system of complex numbers.***

This is a very powerful theorem. Note that although it guarantees that a root exists, it does not tell how to find it. We now develop some theory that can help in finding roots. First, we prove a corollary of the fundamental theorem of algebra.

THEOREM 4

Every polynomial of degree n, where $n > 0$, having complex coefficients, can be factored into n linear factors.

Proof. Let us consider any polynomial of degree n, say $P(x)$. By the fundamental theorem, it has a root r_1. By the factor theorem, $x - r_1$ is a factor of $P(x)$. Thus we know that

$$P(x) = (x - r_1) \cdot Q_1(x),$$

where $Q_1(x)$ is the quotient that would be obtained by dividing $P(x)$ by $x - r_1$. Let the leading coefficient of $P(x)$ be a_n. By considering the actual division process we see that the leading coefficient of $Q_1(x)$ is also a_n and that the degree of $Q_1(x)$ is $n - 1$. Now if the degree of $Q_1(x)$ is greater than 0, then it has a root r_2, and we have

$$Q_1(x) = (x - r_2) \cdot Q_2(x),$$

where the degree of $Q_2(x)$ is $n - 2$ and the leading coefficient is a_n. Thus we have

$$P(x) = (x - r_1)(x - r_2) \cdot Q_2(x).$$

This process can be continued until a quotient $Q_n(x)$ is obtained having degree 0. The leading coefficient will be a_n, so $Q_n(x)$ is actually the constant a_n. We now have the following:

$$P(x) = a_n(x - r_1)(x - r_2)(x - r_3) \cdots (x - r_n).$$

This completes the proof. We have actually shown a little more than what the theorem states. We see that $P(x)$ has been factored with one constant factor a_n and n linear factors having leading coefficient 1.

OBJECTIVES

You should be able to:

[i] Factor polynomials and find their roots and their multiplicities.
[ii] Find a polynomial with specified roots.
[iii] In certain cases, given some of the roots of a polynomial, find such a polynomial and find the rest of its roots.

---

*It is well to keep in mind that when we speak of "roots," or "zeros," of a polynomial $P(x)$, we are also speaking about "roots" or "solutions" of the polynomial equation $P(x) = 0$.

Find the roots of each polynomial and state the multiplicity of each.

1. $P(x) = (x - 5)(x - 5)(x + 6)$

2. $P(x) = 4(x + 7)^2(x - 3)$

3. $P(x) = (x + 2)^3(x^2 - 9)$

4. $P(x) = (x^2 - 7x + 12)^2$

5. $5x^2 - 5$

6. Find a polynomial of degree 3 that has -1, 2, and 5 as roots.

7. Find a polynomial of degree 3 that has -1, 2, and $-5i$ as roots.

8. Find a polynomial of degree 5 with -2 as a root of multiplicity 3 and 0 as a root of multiplicity 2.

9. Find a polynomial of degree 4 with 1 as a root of multiplicity 3 and -5 as a root of multiplicity 1.

[i] Finding Roots by Factoring

To find the roots of a polynomial, we can attempt to factor it.

Examples

1. $x^4 + x^3 - 13x^2 - x + 12$ can be factored as

$$(x - 3)(x + 4)(x + 1)(x - 1),$$

so the roots are 3, -4, -1, and 1.

2. $3x^4 - 15x^3 + 18x^2 + 12x - 24$ can be factored as

$$3(x - 2)(x - 2)(x - 2)(x + 1),$$

so the roots are 2 and -1.

In Example 2 the factor $x - 2$ occurs three times. In a case like this we sometimes say that the root we obtain, 2, has a *multiplicity* of three.

From Theorem 4 and the above examples, we have the following.

THEOREM 5

Every polynomial of degree n, where $n > 0$, has at least one root and at most n roots.*

DO EXERCISES 1–5.

[ii] Finding Polynomials with Given Roots

Given several numbers, we can find a polynomial having them as its roots.

Example 3 Find a polynomial of degree three, having the roots -2, 1, and $3i$.

By Theorem 2, such a polynomial has factors $x + 2$, $x - 1$, and $x - 3i$, so we have

$$P(x) = a_n(x + 2)(x - 1)(x - 3i).$$

The number a_n can be any nonzero number. The simplest polynomial will be obtained if we let it be 1. If we then multiply the factors we obtain

$$P(x) = x^3 + (1 - 3i)x^2 + (-2 - 3i)x + 6i.$$

Example 4 Find a polynomial of degree 5 with -1 as a root of multiplicity 3, 4 as a root of multiplicity 1, and 0 as a root of multiplicity 1.

Proceeding as in Example 3, letting $a_n = 1$, we obtain

$$(x + 1)^3(x - 4)(x - 0), \quad \text{or} \quad x^5 - x^4 - 9x^3 - 11x^2 - 4x.$$

DO EXERCISES 6–9.

---

*Theorem 5 is often stated as follows: "Every polynomial of degree n, where $n > 0$, has *exactly n* roots." This statement is not incompatible with Theorem 5, as it first seems, because to make sense of the statement just quoted, one must take multiplicities into account. Theorem 5, as stated here, is simpler and more straightforward.

[iii] Roots of Polynomials with Real Coefficients

Consider the quadratic equation $x^2 - 2x + 2 = 0$, with real coefficients. Its roots are $1 + i$ and $1 - i$. Note that they are complex conjugates. This generalizes to any polynomial with real coefficients.

THEOREM 6

If a complex number z is a root of a polynomial $P(x)$ of degree greater than or equal to 1 with real coefficients, then its conjugate $\bar{z}$ is also a root. (Nonreal roots occur in conjugate pairs.)

Proof. Let

$$P(x) = a_n x^n + a_{n-1} x^{n-1} + \cdots + a_1 x + a_0,$$

where the coefficients are real numbers. Suppose z is a complex root of $P(x)$. Then $P(z) = 0$, or

$$a_n z^n + a_{n-1} z^{n-1} + \cdots + a_1 z + a_0 = 0.$$

Now let us find the conjugate of each side of the equation. First note that $\bar{0} = 0$, since 0 is a real number. Then we have the following.

$$0 = \bar{0} = \overline{a_n z^n + a_{n-1} z^{n-1} + \cdots + a_1 z + a_0}$$

$$= \overline{a_n z^n} + \overline{a_{n-1} z^{n-1}} + \cdots + \overline{a_1 z} + \overline{a_0} \qquad \text{By Theorem 3, Chapter 12}$$

$$= \overline{a_n} \cdot \overline{z^n} + \overline{a_{n-1}} \cdot \overline{z^{n-1}} + \cdots + \overline{a_1} \cdot \overline{z} + \overline{a_0} \qquad \text{By Theorem 4, Chapter 12}$$

$$= a_n \overline{z^n} + a_{n-1} \overline{z^{n-1}} + \cdots + a_1 \overline{z} + a_0 \qquad \text{By Theorem 6, Chapter 12}$$

$$= a_n \bar{z}^n + a_{n-1} \bar{z}^{n-1} + \cdots + a_1 \bar{z} + a_0. \qquad \text{By Theorem 5, Chapter 12}$$

Thus $P(\bar{z}) = 0$, so $\bar{z}$ is a root of the polynomial.

For Theorem 6, it is essential that the coefficients be real numbers. This can be seen by considering Example 3. In that polynomial the root $3i$ occurs, but its conjugate does not. This can happen because some of the coefficients of the polynomial are not real.

Rational Coefficients

When a polynomial has rational numbers for coefficients, certain irrational roots also occur in pairs, as described in the following theorem.

THEOREM 7

Suppose $P(x)$ is a polynomial with rational coefficients and of degree greater than 0. Then if either of the following is a root, so is the other: $a + c\sqrt{b}$, $a - c\sqrt{b}$, a and c rational, b not a square.

This theorem can be proved in a manner analogous to Theorem 6, but we shall not do it here. The theorem can be used to help in finding roots.

10. Suppose a polynomial of degree 5 with rational coefficients has -4, $7 - 2i$, and $3 + \sqrt{5}$ as roots. Find the other roots.

11. Find a polynomial of lowest degree with rational coefficients that has $2 + \sqrt{3}$ and $1 - i$ as some of its roots.

12. Find a polynomial of lowest degree with real coefficients that has $2i$ and 2 as some of its roots.

13. Find the other roots of $x^4 + x^3 - x^2 + x - 2$, given that i is a root.

Example 5 Suppose a polynomial of degree 6 with rational coefficients has $-2 + 5i$, $-2i$, and $1 - \sqrt{3}$ as some of its roots. Find the other roots.

The other roots are $-2 - 5i$, $2i$, and $1 + \sqrt{3}$. There are no other roots since the degree is 6.

Example 6 Find a polynomial of lowest degree with rational coefficients that has $1 - \sqrt{2}$ and $1 + 2i$ as some of its roots.

The polynomial must also have the roots $1 + \sqrt{2}$ and $1 - 2i$. Thus the polynomial is

$$[x - (1 - \sqrt{2})][x - (1 + \sqrt{2})][x - (1 + 2i)][x - (1 - 2i)],$$

or

$$(x^2 - 2x - 1)(x^2 - 2x + 5),$$

or

$$x^4 - 4x^3 + 8x^2 - 8x - 5.$$

DO EXERCISES 10–12.

Example 7 Let $P(x) = x^4 - 5x^3 + 10x^2 - 20x + 24$. Find the other roots of $P(x)$, given that $2i$ is a root.

Since $2i$ is a root, we know that $-2i$ is also a root. Thus

$$P(x) = (x - 2i)(x + 2i) \cdot Q(x)$$

for some $Q(x)$. Since $(x - 2i)(x + 2i) = x^2 + 4$, we know that

$$P(x) = (x^2 + 4) \cdot Q(x).$$

We find, using division, that $Q(x) = x^2 - 5x + 6$, and since we can factor $x^2 - 5x + 6$, we get

$$P(x) = (x^2 + 4)(x - 2)(x - 3).$$

Thus the other roots are $-2i$, 2, and 3.

DO EXERCISE 13.

EXERCISE SET 13.3

[i] In Exercises 1–4, find the roots of each polynomial, or polynomial equation, and state the multiplicity of each.

1. $(x + 3)^2(x - 1)$

2. $-8(x - 3)^2(x + 4)^3x^4 = 0$

3. $x^3(x - 1)^2(x + 4) = 0$

4. $(x^2 - 5x + 6)^2$

[ii] In Exercises 5–9, find a polynomial of degree 3 with the given numbers as roots.

5. $-2, 3, 5$

6. $2, i, -i$

7. $-3, 2i, -2i$

8. $1 + 4i, 1 - 4i, -1$

9. $\sqrt{2}, -\sqrt{2}, \sqrt{3}$
Are the coefficients rational?

10. Find a polynomial equation of degree 4 with -2 as a root of multiplicity 1, 3 as a root of multiplicity 2, and -1 as a root of multiplicity 1.

[iii] In Exercises 11 and 12, suppose a polynomial or polynomial equation of degree 5 with rational coefficients has the given numbers as roots. Find the other roots.

11. $6, -3 + 4i, 4 - \sqrt{5}$

12. $-2, 3, 4, 1 - i$

In Exercises 13–18, find a polynomial of lowest degree with rational coefficients that has the given numbers as some of its roots.

13. $1 + i, 2$

14. $2 - i, -1$

15. $-4i, 5$

16. $2 - \sqrt{3}, 1 + i$

17. $\sqrt{5}, -3i$

18. $-\sqrt{2}, 4i$

In Exercises 19–22, given that the polynomial or polynomial equation has the given root, find the other roots.

19. $x^4 - 5x^3 + 7x^2 - 5x + 6; -i$

20. $x^4 - 16 = 0; 2i$

21. $x^3 - 6x^2 + 13x - 20 = 0; 4$

22. $x^3 - 8; 2$

☆ ———

In Exercises 23–25, solve the equations. Use synthetic division, the quadratic formula, the theorems of this section, or whatever else you think might help.

23. $x^3 - 4x^2 + x - 4 = 0$

24. $x^3 - x^2 - 7x + 15 = 0$

25. $x^4 - 2x^3 - 2x - 1 = 0$

26. The equation $x^2 + 2ax + b = 0$ has a double root. Find it.

27. What does the fundamental theorem of algebra tell you about the following equation?

$$2 \sin^5 x - 2 \sin^3 x + \sin x = \frac{1}{8}$$

28. Prove that a polynomial with positive coefficients cannot have a positive root.

★ ———

29. Prove that every polynomial of odd degree, with real coefficients, has at least one real root.

30. Prove that there is no polynomial P that defines the cosine function, $P(x) \equiv \cos x$.

31. Prove that every polynomial with real coefficients has a factorization into linear and quadratic factors (with real coefficients).

13.4 RATIONAL ROOTS

[i] Integer Coefficients

It is not always easy to find the roots of a polynomial. However, if a polynomial has integer coefficients, there is a procedure that will yield all the rational roots.

THEOREM 8
———

The rational roots theorem. Let

$$P(x) = a_n x^n + a_{n-1}x^{n-1} + \cdots + a_1 x + a_0,$$

where all the coefficients are integers. Consider a rational number denoted by c/d, where c and d are relatively prime (having no common factor besides 1 and -1). If c/d is a root of $P(x)$, then c is a factor of a_0 and d is a factor of a_n.

OBJECTIVES

You should be able to:

[i] Given a polynomial with integer coefficients, find the rational roots and find the other roots, if possible.

[ii] Do the same for polynomials with rational coefficients.

Proof. Since c/d is a root of $P(x)$, we know that

$$a_n\left(\frac{c}{d}\right)^n + a_{n-1}\left(\frac{c}{d}\right)^{n-1} + \cdots + a_1\left(\frac{c}{d}\right) + a_0 = 0. \qquad (1)$$

We multiply by d^n and get the equation

$$a_n c^n + a_{n-1}c^{n-1}d + \cdots + a_1 cd^{n-1} + a_0 d^n = 0. \qquad (2)$$

Then we have

$$a_n c^n = (-a_{n-1}c^{n-1} - \cdots - a_1 cd^{n-2} - a_0 d^{n-1})d.$$

Note that d is a factor of $a_n c^n$. Now d is not a factor of c because c and d are relatively prime. Thus d is not a factor of c^n. So d is a factor of a_n.

In a similar way we can show from equation (2) that

$$a_0 d^n = (-a_n c^{n-1} - a_{n-1}c^{n-2}d - \cdots - a_1 d^{n-1})c.$$

Thus c is a factor of $a_0 d^n$. Again, c is not a factor of d^n, so it must be a factor of a_0.

Example 1 Let $P(x) = 3x^4 - 11x^3 + 10x - 4$. Find the rational roots of $P(x)$. If possible, find the other roots.

By the rational roots theorem if c/d is a root of $P(x)$, then c must be a factor of -4 and d must be a factor of 3. Thus the possibilities for c and d are

$$c:\;\; 1,\; -1,\; 4,\; -4,\; 2,\; -2; \qquad d:\;\; 1,\; -1,\; 3,\; -3.$$

Then the resulting possibilities for c/d are

$$\frac{c}{d}:\;\; 1,\; -1,\; 4,\; -4,\; \frac{1}{3},\; -\frac{1}{3},\; \frac{4}{3},\; -\frac{4}{3},\; \frac{2}{3},\; -\frac{2}{3},\; 2,\; -2.$$

Of these 12 possibilities, we know that at most 4 of them could be roots because $P(x)$ is of degree 4. To find which are roots we could use substitution, but synthetic division is usually more efficient.

We try 1:

$$
\begin{array}{r|rrrrr}
1 & 3 & -11 & 0 & 10 & -4 \\
  &   & 3 & -8 & -8 & 2 \\
\hline
  & 3 & -8 & -8 & 2 & \!\!\!\!\mid -2.
\end{array}
$$

We try -1:

$$
\begin{array}{r|rrrrr}
-1 & 3 & -11 & 0 & 10 & -4 \\
   &   & -3 & 14 & -14 & 4 \\
\hline
   & 3 & -14 & 14 & -4 & \!\!\!\!\mid 0.
\end{array}
$$

Since $P(1) = -2$, 1 is not a root; but $P(-1) = 0$, so -1 is a root. We use synthetic division again, to see whether -1 is a double root.

$$
\begin{array}{r|rrrr}
-1 & 3 & -14 & 14 & -4 \\
   &   & -3 & 17 & -31 \\
\hline
   & 3 & -17 & 31 & \!\!\!\!\mid -35
\end{array}
$$

It is not. Using the results of the second synthetic division above, we can express $P(x)$ as follows:

$$P(x) = (x + 1)(3x^3 - 14x^2 + 14x - 4).$$

We now use $3x^3 - 14x^2 + 14x - 4$ and check the other possible roots. We try $\frac{2}{3}$:

$$\begin{array}{r|rrr r}
\frac{2}{3} & 3 & -14 & 14 & -4 \\
 & & 2 & -8 & 4 \\
\hline
 & 3 & -12 & 6 & 0.
\end{array}$$

Since $P(\frac{2}{3}) = 0$, $\frac{2}{3}$ is a root. Again using the results of the synthetic division, we can express $P(x)$ as

$$P(x) = (x + 1)\left(x - \frac{2}{3}\right)(3x^2 - 12x + 6).$$

Since the factor $3x^2 - 12x + 6$ is quadratic, we can use the quadratic formula to find that the other roots are $2 + \sqrt{2}$ and $2 - \sqrt{2}$. These are irrational numbers. Thus the rational roots are -1 and $\frac{2}{3}$.

DO EXERCISE 1.

Example 2 Let $P(x) = x^3 + 6x^2 + x + 6$. Find the rational roots of $P(x)$. If possible, find the other roots.

By the rational roots theorem, if c/d is a root of $P(x)$, then c must be a factor of 6 and d must be a factor of 1. Thus the possibilities for c and d are

$$c: \quad 1, -1, 2, -2, 3, -3, 6, -6; \qquad d: \quad 1, -1.$$

Then the resulting possibilities for c/d are

$$\frac{c}{d}: \quad 1, -1, 2, -2, 3, -3, 6, -6.$$

Note that these are the same as the possibilities for c. If the leading coefficient is 1, we need only check the factors of the last coefficient as possibilities for rational roots.

There is another aid in eliminating possibilities for rational roots. Note that all coefficients of $P(x)$ are positive. Thus when any positive number is substituted in $P(x)$, we get a positive value, never 0. Therefore no positive number can be a root. Thus the only possibilities for roots are

$$-1, -2, -3, -6.$$

We try -6:

$$\begin{array}{r|rrr r}
-6 & 1 & 6 & 1 & 6 \\
 & & -6 & 0 & -6 \\
\hline
 & 1 & 0 & 1 & 0.
\end{array}$$

Thus $P(-6) = 0$, so -6 is a root. We could divide again, to determine whether -6 is a double root, but since we can now factor $P(x)$ as a product of a linear and a quadratic polynomial, it is preferable to proceed that way. We have

$$P(x) = (x + 6)(x^2 + 1).$$

Now $x^2 + 1$ has the complex roots i and $-i$. Thus the only rational root of $P(x)$ is -6.

DO EXERCISE 2.

1. Let
$$P(x) = 2x^4 - 7x^3 - 35x^2 + 13x + 3.$$
If c/d is a rational root of $P(x)$, then:
a) What are the possibilities for c?
b) What are the possibilities for d?
c) What are the possibilities for $\frac{c}{d}$?
d) Find the rational roots.
e) If possible, find the other roots.

2. Let $P(x) = x^3 + 7x^2 + 4x + 28$. If c/d is a rational root of $P(x)$, then:
a) What are the possibilities for c?
b) What are the possibilities for d?
c) What are the possibilities for $\frac{c}{d}$?
d) How can you tell without substitution or synthetic division that there are no positive roots?
e) Find the rational roots of $P(x)$.
f) Find the other roots if they exist.

3. Let $P(x) = x^4 + x^2 + 2x + 6$.

  a) How do you know at the outset that this polynomial has no positive roots?

  b) Find the rational roots of $P(x)$.

4. a) Find the rational roots of $x^2 + 3x + 3$.

  b) Can you find the other roots of this polynomial? What are they? Why can you find them?

5. Let

$$P(x) = x^4 - \frac{1}{6}x^3 - \frac{4}{3}x^2 + \frac{1}{6}x + \frac{1}{3}.$$

  a) Which, if any, of the coefficients are *not* integers?

  b) What is the LCM of the denominators?

  c) Multiply by the LCM.

  d) Find the rational roots of the resulting polynomial.

  e) Are they rational roots of $P(x)$? Why?

Example 3 Find the rational roots of $x^4 + 2x^3 + 2x^2 - 4x - 8$.

Since the leading coefficient is 1, the only possibilities for rational roots are the factors of the last coefficient -8:

$$1, -1, 2, -2, 4, -4, 8, -8.$$

But, using substitution or synthetic division, we find that none of the possibilities is a root. We leave it to the student to verify this. Thus there are no rational roots.

The polynomial given in Example 3 has no rational roots. We can approximate the irrational roots of such a polynomial by graphing it and determining the x-intercepts (see Section 13.6).

DO EXERCISES 3 AND 4.

[ii] Rational Coefficients

Suppose some (or all) of the coefficients of a polynomial are rational, but not integers. After multiplying on both sides by the LCM of the denominators, we can then find the rational roots.

Example 4 Let $P(x) = \frac{1}{12}x^3 - \frac{1}{12}x^2 - \frac{2}{3}x + 1$. Find the rational roots of $P(x)$.

The LCM of the denominators is 12. When we multiply on both sides by 12, we get

$$12P(x) = x^3 - x^2 - 8x + 12.$$

This equation is equivalent to the first, and all coefficients on the right are integers. Thus any root of $12P(x)$ is a root of $P(x)$. We leave it to the student to verify that 2 and -3 are the rational roots, in fact the only roots, of $P(x)$.

DO EXERCISE 5.

EXERCISE SET 13.4

[i] In Exercises 1–4, use Theorem 8 to list all *possible* rational roots.

1. $x^5 - 3x^2 + 1$

2. $x^7 + 37x^5 - 6x^2 + 12$

3. $15x^6 + 47x^2 + 2$

4. $10x^{25} + 3x^{17} - 35x + 6$

In Exercises 5–16, find the rational roots, if they exist, of each polynomial or equation. If possible, find the other roots. Then write the equation or polynomial in factored form.

5. $x^3 + 3x^2 - 2x - 6$

6. $x^3 - x^2 - 3x + 3 = 0$

7. $5x^4 - 4x^3 + 19x^2 - 16x - 4 = 0$

8. $3x^4 - 4x^3 + x^2 + 6x - 2$

9. $x^4 - 3x^3 - 20x^2 - 24x - 8$

10. $x^4 + 5x^3 - 27x^2 + 31x - 10$

11. $x^3 + 3x^2 - x - 3 = 0$

12. $x^3 + 5x^2 - x - 5$

13. $x^3 + 8$

14. $x^3 - 8 = 0$

[ii]

15. $\frac{1}{3}x^3 - \frac{1}{2}x^2 - \frac{1}{6}x + \frac{1}{6}$

16. $\frac{2}{3}x^3 - \frac{1}{2}x^2 + \frac{2}{3}x - \frac{1}{2}$

[i] In Exercises 17–24, find only the rational roots.

17. $x^4 + 32$

18. $x^6 + 8 = 0$

19. $x^3 - x^2 - 4x + 3 = 0$

20. $2x^3 + 3x^2 + 2x + 3$

21. $x^4 + 2x^3 + 2x^2 - 4x - 8 = 0$

22. $x^4 + 6x^3 + 17x^2 + 36x + 66 = 0$

23. $x^5 - 5x^4 + 5x^3 + 15x^2 - 36x + 20$

24. $x^5 - 3x^4 - 3x^3 + 9x^2 - 4x + 12$

☆

25. The volume of a cube is 64 cm^3. Find the length of a side. [*Hint:* Solve $x^3 - 64 = 0$.]

26. The volume of a cube is 125 cm^3. Find the length of a side.

27. An open box of volume 48 cm^3 can be made from a piece of tin 10 cm on a side by cutting a square from each corner and folding up the edges. What is the length of a side of the squares?

28. An open box of volume 500 cm^3 can be made from a piece of tin 20 cm on a side by cutting a square from each corner and folding up the edges. What is the length of a side of the squares?

★

29. Show that $\sqrt{5}$ is irrational, by considering the equation $x^2 - 5 = 0$.

30. Generalize the result of Exercise 29 to find which integers have rational square roots.

13.5 FURTHER HELPS IN FINDING ROOTS

[i] Descartes' Rule of Signs

A rule that helps determine the number of positive real roots of a polynomial is due to Descartes. To use the rule we must have the polynomial arranged in descending or ascending order, with no zero terms written in. Then we determine the number of *variations of sign,* that is, the number of times, in going through the polynomial, that successive coefficients are of different sign.

Example 1 Determine the number of variations of sign in the polynomial $2x^6 - 3x^2 + x + 4$.

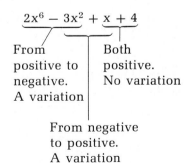

$$2x^6 - 3x^2 + x + 4$$

From positive to negative. A variation

Both positive. No variation

From negative to positive. A variation

The number of variations of sign is two.

DO EXERCISES 1 AND 2.

OBJECTIVES

You should be able to:

[i] Use Descartes' rule of signs to find information about the number of real roots of a polynomial with real coefficients.

[ii] Find upper and lower bounds of the roots of a polynomial with real coefficients, using synthetic division.

Determine the number of variations of sign in each polynomial.

1. $3x^5 - 2x^3 - x^2 + x - 2$

2. $4p^7 + 6p^5 + 2p^3 - 5p^2 + 3$

In each case, what does Descartes' rule of signs tell you about the number of positive real roots?

3. $5x^3 - 4x - 5$

4. $6p^6 - 5p^4 + 3p^3 - 7p^2 + p - 2$

5. $3x^2 - 2x + 4$

We now state Descartes' rule, without proof.

THEOREM 9

Descartes' rule of signs. **The number of positive real roots of a polynomial with real coefficients is either**

1. **The same as the number of its variations of sign, or**
2. **Less than the number of its variations of sign by a positive even integer.**

A root of multiplicity m must be counted m times.

Examples In each case, what does Descartes' rule of signs tell you about the number of positive real roots?

2. $2x^5 - 5x^2 + 3x + 6$

The number of variations of sign is two. Therefore the number of positive real roots either is 2 or is less than 2 by 2, 4, 6, etc. Thus the number of positive roots is either 2 or 0, a negative number of roots having no meaning.

3. $5x^4 - 3x^3 + 7x^2 - 12x + 4$

There are four variations of sign. Thus the number of positive real roots is either

$$4 \quad \text{or} \quad 4 - 2 \quad \text{or} \quad 4 - 4.$$

That is, the number of roots is 4, 2, or 0.

4. $6x^5 - 2x - 5 = 0$

The number of variations of sign is 1. Therefore there is exactly one positive real root.

DO EXERCISES 3–5.

Negative Roots

Descartes' rule can also be used to help determine the number of negative roots of a polynomial. Consider the following graphs of a polynomial equation $y = P(x)$ and its reflection across the y-axis, that is, $y = P(-x)$. The points at which the graphs cross the x-axis are the roots of the polynomials.

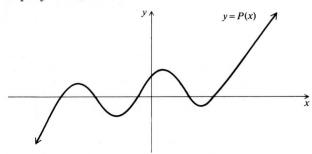

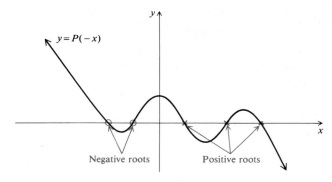

From the graphs we see that the number of positive roots of $P(-x)$ is the same as the number of negative roots of $P(x)$.

THEOREM 10

Corollary to Descartes' rule of signs. The number of negative real roots of a polynomial $P(x)$ with real coefficients is either

1. The number of variations of sign of $P(-x)$, or
2. Less than the number of variations of sign of $P(-x)$ by a positive even integer.

To apply Theorem 10, we construct $P(-x)$ by replacing x by $-x$ wherever it occurs and then count the variations of sign.

Examples In each case, what does Descartes' rule of signs tell you about the number of negative real roots?

5. $5x^4 + 3x^3 + 7x^2 + 12x + 4$

$$P(-x) = 5(-x)^4 + 3(-x)^3 + 7(-x)^2 + 12(-x) + 4 \quad \text{Replacing}$$
$$\text{x by } -x$$
$$= 5x^4 \quad - 3x^3 \quad + 7x^2 \quad - 12x \quad + 4 \quad \text{Simplifying}^*$$

There are four variations of sign, so the number of negative roots is either 4 or 2 or 0.

6. $2x^5 - 5x^2 + 3x + 6$

$$P(-x) = 2(-x)^5 - 5(-x)^2 + 3(-x) + 6 \quad \text{Replacing}$$
$$\text{x by } -x$$
$$= -2x^5 \quad - 5x^2 \quad - 3x \quad + 6 \quad \text{Simplifying}$$

There is one variation of sign, so there is exactly one negative root.

DO EXERCISES 6-8.

———————————
*It is simple to construct $P(-x)$ mechanically. If $P(x)$ is in descending order with no missing terms, we change the sign of every second term beginning with the second term from the right.

In each case, what does Descartes' rule of signs tell you about the number of negative real roots?

6. $5x^3 - 4x - 5$

7. $6p^6 - 5p^4 + 3p^3 - 7p^2 + p - 2$

8. $3x^2 - 2x + 4$

Change signs.

[ii] Upper Bounds on Roots

Recall that when a polynomial $P(x)$ is divided by $x - a$ we obtain a quotient $Q(x)$ and a remainder R, related as follows:

$$P(x) = (x - a) \cdot Q(x) + R.$$

Let us consider the case in which a is positive, R is positive or zero, and all the coefficients of $Q(x)$ are nonnegative. Then $P(b)$ will be positive for any positive number b that is greater than a. We show this as follows.

$$P(b) \quad = \quad \underbrace{(b - a)} \cdot \underbrace{Q(b)} \quad + \quad R$$

positive   positive   nonnegative

$b - a > 0$ because we assumed that $b > a$;

$Q(b) > 0$ because all coefficients are nonnegative and the number b we are substituting is positive. (We know that $Q(x)$ is not the zero polynomial, so it has some nonzero coefficients.)

R is assumed to be nonnegative.

This shows that there can be no root of $P(x)$ greater than the positive number a. If there were such a root r it would make $P(r) = 0$ and this cannot occur, because all values of $P(x)$ are positive when x is greater than a. In terms of synthetic division, this tells us the following.

If this $a\rfloor$ ☐ ☐ ☐ ☐   ☐
number is
positive   ☐ ☐ ☐   ☐

☐ ☐ ☐ ☐ R

and all of the numbers in the bottom row are nonnegative,

then there is no root greater than a. In other words, the number a is an *upper bound* to all the roots of $P(x)$. Of course, if the number R is 0, we know that a is actually a root, as well as an upper bound. We state this result as a theorem.

THEOREM 11

If when a polynomial is divided by $x - a$, where a is positive, the remainder and all coefficients of the quotient are nonnegative, the number a is an upper bound to the roots of the polynomial (all of the roots are less than or equal to a).

Example 7 Determine an upper bound to the roots of

$$3x^4 - 11x^3 + 10x - 4.$$

We choose some *positive* number a and divide by $x - a$, using synthetic division. Let's try 1 for a.

$$
\begin{array}{r|rrrrr}
1 & 3 & -11 & 0 & 10 & -4 \\
  &   & 3   & -8 & -8 & 2 \\
\hline
  & 3 & -8 & -8 & 2 & -2 \\
\end{array}
$$

Since some of the coefficients are negative the theorem does not guarantee that 1 is an upper bound. We choose a larger number, this time 4.

$$
\begin{array}{r|rrrrr}
4 & 3 & -11 & 0 & 10 & -4 \\
  &   & 12 & 4 & 16 & 104 \\
\hline
  & 3 & 1   & 4 & 26 & 100 \\
\end{array}
$$

⌈One reason for choosing 4 is that we know *this* number must be at least 11, in order for the addition to give a nonnegative result.⌋

Since there are no negative numbers in the bottom row, 4 is an upper bound.

CAUTION! **Remember, in applying Theorem 11 the number a must be a *positive* number.**

DO EXERCISES 9 AND 10.

Lower Bounds

A corollary of the above theorem can be used to find lower bounds of roots of a polynomial. Consider the following graphs.

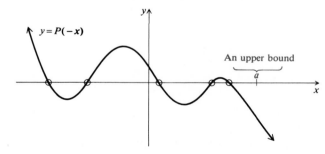

Here we have an upper bound, a, to the roots of the polynomial $P(-x)$. Now let us consider the reflection across the y-axis, $y = P(x)$.

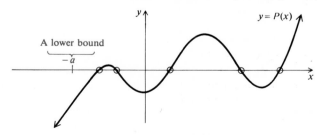

It is easy to see that the negative number $-a$ is a lower bound to the roots of $P(x)$.

In each case, determine an upper bound to the roots.

9. $5x^4 - 18x^2 + 3x - 2$

10. $x^3 - 75x^2 + 3$

THEOREM 12

> The number $-a$, where a is positive, is a lower bound to the roots of the polynomial $P(x)$, if when $P(-x)$ is divided by $x - a$ the remainder and all coefficients of the quotient are nonnegative.

Example 8 Determine a lower bound to the roots of $P(x)$, where

$$P(x) = 3x^4 + 11x^3 + 10x - 4.$$

We first construct $P(-x)$ by replacing x by $-x$:

$$3(-x)^4 + 11(-x)^3 + 10(-x) - 4,$$
$$3x^4 - 11x^3 - 10x - 4. \quad \text{Simplifying}$$

Next, we use synthetic division, dividing by x $-$ 4:*

$$\begin{array}{r|rrrrr} 4 & 3 & -11 & 0 & -10 & -4 \\ & & 12 & 4 & 16 & 24 \\ \hline & 3 & 1 & 4 & 6 & 20 \end{array}$$

Since none of the numbers in the bottom row is negative, 4 is an upper bound to the roots of $P(-x)$. Thus -4 is a lower bound to the roots of $3x^4 + 11x^3 + 10x - 4$, which is $P(x)$.

Example 9 Determine a lower bound to the roots of $P(x)$, where

$$P(x) = 5x^3 - 12x^2 + 2x + 3.$$

We construct $P(-x)$ by replacing x by $-x$ (or by simply changing the sign of every other term, beginning at the second from the right).

$$P(-x) = 5(-x)^3 - 12(-x)^2 + 2(-x) + 3$$
$$= -5x^3 - 12x^2 - 2x + 3.$$

We try synthetic division.

$$\begin{array}{r|rrrr} a & -5 & -12 & -2 & 3 \\ & & b & & \\ \hline & -5 & c & & \end{array}$$

Recall that the number a must be positive in this procedure. It is clear that no positive number a will give us a positive number in position b. Therefore it will be impossible to get a nonnegative number in position c.

In a case like this we will use $-P(-x)$ instead of $P(-x)$, since the two polynomials have the same roots. To change to $-P(-x)$ all we need do is change the sign of every coefficient in $P(-x)$. Then we

---

*To use a shortcut, we can set up the synthetic division using $P(x)$.

$$\begin{array}{r|rrrrr} 4 & 3 & 11 & 0 & 10 & -4 \end{array}$$

Then we change the sign of every second number, beginning at the second from the right. Then divide.

$$\begin{array}{r|rrrrr} 4 & 3 & \boxed{-11} & 0 & \boxed{-10} & -4 \end{array}$$

Change signs.

proceed as before. We try 1 for a.

$$
\begin{array}{r|rrrr}
1 & 5 & 12 & 2 & -3 \\
& & 5 & 17 & 19 \\
\hline
& 5 & 17 & 19 & \big|\ 16
\end{array}
$$

The number -1 is a lower bound.

In determining bounds for roots, if the leading coefficient of $P(x)$ is negative, we simply convert to $-P(x)$ by changing the sign of every term. Then we proceed as before.

DO EXERCISES 11 AND 12.

A Shortcut

On p. 503 you were cautioned to use only positive numbers when seeking upper bounds. Likewise, in looking for lower bounds we have used only positive numbers, in accordance with Theorem 11. In looking for lower bounds there is a way to use negative numbers in the synthetic division that may save a small amount of time. We illustrate, using the polynomial of Example 8. The following is what we have already done.

$$
\begin{array}{r|rrrrr}
4 & 3 & -11 & 0 & -10 & -4 \\
& & 12 & 4 & 16 & 24 \\
\hline
& 3 & 1 & 4 & 6 & \big|\ 20
\end{array}
$$
←——— This is $P(-x)$.

Now let us use -4 with $P(x)$ and compare.

$$
\begin{array}{r|rrrrr}
-4 & 3 & 11 & 0 & 10 & -4 \\
& & -12 & 4 & -16 & 24 \\
\hline
& 3 & -1 & 4 & -6 & \big|\ 20
\end{array}
$$
←——— This is $P(x)$.

Note that the only difference is a change of sign in the indicated columns. This illustrates the following corollary to Theorem 12.

THEOREM 13

When a polynomial is divided by $x - a$, where a is negative, a will be a lower bound to the roots if in the result of the synthetic division the first number is nonnegative, the second nonpositive, the third nonnegative, the fourth nonpositive, and so on.

Example 10 Determine a lower bound to the roots of $P(x)$, where

$$P(x) = 5x^3 - 12x^2 + 2x + 3.$$

We try -2:

$$
\begin{array}{r|rrrr}
-2 & 5 & -12 & 2 & 3 \\
& & -10 & 44 & -92 \\
\hline
& 5 & -22 & 46 & \big|\ -89.
\end{array}
$$

Since the odd-numbered coefficients (from left to right) are nonnegative and the even-numbered ones are nonpositive, -2 is a lower bound to the roots.

DO EXERCISE 13.

Determine a lower bound to the roots.

11. $5x^4 + 18x^3 + 3x - 2$

12. $4x^3 + 7x^2 + 3x + 5$

13. Determine a lower bound to the roots of $5x^4 + 18x^3 - 3x - 3$.

In each case, what does Descartes' rule of signs tell you about the roots of the polynomial? Find upper and lower bounds to the real roots.

14. $x^4 - 6x^3 + 7x^2 + 6x - 2$

15. $x^4 + 2x^2 + x - 2$

Example 11 What does Descartes' rule of signs tell you about the roots of $P(x)$? Find upper and lower bounds to the real roots.

$$P(x) = 4x^4 + 3x^3 + x - 1$$

a) There is one variation of sign, so there is just one real positive root.

b) $P(-x) = 4(-x)^4 + 3(-x)^3 + (-x) - 1$
$\qquad\quad = 4x^4 \quad\; - 3x^3 \quad\; - x \quad\; - 1$
There is just one variation of sign, so there is just one negative root.

c) Since the equation is of degree four it has four roots. Therefore, there are two nonreal roots.

d) We look for an upper bound.

$$
\begin{array}{r|rrrr}
1 & 4 & 3 & 0 & 1 & -1 \\
  &   & 4 & 7 & 7 & 8 \\
\hline
  & 4 & 7 & 7 & 8 & \big|\; 7
\end{array}
$$

The number 1 is an upper bound. Let's try $\frac{1}{4}$.

$$
\begin{array}{r|rrrrr}
\frac{1}{4} & 4 & 3 & 0 & 1 & -1 \\
  &   & 1 & 1 & \frac{1}{4} & \frac{5}{16} \\
\hline
  & 4 & 4 & 1 & \frac{5}{4} & \big|\; -\frac{11}{16}
\end{array}
$$

We do not know whether or not $\frac{1}{4}$ is an upper bound,* but we look no further.

e) We look for a lower bound, using Theorem 12.

$$
\begin{array}{r|rrrrr}
-1 & 4 & 3 & 0 & 1 & -1 \\
   &   & -4 & 1 & -1 & 0 \\
\hline
   & 4 & -1 & 1 & 0 & \big|\; -1
\end{array}
$$

We do not know whether or not -1 is a lower bound.

$$
\begin{array}{r|rrrrr}
-2 & 4 & 3 & 0 & 1 & -1 \\
   &   & -8 & 10 & -20 & 38 \\
\hline
   & 4 & -5 & 10 & -19 & \big|\; 37
\end{array}
$$

The number -2 is a lower bound and not a root (because $R \neq 0$).

To summarize, we have learned that $P(x)$ has two real roots, one between 0 and 1 and one between -2 and 0. It also has two nonreal roots.

From the information obtained in Example 11, we can actually say a little more. In (d) we found that $P(1) = 7$ (a positive number) and $P(\frac{1}{4}) = -\frac{11}{16}$ (a negative number). Thus $P(x)$ must be 0 for some number between $\frac{1}{4}$ and 1. One of the roots is therefore between $\frac{1}{4}$ and 1.

DO EXERCISES 14 AND 15.

---

*The converse of Theorem 10 is not true. That is, if the positive number a is an upper bound to the roots, it is not necessarily true that when we divide by $x - a$ all of the coefficients of the quotient and the remainder will be nonnegative. Thus when we divide and find negative numbers in the bottom row, this does not mean that the number a is *not* an upper bound. We simply do not know whether it is or not.

EXERCISE SET 13.5

[i] What does Descartes' rule of signs tell you about the number of positive real roots?

1. $3x^5 - 2x^2 + x - 1$

2. $5x^6 - 3x^3 + x^2 - x$

3. $6x^7 + 2x^2 + 5x + 4 = 0$

4. $-3x^5 - 7x^3 - 4x - 5 = 0$

5. $3p^{18} + 2p^4 - 5p^2 + p + 3$

6. $5t^{12} - 7t^4 + 3t^2 + t + 1$

What does Descartes' rule of signs tell you about the number of negative real roots?

7. $3x^5 - 2x^2 + x - 1$

8. $5x^6 - 3x^3 + x^2 - x$

9. $6x^7 + 2x^2 + 5x + 4 = 0$

10. $-3x^5 - 7x^3 - 4x - 5 = 0$

11. $3p^{18} + 2p^3 - 5p^2 + p + 3$

12. $5t^{11} - 7t^4 + 3t^2 + t + 1$

[ii] Determine an upper bound to the roots.

13. $3x^4 - 15x^2 + 2x - 3$

14. $4x^4 - 14x^2 + 4x - 2$

15. $6x^3 - 17x^2 - 3x - 1$

16. $5x^3 - 15x^2 + 5x - 4$

Determine a lower bound to the roots.

17. $3x^4 - 15x^3 + 2x - 3$

18. $4x^4 - 17x^3 + 3x - 2$

19. $6x^3 + 15x^2 + 3x - 1$

20. $6x^3 + 12x^2 + 5x - 3$

[i], [ii] What does Descartes' rule of signs tell you about the roots of the polynomial or equation? Find upper and lower bounds to the real roots.

21. $x^4 - 2x^2 + 12x - 8$

22. $x^4 - 6x^2 + 20x - 24$

23. $x^4 - 2x^2 - 8 = 0$

24. $3x^4 - 5x^2 - 4 = 0$

25. $x^4 - 9x^2 - 6x + 4$

26. $x^4 - 21x^2 + 4x + 6$

27. $x^4 + 3x^2 + 2 = 0$

28. $x^4 + 5x^2 + 6 = 0$

☆

29. Prove that for n a positive even integer, $x^n - 1$ has only two real roots.

30. Prove that for n an odd positive integer, $x^n - 1$ has only one real root.

★

31. Show that $x^4 + ax^2 + bx - c$, where a, b, and c are positive, has just two nonreal roots.

32. In the three-body problem that arises in astronomy the following equation occurs:

$$r^5 + (3 - k)r^4 + (3 - 2k)r^3 - kr^2 - 2kr - k = 0,$$

where $0 < k < 1$. Show that this equation has just one positive real solution.

13.6 GRAPHS OF POLYNOMIAL FUNCTIONS

[i] Graphs of first-degree polynomial functions are lines; graphs of second-degree, or quadratic, functions are parabolas. We now consider polynomials of higher degree. We begin with some general principles to be kept in mind while graphing. We consider only polynomials with real coefficients.

1. Every polynomial function is a continuous function, whose domain is the set of all real numbers. The graph of any function must pass

OBJECTIVES

You should be able to:

[i] Graph polynomial functions.

[ii] Approximate roots of polynomial equations by graphing.

the vertical line test, so no polynomial function can have a graph like the following.

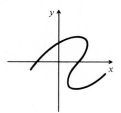

2. Unless a polynomial function is linear, no part of its graph is straight. A common error is to make a graph look like the black one below, rather than the colored one below.

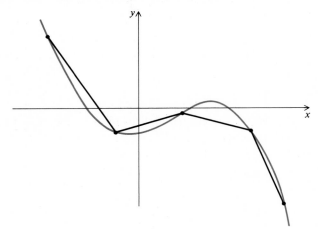

3. A polynomial of degree n cannot have more than n real roots. This means that the graph cannot cross the x-axis more than n times, so we know something about how the curve "wiggles." Third-degree, or *cubic*, functions have graphs like the following.

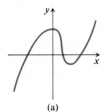

(a)

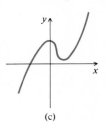

(b)   (c)

In (a) the graph crosses the axis three times, so there are three real roots. In (b) and (c) there is only one x-intercept, so there is only one real root in each case. The graph of a cubic cannot look like the following, because there would be a possibility that it might cross the x-axis more than three times.

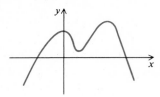

Graphs of fourth-degree, or *quartic,* polynomials look like these.

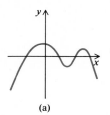

(a)

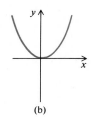

(b)

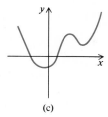

(c)

In (a) there are four real roots, in (b) there is one, and in (c) there are two.

Multiple roots occur at points like the following.

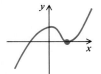

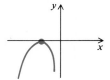

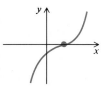

When a graph looks like the one below, missing the x-axis at one of its opportunities, a pair of nonreal roots occurs. There is no easy way to find them from the graph.

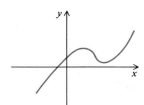

4. The leading term of a polynomial tells a lot about how the graph looks for values of x with large absolute value. This is so because for such values of x the contributions of the other terms are relatively minor. First, let us suppose that the coefficient is positive. Then for large positive values of x, the function value will be positive and increasing. Thus we know that as we move far to the right the graph looks like (a) below.

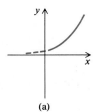

(a)

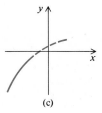

(b)

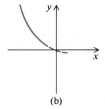

(c)

If the exponent of the leading term is even, the same thing happens as we move far to the left, as in (b). If the exponent is odd, then the function values are negative and they decrease as we move far to the left, as in (c) above. If the leading coefficient is negative, the graphs will of course be reflected across the x-axis.

To graph polynomials, keep in mind previously established results and proceed somewhat as follows.

TO GRAPH POLYNOMIALS

a) **Look at the degree of the polynomial and its leading coefficient. This gives a lot of information about the general shape of the graph.**

b) **Look for symmetries, as covered in Chapter 3.**

c) **Make a table of values using synthetic division.**

d) **Find the *y*-intercept and as many *x*-intercepts as possible (the latter are roots of the polynomial). In doing this, recall the theorems about roots, including Descartes' rule of signs.**

e) **Plot the points and connect them appropriately.**

Example 1 Graph $P(x) = 2x^3 - x + 2$.

a) This polynomial is of degree 3 with leading coefficient positive. Thus as we move far to the right, function values will increase beyond bound. As we move far left, they will be negative and decrease beyond bound.

    The curve will have the general shape of a cubic, one of those shown on p. 508.

b) The function is not odd or even. However, $2x^3 - x$ is an odd function, with the origin as a point of symmetry. $P(x)$ is a translation of this upward 2 units; hence the point $(0, 2)$ is a point of symmetry.

c) We make a table of values. The first entry is merely the *y*-intercept. The other rows contain the bottom line in synthetic division (which is really all we need to write ordinarily).

	2	0	−1	2	(x, y)
0				2	(0, 2)
1	2	2	1	3	(1, 3)
2	2	4	7	16	(2, 16)
−1	2	−2	1	1	(−1, 1)
−2	2	−4	7	−12	(−2, −12)

The *y*-intercept

Since all numbers are positive, 1 is an upper bound to the roots.

Since signs alternate, −2 is a lower bound to the roots.

d) Because we know the curve is symmetric with respect to the point $(0, 2)$, we do not really need to calculate $P(x)$ for negative values of x. We can also see from this table that the graph will cross the x-axis between −1 and −2. This is because $P(-1)$ is positive (and the graph there is above the axis), whereas $P(-2)$ is negative (and the graph there is below the axis). Somewhere between −1 and −2, then, $P(x)$ must be 0. Thus there is a real root of $P(x)$ between −1 and −2.

By Descartes' rule of signs, there are either 2 or 0 positive real roots. From our table of values, we see that $P(x)$ is positive and increasing rapidly. These facts would lead us to *suspect* that there are *no* positive roots. If there are any, they must be between 0 and 1 because 1 is an upper bound. Now $P(-x) = -2x^3 + x + 2$, so Descartes' rule tells us there is just one negative root. Thus the one we have already eyeballed is the only one. We also know that if there are no positive roots, there will be two nonreal ones, and they will be conjugates of each other.

e) We now plot points and connect them appropriately.*

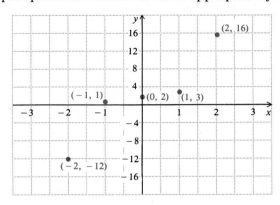

In this example, it is not easy to see how to draw the curve. It might look like any of the following so far as we can tell from the points plotted.

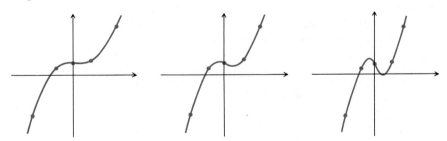

We therefore need to plot some more points. Because of symmetry, it will be sufficient to consider positive values of x.

	2	0	−1	2	(x, y)
0.1	2	0.2	0.98	1.9	(0.1, 1.9)
0.3	2	0.6	−0.82	1.75	(0.3, 1.75)
0.5	2	1.0	−0.5	1.75	(0.5, 1.75)
0.7	2	1.4	−0.02	1.99	(0.7, 1.99)
0.9	2	1.8	0.62	2.56	(0.9, 2.56)

*In this graph we are using different scales on the x-axis and the y-axis. We are reluctant to do so, but it is sometimes necessary.

1. Graph.

$$P(x) = x^3 - 4x^2 - 3x + 12$$

We now have enough points to determine the shape of the graph.

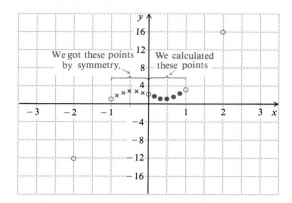

We can now draw the graph confidently.

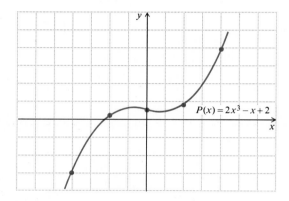

DO EXERCISE 1.

[ii] Solving Equations

Whenever we find the roots, or zeros, of a function $P(x)$ we have solved the equation $P(x) = 0$. One means of finding approximate solutions of polynomial equations is by graphing. We graph the function $y = P(x)$ and note where the graph crosses the x-axis. In Example 1, there is a real solution of the equation $2x^3 - x + 2 = 0$ at about -1.1.

We can approximate roots more precisely by doing further calculation.

▦ **Example 2** Find a better approximation of the solution of $2x^3 - x + 2 = 0$, which is near -1.1.

We know that $P(-1) = 1$ and $P(-2) = -12$, so there must be a root between -1 and -2. To find a better approximation we use synthetic division.

	2	0	−1	2
−1.1	2	−2.2	1.42	0.44
−1.2	2	−2.4	1.88	−0.26
.	.	.	.	.
.	.	.	.	.
−1.15	2	−2.30	1.65	0.11
−1.16	2	−2.32	1.69	0.04
−1.17	2	−2.34	1.74	−0.03

Since $P(-1.1) > 0$, there is no root between −1 and −1.1.

Since $P(-1.2) < 0$ and $P(-1.1) > 0$, there is a root between −1.1 and −1.2. We'll try −1.15, halfway between.

We have a root between −1.16 and −1.17.

2. a) Graph $P(x) = x^3 - 3x^2 + 1$.
b) Use the graph to approximate the roots to tenths.

We now have an approximation to hundredths, −1.16. In this manner the approximation can be made as close as we please.

DO EXERCISE 2.

EXERCISE SET 13.6

[i] In Exercises 1–6, graph.

1. $P(x) = x^3 - 3x^2 - 2x - 6$

2. $P(x) = x^3 + 4x^2 - 3x - 12$

3. $P(x) = 2x^4 + x^3 - 7x^2 - x + 6$

4. $P(x) = 3x^4 + 5x^3 + 5x^2 - 5x - 6$

5. $P(x) = x^5 - 2x^4 - x^3 + 2x^2$

6. $P(x) = x^5 + 4x^4 - 5x^3 - 14x^2 - 8x$

[ii] In Exercises 7–16, graph the corresponding polynomial functions, in order to find approximate solutions of the equations.

7. $x^3 - 3x - 2 = 0$

8. $x^3 - 3x^2 + 3 = 0$

9. $x^3 - 3x - 4 = 0$

10. $x^3 - 3x^2 + 5 = 0$

11. $x^4 + x^2 + 1 = 0$

12. $x^4 + 2x^2 + 2 = 0$

13. $x^4 - 6x^2 + 8 = 0$

14. $x^4 - 4x^2 + 2 = 0$

15. $x^5 + x^4 - x^3 - x^2 - 2x - 2 = 0$

16. $x^5 - 2x^4 - 2x^3 + 4x^2 - 3x + 6 = 0$

17. ▦ The following equation has a solution between 0 and 1. Approximate it, to hundredths.

$$2x^5 + 2x^3 - x^2 - 1 = 0$$

18. ▦ The following equation has a solution between 1 and 2. Approximate it, to hundredths.

$$x^4 - 2x^3 - 3x^2 + 4x + 2$$

In each of the following, graph and then approximate the irrational roots, to hundredths.

19. ▦ $P(x) = x^3 - 2x^2 - x + 4$

20. ▦ $P(x) = x^3 - 4x^2 + x + 3$

★

21. The equation $\log x = \sin x$ has a solution between 2 and 3. Find it, to the nearest tenth.

22. ▦ Graph $P(x) = 5.8x^4 - 2.3x^2 - 6.1$.

23. *Nested evaluation.* A procedure for evaluating a polynomial is as follows. Given a polynomial, such as $3x^4 - 5x^3 + 4x^2 - 5$, successively factor out x, as shown:

$$x\big(x(x(3x - 5) + 4)\big) - 5.$$

Given a value for x, substitute it in the innermost parentheses and work your way out, at each step multiplying, then adding or subtracting. Show that this process is identical to synthetic division.

OBJECTIVE

You should be able to:

[i] Graph rational functions.

13.7 RATIONAL FUNCTIONS

[i] A *rational function* is a function definable as the quotient of two polynomials. Here are some examples:

$$y = \frac{x^2 + 3x - 5}{x + 4}, \qquad y = \frac{5}{x^2 + 3}, \qquad y = \frac{3x^5 - 5x + 2}{4}.$$

DEFINITION

A *rational function* **is a function that can be defined as** $y = P(x)/Q(x)$, **where** $P(x)$ **and** $Q(x)$ **are polynomials having no common factors other than 1 and** -1, **and with** $Q(x)$ **not the zero polynomial.**

Polynomial functions are themselves special kinds of rational functions, since $Q(x)$ can be the polynomial 1. Here we are interested in rational functions in which the denominator is not a constant. We begin with the simplest such function.

Example 1 Graph the function $y = 1/x$.

a) Note that the domain of this function consists of all real numbers except 0.

b) For nonzero x or y, the above equation is equivalent to $xy = 1$. Now it is easy to see that the graph is symmetric with respect to the line $y = x$ (because interchanging x and y produces an equivalent equation).

c) The graph is also symmetric with respect to the origin (because replacing x with $-x$ and y with $-y$ produces an equivalent equation).

d) Now we find some values, keeping in mind the two symmetries.

x	1	2	3	4	5
y	1	$\frac{1}{2}$	$\frac{1}{3}$	$\frac{1}{4}$	$\frac{1}{5}$

e) We plot these points. We use one symmetry to get other points in the first quadrant. We use the other symmetry to get the points in the third quadrant.

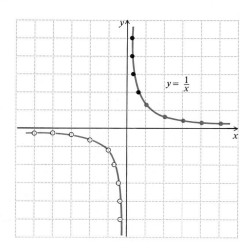

$$y = \frac{1}{x}$$

The points indicated by ● are obtained from the table. Those marked ● are obtained by reflection across the line $y = x$. Following this, the points marked ○ are obtained by reflection across the origin.

Asymptotes

Note that this curve does not touch either axis, but comes very close. As $|x|$ becomes very large the curve comes very near to the x-axis. In fact, we can find points as close to the x-axis as we please by choosing x large enough. We say that the curve approaches the x-axis *asymptotically,* and we say that the x-axis is an *asymptote* to the curve. The y-axis is also an asymptote to this curve.

Using the ideas of transformations we can easily graph certain variations of the above function.

Example 2 $y = -1/x$ is a reflection across the x-axis (or the y-axis; the result is the same).

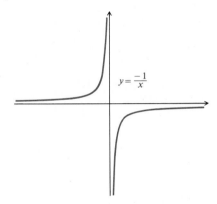

Example 3 $y = 1/(x - 2)$ is a translation of $y = 1/x$ two units to the right.

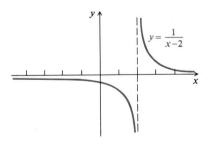

DO EXERCISES 1–4.

Example 4 Graph the function $y = 1/x^2$.

a) Note that this function is defined for all x except 0. Therefore the line $x = 0$ is an asymptote.

b) As $|x|$ gets very large, y approaches 0. Therefore the x-axis is also an asymptote, for both positive and negative values of x.

c) This function is even. Therefore it is symmetric with respect to the y-axis.

Graph these equations. (Use graph paper.)

1. $y = \dfrac{1}{x + 5}$

2. $y = \dfrac{3x + 1}{x}$

$\left(Hint:\ \dfrac{3x + 1}{x} = 3 + \dfrac{1}{x}.\right)$

3. $y = \dfrac{2}{x}$

4. $y = \dfrac{1}{3x + 15}$

[*Hint:* $3x + 15 = 3(x + 5)$.] Compare with Exercise 1.

516

POLYNOMIALS AND RATIONAL FUNCTIONS

Graph.

5. $y = -\dfrac{1}{x^2}$

d) All function values are positive. Therefore the entire graph is above the x-axis.

e) With this much information, we can already sketch a rough graph of the function. However, a table of values will help.

x	1	2	3	4	$\frac{1}{2}$	$\frac{1}{3}$
y	1	$\frac{1}{4}$	$\frac{1}{9}$	$\frac{1}{16}$	4	9

The graph is as follows. Points marked ● are obtained from the table. Points marked ● are obtained by reflection across the y-axis.

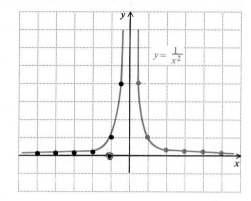

6. $y = \dfrac{-3x^2 + 1}{x^2}$

(*Hint:* Divide numerator by denominator.)

DO EXERCISES 5–8.

7. $y = \dfrac{1}{(x - 3)^2}$

Occurrence of Asymptotes

It is important in graphing rational functions to determine where the asymptotes, if any, occur. Vertical asymptotes are easy to locate when a denominator is factored. The x-values that make a denominator 0 but do not make the numerator 0 are those of the vertical asymptotes.

Examples Determine the vertical asymptotes.

8. Find the vertical asymptotes.

$y = \dfrac{x + 3}{x^3 - x^2 - 6x}$

5. $y = \dfrac{3x - 2}{x(x - 5)(x + 3)}$

The vertical asymptotes are the lines $x = 0$, $x = 5$, and $x = -3$.

6. $y = \dfrac{x - 2}{x^3 - x}$

We factor the denominator:

$$x^3 - x = x(x + 1)(x - 1).$$

The vertical asymptotes are $x = 0$, $x = -1$, and $x = 1$.

Horizontal asymptotes occur when the degree of the numerator is the same as or less than that of the denominator. Let us first consider a function for which the degree of the numerator is less than that of the denominator:

$$y = \dfrac{2x + 3}{x^3 - 2x^2 + 4}.$$

We shall multiply by $\dfrac{1/x^3}{1/x^3}$, to obtain

$$y = \frac{\dfrac{2}{x^2} + \dfrac{3}{x^3}}{1 - \dfrac{2}{x} + \dfrac{4}{x^3}}.$$

Let us now consider what happens to the function values as $|x|$ becomes very large. Each expression with x in its denominator takes on smaller and smaller values, approaching 0. Thus the numerator approaches 0 and the denominator approaches 1; hence the entire expression takes on values closer and closer to 0. Therefore, the x-axis is an asymptote. Whenever the degree of a numerator is less than that of the denominator, the x-axis will be an asymptote.

DO EXERCISE 9.

Next, we consider a function for which the numerator and denominator have the same degree:

$$y = \frac{3x^2 + 2x - 4}{2x^2 - x + 1} = \frac{3x^2 + 2x - 4}{2x^2 - x + 1} \cdot \frac{\dfrac{1}{x^2}}{\dfrac{1}{x^2}}$$

$$= \frac{3 + \dfrac{2}{x} - \dfrac{4}{x^2}}{2 - \dfrac{1}{x} + \dfrac{1}{x^2}}.$$

As $|x|$ gets very large the numerator approaches 3 and the denominator approaches 2. Therefore the function values get very close to $\frac{3}{2}$, and thus the line $y = \frac{3}{2}$ is an asymptote. From this example, we can see that the asymptote in such cases can be determined by dividing the leading coefficients of the two polynomials.

Examples Determine the horizontal asymptotes.

7. $y = \dfrac{5x^3 - x^2 + 7}{3x^3 + x - 10}$

The line $y = \frac{5}{3}$ is an asymptote.

8. $y = \dfrac{-7x^4 - 10x^2 + 1}{11x^4 + x - 2}$

The line $y = -\frac{7}{11}$ is an asymptote.

DO EXERCISES 10 AND 11.

There are asymptotes that are neither vertical nor horizontal. They are called *oblique*, and they occur when the degree of the numerator is greater than that of the denominator by 1. To find the asymptote we divide the numerator by the denominator. Consider

$$y = \frac{2x^2 - 3x - 1}{x - 2}.$$

9. Which of the following have the x-axis for an asymptote?

a) $y = \dfrac{3x^4 - x^2 + 4}{18x^4 - x^3 + 44}$

b) $y = \dfrac{17x^2 + 14x - 5}{x^3 - 2x - 1}$

c) $y = \dfrac{135x^5 - x^2}{x^7}$

Find the horizontal asymptotes.

10. $y = \dfrac{3x^3 + 4x - 9}{6x^3 - 7x^2 + 3}$

11. $y = \dfrac{9x^4 - 7x^2 - 9}{3x^4 + 7x^2 + 9}$

Find the oblique asymptotes.

12. $y = \dfrac{3x^2 - 7x + 2}{x - 2}$

13. $y = \dfrac{5x^3 + 2x + 1}{x^2 - 4}$

When we divide the numerator by the denominator we obtain a quotient of $2x + 1$ and a remainder of 1. Thus,

$$y = 2x + 1 + \frac{1}{x - 2}.$$

Now we can see that when $|x|$ becomes very large, $1/(x - 2)$ approaches 0, and the y-values thus approach $2x + 1$. This means that the graph comes closer and closer to the straight line $y = 2x + 1$.

Example 9 Find and draw the asymptotes of the function

$$y = \frac{2x^2 - 11x - 10}{x - 4}.$$

a) We first note that $x = 4$ is a vertical asymptote.

b) We divide, to obtain $y = 2x - 3 - \dfrac{22}{x - 4}$.

Thus the line $y = 2x - 3$ is an asymptote.

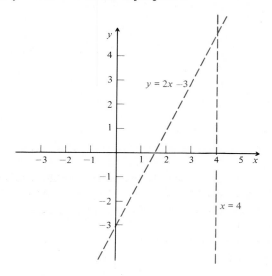

DO EXERCISES 12 AND 13.

The following summarizes the conditions under which asymptotes occur.

ASYMPTOTES OF A RATIONAL FUNCTION OCCUR AS FOLLOWS.

1. *Vertical asymptotes.* **Have the same *x*-values that make the denominator zero, but the numerator nonzero.**

2. *The x-axis an asymptote.* **When the degree of the denominator is greater than that of the numerator.**

3. *Horizontal asymptotes other than the x-axis.* **Occur when numerator and denominator have the same degree.**

4. *Oblique asymptotes.* **Occur when the degree of the numerator is greater than that of the denominator by 1.**

Zeros or Roots

Zeros, or roots, of a rational function occur when the numerator is 0 but the denominator is not 0. The zeros of a function occur at points where the graph crosses the x-axis. Therefore knowing the zeros helps in making a graph. If the numerator can be factored, the zeros are easy to determine.

Example 10 Find the zeros of the function $y = \dfrac{x^3 - x^2 - 6x}{x^2 - 3x + 2}$.

We factor numerator and denominator:

$$y = \frac{x(x + 2)(x - 3)}{(x - 1)(x - 2)}.$$

The values making the numerator zero are 0, -2, and 3. Since none of these makes the denominator zero, they are the zeros of the function.

DO EXERCISES 14 AND 15.

The following is an outline of a procedure to be followed in graphing rational functions.

TO GRAPH A RATIONAL FUNCTION

a) **Determine any symmetries.**

b) **Determine any horizontal or oblique asymptotes and sketch them.**

c) **Factor the denominator and the numerator.**
   i) **Determine any vertical asymptotes and sketch them.**
   ii) **Determine the zeros, if possible, and plot them.**

d) **Make a table, showing where the function values are positive and where negative.**

e) **Make a table of values and plot them.**

f) **Sketch the curve.**

Example 11 Graph $y = \dfrac{x^3 - x^2 - 6x}{x^2 - 3x + 2}$.

We follow the outline above.

a) The function is neither even nor odd and no symmetries are apparent.

b) The degree of the numerator is one greater than that of the denominator. Dividing numerator by denominator will show that the line $y = x + 2$ is an oblique asymptote.

c) The numerator and denominator are easily factorable. We have

$$y = \frac{x(x + 2)(x - 3)}{(x - 1)(x - 2)}.$$

The zeros are 0, -2, and 3 and there are vertical asymptotes $x = 1$ and $x = 2$.

Find the zeros of these functions.

14. $y = \dfrac{x(x - 3)(x + 5)}{(x + 2)(x - 7)}$

15. $y = \dfrac{x^3 + 2x^2 - 3x}{x^2 + 5}$

d) The zeros and asymptotes divide the x-axis into intervals in a natural way, pictured as follows. We tabulate signs in these intervals.

Interval	x + 2	x	x − 1	x − 2	x − 3	y
x < −2	−	−	−	−	−	−
−2 < x < 0	+	−	−	−	−	+
0 < x < 1	+	+	−	−	−	−
1 < x < 2	+	+	+	−	−	+
2 < x < 3	+	+	+	+	−	−
3 < x	+	+	+	+	+	+

e) We next make a table of function values and plot them.

x	$\frac{1}{2}$	$\frac{3}{2}$	$\frac{5}{2}$	4	5	−1	−3	−5
y	$-\frac{5}{2}$	63	$-\frac{15}{2}$	4	$\frac{35}{6}$	$\frac{2}{3}$	$-\frac{9}{5}$	$-\frac{20}{7}$

f) Using all available information, we draw the graph.

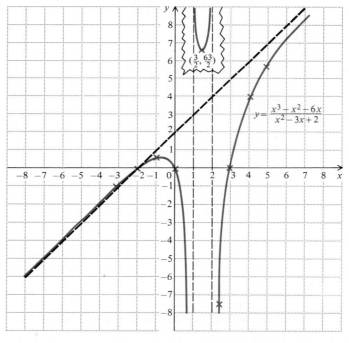

$$y = \frac{x^3 - x^2 - 6x}{x^2 - 3x + 2}$$

The lower left-hand part of the graph is as shown on the following page. The curve crosses the oblique asymptote at $(-2, 0)$, and then, moving to the left, comes back close to the asymptote from above.*

---

*Mathematical folklore often has it that no curve ever crosses its asymptote. This example shows the folklore to be false.

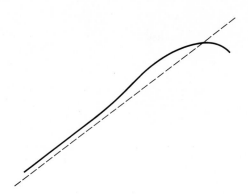

16. Graph this function.

$$y = \frac{x^2 - 4}{x^2 - x - 12}$$

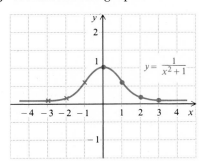

Example 12 Graph $y = \frac{1}{x^2 + 1}$.

We follow the outline above.

a) The function is even, so the graph is symmetric with respect to the y-axis.

b) The degree of the denominator is greater than that of the numerator. Thus the x-axis is an asymptote.

c) The denominator is not factorable. Neither is the numerator. Therefore there are no vertical asymptotes and there are no zeros.

d) The numerator is a positive constant and the denominator can never be negative, so this function has only positive values. Thus no table of signs is necessary.

e) We tabulate some values.   **f)** We sketch the graph.

x	0	1	2	3
y	1	$\frac{1}{2}$	$\frac{1}{5}$	$\frac{1}{10}$

DO EXERCISE 16.

EXERCISE SET 13.7

[i] Graph these functions.

1. $y = \dfrac{1}{x - 3}$

2. $y = \dfrac{1}{x - 5}$

3. $y = \dfrac{-2}{x - 5}$

4. $y = \dfrac{-3}{x - 3}$

5. $y = \dfrac{2x + 1}{x}$

6. $y = \dfrac{3x - 1}{x}$

7. $y = \dfrac{1}{(x - 2)^2}$

8. $y = \dfrac{-2}{(x - 3)^2}$

9. $y = \dfrac{2}{x^2}$

10. $y = \dfrac{1}{3x^2}$

11. $y = \dfrac{1}{x^2 + 3}$

12. $y = \dfrac{-1}{x^2 + 2}$

13. $y = \dfrac{x - 1}{x + 2}$

14. $y = \dfrac{x - 2}{x + 1}$

15. $y = \dfrac{3x}{x^2 + 5x + 4}$

16. $y = \dfrac{x + 3}{2x^2 - 5x - 3}$

17. $y = \dfrac{x^2 - 4}{x - 1}$

18. $y = \dfrac{x^2 - 9}{x + 1}$

19. $y = \dfrac{x^2 + x - 2}{2x^2 + 1}$

20. $y = \dfrac{x^2 - 2x - 3}{3x^2 + 2}$

21. $y = \dfrac{x - 1}{x^2 - 2x - 3}$

22. $y = \dfrac{x + 2}{x^2 + 2x - 15}$

23. $y = \dfrac{x + 2}{(x - 1)^3}$

24. $y = \dfrac{x - 3}{(x + 1)^3}$

25. $y = \dfrac{x^3 + 1}{x}$

26. $y = \dfrac{x^3 - 1}{x}$

27. $y = \dfrac{x^3 + 2x^2 - 15x}{x^2 - 5x - 14}$

28. $y = \dfrac{x^3 + 2x^2 - 3x}{x^2 - 25}$

☆

29. $y = \dfrac{1}{x^2 + 3} - 5$

30. $y = \dfrac{-1}{x^2 + 2} + 4$

31. $y = \dfrac{1}{|x + 2|}$

32. $y = \left|\dfrac{1}{x} - 2\right|$

33. $y = \left|\dfrac{1}{x - 2} - 3\right|$

34. $y = \dfrac{3x}{|x^2 + 5x + 4|}$

CHAPTER 13 REVIEW

[13.2, ii] **1.** Find the remainder when $x^4 + 3x^3 + 3x^2 + 3x + 2$ is divided by $x + 2$.

[13.2, i] **2.** Use synthetic division to find the quotient and remainder:

$$(2x^4 - 6x^3 + 7x^2 - 5x + 1) \div (x + 2).$$

[13.3, ii] **3.** Find a polynomial of degree 3 with roots 0, 1, and 2.

4. Find a polynomial of lowest degree having roots 1 and -1, and having 2 as a root of multiplicity 2 and -3 as a root of multiplicity 3.

5. Find a complete factorization of $x^3 - 1$.

[13.2, ii] **6.** Use synthetic division to find $P(3)$:

$$P(x) = 2x^4 - 3x^3 + x^2 - 3x + 7.$$

7. Factor the polynomial $P(x)$. Then solve the equation $P(x) = 0$.

$$P(x) = x^3 + 7x^2 + 7x - 15$$

[13.2, iii] **8.** Determine whether $x + 1$ is a factor of $x^3 + 6x^2 + x + 30$.

9. Find k such that $x + 3$ is a factor of $x^3 + kx^2 + kx - 15$.

[13.3, iii] **10.** The equation $x^4 - 81 = 0$ has $3i$ for a root. Find the other roots.

11. The equation $x^2 - 8x + c = 0$ has a double root. Find it.

12. A polynomial of degree 4 with rational coefficients has roots $-8 - 7i$ and $10 + \sqrt{5}$. Find the other roots.

[13.4, i] **13.** List all possible rational roots of $2x^6 - 12x^4 + 17x^2 + 12$.

[13.5, i] **14.** What does Descartes' rule of signs tell you about the number of positive real roots of $3x^{12} + 3x^4 - 7x^2 + x + 5$?

15. What does Descartes' rule of signs tell you about the number of negative real roots of $6x^8 - 12x^4 + 5x^2 + x + 2$?

[13.5, ii] **16.** Find an upper bound to the roots of $2x^4 - 7x^2 + 2x - 1$.

17. Find a lower bound to the roots of $12x^3 + 24x^2 + 10x - 6$.

[13.6, ii] **18.** The equation $x^5 + x^4 - x^3 - x^2 - 2x - 2$ has a root between 1 and 2. Approximate this root to hundredths.

19. Graph $P(x) = x^3 - 3x^2 + 3$. Use the graph to approximate the irrational roots.

[13.7, i] **20.** Graph $y = \dfrac{x^2 + x - 6}{x^2 - x - 20}$.

14

EQUATIONS OF SECOND DEGREE AND THEIR GRAPHS

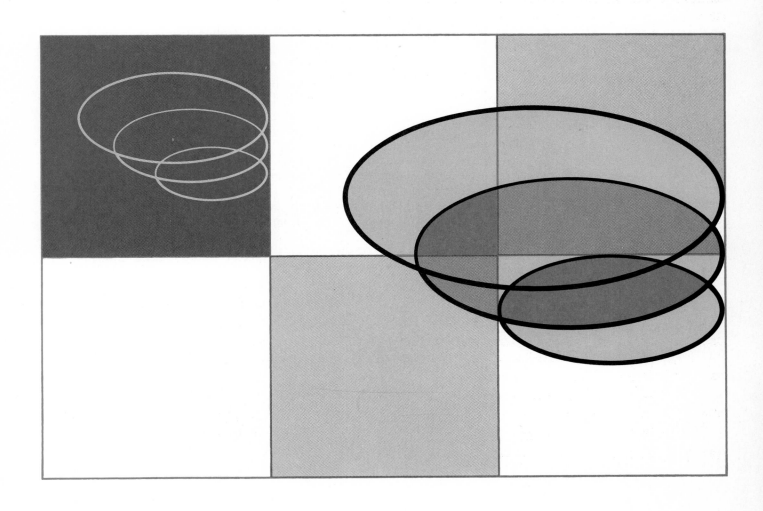

OBJECTIVES

You should be able to:

[i] Graph equations whose graph consists of a union of lines.
[ii] Graph equations whose graph consists of a single point.
[iii] Given the center and radius of a circle, find an equation for the circle.
[iv] Find the center and radius of a circle whose equation is

$$(x - h)^2 + (y - k)^2 = r^2$$

and graph the circle.

[v] Given an equation of a circle that is not in standard form, complete the square and find its center and radius.
[vi] Given the center and a point that a circle passes through, find an equation of the circle.

14.1 CONIC SECTIONS

Cones

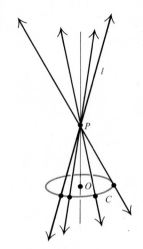 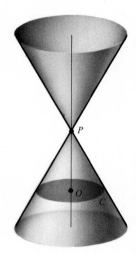

Suppose C is a circle with center O and P is a point not in the same plane as C such that the line $\overleftrightarrow{OP}$ is perpendicular to the plane of the circle C. The set of points on all lines through P and a point of the circle form a *right circular cone* (or *conical surface*). Any line contained in the surface is called a *surface element*. Note that there are two parts or *nappes* of a cone. Point P is called the *vertex* and line $\overleftrightarrow{OP}$ is called the *axis*.

Conic Sections

The nonempty intersection of any plane with a cone is a *conic section*. Some conic sections are shown below and at the top of the following page.

In this chapter we study equations of second degree and their graphs. Graphs of most such equations are conic sections.

(a) Ellipse (b) Parabola (c) Hyperbola

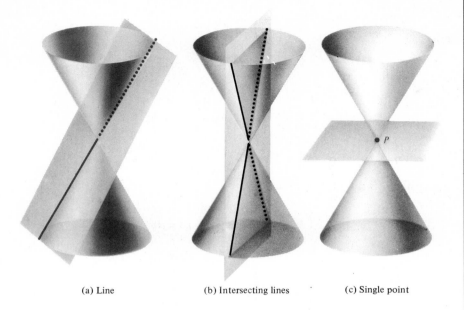

(a) Line (b) Intersecting lines (c) Single point

[i] Lines

Some second-degree equations have graphs consisting of a line. Some have graphs consisting of two lines. The graphs are conic sections except when the two lines are parallel.

Example 1 Graph $3x^2 + 2xy - y^2 = 0$.

We factor and use the principle of zero products:

$$(3x - y)(x + y) = 0$$
$$3x - y = 0 \quad \text{or} \quad x + y = 0$$
$$y = 3x \quad \text{or} \quad y = -x.$$

The graph consists of two intersecting lines.

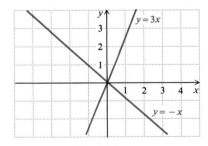

Example 2 Graph $(y - x)(y - x - 1) = 0$.

$$(y - x)(y - x - 1) = 0$$
$$y - x = 0 \quad \text{or} \quad y - x - 1 = 0 \qquad \text{Using the principle of}$$
$$\text{zero products}$$
$$y = x \quad \text{or} \quad y = x + 1$$

Graph. (Use graph paper.)

1. $x^2 - 4y^2 = 0$

2. $y^2 = 4$

3. $y^2 + 9x^2 = 6xy$

4. $x^2 + 2y^2 = 0$

5. $x^2 + y^2 = -3$

The graph consists of two parallel lines.

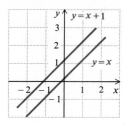

Example 3 Graph $y^2 + x^2 = 2xy$.

$$y^2 - 2xy + x^2 = 0$$
$$(y - x)(y - x) = 0$$
$$y - x = 0 \quad \text{or} \quad y - x = 0$$
$$y = x \quad \text{or} \quad y = x$$

The graph consists of a single line.

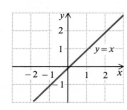

Generally a second-degree equation with 0 on one side and a factorable expression on the other has a graph consisting of one or two lines. A special case is $xy = 0$, in which the graph consists of the coordinate axes.

DO EXERCISES 1–3.

[ii] Single Points or No Points

When a plane intersects only the vertex of a cone, the result is a single point. The following is an equation for such a conic section.

Example 4 Graph $x^2 + 4y^2 = 0$.

The expression $x^2 + 4y^2$ is not factorable in the real-number system. The only real-number solution of the equation is $(0, 0)$.

Example 5 Graph $3x^2 + 7y^2 = -2$.

Since squares of numbers are never negative, the left side of the equation can never be negative. The equation has no real-number solutions, hence has no graph.

DO EXERCISES 4 AND 5.

Circles

Some equations of second degree have graphs that are circles. Circles are defined as follows.

DEFINITION

A *circle* is the locus or set of all points in a plane that are at a fixed distance from a fixed point in that plane.

When a plane intersects a cone as shown, perpendicular to the axis of the cone, a circle is formed. We shall prove this later.

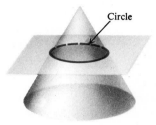

[iii] **Equations of Circles**

We first obtain an equation for a circle centered at the origin.

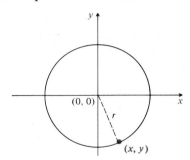

We let r represent the radius. If a point (x, y) is on the circle, then by the definition of a circle and the distance formula we have

$$r^2 = x^2 + y^2.$$

We have shown that if a point (x, y) is on the circle, then $x^2 + y^2 = r^2$. We now prove the converse, i.e., if a point (x, y) satisfies the equation $x^2 + y^2 = r^2$, then it is on the circle. We assume that (x, y) satisfies $x^2 + y^2 = r^2$. Then we have

$$(x - 0)^2 + (y - 0)^2 = r^2.$$

Taking the principal square root, we obtain

$$\sqrt{(x - 0)^2 + (y - 0)^2} = r.$$

Thus the distance from (x, y) to $(0, 0)$ is r, and the point (x, y) is on the circle. The two parts of this proof together show that the equation $x^2 + y^2 = r^2$ gives *all* the points of the circle *and no others*.

6. Find an equation of the circle having center $(-3, 7)$ and radius 5.

When a circle is translated so that its center is the point (h, k), an equation for it can be found by replacing x by $x - h$ and y by $y - k$.

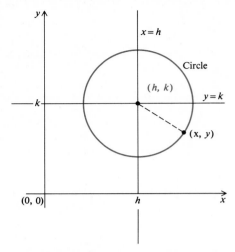

The equation, in standard form, of a circle with center (h, k) and radius r is

$$(x - h)^2 + (y - k)^2 = r^2.$$

Example 6 Find an equation of the circle having center $(4, -5)$ and radius 6.

Using the standard form, we obtain

$$(x - 4)^2 + [y - (-5)]^2 = 6^2,$$

or

$$(x - 4)^2 + (y + 5)^2 = 36.$$

DO EXERCISE 6.

We now prove that a conic section obtained when the plane is perpendicular to the axis of a cone is actually a circle.

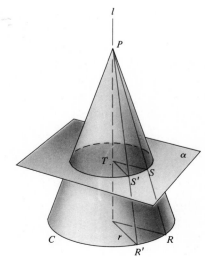

Consider a plane α that cuts a right circular cone perpendicular to its axis, and of course parallel to its base. Then look at the cross sections on any two planes containing the axis of the cone l. Angles PTS and PTS' are right angles because α is perpendicular to l. Thus $\triangle PTS \simeq \triangle PTS'$. Then $\overline{TS} \simeq \overline{TS'}$, for any two points S and S' on the intersection of the cone and the plane α. Thus the intersection consists of all points that are a distance TS from T and is thus a circle, centered at T.

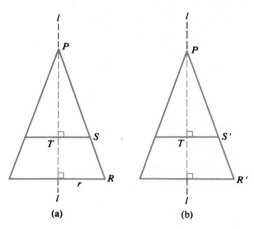

(a)   (b)

7. Find the center and radius of

$$(x + 1)^2 + (y - 3)^2 = 4.$$

Then graph the circle.

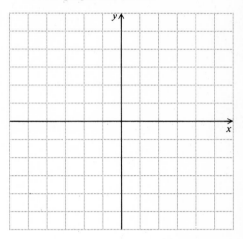

[iv] Finding Center and Radius

Example 7 Find the center and radius of $(x - 2)^2 + (y + 3)^2 = 16$. Then graph the circle.

We may first write standard form: $(x - 2)^2 + [y - (-3)]^2 = 4^2$. Then the *center* is $(2, -3)$ and the *radius* is 4. The graph is then easy to draw, as shown, using a compass.

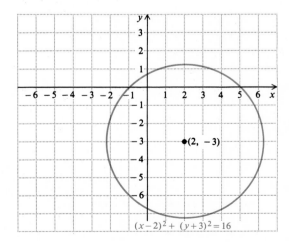

DO EXERCISE 7.

8. Find the center and radius of the circle $x^2 + y^2 - 14x + 4y - 11 = 0$.

9. Find an equation of the circle with center $(-1, 4)$ that passes through $(3, -1)$.

[v] Standard Form by Completing the Square

Completing the square allows us to find the standard form for the equation of a circle.

Example 8 Find the center and radius of the circle

$$x^2 + y^2 + 8x - 2y + 15 = 0.$$

We complete the square twice to get standard form:

$$(x^2 + 8x + \quad) + (y^2 - 2y + \quad) = -15.$$

We take half the coefficient of the x-term and square it, obtaining 16. We add $16 - 16$ in the first parentheses. Similarly, we add $1 - 1$ in the second parentheses.

$$(x^2 + 8x + 16 - 16) + (y^2 - 2y + 1 - 1) = -15$$

Next we do some rearranging and factoring:

$$(x^2 + 8x + 16) + (y^2 - 2y + 1) - 16 - 1 = -15$$
$$(x + 4)^2 + (y - 1)^2 = 2. \quad \text{This is standard form.}$$

The center is $(-4, 1)$ and the radius is $\sqrt{2}$.

DO EXERCISE 8.

[vi] Circle Determined by Center and One Point

Example 9 Find an equation of a circle with center $(-2, -3)$ that passes through the point $(1, 1)$.

Since $(-2, -3)$ is the center, we have

$$(x + 2)^2 + (y + 3)^2 = r^2.$$

The circle passes through $(1, 1)$. We find r by substituting in the above equation:

$$(1 + 2)^2 + (1 + 3)^2 = r^2$$
$$9 + 16 = r^2$$
$$25 = r^2$$
$$5 = r.$$

Then $(x + 2)^2 + (y + 3)^2 = 25$ is an equation of the circle.

DO EXERCISE 9.

EXERCISE SET 14.1

[i], [ii] Graph.

1. $x^2 - y^2 = 0$

2. $x^2 - 9y^2 = 0$

3. $3x^2 + xy - 2y^2 = 0$

4. $x^2 - xy - 2y^2 = 0$

5. $2x^2 + y^2 = 0$

6. $5x^2 + y^2 = -3$

[iii] Find an equation of a circle with center and radius as given.

7. Center: $(0, 0)$
Radius: 7

8. Center: $(-2, 7)$
Radius: $\sqrt{5}$

[iv] Find the center and radius of each circle. Then graph the circle.

9. $(x + 1)^2 + (y + 3)^2 = 4$

10. $(x - 2)^2 + (y + 3)^2 = 1$

Find the center and radius of each circle.

11. $(x - 8)^2 + (y + 3)^2 = 40$

12. $(x + 5)^2 + (y - 1)^2 = 75$

[v]

13. $x^2 + y^2 + 8x - 6y - 15 = 0$

14. $x^2 + y^2 + 25x + 10y + 12 = 0$

15. $x^2 + y^2 - 4x = 0$

16. $x^2 + y^2 + 10y - 75 = 0$

17. ▦ $x^2 + y^2 + 8.246x - 6.348y - 74.35 = 0$

18. ▦ $x^2 + y^2 + 25.074x + 10.004y + 12.054 = 0$

19. $9x^2 + 9y^2 = 1$

20. $16x^2 + 16y^2 = 1$

[vi] Find an equation of a circle satisfying the given conditions.

21. Center $(0, 0)$, passing through $(-3, 4)$

22. Center $(3, -2)$, passing through $(11, -2)$

☆ _____

Find an equation of a circle satisfying the given conditions.

23. Center $(2, 4)$ and tangent (touching at one point) to the x-axis

24. Center $(-3, -2)$ and tangent to the y-axis

25. ▦ A circular swimming pool is being constructed in the corner of a lot as shown. In laying out the pool the contractor wishes to know the distances a_1 and a_2. Find them.

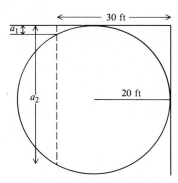

26. Find an equation of a circle such that the endpoints of a diameter are $(5, -3)$ and $(-3, 7)$.

27. a) Graph $x^2 + y^2 = 4$. Is this relation a function?

b) Solve $x^2 + y^2 = 4$ for y.

c) Graph $y = \sqrt{4 - x^2}$ and determine whether it is a function. Find the domain and range.

d) Graph $y = -\sqrt{4 - x^2}$ and determine whether it is a function. Find the domain and range.

28. Show that the following equation is an equation of a circle with center (h, k) and radius r.

$$\begin{vmatrix} x - h & -(y - k) \\ y - k & x - h \end{vmatrix} = r^2$$

Determine whether each of the following points lies on the *unit circle* $x^2 + y^2 = 1$.

29. $(0, -1)$

30. $\left(\dfrac{\sqrt{3}}{2}, -\dfrac{1}{2}\right)$

31. $(\sqrt{2} + \sqrt{3}, 0)$

32. $\left(\dfrac{\pi}{4}, \dfrac{4}{\pi}\right)$

33. $(\cos \theta, \sin \theta)$ **34.** $(\sin \phi, \cos \phi)$ **35.** $(\tan \theta, \sec \theta)$ **36.** $(\sec \theta, \tan \theta)$

37. Prove that $\angle ABC$ is a right angle. Assume point B is on the circle whose radius is a and whose center is at the origin. (*Hint:* Use slopes and an equation of the circle.)

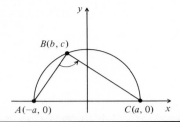

OBJECTIVES

You should be able to:

[i] Given an equation of an ellipse centered at the origin, find the vertices and foci, and graph the ellipse.

[ii] Given an equation of an ellipse centered at other than the origin, complete the square if necessary and then find the vertices and foci, and graph the ellipse.

14.2 ELLIPSES

Some equations of second degree have graphs that are ellipses. Ellipses are defined as follows.

DEFINITION

An *ellipse* is the locus or set of all points P in a plane such that the sum of the distances from P to two fixed points F_1 and F_2 in the plane is constant. F_1 and F_2 are called *foci* (singular, *focus*) of the ellipse.

When a plane intersects a cone as shown, not perpendicular to the axis of the cone, an ellipse is formed. See the figure on p. 524. We shall prove this later.

Here is a way to draw an ellipse. Stick two tacks in a piece of cardboard. These will be the foci F_1 and F_2. Attach a piece of string to the tacks. The length of the string will be the constant sum of the distances from the foci to points on the ellipse. Take a pencil and pull the string tight. Now swing the pencil around, keeping the string tight.

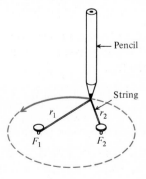

[i] Equations of Ellipses

We first consider an equation of an ellipse whose center is at the origin and whose foci lie on one of the coordinate axes.

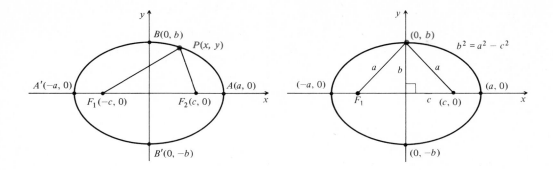

Suppose we have an ellipse with foci $F_1(-c, 0)$ and $F_2(c, 0)$. If $P(x, y)$ is a point on the ellipse, then $F_1P + F_2P$ is the given constant distance. We will call it $2a$:

$$F_1P + F_2P = 2a.$$

By the distance formula,

$$\sqrt{(x + c)^2 + y^2} + \sqrt{(x - c)^2 + y^2} = 2a,$$

or

$$\sqrt{(x + c)^2 + y^2} = 2a - \sqrt{(x - c)^2 + y^2}.$$

Squaring, we get

$$x^2 + 2cx + c^2 + y^2 = 4a^2 - 4a\sqrt{(x - c)^2 + y^2} + x^2 - 2cx + c^2 + y^2,$$

or

$$-4a^2 + 4cx = -4a\sqrt{(x - c)^2 + y^2},$$
$$-a^2 + cx = -a\sqrt{(x - c)^2 + y^2}.$$

Squaring again, we get

$$a^4 - 2a^2cx + c^2x^2 = a^2x^2 - 2cxa^2 + a^2c^2 + a^2y^2,$$

or

$$x^2(a^2 - c^2) + a^2y^2 = a^2(a^2 - c^2).$$

It follows from when P is at $(0, b)$ that $b^2 = a^2 - c^2$. Substituting b^2 for $a^2 - c^2$ in the last equation, we have the equation of the ellipse $b^2x^2 + a^2y^2 = a^2b^2$, or the following.

$$\frac{x^2}{a^2} + \frac{y^2}{b^2} = 1 \qquad \text{**Standard form of an equation of an ellipse, center at the origin**}$$

We have proved that if a point is on the ellipse, then its coordinates satisfy this equation. We also need to know the converse, that is, if the coordinates of a point satisfy this equation, then the point is on the ellipse. The proof of the latter will be omitted here. In the above, the longer axis of symmetry $\overline{A'A}$ is called the *major axis*. The shorter axis of symmetry $\overline{B'B}$ is called the *minor axis*. The intersection of these axes is called the *center*. The points A, A', B, and B' are called *vertices*. If the center of an ellipse is at the origin, the vertices are also the intercepts.

534

1. Suppose a circle has been distorted to get the ellipse below.

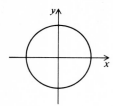

　a) Has the circle been stretched or shrunk in the x-direction?

　b) Has the circle been stretched or shrunk in the y-direction?

For each ellipse find the vertices and the foci, and draw a graph.

2. $x^2 + 9y^2 = 9$

3. $9x^2 + 25y^2 = 225$

4. $2x^2 + 4y^2 = 8$

Ellipses as Stretched Circles

　Let us consider a unit circle centered at the origin:

$$x^2 + y^2 = 1.$$

If we replace x by x/a and y by y/b we get an equation of an ellipse:

$$\left(\frac{x}{a}\right)^2 + \left(\frac{y}{b}\right)^2 = 1 \quad \text{or} \quad \frac{x^2}{a^2} + \frac{y^2}{b^2} = 1.$$

It follows that an ellipse is a circle transformed by a stretch or shrink in the x-direction and also in the y-direction. If $a = 2$, for example, the unit circle is stretched in the x-direction by a factor of 2. In any case, the x-intercepts become a and $-a$ and the y-intercepts become b and $-b$.

Example 1 For the ellipse $x^2 + 16y^2 = 16$, find the vertices and the foci. Then graph the ellipse.

a) We first multiply by $\frac{1}{16}$ to find standard form:

$$\frac{x^2}{16} + \frac{y^2}{1} = 1 \quad \text{or} \quad \frac{x^2}{4^2} + \frac{y^2}{1^2} = 1.$$

Thus $a = 4$ and $b = 1$. Then two of the vertices are $(-4, 0)$ and $(4, 0)$. These are also x-intercepts. The other vertices are $(0, 1)$ and $(0, -1)$. These are also y-intercepts. Since we know that $c^2 = a^2 - b^2$, we have $c^2 = 16 - 1$, so $c = \sqrt{15}$ and the foci are $(-\sqrt{15}, 0)$ and $(\sqrt{15}, 0)$.

b) The graph is as follows.

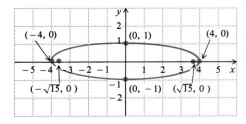

DO EXERCISES 1–4.

Example 2 Graph this ellipse and its foci: $9x^2 + 2y^2 = 18$.

a) We first multiply by $\frac{1}{18}$:

$$\frac{x^2}{2} + \frac{y^2}{9} = 1 \quad \text{or} \quad \frac{x^2}{(\sqrt{2})^2} + \frac{y^2}{3^2} = 1.$$

Thus $a = \sqrt{2}$ and $b = 3$.

b) Since $b > a$ the foci are on the y-axis and the major axis lies along the y-axis. To find c in this case we proceed as follows:

$$c^2 = b^2 - a^2 = 9 - 2 = 7$$
$$c = \sqrt{7}.$$

c) The graph is as follows.

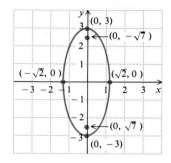

DO EXERCISES 5–7.

[ii] **Standard Form by Completing the Square**

If the center of an ellipse is not at the origin but at some point (h, k), then the standard form of the equation is as follows.

$$\frac{(x - h)^2}{a^2} + \frac{(y - k)^2}{b^2} = 1 \qquad \begin{array}{l}\textbf{Standard form of an equation} \\ \textbf{of an ellipse, center at } (h, k)\end{array}$$

Example 3 For the ellipse

$$16x^2 + 4y^2 + 96x - 8y + 84 = 0,$$

find the center, vertices, and foci. Then graph the ellipse.

a) We first complete the square to get standard form:

$$16(x^2 + 6x + \quad) + 4(y^2 - 2y + \quad) = -84$$
$$16(x^2 + 6x + 9 - 9) + 4(y^2 - 2y + 1 - 1) = -84$$
$$16(x^2 + 6x + 9) + 4(y^2 - 2y + 1) = -84 + 144 + 4$$
$$16(x + 3)^2 + 4(y - 1)^2 = 64$$
$$\frac{(x + 3)^2}{2^2} + \frac{(y - 1)^2}{4^2} = 1.$$

The center is $(-3, 1)$, $a = 2$, and $b = 4$.

b) The vertices of $x^2/2^2 + y^2/4^2 = 1$ are $(2, 0)$, $(-2, 0)$, $(0, 4)$, and $(0, -4)$; and since $c^2 = 16 - 4 = 12$, $c = 2\sqrt{3}$, and its foci are $(0, 2\sqrt{3})$ and $(0, -2\sqrt{3})$.

c) Then the vertices and foci of the translated ellipse are found by translation in the same way in which the center has been translated. Thus the vertices are

$$(-3 + 2, 1), \quad (-3 - 2, 1), \quad (-3, 1 + 4), \quad (-3, 1 - 4),$$

or

$$(-1, 1), \quad (-5, 1), \quad (-3, 5), \quad (-3, -3).$$

The foci are $(-3, 1 + 2\sqrt{3})$ and $(-3, 1 - 2\sqrt{3})$.

For each ellipse find the center, the vertices, and the foci, and draw a graph.

5. $9x^2 + y^2 = 9$

6. $25x^2 + 9y^2 = 225$

7. $4x^2 + 2y^2 = 8$

For each ellipse find the center, the vertices, and the foci, and graph the ellipse.

8. $25x^2 + 9y^2 + 150x - 36y + 260 = 0$

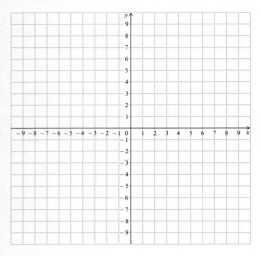

9. $9x^2 + 25y^2 - 36x + 150y + 260 = 0$

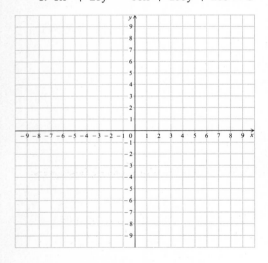

d) The graph is as follows.

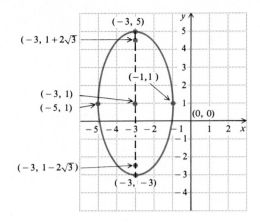

DO EXERCISES 8 AND 9.

Applications

Ellipses have many applications. Earth satellites travel in elliptical orbits. The planets travel around the sun in elliptical orbits with the sun at one focus.

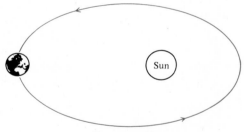

Planetary orbit

An interesting attraction found in museums is the *whispering gallery*. It is elliptical. Persons with their heads at the foci can whisper and hear each other clearly, while persons at other positions cannot hear them. This happens because sound waves emanating from one focus are reflected to the other focus, being concentrated there.

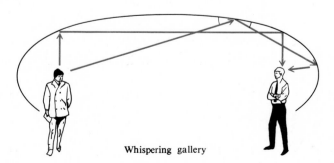

Whispering gallery

A Conic Section

We now prove that a conic section obtained when the plane is not perpendicular to the axis of a cone is actually an ellipse.

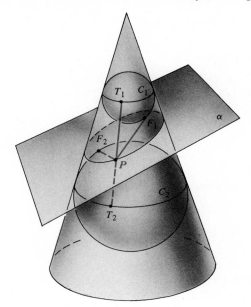

Consider a cone cut by a plane α not perpendicular to the axis. Now consider two spheres tangent to the cones at circles C_1 and C_2 and to plane α at points F_1 and F_2, respectively. Then consider any point P on the intersection of α with the cone. $\overline{PF_2}$ is tangent to the large sphere at F_2. $\overline{PT_2}$, a segment on an element of the cone, is tangent to the large sphere at T_2. Hence $PF_2 = PT_2$.

Similarly, $\overline{PF_1}$ and $\overline{PT_1}$ are tangent to the small sphere, hence $PF_1 = PT_1$. Thus $PF_1 + PF_2 = PT_1 + PT_2$. But $PT_1 + PT_2$ is constant, being the distance between the circles C_1 and C_2. Therefore, for all points P in the intersection of α with the cone, $PF_1 + PF_2$ is constant. The curve is thus an ellipse.

EXERCISE SET 14.2

[i] For each ellipse find the vertices and the foci, and draw a graph.

1. $\dfrac{x^2}{4} + \dfrac{y^2}{1} = 1$

2. $\dfrac{x^2}{1} + \dfrac{y^2}{4} = 1$

3. $16x^2 + 9y^2 = 144$

4. $9x^2 + 16y^2 = 144$

5. $2x^2 + 3y^2 = 6$

6. $5x^2 + 7y^2 = 35$

7. $4x^2 + 9y^2 = 1$

8. $25x^2 + 16y^2 = 1$

[ii] Find the center, vertices, and foci, and draw a graph.

9. $\dfrac{(x-1)^2}{4} + \dfrac{(y-2)^2}{1} = 1$

10. $\dfrac{(x-1)^2}{1} + \dfrac{(y-2)^2}{4} = 1$

11. $\dfrac{(x+3)^2}{25} + \dfrac{(y-2)^2}{16} = 1$

12. $\dfrac{(x-2)^2}{25} + \dfrac{(y+3)^2}{16} = 1$

13. $3(x+2)^2 + 4(y-1)^2 = 192$

14. $4(x-5)^2 + 3(y-5)^2 = 192$

15. $4x^2 + 9y^2 - 16x + 18y - 11 = 0$

16. $x^2 + 2y^2 - 10x + 8y + 29 = 0$

17. $4x^2 + y^2 - 8x - 2y + 1 = 0$

18. $9x^2 + 4y^2 + 54x - 8y + 49 = 0$

For each ellipse find the center and vertices.

19. ▦ $4x^2 + 9y^2 - 16.025x + 18.0927y - 11.346 = 0$

20. ▦ $9x^2 + 4y^2 + 54.063x - 8.016y + 49.872 = 0$

☆ ───

Find equations of the ellipses with the following vertices. (*Hint:* Graph the vertices.)

21. $(2, 0)$, $(-2, 0)$, $(0, 3)$, $(0, -3)$

22. $(1, 0)$, $(-1, 0)$, $(0, 4)$, $(0, -4)$

23. $(1, 1)$, $(5, 1)$, $(3, 6)$, $(3, -4)$

24. $(-1, -1)$, $(-1, 5)$, $(-3, 2)$, $(1, 2)$

Find equations of the ellipses satisfying the given conditions.

25. Center at $(-2, 3)$ with major axis of length 4 and parallel to the y-axis, minor axis of length 1.

26. Vertices $(3, 0)$ and $(-3, 0)$ and containing the point $\left(2, \frac{22}{3}\right)$.

27. a) Graph $9x^2 + y^2 = 9$. Is this relation a function?

b) Solve $9x^2 + y^2 = 9$ for y.

c) Graph $y = 3\sqrt{1 - x^2}$ and determine whether it is a function. Find the domain and range.

d) Graph $y = -3\sqrt{1 - x^2}$ and determine whether it is a function. Find the domain and range.

28. Describe the graph of $\frac{x^2}{a^2} + \frac{y^2}{b^2} = 1$ when $a^2 = b^2$.

29. ▦ Draw a large-scale precise graph of $\frac{x^2}{25} + \frac{y^2}{16} = 1$ by calculating and plotting a large number of points.

30. The maximum distance of the earth from the sun is 9.3×10^7 miles. The minimum distance is 9.1×10^7 miles. The sun is at one focus of the elliptical orbit. Find the distance from the sun to the other focus.

31. A point moves so that its x-coordinate is given by $x = a \cos t$ and its y-coordinate is given by $y = b \sin t$, where a and b are constants and t is time. Show that the path of the point is an ellipse.

★ ───

32. The unit square on the left is transformed to the rectangle on the right by a stretch or shrink in the x-direction and a stretch or shrink in the y-direction.

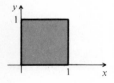 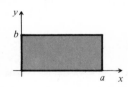

a) Use the above result to develop a formula for the area of the ellipse $\frac{x^2}{a^2} + \frac{y^2}{b^2} = 1$. (*Hint:* The area of the circle $x^2 + y^2 = r^2$ is $\pi \cdot r \cdot r$.)

b) Use the result of (a) to find the area of the ellipse $\frac{x^2}{16} + \frac{y^2}{25} = 1$.

c) Use the result of (a) to find the area of the ellipse $\frac{x^2}{4} + \frac{y^2}{3} = 1$.

33. The toy pictured here is called a "vacuum grinder." It consists of a rod hinged to two blocks A and B that slide in perpendicular grooves. One grasps the knob at C and grinds. Determine (and prove) whether or not the path of the handle C is an ellipse.

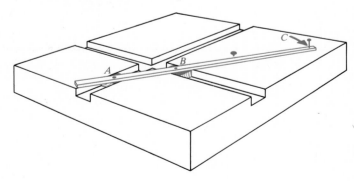

14.3 HYPERBOLAS

Some equations of second degree have graphs that are hyperbolas. Hyperbolas are defined as follows.

DEFINITION

A *hyperbola* is a locus or set of all points P in a plane such that the absolute value of the difference of the distances from P to two fixed points F_1 and F_2 in the plane is constant. The points F_1 and F_2 are called *foci* (singular, *focus*) and the midpoint of the segment joining them is called the *center*.

When a plane intersects a cone as shown, parallel to the axis of the cone, a hyperbola is formed. See p. 524. Note that a hyperbola has two parts. These are called *branches*.

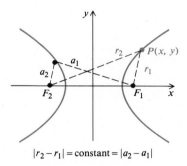

$$|r_2 - r_1| = \text{constant} = |a_2 - a_1|$$

[i] Equations of Hyperbolas

Now let us find equations for hyperbolas. We first consider an equation of a hyperbola whose center is at the origin and whose foci lie on one of the coordinate axes. Suppose we have a hyperbola as shown with foci $F_1(c, 0)$ and $F_2(-c, 0)$ on the x-axis. We consider a point $P(x, y)$ in the first quadrant. The proof for the other quadrants is similar to the proof that follows.

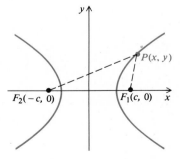

We know that $PF_2 > PF_1$, so $PF_2 - PF_1 > 0$ and $|PF_2 - PF_1| = PF_2 - PF_1$. Let the constant difference be $2a$. Then

$$|PF_2 - PF_1| = PF_2 - PF_1 = 2a.$$

In the triangle F_2PF_1, $PF_2 - PF_1 < F_1F_2$, or $2a < 2c$; therefore $a < c$.

OBJECTIVES

You should be able to:

[i] Given an equation of a hyperbola, center at the origin, find the vertices, foci, and asymptotes, and graph.

[ii] Given an equation of a hyperbola, center not at the origin, complete the square if necessary, then find the center, vertices, foci, and asymptotes, and graph.

[iii] Graph hyperbolas having an equation $xy = c$.

Using the distance formula, we have

$$\sqrt{(x + c)^2 + y^2} - \sqrt{(x - c)^2 + y^2} = 2a,$$

or

$$\sqrt{(x + c)^2 + y^2} = 2a + \sqrt{(x - c)^2 + y^2}.$$

Squaring, we get

$$x^2 + 2xc + c^2 + y^2 = 4a^2 + 4a\sqrt{(x - c)^2 + y^2} + x^2 - 2xc + c^2 + y^2,$$

which simplifies to

$$4cx - 4a^2 = 4a\sqrt{(x - c)^2 + y^2},$$

or

$$cx - a^2 = a\sqrt{(x - c)^2 + y^2}.$$

Squaring again, we get

$$c^2x^2 - 2a^2cx + a^4 = a^2x^2 - 2a^2cx + a^2c^2 + a^2y^2,$$

or

$$x^2(c^2 - a^2) - a^2y^2 = a^2(c^2 - a^2).$$

Since $c > a$, $c^2 > a^2$, so $c^2 - a^2$ is positive. We represent $c^2 - a^2$ by b^2. The previous equation then becomes

$$x^2b^2 - a^2y^2 = a^2b^2,$$

or the following.

$$\frac{x^2}{a^2} - \frac{y^2}{b^2} = 1$$
     Standard equation of a hyperbola, center at the origin, foci on the x-axis

---

We have shown that if a point is on the hyperbola it satisfies this equation. We also need to know the converse: If a point satisfies the equation, then it is on the hyperbola. We omit the proof.

The following figure is a hyperbola.

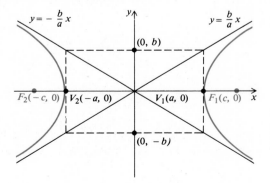

Points $V_1(a, 0)$ and $V_2(-a, 0)$ are called the *vertices,* and the line segment $\overline{V_1V_2}$ is called the *transverse axis.* The line segment from $(0, b)$ to $(0, -b)$ is called the *conjugate axis.* Note that when we replace either or both of x by $-x$ and/or y by $-y$, we get an equivalent equation. Thus the hyperbola is symmetric with respect to the origin, and the x- and y-axes are lines of symmetry.

The lines $y = (b/a)x$ and $y = -(b/a)x$ are *asymptotes*. They have slopes b/a and $-b/a$.

Example 1 For the hyperbola $9x^2 - 16y^2 = 144$, find the vertices, the foci, and the asymptotes. Then graph the hyperbola.

a) We first multiply by $\frac{1}{144}$ to find the standard form:

$$\frac{x^2}{16} - \frac{y^2}{9} = 1.$$

Thus $a = 4$ and $b = 3$. The vertices are $(4, 0)$ and $(-4, 0)$. Since $b^2 = c^2 - a^2$, $c = \sqrt{a^2 + b^2} = \sqrt{4^2 + 3^2} = 5$. Thus the foci are $(5, 0)$ and $(-5, 0)$. The asymptotes are $y = \frac{3}{4}x$ and $y = -\frac{3}{4}x$.

b) To graph the hyperbola it is helpful to first graph the asymptotes. An easy way to do this is to draw the rectangle shown below. Then draw the branches of the hyperbola outward from the vertices toward the asymptotes.

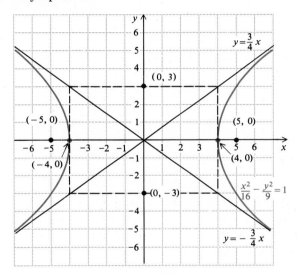

Why are $y = (b/a)x$ and $y = -(b/a)x$ asymptotes? To answer this we solve $x^2/16 - y^2/9 = 1$ for y^2:

$$-16y^2 = 144 - 9x^2$$
$$16y^2 = 9x^2 - 144$$
$$y^2 = \tfrac{1}{16}(9x^2 - 144) = \frac{9x^2 - 144}{16}.$$

From this last equation, we see that as $|x|$ gets larger the term -144 is very small compared to $9x^2$, so y^2 gets close to $9x^2/16$. That is, when $|x|$ is large,

$$y^2 \approx \frac{9x^2}{16},$$

so $\qquad\qquad y \approx |\tfrac{3}{4}x| \qquad$ or $\qquad y \approx \pm\tfrac{3}{4}x.$

Thus the lines $y = \frac{3}{4}x$ and $y = -\frac{3}{4}x$ are asymptotes.

DO EXERCISES 1 AND 2.

For each hyperbola find the vertices, the foci, and the asymptotes. Then draw a graph.

1. $4x^2 - 9y^2 = 36$

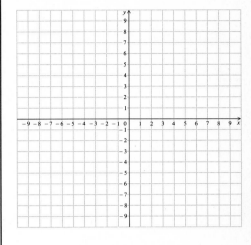

2. $x^2 - y^2 = 16$

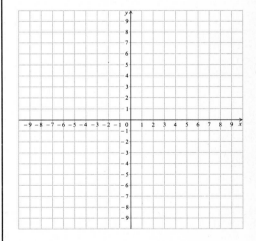

3. Find the vertices, the foci, and the asymptotes. Then draw a graph.

$$9y^2 - 25x^2 = 225$$

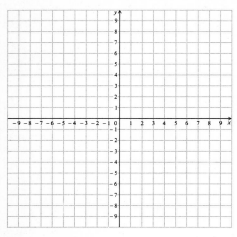

4. Find the vertices, the foci, and the asymptotes. Then draw a graph.

$$y^2 - x^2 = 25$$

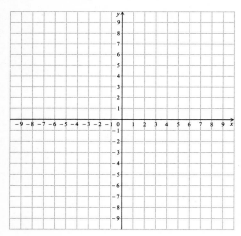

The foci of a hyperbola can be on the y-axis. In that case, the equation is as follows.

$\dfrac{y^2}{b^2} - \dfrac{x^2}{a^2} = 1$	**Standard equation of a hyperbola, center at the origin, foci on the y-axis**

In this case the slopes of the asymptotes are still $\pm b/a$ and it is still true that $c^2 = a^2 + b^2$. There are now y-intercepts and they are $\pm b$.

Example 2 For the hyperbola $25y^2 - 16x^2 = 400$, find the vertices, the foci, and the asymptotes. Then draw a graph.

a) We first multiply by $\frac{1}{400}$ to find the standard form:

$$\frac{y^2}{16} - \frac{x^2}{25} = 1.$$

Thus $a = 5$ and $b = 4$. The vertices are $(0, 4)$ and $(0, -4)$. Since $c = \sqrt{4^2 + 5^2} = \sqrt{41}$, the foci are $(0, \sqrt{41})$ and $(0, -\sqrt{41})$. The asymptotes are $y = \frac{4}{5}x$ and $y = -\frac{4}{5}x$.

b) The graph is as shown.

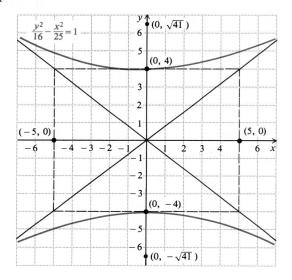

DO EXERCISE 3.

[ii] Standard Form by Completing the Square

If the center of a hyperbola is not at the origin but at some point (h, k), then the standard equation is one of the following.

$\dfrac{(x - h)^2}{a^2} - \dfrac{(y - k)^2}{b^2} = 1$	**Hyperbola, transverse axis parallel to the x-axis**
$\dfrac{(y - k)^2}{b^2} - \dfrac{(x - h)^2}{a^2} = 1$	**Hyperbola, transverse axis parallel to the y-axis**

Example 3 For the hyperbola

$$4x^2 - y^2 + 24x + 4y + 28 = 0,$$

find the center, the vertices, the foci, and the asymptotes. Then graph the hyperbola.

a) We complete the square to find standard form:

$$4(x^2 + 6x + \quad) - (y^2 - 4y + \quad) = -28$$
$$4(x^2 + 6x + 9 - 9) - (y^2 - 4y + 4 - 4) = -28$$
$$4(x^2 + 6x + 9) - (y^2 - 4y + 4) = -28 + 36 - 4$$
$$4(x + 3)^2 - (y - 2)^2 = 4$$
$$\frac{(x + 3)^2}{1} - \frac{(y - 2)^2}{4} = 1.$$

The center is $(-3, 2)$.

b) Consider $x^2/1 - y^2/4 = 1$. We have $a = 1$ and $b = 2$. The vertices of this hyperbola are $(1, 0)$ and $(-1, 0)$. Also, $c = \sqrt{1^2 + 2^2} = \sqrt{5}$, so the foci are $(\sqrt{5}, 0)$ and $(-\sqrt{5}, 0)$. The asymptotes are $y = 2x$ and $y = -2x$.

c) The vertices, foci, and asymptotes of the translated hyperbola are found in the same way in which the center has been translated. The vertices are $(-3 + 1, 2)$, $(-3 - 1, 2)$, or $(-2, 2)$, $(-4, 2)$. The foci are $(-3 + \sqrt{5}, 2)$ and $(-3 - \sqrt{5}, 2)$. The asymptotes are

$$y - 2 = 2(x + 3) \qquad \text{and} \qquad y - 2 = -2(x + 3).$$

d) The graph is as follows.

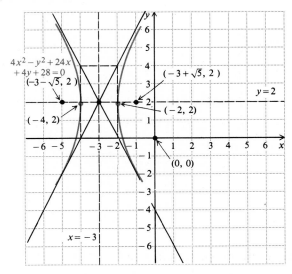

DO EXERCISES 4–6. (NOTE THAT EXERCISE 4 IS ON THE PRECEDING PAGE.)

[iii] Asymptotes on the Coordinate Axes

If a hyperbola has its center at the origin and its axes at 45° to the coordinate axes, it has a simple equation. The coordinate axes are its asymptotes.

For each hyperbola, find the center, the vertices, the foci, and the asymptotes. Then draw a graph.

5. $4x^2 - 25y^2 - 8x - 100y - 196 = 0$

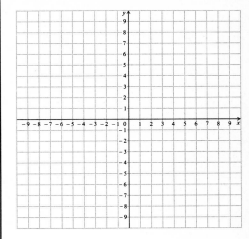

6. $\dfrac{(y - 2)^2}{9} - \dfrac{(x + 1)^2}{16} = 1$

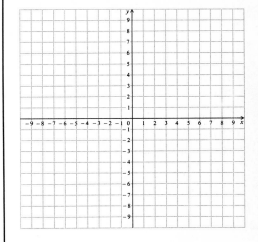

7. Graph.

$$xy = 3$$

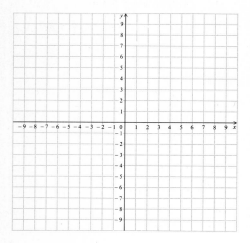

| $xy = c,$ c a nonzero constant | **Hyperbola, asymptotes the coordinate axes** |

If c is positive, the branches of the hyperbola lie in the first and third quadrants. If c is negative, the branches lie in the second and fourth quadrants. In either case the asymptotes are the x-axis and the y-axis.

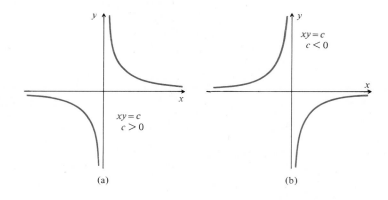

(a)   (b)

DO EXERCISES 7 AND 8.

8. Graph.

$$xy = -12$$

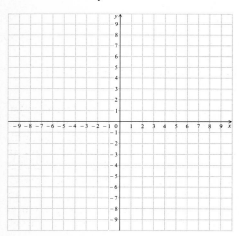

Applications

Hyperbolas also have many applications. A jet breaking the sound barrier creates a sonic boom whose wave front has the shape of a cone. The cone intersects the ground in one branch of a hyperbola.

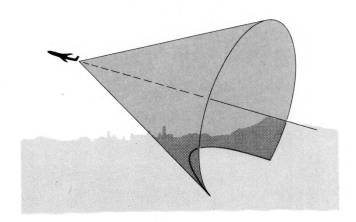

Some comets travel in hyperbolic orbits. A cross section of an amphitheater may be half of one branch of a hyperbola.

EXERCISE SET 14.3

[i] For each hyperbola find the center, the vertices, the foci, and the asymptotes, and graph the hyperbola.

1. $\dfrac{x^2}{9} - \dfrac{y^2}{1} = 1$ **2.** $\dfrac{x^2}{1} - \dfrac{y^2}{9} = 1$ **3.** $\dfrac{(x-2)^2}{9} - \dfrac{(y+5)^2}{1} = 1$ **4.** $\dfrac{(x-2)^2}{1} - \dfrac{(y+5)^2}{9} = 1$

5. $\dfrac{(y+3)^2}{4} - \dfrac{(x+1)^2}{16} = 1$ **6.** $\dfrac{(y+3)^2}{25} - \dfrac{(x+1)^2}{16} = 1$ **7.** $x^2 - 4y^2 = 4$ **8.** $4x^2 - y^2 = 4$

9. $4y^2 - x^2 = 4$ **10.** $y^2 - 4x^2 = 4$ **11.** $x^2 - y^2 = 2$ **12.** $x^2 - y^2 = 3$

13. $x^2 - y^2 = \dfrac{1}{4}$ **14.** $x^2 - y^2 = \dfrac{1}{9}$

[ii]

15. $x^2 - y^2 - 2x - 4y - 4 = 0$ **16.** $4x^2 - y^2 + 8x - 4y - 4 = 0$

17. $36x^2 - y^2 - 24x + 6y - 41 = 0$ **18.** $9x^2 - 4y^2 + 54x + 8y + 45 = 0$

[iii] Graph.

19. $xy = 1$ **20.** $xy = -4$ **21.** $xy = -8$ **22.** $xy = 3$

☆ ──────────────────────────────

23. ▦ Find the center, the vertices, and the asymptotes.
$$x^2 - y^2 - 2.046x - 4.088y - 4.228 = 0$$

Find an equation of a hyperbola having:

24. Vertices at $(1, 0)$ and $(-1, 0)$; and foci at $(2, 0)$ and $(-2, 0)$.

25. Asymptotes $y = \dfrac{3}{2}x$ and $y = -\dfrac{3}{2}x$ and one vertex $(2, 0)$.

26. a) Graph $x^2 - 4y^2 = 4$. Is this relation a function?
 b) Solve $x^2 - 4y^2 = 4$ for y.
 c) Graph $y = \frac{1}{2}\sqrt{x^2 - 4}$ and determine whether it is a function. Find the domain and range.
 d) Graph $y = -\frac{1}{2}\sqrt{x^2 - 4}$ and determine whether it is a function. Find the domain and range.

27. Show that the equation

$$\begin{vmatrix} \dfrac{x-h}{a} & \dfrac{y-k}{b} \\[2mm] \dfrac{y-k}{b} & \dfrac{x-h}{a} \end{vmatrix} = 1$$

is an equation of a hyperbola with center (h, k).

★ ──────────────────────────────

28. A rifle at A fires a bullet which hits a target at B. A person at C hears the sound of the rifle shot and the sound of the bullet hitting the target simultaneously. Describe the set of all such points C.

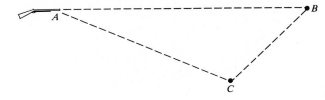

29. In a navigation system called Loran, a radio transmitter at M (the *master* station) sends out pulses. Each pulse triggers another transmitter at S (the *slave* station), which then also transmits a pulse. A ship or airplane at A receives pulses from both M and S and a device measures the difference in the time at which they arrive at A. Knowing this difference in time, the navigator can locate the vessel as being somewhere along a curve predrawn on a chart. What is the shape of that curve?

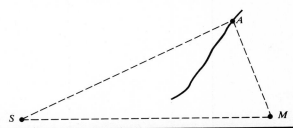

14.4 PARABOLAS

Some equations of second degree have graphs that are parabolas. Parabolas are defined as follows.

DEFINITION

A *parabola* is a locus or set of all points P in a plane equidistant from a fixed line and a fixed point in the plane. The fixed line is called the *directrix* and the fixed point is called the *focus*.

When a plane intersects a cone parallel to an element of the cone, as shown on p. 524, a parabola is formed.

[i] Equations of Parabolas

Now let us find equations for parabolas. Given the focus F and the directrix l, we place coordinate axes as shown. The y-axis contains F and is perpendicular to l. The x-axis is halfway between F and l. We shall call the distance from F to the x-axis p. Then F has coordinates $(0, p)$ and l has the equation $y = -p$.

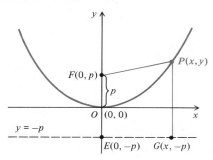

Let $P(x, y)$ be any point of the parabola and consider $\overline{PG}$ perpendicular to the line $y = -p$. The coordinates of G are $(x, -p)$. By definition of a parabola,

$$PF = PG.$$

Then using the distance formula, we have

$$\sqrt{(x - 0)^2 + (y - p)^2} = \sqrt{(x - x)^2 + (y + p)^2}.$$

Squaring, we get

$$x^2 + y^2 - 2py + p^2 = y^2 + 2py + p^2$$
$$x^2 = 4py.$$

Thus we have the following.

$$x^2 = 4py$$

Standard equation of a parabola with focus at $(0, p)$ and directrix $y = -p$. The vertex is $(0, 0)$ and the y-axis is the only line of symmetry.

We have shown that if $P(x, y)$ is on the parabola, then its coordinates satisfy this equation. The converse is also true, but we omit the proof.

Note that if $p > 0$, as above, the graph opens upward. If $p < 0$, the graph opens downward and the focus and directrix exchange sides of the x-axis.

The inverse of the above parabola is described as follows.

$$y^2 = 4px$$

Standard equation of a parabola with focus at $(p, 0)$ and directrix $x = -p$. The vertex is $(0, 0)$ and the x-axis is the only line of symmetry.

Example 1 For the parabola $y = x^2$, find the vertex, the focus, and the directrix, and draw a graph.

We first write

$$x^2 = 4py$$

$$x^2 = 4\left(\frac{1}{4}\right)y.$$

Vertex: $(0, 0)$

Focus: $\left(0, \frac{1}{4}\right)$

Directrix: $y = -\frac{1}{4}$

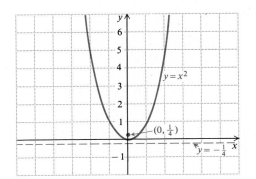

Example 2 For the parabola $y^2 = -12x$, find the vertex, the focus, and the directrix, and draw a graph.

We first write

$$y^2 = 4px$$

$$y^2 = 4(-3)x.$$

Vertex: $(0, 0)$

Focus: $(-3, 0)$

Directrix: $x = -(-3)$

$\qquad = 3$

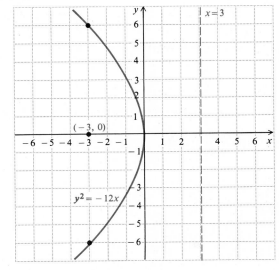

DO EXERCISES 1–3.

For each parabola find the vertex, the focus, and the directrix, and draw a graph.

1. $8y = x^2$

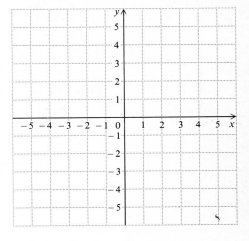

2. $y = 2x^2$

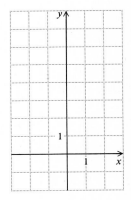

3. $y^2 = -6x$

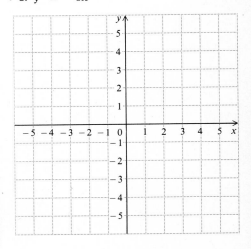

Find an equation of a parabola satisfying the given conditions.

4. Focus $(3, 0)$, directrix $x = -3$

5. Focus $\left(0, \dfrac{1}{2}\right)$, directrix $y = -\dfrac{1}{2}$

6. Focus $(-6, 0)$, directrix $x = 6$

7. Focus $(0, -1)$, directrix $y = 1$

[ii] Focus and Directrix Known

Example 3 Find an equation of a parabola with focus $(5, 0)$ and directrix $x = -5$.

The focus is on the x-axis and $x = -5$ is the directrix, so the line of symmetry is the x-axis. Thus the equation is of the type

$$y^2 = 4px.$$

Since $p = 5$, the equation is $y^2 = 20x$.

Example 4 Find an equation of a parabola with focus $(0, -7)$ and directrix $y = 7$.

The focus is on the y-axis and $y = 7$ is the directrix, so the line of symmetry is the y-axis. Thus the equation is of the type

$$x^2 = 4py.$$

Since $p = -7$, we obtain $x^2 = -28y$.

DO EXERCISES 4–7.

[iii] Standard Form by Completing the Square

If a parabola is translated so that its vertex is (h, k) and its axis of symmetry is parallel to the y-axis, it has an equation as follows:

$$(x - h)^2 = 4p(y - k),$$

where the vertex is (h, k), the focus is $(h, k + p)$, and the directrix is $y = k - p$.

If a parabola is translated so that its vertex is (h, k) and its axis of symmetry is parallel to the x-axis, it has an equation as follows:

$$(y - k)^2 = 4p(x - h),$$

where the vertex is (h, k), the focus is $(h + p, k)$, and the directrix is $x = h - p$.

Example 5 For the parabola

$$x^2 + 6x + 4y + 5 = 0,$$

find the vertex, the focus, and the directrix, and graph the parabola.

We complete the square:

$$
\begin{aligned}
x^2 + 6x \qquad\quad &= -4y - 5 \\
x^2 + 6x + 9 - 9 &= -4y - 5 \\
x^2 + 6x + 9 &= -4y + 4 \\
(x + 3)^2 &= -4y + 4 = 4(-1)(y - 1).
\end{aligned}
$$

Vertex: $(-3, 1)$

Focus: $\big(-3, 1 + (-1)\big)$ or $(-3, 0)$

Directrix: $y = 2$

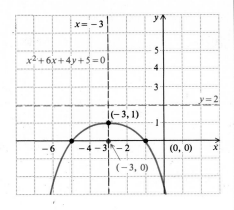

$x = -3$

$x^2 + 6x + 4y + 5 = 0$

$y = 2$

$(-3, 1)$

$(0, 0)$

$(-3, 0)$

For each parabola find the vertex, the focus, and the directrix, and graph the parabola.

8. $x^2 + 2x - 8y - 3 = 0$

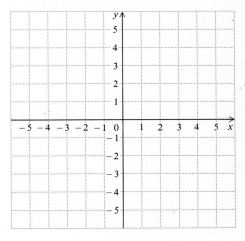

Example 6 For the parabola

$$y^2 + 6y - 8x - 31 = 0,$$

find the vertex, the focus, and the directrix, and draw a graph.

We complete the square:

$$y^2 + 6y \qquad\quad = 8x + 31$$
$$y^2 + 6y + 9 - 9 = 8x + 31$$
$$y^2 + 6y + 9 = 8x + 40$$
$$(y + 3)^2 = 8x + 40 = 8(x + 5) = 4(2)(x + 5).$$

Vertex: $(-5, -3)$

Focus: $(-5 + 2, -3)$ or
   $(-3, -3)$

Directrix: $x = -5 - 2 = -7$

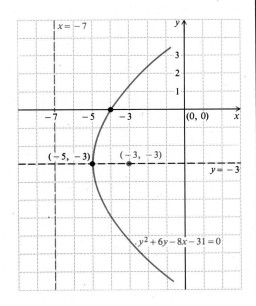

$x = -7$

$(0, 0)$

$(-5, -3)$ $(-3, -3)$

$y = -3$

$y^2 + 6y - 8x - 31 = 0$

9. $y^2 + 2y + 4x - 7 = 0$

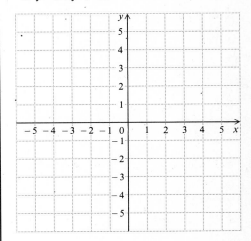

DO EXERCISES 8 AND 9.

Applications

   Parabolas have many applications. Cross sections of headlights are parabolas. The bulb is located at the focus. All light from that point is reflected outward, parallel to the axis of symmetry.

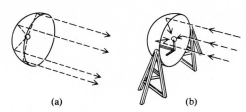

(a)   (b)

Radar and radio antennas may have cross sections that are parabolas. Incoming radio waves are reflected and concentrated at the focus. Cables hung between structures to form suspension bridges form parabolas. When a cable supports only its own weight it does not form a parabola, but rather a curve called a *catenary*.

EXERCISE SET 14.4

[i] For each parabola find the vertex, the focus, and the directrix, and graph the parabola.

1. $x^2 = 8y$

2. $x^2 = 16y$

3. $y^2 = -6x$

4. $y^2 = -2x$

5. $x^2 - 4y = 0$

6. $y^2 + 4x = 0$

7. $y = 2x^2$

8. $y = \frac{1}{2}x^2$

[ii] Find an equation of a parabola satisfying the given conditions.

9. Focus $(4, 0)$, directrix $x = -4$

10. Focus $\left(0, \frac{1}{4}\right)$, directrix $y = -\frac{1}{4}$

11. Focus $(-\sqrt{2}, 0)$, directrix $x = \sqrt{2}$

12. Focus $(0, -\pi)$, directrix $y = \pi$

13. Focus $(3, 2)$, directrix $x = -4$

14. Focus $(-2, 3)$, directrix $y = -3$

[iii] Find the vertex, focus, and directrix, and graph.

15. $(x + 2)^2 = -6(y - 1)$

16. $(y - 3)^2 = -20(x + 2)$

17. $x^2 + 2x + 2y + 7 = 0$

18. $y^2 + 6y - x + 16 = 0$

19. $x^2 - y - 2 = 0$

20. $x^2 - 4x - 2y = 0$

21. $y = x^2 + 4x + 3$

22. $y = x^2 + 6x + 10$

23. $4y^2 - 4y - 4x + 24 = 0$

24. $4y^2 + 4y - 4x - 16 = 0$

For each parabola find the vertex, the focus, and the directrix.

25. ▦ $x^2 = 8056.25y$

26. ▦ $y^2 = -7645.88x$

☆ ───────────────────────────────────

27. Graph each of the following, using the same set of axes.

$$x^2 - y^2 = 0, \quad x^2 - y^2 = 1,$$
$$x^2 + y^2 = 1, \quad\quad y = x^2.$$

29. Find equations of the following parabola: Line of symmetry parallel to the y-axis, vertex $(-1, 2)$, and passing through $(-3, 1)$.

28. Graph each of the following using the same set of axes.

$$x^2 - 4y^2 = 0, \quad x^2 - 4y^2 = 1,$$
$$x^2 + 4y^2 = 1, \quad\quad x = 4y^2.$$

30. a) Graph $(y - 3)^2 = -20(x + 1)$. Is this relation a function?

b) In general, is $(y - k)^2 = 4p(x - h)$ a function?

31. Show that the equation

$$\begin{vmatrix} y - k & x - h \\ 4p & y - k \end{vmatrix} = 0$$

is an equation of a parabola with vertex (h, k).

32. The cables of a suspension bridge are 50 ft above the roadbed at the ends of the bridge and 10 ft above it in the center of the bridge. The roadbed is 200 ft long. Vertical cables are to be spaced every 20 ft along the bridge. Calculate the lengths of these vertical cables.

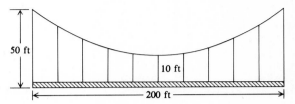

★

33. Prove that when a cable supports a load distributed uniformly horizontally, it hangs in the shape of a parabola. (*Hint:* Proceed as follows.)

a) Place a coordinate system as shown, with the origin at the lowest point of the cable.

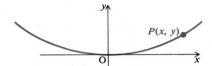

b) For a point $P(x, y)$ on the cable write an equation of rotational equilibrium. The forces involved are the tensions in the cable, F_1 and F_2, and the weight supported, W (which is a function of x). The weight of the cable is essentially neglected. Use point P as the center of rotation.

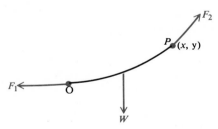

c) Solve for y.

14.5 SYSTEMS OF FIRST-DEGREE AND SECOND-DEGREE EQUATIONS

When we studied systems of linear equations, we solved them both graphically and algebraically. Here we study systems in which one equation is of first degree and one is of second degree. We will use graphical and then algebraic methods of solving.

[i] Solving Graphically

We consider a system of equations, an equation of a circle and an equation of a line. Let us think about the possible ways in which a circle and a line can intersect. The three possibilities are shown in the figure. For L_1 there is no point of intersection, hence the system of

OBJECTIVES

You should be able to:

[i] Solve a system of one first-degree and one second-degree equation graphically or using the substitution method.

[ii] Solve applied problems involving the solution of one first-degree and one second-degree equation.

Solve these systems graphically.

1. $x^2 + y^2 = 25$, $y - x = -1$

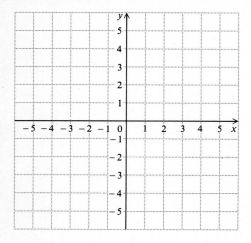

2. $y = x^2 - 2x - 1$, $y = x + 3$

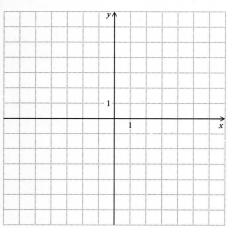

3. $y = \dfrac{x^2}{4}$, $x + 2y = 4$

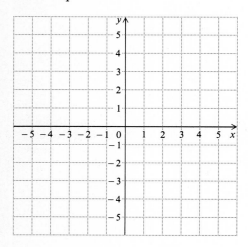

equations has no real solution. For L_2 there is one point of intersection, hence one real solution. For L_3 there are two points of intersection, hence two real solutions.

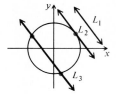

Example 1 Solve this system graphically:

$$x^2 + y^2 = 25,$$
$$3x - 4y = 0.$$

We graph the two equations, using the same axes. The points of intersection have coordinates that must satisfy both equations. The solutions seem to be $(4, 3)$ and $(-4, -3)$. We check.

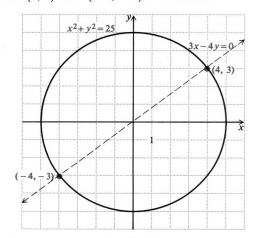

Check: For $(-4, -3)$. You should do the check for $(4, 3)$.

$x^2 + y^2 = 25$	
$(-4)^2 + (-3)^2$	25
$16 + 9$	
25	

$3x - 4y = 0$	
$3(-4) - 4(-3)$	0
$-12 + 12$	
0	

DO EXERCISES 1–3.

Solving by Substitution

Remember that we used the *addition* and *substitution* methods to solve systems of linear equations. In solving systems where one equation is of first degree and one is of second degree, it is preferable to use the *substitution* method.

Example 2 Solve this system algebraically:

$$x^2 + y^2 = 25, \tag{1}$$
$$3x - 4y = 0. \tag{2}$$

First solve the linear equation (2) for x:

$$x = \frac{4}{3}y.$$

Then substitute $\frac{4}{3}y$ for x in equation (1) and solve for y:

$$\left(\frac{4}{3}y\right)^2 + y^2 = 25$$

$$\frac{16}{9}y^2 + y^2 = 25$$

$$\frac{16}{9}y^2 + \frac{9}{9}y^2 = 25$$

$$\frac{25}{9}y^2 = 25$$

$$\frac{9}{25} \cdot \frac{25}{9}y^2 = \frac{9}{25} \cdot 25$$

$$y^2 = 9$$

$$y = \pm 3.$$

Now substitute these numbers into the linear equation and solve for x:

$$x = \frac{4}{3}(3), \quad \text{or} \quad 4;$$

$$x = \frac{4}{3}(-3), \quad \text{or} \quad -4.$$

The pairs $(4, 3)$ and $(-4, -3)$ check, hence are solutions.

DO EXERCISES 4–6.

Sometimes an equation may not take on a familiar form like those studied previously in this chapter. We can still rely on the substitution method for solution.

Example 3 Solve the system

$$y + 3 = 2x, \tag{1}$$
$$x^2 + 2xy = -1. \tag{2}$$

First solve the linear equation (1) for y:

$$y = 2x - 3.$$

Then substitute $2x - 3$ for y in equation (2) and solve for x:

$$x^2 + 2x(2x - 3) = -1$$
$$x^2 + 4x^2 - 6x = -1$$
$$5x^2 - 6x + 1 = 0$$
$$(5x - 1)(x - 1) = 0$$
$$5x - 1 = 0 \quad \text{or} \quad x - 1 = 0$$

$$x = \frac{1}{5} \quad \text{or} \quad x = 1.$$

4. Solve the system in Exercise 1 algebraically.

5. Solve the system in Exercise 2 algebraically.

6. Solve the system in Exercise 3 algebraically.

7. Solve the system.

$$y + 3x = 1,$$
$$x^2 - 2xy = 5$$

8. The difference of two numbers is 4 and the difference of their squares is 72. What are the numbers?

9. The perimeter of a rectangular field is 34 ft and the length of a diagonal is 13 ft. Find the dimensions of the field.

Now substitute these numbers into the linear equation and solve for y:

$$y = 2\left(\frac{1}{5}\right) - 3, \quad \text{or} \quad -\frac{13}{5};$$
$$y = 2(1) - 3, \quad \text{or} \quad -1.$$

The pairs $\left(\frac{1}{5}, -\frac{13}{5}\right)$ and $(1, -1)$ check and are solutions.

DO EXERCISE 7.

[ii] **Applied Problems**

Example 4 The perimeter of a rectangular field is 204 m and the area is 2565 m². Find the dimensions of the field.

We first translate the conditions of the problem to equations, using w for the width and l for the length:

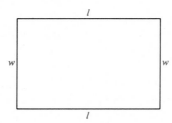

$$\text{Perimeter: } 2w + 2l = 204$$
$$\text{Area: } lw = 2565$$

Now we solve the system

$$2w + 2l = 204,$$
$$lw = 2565$$

and get the solution $(45, 57)$. Now check in the original problem: The perimeter is $2 \cdot 45 + 2 \cdot 57$, or 204. The area is $45 \cdot 57$, or 2565. The numbers check, so the answer is $l = 57$ m, $w = 45$ m.

DO EXERCISES 8 AND 9.

EXERCISE SET 14.5

[i] Solve each system graphically. Then solve algebraically.

1. $x^2 + y^2 = 25,$
$y - x = 1$

2. $x^2 + y^2 = 100,$
$y - x = 2$

3. $y^2 - x^2 = 9,$
$2x - 3 = y$

4. $x + y = -6,$
$xy = -7$

5. $4x^2 + 9y^2 = 36,$
$3y + 2x = 6$

6. $9x^2 + 4y^2 = 36,$
$3x + 2y = 6$

7. $y^2 = x + 3,$
$2y = x + 4$

8. $y = x^2,$
$3x = y + 2$

Solve.

9. $x^2 + 4y^2 = 25,$
$x + 2y = 7$

10. $y^2 - x^2 = 16,$
$2x - y = 1$

11. $x^2 - xy + 3y^2 = 27,$
$x - y = 2$

12. $2y^2 + xy + x^2 = 7,$
  $x - 2y = 5$

13. $3x + y = 7,$
  $4x^2 + 5y = 56$

14. $2y^2 + xy = 5,$
  $4y + x = 7$

[ii] **Applied problems**

15. The sum of two numbers is 14 and the sum of their squares is 106. What are the numbers?

16. The sum of two numbers is 15 and the difference of their squares is also 15. What are the numbers?

17. A rectangle has perimeter 28 cm and the length of a diagonal is 10 cm. What are its dimensions?

18. A rectangle has perimeter 6 m and the length of a diagonal $\sqrt{5}$ m. What are its dimensions?

19. A rectangle has area 20 in^2 and perimeter 18 in. Find its dimensions.

20. A rectangle has area 2 yd^2 and perimeter 6 yd. Find its dimensions.

[i] Solve.

21. ▦ $x^2 + y^2 = 19,380,510.36,$
  $27,942.25x - 6.125y = 0$

22. ▦ $2x + 2y = 1660,$
  $xy = 35,325$

☆ ————————————————————————————

23. Given the area A and the perimeter P of a rectangle, show that the length L and the width W are given by the formulas

$$L = \frac{1}{4}(P + \sqrt{P^2 - 16A}),$$

$$W = \frac{1}{4}(P - \sqrt{P^2 - 16A})$$

24. Show that a hyperbola does not intersect its asymptotes. That is, solve the system

$$\frac{x^2}{a^2} - \frac{y^2}{b^2} = 1,$$

$$y = \frac{b}{a}x \quad \left(\text{or, } y = -\frac{b}{a}x\right)$$

25. Find an equation of a circle that passes through the points (2, 4) and (3, 3) and whose center is on the line $3x - y = 3$.

26. Find an equation of a circle that passes through the points (7, 3) and (5, 5) and whose center is on the line $y - 4x = 1$.

14.6 SYSTEMS OF SECOND-DEGREE EQUATIONS

We now consider systems of two second-degree equations. The following figure shows ways in which a circle and a hyperbola can intersect.

4 real solutions

3 real solutions

2 real solutions

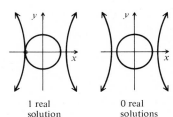

1 real solution

0 real solutions

OBJECTIVES

You should be able to:

[i] Solve systems of second-degree equations graphically.

[ii] Use the addition method to solve systems of equations like

$$2x^2 + 5y^2 = 20,$$
$$3x^2 - y^2 = -1.$$

[iii] Use the substitution method to solve systems of equations like

$$x^2 + 4y^2 = 20,$$
$$xy = 4.$$

[iv] Solve applied problems involving systems of two second-degree equations.

Solve these systems graphically.

1. $x^2 + y^2 = 4,$

$\dfrac{x^2}{4} - \dfrac{y^2}{4} = 1$

2. $x^2 + y^2 = 16,$

$\dfrac{x^2}{16} + \dfrac{y^2}{9} = 1$

3. $x^2 + y^2 = 4,$

$\dfrac{x^2}{25} + \dfrac{y^2}{4} = 1$

[i] Solving Graphically

Example 1 Solve this system graphically:

$$x^2 + y^2 = 25,$$
$$\frac{x^2}{25} - \frac{y^2}{25} = 1.$$

We graph the two equations using the same axes. The points of inter-section have coordinates that must satisfy both equations; the solutions seem to be $(5, 0)$ and $(-5, 0)$.

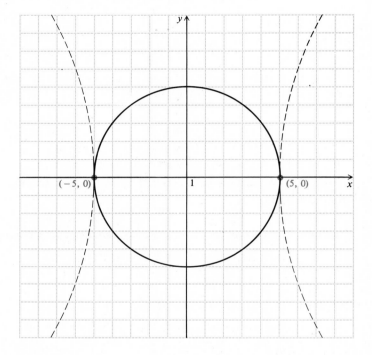

Check: Since $(5)^2 = 25$ and $(-5)^2 = 25$, we can do both checks at once.

$$
\begin{array}{c|c}
x^2 + y^2 = & 25 \\
\hline
(\pm 5)^2 + 0^2 & 25 \\
25 + 0 & \\
25 &
\end{array}
\qquad
\begin{array}{c|c}
\dfrac{x^2}{25} - \dfrac{y^2}{25} = 1 \\
\hline
\dfrac{(\pm 5)^2}{25} - \dfrac{0^2}{25} & 1 \\
\dfrac{25}{25} - 0 & \\
1 &
\end{array}
$$

DO EXERCISES 1–3.

[ii] The Addition Method

To solve systems of two second-degree equations we can use either the substitution or the addition method.

Example 2 Solve this system:

$$2x^2 + 5y^2 = 22, \quad (1)$$
$$3x^2 - y^2 = -1. \quad (2)$$

Here we use the addition method:

$$2x^2 + 5y^2 = 22.$$

From the second equation, we get

$$\underline{15x^2 - 5y^2 = -5} \qquad \text{Multiplying by 5}$$
$$17x^2 = 17 \qquad \text{Adding}$$
$$x^2 = 1$$
$$x = \pm 1.$$

If $x = 1$, $x^2 = 1$, and if $x = -1$, $x^2 = 1$, so substituting 1 or -1 for x in equation (2) we have

$$3 \cdot 1^2 - y^2 = -1$$
$$y^2 = 4$$
$$y = \pm 2.$$

Thus, if $x = 1$, $y = 2$ or $y = -2$, and if $x = -1$, $y = 2$ or $y = -2$. The possible solutions are $(1, 2)$, $(1, -2)$, $(-1, 2)$, and $(-1, -2)$.

Check: Since $(2)^2 = 4$, $(-2)^2 = 4$, $(1)^2 = 1$, and $(-1)^2 = 1$, we can check all four pairs at one time.

$2x^2 + 5y^2 = 22$		$3x^2 - y^2 = -1$	
$2(\pm 1)^2 + 5(\pm 2)^2$	22	$3(\pm 1)^2 - (\pm 2)^2$	-1
$2 + 20$		$3 - 4$	
22		-1	

DO EXERCISE 4.

[iii] The Substitution Method

Example 3 Solve the system

$$x^2 + 4y^2 = 20, \quad (1)$$
$$xy = 4. \quad (2)$$

Here we use the substitution method. First solve equation (2) for y:

$$y = \frac{4}{x}.$$

Then substitute $4/x$ for y in equation (1) and solve for x:

$$x^2 + 4\left(\frac{4}{x}\right)^2 = 20$$
$$x^2 + \frac{64}{x^2} = 20$$
$$x^4 + 64 = 20x^2 \qquad \text{Multiplying by } x^2$$
$$x^4 - 20x^2 + 64 = 0$$
$$u^2 - 20u + 64 = 0 \qquad \text{Letting } u = x^2$$
$$(u - 16)(u - 4) = 0.$$

4. Solve this system.

$$2y^2 - 3x^2 = 6,$$
$$5y^2 + 2x^2 = 53$$

5. Solve this system.
$$x^2 + xy + y^2 = 19,$$
$$xy = 6$$

6. The area of a rectangle is 2 ft^2 and the length of a diagonal is $\sqrt{5}$ ft. Find the dimensions of the rectangle.

Then $x^2 = 4$ or $x^2 = -4$ or $x = 2$ or $x = -2$. Since $y = 4/x$, if $x = 4$, $y = 1$; if $x = -4$, $y = -1$; if $x = 2$, $y = 2$; if $x = -2$, $y = -2$. The solutions are $(4, 1)$, $(-4, -1)$, $(2, 2)$, and $(-2, -2)$.

DO EXERCISE 5.

[iv] An Applied Problem

Example 4 The area of a rectangle is 300 yd^2 and the length of a diagonal is 25 yd. Find the dimensions.

First make a drawing.

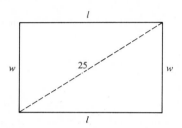

We use l for the length and w for the width and translate to equations.

From the Pythagorean theorem: $l^2 + w^2 = 25^2$
Area: $lw = 300$

Now we *solve* the system

$$l^2 + w^2 = 625,$$
$$lw = 300,$$

and get the solutions $(15, 20)$ and $(-15, -20)$. Now we check in the original problem: $15^2 + 20^2 = 25^2$ and $15 \cdot 20 = 300$, so $(15, 20)$ is a solution of the problem. Lengths of sides cannot be negative so $(-15, -20)$ is not a solution. The answer is $l = 20$ yd and $w = 15$ yd.

DO EXERCISE 6.

EXERCISE SET 14.6

[i], [ii], [iii] Solve each system graphically. Then solve algebraically.

1. $x^2 + y^2 = 25,$
   $y^2 = x + 5$

2. $y = x^2,$
   $x = y^2$

3. $x^2 + y^2 = 9,$
   $x^2 - y^2 = 9$

4. $y^2 - 4x^2 = 4,$
   $4x^2 + y^2 = 4$

5. $x^2 + y^2 = 25,$
   $xy = 12$

6. $x^2 - y^2 = 16,$
   $x + y^2 = 4$

7. $x^2 + y^2 = 4,$
   $16x^2 + 9y^2 = 144$

8. $x^2 + y^2 = 25,$
   $25x^2 + 16y^2 = 400$

[ii], [iii] Solve.

9. $x^2 + y^2 = 16,$
   $y^2 - 2x^2 = 10$

10. $x^2 + y^2 = 14,$
   $x^2 - y^2 = 4$

11. $x^2 + y^2 = 5,$
   $xy = 2$

12. $x^2 + y^2 = 20,$
   $xy = 8$

13. $x^2 + y^2 = 13,$
   $xy = 6$

14. $x^2 + 4y^2 = 20,$
   $xy = 4$

15. $x^2 + y^2 + 6y + 5 = 0,$
   $x^2 + y^2 - 2x - 8 = 0$

16. $2xy + 3y^2 = 7,$
   $3xy - 2y^2 = 4$

17. ▦　$18.465x^2 + 788.723y^2 = 6408,$
　　　$106.535x^2 - 788.723y^2 = 2692$

18. ▦　$0.319x^2 + 2688.7y^2 = 56,548,$
　　　$0.306x^2 - 2688.7y^2 = 43,452$

[iv]

19. Find two numbers whose product is 156 if the sum of their squares is 313.

20. Find two numbers whose product is 60 if the sum of their squares is 136.

21. The area of a rectangle is $\sqrt{3}\,\text{m}^2$ and the length of a diagonal is 2 m. Find the dimensions.

22. The area of a rectangle is $\sqrt{2}\,\text{m}^2$ and the length of a diagonal is $\sqrt{3}\,\text{m}$. Find the dimensions.

23. A garden contains two square peanut beds. Find the length of each bed if the sum of their areas is 832 ft², and the difference of their areas is 320 ft².

24. A certain amount of money saved for 1 yr at a certain interest rate yielded $7.50. If the principal had been $25 more and the interest rate 1% less, the interest would have been the same. Find the principal and the rate.

☆ ──

25. Find an equation of the circle that passes through the points (4, 6), (−6, 2), and (1, −3).

26. Find an equation of the circle that passes through the points (2, 3), (4, 5), and (0, −3).

★ ──

27. Find k such that the following equations have two roots in common.

$$(x - 2)^4 - (x - 2) = 0,$$
$$x^2 - kx + k = 0$$

CHAPTER 14 REVIEW

[14.1, i]　　**1.** Graph $2x^2 - 3xy - 2y^2 = 0$.

[14.1, iii]　　**2.** Find an equation of the circle with center $(-2, 6)$ and radius $\sqrt{13}$.

[14.1, vi]　　**3.** Find an equation of the circle having its center at (3, 4) and passing through the origin.

[14.1, v]　　**4.** Find the center and radius of the circle $2x^2 + 2y^2 - 3x - 5y + 3 = 0$.

[14.1, vi]　　**5.** Find an equation of the circle having a diameter with endpoints $(-3, 5)$ and (7, 3).

[14.4, ii]　　**6.** Find an equation of the parabola with directrix $y = \frac{3}{2}$ and focus $(0, -\frac{3}{2})$.

[14.4, i]　　**7.** Find the focus, vertex, and directrix of the parabola $y^2 = -12x$.

[14.2, ii]　　**8.** Find the center, vertices, and foci of the ellipse $16x^2 + 25y^2 - 64x + 50y - 311 = 0$. Then graph the ellipse.

9. Find an equation of the ellipse having vertices (3, 0) and (0, 4) and centered at the origin.

[14.3, ii]　　**10.** Find the center, vertices, foci, and asymptotes of the hyperbola $x^2 - 2y^2 + 4x + y - \frac{1}{8} = 0$.

[14.6, ii]　　**11.** Solve.
　　　$x^2 - 16y = 0,$
　　　$x^2 - y^2 = 64$

[14.6, ii]　　**12.** Solve.
　　　$4x^2 + 4y^2 = 65,$
　　　$6x^2 - 4y^2 = 25$

[14.5, ii]　　**13.** The sum of two numbers is 11 and the sum of their squares is 65. Find the numbers.

[14.6, iv]　　**14.** The sides of a triangle are 8, 10, and 14. Find the altitude to the longest side.

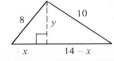

15

SEQUENCES, SERIES, AND MATHEMATICAL INDUCTION

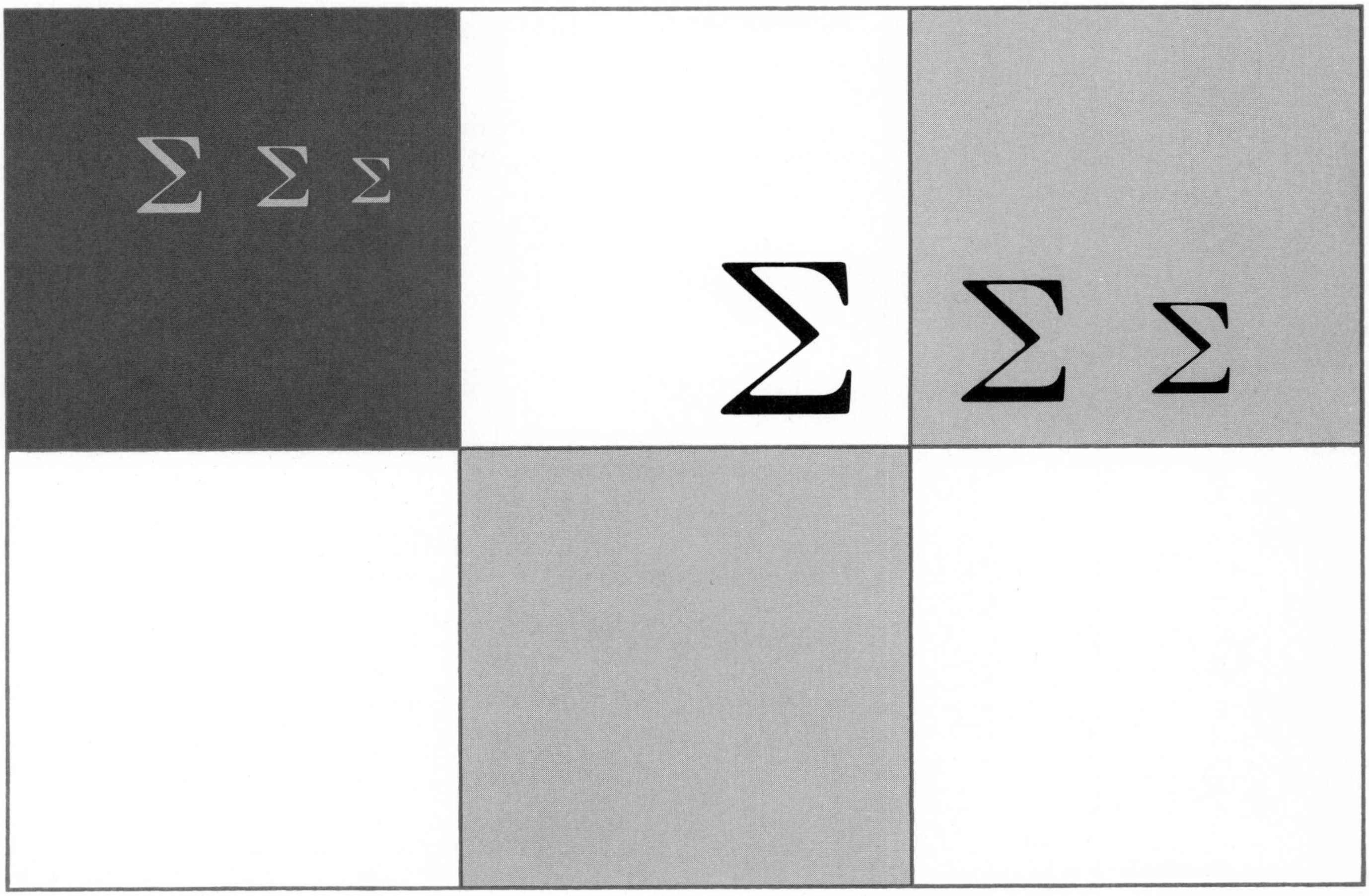

$\Sigma \Sigma \Sigma$

$\Sigma \quad \Sigma \Sigma$

OBJECTIVES

You should be able to:

[i] Given a formula for the nth term (general term) of a sequence, find any term in the sequence, given a value for n.

[ii] Given a sequence, look for a pattern and try to guess a rule or formula for the general term.

[iii] Given a recursively defined sequence, construct its terms.

[iv] Find sums of sequences.

[v] Convert between sigma (Σ) notation and other notation for the sum of a sequence.

1. A sequence is given by

$$a(n) = 2n - 1.$$

a) Find the first 3 terms.
b) Find the 34th term.

2. A sequence is given by

$$a_n = \frac{(-1)^n}{n - 1}, \quad n \geq 2.$$

a) Find the first 4 terms.
b) Find the 48th term.

15.1 SEQUENCES

[i] A sequence is a set in which the order of the elements is specified. It can be defined precisely as a certain kind of function.

DEFINITION

A *sequence* **is a function having for its domain some set of natural numbers.**

As an example, consider the sequence given by

$$a(n) = 2^n, \quad \text{or} \quad a_n = 2^n.*$$

Some of the function values (also known as *terms* of the sequence) are as follows:

$$a_1 = 2^1 = 2,$$
$$a_2 = 2^2 = 4,$$
$$a_3 = 2^3 = 8,$$
$$a_6 = 2^6 = 64.$$

The first term of the sequence is a_1, the fifth term is a_5, and the nth term or *general* term is a_n. This sequence can also be denoted in the following ways.

a) $2, 4, 8, \ldots$ **b)** $2, 4, 8, \ldots, 2^n, \ldots$

Example 1 Find the first 4 terms and the 57th term of the sequence whose general term is given by $a_n = (-1)^n/(n + 1)$.

$$a_1 = \frac{(-1)^1}{1 + 1} = -\frac{1}{2}, \quad a_2 = \frac{(-1)^2}{2 + 1} = \frac{1}{3},$$
$$a_3 = \frac{(-1)^3}{3 + 1} = -\frac{1}{4}, \quad a_4 = \frac{(-1)^4}{4 + 1} = \frac{1}{5},$$
$$a_{57} = \frac{(-1)^{57}}{57 + 1} = \frac{-1}{58}.$$

[ii] **Finding a General Term**

When a sequence is described merely by naming the first few terms, we do not know for sure what the general term is, but the reader is expected to make a guess by looking for a pattern.

DO EXERCISES 1 AND 2.

Examples For each sequence, make a guess at the general term.

2. $1, 4, 9, 16, 25, \ldots$
These are squares of numbers, so the general term may be n^2.

3. $\sqrt{1}, \sqrt{2}, \sqrt{3}, \sqrt{4}, \ldots$
These are square roots of numbers, so the general term may be $\sqrt{n}$.

---

*The notation a_n means the same as $a(n)$, but is more commonly used with sequences.

4. $-1, 2, -4, 8, -16, \ldots$

These are powers of 2 with alternating signs, so the general term may be $(-1)^n 2^{n-1}$.

5. $2, 4, 8, \ldots$

If we see the pattern of powers of 2, we will see 16 as the next term and guess 2^n for the general term. If we see that we can get the second term by adding 2, can get the third term by adding 4, and get the next term by adding 6, and so on, we will see 14 as the next term. The general term is then $n^2 - n + 2$.

DO EXERCISES 3–7.

[iii] Recursive Definitions

A sequence may be defined by a *recursive definition*. Such a definition lists the first term and tells how to get the $(k + 1)$st term from the kth term.

Example 6 Find the first five terms of the sequence defined by

$$a_1 = 5,$$
$$a_{k+1} = 2a_k - 3, \quad \text{for } k \geq 1.$$

$$a_1 = 5,$$
$$a_2 = 2a_1 - 3 = 2 \cdot 5 - 3 = 7,$$
$$a_3 = 2a_2 - 3 = 2 \cdot 7 - 3 = 11,$$
$$a_4 = 2a_3 - 3 = 2 \cdot 11 - 3 = 19,$$
$$a_5 = 2a_4 - 3 = 2 \cdot 19 - 3 = 35.$$

DO EXERCISE 8.

[iv] Sums

With any sequence there is associated another sequence, obtained by taking sums of terms of the given sequence. The general term of this sequence of "partial sums" is denoted S_n. Consider the sequence

$$3, 5, 7, 9, \ldots, 2n + 1.$$

We construct some terms of the sequence of partial sums.

$S_1 = 3$ This is the first term of the given sequence.
$S_2 = 3 + 5 = 8$ This is the sum of the first two terms.
$S_3 = 3 + 5 + 7 = 15$ The sum of the first three terms
$S_4 = 3 + 5 + 7 + 9 = 24$ The sum of the first four terms

A sum like any of the above is also known as a *series*.

Example 7 Find the sum of the sequence $-2, 4, -6, 8, -10$.

$$S = -2 + 4 + (-6) + 8 + (-10)$$
$$= -6$$

DO EXERCISE 9.

For each sequence try to find a rule for finding the general term, or the nth term. Answers may vary.

3. $2, 4, 6, 8, 10, \ldots$

4. $1, 2, 3, 4, 5, 6, \ldots$

5. $1, 8, 27, 64, 125, \ldots$

6. $x, \dfrac{x^2}{2}, \dfrac{x^3}{3}, \dfrac{x^4}{4}, \dfrac{x^5}{5}, \ldots$

7. $1, 2, 4, 8, 16, 32, \ldots$

8. Find the first five terms of each recursively defined sequence.

    a) $a_1 = 2$
       $a_{k+1} = 3a_k - 4, \quad k \geq 1$
    b) $a_1 = -3$
       $a_{k+1} = (-1) \cdot a_k^2, \quad k \geq 1$

9. Find the sum of the sequence

$$\frac{1}{2}, \frac{1}{4}, \frac{1}{8}, \frac{1}{16}, \frac{1}{32}.$$

Name each series without using Σ.

10. $\displaystyle\sum_{k=1}^{3} 2 + \frac{1}{k}$

11. $\displaystyle\sum_{k=0}^{4} 5^k$

12. $\displaystyle\sum_{k=8}^{11} k^3$

Name each series with sigma notation.

13. $2 + 4 + 6 + 8 + 10$

14. $1 + 8 + 27 + 64$

15. $x + \dfrac{x^2}{2} + \dfrac{x^3}{3} + \dfrac{x^4}{4} + \dfrac{x^5}{5} + \dfrac{x^6}{6}$

16. $2 + 3 + 4 + 5$

[v] Sigma (Σ) Notation

Sums, or series, like those above can be handily denoted when the general term is known, using the Greek letter Σ (sigma).* The sum of the first four terms (S_4) of the above sequence can be named as follows:

$$\sum_{k=1}^{4} (2k + 1).$$

This is read "the sum as k goes from 1 to 4 of $(2k + 1)$."

Examples Rename the following sums without using sigma notation.

8. $\displaystyle\sum_{k=1}^{5} k^2 = 1^2 + 2^2 + 3^2 + 4^2 + 5^2$

9. $\displaystyle\sum_{k=1}^{4} (-1)^k(2k) = (-1)^1(2 \cdot 1) + (-1)^2(2 \cdot 2)$
$$+ (-1)^3(2 \cdot 3) + (-1)^4(2 \cdot 4)$$
$$= -2 + 4 - 6 + 8$$

10. $\displaystyle\sum_{k=10}^{12} \log k = \log 10 + \log 11 + \log 12$

To find sigma notation for a sum we must find a formula for the general term.

Examples Write sigma notation for these sums.

11. $x + x^2 + x^3 + x^4 + x^5 + x^6 = \displaystyle\sum_{k=1}^{6} x^k$

12. $\sin \dfrac{\pi}{2} - \sin \dfrac{3\pi}{2} + \sin \dfrac{5\pi}{2} - \sin \dfrac{7\pi}{2} = \displaystyle\sum_{k=1}^{4} (-1)^{k+1} \sin \left(\dfrac{2k - 1}{2} \right) \pi$

DO EXERCISES 10–16.

EXERCISE SET 15.1

[i] In each of the following the general, or nth, term of a sequence is given. In each case find the first 4 terms, the 10th term, and the 15th term.

1. $a_n = 3n + 1$ **2.** $a_n = (n - 1)(n - 2)(n - 3)$ **3.** $a_n = \dfrac{1}{n}$

4. $a_n = \dfrac{(-1)^n}{n}$ **5.** $a_n = \dfrac{n}{n + 1}$ **6.** $a_n = \left(-\dfrac{1}{2} \right)^{n-1}$

[ii] For each of the following sequences find the general, or nth, term. Answers may vary.

7. $1, 3, 5, 7, 9, \ldots$ **8.** $3, 9, 27, 81, 243, \ldots$

---

*The letter Σ corresponds to S, for "sum."

9. $\frac{2}{3}, \frac{3}{4}, \frac{4}{5}, \frac{5}{6}, \frac{6}{7}, \ldots$

10. $\sqrt{2}, \sqrt{4}, \sqrt{6}, \sqrt{8}, \sqrt{10}, \ldots$

11. $\sqrt{3}, 3, 3\sqrt{3}, 9, 9\sqrt{3}, \ldots$

12. $1 \cdot 2, 2 \cdot 3, 3 \cdot 4, 4 \cdot 5, \ldots$

13. $\log 1, \log 10, \log 100, \log 1000, \ldots$

14. $2, 4, 6, 8, 10, 12, 14, 16$

[iii] Find the first four terms of each recursively defined sequence.

15. $a_1 = 4, a_{k+1} = 1 + \dfrac{1}{a_k}$

16. $a_1 = 16, a_{k+1} = \sqrt{a_k}$

17. $a_1 = 64, a_{k+1} = \sqrt{a_k}$

18. $a_1 = e^Q, a_{k+1} = \ln a_k$

For each sequence find the associated sum.

19. $1, 2, 3, 4, 5, 6, 7, \ldots$. Find S_7.

20. $1, -3, 5, -7, 9, -11, \ldots$. Find S_8.

21. $2, 4, 6, 8, \ldots$. Find S_5.

22. $1, \dfrac{1}{4}, \dfrac{1}{9}, \dfrac{1}{16}, \dfrac{1}{25}, \ldots$. Find S_5.

[v] Rename without using Σ notation.

23. $\displaystyle\sum_{k=1}^{5} \frac{1}{2k}$

24. $\displaystyle\sum_{k=1}^{6} \frac{1}{2k+1}$

25. $\displaystyle\sum_{k=0}^{5} 2^k$

26. $\displaystyle\sum_{k=4}^{7} \sqrt{2k-1}$

27. $\displaystyle\sum_{k=7}^{10} \log k$

28. $\displaystyle\sum_{k=0}^{4} \pi k$

[iv], [v] Write Σ notation for each sum. Then compute the sum.

29. $\dfrac{1}{1^2} + \dfrac{1}{2^2} + \dfrac{1}{3^2} + \dfrac{1}{4^2}$

30. $\dfrac{1}{2} + \dfrac{2}{4} + \dfrac{3}{8} + \dfrac{4}{16}$

31. $-2 + 4 - 8 + 16 - 32 + 64$

32. $\dfrac{1}{2^1} - \dfrac{2}{2^2} + \dfrac{3}{2^3} - \dfrac{4}{2^4} + \dfrac{5}{2^5}$

☆ ───────────────────────────────

Find the first five terms of each sequence.

33. $a_n = \dfrac{1}{2n} \log 1000^n$

34. $a_n = i^n, i = \sqrt{-1}$

35. $a_n = \sin \dfrac{n\pi}{2}$

36. $a_n = \ln (1 \cdot 2 \cdot 3 \cdots n)$

37. $a_n = \text{Cos}^{-1} (-1)^n$

38. $a_n = |\cos (x + \pi n)|$

Find decimal notation, rounded to six decimal places, for the sum of the first six terms of each sequence.

39. ▦ $a_n = \left(1 + \dfrac{1}{n}\right)^n$

40. ▦ $a_n = \sqrt{n+1} - \sqrt{n}$

41. ▦ $a_1 = 2, a_{k+1} = \sqrt{1 + \sqrt{a_k}}$

42. ▦ $a_1 = 2, a_{k+1} = \dfrac{1}{2}\left(a_k + \dfrac{2}{a_k}\right)$

★ ───────────────────────────────

For each sequence, find a formula for S_n.

43. $a_n = \ln n$

44. $a_n = \dfrac{1}{n} - \dfrac{1}{n+1}$

OBJECTIVES

You should be able to:

[i] Use the formula of Theorem 1 to find, for any arithmetic sequence,

a) the nth term when n is given;
b) n, when an nth term is given.
c) Given two terms of a sequence, find the common difference and construct the sequence.

[ii] Use Theorems 2 and 3 for arithmetic sequences to find the sum of any arithmetic sequence.

[iii] Insert arithmetic means between two numbers.

1. Find the 13th term of the sequence 2, 6, 10, 14, ...

2. In the sequence given in Exercise 1, what term would be 298? That is, what is n if $a_n = 298$?

15.2 ARITHMETIC SEQUENCES AND SERIES

[i] Arithmetic Sequences

Consider this sequence:

$$2, \ 5, \ 8, \ 11, \ 14, \ 17, \ \ldots .$$

Note that adding 3 to any term produces the following term. In other words, the difference between any term and the preceding one is 3. Sequences in which the difference between successive terms is constant are called *arithmetic sequences*.*

We now find a formula for the general, or nth, term of any arithmetic sequence. Let us denote the difference between successive terms (called the *common difference*) by d, and write out the first few terms.

$$a_1$$
$$a_2 = a_1 + d$$
$$a_3 = a_2 + d = (a_1 + d) + d = a_1 + 2d$$
$$a_4 = a_3 + d = (a_1 + 2d) + d = a_1 + 3d$$

Generalizing, we obtain the following.

THEOREM 1

The nth term of an arithmetic sequence is given by

$$a_n = a_1 + (n - 1)d,$$

where d is the common difference between successive terms.

Example 1 Find the 14th term of the arithmetic sequence 4, 7, 10, 13, ... First note that $a_1 = 4$, $d = 3$, and $n = 14$. Using the formula of Theorem 1, we obtain

$$a_{14} = 4 + (14 - 1) \cdot 3$$
$$= 4 + 39$$
$$= 43.$$

DO EXERCISE 1.

Example 2 In the sequence given in Example 1, which term is 301? That is, what is n if $a_n = 301$?

We substitute into the formula of Theorem 1 and solve for n:

$$a_n = a_1 + (n - 1)d$$
$$301 = 4 + (n - 1) \cdot 3$$
$$301 = 4 + 3n - 3$$
$$300 = 3n$$
$$100 = n.$$

The 100th term is 301.

DO EXERCISE 2.

---
*Sometimes called arithmetic *progressions*.

Given two terms and their places in an arithmetic sequence, we can construct the sequence.

Example 3 The third term of an arithmetic sequence is 8 and the sixteenth term is 47. Find a_1 and d and construct the sequence.

We know that $a_3 = 8$ and $a_{16} = 47$. Thus we would have to add d, thirteen times, to get from 8 to 47. That is,

$$13d = 47 - 8 = 39.$$

Solving, we obtain

$$d = 3.$$

Since $a_3 = 8$, we subtract d twice to get to a_1. Thus

$$a_1 = 2.$$

The sequence is 2, 5, 8, 11, ...

DO EXERCISE 3.

3. The 7th term of an arithmetic sequence is 79 and the 13th term is 151. Find a_1 and d. Construct the sequence.

[ii] Sums

We now look for a formula for the sum of the first n terms of an arithmetic sequence. Let us denote the terms as follows:

$$a_1, (a_1 + d), (a_1 + 2d), \ldots, (a_n - 2d), (a_n - d), a_n.$$

Then the sum of the first n terms is given by

$$S_n = a_1 + (a_1 + d) + (a_1 + 2d) + \cdots + (a_n - 2d) + (a_n - d) + a_n.$$

Reversing the right side, we obtain

$$S_n = a_n + (a_n - d) + (a_n - 2d) + \cdots + (a_1 + 2d) + (a_1 + d) + a_1.$$

Adding these two equations, we obtain

$$2S_n = (a_1 + a_n) + (a_1 + a_n) + \cdots + (a_1 + a_n),$$

a total of n terms on the right. Thus $2S_n = n(a_1 + a_n)$. Dividing by 2 gives us the formula we seek.

THEOREM 2

The sum of the first n terms of an arithmetic sequence is given by

$$S_n = \frac{n}{2}(a_1 + a_n).$$

We now derive a variant of this formula, which we can use in case a_n is not known. We substitute the expression for a_n given in Theorem 1, $a_n = a_1 + (n - 1)d$, into the formula of Theorem 2:

$$S_n = \frac{n}{2}(a_1 + [a_1 + (n - 1)d]).$$

4. Find the sum of the first 200 natural numbers.

We now have the following.

The sum of the first n terms of an arithmetic sequence is given by

$$S_n = \frac{n}{2}[2a_1 + (n-1)d].$$

Example 4 Find the sum of the first 100 natural numbers.

The series is $1 + 2 + 3 + \cdots + 100$, so $a_1 = 1$, $a_n = 100$, and $n = 100$. We substitute into the formula of Theorem 2:

$$S_n = \frac{n}{2}(a_1 + a_n)$$

$$S_{100} = \frac{100}{2}(1 + 100) = 5050.$$

5. Find the sum of the first 15 terms of the arithmetic sequence

$$1, 3, 5, 7, 9, \ldots.$$

Example 5 Find the sum of the first 14 terms of the arithmetic sequence $2, 5, 8, 11, 14, \ldots.$

We note that $a_1 = 2$, $d = 3$, and $n = 14$. We use the formula of Theorem 3:

$$S_n = \frac{n}{2}[2a_1 + (n-1)d]$$

$$S_{14} = \frac{14}{2} \cdot [2 \cdot 2 + (14-1)3]$$

$$= 7 \cdot [4 + 13 \cdot 3]$$

$$= 7 \cdot 43$$

$$S_{14} = 301.$$

Example 6 Find the sum $\sum_{k=1}^{13} (4k + 5)$.

6. Find the sum.

$$\sum_{k=1}^{10} (9k - 4)$$

It is helpful to write out a few terms:

$$9 + 13 + 17 + \cdots.$$

We see that the sequence is arithmetic, with $a_1 = 9$, $d = 4$, and $n = 13$. We use the formula of Theorem 3:

$$S_n = \frac{n}{2}[2a_1 + (n-1)d]$$

$$S_{13} = \frac{13}{2}[2 \cdot 9 + (13-1)4]$$

$$= \frac{13}{2}[18 + 12 \cdot 4]$$

$$= \frac{13}{2} \cdot 66$$

$$S_{13} = 429.$$

DO EXERCISES 4–6.

Example 7 A family saves money in an arithmetic sequence. They save $500 one year, $600 the next, $700 the next, and so on, for 10 years. How much do they save in all, exclusive of interest?

The amount saved is given by the series

$$\$500 + \$600 + \$700 + \cdots.$$

We have $a_1 = 500$, $n = 10$, and $d = 100$. We use the formula of Theorem 3:

$$S_n = \frac{n}{2}[2a_1 + (n-1)d]$$

$$S_{10} = \frac{10}{2}[2 \cdot 500 + (10-1)100]$$

$$S_{10} = 9500.$$

They save $9500.

DO EXERCISES 7 AND 8.

[iii] Arithmetic Means

If p, m, and q form an arithmetic sequence, it can be shown (see Exercise 34) that $m = (p + q)/2$. We call m the *arithmetic mean* of p and q. Given two numbers p and q, if we find k other numbers $m_1, m_2, m_3, \ldots, m_k$ such that the following is an arithmetic sequence,

$$p, m_1, m_2, \ldots, m_k, q,$$

we say that we have "inserted k arithmetic means between p and q."

Example 8 Insert three arithmetic means between 4 and 13.

We look for m_1, m_2, and m_3 such that 4, m_1, m_2, m_3, 13 is an arithmetic sequence. In this case, $a_1 = 4$, $n = 5$, and $a_5 = 13$. We use the formula of Theorem 1:

$$a_n = a_1 + (n-1)d$$
$$13 = 4 + (5-1)d.$$

Then $d = 2\frac{1}{4}$, so we have

$$m_1 = a_1 + d$$
$$= 4 + 2\frac{1}{4} = 6\frac{1}{4},$$
$$m_2 = m_1 + d$$
$$= 6\frac{1}{4} + 2\frac{1}{4} = 8\frac{1}{2},$$
$$m_3 = m_2 + d$$
$$= 8\frac{1}{2} + 2\frac{1}{4} = 10\frac{3}{4}.$$

DO EXERCISE 9.

7. Find the sum of all two-digit numbers divisible by 3.

8. A cheerleader pyramid has 15 girls on the bottom row, 14 on the next row, and so on, until there is just one girl on top. How many cheerleaders are in the pyramid?

9. Insert three arithmetic means between 4 and 16.

EXERCISE SET 15.2

[i] Exercises 1–10 refer to arithmetic sequences.

1. Find the 12th term of

$$2, 6, 10, \ldots.$$

2. Find the 11th term of

$$0.07, 0.12, 0.17, \ldots.$$

3. In the sequence of Exercise 1, which term is 106?

4. In the sequence of Exercise 2, which term is 1.67?

5. If $a_1 = 5$ and $d = 6$, what is a_{17}?

6. If $a_1 = 14$ and $d = -3$, what is a_{20}?

7. If $d = 4$ and $a_8 = 33$, what is a_1?

8. If $a_1 = 8$ and $a_{11} = 26$, what is d?

9. If $a_1 = 5$, $d = -3$, and $a_n = -76$, what is n?

10. If $a_1 = 25$, $d = -14$, and $a_n = -507$, what is n?

11. In an arithmetic sequence $a_{17} = -40$ and $a_{28} = -73$. Find a_1 and d. Find the first 5 terms of the sequence.

12. A bomb drops from an airplane and falls 16 ft the first sec, 48 ft the second sec, and so on, forming an arithmetic sequence. How many feet will the bomb fall during the 20th sec?

[ii]

13. Find the sum of the first 20 terms of the sequence

$$5, 8, 11, 14, \ldots.$$

14. Find the sum of the first 15 terms of the sequence

$$5, \frac{55}{7}, \frac{75}{7}, \frac{95}{7}, \ldots.$$

15. Find the sum of the odd numbers from 1 to 99, inclusive.

16. Find the sum of the multiples of 7, from 7 to 98 inclusive.

17. An arithmetic sequence has $a_1 = 2$, $d = 5$, and $n = 20$. What is S_{20}?

18. Find the sum of all multiples of 4 that are between 14 and 523.

19. Find the sum $\displaystyle\sum_{k=1}^{16} (7k - 76)$.

20. Find the sum $\displaystyle\sum_{k=1}^{12} (6k - 3)$.

21. How many poles will be in a pile of telephone poles if there are 30 in the first layer, 29 in the second, and so on, until there is one in the last layer?

22. If a student saved 10¢ on October 1, 20¢ on October 2, 30¢ on October 3, etc., how much would be saved during October? (October has 31 days.)

[iii]

23. Insert four arithmetic means between 4 and 13.

24. Insert three arithmetic means between 5 and -3.

☆ _____

25. Find a formula for the sum of the first n odd natural numbers.

26. Find three numbers in an arithmetic sequence such that the sum of the first and third is 10 and the product of the first and second is 15.

27. The zeros of this polynomial form an arithmetic sequence. Find them.

$$x^4 + 4x^3 - 84x^2 - 176x + 640$$

28. Insert enough arithmetic means between 1 and 50 so the sum of the resulting arithmetic series will be 459.

29. Suppose that the lengths of the sides of a right triangle form an arithmetic sequence. Prove that the triangle is similar to a right triangle whose sides have lengths 3, 4, 5.

(*Straight-line depreciation.*) A company buys an office machine for $5200 on January 1 of a given year. It is expected to last for 8 years, at which time its *trade-in*, or *salvage*, *value* will be $1100. If the company figures the decline in value to be the same each year, then the *book value*, or *salvage value*, after n years, $0 \leq n \leq 8$, is given by an arithmetic sequence:

$$V_n = C - n\left(\frac{C - S}{L}\right),$$

where $C =$ the original cost of the item ($5200), $L =$ years of expected life (8), and $S =$ the salvage value ($1100).

30. Find the general term for the straight-line depreciation of the office machine.

31. Find the salvage value after 0 years, 1 year, 2 years, 3 years, 4 years, 7 years, 8 years.

32. Find the first 10 terms of the arithmetic sequence for which $a_1 = \$8760$ and $d = -\$798.23$.

33. Find the sum of the first ten terms of the sequence given in Exercise 32.

★ ───────────────────────────────

34. Prove that an expression for the nth term of an arithmetic sequence defines a linear function.

35. Prove that if p, m, and q form an arithmetic sequence, then

$$m = \frac{p + q}{2}.$$

15.3 GEOMETRIC SEQUENCES AND SERIES

[i] Geometric Sequences

Consider this sequence:

$$3, 6, 12, 24, 48, 96, \ldots.$$

Note that multiplying any term by 2 produces the following term. In other words, the ratio of any term and the preceding one is 2. Sequences in which the ratio of successive terms is constant but different from 1 are called *geometric sequences*.*

We now find a formula for the general, or nth, term of any geometric sequence. Let us denote the ratio of successive terms (called the *common ratio*) by r, and write out the first few terms:

$$a_1,$$
$$a_2 = a_1 r,$$
$$a_3 = a_2 r = (a_1 r)r = a_1 r^2,$$
$$a_4 = a_3 r = (a_1 r^2)r = a_1 r^3$$

Generalizing, we obtain the following.

THEOREM 4
═══════════════════════════════════

The nth term of a geometric sequence is given by

$$a_n = a_1 r^{n-1},$$

where r is the common ratio of successive terms.
═══════════════════════════════════

Example 1 Find the 6th term of the geometric sequence 4, 20, 100, ...

First note that $a_1 = 4$, $n = 6$, and $r = 5$. We use the formula given in Theorem 4:

$$a_n = a_1 r^{n-1}$$
$$a_6 = 4 \cdot 5^{6-1} = 4 \cdot 5^5$$
$$= 12{,}500.$$

DO EXERCISE 1.

───────────────

*Sometimes called geometric *progressions*.

OBJECTIVES

You should be able to:

[i] Use Theorem 4 for geometric sequences to find a given term.
[ii] Find sums of geometric sequences.

1. Find the 8th term of the geometric sequence

$$2, 4, 8, 16, \ldots.$$

2. Find the 6th term of the geometric sequence

$$3, \; -15, \; 75, \; \ldots .$$

Example 2 Find the 11th term of the geometric sequence 64, -32, 16, -8, $\ldots$.

Note that $a_1 = 64$, $n = 11$, and $r = -\frac{1}{2}$. We use the formula given in Theorem 4:

$$a_n = a_1 r^{n-1}$$
$$a_{11} = 64 \cdot \left(-\frac{1}{2}\right)^{11-1}$$
$$= 64 \cdot \left(-\frac{1}{2}\right)^{10}$$
$$= 2^6 \cdot \frac{1}{2^{10}}$$
$$= \frac{1}{2^4} \quad \text{or} \quad \frac{1}{16}.$$

DO EXERCISE 2.

Example 3 A college student borrows \$600 at 12% interest compounded annually. The loan is paid off at the end of 3 years. How much is paid back?

For principal P, at 12% interest, $P + 0.12P$ will be owed at the end of a year. This is also $1.12P$, and is the principal for the second year, so at the end of that year $1.12(1.12P)$ is owed. Thus the principal at the beginnings of successive years is

$$P, \; 1.12P, \; 1.12^2P, \; 1.12^3P, \quad \text{and so on.}$$

We have a geometric sequence with $a_1 = P = 600$, $n = 4$, and $r = 1.12$. We use the formula given in Theorem 4:

$$a_n = a_1 r^{n-1}$$
$$a_4 = 600 \cdot (1.12)^{4-1}$$
$$= 600(1.12)^3$$
$$= 600(1.404928)$$
$$= 842.96.$$

The amount paid back is \$842.96.

3. A college student borrows \$400 at 6% interest compounded annually. She pays off the loan at the end of 3 yr. How much does she pay?

Note that the equation $a_4 = \$600(1.12)^3$, used in Example 3, conforms to the compound interest formula $A = P(1 + i)^t$, developed in Chapter 2. If we consult that development we see that the numbers A_1, A_2, A_3, and so on form a geometric sequence, where $A_1 = P$, $A_2 = P(1 + i)$, $r = 1 + i$, and $n - 1 = t$. (Note that "after t years" and "at the beginning of the nth year" have the same meaning here.)

DO EXERCISE 3.

[ii] Sums

We now look for a formula for the sum of the first n terms of a geometric sequence.* Let us denote the terms as follows:

$$a_1,\ a_1r,\ a_1r^2,\ a_1r^3,\ \ldots,\ a_1r^{n-1}.$$

Then the sum of the first n terms is given by

$$S_n = a_1 + a_1r + a_1r^2 + \cdots + a_1r^{n-2} + a_1r^{n-1}.$$

We next multiply on both sides of this equation by $-r$, obtaining

$$-rS_n = -a_1r - a_1r^2 - \cdots - a_1r^{n-1} - a_1r^n.$$

We now add the two equations, obtaining

$$S_n - rS_n = a_1 - a_1r^n.$$

Solving for S_n gives us the formula we seek.

THEOREM 5

The sum of the first n terms of a geometric sequence is given by

$$S_n = \frac{a_1 - a_1r^n}{1 - r}, \quad r \neq 1.$$

We now derive a variant of this formula, which we can use when a_n is known. Using Theorem 4 we substitute a_n for a_1r^{n-1} into the formula of Theorem 5. We then have the following.

THEOREM 6

The sum of the first n terms of a geometric sequence is given by

$$S_n = \frac{a_1 - ra_n}{1 - r}, \quad r \neq 1.$$

Example 4 Find the sum of the first 6 terms of the geometric sequence 3, 6, 12, 24, ...

Note that $a_1 = 3$, $n = 6$, and $r = 2$. We use the formula given in Theorem 5:

$$S_n = \frac{a_1 - a_1r^n}{1 - r}$$

$$S_6 = \frac{3 - 3 \cdot 2^6}{1 - 2}$$

$$= \frac{3 - 192}{-1}, \quad \text{or} \quad 189.$$

---

*It is not uncommon to see or hear the redundant terminology "sum of a geometric series."

4. Find the sum of the first 6 terms of the geometric sequence

$$3, 15, 75, 375, \ldots.$$

5. Find the sum of the first 10 terms of the geometric sequence

$$2, -1, \frac{1}{2}, -\frac{1}{4}, \ldots.$$

6. Find the sum.

$$\sum_{k=1}^{5} 3^k$$

7. In Example 3, how much would you earn in 20 days?

▦ **Example 5** Find the sum $\displaystyle\sum_{k=1}^{11} (0.3)^k$.

This is a geometric series. The first term is 0.3, $r = 0.3$, and $n = 11$. We use the formula given in Theorem 5:

$$S_n = \frac{a_1 - a_1 r^n}{1 - r}$$

$$S_{11} = \frac{0.3 - 0.3 \cdot (0.3)^{11}}{1 - 0.3}$$

$$= \frac{0.3 - (0.3)^{12}}{0.7}$$

$$= 0.42857+.$$

DO EXERCISES 4–6.

▦ **Example 6** Someone offers you a job for 30 days, the pay being 1¢ the first day, 2¢ the second day, and doubled every day thereafter. Would you take the job? Calculate your pay for the 30 days.

Your pay is the sum of the geometric sequence (in dollars)

$$0.01, \; 2(0.01), \; 2^2(0.01), \; 2^3(0.01), \ldots, \; 2^{29}(0.01).$$

We note that $a_1 = 0.01$, $n = 30$, and $r = 2$. We use the formula given in Theorem 5:

$$S_n = \frac{a_1 - a_1 r^n}{1 - r}$$

$$S_{30} = \frac{0.01 - 0.01(2)^{30}}{1 - 2}$$

$$= \frac{0.01(1 - 2^{30})}{-1} = \frac{0.01(1 - 1073741824)}{-1}$$

$$= 10,737,418.23.$$

Your pay: $10,737,418.23.

DO EXERCISE 7.

EXERCISE SET 15.3

[i]

1. Find the 10th term of the geometric sequence

$$\frac{8}{243}, \frac{4}{81}, \frac{2}{27}, \ldots.$$

[ii]

3. Find the sum of the first 8 terms of the geometric sequence

$$5, 10, 20, \ldots.$$

2. Find the 5th term of the geometric sequence

$$2, -10, 50, \ldots.$$

4. Find the sum of the first 6 terms of the geometric sequence

$$16, -8, 4, \ldots.$$

5. Find the sum of the first 7 terms of the geometric sequence

$$\frac{1}{18}, \; -\frac{1}{6}, \; \frac{1}{2}, \; \dots.$$

6. Find the sum of the geometric sequence

$$-8, \; 4, \; -2, \; \dots -\frac{1}{32}.$$

7. Find the sum.

$$\sum_{k=1}^{6} \left(\frac{1}{2}\right)^{k-1}$$

8. Find the sum.

$$\sum_{k=1}^{10} 2^k$$

9. A college student borrows $800 at 8% interest compounded annually. She pays off the loan in full at the end of 2 yr. How much does she pay?

10. A college student borrows $1000 at 8% interest compounded annually. He pays off the loan in full at the end of 4 yr. How much does he pay?

11. A ping-pong ball is dropped from a height of 16 ft and always rebounds $\frac{1}{4}$ of the distance of the previous fall. What distance does it rebound the 6th time?

12. Gaintown has a population of 100,000 now and the population is increasing 10% every year. What will be the population in 5 yr?

13. ▦ Find the sum of the first 5 terms of the geometric sequence

$$\$1000, \; \$1000(1.08), \; \$1000(1.08)^2, \; \dots.$$

Round to the nearest cent.

14. ▦ Find the sum of the first 6 terms of the geometric sequence

$$\$200, \; \$200(1.13), \; \$200(1.13)^2, \; \dots.$$

Round to the nearest cent.

☆ ──────────────────────────────

For Exercises 15–17, assume that $a_1, a_2, a_3, \dots,$ is a geometric sequence.

15. Show that

$$a_1^2, \; a_2^2, \; a_3^2, \; \dots,$$

is a geometric sequence.

16. Show that

$$a_1^{-3}, \; a_2^{-3}, \; a_3^{-3}, \; \dots,$$

is a geometric sequence.

17. Show that

$$\ln a_1, \; \ln a_2, \; \ln a_3, \; \dots,$$

is an arithmetic sequence.

18. Show that

$$5^{a_1}, \; 5^{a_2}, \; 5^{a_3}, \; \dots,$$

is a geometric sequence, if $a_1, a_2, a_3, \dots,$ is an arithmetic sequence.

19. Show that if the positive numbers p, G, q form a geometric sequence

$$G = \sqrt{pq},$$

G is called the *geometric mean* of the numbers p and q.

20. Find the geometric mean between each pair of numbers.

a) 4 and 9

b) 2 and 6

c) $\frac{1}{2}$ and $\frac{1}{3}$

d) $\sqrt{5} + \sqrt{2}$ and $\sqrt{5} - \sqrt{2}$

21. ▦ A piece of paper is 0.01 in. thick. It is folded in such a way that its thickness is doubled each time for 20 times. How thick is the result?

22. ▦ A superball dropped from the top of the Washington Monument (556 ft high) always rebounds $\frac{3}{4}$ of the distance of the previous fall. How far up and down has it traveled when it hits the ground for the sixth time?

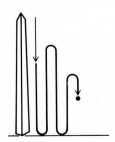

23. (*An annuity.*) A person decides to save money in a savings account for retirement. At the beginning of each year $1000 is invested at 8% compounded annually. How much is in the retirement fund at the end of 40 years?

24. A square has sides 1 meter in length. Another square is formed whose vertices are the midpoints of the original square. Inside this square another is formed whose vertices are the midpoints of the second square, and so on.

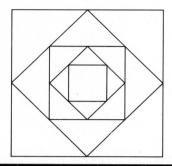

a) Find the first five terms of the sequence of areas of the squares.

b) Find the nth term.

c) Find a formula for the sum of the areas of the first n squares.

OBJECTIVES

You should be able to:

[i] Determine whether the sum of an infinite geometric sequence exists, and if so use the formula of Theorem 7 to find it.

[ii] Convert from repeating decimal notation to fractional notation.

15.4 INFINITE GEOMETRIC SEQUENCES AND SERIES

Recall the definition of a *sequence* as a function having for its domain some set of natural numbers. If that domain is an infinite (unending) set of natural numbers, then the sequence is called an *infinite sequence*. In the examples that follow, the domain will be the set of *all* natural numbers, even though this is not a requirement of the definition.

Let us look at two examples of infinite geometric sequences and their sequences of partial sums.

A. $2, 4, 8, 16, \ldots, 2^n, \ldots$

$$S_1 = 2$$
$$S_2 = 6$$
$$S_3 = 14$$
$$S_4 = 30$$
$$S_5 = 62$$

The terms of the sequence S_n get larger and larger, without bound. Now let us look at the next sequence.

B. $\dfrac{1}{2}, \dfrac{1}{4}, \dfrac{1}{8}, \dfrac{1}{16}, \ldots, \dfrac{1}{2^n}, \ldots$

$$S_1 = \frac{1}{2}$$

$$S_2 = \frac{1}{2} + \frac{1}{4} = \frac{3}{4}$$

$$S_3 = S_2 + \frac{1}{8} = \frac{3}{4} + \frac{1}{8} = \frac{7}{8}$$

$$S_4 = S_3 + \frac{1}{16} = \frac{15}{16}$$

$$S_5 = S_4 + \frac{1}{32} = \frac{31}{32}$$

It appears that we can write a formula for S_n as follows:

$$S_n = \frac{2^n - 1}{2^n}.$$

Later (Example 5, p. 583) we will prove that this is correct. In this formula the numerator is less than the denominator for all values of n. Yet as n increases the numerator gets very close to the denominator. Thus the values of S_n approach the number 1 more and more closely. We say that S_n "approaches 1 as a limit."

[i] Sums

The sequence (B) in the above example is infinite. For any finite number of terms, the sum is given by

$$S_n = \sum_{k=1}^{n} \frac{1}{2^k} = \frac{2^n - 1}{2^n}.$$

Since S_n approaches a limit, we *define* that limit to be the "sum" of the infinite sequence of terms, and use the following notation:*

$$S_\infty = \sum_{k=1}^{\infty} \frac{1}{2^k} = 1.$$

Not all infinite sequences have sums (for example, sequence (A) above). In those cases the notation $\sum_{k=1}^{\infty} a_k$ is meaningless. An infinite sum is also called an *infinite series*.

We now find a formula for the "sum" of an infinite geometric sequence, by looking at values of S_n as n becomes very large. Recall that

$$S_n = \frac{a_1 - a_1 r^n}{1 - r}$$

or, equivalently,

$$S_n = \frac{a_1}{1 - r} - \frac{a_1 r^n}{1 - r}.$$

If $r = 1$, the formula does not hold, since 1 is a nonsensible replacement for r. However, in that case the sequence is not geometric. For any other value of r, the first fractional expression can be evaluated. It remains to look at the second fractional expression as n becomes large.

If $r > 1$, r_n will become very large and the second fractional expression will become large, hence no sum exists. If r is negative and $|r| > 1$, then r^n will alternate between positive and negative numbers having larger and larger absolute value. In either case, S_n will not approach a limit.

---

* $\sum_{k=1}^{\infty} 1/2^k$ is read "the sum, as k goes from 1 to infinity, of $1/2^k$."

Determine which geometric sequences have sums.

1. 4, 16, 64, . . .

2. 5, −30, 180, . . .

3. $1, \dfrac{1}{3}, \dfrac{1}{9}, \dfrac{1}{27}, \cdots$

Find these infinite sums. The series are geometric.

4. $1 + \dfrac{1}{3} + \dfrac{1}{9} + \dfrac{1}{27} + \cdots$

5. $\displaystyle\sum_{n=0}^{\infty} \dfrac{1}{2^n}$

6. $4 - 1 + \dfrac{1}{4} - \dfrac{1}{16}, \cdots$

If $|r| < 1$, then the values of r^n approach 0 as n becomes very large, hence the second fractional expression also approaches 0, and we have the following.

THEOREM 7

An infinite geometric sequence has a sum if and only if $|r| < 1$. That sum is given by

$$S_{\infty} = \frac{a_1}{1 - r}.$$

Example 1 Find the sum of the infinite geometric sequence $5, \frac{5}{2}, \frac{5}{4}, \frac{5}{8}, \ldots,$ if it exists. That is, find $\displaystyle\sum_{k=1}^{\infty} 5/2^{k-1}$.

Note that $a_1 = 5$. The sum exists because $r = \frac{1}{2}$, hence $|r| < 1$. We use the formula of Theorem 7:

$$S_{\infty} = \frac{a_1}{1 - r} = \frac{5}{1 - \dfrac{1}{2}} = 10.$$

DO EXERCISES 1–6.

[ii] Repeating Decimals

Any repeating decimal represents a geometric series. For example,

$$0.33333\ldots = 0.3 + 0.03 + 0.003 + 0.0003 + \cdots.$$

In this series the common ratio is 0.1. It is a common series and the sum is known to be $\frac{1}{3}$. Consider the following:

$$0.35\overline{3535}^* = 0.35 + 0.0035 + 0.000035 + \cdots.$$

In this case the common ratio is 0.01. In any repeating decimal the common ratio will be 0.1, 0.01, 0.001, or 10^{-n} for some natural number n. Thus in any repeating decimal the common ratio is less than 1, so the infinite sum exists.

Example 2 Find fractional notation for 0.272727.

We can think of this as $0.27 + 0.0027 + 0.000027 + 0.00000\overline{27} + \cdots$. Note that $a_1 = 0.27$ and that $r = 0.01$. We use the formula given in Theorem 7:

$$S_{\infty} = \frac{a_1}{1 - r}$$

$$S_{\infty} = \frac{0.27}{1 - 0.01} = \frac{0.27}{0.99} = \frac{3}{11}.$$

Fractional notation for $0.27\overline{27}$ is $\frac{3}{11}$. This can be checked by dividing 3 by 11.

---

*The bar indicates the repeating cycle.

Example 3 Find fractional notation for $0.416\overline{6}$.

We first split this up into a repeating and a nonrepeating part as follows: $0.41 + 0.0066\overline{6}$. The repeating part can be considered as $0.006 + 0.0006 + 0.00006 + 0.000006 + \cdots$. Thus the repeating part is a geometric series with $a_1 = 0.006$ and $r = 0.1$. We find fractional notation for this sum first:

$$S_\infty = \frac{a_1}{1 - r}$$

$$= \frac{0.006}{1 - 0.1}$$

$$= \frac{0.006}{0.9}$$

$$= \frac{6}{900}.$$

The number in question is then $0.41 + \frac{6}{900}$ or $\frac{41}{100} + \frac{6}{900}$. Adding and simplifying, we obtain $\frac{5}{12}$. The reader should check this by dividing.

DO EXERCISES 7 AND 8.

Example 4 (*The economic multiplier.*) Suppose that a change in the tax laws reduces the tax paid by a family by $1000. Each family then has an extra $1000 to spend. Let us suppose that they spend 90% of this amount, that the people who get it spend 90% of it, the people who get that spend 90% of it, and so on.

According to certain economic theory, the money that this effectively puts into the economy can be calculated as the sum of an infinite geometric sequence, as follows:

$$\$1000 + \$1000(0.90) + \$1000(0.90)^2 + \$1000(0.90)^3 + \cdots.$$

Using the formula of Theorem 7 we find this amount to be

$$\frac{\$1000}{1 - 0.90}, \quad \text{or} \quad \$10,000.$$

DO EXERCISE 9.

Find fractional notation for each number.

7. $0.45\overline{45}$

8. $5.36\overline{36}$

9. Some states gave veterans a bonus after World War II. In one such state, the bonus was $500. Assuming that each veteran spent 85% of his bonus and that each person receiving the money thus put into circulation also spent 85% of it, how much money was effectively put into the economy?

EXERCISE SET 15.4

[i] Determine which of the following infinite geometric sequences have sums.

1. 5, 10, 20, 40, . . .

2. 16, 8, 4, 2, . . .

3. $6, 2, \frac{2}{3}, \frac{2}{9}, \ldots$

4. $2, -4, 8, -16, 32, \ldots$

5. 1, 0.1, 0.01, 0.001, 0.0001, . . .

6. $-\frac{5}{3}, -\frac{10}{9}, -\frac{20}{27}, \ldots$

7. $1, -\frac{1}{5}, \frac{1}{25}, -\frac{1}{125}, \ldots$

8. $6, \frac{42}{5}, \frac{294}{25}, \ldots$

Find these infinite sums. (The series are geometric.)

9. $4 + 2 + 1 + \cdots$

10. $7 + 3 + \frac{9}{7} + \cdots$

11. $25 + 20 + 16 + \cdots$

12. $12 + 9 + \dfrac{27}{4} + \cdots$

13. $\displaystyle\sum_{k=1}^{\infty} \dfrac{1}{2^{k-1}}$

14. $\displaystyle\sum_{k=1}^{\infty} \dfrac{8}{3}\left(\dfrac{1}{2}\right)^{k-1}$

15. $\displaystyle\sum_{k=1}^{\infty} 16(0.1)^{k-1}$

16. $\displaystyle\sum_{k=1}^{\infty} 4(0.6)^{k-1}$

17. ▦ $\$1000(1.08)^{-1} + \$1000(1.08)^{-2} + \$1000(1.08)^{-3} + \cdots$

18. ▦ $\$500(1.12)^{-1} + \$500(1.12)^{-2} + \$500(1.12)^{-3} + \cdots$

[ii] Find fractional notation for these numbers.

19. $0.7777\overline{7}$

20. $0.5333\overline{3}$

21. $0.2121\overline{21}$

22. $0.64444\overline{4}$

23. $5.1515\overline{15}$

24. $0.4125\overline{125}$

25. The government makes an $8,000,000,000 expenditure for a new type of aircraft. If 85% of this gets spent again, and 85% of that gets spent again, and so on, what is the total effect on the economy?

26. Repeat Exercise 25 for $9,400,000,000 and 99%.

☆ ───────────────────────────────

27. (*Advertising effect.*) A company is marketing a new product in a city of 5,000,000 people. They plan an advertising campaign that they think will induce 40% of the people to buy the product. They estimate that if those people like the product, they will induce 40% (of the 40% of 5,000,000) more to buy the product, and those will induce 40%, and so on. In all, how many people will buy the product as a result of the advertising campaign? What percentage of the population is this?

28. How far (up and down) will a ball travel before stopping if it is dropped from a height of 12 ft, and each rebound is $\frac{1}{3}$ of the previous distance? (*Hint:* Use an infinite geometric series.)

★ ───────────────────────────────

29. ▦ The function e^x is given by the infinite series

$$e^x = 1 + x + \dfrac{x^2}{2!} + \dfrac{x^3}{3!} + \cdots$$

($n!$ is read "n factorial," and is defined to be the product $1 \cdot 2 \cdot 3 \cdot \cdots \cdot n$). Approximate e using the first 6 terms of the series.

30. ▦ An infinite sequence is defined recursively by

$$a_1 = 2, \quad a_{k+1} = \dfrac{1}{2}\left(a_k + \dfrac{2}{a_k}\right).$$

Find decimal notation, rounded to six decimal places, for S_1, S_2, S_3, S_4, S_5, and S_6. Make a conjecture about the value of S_∞.

31. ▦ The infinite sequence

$$2, \; \dfrac{1}{2}, \; \dfrac{1}{2 \cdot 3}, \; \dfrac{1}{2 \cdot 3 \cdot 4}, \; \dfrac{1}{2 \cdot 3 \cdot 4 \cdot 5}, \; \dfrac{1}{2 \cdot 3 \cdot 4 \cdot 5 \cdot 6}, \; \cdots$$

is not geometric, but does have a sum. Find S_1, S_2, S_3, S_4, S_5, and S_6. Make a conjecture about the value of S_∞.

━━━━━━━━━━━━━━━━━━━━━━━━━━━━━━━━

OBJECTIVES

You should be able to:

[i] List the statements of an infinite sequence that is defined by a formula.

[ii] Do proofs by mathematical induction.

15.5 MATHEMATICAL INDUCTION

[i] Sequences of Statements

Infinite sequences of statements occur often in mathematics. In an infinite sequence of statements there is of course a statement for each natural number. For example, consider the sentence

For x between 0 and 1, $0 < x^n < 1$.

Let us think of this as $S(n)$ or S_n. Substituting natural numbers for n gives a sequence of statements. We list a few of them.

Statement 1 (S_1):*   For x between 0 and 1, $0 < x^1 < 1$.

Statement 2 (S_2):   For x between 0 and 1, $0 < x^2 < 1$.

S_3:   For x between 0 and 1, $0 < x^3 < 1$.

S_4:   For x between 0 and 1, $0 < x^4 < 1$.

Example 1 List the first few statements in the sequence obtainable from $\log n < n$.

This time S_n is "$\log n < n$.

$$S_1: \quad \log 1 < 1$$
$$S_2: \quad \log 2 < 2$$
$$S_3: \quad \log 3 < 3$$

Many sequences of statements concern sums.

Example 2 List the first few statements in the sequence obtainable from

$$1 + 3 + 5 + \cdots + (2n - 1) = n^2.$$

This time the entire equation is the statement S_n.

$$S_1: \quad 1 = 1^2$$
$$S_2: \quad 1 + 3 = 2^2$$
$$S_3: \quad 1 + 3 + 5 = 3^2$$
$$S_4: \quad 1 + 3 + 5 + 7 = 4^2$$

DO EXERCISES 1 AND 2.

[ii] Proving Infinite Sequences of Statements

The method of proof of this section, called *mathematical induction,* allows us to prove infinite sequences of statements. They are usually verbalized somewhat as follows:

For all natural numbers n, S_n,

where S_n is some sentence such as those in the preceding examples. Of course, we cannot prove each statement of an infinite sequence individually. Instead, we try to show that whenever S_k holds, then S_{k+1} will have to hold. We abbreviate this as $S_k \rightarrow S_{k+1}$. (This is read "S_k *implies* S_{k+1}.") If we can establish that this holds for all natural numbers k, we then have the following.

$S_1 \rightarrow S_2$ Meaning whenever S_1 holds, S_2 must hold.

$S_2 \rightarrow S_3$ Meaning whenever S_2 holds, S_3 must hold.

$S_3 \rightarrow S_4$ Meaning whenever S_3 holds, S_4 must hold.

and so on, indefinitely.

---

*Note that S_1, S_2, and so on do *not* represent sums in this context.

1. List the first five statements in the sequence obtainable from

$$n^2 + 1 > n + 1.$$

2. List the first five statements in the sequence obtainable from

$$1 + 2 + \cdots + n = \frac{n(n + 1)}{2}.$$

At this stage we do not yet know whether there is *any k* for which S_k holds. All we know is that *if* S_k holds, then S_{k+1} must hold. So we now show that S_k holds for some k, usually $k = 1$. We are then in good condition because we have the following.

S_1 is true.	We have verified, or proved, this.
$S_1 \to S_2$	This means that whenever S_1 holds,
Therefore S_2 is true.	S_2 must hold.
$S_2 \to S_3$	This means that whenever S_2 holds,
Therefore S_3 is true,	S_3 must hold.
and so on.	

We conclude that S_n is true for all natural numbers n.*

We now state the principle of mathematical induction.

Principle of Mathematical Induction

We can prove an infinite sequence of statements S_n by showing the following.

a) S_1 is true. (This is called the *basis* step.)

b) For all natural numbers k, $S_k \to S_{k+1}$. (This is called the *induction* step.)

In learning to do proofs by mathematical induction, it is helpful to first write out S_n, S_1, S_k, and S_{k+1}. This helps to identify what is to be assumed and what is to be deduced.

Example 3 Prove: For every natural number n, $n < 2^n$.

We first list S_n, S_1, S_k, and S_{k+1}.

$$S_n: \quad n < 2^n$$
$$S_1: \quad 1 < 2^1$$
$$S_k: \quad k < 2^k$$
$$S_{k+1}: \quad k + 1 < 2^{k+1}$$

a) *Basis step.* S_1 as listed is obviously true.

b) *Induction step.* We assume S_k as hypothesis and try to show that it implies S_{k+1}. For any k,

$$k < 2^k \qquad \text{By hypothesis (this is } S_k\text{)}$$
$$2k < 2 \cdot 2^k \qquad \text{Multiplying on both sides by 2}$$
$$2k < 2^{k+1} \qquad \text{Adding exponents on the right}$$

Now, since k is any natural number,

$$1 \le k$$
$$k + 1 \le k + k \qquad \text{Adding } k \text{ on both sides}$$
$$k + 1 \le 2k.$$

---

*This is reminiscent of knocking over dominoes. The first one hits the second one, the second hits the third, and so on.

Thus we have

$$k + 1 \leq 2k < 2^{k+1}$$

and

$$k + 1 < 2^{k+1}. \qquad \text{This is } S_{k+1}.$$

We have now shown that $S_k \rightarrow S_{k+1}$ for any natural number k. Thus the induction step and the proof are complete. We have proved that for every natural number n, $n < 2^n$.

DO EXERCISE 3.

Example 4 Prove: For every natural number n,

$$1 + 3 + 5 + 7 + \cdots + (2n - 1) = n^2.$$

We first list S_n, S_1, S_k, and S_{k+1}.

S_n: $1 + 3 + 5 + \cdots + (2n - 1) = n^2$
S_1: $1 = 1^2$
S_k: $1 + 3 + 5 + \cdots + (2k - 1) = k^2$
S_{k+1}: $1 + 3 + 5 + \cdots + (2k - 1) + [2(k + 1) - 1] = (k + 1)^2$

a) *Basis step.* S_1 as listed is obviously true.

b) *Induction step.* We assume S_k as hypothesis and try to show that it implies S_{k+1}.

$$1 + 3 + 5 + \cdots + (2k - 1) = k^2 \quad \text{for any natural number } k.$$

This is hypothesis (it is S_k). Let us add $[2(k + 1) - 1]$ on both sides. This gives us

$$
\begin{aligned}
1 + 3 + \cdots + (2k - 1) + [2(k + 1) - 1] &= k^2 + [2(k + 1) - 1] \\
&= k^2 + 2k + 1 \\
&= (k + 1)^2.
\end{aligned}
$$

We have arrived at S_{k+1}. Thus we have shown that for all natural numbers k, $S_k \rightarrow S_{k+1}$. This completes the induction step. The proof is complete. We have proved that the equation is true for all natural numbers n.

DO EXERCISE 4.

Example 5 Prove: $\displaystyle\sum_{p=1}^{n} \frac{1}{2^p} = \frac{2^n - 1}{2^n}$, for all natural numbers n. In other words, prove that for all natural numbers n,

$$\frac{1}{2} + \frac{1}{4} + \frac{1}{8} + \cdots + \frac{1}{2^n} = \frac{2^n - 1}{2^n}.$$

3. Prove that if x is a number greater than 1, then for any natural number n,

$$x \leq x^n.$$

4. Consider

$$2 + 4 + 6 + \cdots + 2n = n(n + 1).$$

a) List S_1 and S_2.
b) List S_k.
c) List S_{k+1}.
d) Complete the basis step, i.e., verify that S_1 is true.
e) Complete the proof that the formula holds for all n, by proving that S_k implies S_{k+1} for all natural numbers k.

5. Prove that

$$\sum_{p=1}^{n} (3p - 1) = \frac{n(3n + 1)}{2}$$

for all natural numbers n.

We first list S_n, S_1, S_k, and S_{k+1}.

S_n: $\displaystyle\sum_{p=1}^{n} \frac{1}{2^p} = \frac{1}{2} + \frac{1}{4} + \frac{1}{8} + \cdots + \frac{1}{2^n} = \frac{2^n - 1}{2^n}$

S_1: $\dfrac{1}{2} = \dfrac{2^1 - 1}{2^1}$

S_k: $\displaystyle\sum_{p=1}^{k} \frac{1}{2^p} = \frac{1}{2} + \frac{1}{4} + \cdots + \frac{1}{2^k} = \frac{2^k - 1}{2^k}$

S_{k+1}: $\displaystyle\sum_{p=1}^{k+1} \frac{1}{2^p} = \frac{1}{2} + \frac{1}{4} + \frac{1}{8} + \cdots + \frac{1}{2^k} + \frac{1}{2^{k+1}} = \frac{2^{k+1} - 1}{2^{k+1}}$

a) *Basis step.*

$$\frac{2^1 - 1}{2^1} = \frac{2 - 1}{2} = \frac{1}{2}, \quad \text{so } S_1 \text{ is true.}$$

b) *Induction step.* Using S_k as hypothesis, we have, for all natural numbers k,

$$\frac{1}{2} + \frac{1}{4} + \frac{1}{8} + \cdots + \frac{1}{2^k} = \frac{2^k - 1}{2^k}.$$

Let us add $1/2^{k+1}$ on both sides. Then we have

$$\frac{1}{2} + \frac{1}{4} + \cdots + \frac{1}{2^k} + \frac{1}{2^{k+1}} = \frac{2^k - 1}{2^k} + \frac{1}{2^{k+1}} = \frac{2^k - 1}{2^k} \cdot \frac{2}{2} + \frac{1}{2^{k+1}}$$

$$= \frac{(2^k - 1) \cdot 2 + 1}{2^{k+1}} = \frac{2^{k+1} - 1}{2^{k+1}}.$$

We have arrived at S_{k+1}. Thus we have shown that for all natural numbers k, $S_k \to S_{k+1}$. This completes the induction and the proof is complete. We have proved that the equation is true for all natural numbers n.

DO EXERCISE 5.

EXERCISE SET 15.5

[i] List the first five statements in the sequence obtainable from each of the following.

1. $n^2 < n^3$

2. $n^2 + n + 40$ is prime

3. A polygon of n sides has $\dfrac{n(n - 3)}{2}$ diagonals.

4. The sum of the angles of a polygon of n sides is $(n - 2) \cdot 180°$.

[ii] In Exercises 5–22, use mathematical induction to prove each, for every natural number n.

5. $1 + 2 + 3 + \cdots + n = \dfrac{n(n + 1)}{2}$

6. $4 + 8 + 12 + \cdots + 4n = 2n(n + 1)$

7. $1 + 5 + 9 + \cdots + (4n - 3) = n(2n - 1)$

8. $3 + 6 + 9 + \cdots + 3n = \dfrac{3n(n + 1)}{2}$

9. $\dfrac{1}{1 \cdot 2} + \dfrac{1}{2 \cdot 3} + \cdots + \dfrac{1}{n(n + 1)} = \dfrac{n}{n + 1}$

10. $2 + 4 + 8 + \cdots + 2^n = 2(2^n - 1)$

11. $n < n + 1$

12. $2 \leq 2^n$

13. $3^n < 3^{n+1}$

14. $2n \leq 2^n$

15. $1^3 + 2^3 + 3^3 + \cdots + n^3 = \dfrac{n^2(n+1)^2}{4}$

16. $\dfrac{1}{1 \cdot 2 \cdot 3} + \dfrac{1}{2 \cdot 3 \cdot 4} + \dfrac{1}{3 \cdot 4 \cdot 5} + \cdots + \dfrac{1}{n(n+1)(n+2)} = \dfrac{n(n+3)}{4(n+1)(n+2)}$

☆ ——————————————————————————————————————

17. $\left(1 + \dfrac{1}{1}\right)\left(1 + \dfrac{1}{2}\right)\left(1 + \dfrac{1}{3}\right) \cdots \left(1 + \dfrac{1}{n}\right) = n + 1$

18. $a_1 + (a_1 + d) + (a_2 + d) + \cdots + [a_1 + (n-1)d] = \dfrac{n}{2}[2a_1 + (n-1)d]$ [This is a formula for the sum of the first n terms of an *arithmetic* sequence (Theorem 3)].

19. $a_1 + a_1r + a_1r^2 + \cdots + a_1r^{n-1} = \dfrac{a_1 - a_1r^n}{1 - r}$ [This is a formula for the sum of the first n terms of a *geometric* sequence (Theorem 5)].

20. $\cos n\pi = (-1)^n$
[*Hint:* Use the identity for $\cos(\alpha + \beta)$.]

21. (*DeMoivre's theorem.*) For any angle θ and any nonnegative number r, and $i^2 = -1$,

$$[r(\cos\theta + i\sin\theta)]^n = r^n(\cos n\theta + i\sin n\theta).$$

22. $|\sin(nx)| \leq n|\sin x|$

In Exercises 23–25, prove using mathematical induction.

23. For every natural number $n \geq 2$,

$$\cos x \cdot \cos 2x \cdot \cos 4x \cdots \cos 2^{n-1}x = \dfrac{\sin 2^n x}{2^n \sin x}.$$

24. For every natural number $n \geq 2$,

$$\log_a(b_1 b_2 \cdots b_n) = \log_a b_1 + \log_a b_2 + \cdots + \log_a b_n.$$

25. For every natural number $n \geq 2$,

$$\left(1 - \dfrac{1}{2^2}\right)\left(1 - \dfrac{1}{3^2}\right) \cdots \left(1 - \dfrac{1}{n^2}\right) = \dfrac{n+1}{2n}.$$

Prove the following for any complex numbers $z_1, \ldots, z_n$, where $i^2 = -1$ and $\bar{z}$ is the conjugate of z (see Section 12.2).

26. $\overline{z^n} = \bar{z}^n$

27. $\overline{z_1 + z_2 + \cdots + z_n} = \bar{z}_1 + \bar{z}_2 + \cdots + \bar{z}_n$

28. $\overline{z_1 \cdot z_2 \cdots z_n} = \bar{z}_1 \cdot \bar{z}_2 \cdots \bar{z}_n$

29. i^n is either 1, −1, i, or −i.

For any integers a and b, b is a factor of a if there exists an integer c such that $a = bc$. Prove the following for any natural number n.

30. 3 is a factor of $n^3 + 2n$.

31. 2 is a factor of $n^2 + n$.

★ ——————————————————————————————————————

32. For every natural number $n \geq 2$,

$$\dfrac{1}{\sqrt{1}} + \dfrac{1}{\sqrt{2}} + \dfrac{1}{\sqrt{3}} + \cdots + \dfrac{1}{\sqrt{n}} > \sqrt{n}.$$

33. (*The Tower of Hanoi problem.*) There are three pegs on a board. On one peg are n disks, each smaller than the one on which it rests. The problem is to move this pile of disks to another peg. The final order must be the same, but you can move only one disk at a time and you can never place a larger disk on a smaller one.

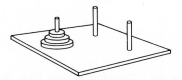

a) What is the *least* number of moves it takes to move 3 disks?

b) What is the *least* number of moves it takes to move 4 disks?

c) What is the *least* number of moves it takes to move 2 disks?

d) What is the *least* number of moves it takes to move 1 disk?

e) Conjecture a formula for the *least* number of moves it takes to move n disks. Prove it by mathematical induction.

CHAPTER 15 REVIEW

[15.2, i]   **1.** Find the 10th term in the arithmetic sequence $\frac{3}{4}, \frac{13}{12}, \frac{17}{12}, \ldots$.

2. Find the 6th term in the arithmetic sequence $a - b, a, a + b, \ldots$.

[15.2, ii]   **3.** Find the sum of the first 18 terms of the arithmetic sequence 4, 7, 10, ...

4. Find the sum of the first 30 positive integers.

[15.2, i]   **5.** The first term of an arithmetic sequence is 5. The 17th term is 53. Find the 3rd term.

6. The common difference in an arithmetic sequence is 3. The 10th term is 23. Find the first term.

[15.3, i]   **7.** For a geometric sequence, $a_1 = -2$, $r = 2$, and $a_n = -64$. Find n and S_n.

8. For a geometric sequence, $r = \frac{1}{2}$, $n = 5$, and $S_n = \frac{31}{2}$. Find a_1 and a_n.

9. Write the first 3 terms of the infinite geometric sequence with $S_\infty = \frac{3}{11}$ and $r = 0.01$.

10. Write the first 3 terms of the infinite geometric sequence with $r = -\frac{1}{3}$ and $S_\infty = \frac{3}{8}$.

[15.4, ii]   **11.** Find fractional notation for $2.\overline{13}$.

[15.2, i]   **12.** Insert four arithmetic means between 5 and 9.

13. Find the geometric mean between 5 and 9.

[15.3, ii]   **14.** A golf ball is dropped from a height of 30 ft to the pavement, and the rebound is $\frac{1}{4}$ the distance it drops. If, after each descent, it continues to rebound $\frac{1}{4}$ the distance dropped, what is the total distance the ball has traveled when it reaches the pavement on its 10th descent?

[15.2, ii]   **15.** You receive 10¢ on the first day of the year, 12¢ on the 2nd day, 14¢ on the 3rd day, and so on. How much will you receive on the 365th day? What is the sum of all these 365 gifts?

[15.3, i]   **16.** The present population of a city is 30,000. Its population is supposed to double every 10 yr. What will its population be at the end of 80 yr?

[15.4, i]   **17.** The sides of a square are each 16 in. long. A second square is inscribed by joining the midpoints of the sides, successively. In the second square we repeat the process, inscribing a third square. If this process is continued indefinitely, what is the sum of the perimeters of all of the squares? (*Hint:* Use an infinite geometric series.)

18. A pendulum is moving back and forth in such a way that it traverses an arc 10 cm in length and thereafter arcs that are 4/7 the length of the previous arc. What is the sum of the arc lengths that the pendulum traverses?

[15.5, ii]   Use mathematical induction.

19. Prove: For all natural numbers n,

$$1 + 4 + 7 + \cdots + (3n - 2) = \frac{n(3n - 1)}{2}.$$

20. Prove: For all natural numbers n,

$$1 + 3 + 3^2 + \cdots + 3^{n-1} = \frac{3^n - 1}{2}.$$

21. Prove: For every natural number $n \geq 2$,

$$\left(1 - \frac{1}{2}\right) \cdot \left(1 - \frac{1}{3}\right) \cdots \left(1 - \frac{1}{n}\right) = \frac{1}{n}.$$

22. Explain why the following cannot be proved by mathematical induction. For every natural number n,

a) $3 + 5 + \cdots + (2n + 1) = (n + 1)^2$;

b) $1 + 3 + \cdots + (2n - 1) = n^2 + 3$.

23. Suppose a and b are geometric sequences. Prove that the sequence c is a geometric sequence where $c_n = a_n b_n$.

24. Suppose a is an arithmetic sequence. Prove that c is a geometric sequence, where $c_n = b^{a_n}$, for some positive number b.

25. Suppose a is an arithmetic sequence. Under what conditions is b an arithmetic sequence where:

   a) $b_n = |a_n|$?   b) $b_n = a_n + 8$?   c) $b_n = 7a_n$?

   d) $b_n = \dfrac{1}{a_n}$?   e) $b_n = \log a_n$?   f) $b_n = a_n^3$?

26. The zeros of this polynomial form an arithmetic sequence. Find them.

$$x^4 - 4x^3 - 4x^2 + 16x$$

[15.1, iii]   27. Find the first four terms of this recursively defined sequence.

$$a_1 = 5, \ a_{k+1} = 2a_k^2 + 1$$

[15.1, v]   28. Write Σ notation for this sequence.

$$0, \ 3, \ 8, \ 15, \ 24, \ 35, \ 48$$

16
COMBINATORICS AND PROBABILITY

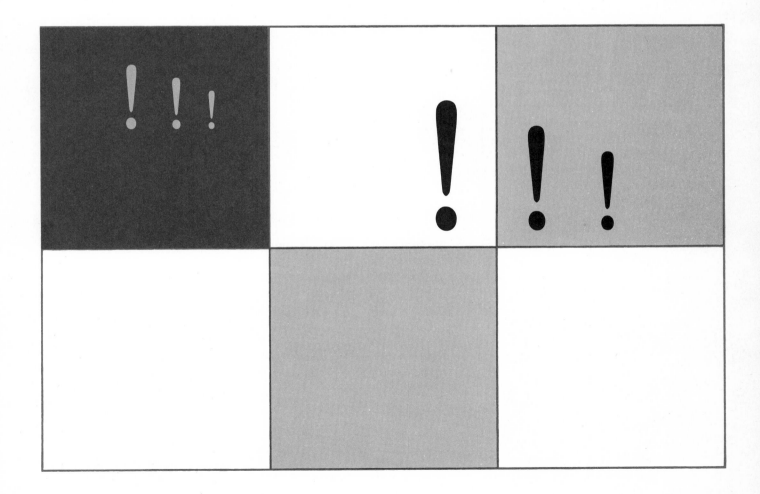

OBJECTIVE

You should be able to:

[i] Evaluate factorial and permutation notation and solve related counting problems.

16.1 PERMUTATIONS

Combinatorics

We shall develop means of determining the number of ways a set of objects can be arranged or combined, certain objects can be chosen, or a succession of events can occur. The study of such things is called *combinatorics*.

Example 1 How many 3-letter code symbols can be formed with the letters A, B, and C *without* repetition?

Consider placing letters in these frames.

We can select any of the 3 letters for the first letter in the symbol. Once this letter has been selected, the second is selected from the remaining 2 letters. Then the third is already determined since there is only 1 letter left. The possibilities can be arrived at with a *tree diagram*.

$$
\begin{array}{c}
A \diagup \begin{array}{c} B-C \\ C-B \end{array} \\[4pt]
B \diagup \begin{array}{c} A-C \\ C-A \end{array} \\[4pt]
C \diagup \begin{array}{c} A-B \\ B-A \end{array}
\end{array}
$$

There are $3 \cdot 2 \cdot 1$ possibilities. The set of all of them is as follows:

$$\{ABC, ACB, BAC, BCA, CAB, CBA\}.$$

Suppose we perform an experiment such as selecting letters (as in the preceding example), flipping a coin, or drawing a card. The results are called *outcomes*. An *event* is a set of outcomes. When several events occur together, we say the event is *compound*. The following principle is concerned with compound events.

Fundamental Counting Principle

In a compound event in which the first event can occur independently in n_1 ways, the second can occur in n_2 ways, and so on, the kth event in n_k ways, the total number of ways the compound event can occur is

$$n_1 \cdot n_2 \cdot n_3 \cdots n_k.$$

Example 2 How many 3-letter code symbols can be formed with the letters A, B, and C *with* repetition?

There are 3 choices for the first letter, and since we allow repetition, 3 choices for the second, and 3 for the third. Thus by the fundamental counting principle there are $3 \cdot 3 \cdot 3$, or 27, choices.

DO EXERCISES 1–3.

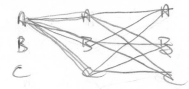

Permutations

DEFINITION

A *permutation* of a set of n objects is an ordered arrangement of all n objects.

Consider, for example, a set of 4 objects

$$\{A, B, C, D\}.$$

To find the number of ordered arrangements of the set, we select a first letter; there are 4 choices. Then we select a second letter; there are 3 choices. Then we select a third letter; there are 2 choices. Finally there is 1 choice for the last selection. Thus by the fundamental counting principle, there are $4 \cdot 3 \cdot 2 \cdot 1$, or 24, permutations of a set of 4 objects.

DO EXERCISE 4.

We can find a formula for the total number of permutations of all objects in a set of n objects. We have n choices for the first selection, $n - 1$ for the second, $n - 2$ for the third, and so on. For the nth selection there is only 1 choice.

THEOREM 1

The total number of permutations of a set of n objects, denoted $_nP_n$, is given by

$$_nP_n = n(n - 1)(n - 2) \cdots (3)(2)(1).$$

Example 3 Find $_4P_4$ and $_7P_7$.

a) $_4P_4 = 4 \cdot 3 \cdot 2 \cdot 1 = 24$

b) $_7P_7 = 7 \cdot 6 \cdot 5 \cdot 4 \cdot 3 \cdot 2 \cdot 1 = 5040$

Example 4 In how many ways can 9 letters be placed in 9 mailboxes, one letter to a box?

$$_9P_9 = 9 \cdot 8 \cdot 7 \cdot 6 \cdot 5 \cdot 4 \cdot 3 \cdot 2 \cdot 1 = 362{,}880$$

DO EXERCISES 5–10.

1. How many 3-digit numbers can be named using all the digits 5, 6, 7 without repetition? with repetition?

2. A man is planning a date. He will first put on one of 4 suits, then call one of 2 girlfriends, and then select one of 5 restaurants. How many possible arrangements can he make for his date (assuming either girl would accept)?

3. In how many ways can 5 different cars be parked in a row in a parking lot?

4. How many permutations are there of a set of 5 objects? Consider a set $\{A, B, C, D, E\}$.

Compute.

5. $_3P_3$

6. $_5P_5$

7. $_6P_6$

8. In how many different ways can 4 horses be lined up for a race?

9. In how many different ways can 6 people line up at a ticket window?

10. In how many ways can the 9-man batting order of a baseball team be made up, if you assume the pitcher bats last?

11. Find 9!.

12. Find 8!.

13. Using factorial notation only, represent the number of permutations of 18 objects.

14. Represent each in the form $n(n-1)!$

a) 10!   b) 20!

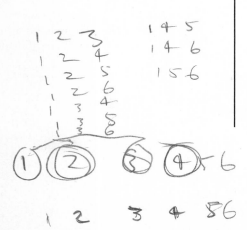

Factorial Notation

Products of successive natural numbers, such as $7 \cdot 6 \cdot 5 \cdot 4 \cdot 3 \cdot 2 \cdot 1$ and $9 \cdot 8 \cdot 7 \cdot 6 \cdot 5 \cdot 4 \cdot 3 \cdot 2 \cdot 1$, are used so often that it is convenient to adopt a notation for them.

For the product $7 \cdot 6 \cdot 5 \cdot 4 \cdot 3 \cdot 2 \cdot 1$ we write 7!, read "7-factorial."

DEFINITION

$$n! = n(n-1)(n-2) \cdots (3)(2)(1).$$

Here are some examples.

$$
\begin{aligned}
7! &= 7 \cdot 6 \cdot 5 \cdot 4 \cdot 3 \cdot 2 \cdot 1 = 5040 \\
6! &= 6 \cdot 5 \cdot 4 \cdot 3 \cdot 2 \cdot 1 = 720 \\
5! &= 5 \cdot 4 \cdot 3 \cdot 2 \cdot 1 = 120 \\
4! &= 4 \cdot 3 \cdot 2 \cdot 1 = 24 \\
3! &= 3 \cdot 2 \cdot 1 = 6 \\
2! &= 2 \cdot 1 = 2 \\
1! &= 1 = 1
\end{aligned}
$$

DO EXERCISES 11 AND 12.

We also define 0! to be 1. We do this so that certain formulas and theorems can be stated concisely.

We can now simplify the formula of Theorem 1 as follows.

$$_nP_n = n!$$

DO EXERCISE 13.

Note that $8! = 8 \cdot 7!$ We can see this as follows: By definition of factorial notation,

$$
\begin{aligned}
8! &= 8 \cdot 7 \cdot 6 \cdot 5 \cdot 4 \cdot 3 \cdot 2 \cdot 1 \\
&= 8 \cdot (7 \cdot 6 \cdot 5 \cdot 4 \cdot 3 \cdot 2 \cdot 1) \\
&= 8 \cdot 7!.
\end{aligned}
$$

Generalizing, we get the following.

For any natural number n, $n! = n(n-1)!$

By using this result repeatedly, we can further manipulate factorial notation.

Example 5 Rewrite 7! with a factor of 5!.

$$7! = 7 \cdot 6 \cdot 5!$$

DO EXERCISE 14.

Permutations of n Objects Taken r at a Time

Consider a set of 6 objects. How many ordered arrangements are there having three members? We can select the first object in 6 ways.

There are then 5 choices for the second and then 4 choices for the third. By the fundamental counting principle, there are then $6 \cdot 5 \cdot 4$ ways to construct the subset. In other words, there are $6 \cdot 5 \cdot 4$ permutations of a set of 6 objects taken three at a time. Note that

$$6 \cdot 5 \cdot 4 = \frac{6 \cdot 5 \cdot 4 \cdot 3 \cdot 2 \cdot 1}{3 \cdot 2 \cdot 1} \quad \text{or} \quad \frac{6!}{3!}.$$

DEFINITION

A *permutation* of a set of n objects taken r at a time is an ordered arrangement of r objects taken from the set.

Consider a set of n objects and the selecting of an ordered arrangement of r objects. The first object can be selected in n ways. The second can be selected in $n - 1$ ways, and so on. The rth can be selected in $n - (r - 1)$ ways. By the fundamental counting principle, the total number of permutations is

$$n(n - 1)(n - 2) \cdots [n - (r - 1)].$$

We now multiply by 1:

$$n(n - 1)(n - 2) \cdots [n - (r - 1)] \frac{(n - r)!}{(n - r)!}$$

$$= \frac{n(n - 1)(n - 2)(n - 3) \cdots [n - (r - 1)](n - r)!}{(n - r)!}.$$

The numerator is now the product of all natural numbers from n to 1, hence is $n!$. Thus the total number of permutations is

$$\frac{n!}{(n - r)!}.$$

This gives us the following theorem.

THEOREM 2

The number of permutations of a set of n objects taken r at a time, denoted $_nP_r$, is given by

$$_nP_r = n(n - 1)(n - 2) \cdots [n - (r - 1)] \qquad (1)$$

$$= \frac{n!}{(n - r)!}. \qquad (2)$$

Formula (1) is most useful in application, but formula (2) will be important in a later development.

Example 6 Compute $_6P_4$ using both formulas of Theorem 2.

Using formula (1), we have

$$_6P_4 = \underbrace{6 \cdot 5 \cdot 4 \cdot 3}_{} \quad \text{Note that the 6 in } _6P_4 \text{ shows where to start and the 4 shows how many factors there are.}$$

$$= 360.$$

15. Compute $_7P_3$.

16. Compute.

a) $_{10}P_4$

b) $_8P_2$

c) $_{11}P_5$

d) $_nP_1$

e) $_nP_2$

17. In how many ways can a 5-man starting unit be selected from a 12-man basketball squad and arranged in a straight line?

18. A professor wants to write a 6-item test from a pool of 10 questions. In how many ways can he do this?

19. How many 7-digit numbers can be named, without repetition, using the digits 2, 3, 4, 5, 6, 7, and 8, if the even digits come first?

Using formula (2) of Theorem 2, we have

$$_6P_4 = \frac{6!}{(6-4)!} = \frac{6!}{2!} = \frac{6 \cdot 5 \cdot 4 \cdot 3 \cdot 2 \cdot 1}{2 \cdot 1} = 6 \cdot 5 \cdot 4 \cdot 3 = 360.$$

DO EXERCISES 15 AND 16.

Example 7 In how many ways can letters of the set {A, B, C, D, E, F, G} be arranged to form code words of (a) 7 letters? (b) 5 letters? (c) 4 letters? (d) 2 letters?

a) $_7P_7 = 7 \cdot 6 \cdot 5 \cdot 4 \cdot 3 \cdot 2 \cdot 1 = 5040$

b) $_7P_5 = 7 \cdot 6 \cdot 5 \cdot 4 \cdot 3 \qquad = 2520$

c) $_7P_4 = 7 \cdot 6 \cdot 5 \cdot 4 \qquad\quad = 840$

d) $_7P_2 = 7 \cdot 6 \qquad\qquad\quad = 42$

Example 8 A baseball manager arranges his batting order as follows: The 4 infielders will bat first, then the outfielders, catcher, and pitcher will follow, not necessarily in that order. How many batting orders are possible?

The infielders can bat in 4! different ways; the rest in 5! different ways. Then by the fundamental counting principle we have $_4P_4 \cdot {_5P_5} = 4! \cdot 5!$, or 2880, possible batting orders.

DO EXERCISES 17–19.

Permutations of Sets with Nondistinguishable Objects

Consider a set of 7 marbles, 4 of which are colored and 3 of which are black. Although the marbles are all different, when they are lined up, one black marble will look just like any other black marble. In this sense we say that the marbles are nondistinguishable.

We know that there are 7! permutations of this set. Many of them will look alike, however. We develop a formula for finding the number of distinguishable permutations.

Consider a set of n objects in which n_1 are of one kind, n_2 are of a second kind, ..., n_k are of the kth kind. By Theorem 1 the total number of permutations of the set is $n!$. Let P be the number of distinguishable permutations. For each of these P permutations there are $n_1!$ actual permutations, obtained by permuting the objects of the first kind. For each of these $P_n!$ permutations there are $n_2!$ actual permutations, obtained by permuting the objects of the second kind. And so on. By the fundamental counting principle, the total number of actual permutations is

$$P \cdot n_1! \cdot n_2! \cdots n_k!.$$

Then we have $P \cdot n_1! \cdot n_2! \cdots n_k! = n!$. Solving for P, we obtain

$$P = \frac{n!}{n_1! n_2! \cdots n_k!}.$$

This proves the following theorem.

THEOREM 3

For a set of n objects in which n_1 are of one kind, n_2 are of another kind, ..., n_k are of a kth kind, the number of distinguishable permutations is

$$\frac{n!}{n_1! \cdot n_2! \cdots n_k!}.$$

Example 9 In how many distinguishable ways can the letters of the word CINCINNATI be arranged?

Note: There are 2 C's, 3 I's, 3 N's, 1 A, and 1 T, for a total of 10. Thus

$$P = \frac{10!}{2! \cdot 3! \cdot 3! \cdot 1! \cdot 1!}, \quad \text{or} \quad 50,400.$$

DO EXERCISES 20–22.

Circular Arrangements

Suppose we arrange the 4 letters A, B, C, and D in a circular arrangement.

Note that the arrangements ABCD, BCDA, CDAB, and DABC are not distinguishable. For each circular arrangement there are 4 distinguishable arrangements on a line. Thus if there are P circular arrangements, these yield $4 \cdot P$ arrangements on a line which we know is 4!. Thus $4 \cdot P = 4!$, so $P = 4!/4$, or 3!. To generalize:

THEOREM 4

The number of distinct (different) circular arrangements of n objects is $(n-1)!$.

Example 10 In how many ways can 9 different foods be arranged around a lazy Susan?

$$(9-1)! = 8! = 8 \cdot 7 \cdot 6 \cdot 5 \cdot 4 \cdot 3 \cdot 2 \cdot 1 = 40,320$$

DO EXERCISES 23 AND 24.

20. In how many distinguishable ways can the letters of the word MISSISSIPPI be arranged?

21. How many 6-digit numbers can be named with all the digits 3, 3, 3, 4, 4, and 5?

22. How many vertical signal flag arrangements can be formed with 3 solid red, 3 solid green, and 2 solid yellow flags?

23. In how many ways can 7 men be arranged around a round table?

24. In how many ways can the numbers on a clock face be arranged?

25. How many 5-letter code symbols can be formed by repeated use of the letters of the alphabet?

Repeated Use of the Same Object

Example 11 How many 5-letter code symbols can be formed with the letters A, B, C, and D if we allow a letter to occur more than once?

We have five spaces:

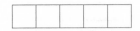

We can select the first letter in 4 ways, the second in 4 ways, and so on. Thus there are 4^5, or 1024, arrangements. Generalizing we have the following.

THEOREM 5

> The number of distinct arrangements of n objects taken r at a time, allowing repetition, is n^r.

DO EXERCISE 25.

EXERCISE SET 16.1

[i] Evaluate.

1. $_4P_3$

2. $_7P_5$

3. $_{10}P_7$

4. $_{10}P_3$

Table 5 at the back of the book can help with many of these problems.

5. How many 5-digit numbers can be named using the digits 5, 6, 7, 8, and 9 without repetition? with repetition?

6. How many 4-digit numbers can be named using the digits 2, 3, 4, and 5 without repetition? with repetition?

7. In how many ways can 5 students be arranged in a straight line? in a circle?

8. In how many ways can 7 athletes be arranged in a straight line? in a circle?

9. In how many distinguishable ways can the letters of the word PEACE be arranged?

10. In how many distinguishable ways can the letters of the word GROOVY be arranged?

11. How many 7-digit phone numbers can be formed with the digits 0, 1, 2, 3, 4, 5, 6, 7, 8, and 9, assuming that no digit is used more than once and the first digit is not 0?

12. A program is planned and is to have 5 rock numbers and 4 speeches. In how many ways can this be done if a rock number and a speech are to alternate and the rock numbers come first?

13. Suppose the expression $a^2b^3c^4$ is rewritten without exponents. In how many ways can this be done?

14. Suppose the expression a^3bc^2 is rewritten without exponents. In how many ways can this be done?

15. In how many ways could King Arthur and his 12 knights sit at his Round Table?

16. In how many ways can 4 people be seated at a bridge table?

17. A penny, nickel, dime, quarter, and half dollar (if you have one) are arranged in a straight line. (a) Considering just the coins, in how many ways can they be lined up? (b) Considering the coins and heads and tails, in how many ways can they be lined up?

18. A penny, nickel, dime, and quarter are arranged in a straight line. (a) Considering just the coins, in how many ways can they be lined up? (b) Considering the coins and heads and tails, in how many ways can they be lined up?

19. ▦ Compute $_{52}P_4$.

20. ▦ Compute $_{50}P_5$.

21. A professor is going to grade her 24 students on a curve. She will give 3 A's, 5 B's, 9 C's, 4 D's, and 3 F's. In how many ways can she do this?

22. A professor is planning to grade his 20 students on a curve. He will give 2 A's, 5 B's, 8 C's, 3 D's, and 2 F's. In how many ways can he do this?

23. ▦ How many distinguishable code symbols can be formed from the letters of the word MATH? BUSINESS? PHILOSOPHICAL?

24. ▦ How many distinguishable code symbols can be formed from the letters of the word ORANGE? BIOLOGY? MATHEMATICS?

25. A state forms its license plates by first listing a number that corresponds to the county in which the car owner lives (the names of the counties are alphabetized and the number is its location in the order). Then the plate lists a letter of the alphabet and this is followed by a number from 1 to 9999. How many such plates are possible if there are 80 counties?

```
29 B 7480
   INDIANA
```

26. How many code symbols can be formed using 4 out of 5 letters of A, B, C, D, E if the letters

a) are not repeated?

b) can be repeated?

c) are not repeated but must begin with D?

d) are not repeated but must end with DE?

☆

Solve for n.

27. $_nP_5 = 7 \cdot _nP_4$

28. $_nP_4 = 8 \cdot _{n-1}P_3$

29. $_nP_5 = 9 \cdot _{n-1}P_4$

30. $_nP_4 = 8 \cdot _nP_3$

★

---

31. In a single-elimination sports tournament consisting of n teams, a team is eliminated when it loses one game. How many games are required to complete the tournament?

32. In a double-elimination softball tournament consisting of n teams, a team is eliminated when it loses two games. At most how many games are required to complete the tournament?

33. Find a formula for the total number of arrangements of n keys on a key ring.

16.2 COMBINATIONS

[i] Permutations of a set are ordered arrangements. *Combinations* of a set are simply subsets and are not considered with regard to order.

Example 1 How many combinations are there of the set {A, B, C, D} taken 3 at a time?

The combinations are the subsets {A, B, C}, {A, C, D}, {B, C, D}, {A, B, D}. There are four of them.

DEFINITION

> The number of combinations taken r at a time from a set of n objects, denoted $_nC_r$, is the number of subsets containing exactly r objects.

DO EXERCISE 1.

We develop a formula for finding $_nC_r$, but we first define some convenient symbolism.

DEFINITION

$$\binom{n}{r} \text{ is defined to mean } \frac{n!}{r!(n-r)!}.$$

The notation $\binom{n}{r}$ is read "n over r," or (for reasons we see later) "binomial coefficient n over r."

OBJECTIVE

You should be able to:

[i] Evaluate combination notation and solve related problems.

1. Consider the set

$$\{A, B, C, D\}.$$

How many combinations are there taken

a) 2 at a time?

b) 4 at a time?

c) 1 at a time?

d) 0 at a time?

2. Consider the set of 5 objects

$$\{A, B, C, D, E\}.$$

a) Calculate the number of permutations of this set taken 3 at a time.

b) List these permutations.

c) Calculate the number of combinations of this set taken 3 at a time.

d) List these combinations.

3. Compute.

a) $\binom{10}{8}$ b) $\binom{10}{2}$

c) $_7C_5$ d) $_7C_7$

4. Compute $\binom{9}{7}$ and $\binom{9}{2}$.

> **CAUTION!**
>
> $\binom{n}{r}$ does *not* mean $n \div r$, or $\dfrac{n}{r}$.

Example 2 Evaluate $\binom{5}{2}$.

$$\binom{5}{2} = \frac{5!}{2!(5-2)!} = \frac{5!}{2!3!} = \frac{5 \cdot 4 \cdot 3!}{2!3!} = \frac{5 \cdot 4}{2} = 10$$

We now find a formula for $_nC_r$. Consider a set of n objects. The number of subsets with r members is $_nC_r$. Each subset with r members can be arranged in $r!$ ways. By the fundamental counting principle we have $_nC_r \cdot r!$ ordered subsets with r members. This gives us

$$_nP_r = {_nC_r} \cdot r!.$$

Then

$$_nC_r = \frac{_nP_r}{r!}.$$

Since

$$_nP_r = \frac{n!}{(n-r)!}, \qquad \text{Theorem 2}$$

we get

$$_nC_r = \frac{n!}{r!(n-r)!},$$

or

$$_nC_r = \binom{n}{r}.$$

This proves the following.

THEOREM 6

The number of combinations of a set of n objects taken r at a time is given by

$$_nC_r = \binom{n}{r} = \frac{n!}{r!(n-r)!}.$$

Example 3 Compute $_7C_4$.

By Theorem 6 we have $_7C_4 = \binom{7}{4}$.

$$\binom{7}{4} = \frac{7!}{4!3!} = \frac{7 \cdot 6 \cdot 5 \cdot 4!}{3! \cdot 4!} = \frac{7 \cdot 6 \cdot 5}{3 \cdot 2} = 35$$

DO EXERCISES 2–4.

Example 4 How many committees can be formed from a set of 5 governors and 7 senators if each committee contains 3 governors and 4 senators?

The 3 governors can be selected in $_5C_3$ ways and the 4 senators can be selected in $_7C_4$ ways. Using the fundamental counting principle, it follows that the number of possible committees is

$$_5C_3 \cdot {_7C_4} = 10 \cdot 35 = 350.$$

DO EXERCISES 5 AND 6.

5. In how many ways can a 5-man starting unit be selected from a 12-man basketball squad?

6. A committee is to be chosen from 12 men and 8 women and is to consist of 3 men and 2 women. How many committees can be formed?

EXERCISE SET 16.2

[i] Compute. Table 5 at the back of the book or a calculator can be of help with these problems.

1. $_9C_5$

2. $_{14}C_2$

3. $\binom{50}{2}$

4. $\binom{40}{3}$

5. $_nC_3$

6. $_nC_2$

7. There are 23 students in a fraternity. How many sets of 4 officers can be selected?

8. How many basketball games can be played in a 9-team league if each team plays all other teams once? twice?

9. On a test a student is to select 6 out of 10 questions. In how many ways can he do this?

10. On a test a student is to select 7 out of 11 questions. In how many ways can she do this?

11. How many lines are determined by 8 points, no 3 of which are collinear? How many triangles are determined by the same points if no 4 are coplanar?

12. How many lines are determined by 7 points, no 3 of which are collinear? How many triangles are determined by the same points if no 4 are coplanar?

13. Of the first 10 questions on a test, a student must answer 7. On the second 5 questions, she must answer 3. In how many ways can this be done?

14. Of the first 8 questions on a test, a student must answer 6. On the second 4 questions, he must answer 3. In how many ways can this be done?

15. Suppose the Senate of the United States consisted of 58 Democrats and 42 Republicans. How many committees consisting of 6 Democrats and 4 Republicans could be formed? You do not need to simplify the expression.

16. Suppose the Senate of the United States consisted of 63 Republicans and 37 Democrats. How many committees consisting of 8 Republicans and 12 Democrats could be formed? You need not simplify the expression.

17. How many 5-card poker hands consisting of 3 aces and 2 cards that are not aces are possible with a 52-card deck? See Section 16.6 for a description of a 52-card deck.

18. How many 5-card poker hands consisting of 2 kings and 3 cards that are not kings are possible with a 52-card deck?

☆ ——————————————————————————————

19. How many 5-card poker hands are possible with a 52-card deck?

20. ▦ How many 13-card bridge hands are possible with a 52-card deck?

21. There are 8 points on a circle. How many triangles can be inscribed with these points as vertices?

22. There are n points on a circle. How many quadrilaterals can be inscribed with these points as vertices?

23. A set of 5 parallel lines crosses another set of 8 parallel lines at angles that are not right angles. How many parallelograms are formed?

24. Prove: For any natural numbers n and $r \leq n$,

$$\binom{n}{r} = \binom{n}{n-r}.$$

Solve for n.

25. $\binom{n+1}{3} = 2 \cdot \binom{n}{2}$

26. $\binom{n}{n-2} = 6$

27. $\binom{n+2}{4} = 6 \cdot \binom{n}{2}$

28. $\binom{n}{3} = 2 \cdot \binom{n-1}{2}$

★ _____

29. How many line segments are determined by the 5 vertices of a pentagon? Of these, how many are diagonals?

30. How many line segments are determined by the 6 vertices of a hexagon? Of these, how many are diagonals?

31. How many line segments are determined by the n vertices of an n-gon? Of these, how many are diagonals? Use mathematical induction to prove the result for the diagonals.

32. Prove: For any natural numbers n and $r \leq n$,

$$\binom{n}{r-1} + \binom{n}{r} = \binom{n+1}{r}.$$

OBJECTIVES

You should be able to:

[i] Expand a binomial power and find the rth term of such an expansion.

[ii] Find the total number of subsets of a set of n objects.

16.3 BINOMIAL SERIES

[i] Expanding Powers of Binomials

Consider the following powers of $a + b$, where $a + b$ is any binomial. Look for patterns.

$$
\begin{aligned}
(a + b)^0 &= 1 \\
(a + b)^1 &= a + b \\
(a + b)^2 &= a^2 + 2ab + b^2 \\
(a + b)^3 &= a^3 + 3a^2b + 3ab^2 + b^3 \\
(a + b)^4 &= a^4 + 4a^3b + 6a^2b^2 + 4ab^3 + b^4 \\
(a + b)^5 &= a^5 + 5a^4b + 10a^3b^2 + 10a^2b^3 + 5ab^4 + b^5
\end{aligned}
$$

Each expansion is a polynomial that can be regarded as a series. There are some patterns to be noted in the expansions.

1. In each term, the sum of the exponents is n.

2. The exponents of a start with n and decrease, to 0. The last term has no factor of a. The first term has no factor of b. The exponents of b start in the second term with 1 and increase to n.

3. There is one more term than the degree of the polynomial. The expansion of $(a + b)^n$ has $n + 1$ terms.

We now find a way to determine the coefficients. Let us consider the nth power of a binomial $(a + b)$:

$$(a + b)^n = \underbrace{(a + b)(a + b)(a + b) \cdots (a + b)}_{n \text{ factors}}.$$

When we multiply, we will find all possible products of a's and b's. For example, when we multiply all the first terms we will get n factors of a, or a^n. Thus the first term in the expansion is a^n, and the first coefficient is 1. Similarly, the coefficient of the last term is 1.

To get a term such as the $a^{n-r}b^r$ term, we will take a's from $n - r$ factors and b's from r factors. Thus we take n objects, $n - r$ of them a's and r of them b's. The number of ways in which we can do this is

$$\frac{n!}{(n - r)!r!}.$$

This is $\binom{n}{r}$. Thus the $(r + 1)$st term in the expansion is $\binom{n}{r} a^{n-r}b^r$. We now have a theorem.

1. Expand $(x^2 - 1)^5$.

THEOREM 7

The binomial theorem. For any binomial $(a + b)$ and any natural number n,

$$(a + b)^n = \binom{n}{0}a^n + \binom{n}{1}a^{n-1}b + \binom{n}{2}a^{n-2}b^2 + \cdots + \binom{n}{n}b^n.$$

The binomial theorem can be proved by mathematical induction but we shall not do that here.

Sigma notation for a binomial series is as follows:

$$(a + b)^n = \sum_{r=0}^{n} \binom{n}{r} a^{n-r}b^r.$$

Because of Theorem 7, $\binom{n}{r}$ is called a *binomial coefficient*. It should now be apparent why 0! is defined to be 1. In the binomial expansion we want $\binom{n}{0}$ to equal 1 and we also want the definition

$$\binom{n}{r} = \frac{n!}{(n - r)!r!}$$

to hold for all whole numbers n and r. Thus we must have

$$\binom{n}{0} = \frac{n!}{(n - 0)!0!}$$
$$= \frac{n!}{n!0!} = 1.$$

This will be satisfied if 0! is defined to be 1.

Example 1 Expand $(x^2 - 2y)^5$.

Note that $a = x^2$, $b = -2y$, and $n = 5$. Then using the binomial theorem we have

$$(x^2 - 2y)^5 = \binom{5}{0}(x^2)^5 + \binom{5}{1}(x^2)^4(-2y) + \binom{5}{2}(x^2)^3(-2y)^2$$
$$+ \binom{5}{3}(x^2)^2(-2y)^3 + \binom{5}{4}x^2(-2y)^4 + \binom{5}{5}(-2y)^5$$
$$= \frac{5!}{0!5!}x^{10} + \frac{5!}{1!4!}x^8(-2y) + \frac{5!}{2!3!}x^6(-2y)^2$$
$$+ \frac{5!}{3!2!}x^4(-2y)^3 + \frac{5!}{4!1!}x^2(-2y)^4 + \frac{5!}{5!0!}(-2y)^5$$
$$= x^{10} - 10x^8y + 40x^6y^2 - 80x^4y^3 + 80x^2y^4 - 32y^5.$$

DO EXERCISE 1.

2. Expand $\left(2x + \dfrac{1}{y}\right)^4$.

3. Expand $(x - \sqrt{2})^6$.

4. Find the 4th term of $(x - 3)^8$.

5. Find the 5th term of $(x - 3)^8$.

6. Find the 6th term of $(y + 2)^{10}$.

7. Find the 1st term of $(3x + 5)^5$.

Example 2 Expand $(2/x + 3\sqrt{x})^4$.

Note that $a = 2/x$, $b = 3\sqrt{x}$, and $n = 4$. Then using the binomial theorem, we have

$$\left(\frac{2}{x} + 3\sqrt{x}\right)^4 = \binom{4}{0}\cdot\left(\frac{2}{x}\right)^4 + \binom{4}{1}\cdot\left(\frac{2}{x}\right)^3(3\sqrt{x}) + \binom{4}{2}\cdot\left(\frac{2}{x}\right)^2(3\sqrt{x})^2$$

$$+ \binom{4}{3}\left(\frac{2}{x}\right)(3\sqrt{x})^3 + \binom{4}{4}(3\sqrt{x})^4$$

$$= \frac{4!}{0!4!}\cdot\frac{16}{x^4} + \frac{4!}{1!3!}\cdot\frac{8}{x^3}3\sqrt{x} + \frac{4!}{2!2!}\cdot\frac{4}{x^2}\cdot9x$$

$$+ \frac{4!}{3!1!}\cdot\frac{2}{x}\cdot27x^{3/2} + \frac{4!}{4!0!}\cdot81x^2$$

$$= \frac{16}{x^4} + \frac{96}{x^{5/2}} + \frac{216}{x} + 216\sqrt{x} + 81x^2.$$

DO EXERCISES 2 AND 3.

Example 3 Find the 7th term of $(4x - y^3)^9$.

We let $r = 6$, $n = 9$, $a = 4x$, and $b = -y^3$ in the formula $\binom{n}{r}a^{n-r}b^r$. Then

$$\binom{9}{6}(4x)^3(-y^3)^6 = \frac{9!}{6!3!}(4x)^3(-y^3)^6$$

$$= \frac{9\cdot8\cdot7\cdot6!}{3!\cdot6!}64x^3y^{18}$$

$$= 5376x^3y^{18}.$$

DO EXERCISES 4–7.

Subsets

Suppose a set has n objects. The number of subsets containing r members is $\binom{n}{r}$, by Theorem 6. The total number of subsets of a set is the number with 0 elements, plus the number with 1 element, plus the number with 2 elements, and so on. The total number of subsets of a set with n members is

$$\binom{n}{0} + \binom{n}{1} + \binom{n}{2} + \cdots + \binom{n}{n}.$$

Now let us expand $(1 + 1)^n$:

$$(1 + 1)^n = \binom{n}{0} + \binom{n}{1} + \binom{n}{2} + \cdots + \binom{n}{n}.$$

Thus the total number of subsets is $(1 + 1)^n$ or 2^n. We have proved the following theorem.

THEOREM 8

The total number of subsets of a set with n members is 2^n.

Example 4 The set {A, B, C, D, E} has how many subsets?

The set has 5 members, so the number of subsets is 2^5, or 32.

Example 5 A fast-food restaurant can make hamburgers 256 ways using combinations of 8 seasonings. Show why.

The total number of combinations is

$$\binom{8}{0} + \binom{8}{1} + \cdots + \binom{8}{8} = 2^8$$
$$= 256.$$

DO EXERCISES 8 AND 9.

8. How many subsets are there in the set {a, b, c, d, e, f}?

9. How many subsets are there of the set of all states of the United States?

EXERCISE SET 16.3

[i] Find the indicated term of the binomial expression.

1. 3rd, $(a + b)^6$

2. 6th, $(x + y)^7$

3. 12th, $(a - 2)^{14}$

4. 11th, $(x - 3)^{12}$

5. 5th, $(2x^3 - \sqrt{y})^8$

6. 4th, $\left(\frac{1}{b^2} + \frac{b}{3}\right)^7$

7. Middle, $(2u - 3v^2)^{10}$

8. Middle two, $(\sqrt{x} + \sqrt{3})^5$

Expand.

9. $(m + n)^5$

10. $(a - b)^4$

11. $(x^2 - 3y)^5$

12. $(3c - d)^6$

13. $(x^{-2} + x^2)^4$

14. $\left(\frac{1}{\sqrt{x}} - \sqrt{x}\right)^6$

15. $(1 - 1)^n$

16. $(1 + 3)^n$

17. $(\sqrt{2} + 1)^6 - (\sqrt{2} - 1)^6$

18. $(1 - \sqrt{2})^4 + (1 + \sqrt{2})^4$

19. $(\sqrt{3} - t)^4$

20. $(\sqrt{5} + t)^6$

[ii] Determine the number of subsets:

21. Of a set of 7 members.

22. Of a set of 6 members.

23. ▦ Of the set of letters of the English alphabet, which contains 26 letters.

24. ▦ Of the set of letters of the Greek alphabet, which contains 24 letters.

☆

Expand.

25. $(\sqrt{2} - i)^4$, where $i^2 = -1$

26. $(1 + i)^6$, where $i^2 = -1$

27. $(\sin t - \csc t)^7$

28. $(\tan \theta + \cot \theta)^{11}$

29. Find a formula for $(a - b)^n$. Use sigma notation.

30. Expand and simplify $\frac{(x + h)^n - x^n}{h}$. Use sigma notation.

Solve for x.

31. $\sum_{r=0}^{8} \binom{8}{r} x^{8-r} 3^r = 0$

32. $\sum_{r=0}^{4} \binom{4}{r} 5^{4-r} x^r = 64$

33. $\displaystyle\sum_{r=0}^{5}\binom{5}{r}(-1)^r x^{5-r} 3^r = 32$

34. $\displaystyle\sum_{r=0}^{9}\binom{9}{r}\sin^r x = 0$

35. A money clip contains one each of the following bills: \$1, \$2, \$5, \$10, \$20, \$50, and \$100. How many different sums of money can be formed using the bills?

★ ───

Simplify.

36. $\displaystyle\sum_{r=0}^{23}\binom{23}{r}(\log_a x)^{23-r}(\log_a t)^r$

37. $\ln\left[\displaystyle\sum_{r=0}^{15}(\cos t)^{30-2r}(\sin t)^{2r}\right]$

38. Use mathematical induction and the property $\binom{k}{r-1} + \binom{k}{r} = \binom{k+1}{r}$ to prove the binomial theorem.

───

OBJECTIVE

You should be able to:

[i] Compute the probability of a simple event.

16.4 PROBABILITY

We say that when a coin is tossed the chances that it will fall heads are 1 out of 2, or that the *probability* that it will fall heads is $\frac{1}{2}$. Of course this does not mean that if a coin is tossed ten times it will necessarily fall heads exactly five times. If the coin is tossed a great number of times, however, it will fall heads very nearly half of them.

Experimental and Theoretical Probability

If we toss a coin a large number of times, say 1000, and count the number of heads, we can determine the probability. If there are 503 heads, we would calculate the probability to be

$$\frac{503}{1000} \quad \text{or} \quad 0.503.$$

This is an *experimental* determination of probability.

If we consider a coin and reason that it is just as likely to fall heads as tails we would calculate the probability to be $\frac{1}{2}$. This is a *theoretical* determination of probability. You might ask "What is the *true* probability?" In fact there is none. Experimentally we can determine probabilities within certain limits. These may or may not agree with what we obtain theoretically.

[i] Computing Theoretical Probabilities

We need some terminology before we can continue. Suppose we perform an experiment such as flipping a coin, throwing a dart, drawing a card from a deck, or checking an item off an assembly line for quality. The results of an experiment are called *outcomes*. The set of all possible outcomes is called the *sample space*. An *event* is a set of outcomes, that is, a subset of the sample space. For example, for the experiment "throwing a dart," suppose the dartboard is as follows.

Then one event is

{hitting red},   (An event)

which is a subset of the sample space

{red, white, gray},   (Sample space)

assuming the dart must hit the target somewhere.

We denote the probability that an event *E* occurs *P*(*E*). For example, "getting a head" may be denoted by H. Then *P*(H) represents the probability of getting a head. When the outcomes of an experiment all have the same probability of occurring, we say that they are *equally likely*. To see the distinction between events that are equally likely and those that are not, consider the dartboards.

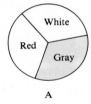
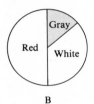

A   B

For dartboard A, the events *hitting red, hitting white,* and *hitting gray* are equally likely, but for board B they are not. A sample space that can be expressed as a union of equally likely events can allow us to calculate probabilities of other events.

Principle P

If an event E can occur *m* ways out of *n* possible equally likely outcomes of a sample space *S*, the probability of that event is given by

$$P(\text{E}) = \frac{m}{n}.$$

A die (pl., dice) is a cube, with six faces, each containing a number of dots from 1 to 6.

Example 1 What is the probability of rolling a 3 on a die?

On a fair die there are 6 equally likely outcomes and there is 1 way to get a 3. By Principle *P*, $P(3) = \frac{1}{6}$.

1. What is the probability of rolling a prime number on a die?

Example 2 What is the probability of rolling an even number on a die?

The event is getting an *even* number. It can occur in 3 ways (getting 2, 4, or 6). The number of equally likely outcomes is 6. By Principle P, $P(\text{even}) = \frac{3}{6}$ or $\frac{1}{2}$.

DO EXERCISE 1.

We use a number of examples related to a standard bridge deck of 52 cards. Such a deck is made up as shown in the following figure.

A DECK OF 52 CARDS:

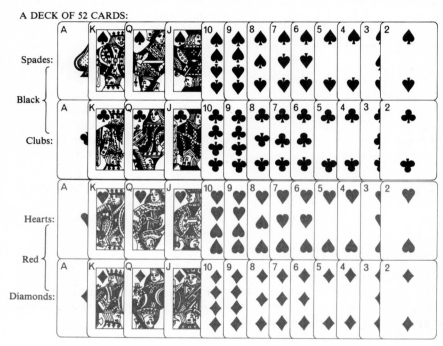

2. Suppose we draw a card from a well-shuffled deck of 52 cards.

   a) What is the probability of drawing a king?

   b) What is the probability of drawing a spade?

   c) What is the probability of drawing a black card?

   d) What is the probability of drawing a jack or a queen?

Example 3 What is the probability of drawing an ace from a well-shuffled deck of 52 cards?

Since there are 52 outcomes (cards in the deck) and they are equally likely (from a well-shuffled deck) and there are 4 ways to obtain an ace, by Principle P we have

$$P(\text{drawing an ace}) = \frac{4}{52}, \quad \text{or} \quad \frac{1}{13}.$$

3. Suppose we select, without looking, one marble from a bag containing 5 red marbles and 6 green marbles. What is the probability of selecting a green marble?

Example 4 Suppose we select, without looking, one marble from a bag containing 3 red marbles and 4 green marbles. What is the probability of selecting a red marble?

There are 7 equally likely ways of selecting any marble, and since the number of ways of getting a red marble is 3,

$$P(\text{selecting a red marble}) = \frac{3}{7}.$$

DO EXERCISES 2 AND 3.

The following are some results that follow from Principle P.

THEOREM 9

If an event E cannot occur, then $P(E) = 0$.

For example in coin tossing, the event that a coin would land on its edge has probability 0.

THEOREM 10

If an event E is certain to occur (every trial is a success), then $P(E) = 1$.

For example in coin tossing, the event that a coin falls either heads or tails has probability 1.
In general,

THEOREM 11

The probability that an event E will occur is a number from 0 to 1:

$$0 \le P(E) \le 1.$$

DO EXERCISES 4 AND 5.

Example 5 Suppose 2 cards are drawn from a well-shuffled deck of 52 cards. What is the probability that both of them are spades?

The number n of ways of drawing 2 cards from a deck of 52 is $_{52}C_2$. Now 13 of the 52 cards are spades, so the number m of ways of drawing 2 spades is $_{13}C_2$. Thus,

$$P(\text{getting 2 spades}) = \frac{m}{n} = \frac{_{13}C_2}{_{52}C_2} = \frac{78}{1326} = \frac{1}{17}.$$

Example 6 Suppose 2 people are selected at random from a group that consists of 6 men and 4 women. What is the probability that both of them are women?

The number of ways of selecting 2 people from a group of 10 is $_{10}C_2$. The number of ways of selecting 2 women from a group of 4 is $_4C_2$. Thus the probability of selecting 2 women from the group of 10 is P, where

$$P = \frac{_4C_2}{_{10}C_2} = \frac{6}{45} = \frac{2}{15}.$$

Example 7 Suppose 3 people are selected at random from a group that consists of 6 men and 4 women. What is the probability that 1 man and 2 women are selected?

The number of ways of selecting 3 people from a group of 10 is $_{10}C_3$. One man can be selected in $_6C_1$ ways, and 2 women can be selected in

4. On a single roll of a die, what is the probability of getting a 7?

5. On a single roll of a die, what is the probability of getting a 1, 2, 3, 4, 5, or 6?

6. Suppose 3 cards are drawn from a well-shuffled deck of 52 cards. What is the probability that all three of them are spades?

7. Suppose 2 people are selected at random from a group that consists of 8 men and 6 women. What is the probability that both of them are women?

8. Suppose 3 people are selected at random from a group that consists of 8 men and 6 women. What is the probability that 2 men and 1 woman are selected?

9. What is the probability of getting a total of 7 on a roll of a pair of dice?

$_4C_2$ ways. By the fundamental counting principle the number of ways of selecting 1 man and 2 women is $_6C_1 \cdot {_4C_2}$. Thus the probability is

$$\frac{_6C_1 \cdot {_4C_2}}{_{10}C_3}, \quad \text{or} \quad \frac{3}{10}.$$

DO EXERCISES 6–8.

Example 8 What is the probability of getting a total of 8 on a roll of a pair of dice? (Assume the dice are different, say one red and one black.)

On each die there are 6 possible outcomes. The outcomes are paired so there are $6 \cdot 6$, or 36, possible ways in which the two can fall.

Red die

6	(1, 6)	(2, 6)	(3, 6)	(4, 6)	(5, 6)	(6, 6)
5	(1, 5)	(2, 5)	(3, 5)	(4, 5)	(5, 5)	(6, 5)
4	(1, 4)	(2, 4)	(3, 4)	(4, 4)	(5, 4)	(6, 4)
3	(1, 3)	(2, 3)	(3, 3)	(4, 3)	(5, 3)	(6, 3)
2	(1, 2)	(2, 2)	(3, 2)	(4, 2)	(5, 2)	(6, 2)
1	(1, 1)	(2, 1)	(3, 1)	(4, 1)	(5, 1)	(6, 1)

$\qquad$ 1 $\qquad$ 2 $\qquad$ 3 $\qquad$ 4 $\qquad$ 5 $\qquad$ 6 Black die

The pairs that total 8 are as shown. Thus there are 5 possible ways of getting a total of 8, so the probability is $\frac{5}{36}$.

DO EXERCISE 9.

Origin and Use of Probability

A desire to calculate odds in games of chance gave rise to the theory of probability. Today the theory of probability and its closely related field, mathematical statistics, have many applications, most of them not related to games of chance. Opinion polls, with such uses as predicting elections, are a familiar example. Quality control, in which a prediction about the percentage of faulty items manufactured is made without testing them all, is an important application, among many, in business. Still other applications are in the areas of genetics, medicine, and the kinetic theory of gases.

EXERCISE SET 16.4

[i] Suppose we draw a card from a well-shuffled deck of 52 cards.

1. How many equally likely outcomes are there?

2. What is the probability drawing a queen?

3. What is the probability of drawing a heart?

4. What is the probability of drawing a club?

5. What is the probability of drawing a 4?

6. What is the probability of drawing a red card?

7. What is the probability of drawing a black card?

9. What is the probability of drawing a 9 or a king?

8. What is the probability of drawing an ace or a deuce?

Suppose we select, without looking, one marble from a bag containing 4 red marbles and 10 green marbles.

10. What is the probability of selecting a red marble?

12. What is the probability of selecting a purple marble?

11. What is the probability of selecting a green marble?

13. What is the probability of selecting a white marble?

Suppose 4 cards are drawn from a well-shuffled deck of 52 cards.

14. What is the probability that all 4 are spades?

15. What is the probability that all 4 are hearts?

16. If 4 marbles are drawn at random all at once from a bag containing 8 white marbles and 6 black marbles, what is the probability that 2 will be white and 2 will be black?

17. From a group of 8 men and 7 women, a committee of 4 is chosen. What is the probability that 2 men and 2 women will be chosen?

18. What is the probability of getting a total of 6 on a roll of a pair of dice?

19. What is the probability of getting a total of 3 on a roll of a pair of dice?

20. What is the probability of getting snake eyes (a total of 2) on a roll of a pair of dice?

21. What is the probability of getting box-cars (a total of 12) on a roll of a pair of dice?

22. From a bag containing 5 nickels, 8 dimes, and 7 quarters, 5 coins are drawn at random all at once. What is the probability of getting 2 nickels, 2 dimes, and 1 quarter?

23. From a bag containing 6 nickels, 10 dimes, and 4 quarters, 6 coins are drawn at random all at once. What is the probability of getting 3 nickels, 2 dimes, and 1 quarter?

Roulette   A roulette wheel contains slots numbered 00, 0, 1, 2, 3, . . . , 35, 36. Eighteen of the slots numbered 1 through 36 are colored red and eighteen are colored black. The 00 and 0 slots are uncolored. The wheel is spun, and a ball is rolled around the rim until it falls into a slot. What is the probability that the ball falls in

24. a black slot?

25. a red slot?

26. a red or black slot?

27. the 00 slot?

28. the 0 slot?

29. either the 00 or 0 slot? (Here the house always wins.)

30. an odd-numbered slot?

☆ ──

Five-card poker hands and probabilities   In part (a) of each problem, give a reasoned expression as well as the answer. Read all the problems before beginning.

31. ▦ How many 5-card poker hands can be dealt from a standard 52-card deck?

32. ▦ A *royal flush* consists of a 5-card hand with A-K-Q-J-10 of the same suit.

  a) How many royal flushes are there?

  b) What is the probability of getting a royal flush?

33. ▦ A *straight flush* consists of five cards in sequence in the same suit, but excludes royal flushes. An ace can be used either high, after a king, or low, before a two.

  a) How many straight flushes are there?

  b) What is the probability of getting a straight flush?

34. ▦ *Four of a kind* is a 5-card hand in which 4 of the cards are of the same denomination, such as J-J-J-J-6, 7-7-7-7-A, or 2-2-2-2-5.

  a) How many are there?

  b) What is the probability of getting four of a kind?

★ ──

35. ▦ A *full house* consists of a pair and three of a kind, such as Q-Q-Q-4-4.

  a) How many are there?

  b) What is the probability of getting a full house?

36. ▦ A *pair* is a 5-card hand in which just 2 of the cards are the same denomination, such as Q-Q-8-A-3.

  a) How many are there?

  b) What is the probability of getting a pair?

37. ▦ *Three of a kind* consists of a 5-card hand in which 3 of the cards are the same denomination, such as Q-Q-Q-10-7.

a) How many are there?

b) What is the probability of getting three of a kind?

39. ▦ *Two pairs* is a hand such as Q-Q-3-3-A.

a) How many are there?

b) What is the probability of getting two pairs?

38. ▦ A *flush* is a 5-card hand in which all the cards are the same suit, but not all in sequence (not a straight flush or royal flush).

a) How many are there?

b) What is the probability of getting a flush?

40. ▦ A *straight* is any five cards in sequence, but not in the same suit—for example, 4 of spades, 5 of spades, 6 of diamonds, 7 of hearts, and 8 of clubs.

a) How many are there?

b) What is the probability of getting a straight?

CHAPTER 16 REVIEW

[16.1, i] **1.** In how many ways can 6 books be arranged on a shelf?

2. If 9 different signal flags are available, how many different displays are possible using 4 flags in a row?

[16.2, i] **3.** The winner of a contest can choose any 8 of 15 prizes. How many different selections can be made?

[16.1, i] **4.** The Greek alphabet contains 24 letters. How many fraternity or sorority names can be formed using 3 different letters?

5. In how many distinguishable ways can the letters of the word TENNESSEE be arranged?

6. A manufacturer of houses has one floor plan but achieves variety by having 3 different colored roofs, 4 different ways of attaching the garage, and 3 different types of entrance. Find the number of different houses that can be produced.

7. How many code symbols can be formed using 5 out of 6 of the letters of G, H, I, J, K, L if the letters

a) are not repeated?   b) can be repeated?

c) are not repeated but must begin with K?   d) are not repeated but must end with IGH?

8. In how many different ways can 7 people be seated at a round table?

[16.3, i] **9.** Find the 4th term of $(a + x)^{12}$.

10. Find the 12th term of $(a + x)^{18}$. Do not multiply out the factorials.

Expand.

11. $(m + n)^7$

12. $(x^2 + 3y)^4$

13. $(5i + 1)^6$, where $i^2 = -1$

14. $(\cos t + \sec t)^8$

[16.4, i] **15.** What is the probability of rolling a 10 on a roll of a pair of dice? on a roll of one die?

16. From a deck of 52 cards 1 card is drawn. What is the probability that it is a club?

17. From a deck of 52 cards 3 are drawn at random without replacement. What is the probability that 2 are aces and 1 is a king?

Simplify.

18. $\displaystyle\sum_{r=0}^{10} (-1)^r (\log x)^{10-r} (\log y)^r$

Solve for n.

19. $\dbinom{n}{n-1} = 36$

20. $26 \cdot \dbinom{n}{1} = \dbinom{n}{3}$

TABLES

Table 1 Powers, Roots, and Reciprocals

n	n^2	n^3	$\sqrt{n}$	$\sqrt[3]{n}$	$\sqrt{10n}$	$\frac{1}{n}$	n	n^2	n^3	$\sqrt{n}$	$\sqrt[3]{n}$	$\sqrt{10n}$	$\frac{1}{n}$
1	1	1	1.000	1.000	3.162	1.0000	51	2,601	132,651	7.141	3.708	22.583	.0196
2	4	8	1.414	1.260	4.472	.5000	52	2,704	140,608	7.211	3.733	22.804	.0192
3	9	27	1.732	1.442	5.477	.3333	53	2,809	148,877	7.280	3.756	23.022	.0189
4	16	64	2.000	1.587	6.325	.2500	54	2,916	157,464	7.348	3.780	23.238	.0185
5	25	125	2.236	1.710	7.071	.2000	55	3,025	166,375	7.416	3.803	23.452	.0182
6	36	216	2.449	1.817	7.746	.1667	56	3,136	175,616	7.483	3.826	23.664	.0179
7	49	343	2.646	1.913	8.367	.1429	57	3,249	185,193	7.550	3.849	23.875	.0175
8	64	512	2.828	2.000	8.944	.1250	58	3,364	195,112	7.616	3.871	24.083	.0172
9	81	729	3.000	2.080	9.487	.1111	59	3,481	205,379	7.681	3.893	24.290	.0169
10	100	1,000	3.162	2.154	10.000	.1000	60	3,600	216,000	7.746	3.915	24.495	.0167
11	121	1,331	3.317	2.224	10.488	.0909	61	3,721	226,981	7.810	3.936	24.698	.0164
12	144	1,728	3.464	2.289	10.954	.0833	62	3,844	238,328	7.874	3.958	24.900	.0161
13	169	2,197	3.606	2.351	11.402	.0769	63	3,969	250,047	7.937	3.979	25.100	.0159
14	196	2,744	3.742	2.410	11.832	.0714	64	4,096	262,144	8.000	4.000	25.298	.0156
15	225	3,375	3.873	2.466	12.247	.0667	65	4,225	274,625	8.062	4.021	25.495	.0154
16	256	4,096	4.000	2.520	12.648	.0625	66	4,356	287,496	8.124	4.041	25.690	.0152
17	289	4,913	4.123	2.571	13.038	.0588	67	4,489	300,763	8.185	4.062	25.884	.0149
18	324	5,832	4.243	2.621	13.416	.0556	68	4,624	314,432	8.246	4.082	26.077	.0147
19	361	6,859	4.359	2.668	13.784	.0526	69	4,761	328,509	8.307	4.102	26.268	.0145
20	400	8,000	4.472	2.714	14.142	.0500	70	4,900	343,000	8.367	4.121	26.458	.0143
21	441	9,261	4.583	2.759	14.491	.0476	71	5,041	357,911	8.426	4.141	26.646	.0141
22	484	10,648	4.690	2.802	14.832	.0455	72	5,184	373,248	8.485	4.160	26.833	.0139
23	529	12,167	4.796	2.844	15.166	.0435	73	5,329	389,017	8.544	4.179	27.019	.0137
24	576	13,824	4.899	2.884	15.492	.0417	74	5,476	405,224	8.602	4.198	27.203	.0135
25	625	15,625	5.000	2.924	15.811	.0400	75	5,625	421,875	8.660	4.217	27.386	.0133
26	676	17,576	5.099	2.962	16.125	.0385	76	5,776	438,976	8.718	4.236	27.568	.0132
27	729	19,683	5.196	3.000	16.432	.0370	77	5,929	456,533	8.775	4.254	27.749	.0130
28	784	21,952	5.292	3.037	16.733	.0357	78	6,084	474,552	8.832	4.273	27.928	.0128
29	841	24,389	5.385	3.072	17.029	.0345	79	6,241	493,039	8.888	4.291	28.107	.0127
30	900	27,000	5.477	3.107	17.321	.0333	80	6,400	512,000	8.944	4.309	28.284	.0125
31	961	29,791	5.568	3.141	17.607	.0323	81	6,561	531,441	9.000	4.327	28.460	.0123
32	1,024	32,768	5.657	3.175	17.889	.0312	82	6,724	551,368	9.055	4.344	28.636	.0122
33	1,089	35,937	5.745	3.208	18.166	.0303	83	6,889	571,787	9.110	4.362	28.810	.0120
34	1,156	39,304	5.831	3.240	18.439	.0294	84	7.056	592,704	9.165	4.380	28.983	.0119
35	1,225	42,875	5.916	3.271	18.708	.0286	85	7,225	614,125	9.220	4.397	29.155	.0118
36	1,296	46,656	6.000	3.302	18.974	.0278	86	7,396	636,056	9.274	4.414	29.326	.0116
37	1,369	50,653	6.083	3.332	19.235	.0270	87	7,569	658,503	9.327	4.431	29.496	.0115
38	1,444	54,872	6.164	3.362	19.494	.0263	88	7,744	681,472	9.381	4.448	29.665	.0114
39	1,521	59,319	6.245	3.391	19.748	.0256	89	7,921	704,969	9.434	4.465	29.833	.0112
40	1,600	64,000	6.325	3.420	20.000	.0250	90	8,100	729,000	9.487	4.481	30.000	.0111
41	1,681	68,921	6.403	3.448	20.248	.0244	91	8,281	753,571	9.539	4.498	30.166	.0110
42	1,764	74,088	6.481	3.476	20.494	.0238	92	8,464	778,688	9.592	4.514	30.332	.0109
43	1,849	79,507	6.557	3.503	20.736	.0233	93	8,649	804,357	9.644	4.531	30.496	.0108
44	1,936	85,184	6.633	3.530	20.976	.0227	94	8,836	830,584	9.695	4.547	30.659	.0106
45	2,025	91,125	6.708	3.557	21.213	.0222	95	9,025	857,375	9.747	4.563	30.822	.0105
46	2,116	97,336	6.782	3.583	21.448	.0217	96	9,216	884,736	9.798	4.579	30.984	.0104
47	2,209	103,823	6.856	3.609	21.679	.0213	97	9,409	912,673	9.849	4.595	31.145	.0103
48	2,304	110,592	6.928	3.634	21.909	.0208	98	9,604	941,192	9.899	4.610	31.305	.0102
49	2,401	117,649	7.000	3.659	22.136	.0204	99	9,801	970,299	9.950	4.626	31.464	.0101
50	2,500	125,000	7.071	3.684	22.361	.0200	100	10,000	1,000,000	10.000	4.642	31.623	.0100

Table 2 Common Logarithms

x	0	1	2	3	4	5	6	7	8	9
1.0	.0000	.0043	.0086	.0128	.0170	.0212	.0253	.0294	.0334	.0374
1.1	.0414	.0453	.0492	.0531	.0569	.0607	.0645	.0682	.0719	.0755
1.2	.0792	.0828	.0864	.0899	.0934	.0969	.1004	.1038	.1072	.1106
1.3	.1139	.1173	.1206	.1239	.1271	.1303	.1335	.1367	.1399	.1430
1.4	.1461	.1492	.1523	.1553	.1584	.1614	.1644	.1673	.1703	.1732
1.5	.1761	.1790	.1818	.1847	.1875	.1903	.1931	.1959	.1987	.2014
1.6	.2041	.2068	.2095	.2122	.2148	.2175	.2201	.2227	.2253	.2279
1.7	.2304	.2330	.2355	.2380	.2405	.2430	.2455	.2480	.2504	.2529
1.8	.2553	.2577	.2601	.2625	.2648	.2672	.2695	.2718	.2742	.2765
1.9	.2788	.2810	.2833	.2856	.2878	.2900	.2923	.2945	.2967	.2989
2.0	.3010	.3032	.3054	.3075	.3096	.3118	.3139	.3160	.3181	.3201
2.1	.3222	.3243	.3263	.3284	.3304	.3324	.3345	.3365	.3385	.3404
2.2	.3424	.3444	.3464	.3483	.3502	.3522	.3541	.3560	.3579	.3598
2.3	.3617	.3636	.3655	.3674	.3692	.3711	.3729	.3747	.3766	.3784
2.4	.3802	.3820	.3838	.3856	.3874	.3892	.3909	.3927	.3945	.3962
2.5	.3979	.3997	.4014	.4031	.4048	.4065	.4082	.4099	.4116	.4133
2.6	.4150	.4166	.4183	.4200	.4216	.4232	.4249	.4265	.4281	.4298
2.7	.4314	.4330	.4346	.4362	.4378	.4393	.4409	.4425	.4440	.4456
2.8	.4472	.4487	.4502	.4518	.4533	.4548	.4564	.4579	.4594	.4609
2.9	.4624	.4639	.4654	.4669	.4683	.4698	.4713	.4728	.4742	.4757
3.0	.4771	.4786	.4800	.4814	.4829	.4843	.4857	.4871	.4886	.4900
3.1	.4914	.4928	.4942	.4955	.4969	.4983	.4997	.5011	.5024	.5038
3.2	.5051	.5065	.5079	.5092	.5105	.5119	.5132	.5145	.5159	.5172
3.3	.5185	.5198	.5211	.5224	.5237	.5250	.5263	.5276	.5289	.5307
3.4	.5315	.5328	.5340	.5353	.5366	.5378	.5391	.5403	.5416	.5428
3.5	.5441	.5453	.5465	.5478	.5490	.5502	.5514	.5527	.5539	.5551
3.6	.5563	.5575	.5587	.5599	.5611	.5623	.5635	.5647	.5658	.5670
3.7	.5682	.5694	.5705	.5717	.5729	.5740	.5752	.5763	.5775	.5786
3.8	.5798	.5809	.5821	.5832	.5843	.5855	.5866	.5877	.5888	.5899
3.9	.5911	.5922	.5933	.5944	.5955	.5966	.5977	.5988	.5999	.6010
4.0	.6021	.6031	.6042	.6053	.6064	.6075	.6085	.6096	.6107	.6117
4.1	.6128	.6138	.6149	.6160	.6170	.6180	.6191	.6201	.6212	.6222
4.2	.6232	.6243	.6253	.6263	.6274	.6284	.6294	.6304	.6314	.6325
4.3	.6335	.6345	.6355	.6365	.6375	.6385	.6395	.6405	.6415	.6425
4.4	.6435	.6444	.6454	.6464	.6474	.6484	.6493	.6503	.6513	.6522
4.5	.6532	.6542	.6551	.6561	.6571	.6580	.6590	.6599	.6609	.6618
4.6	.6628	.6637	.6646	.6656	.6665	.6675	.6684	.6693	.6702	.6712
4.7	.6721	.6730	.6739	.6749	.6758	.6767	.6776	.6785	.6794	.6803
4.8	.6812	.6821	.6830	.6839	.6848	.6857	.6866	.6875	.6884	.6893
4.9	.6902	.6911	.6920	.6928	.6937	.6946	.6955	.6964	.6972	.6981
5.0	.6990	.6998	.7007	.7016	.7024	.7033	.7042	.7050	.7059	.7067
5.1	.7076	.7084	.7093	.7101	.7110	.7118	.7126	.7135	.7143	.7152
5.2	.7160	.7168	.7177	.7185	.7193	.7202	.7210	.7218	.7226	.7235
5.3	.7243	.7251	.7259	.7267	.7275	.7284	.7292	.7300	.7308	.7316
5.4	.7324	.7332	.7340	.7348	.7356	.7364	.7372	.7380	.7388	.7396
x	0	1	2	3	4	5	6	7	8	9

Table 2 (continued)

x	0	1	2	3	4	5	6	7	8	9
5.5	.7404	.7412	.7419	.7427	.7435	.7443	.7451	.7459	.7466	.7474
5.6	.7482	.7490	.7497	.7505	.7513	.7520	.7528	.7536	.7543	.7551
5.7	.7559	.7566	.7574	.7582	.7589	.7597	.7604	.7612	.7619	.7627
5.8	.7634	.7642	.7649	.7657	.7664	.7672	.7679	.7686	.7694	.7701
5.9	.7709	.7716	.7723	.7731	.7738	.7745	.7752	.7760	.7767	.7774
6.0	.7782	.7789	.7796	.7803	.7810	.7818	.7825	.7832	.7839	.7846
6.1	.7853	.7860	.7868	.7875	.7882	.7889	.7896	.7903	.7910	.7917
6.2	.7924	.7931	.7938	.7945	.7952	.7959	.7966	.7973	.7980	.7987
6.3	.7993	.8000	.8007	.8014	.8021	.8028	.8035	.8041	.8048	.8055
6.4	.8062	.8069	.8075	.8082	.8089	.8096	.8102	.8109	.8116	.8122
6.5	.8129	.8136	.8142	.8149	.8156	.8162	.8169	.8176	.8182	.8189
6.6	.8195	.8202	.8209	.8215	.8222	.8228	.8235	.8241	.8248	.8254
6.7	.8261	.8267	.8274	.8280	.8287	.8293	.8299	.8306	.8312	.8319
6.8	.8325	.8331	.8338	.8344	.8351	.8357	.8363	.8370	.8376	.8382
6.9	.8388	.8395	.8401	.8407	.8414	.8420	.8426	.8432	.8439	.8445
7.0	.8451	.8457	.8463	.8470	.8476	.8482	.8488	.8494	.8500	.8506
7.1	.8513	.8519	.8525	.8531	.8537	.8543	.8549	.8555	.8561	.8567
7.2	.8573	.8579	.8585	.8591	.8597	.8603	.8609	.8615	.8621	.8627
7.3	.8633	.8639	.8645	.8651	.8657	.8663	.8669	.8675	.8681	.8686
7.4	.8692	.8698	.8704	.8710	.8716	.8722	.8727	.8733	.8739	.8745
7.5	.8751	.8756	.8762	.8768	.8774	.8779	.8785	.8791	.8797	.8802
7.6	.8808	.8814	.8820	.8825	.8831	.8837	.8842	.8848	.8854	.8859
7.7	.8865	.8871	.8876	.8882	.8887	.8893	.8899	.8904	.8910	.8915
7.8	.8921	.8927	.8932	.8938	.8943	.8949	.8954	.8960	.8965	.8971
7.9	.8976	.8982	.8987	.8993	.8998	.9004	.9009	.9015	.9020	.9025
8.0	.9031	.9036	.9042	.9047	.9053	.9058	.9063	.9069	.9074	.9079
8.1	.9085	.9090	.9096	.9101	.9106	.9112	.9117	.9122	.9128	.9133
8.2	.9138	.9143	.9149	.9154	.9159	.9165	.9170	.9175	.9180	.9186
8.3	.9191	.9196	.9201	.9206	.9212	.9217	.9222	.9227	.9232	.9238
8.4	.9243	.9248	.9253	.9258	.9263	.9269	.9274	.9279	.9284	.9289
8.5	.9294	.9299	.9304	.9309	.9315	.9320	.9325	.9330	.9335	.9340
8.6	.9345	.9350	.9555	.9360	.9365	.9370	.9375	.9380	.9385	.9390
8.7	.9395	.9400	.9405	.9410	.9415	.9420	.9425	.9430	.9435	.9440
8.8	.9445	.9450	.9455	.9460	.9465	.9469	.9474	.9479	.9484	.9489
8.9	.9494	.9499	.9504	.9509	.9513	.9518	.9523	.9528	.9533	.9538
9.0	.9542	.9547	.9552	.9557	.9562	.9566	.9571	.9576	.9581	.9586
9.1	.9590	.9595	.9600	.9605	.9609	.9614	.9619	.9624	.9628	.9633
9.2	.9638	.9643	.9647	.9652	.9657	.9661	.9666	.9671	.9675	.9680
9.3	.9685	.9689	.9694	.9699	.9703	.9708	.9713	.9717	.9722	.9727
9.4	.9731	.9736	.9741	.9745	.9750	.9754	.9759	.9763	.9768	.9773
9.5	.9777	.9782	.9786	.9791	.9795	.9800	.9805	.9809	.9814	.9818
9.6	.9823	.9827	.9832	.9836	.9841	.9845	.9850	.9854	.9859	.9863
9.7	.9868	.9872	.9877	.9881	.9886	.9890	.9894	.9899	.9903	.9908
9.8	.9912	.9917	.9921	.9926	.9930	.9934	.9939	.9943	.9948	.9952
9.9	.9956	.9961	.9965	.9969	.9974	.9978	.9983	.9987	.9991	.9996
x	0	1	2	3	4	5	6	7	8	9

Table 3   Values of Trigonometric Functions

Degrees	Radians	Sin	Cos	Tan	Cot	Sec	Csc		
0° 00'	.0000	.0000	1.0000	.0000	——	1.000	——	1.5708	**90° 00'**
10	029	029	000	029	343.8	000	343.8	679	50
20	058	058	000	058	171.9	000	171.9	650	40
30	.0087	.0087	1.0000	.0087	114.6	1.000	114.6	1.5621	30
40	116	116	.9999	116	85.94	000	85.95	592	20
50	145	145	999	145	68.75	000	68.76	563	10
1° 00'	.0175	.0175	.9998	.0175	57.29	1.000	57.30	1.5533	**89° 00'**
10	204	204	998	204	49.10	000	49.11	504	50
20	233	233	997	233	42.96	000	42.98	475	40
30	.0262	.0262	.9997	.0262	38.19	1.000	38.20	1.5446	30
40	291	291	996	291	34.37	000	34.38	417	20
50	320	320	995	320	31.24	001	31.26	388	10
2° 00'	.0349	.0349	.9994	.0349	28.64	1.001	28.65	1.5359	**88° 00'**
10	378	378	993	378	26.43	001	26.45	330	50
20	407	407	992	407	24.54	001	24.56	301	40
30	.0436	.0436	.9990	.0437	22.90	1.001	22.93	1.5272	30
40	465	465	989	466	21.47	001	21.49	243	20
50	495	494	988	495	20.21	001	20.23	213	10
3° 00'	.0524	.0523	.9986	.0524	19.08	1.001	19.11	1.5184	**87° 00'**
10	553	552	985	553	18.07	002	18.10	155	50
20	582	581	983	582	17.17	002	17.20	126	40
30	.0611	.0610	.9981	.0612	16.35	1.002	16.38	1.5097	30
40	640	640	980	641	15.60	002	15.64	068	20
50	669	669	978	670	14.92	002	14.96	039	10
4° 00'	.0698	.0698	.9976	.0699	14.30	1.002	14.34	1.5010	**86° 00'**
10	727	727	974	729	13.73	003	13.76	981	50
20	756	756	971	758	13.20	003	13.23	952	40
30	.0785	.0785	.9969	.0787	12.71	1.003	12.75	1.4923	30
40	814	814	967	816	12.25	003	12.29	893	20
50	844	843	964	846	11.83	004	11.87	864	10
5° 00'	.0873	.0872	.9962	.0875	11.43	1.004	11.47	1.4835	**85° 00'**
10	902	901	959	904	11.06	004	11.10	806	50
20	931	931	957	934	10.71	004	10.76	777	40
30	.0960	.0958	.9954	.0963	10.39	1.005	10.43	1.4748	30
40	989	987	951	992	10.08	005	10.13	719	20
50	.1018	.1016	948	.1022	9.788	005	9.839	690	10
6° 00'	.1047	.1045	.9945	.1051	9.514	1.006	9.567	1.4661	**84° 00'**
10	076	074	942	080	9.255	006	9.309	632	50
20	105	103	939	110	9.010	006	9.065	603	40
30	.1134	.1132	.9936	.1139	8.777	1.006	8.834	1.4573	30
40	164	161	932	169	8.556	007	8.614	544	20
50	193	190	929	198	8.345	007	8.405	515	10
7° 00'	.1222	.1219	.9925	.1228	8.144	1.008	8.206	1.4486	**83° 00'**
10	251	248	922	257	7.953	008	8.016	457	50
20	280	276	918	287	7.770	008	7.834	428	40
30	.1309	.1305	.9914	.1317	7.596	1.009	7.661	1.4399	30
40	338	334	911	346	7.429	009	7.496	370	20
50	367	363	907	376	7.269	009	7.337	341	10
8° 00'	.1396	.1392	.9903	.1405	7.115	1.010	7.185	1.4312	**82° 00'**
10	425	421	899	435	6.968	010	7.040	283	50
20	454	449	894	465	6.827	011	6.900	254	40
30	.1484	.1478	.9890	.1495	6.691	1.011	6.765	1.4224	30
40	513	507	886	524	6.561	012	6.636	195	20
50	542	536	881	554	6.435	012	6.512	166	10
9° 00'	.1571	.1564	.9877	.1584	6.314	1.012	6.392	1.4137	**81° 00'**
		Cos	Sin	Cot	Tan	Csc	Sec	Radians	Degrees

Table 3 (continued)

Degrees	Radians	Sin	Cos	Tan	Cot	Sec	Csc		
9° 00′	.1571	.1564	.9877	.1584	6.314	1.012	6.392	1.4137	81° 00′
10	600	593	872	614	197	013	277	108	50
20	629	622	868	644	084	013	166	079	40
30	.1658	.1650	.9863	.1673	5.976	1.014	6.059	1.4050	30
40	687	679	858	703	871	014	5.955	1.4021	20
50	716	708	853	733	769	015	855	992	10
10° 00′	.1745	.1736	.9848	.1763	5.671	1.015	5.759	1.3963	80° 00′
10	774	765	843	793	576	016	665	934	50
20	804	794	838	823	485	016	575	904	40
30	.1833	.1822	.9833	.1853	5.396	1.017	5.487	1.3875	30
40	862	851	827	883	309	018	403	846	20
50	891	880	822	914	226	018	320	817	10
11° 00′	.1920	.1908	.9816	.1944	5.145	1.019	5.241	1.3788	79° 00′
10	949	937	811	974	066	019	164	759	50
20	978	965	805	.2004	4.989	020	089	730	40
30	.2007	.1994	.9799	.2035	4.915	1.020	5.016	1.3701	30
40	036	.2022	793	065	843	021	4.945	672	20
50	065	051	787	095	773	022	876	643	10
12° 00′	.2094	.2079	.9781	.2126	4.705	1.022	4.810	1.3614	78° 00′
10	123	108	775	156	638	023	745	584	50
20	153	136	769	186	574	024	682	555	40
30	.2182	.2164	.9763	.2217	4.511	1.024	4.620	1.3526	30
40	211	193	757	247	449	025	560	497	20
50	240	221	750	278	390	026	502	468	10
13° 00′	.2269	.2250	.9744	.2309	4.331	1.026	4.445	1.3439	77° 00′
10	298	278	737	339	275	027	390	410	50
20	327	306	730	370	219	028	336	381	40
30	.2356	.2334	.9724	.2401	4.165	1.028	4.284	1.3352	30
40	385	363	717	432	113	029	232	323	20
50	414	391	710	462	061	030	182	294	10
14° 00′	.2443	.2419	.9703	.2493	4.011	1.031	4.134	1.3265	76° 00′
10	473	447	696	524	3.962	·031	086	235	50
20	502	476	689	555	914	032	039	206	40
30	.2531	.2504	.9681	.2586	3.867	1.033	3.994	1.3177	30
40	560	532	674	617	821	034	950	148	20
50	589	560	667	648	776	034	906	119	10
15° 00′	.2618	.2588	.9659	.2679	3.732	1.035	3.864	1.3090	75° 00′
10	647	616	652	711	689	036	822	061	50
20	676	644	644	742	647	037	782	032	40
30	.2705	.2672	.9636	.2773	3.606	1.038	3.742	1.3003	30
40	734	700	628	805	566	039	703	974	20
50	763	728	621	836	526	039	665	945	10
16° 00′	.2793	.2756	.9613	.2867	3.487	1.040	3.628	1.2915	74° 00′
10	822	784	605	899	450	041	592	886	50
20	851	812	596	931	412	042	556	857	40
30	.2880	.2840	.9588	.2962	3.376	1.043	3.521	1.2828	30
40	909	868	580	994	340	044	487	799	20
50	938	896	572	.3026	305	045	453	770	10
17° 00′	.2967	.2924	.9563	.3057	3.271	1.046	3.420	1.2741	73° 00′
10	996	952	555	089	237	047	388	712	50
20	.3025	979	546	121	204	048	356	683	40
30	.3054	.3007	.9537	.3153	3.172	1.049	3.326	1.2654	30
40	083	035	528	185	140	049	295	625	20
50	113	062	520	217	108	050	265	595	10
18° 00′	.3142	.3090	.9511	.3249	3.078	1.051	3.236	1.2566	72° 00′
		Cos	Sin	Cot	Tan	Csc	Sec	Radians	Degrees

(Continued)

Table 3 (continued)

Degrees	Radians	Sin	Cos	Tan	Cot	Sec	Csc		
18° 00'	.3142	.3090	.9511	.3249	3.078	1.051	3.236	1.2566	**72° 00'**
10	171	118	502	281	047	052	207	537	50
20	200	145	492	314	018	053	179	508	40
30	.3229	.3173	.9483	.3346	2.989	1.054	3.152	1.2479	30
40	258	201	474	378	960	056	124	450	20
50	287	228	465	411	932	057	098	421	10
19° 00'	.3316	.3256	.9455	.3443	2.904	1.058	3.072	1.2392	**71° 00'**
10	345	283	446	476	877	059	046	363	50
20	374	311	436	508	850	060	021	334	40
30	.3403	.3338	.9426	.3541	2.824	1.061	2.996	1.2305	30
40	432	365	417	574	798	062	971	275	20
50	462	393	407	607	773	063	947	246	10
20° 00'	.3491	.3420	.9397	.3640	2.747	1.064	2.924	1.2217	**70° 00'**
10	520	448	387	673	723	065	901	188	50
20	549	475	377	706	699	066	878	159	40
30	.3578	.3502	.9367	.3739	2.675	1.068	2.855	1.2130	30
40	607	529	356	772	651	069	833	101	20
50	636	557	346	805	628	070	812	072	10
21° 00'	.3665	.3584	.9336	.3839	2.605	1.071	2.790	1.2043	**69° 00'**
10	694	611	325	872	583	072	769	1.2014	50
20	723	638	315	906	560	074	749	985	40
30	.3752	.3665	.9304	.3939	2.539	1.075	2.729	1.1956	30
40	782	692	293	973	517	076	709	926	20
50	811	719	283	.4006	496	077	689	897	10
22° 00'	.3840	.3746	.9272	.4040	2.475	1.079	2.669	1.1868	**68° 00'**
10	869	773	261	074	455	080	650	839	50
20	898	800	250	108	434	081	632	810	40
30	.3927	.3827	.9239	.4142	2.414	1.082	2.613	1.1781	30
40	956	854	228	176	394	084	595	752	20
50	985	881	216	210	375	085	577	723	10
23° 00'	.4014	.3907	.9205	.4245	2.356	1.086	2.559	1.1694	**67° 00'**
10	043	934	194	279	337	088	542	665	50
20	072	961	182	314	318	089	525	636	40
30	.4102	.3987	.9171	.4348	2.300	1.090	2.508	1.1606	30
40	131	.4014	159	383	282	092	491	577	20
50	160	041	147	417	264	093	475	548	10
24° 00'	.4189	.4067	.9135	.4452	2.246	1.095	2.459	1.1519	**66° 00'**
10	218	094	124	487	229	096	443	490	50
20	247	120	112	522	211	097	427	461	40
30	.4276	.4147	.9100	.4557	2.194	1.099	2.411	1.1432	30
40	305	173	088	592	177	100	396	403	20
50	334	200	075	628	161	102	381	374	10
25° 00'	.4363	.4226	.9063	.4663	2.145	1.103	2.366	1.1345	**65° 00'**
10	392	253	051	699	128	105	352	316	50
20	422	279	038	734	112	106	337	286	40
30	.4451	.4305	.9026	.4770	2.097	1.108	2.323	1.1257	30
40	480	331	013	806	081	109	309	228	20
50	509	358	001	841	066	111	295	199	10
26° 00'	.4538	.4384	.8988	.4877	2.050	1.113	2.281	1.1170	**64° 00'**
10	567	410	975	913	035	114	268	141	50
20	596	436	962	950	020	116	254	112	40
30	.4625	.4462	.8949	.4986	2.006	1.117	2.241	1.1083	30
40	654	488	936	.5022	1.991	119	228	054	20
50	683	514	923	059	977	121	215	1.1025	10
27° 00'	.4712	.4540	.8910	.5095	1.963	1.122	2.203	1.0996	**63° 00'**
		Cos	Sin	Cot	Tan	Csc	Sec	Radians	Degrees

Table 3 (continued)

Degrees	Radians	Sin	Cos	Tan	Cot	Sec	Csc		
27° 00'	.4712	.4540	.8910	.5095	1.963	1.122	2.203	1.0996	**63° 00'**
10	741	566	897	132	949	124	190	966	50
20	771	592	884	169	935	126	178	937	40
30	.4800	.4617	.8870	.5206	1.921	1.127	2.166	1.0908	30
40	829	643	857	243	907	129	154	879	20
50	858	669	843	280	894	131	142	850	10
28° 00'	.4887	.4695	.8829	.5317	1.881	1.133	2.130	1.0821	**62° 00'**
10	916	720	816	354	868	134	118	792	50
20	945	746	802	392	855	136	107	763	40
30	.4974	.4772	.8788	.5430	1.842	1.138	2.096	1.0734	30
40	.5003	797	774	467	829	140	085	705	20
50	032	823	760	505	816	142	074	676	10
29° 00'	.5061	.4848	.8746	.5543	1.804	1.143	2.063	1.0647	**61° 00'**
10	091	874	732	581	792	145	052	617	50
20	120	899	718	619	780	147	041	588	40
30	.5149	.4924	.8704	.5658	1.767	1.149	2.031	1.0559	30
40	178	950	689	696	756	151	020	530	20
50	207	975	675	735	744	153	010	501	10
30° 00'	.5236	.5000	.8660	.5774	1.732	1.155	2.000	1.0472	**60° 00'**
10	265	025	646	812	720	157	1.990	443	50
20	294	050	631	851	709	159	980	414	40
30	.5323	.5075	.8616	.5890	1.698	1.161	1.970	1.0385	30
40	352	100	601	930	686	163	961	356	20
50	381	125	587	969	675	165	951	327	10
31° 00'	.5411	.5150	.8572	.6009	1.664	1.167	1.942	1.0297	**59° 00'**
10	440	175	557	048	653	169	932	268	50
20	469	200	542	088	643	171	923	239	40
30	.5498	.5225	.8526	.6128	1.632	1.173	1.914	1.0210	30
40	527	250	511	168	621	175	905	181	20
50	556	275	496	208	611	177	896	152	10
32° 00'	.5585	.5299	.8480	.6249	1.600	1.179	1.887	1.0123	**58° 00'**
10	614	324	465	289	590	181	878	094	50
20	643	348	450	330	580	184	870	065	40
30	.5672	.5373	.8434	.6371	1.570	1.186	1.861	1.0036	30
40	701	398	418	412	560	188	853	1.0007	20
50	730	422	403	453	550	190	844	977	10
33° 00'	.5760	.5446	.8387	.6494	1.540	1.192	1.836	.9948	**57° 00'**
10	789	471	371	536	530	195	828	919	50
20	818	495	355	577	520	197	820	890	40
30	.5847	.5519	.8339	.6619	1.511	1.199	1.812	.9861	30
40	876	544	323	661	501	202	804	832	20
50	905	568	307	703	1.492	204	796	803	10
34° 00'	.5934	.5592	.8290	.6745	1.483	1.206	1.788	.9774	**56° 00'**
10	963	616	274	787	473	209	781	745	50
20	992	640	258	830	464	211	773	716	40
30	.6021	.5664	.8241	.6873	1.455	1.213	1.766	.9687	30
40	050	688	225	916	446	216	758	657	20
50	080	712	208	959	437	218	751	628	10
35° 00'	.6109	.5736	.8192	.7002	1.428	1.221	1.743	.9599	**55° 00'**
10	138	760	175	046	419	223	736	570	50
20	167	783	158	089	411	226	729	541	40
30	.6196	.5807	.8141	.7133	1.402	1.228	1.722	.9512	30
40	225	831	124	177	393	231	715	483	20
50	254	854	107	221	385	233	708	454	10
36° 00'	.6283	.5878	.8090	.7265	1.376	1.236	1.701	.9425	**54° 00'**
		Cos	Sin	Cot	Tan	Csc	Sec	Radians	Degrees

(Continued)

618TABLES**Table 3**(continued)

Degrees	Radians	Sin	Cos	Tan	Cot	Sec	Csc		
36° 00'	.6283	.5878	.8090	.7265	1.376	1.236	1.701	.9425	**54° 00'**
10	312	901	073	310	368	239	695	396	50
20	341	925	056	355	360	241	688	367	40
30	.6370	.5948	.8039	.7400	1.351	1.244	1.681	.9338	30
40	400	972	021	445	343	247	675	308	20
50	429	995	004	490	335	249	668	279	10
37° 00'	.6458	.6018	.7986	.7536	1.327	1.252	1.662	.9250	**53° 00'**
10	487	041	969	581	319	255	655	221	50
20	516	065	951	627	311	258	649	192	40
30	.6545	.6088	.7934	.7673	1.303	1.260	1.643	.9163	30
40	574	111	916	720	295	263	636	134	20
50	603	134	898	766	288	266	630	105	10
38° 00'	.6632	.6157	.7880	.7813	1.280	1.269	1.624	.9076	**52° 00'**
10	661	180	862	860	272	272	618	047	50
20	690	202	844	907	265	275	612	.9018	40
30	.6720	.6225	.7826	.7954	1.257	1.278	1.606	.8988	30
40	749	248	808	.8002	250	281	601	959	20
50	778	271	790	050	242	284	595	930	10
39° 00'	.6807	.6293	.7771	.8098	1.235	1.287	1.589	.8901	**51° 00'**
10	836	316	753	146	228	290	583	872	50
20	865	338	735	195	220	293	578	843	40
30	.6894	.6361	.7716	.8243	1.213	1.296	1.572	.8814	30
40	923	383	698	292	206	299	567	785	20
50	952	406	679	342	199	302	561	756	10
40° 00'	.6981	.6428	.7660	.8391	1.192	1.305	1.556	.8727	**50° 00'**
10	.7010	450	642	441	185	309	550	698	50
20	039	472	623	491	178	312	545	668	40
30	.7069	.6494	.7604	.8541	1.171	1.315	1.540	.8639	30
40	098	517	585	591	164	318	535	610	20
50	127	539	566	642	157	322	529	581	10
41° 00'	.7156	.6561	.7547	.8693	1.150	1.325	1.524	.8552	**49° 00'**
10	185	583	528	744	144	328	519	523	50
20	214	604	509	796	137	332	514	494	40
30	.7243	.6626	.7490	.8847	1.130	1.335	1.509	.8465	30
40	272	648	470	899	124	339	504	436	20
50	301	670	451	952	117	342	499	407	10
42° 00'	.7330	.6691	.7431	.9004	1.111	1.346	1.494	.8378	**48° 00'**
10	359	713	412	057	104	349	490	348	50
20	389	734	392	110	098	353	485	319	40
30	.7418	.6756	.7373	.9163	1.091	1.356	1.480	.8290	30
40	447	777	353	217	085	360	476	261	20
50	476	799	333	271	079	364	471	232	10
43° 00'	.7505	.6820	.7314	.9325	1.072	1.367	1.466	.8203	**47° 00'**
10	534	841	294	380	066	371	462	174	50
20	563	862	274	435	060	375	457	145	40
30	.7592	.6884	.7254	.9490	1.054	1.379	1.453	.8116	30
40	621	905	234	545	048	382	448	087	20
50	650	926	214	601	042	386	444	058	10
44° 00'	.7679	.6947	.7193	.9657	1.036	1.390	1.440	.8029	**46° 00'**
10	709	967	173	713	030	394	435	999	50
20	738	988	153	770	024	398	431	970	40
30	.7767	.7009	.7133	.9827	1.018	1.402	1.427	.7941	30
40	796	030	112	884	012	406	423	912	20
50	825	050	092	942	006	410	418	883	10
45° 00'	.7854	.7071	.7071	1.000	1.000	1.414	1.414	.7854	**45° 00'**
		Cos	Sin	Cot	Tan	Csc	Sec	Radians	Degrees

Table 4 Exponential Functions

x	e^x	e^{-x}	x	e^x	e^{-x}	x	e^x	e^{-x}
0.00	1.0000	1.0000	0.55	1.7333	0.5769	3.6	36.598	0.0273
0.01	1.0101	0.9900	0.60	1.8221	0.5488	3.7	40.447	0.0247
0.02	1.0202	0.9802	0.65	1.9155	0.5220	3.8	44.701	0.0224
0.03	1.0305	0.9704	0.70	2.0138	0.4966	3.9	49.402	0.0202
0.04	1.0408	0.9608	0.75	2.1170	0.4724	4.0	54.598	0.0183
0.05	1.0513	0.9512	0.80	2.2255	0.4493	4.1	60.340	0.0166
0.06	1.0618	0.9418	0.85	2.3396	0.4274	4.2	66.686	0.0150
0.07	1.0725	0.9324	0.90	2.4596	0.4066	4.3	73.700	0.0136
0.08	1.0833	0.9231	0.95	2.5857	0.3867	4.4	81.451	0.0123
0.09	1.0942	0.9139	1.0	2.7183	0.3679	4.5	90.017	0.0111
0.10	1.1052	0.9048	1.1	3.0042	0.3329	4.6	99.484	0.0101
0.11	1.1163	0.8958	1.2	3.3201	0.3012	4.7	109.95	0.0091
0.12	1.1275	0.8869	1.3	3.6693	0.2725	4.8	121.51	0.0082
0.13	1.1388	0.8781	1.4	4.0552	0.2466	4.9	134.29	0.0074
0.14	1.1503	0.8694	1.5	4.4817	0.2231	5	148.41	0.0067
0.15	1.1618	0.8607	1.6	4.9530	0.2019	6	403.43	0.0025
0.16	1.1735	0.8521	1.7	5.4739	0.1827	7	1096.6	0.0009
0.17	1.1853	0.8437	1.8	6.0496	0.1653	8	2981.0	0.0003
0.18	1.1972	0.8353	1.9	6.6859	0.1496	9	8103.1	0.0001
0.19	1.2092	0.8270	2.0	7.3891	0.1353	10	22026	0.00005
0.20	1.2214	0.8187	2.1	8.1662	0.1225	11	59874	0.00002
0.21	1.2337	0.8106	2.2	9.0250	0.1108	12	162,754	0.000006
0.22	1.2461	0.8025	2.3	9.9742	0.1003	13	442,413	0.000002
0.23	1.2586	0.7945	2.4	11.023	0.0907	14	1,202,604	0.0000008
0.24	1.2712	0.7866	2.5	12.182	0.0821	15	3,269,017	0.0000003
0.25	1.2840	0.7788	2.6	13.464	0.0743			
0.26	1.2969	0.7711	2.7	14.880	0.0672			
0.27	1.3100	0.7634	2.8	16.445	0.0608			
0.28	1.3231	0.7558	2.9	18.174	0.0550			
0.29	1.3364	0.7483	3.0	20.086	0.0498			
0.30	1.3499	0.7408	3.1	22.198	0.0450			
0.35	1.4191	0.7047	3.2	24.533	0.0408			
0.40	1.4918	0.6703	3.3	27.113	0.0369			
0.45	1.5683	0.6376	3.4	29.964	0.0334			
0.50	1.6487	0.6065	3.5	33.115	0.0302			

Table 5 Factorials and Large Powers of 2

n	n!	2^n
0	1	1
1	1	2
2	2	4
3	6	8
4	24	16
5	120	32
6	720	64
7	5040	128
8	40,320	256
9	362,880	512
10	3,628,800	1024
11	39,916,800	2048
12	479,001,600	4096
13	6,227,020,800	8192
14	87,178,291,200	16,384
15	1,307,674,368,000	32,768
16	20,922,789,888,000	65,536
17	355,687,428,096,000	131,072
18	6,402,373,705,728,000	262,144
19	121,645,100,408,832,000	524,288
20	2,432,902,008,176,640,000	1,048,576

Table 6 Tables of Measures

LENGTH

1 *kilometer* (km)	=	1000 meters (m)
1 *hectometer* (hm)	=	100 meters
1 *dekameter* (dam)	=	10 meters
1 *decimeter* (dm)	=	0.1 meter
1 *centimeter* (cm)	=	0.01 meter
1 *millimeter* (mm)	=	0.001 meter

MASS OR WEIGHT

1 *kilogram* (kg)	=	1000 grams (g)
1 *hectogram* (hg)	=	100 grams
1 *dekagram* (dag)	=	10 grams
1 *decigram* (dg)	=	0.1 gram
1 *centigram* (dg)	=	0.01 gram
1 metric ton (MT or t)	=	1000 kilograms

AREA

1 *hectare* (ha) = 100 are (a), or 10,000 sq m (m²)
1 *are* (a) = 100 sq m (m²)
1 *centare* (ca) = 0.01 are, or 1 m²
The word "are" is pronounced "AIR."

VOLUME

1000 cubic centimeters (cc or cm³) = 1 liter (L)
1 cubic centimeter (cc) = 1 milliliter (mL)
1 mL of water weighs 1 g
1 stere = 1 cubic meter

Metric-American Conversions (Approximate)

LENGTH

1 m = 39.37 in. = 3.3 ft
1 in. = 2.54 cm
1 km = 0.62 mi
1 mi = 1.6 km
1 cm = 3/8 in.

MASS OR WEIGHT

1 kg = 2.2 lb
1 MT = 1.1 tons
1 lb = 454 g
1 oz = 28 g

AREA

1 hectare = 2.47 acres
1 are = 120 sq yd

VOLUME

1 liter = 1.057 qt = 2.1 pt
1 cup = 240 mL
1 ounce (liquid) = 30 mL
1 gallon = 3.78 liters
1 tablespoon = 15 mL
1 teaspoon = 5 mL

Table 7 Natural Logarithms (ln x)

x	0.00	0.01	0.02	0.03	0.04	0.05	0.06	0.07	0.08	0.09
1.0	0.0000	0.0100	0.0198	0.0296	0.0392	0.0488	0.0583	0.0677	0.0770	0.0862
1.1	0.0953	0.1044	0.1133	0.1222	0.1310	0.1398	0.1484	0.1570	0.1655	0.1740
1.2	0.1823	0.1906	0.1989	0.2070	0.2151	0.2231	0.2311	0.2390	0.2469	0.2546
1.3	0.2624	0.2700	0.2776	0.2852	0.2927	0.3001	0.3075	0.3148	0.3221	0.3293
1.4	0.3365	0.3436	0.3507	0.3577	0.3646	0.3716	0.3784	0.3853	0.3920	0.3988
1.5	0.4055	0.4121	0.4187	0.4253	0.4318	0.4383	0.4447	0.4511	0.4574	0.4637
1.6	0.4700	0.4762	0.4824	0.4886	0.4947	0.5008	0.5068	0.5128	0.5188	0.5247
1.7	0.5306	0.5365	0.5423	0.5481	0.5539	0.5596	0.5653	0.5710	0.5766	0.5822
1.8	0.5878	0.5933	0.5988	0.6043	0.6098	0.6152	0.6206	0.6259	0.6313	0.6366
1.9	0.6419	0.6471	0.6523	0.6575	0.6627	0.6678	0.6729	0.6780	0.6831	0.6881
2.0	0.6931	0.6981	0.7031	0.7080	0.7130	0.7178	0.7227	0.7275	0.7324	0.7372
2.1	0.7419	0.7467	0.7514	0.7561	0.7608	0.7655	0.7701	0.7747	0.7793	0.7839
2.2	0.7885	0.7930	0.7975	0.8020	0.8065	0.8109	0.8154	0.8198	0.8242	0.8286
2.3	0.8329	0.8372	0.8416	0.8459	0.8502	0.8544	0.8587	0.8629	0.8671	0.8713
2.4	0.8755	0.8796	0.8838	0.8879	0.8920	0.8961	0.9002	0.9042	0.9083	0.9123
2.5	0.9163	0.9203	0.9243	0.9282	0.9322	0.9361	0.9400	0.9439	0.9478	0.9517
2.6	0.9555	0.9594	0.9632	0.9670	0.9708	0.9746	0.9783	0.9821	0.9858	0.9895
2.7	0.9933	0.9969	1.0006	1.0043	1.0080	1.0116	1.0152	1.0188	1.0225	1.0260
2.8	1.0296	1.0332	1.0367	1.0403	1.0438	1.0473	1.0508	1.0543	1.0578	1.0613
2.9	1.0647	1.0682	1.0716	1.0750	1.0784	1.0818	1.0852	1.0886	1.0919	1.0953
3.0	1.0986	1.1019	1.1053	1.1086	1.1119	1.1151	1.1184	1.1217	1.1249	1.1282
3.1	1.1314	1.1346	1.1378	1.1410	1.1442	1.1474	1.1506	1.1537	1.1569	1.1600
3.2	1.1632	1.1663	1.1694	1.1725	1.1756	1.1787	1.1817	1.1848	1.1878	1.1909
3.3	1.1939	1.1970	1.2000	1.2030	1.2060	1.2090	1.2119	1.2149	1.2179	1.2208
3.4	1.2238	1.2267	1.2296	1.2326	1.2355	1.2384	1.2413	1.2442	1.2470	1.2499
3.5	1.2528	1.2556	1.2585	1.2613	1.2641	1.2669	1.2698	1.2726	1.2754	1.2782
3.6	1.2809	1.2837	1.2865	1.2892	1.2920	1.2947	1.2975	1.3002	1.3029	1.3056
3.7	1.3083	1.3110	1.3137	1.3164	1.3191	1.3218	1.3244	1.3271	1.3297	1.3324
3.8	1.3350	1.3376	1.3403	1.3429	1.3455	1.3481	1.3507	1.3533	1.3558	1.3584
3.9	1.3610	1.3635	1.3661	1.3686	1.3712	1.3737	1.3762	1.3788	1.3813	1.3838
4.0	1.3863	1.3888	1.3913	1.3938	1.3962	1.3987	1.4012	1.4036	1.4061	1.4085
4.1	1.4110	1.4134	1.4159	1.4183	1.4207	1.4231	1.4255	1.4279	1.4303	1.4327
4.2	1.4351	1.4375	1.4398	1.4422	1.4446	1.4469	1.4493	1.4516	1.4540	1.4563
4.3	1.4586	1.4609	1.4633	1.4656	1.4679	1.4702	1.4725	1.4748	1.4770	1.4793
4.4	1.4816	1.4839	1.4861	1.4884	1.4907	1.4929	1.4952	1.4974	1.4996	1.5019
4.5	1.5041	1.5063	1.5085	1.5107	1.5129	1.5151	1.5173	1.5195	1.5217	1.5239
4.6	1.5261	1.5282	1.5304	1.5326	1.5347	1.5369	1.5390	1.5412	1.5433	1.5454
4.7	1.5476	1.5497	1.5518	1.5539	1.5560	1.5581	1.5602	1.5623	1.5644	1.5665
4.8	1.5686	1.5707	1.5728	1.5748	1.5769	1.5790	1.5810	1.5831	1.5851	1.5872
4.9	1.5892	1.5913	1.5933	1.5953	1.5974	1.5994	1.6014	1.6034	1.6054	1.6074
5.0	1.6094	1.6114	1.6134	1.6154	1.6174	1.6194	1.6214	1.6233	1.6253	1.6273
5.1	1.6292	1.6312	1.6332	1.6351	1.6371	1.6390	1.6409	1.6429	1.6448	1.6467
5.2	1.6487	1.6506	1.6525	1.6544	1.6563	1.6582	1.6601	1.6620	1.6639	1.6658
5.3	1.6677	1.6696	1.6715	1.6734	1.6752	1.6771	1.6790	1.6808	1.6827	1.6845
5.4	1.6864	1.6882	1.6901	1.6919	1.6938	1.6956	1.6974	1.6993	1.7011	1.7029
5.5	1.7047	1.7066	1.7084	1.7102	1.7120	1.7138	1.7156	1.7174	1.7192	1.7210
5.6	1.7228	1.7246	1.7263	1.7281	1.7299	1.7317	1.7334	1.7352	1.7370	1.7387
5.7	1.7405	1.7422	1.7440	1.7457	1.7475	1.7492	1.7509	1.7527	1.7544	1.7561
5.8	1.7579	1.7596	1.7613	1.7630	1.7647	1.7664	1.7682	1.7699	1.7716	1.7733
5.9	1.7750	1.7766	1.7783	1.7800	1.7817	1.7834	1.7851	1.7867	1.7884	1.7901

Table 7   (continued)

x	0.00	0.01	0.02	0.03	0.04	0.05	0.06	0.07	0.08	0.09
6.0	1.7918	1.7934	1.7951	1.7967	1.7984	1.8001	1.8017	1.8034	1.8050	1.8066
6.1	1.8083	1.8099	1.8116	1.8132	1.8148	1.8165	1.8181	1.8197	1.8213	1.8229
6.2	1.8245	1.8262	1.8278	1.8294	1.8310	1.8326	1.8342	1.8358	1.8374	1.8390
6.3	1.8406	1.8421	1.8437	1.8453	1.8469	1.8485	1.8500	1.8516	1.8532	1.8547
6.4	1.8563	1.8579	1.8594	1.8610	1.8625	1.8641	1.8656	1.8672	1.8687	1.8703
6.5	1.8718	1.8733	1.8749	1.8764	1.8779	1.8795	1.8810	1.8825	1.8840	1.8856
6.6	1.8871	1.8886	1.8901	1.8916	1.8931	1.8946	1.8961	1.8976	1.8991	1.9006
6.7	1.9021	1.9036	1.9051	1.9066	1.9081	1.9095	1.9110	1.9125	1.9140	1.9155
6.8	1.9169	1.9184	1.9199	1.9213	1.9228	1.9242	1.9257	1.9272	1.9286	1.9301
6.9	1.9315	1.9330	1.9344	1.9359	1.9373	1.9387	1.9402	1.9416	1.9430	1.9445
7.0	1.9459	1.9473	1.9488	1.9502	1.9516	1.9530	1.9544	1.9559	1.9573	1.9587
7.1	1.9601	1.9615	1.9629	1.9643	1.9657	1.9671	1.9685	1.9699	1.9713	1.9727
7.2	1.9741	1.9755	1.9769	1.9782	1.9796	1.9810	1.9824	1.9838	1.9851	1.9865
7.3	1.9879	1.9892	1.9906	1.9920	1.9933	1.9947	1.9961	1.9974	1.9988	2.0001
7.4	2.0015	2.0028	2.0042	2.0055	2.0069	2.0082	2.0096	2.0109	2.0122	2.0136
7.5	2.0149	2.0162	2.0176	2.0189	2.0202	2.0215	2.0229	2.0242	2.0255	2.0268
7.6	2.0282	2.0295	2.0308	2.0321	2.0334	2.0347	2.0360	2.0373	2.0386	2.0399
7.7	2.0412	2.0425	2.0438	2.0451	2.0464	2.0477	2.0490	2.0503	2.0516	2.0528
7.8	2.0541	2.0554	2.0567	2.0580	2.0592	2.0605	2.0618	2.0631	2.0643	2.0665
7.9	2.0669	2.0681	2.0694	2.0707	2.0719	2.0732	2.0744	2.0757	2.0769	2.0782
8.0	2.0794	2.0807	2.0819	2.0832	2.0844	2.0857	2.0869	2.0882	2.0894	2.0906
8.1	2.0919	2.0931	2.0943	2.0956	2.0968	2.0980	2.0992	2.1005	2.1017	2.1029
8.2	2.1041	2.1054	2.1066	2.1078	2.1090	2.1102	2.1114	2.1126	2.1133	2.1150
8.3	2.1163	2.1175	2.1187	2.1199	2.1211	2.1223	2.1235	2.1247	2.1258	2.1270
8.4	2.1282	2.1294	2.1306	2.1318	2.1330	2.1342	2.1353	2.1365	2.1377	2.1389
8.5	2.1401	2.1412	2.1424	2.1436	2.1448	2.1459	2.1471	2.1483	2.1494	2.1506
8.6	2.1518	2.1529	2.1541	2.1552	2.1564	2.1576	2.1587	2.1599	2.1610	2.1622
8.7	2.1633	2.1645	2.1656	2.1668	2.1679	2.1691	2.1702	2.1713	2.1725	2.1736
8.8	2.1748	2.1759	2.1770	2.1782	2.1793	2.1804	2.1815	2.1827	2.1838	2.1849
8.9	2.1861	2.1872	2.1883	2.1894	2.1905	2.1917	2.1928	2.1939	2.1950	2.1961
9.0	2.1972	2.1983	2.1994	2.2006	2.2017	2.2028	2.2039	2.2050	2.2061	2.2072
9.1	2.2083	2.2094	2.2105	2.2116	2.2127	2.2138	2.2148	2.2159	2.2170	2.2181
9.2	2.2192	2.2203	2.2214	2.2225	2.2235	2.2246	2.2257	2.2268	2.2279	2.2289
9.3	2.2300	2.2311	2.2322	2.2332	2.2343	2.2354	2.2364	2.2375	2.2386	2.2396
9.4	2.2407	2.2418	2.2428	2.2439	2.2450	2.2460	2.2471	2.2481	2.2492	2.2502
9.5	2.2513	2.2523	2.2534	2.2544	2.2555	2.2565	2.2576	2.2586	2.2597	2.2607
9.6	2.2618	2.2628	2.2638	2.2649	2.2659	2.2670	2.2680	2.2690	2.2701	2.2711
9.7	2.2721	2.2732	2.2742	2.2752	2.2762	2.2773	2.2783	2.2793	2.2803	2.2814
9.8	2.2824	2.2834	2.2844	2.2854	2.2865	2.2875	2.2885	2.2895	2.2905	2.2915
9.9	2.2925	2.2935	2.2946	2.2956	2.2966	2.2976	2.2986	2.2996	2.3006	2.3016

Examples.

$$\ln 96{,}700 = \ln 9.67 + 4\ln 10$$
$$= 2.2690 + 9.2103$$
$$= 11.4793.$$

$$\ln 0.00967 = \ln 9.67 - 3\ln 10$$
$$= 2.2690 - 6.9078$$
$$= -4.6388.$$

$\ln 10$	= 2.3026	$7 \ln 10$	= 16.1181
$2 \ln 10$	= 4.6052	$8 \ln 10$	= 18.4207
$3 \ln 10$	= 6.9078	$9 \ln 10$	= 20.7233
$4 \ln 10$	= 9.2103	$10 \ln 10$	= 23.0259
$5 \ln 10$	= 11.5129	$11 \ln 10$	= 25.3284
$6 \ln 10$	= 13.8155	$12 \ln 10$	= 27.6310

Note: Adapted from *Functional Approach to Precalculus*, 2nd ed., Mustafa A. Munem and James P. Yizze (New York, NY: Worth Publishers, Inc., © 1974), pp. 500–501. Reproduced by permission of the publisher.

ANSWERS

CHAPTER 1

Margin Exercises, Section 1.1

1. 1, 19   **2.** 0, 1, 19   **3.** -6, 0, 1, 19   **4.** All of them   **5.** Rational   **6.** Rational   **7.** Rational   **8.** Rational
9. Irrational   **10.** Irrational   **11.** 2   **12.** $\sqrt{3}$ **13.** 11.3   **14.** $\frac{3}{4}$ **15.** -12 **16.** -4.7 **17.** $-\frac{4}{5}$ **18.** -0.2 **19.** 5
20. 0   **21.** -6, -6 **22.** 8, 8   **23.** 3.4, 3.4   **24.** 2, -4 **25.** -24 **26.** $\frac{21}{25}$ **27.** 144   **28.** 1.3   **29.** 17   **30.** $-\frac{11}{5}$
31. -13 **32.** 4   **33.** -3 **34.** $-\frac{8}{3}$ **35.** 2

Exercise Set 1.1, p. 10

1. 3, 14   **3.** $\sqrt{3}$, $-\sqrt{7}$, $\sqrt[3]{2}$ **5.** -6, 0, 3, -2, 14   **7.** Rational   **9.** Rational   **11.** Rational   **13.** Irrational
15. Irrational   **17.** Irrational   **19.** Rational   **21.** Irrational   **23.** 12   **25.** 47   **27.** 7, 7   **29.** -57, -57 **31.** -87
33. -16 **35.** -10.3 **37.** $\frac{39}{10}$ **39.** 28   **41.** -49.2 **43.** 210   **45.** $-\frac{833}{5}$ **47.** 5   **49.** $-\frac{1}{7}$ **51.** $-\frac{3}{49}$ **53.** 25   **55.** -4
57. 18   **59.** -11.6 **61.** $-\frac{7}{2}$ **63.** (a) 1.96, 1.9881, 1.999396, 1.999962, 1.999990; (b) $\sqrt{2}$ **65.** Identity $(+)$
67. Distributive   **69.** Commutativity $(+)$ **71.** Associativity $(\times)$ **73.** Inverse $(\times)$
75. $7 - 5 \neq 5 - 7$; $7 - 5 = 2$, $5 - 7 = -2$ **77.** $16 \div (4 \div 2) \neq (16 \div 4) \div 2$; $16 \div (4 \div 2) = 8$, $(16 \div 4) \div 2 = 2$

Margin Exercises, Section 1.2

1. 8^4 **2.** x^3 **3.** $(4y)^4$ **4.** $3 \cdot 3 \cdot 3 \cdot 3$, or 81   **5.** $5x \cdot 5x \cdot 5x \cdot 5x$, or $625x^4$ **6.** $(-5)(-5)(-5)(-5)$, or 625

7. $-[5 \cdot 5 \cdot 5 \cdot 5]$, or -625 **8.** 1   **9.** $25y^2$ **10.** $-8x^3$ **11.** 4^{-3} **12.** $\dfrac{1}{10^4}$, or $\dfrac{1}{10 \cdot 10 \cdot 10 \cdot 10}$, or $\dfrac{1}{10{,}000}$

13. $\dfrac{1}{4^3}$, $\dfrac{1}{4 \cdot 4 \cdot 4}$, $\dfrac{1}{64}$ **14.** 8^4 **15.** y^5 **16.** $-18x^{11}$ **17.** $-75x^{-14}$ **18.** $-10x^{-12}y^2$ **19.** $60y$ **20.** 4^3 **21.** 5^6 **22.** 10^{-14}
23. 9^{-6} **24.** y^{11} **25.** $5y^{-1}$ **26.** $-2x^{-3}y^9$ **27.** 3^{49} **28.** 8^{-14} **29.** y^{-28} **30.** $8x^3y^3$ **31.** $16x^{-4}y^{14}$ **32.** $\frac{1}{27}x^{-12}y^{-6}$
33. $1000x^{-12}y^{21}z^{-6}$ **34.** 4.65×10^5 **35.** 3.789×10^3 **36.** 1.45×10^{-4} **37.** 6.7×10^{-10} **38.** 0.0000467
39. 7,894,000,000,000

Exercise Set 1.2, pp. 14–16

1. 2^{-1} **3.** 1   **5.** 4^3 **7.** $6x^5$ **9.** $15a^{-1}b^5$ **11.** $72x^5$ **13.** $-18x^7yz$ **15.** b^3 **17.** x^3y^{-3} **19.** 1   **21.** $3ab^2$
23. $\frac{4}{7}xyz^{-5}$ **25.** $8a^3b^6$ **27.** $16x^{12}$ **29.** $-16x^{12}$ **31.** $36a^4b^6c^2$ **33.** $\frac{1}{25}c^2d^4$ **35.** 1   **37.** 32   **39.** $\frac{27}{4}a^8b^{-10}c^{18}$ **41.** $\frac{3}{4}xy$
43. 5.8×10^7 **45.** 3.65×10^5 **47.** 2.7×10^{-6} **49.** 2.7×10^{-2} **51.** 9.1×10^{11} **53.** 400,000   **55.** 0.0062
57. 7,690,000,000,000   **59.** 9,460,000,000,000   **61.** -25, 25   **63.** -1.1664, 1.1664   **65.** $1044.65   **67.** In $(x^3)^2$ exponents
were added instead of multiplied; x^{10} **69.** In 2^3 the base 2, and the exponent 3, were multiplied. In $(x^{-4})^3$ the
exponents were added. In $(y^6)^3$ the exponents were subtracted. In $(z^3)^3$ the exponents were added; $8x^{-12}y^{18}z^9$

Margin Exercises, Section 1.3

1. 8, 6, 4, 9, 0; 9   **2.** 4, 4, 5, 6, 0; 6   **3.** $9x^3y^2 - 2x^2y^3$ **4.** $7xy^2 - 2x^2y$ **5.** $3x^4 \cdot \sqrt{y} + 2$ **6.** $-4x^3 + 2x^2 - 4x - \frac{3}{2}$
7. $5p^2q^4 + p^2q^2 - 6pq^2 - 3q + 5$ **8.** $-5x^2t^2 + 4xy^2t + 3xt - 6x + 5$ **9.** $3x^2y - 5xy + 7x - 4y - 2$
10. $8xy^4 - 9xy^2 + 4x^2 + 2y - 7$ **11.** $7x^2y - 9x^3y^2 + 5x^2y^3 - x^2y^2 + 9y$

Exercise Set 1.3, p. 19

1. 4, 3, 2, 1, 0; 4   **3.** 3, 6, 6, 0; 6   **5.** 5, 6, 2, 1, 0; 6   **7.** $3x^2y - 5xy^2 + 7xy + 2$ **9.** $-10pq^2 - 5p^2q + 7pq - 4p + 2q + 3$
11. $3x + 2y - 2z - 3$ **13.** $5x\sqrt{y} - 4y\sqrt{x} - \frac{2}{5}$ **15.** $-5x^3 + 7x^2 - 3x + 6$ **17.** $-2x^2 + 6x - 2$ **19.** $6a - 5b - 2c + 4d$
21. $x^4 - 3x^3 - 4x^2 + 9x - 3$ **23.** $9x\sqrt{y} - 3y\sqrt{x} + 9.1$ **25.** $-1.047p^2q - 2.479pq^2 + 8.879pq - 104.144$

Margin Exercises, Section 1.4

1. $3x^3y^2 + 4x^2y^2 - xy^2 + 6y^2$ **2.** $2p^4q^2 + 3p^3q^2 + 3p^2q^2 + 2q^2$ **3.** $2x^3y - 4xy + 3x^3 - 6x$ **4.** $15x^2 - xy - 6y^2$
5. $6xy - 2\sqrt{2}x + 3\sqrt{2}y - 2$ **6.** $16x^2 - 40xy + 25y^2$ **7.** $4y^4 + 24x^2y^3 + 36x^4y^2$ **8.** $16x^2 - 49$ **9.** $25x^4y^2 - 4y^2$
10. $16y^4 - 3$ **11.** $4x^2 + 12x + 9 - 25y^2$ **12.** $25t^2 - 4x^6y^4$ **13.** $x^3 + 3x^2 + 3x + 1$ **14.** $x^3 - 3x^2 + 3x - 1$
15. $t^6 - 9t^4b + 27t^2b^2 - 27b^3$ **16.** $8a^9 - 60a^6b^2 + 150a^3b^4 - 125b^6$

Exercise Set 1.4, p. 23

1. $6x^3 + 4x^2 + 32x - 64$ **3.** $4a^3b^2 - 10a^2b^2 + 3ab^3 + 4ab^2 - 6b^3 + 4a^2b - 2ab + 3b^2$ **5.** $a^3 - b^3$
7. $4x^2 + 8xy + 3y^2$ **9.** $12x^3 + x^2y - \frac{3}{2}xy - \frac{1}{8}y^2$ **11.** $2x^3 - 2\sqrt{2}x^2y - \sqrt{2}xy^2 + 2y^3$ **13.** $4x^2 + 12xy + 9y^2$
15. $4x^4 - 12x^2y + 9y^2$ **17.** $4x^6 + 12x^3y^2 + 9y^4$ **19.** $\frac{1}{4}x^4 - \frac{3}{5}x^2y + \frac{9}{25}y^2$ **21.** $0.25x^2 + 0.70xy^2 + 0.49y^4$
23. $9x^2 - 4y^2$ **25.** $x^4 - y^2z^2$ **27.** $9x^4 - 2$ **29.** $4x^2 + 12xy + 9y^2 - 16$ **31.** $x^4 + 6x^2y + 9y^2 - y^4$ **33.** $x^4 - 1$
35. $16x^4 - y^4$ **37.** $0.002601x^2 + 0.00408xy + 0.0016y^2$ **39.** $2462.0358x^2 - 945.0214x - 38.908$
41. $y^3 + 15y^2 + 75y + 125$ **43.** $m^6 - 6m^4n + 12m^2n^2 - 8n^3$ **45.** A term is missing—found by calculating twice the
product of the terms; the first term is $9a^2$, not $3a^2$, where the 3 was not squared; $9a^2 + 6ab + b^2$ **47.** In step (1) 3x
should be 6x; the 2 and 3 were not multiplied. Similarly, -3 should be -12; the 4 and -3 were not multiplied. In step
(2) $3x - 3$ is not x since they are not like terms; $6x^2 + 6x - 12$

Margin Exercises, Section 1.5

1. $4x^2y(5x + 3)$ **2.** $(p + q)(2x + y + 2)$ **3.** $(4x - 3)(x + 5)$ **4.** $(x - 4)(x + 4)$ **5.** $(5y^2 + 4x)(5y^2 - 4x)$
6. $2(y^2 + 4x^2)(y - 2x)(y + 2x)$ **7.** $(x - \sqrt{3})(x + \sqrt{3})$ **8.** $(3y - 5)^2$ **9.** $(4x + 9y)^2$ **10.** $-3y^2(2x^2 - 5y^3)^2$
11. $(x + 7)(x - 2)$ **12.** $(3x + 2)(x + 1)$ **13.** $3(2x^2y^3 + 5)(x^2y^3 - 4)$ **14.** $(x - 2)(x^2 + 2x + 4)$
15. $(4 - t)(16 + 4t + t^2)$ **16.** $(3x + y)(9x^2 - 3xy + y^2)$ **17.** $(2m + 5t)(4m^2 - 10mt + 25t^2)$
18. $2y(4y^2 - 5x^2)(16y^4 + 20x^2y^2 + 25x^4)$

Exercise Set 1.5, p. 27

1. $3ab(6a - 5b)$ **3.** $(a + c)(b - 2)$ **5.** $(x + 6)(x + 3)$ **7.** $(3x - 5)(3x + 5)$ **9.** $4x(y^2 - z)(y^2 + z)$ **11.** $(y - 3)^2$
13. $(1 - 4x)^2$ **15.** $(2x - \sqrt{5})(2x + \sqrt{5})$ **17.** $(xy - 7)^2$ **19.** $4a(x + 7)(x - 2)$ **21.** $(a + b + c)(a + b - c)$
23. $(x + y - a - b)(x + y + a + b)$ **25.** $5(y^2 + 4x^2)(y - 2x)(y + 2x)$ **27.** $(x + 2)(x^2 - 2x + 4)$
29. $3(x - \frac{1}{2})(x^2 + \frac{1}{2}x + \frac{1}{4})$ **31.** $(x + 0.1)(x^2 - 0.1x + 0.01)$ **33.** $3(z - 2)(z^2 + 2z + 4)$
35. $(a - t)(a + t)(a^2 - at + t^2)(a^2 + at + t^2)$ **37.** $2ab(2a^2 + 3b^2)(4a^4 - 6a^2b^2 + 9b^4)$ **39.** $(x + 4.19524)(x - 4.19524)$
41. $37(x + 0.626y)(x - 0.626y)$ **43.** $h(3x^2 + 3xh + h^2)$

Margin Exercises, Section 1.6

1. $\{9\}$ **2.** $\{0, -1\}$ **3.** $\frac{4}{3}$ **4.** $\frac{17}{2}$ **5.** $-\frac{19}{8}$ **6.** $\emptyset$ **7.** $\emptyset$ **8.** $7, -\frac{3}{2}$ **9.** $5, -4$ **10.** $x > \frac{3}{2}$ **11.** $\frac{22}{13} \leq y$ **12.** $x < 5$
13. $\{x \mid x > \frac{5}{2}\}$ **14.** $\{y \mid y \geq -7\}$ **15.** $\{x \mid x^2 = 5\}$

Exercise Set 1.6, pp. 31–32

1. 12 **3.** -6 **5.** 8 **7.** $\frac{4}{5}$ **9.** 2 **11.** $-\frac{3}{2}$ **13.** -2 **15.** $\frac{3}{2}, \frac{2}{3}$ **17.** 0, 1, -2 **19.** $\frac{2}{3}, -1$ **21.** 4, 1 **23.** $-1, -2$
25. $-\frac{5}{3}, 4, \frac{5}{2}$ **27.** 0, $\frac{1}{4}, -\frac{1}{4}$ **29.** $x > 3$ **31.** $x \geq -\frac{5}{12}$ **33.** $y \geq \frac{22}{13}$ **35.** $x \leq \frac{15}{34}$ **37.** $x < 1$ **39.** $\{x \mid x > 2.5\}$
41. $\{t \mid t^2 = 5\}$ **43.** 0.7892 **45.** $x < -0.7848$

Margin Exercises, Section 1.7

1. All real numbers except 3 **2.** All real numbers except -3 and -4 **3.** $\dfrac{x^2 + 2xy + y^2}{14x^3 - 7x}$ **4.** $\dfrac{x^2 + 2x - 8}{x^2 + 4x + 4}$

5. $\dfrac{x^2 + 5x + 6}{x^2 - 2x - 15}$; all real numbers except 5; all real numbers except 5 and -3

6. $\dfrac{3x + 2}{x + 2}$; all real numbers except 0 and -2; all real numbers except -2

7. $\dfrac{y + 2}{y - 1}$; all real numbers except 1 and -1; all real numbers except 1   **8.** $\dfrac{3x - 3y}{x + y}$ **9.** $\dfrac{2a^2b + 2ab^2}{a - b}$

10. $\dfrac{3x^2 + 4x + 2}{x - 5}$ **11.** $\dfrac{2x^2 + 11}{x - 5}$ **12.** $\dfrac{4x^2 - xy + 4y^2}{2(2x - y)(x - y)}$ **13.** $\dfrac{x - 6}{(x + 4)(x + 6)}$ **14.** $\dfrac{1}{a - x}$ **15.** $\dfrac{b + a}{b - a}$

Exercise Set 1.7, pp. 38–40

1. All numbers except 0, 1   **3.** All numbers except 0, 3, -2 **5.** $\dfrac{x - 2}{x + 3}$; all numbers except -3 **7.** $\dfrac{1}{x - y}$

9. $\dfrac{(x + 5)(2x + 3)}{7x}$ **11.** $\dfrac{a + 2}{a - 5}$ **13.** $m + n$ **15.** $\dfrac{3(x - 4)}{2(x + 4)}$ **17.** $\dfrac{1}{x + y}$ **19.** $\dfrac{x - y - z}{x + y + z}$ **21.** 1   **23.** $\dfrac{y - 2}{y - 1}$

25. $\dfrac{x + y}{2x - 3y}$ **27.** $\dfrac{3x - 4}{x^2 - 4}$ **29.** $\dfrac{3y - 10}{(y - 5)(y + 4)}$ **31.** $\dfrac{4x - 8y}{x^2 - y^2}$ **33.** $\dfrac{3x - 4}{(x - 2)(x - 1)}$ **35.** $\dfrac{5a^2 + 10ab - 4b^2}{(a - b)(a + b)}$

37. $\dfrac{11x^2 - 18x + 8}{(x + 2)(x - 2)^2}$ **39.** 0   **41.** $\dfrac{x + y}{x}$ **43.** $\dfrac{a^2 - 1}{a^2 + 1}$ **45.** $\dfrac{c^2 - 2c + 4}{c}$ **47.** $\dfrac{xy}{x - y}$ **49.** $x - y$ **51.** $\dfrac{x^2 - y^2}{xy}$ **53.** $\dfrac{1 + a}{1 - a}$

55. $\dfrac{b + a}{b - a}$ **57.** $2x + h$ **59.** $3x^2 + 3xh + h^2$ **61.** x^5 **63.** Step (1) uses the wrong reciprocal. The reciprocal of a sum is not the sum of the reciprocals. Step (2) would be correct if step (1) had been. Step (3) would be correct if steps (1) and (2) had been. Step (4) has an improper simplification of the b in the denominator; $\dfrac{12a}{b(4a + 3b)}$

Margin Exercises, Section 1.8

1 $\{x \mid x \geq 2\}$ **2.** $\{x \mid x \geq -3\}$ **3.** All real numbers are sensible replacements.

4. All real numbers are sensible replacements.   **5.** $6|ab|$ **6.** x^8 **7.** $10m^2n^2|n|$ **8.** $\dfrac{2x^2|x|}{y^2}$ **9.** $|x + 2|$

10. $|x| \cdot |y - 2|$ or $|x(y - 2)|$ **11.** $|x + 2|$ **12.** $|x + 4|$ **13.** $-4xy$ **14.** $\sqrt{133}$ **15.** $\sqrt{x^2 - 4y^2}$ **16.** $\sqrt[4]{81}$, or 3

17. $10\sqrt{3}$ **18.** $6|y|$ **19.** $|x + 1|\sqrt{2}$ **20.** $2 \cdot \sqrt[3]{2}$ **21.** $(a + b) \cdot \sqrt[3]{a + b}$ **22.** $\frac{7}{8}$ **23.** $\dfrac{5}{|y|}$ **24.** $\dfrac{\sqrt{35}}{5}$ **25.** $\dfrac{\sqrt[3]{7}}{5}$ **26.** 5

27. $\dfrac{|x|}{5}$ **28.** $\dfrac{2x}{y}$ **29.** 3^{10} **30.** 3^4 **31.** 37.42 mph

Exercise Set 1.8, pp. 46–47

1. $\{x \mid x \geq 3\}$ **3.** $\{x \mid x \leq \frac{3}{4}\}$ **5.** $9|xy|$ **7.** $3a^2|b|$ **9.** 11   **11.** $4|x|$ **13.** $|b + 1|$ **15.** $-3x$ **17.** $|x - 2|$ **19.** 2

21. $6\sqrt{5}$ **23.** $3\sqrt[3]{2}$ **25.** $\dfrac{8|c|\sqrt{2}}{d^2}$ **27.** $3\sqrt{2}$ **29.** $2x^2y\sqrt{6}$ **31.** $3x\sqrt[3]{4y}$ **33.** $2(x + 4)\sqrt[3]{(x + 4)^2}$ **35.** $\sqrt{7b}$ **37.** 2

39. $\dfrac{1}{2x}$ **41.** $\sqrt{a + b}$ **43.** $\dfrac{3a\sqrt{2b}}{4b}$ **45.** $\dfrac{y \cdot \sqrt[3]{20x^2z^2}}{5z^2}$ **47.** $8x^2\sqrt[3]{2}$ **49.** $ab^2x^2y\sqrt{a}$

51. 1.57 sec, 3.14 sec, 8.88126 sec, 11.1016 sec   **53.** $10.124x^2y$ **55.** $\dfrac{0.5933a\sqrt{b}}{b}$ **57.** $h = \dfrac{a}{2}\sqrt{3}$

Margin Exercises, Section 1.9

1. $-6\sqrt{5}$ **2.** $(10y + 7)\sqrt[3]{2y}$ **3.** $-4 - 9\sqrt{6}$ **4.** $\dfrac{\sqrt{3} + \sqrt{5}}{-2}$ **5.** $\dfrac{x - 7\sqrt{x} + 10}{x - 4}$ **6.** $\dfrac{1}{\sqrt{a + 2} + \sqrt{a}}$

7. $\dfrac{x - 5}{x + 2\sqrt{5x} + 5}$

Exercise Set 1.9, pp. 49–50

1. $-12\sqrt{5} - 2\sqrt{2}$ **3.** $19\sqrt[3]{x^2} - 3x$ **5.** $4y\sqrt{3} - 2y\sqrt{6}$ **7.** 1   **9.** $t - 2x\sqrt{t} + x^2$ **11.** $10\sqrt{7}$ **13.** x   **15.** $\dfrac{3(3-\sqrt{5})}{2}$

17. $\dfrac{2\sqrt[3]{6}}{3}$ **19.** $\dfrac{8x - 20\sqrt{xy} - 6x\sqrt{y} + 15y\sqrt{x}}{4x - 25y}$ **21.** $\dfrac{2 - 5a}{6(\sqrt{2} - \sqrt{5a})}$ **23.** $\dfrac{x}{x + 2 - 2\sqrt{x+1}}$ **25.** $\dfrac{a}{3(\sqrt{a+3} + \sqrt{3})}$

27. $\dfrac{(2 + x^2)\sqrt{1+x^2}}{1+x^2}$ **29.** Let $a = 16$ and $b = 9$. Then $\sqrt{a+b} = 25$ and $\sqrt{a} + \sqrt{b} = 7$.

Margin Exercises, Section 1.10

1. $\sqrt{n^3}$, or $n\sqrt{n}$ **2.** $\dfrac{1}{\sqrt[7]{y^6}}$ **3.** 16   **4.** $\dfrac{1}{16}$ **5.** $(5ab)^{4/3}$ or $(5ab)\sqrt[3]{5ab}$ **6.** 8   **7.** $a^{2/3}$ **8.** $4^{1/6}$, $(2^2)^{1/6}$ or $2^{1/3}$

9. $5^{11/6}$ or $5\sqrt[6]{5^5}$ **10.** $\sqrt[4]{a^5}$ or $a\sqrt[4]{a}$ **11.** $\dfrac{1}{\sqrt[5]{x^6}}$ or $\dfrac{1}{x\sqrt[5]{x}}$ **12.** $\sqrt[4]{2^3} + \dfrac{1}{\sqrt[4]{2}}$ or $3\dfrac{\sqrt[4]{2^3}}{2}$ **13.** $\sqrt[6]{200}$ **14.** $\sqrt[6]{x^4y^3z^5}$

15. $\sqrt[4]{(x+y)}$

Exercise Set 1.10, pp. 52–53

1. $\sqrt[4]{x^3}$ **3.** $(\sqrt[4]{16})^3$ or $\sqrt[4]{(2^4)^3}$ or 8   **5.** $\frac{1}{5}$ **7.** $\dfrac{a}{b}\sqrt[4]{ab}$ **9.** $20^{2/3}$ **11.** $13^{5/4}$ **13.** $11^{1/6}$ **15.** $5^{5/6}$ **17.** 4   **19.** $2y^2$

21. $(a^2 + b^2)^{1/3}$ **23.** $3ab^3$ **25.** $\dfrac{m^2n^4}{2}$ **27.** $8a^{4/2}$ or $8a^2$ **29.** $\dfrac{x^{-3}}{3^{-1}b^2}$ or $\dfrac{3}{x^3b^2}$ **31.** $xy^{1/3}$ or $x\sqrt[3]{y}$ **33.** $\sqrt[6]{288}$

35. $\sqrt[12]{x^{11}y^7}$ **37.** $a\sqrt[6]{a^5}$ **39.** $(a+x)\sqrt[12]{(a+x)^{11}}$ **41.** 24.685   **43.** 43.138   **45.** 32.942   **47.** 5.56 ft   **49.** 7.07 ft   **51.** $a^{a/2}$

Margin Exercises, Section 1.11

1. $18{,}600 \dfrac{m}{sec}$ **2.** $0.5 \dfrac{m}{sec}$ **3.** 62 ft   **4.** $\frac{23}{20}$ kg   **5.** $105 \dfrac{cm}{sec}$ **6.** 12 yd   **7.** 80 oz   **8.** $\frac{7}{10}$ **9.** $11.25 \dfrac{\text{in.-lb}}{hr^2}$ **10.** $4\dfrac{lb^2}{m^2}$

11. 1224 in.   **12.** 58,080 ft   **13.** 18,000 sec   **14.** 20 yd   **15.** 36.96 km   **16.** 100 hr   **17.** $176\dfrac{ft}{sec}$ **18.** $0.36\ m^2$ **19.** $50\dfrac{g}{cm^3}$

20. $300\dfrac{\text{¢}}{hr}$

Exercise Set 1.11, p. 56

1. 12 yd   **3.** 48 hr   **5.** 3 g   **7.** 8 m   **9.** $12\ ft^3$ **11.** $\dfrac{7\ kg^2}{10\ m^2}$ **13.** $720\dfrac{\text{lb-mi}^2}{\text{hr}^2\text{-ft}}$ **15.** $\dfrac{15}{2}\dfrac{cm^5\text{-kg}}{sec^3}$ **17.** 6 ft   **19.** 172,800 sec

21. 600 g/cm   **23.** $2{,}160{,}000\ cm^2$ **25.** 150¢/hr   **27.** 6.228 L/hr   **29.** 5,865,696,000,000 mi/yr   **31.** 1621.8 m/min

33. $1638.4\ km^2$ **35.** 7.5 g, 1250 g   **37.** 1600 g   **39.** 15 moles

Chapter 1 Review, pp. 57–58

1. $12, -3, -1, -19, 31, 0$ **2.** 12, 31   **3.** All except $\sqrt{7}, \sqrt[3]{10}$ **4.** All   **5.** $\sqrt{7}, \sqrt[3]{10}$ **6.** 0, 12, 31   **7.** -4 **8.** -16

9. -5 **10.** 30   **11.** -6 **12.** 153   **13.** -3000 **14.** $-\frac{3}{16}$ **15.** $\frac{31}{24}$ **16.** 3,261,000   **17.** 0.00041   **18.** 1.432×10^{-2}

19. 4.321×10^4 **20.** $-14a^{-2}b^7$ **21.** $6x^9y^{-6}z^6$ **22.** 3   **23.** -2 **24.** $\dfrac{b}{a}$ **25.** $\dfrac{x+y}{xy}$ **26.** -4 **27.** $25x^4 - 10x^2\sqrt{2} + 2$

28. $13\sqrt{5}$ **29.** $x^3 + t^3$ **30.** $125a^3 + 300a^2b + 240ab^2 + 64b^3$ **31.** $8xy^4 - 9xy^2 + 4x^2 + 2y - 7$ **32.** $(x^2 - 3)(x + 2)$

33. $3a(2a - 3b^2)(2a + 3b^2)$ **34.** $(x + 12)^2$ **35.** $x(9x - 1)(x + 4)$ **36.** $(2x - 1)(4x^2 + 2x + 1)$

37. $(3x^2 + 5y^2)(9x^4 - 15x^2y^2 + 25y^4)$ **38.** $y^3 \cdot \sqrt[6]{y}$ **39.** $\sqrt[3]{(a+b)^2}$ **40.** $\sqrt[5]{b^7}$ **41.** $\dfrac{m^4n^2}{3}$ **42.** $6, -3$ **43.** -20

44. $-5, 3$ **45.** $-2, 1$ **46.** $y > -2$ **47.** $x \geq 5$ **48.** 3   **49.** $\dfrac{x-5}{(x+3)(x+5)}$ **50.** $\dfrac{x - 2\sqrt{xy} + y}{x - y}$ **51.** $166\frac{2}{3}$ m/min

52. Inverses $(+)$ **53.** Distributive   **54.** Associative $(\times)$ **55.** Commutative $(\times)$

CHAPTER 2

Margin Exercises, Section 2.1

1. Yes   **2.** Yes   **3.** No   **4.** Yes   **5.** No   **6.** No   **7.** Add $5x^2$ **8.** Add $-5x^2$ **9.** $\{0, 5\}$ **10.** $\{-\frac{1}{5}\}$ **11.** $\{0, -\frac{1}{3}, 4\}$
12. $\{1, -1, -\frac{1}{5}\}$ **13.** $\{2, -2, -\frac{1}{3}\}$ **14.** $\emptyset$ **15.** $\emptyset$ **16.** $\{6\}$ **17.** $\{4\}$ **18.** $\{6, -6\}$ **19.** $\{\frac{16}{5}\}$

Exercise Set 2.1, pp. 64–65

1. Yes   **3.** No   **5.** No   **7.** $\{0, 3\}$ **9.** $\{0, 2, -\frac{1}{3}\}$ **11.** $\{\frac{3}{2}, -\frac{2}{3}, 1\}$ **13.** $\{\frac{1}{2}, 0, -3\}$ **15.** $\{-2\}$ **17.** $\{6\}$ **19.** $\emptyset$ **21.** $\emptyset$
23. $\{8, -5\}$ **25.** $\{\frac{5}{3}\}$ **27.** $\{0, 2.1522\}$ **29.** $\{0.94656\}$ **31.** $\{-1, 1, -\frac{1}{5}\}$ **33.** $\{-2, 1, -1\}$
35. (1) Equivalent to (2): (2) not equivalent to (3); (3) equivalent to (4)   **37.** Identity   **39.** Identity   **41.** Not an identity

Margin Exercises, Section 2.2

1. $F = \frac{9}{5}C + 32$ **2.** $r_2 = \dfrac{Rr_1}{r_1 - R}$ **3.** 18%   **4.** 57   **5.** 27   **6.** \$2450   **7.** \$2565.78   **8.** \$2637.93   **9.** 20 ft
10. 36 km/h   **11.** 375 km   **12.** 50 km/h, 60 km/h   **13.** $2\frac{2}{9}$ hr   **14.** Helen: 12 hr; Fran: 6 hr

Exercise Set 2.2, pp. 73–75

1. $w = \dfrac{P - 2l}{2}$ **3.** $I = \dfrac{E}{R}$ **5.** $T_1 = \dfrac{T_2 P_1 V_1}{P_2 V_2}$ **7.** $v_1 = \dfrac{H}{Sm} + v_2$ **9.** $p = \dfrac{Fm}{m - F}$ **11.** $x = \dfrac{5 + ab}{a - b}$ **13.** $x = -\dfrac{a}{9}$
15. 44%   **17.** 6%   **19.** \$14.500; \$16.095   **21.** \$650   **23.** 26°, 130°, 24°   **25.** 68 m, 93 m   **27.** 91%   **29.** 2 cm
31. 810,000   **33.** 12 km/h   **35.** A: 46 mph; B: 58 mph   **37.** 98.3 mi   **39.** $1\frac{34}{71}$ hr   **41.** 6.21 hr
43. (a) \$1137.50; (b) \$1142.23; (c) \$1144.75; (d) \$1147.37; (e) \$1147.40   **45.** 32 mph

Margin Exercises, Section 2.3

1. $\pm\sqrt{3}$ **2.** 0   **3.** $\pm\sqrt{\dfrac{\pi}{3}}$ **4.** $\pm\sqrt{\dfrac{n}{m}}$ **5.** $-4 \pm \sqrt{7}$ **6.** $5 \pm \sqrt{3}$ **7.** $-3, -7$ **8.** $4, (x + 2)^2$ **9.** $9, (x - 3)^2$

10. $\frac{25}{4}, (x + \frac{5}{2})^2$ **11.** $\frac{49}{4}, (x - \frac{7}{2})^2$ **12.** $\frac{9}{64}, (x + \frac{3}{8})^2$ **13.** $\frac{1}{4}, (x - \frac{1}{2})^2$ **14.** $-2 \pm \sqrt{7}$ **15.** 4, 2   **16.** 2, 3   **17.** $\dfrac{-1 \pm \sqrt{7}}{2}$

18. $-1, \frac{1}{4}$ **19.** $\frac{1}{2}, -4$ **20.** $\dfrac{4 \pm \sqrt{31}}{5}$ **21.** One real solution   **22.** No real solutions   **23.** Two real solutions
24. $x^2 + \frac{7}{3}x - \frac{20}{3} = 0$, or $3x^2 + 7x - 20 = 0$ **25.** $x^2 - x - 56 = 0$ **26.** $x^2 - (m + n)x + mn = 0$

Exercise Set 2.3, pp. 80–81

1. $\pm\sqrt{7}$ **3.** $\pm\dfrac{\sqrt{5}}{3}$ **5.** $\pm\sqrt{\dfrac{b}{a}}$ **7.** $7 \pm \sqrt{5}$ **9.** $h \pm \sqrt{a}$ **11.** $-3 \pm \sqrt{5}$ **13.** $\{3, -10\}$ **15.** $\dfrac{2 \pm \sqrt{14}}{5}$ **17.** $-5, \frac{3}{2}$
19. $1, -5$ **21.** $2, -\frac{1}{2}$ **23.** $-1, -\frac{5}{3}$ **25.** $6 \pm \sqrt{33}$ **27.** No real solution   **29.** Two real solutions
31. $x^2 + 2x - 99 = 0$ **33.** $x^2 - 14x + 49 = 0$ **35.** $x^2 - \frac{4}{5}x - \frac{12}{25} = 0$ **37.** $x^2 - \left(\dfrac{c + d}{2}\right)x + \dfrac{cd}{4} = 0$
39. $x^2 - 4\sqrt{2}x + 6 = 0$ **41.** 1.1754, -0.4254 **43.** $2, -\frac{3}{2}$ **45.** $\dfrac{-3 \pm \sqrt{41}}{2}$ **47.** $\frac{3}{2}, \frac{2}{3}$ **49.** $-0.1 \pm \sqrt{0.31}$
51. $\dfrac{-1 \pm \sqrt{1 + 4\sqrt{2}}}{2}$ **53.** $\dfrac{-\sqrt{5} \pm \sqrt{5 + 4\sqrt{3}}}{2}$ **55.** $\dfrac{\sqrt{6} \pm \sqrt{6 + 8\sqrt{10}}}{4}$ **57.** $-2, \frac{3}{4}$ **59.** $\dfrac{1 \pm \sqrt{113}}{2}$ **61.** $3 \pm \sqrt{5}$

Margin Exercises, Section 2.4

1. $r = \sqrt{\dfrac{3V}{\pi h}}$ **2.** $t = \dfrac{-v_0 + \sqrt{v_0^2 + 64S}}{32}$ **3.** 18.75%   **4.** $12 - 2\sqrt{22} \approx 2.619$ ft   **5.** (a) 4.33 sec; (b) 1.87 sec; (c) 44.9 m

Exercise Set 2.4, pp. 84–86

1. $d = \sqrt{\dfrac{kM_1M_2}{F}}$ **3.** $t = \sqrt{\dfrac{2S}{a}}$ **5.** $t = \dfrac{-v_0 \pm \sqrt{v_0^2 - 64s}}{-32}$ **7.** $n = \dfrac{3 + \sqrt{9 + 8d}}{2}$ **9.** $i = -1 + \sqrt{\dfrac{A}{P}}$ **11.** 18.75%

13. 11%   **15.** 9   **17.** 2 ft   **19.** 4.685 cm   **21.** A: 15 mph; B: 20 mph   **23.** (a) 3.91 sec; (b) 1.906 sec; (c) 79.6 m

25. 3.237 cm   **27.** 7   **29.** 12   **31.** $2, -\dfrac{3}{k}$ **33.** $\dfrac{1}{m + n}, \dfrac{-2}{m + n}$ **35.** 11.7%

Margin Exercises, Section 2.5

1. $\emptyset$ **2.** 4   **3.** $\frac{17}{3}$ **4.** 5   **5.** $b = \sqrt{\dfrac{a^2}{A^2 - 1}}$, or $\dfrac{a}{\sqrt{A^2 - 1}}$

Exercise Set 2.5, pp. 88–89

1. $\frac{5}{3}$ **3.** $\pm\sqrt{2}$ **5.** $\emptyset$ **7.** 4   **9.** $\emptyset$ **11.** -6 **13.** $3, -1$ **15.** $\frac{80}{9}$ **17.** 62.4459   **19.** $L = \dfrac{gT^2}{4\pi^2}, g = \dfrac{4L\pi^2}{T^2}$ **21.** 208 mi

23. 14.400 ft   **25.** $5 \pm 2\sqrt{2}$ **27.** $-\frac{8}{9}$ **29.** 2   **31.** $\dfrac{-5 + \sqrt{61}}{18}$

Margin Exercises, Section 2.6

1. (a) 9; (b) $\sqrt{x} = 12 - x$; $(12 - x)^2 = x$; $x = 144 - 24x + x^2$; $0 = x^2 - 25x + 144$; $0 = (x - 9)(x - 16)$. The procedure in (a) was probably easier, since the factoring was easier.   **2.** $\pm\sqrt{\dfrac{5 + \sqrt{3}}{2}}, \pm\sqrt{\dfrac{5 - \sqrt{3}}{2}}$ **3.** $\pm\sqrt{3}, 0$ **4.** $125, -8$

5. 306.25 ft

Exercise Set 2.6, pp. 92–93

1. 1, 81   **3.** $\pm\sqrt{5}$ **5.** -27.8 **7.** 16   **9.** 7, 5, -1, 1   **11.** $1, 4, \dfrac{5 \pm \sqrt{37}}{2}$ **13.** $\pm\sqrt{2 + \sqrt{6}}$ **15.** $-\frac{1}{2}, \frac{1}{3}$ **17.** $-1, 2$

19. $-1 \pm \sqrt{3}, \dfrac{9 \pm \sqrt{89}}{2}$ **21.** $\frac{100}{99}$ **23.** $-\frac{6}{7}$ **25.** 132.66 ft   **27.** 2.0486   **29.** 1, 4

Margin Exercises, Section 2.7

1. $y = 160x$ **2.** 4.5 kg   **3.** 50 volts   **4.** 176,250 tons   **5.** $y = \dfrac{6.4}{x}$ **6.** 7.5 hr   **7.** $\dfrac{A_2}{A_1} = \dfrac{r_2^2}{r_1^2}$ **8.** $y = kx^2; k = 3$

9. $\dfrac{W_2}{W_1} = \dfrac{d_1^2}{d_2^2}$ **10.** $y = \dfrac{9}{x^2}$ **11.** $\dfrac{A_2}{A_1} = \dfrac{b_2 h_2}{b_1 h_1}$ **12.** $y = 7xz$ **13.** $y = 7\dfrac{xz}{w^2}$ **14.** 2 sec   **15.** (a) 128 lb; (b) 4000 mi

Exercise Set 2.7, pp. 98–99

1. $y = \frac{3}{2}x$ **3.** $y = \dfrac{0.0015}{x^2}$ **5.** $y = \dfrac{xz}{w}$ **7.** $y = \dfrac{5}{4}\dfrac{xz}{w^2}$ **9.** y is doubled   **11.** y is multiplied by $\dfrac{1}{n^2}$ **13.** 532,500 tons

15. L is multiplied by 16   **17.** 68.56 m

19. If p varies directly as q, then $p = kq$. Thus, $q = \dfrac{1}{k}p$, so q varies directly as p.   **21.** $\dfrac{\pi}{4}$

3

Chapter 2 Review, pp. 99–100

1. -1 **2.** $3, -\frac{2}{3}, -2$ **3.** $\frac{4}{3}, -2$ **4.** $\dfrac{-3 \pm \sqrt{13}}{2}$ **5.** $\dfrac{3 \pm \sqrt{57}}{6}$ **6.** $\frac{27}{7}$ **7.** $\pm\sqrt{\dfrac{3 \pm \sqrt{5}}{2}}$ **8.** 1   **9.** $\pm\sqrt{3}, 0$

10. $-8, 125$ **11.** 5   **12.** 0, 3   **13.** $8, -2$ **14.** No real solutions   **15.** Two real solutions

16. $x^2 + \frac{5}{2}x - \frac{3}{2} = 0$, or $2x^2 + 5x - 3 = 0$ **17.** $h = \dfrac{v^2}{2g}$ **18.** $t = \dfrac{ab}{a + b}$ **19.** 94%   **20.** $1\frac{1}{3}\,\text{hr}$ **21.** $1\frac{1}{2}\,\text{hr}$ **22.** 60

23. 4.5   **24.** $80\,\text{km/h}$ **25.** 8, 15, 17   **26.** $2 + 2\sqrt{2} \approx 4.8\,\text{km/h}$ **27.** $y = \dfrac{0.5}{x^2}$ **28.** $T = \dfrac{1}{180} \cdot \dfrac{x^2}{p}$ **29.** \$2.27

30. $s = 16t^2$; $7\frac{1}{2}\,\text{sec}$ **31.** $-(a + c)$

CHAPTER 3

Margin Exercises, Section 3.1

1. $\{(d, 1), (d, 2), (e, 1), (e, 2), (f, 1), (f, 2)\}$
2. $\{(1, 1), (1, 2), (1, 3), (1, 4), (2, 1), (2, 2), (2, 3), (2, 4), (3, 1), (3, 2), (3, 3), (3, 4), (4, 1), (4, 2), (4, 3), (4, 4)\}$
3. $\{(1, 1), (2, 2), (3, 3), (4, 4)\}$ **4.** $\{(2, 1), (3, 1), (3, 2), (4, 1), (4, 2), (4, 3)\}$ **5.** Domain $= \{1, 2, 3, 4\}$; range $= \{1, 2, 3, 4\}$
6. Domain $= \{2, 3, 4\}$; range $= \{1, 2, 3\}$ **7.** Domain $= \{1, 2\}$; range $= \{1, 2, 3\}$

Exercise Set 3.1, p. 104

1. $\{(0, a), (0, b), (0, c), (2, a), (2, b), (2, c)\}$ **3.** $\{(-1, 0), (-1, 1), (-1, 2), (0, 1), (0, 2), (1, 2)\}$
5. $\{(-1, -1), (-1, 0), (-1, 1), (-1, 2), (0, 0), (0, 1), (0, 2), (1, 1), (1, 2), (2, 2)\}$ **7.** $\{(-1, -1), (0, 0), (1, 1), (2, 2)\}$
9. Domain $\{0, 1\}$; range $(0, 1, 2)$

Margin Exercises, Section 3.2

1. (a), (b), (c); (d) Domain $= \{-5, -4, 3\}$; range $= \{-2, 2, 3\}$

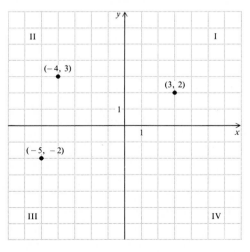

2. Yes   **3.** No   **4.** No   **5.** Yes

6.

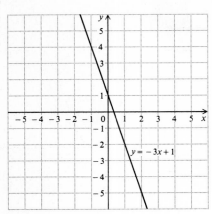

$y = -3x + 1$

7.

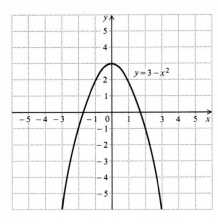

$y = 3 - x^2$

8. The shapes are the same, but this curve opens to the right instead of up.

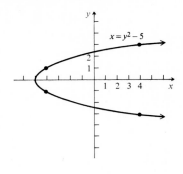

$x = y^2 - 5$

9.

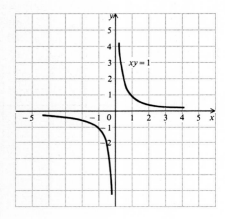

$xy = 1$

10. The shapes are the same, but this graph opens to the right instead of up.

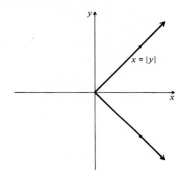

$x = |y|$

11.

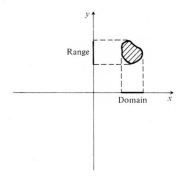

Range

Domain

12.

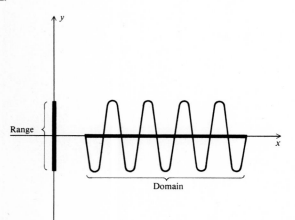

Range

Domain

13.

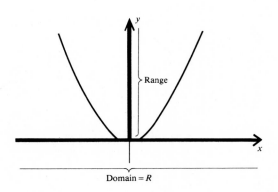

Range

Domain $= R$

Exercise Set 3.2, pp. 110–111

1. Yes   **3.** No   **5.** No

7.

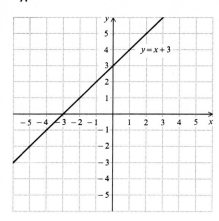

9.

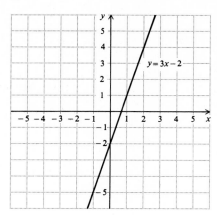

11.

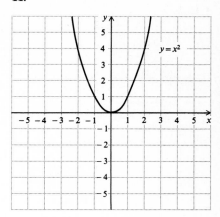

13.

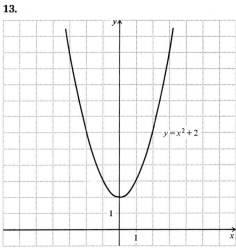

15.

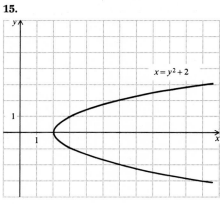

17.

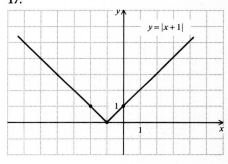

19.

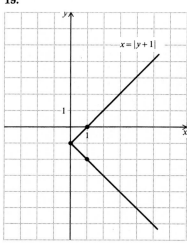

21.

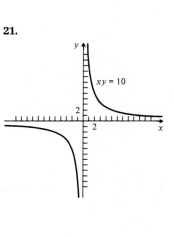

23. Same graphs.

25. Domain $\{x\,|\,2 \leq x \leq 6\}$; range $\{y\,|\,1 \leq y \leq 5\}$ **27.** Horizontal line through $(0, 2)$
29. Line through $(0, 1)$ and $(-1, 0)$ **31.** Line through $(0, 0)$ and $(1, 2)$ **33.** See Exercise 11.
35.

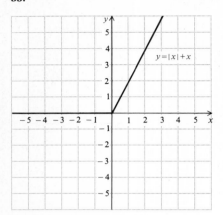

37.

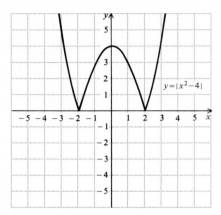

39.

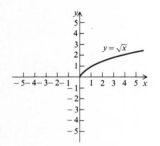

41.

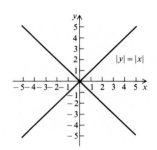

43.

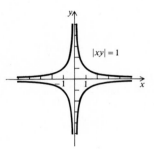

Margin Exercises, Section 3.3

1. (a), (b), (c) **2.** 1, 4, 3, 4; domain $= \{1, 2, 3, 4\}$; range $= \{1, 3, 4\}$ **3.** 1, 3, 4.5
4. (a) 1; (b) 4; (c) 4; (d) $12a^2 + 1$; (e) $3a^2 + 6a + 4$; (f) $6a + 3h$ **5.** $\{x\,|\,x \neq -\frac{4}{3} \text{ and } x \neq -2\}$
6. $\{x\,|\,x \geq -2.5\}$ **7.** All real numbers **8.** $f \circ g(x) = 2x^2 - 2$; $g \circ f(x) = 2x^2 - 8x + 8$
9. $u \circ v(x) = 18x^2 + 24x + 8$; $v \circ u(x) = 6x^2 + 2$

Exercise Set 3.3, pp. 117–119

1. (b), (c), (d) **3.** (a) 0; (b) 1; (c) 57; (d) $5t^2 + 4t$; (e) $5t^2 - 6t + 1$; (f) $10a + 5h + 4$

5. (a) 5; (b) -2; (c) -4; (d) $4|y| + 6y$; (e) $2|a + h| + 3a + 3h$; (f) $\dfrac{2|a + h| + 3h - - 2|a|}{h}$

7. (a) 3.14977; (b) 55.73147; (c) 3178.20675; (d) 1116.70323 **9.** (a) $\frac{2}{3}$; (b) $\frac{10}{9}$; (c) 0; (d) not possible **11.** All real numbers
13. $\{x\,|\,x \neq 0\}$ **15.** $\{x\,|\,x \geq -\frac{4}{7}\}$ **17.** $\{x\,|\,x \neq 2, -2\}$ **19.** $\{x\,|\,x \neq -\frac{3}{4}, 2\}$
21. $f \circ g(x) = 12x^2 - 12x + 5$; $g \circ f(x) = 6x^2 + 3$ **23.** $f \circ g(x) = \dfrac{16}{x^2} - 1$; $g \circ f(x) = \dfrac{2}{4x^2 - 1}$

25. $f \circ g(x) = x^4 - 2x^2 + 2$; $g \circ f(x) = x^4 + 2x^2$ **27.** 0, -3, 3, 2 **29.** $\dfrac{-1}{x(x + h)}$ **31.** $\dfrac{1}{\sqrt{x + h} + \sqrt{x}}$

33. $\{x\,|\,x \neq 2, -1 \text{ and } x \geq -3\}$ **35.** All real numbers
37. Domain of $f \circ g$ is $\{x\,|\,x \neq 0\}$; domain of $g \circ f$ is $\{x\,|\,x \neq \frac{1}{2}, -\frac{1}{2}\}$.

Margin Exercises, Section 3.4

1. (a) $(-3, 2)$; (b) $(4, -5)$

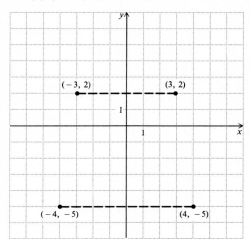

2. (a) $(4, -3)$; (b) $(3, 5)$

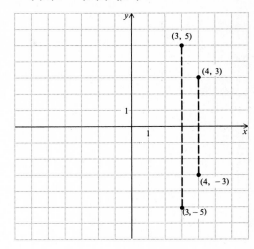

3. x-axis: no; y-axis: yes

6. x-axis: yes; y-axis: yes

9. (a) $(-3, -2)$; (b) $(4, -3)$;
     (c) $(5, 7)$

4. x-axis: yes; y-axis: no

7. a-axis: no; b-axis: no

10. Yes **11.** Yes **12.** Yes **13.** Yes **14.** Yes **15.** No

5. x-axis: yes; y-axis: yes

8. p-axis: no; q-axis; no

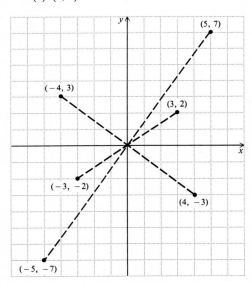

Exercise Set 3.4, p. 125

1. x-axis, no; y-axis, yes; origin, no   **3.** x-axis, no; y-axis, yes; origin, no   **5.** All yes   **7.** All yes   **9.** All no
11. All no   **13.** Yes   **15.** Yes   **17.** Yes   **19.** Yes   **21.** No   **23.** No

25.

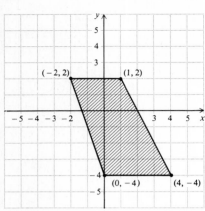

27.

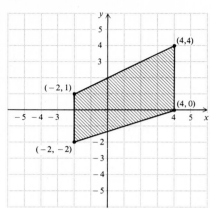

Margin Exercises, Section 3.5

1. (a) Same shape, but the second one is moved up 2 units; (b) same shape as $y = x^2$, but moved down 2 units.
2. Same shape as $y = x^2$, but moved down 3 units.

3.

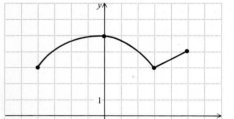

4.

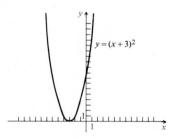

$y = (x + 3)^2$

5.

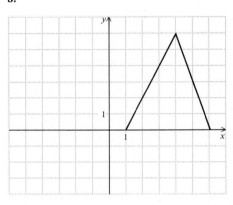

6.

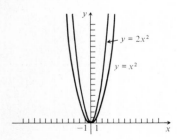

$y = 2x^2$

$y = x^2$

7.

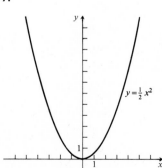

$y = \frac{1}{2} x^2$

8.

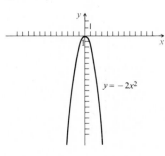

$y = -2x^2$

9.

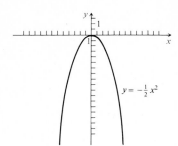

$y = -\frac{1}{2}x^2$

10.

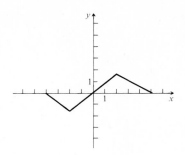

11.

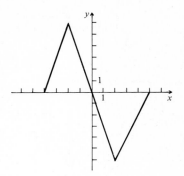

12.

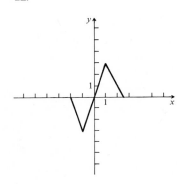

13.

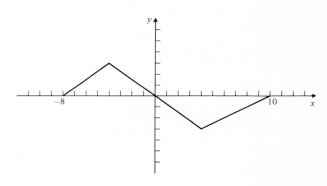

Exercise Set 3.5, pp. 132–134

1. and 3.

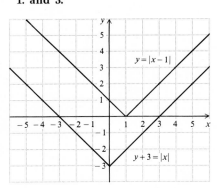

$y = |x - 1|$

$y + 3 = |x|$

5. and 7.

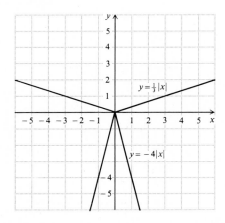

$y = \frac{1}{3}|x|$

$y = -4|x|$

9. and 11.

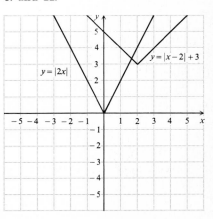

$y = |2x|$

$y = |x - 2| + 3$

13.

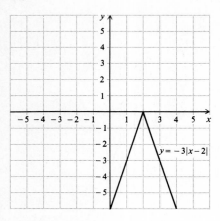

$y = -3|x-2|$

15. and 17.

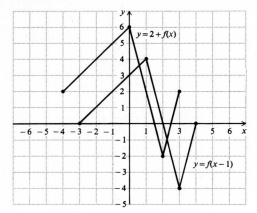

$y = 2 + f(x)$

$y = f(x-1)$

19.

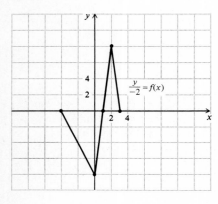

$\dfrac{y}{-2} = f(x)$

21.

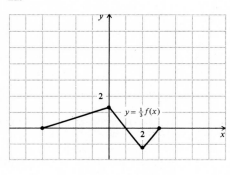

$y = \frac{1}{3} f(x)$

23.

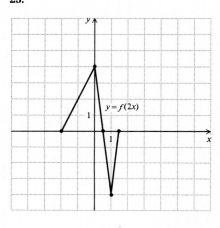

$y = f(2x)$

25.

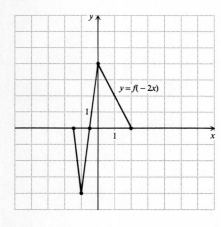

$y = f(-2x)$

27.

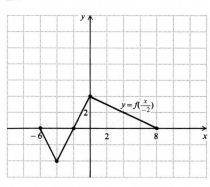

$y = f\left(\frac{x}{-2}\right)$

29.

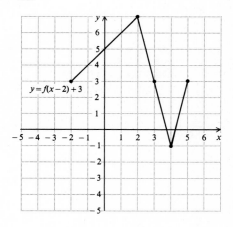

$y = f(x-2) + 3$

31.

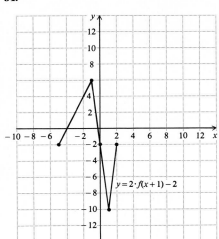

33.

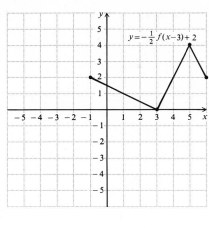

35.

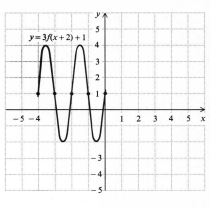

37. The graph is translated 1.8 units to the left, stretched vertically by a factor of $\sqrt{2}$, and reflected across the x-axis.

Margin Exercises, Section 3.6

1. (a) Yes; (b) yes; (c) no; (d) yes　**2.** (a) No; (b) yes; (c) yes; (d) yes　**3.** (a) Odd; (b) odd; (c) even; (d) odd
4. (a) Neither; (b) even; (c) odd; (d) neither; (e) neither　**5.** $f(x) = f(x + 4)$, $f(x) = f(x + 6)$　**6.** (a) Yes; (b) 3
7. $t(x) = t(x + 1)$; t is periodic. It does not have a period, because there is no *smallest p* for which $t(x + p) = t(x)$.
11. (a) $[4, 5\frac{1}{2}]$; (b) $(-3, 0]$; (c) $[-\frac{1}{2}, \frac{1}{2})$; (d) $(-\pi, \pi)$　**12.** (a) No; (b) yes; (c) no; (d) yes; (e) no
13. Where $x = -3, 0, 2$　**14.** (a) Increasing; (b) increasing; (c) decreasing; (d) neither
15. Increasing: $[-3, 0]$; decreasing: $[0, 3]$; there are many answers.
16.

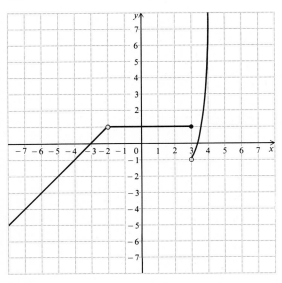

Exercise Set 3.6, pp. 140–145

1. (a) Even; (b) even; (c) odd; (d) neither   **3.** Neither   **5.** Even   **7.** Neither   **9.** Even   **11.** Odd   **13.** Neither
15. Odd   **17.** Even and odd   **19.** Even   **21.** (a) No; (b) yes; (c) yes; (d) no   **23.** 4
25. (a) $(-2, 2)$; (b) $(-5, -1]$; (c) $[c, d]$; (d) $[-5, 1)$ **27.** (a) $(-2, 4)$; (b) $(-\frac{1}{4}, \frac{1}{4}]$; (c) $[7, 10\pi)$; (d) $[-9, -6]$
29. (a) Yes; (b) yes; (c) no; (d) yes; (e) yes   **31.** Where $x = -3$ and $x = 2$
33. (a) Increasing; (b) neither; (c) decreasing; (d) neither
35.   **37.**

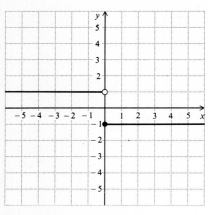   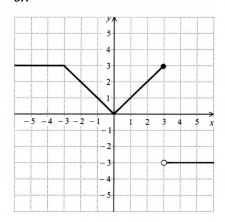

39.

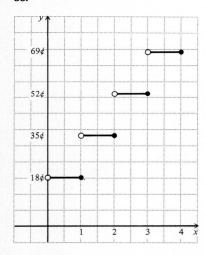

41. Increasing: $[0, 1]$; decreasing: $[-1, 0]$; many possible answers
43. Increasing: (a), (e); decreasing: (b); neither: (c), (d), (f)   **45.** (a) $[2, 3]$; (b) $(0, 9]$; (c) $(-6, 1)$

Margin Exercises, Section 3.7

1. (a) (1, 4) (2, 4) (3, 4) (4, 4)   (b) {(4, 1), (4, 2), (3, 2), (2, 3), (2, 4), (1, 4)}
     (1, 3) (2, 3) (3, 3) (4, 3)   (c) (1, 4) (2, 4) (3, 4) (4, 4)
     (1, 2) (2, 2) (3, 2) (4, 2)   (1, 3) (2, 3) (3, 3) (4, 3)
     (1, 1) (2, 1) (3, 1) (4, 1)   (1, 2) (2, 2) (3, 2) (4, 2)
                                              (1, 1) (2, 1) (3, 1) (4, 1)

2. (a) $x = 3y + 2$; (b) $x = y$; (c) $y^2 + 3x^2 = 4$; (d) $x = 5y^2 + 2$; (e) $x^2 = 4y - 5$; (f) $yx = 5$
3. (d) They are reflections across the line $y = x$.

4. (a) 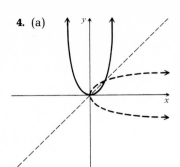   (b)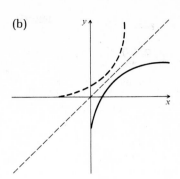

(c)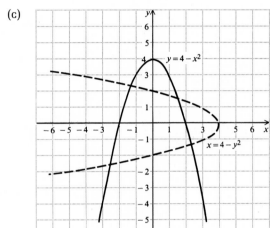

5. Yes   **6.** Yes   **7.** Yes   **8.** No   **9.** Yes   **10.** Yes   **11.** No   **12.** No   **13.** (a), (d)

14. $y = x^2 - 1$ is a function; $x = y^2 - 1$ is not a function.   **15.** $g^{-1}(x) = \dfrac{x + 2}{3}$ **16.** $f^{-1}(x) = \sqrt{x + 1}$ **17.** 5, 5

18. a; a

Exercise Set 3.7, pp. 151–152

1. $x = 4y - 5$ **3.** $y^2 - 3x^2 = 3$ **5.** $x = 3y^2 + 2$ **7.** $yx = 7$

9.   **11.**

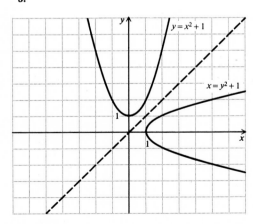   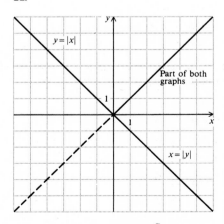

13. No   **15.** Yes   **17.** Yes   **19.** Yes   **21.** No   **23.** Yes   **25.** (a), (c)   **27.** $f^{-1}(x) = \dfrac{x - 5}{2}$ **29.** $f^{-1}(x) = x^2 - 1$

31. 3; -125 **33.** 12,053; $-17,243$ **35.** 1.8   **37.** x-axis: no; y-axis: yes; origin: (no); $y = x$: (no)

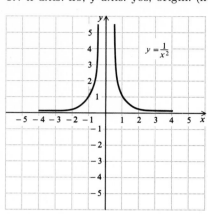

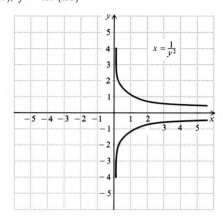

Chapter 3 Review, pp. 152–156

1. $\{(1, 1), (1, 3), (1, 5), (1, 7), (3, 1), (3, 3), (3, 5), (3, 7), (5, 1), (5, 3), (5, 5), (5, 7), (7, 1), (7, 3), (7, 5), (7, 7)\}$

2. Domain $= \{3, 5, 7\}$; range $= \{1, 3, 5, 7\}$

3.

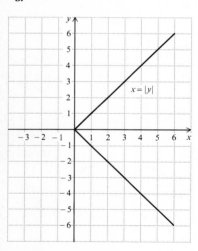

4.

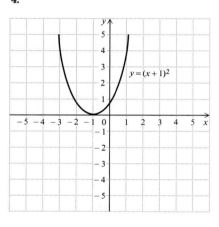

5.

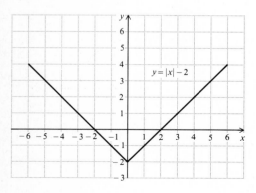

6.

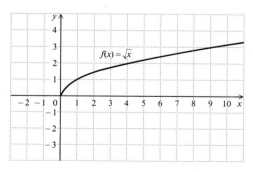

7.

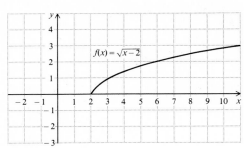

8.

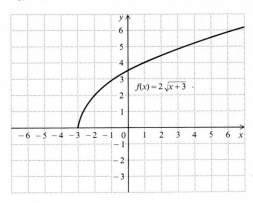

9.

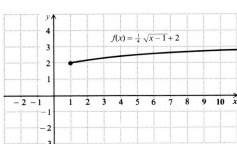

10.

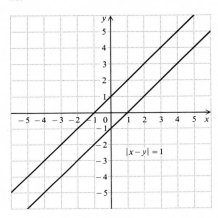

11. (b), (d), (f)   **12.** (a), (b), (d), (g)   **13.** (b), (c), (d), (h)   **14.** (b), (e)   **15.** $x = 3y^2 + 2y - 1$ **16.** $x = \sqrt{y + 2}$
17. (b), (c)   **18.** (d)   **19.** -3 **20.** 9   **21.** $a^2 + 2ah + h^2 - a - h - 3$ **22.** 0   **23.** 4   **24.** $2\sqrt{a + 1}$
25. $g^{-1}(x) = (2x - 4)^2$ **26.** $f^{-1}(x) = \sqrt{x - 2}$ **27.** $\{x \mid x \le \frac{7}{3}\}$ **28.** $\{x \mid x \ne 1, 5\}$ **29.** $\{x \mid x \ne 0, 3, -3\}$

30. $\{x \mid x < 0\}$ **31.** $f \circ g(x) = \dfrac{4}{(3 - 2x)^2}$, $g \circ f(x) = 3 - \dfrac{8}{x^2}$ **32.** $f \circ g(x) = 12x^2 - 4x - 1$, $g \circ f(x) = 6x^2 + 8x - 1$

33. a **34.** t
35. (a) $y = 1 + f(x)$ (b) $y = \frac{1}{2}f(x)$ (c) $y = f(x + 1)$

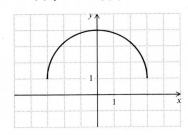
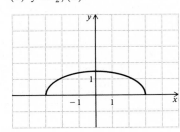
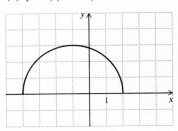

36. (a), (b), (c)   **37.** (e), (f)   **38.** (d)   **39.** (a), (c), (d)   **40.** 2   **41.** (a) Yes; (b) no   **42.** $[-\pi, 2\pi]$ **43.** (0, 1]
44. (c), (d)   **45.** (a)   **46.** (b)

47.

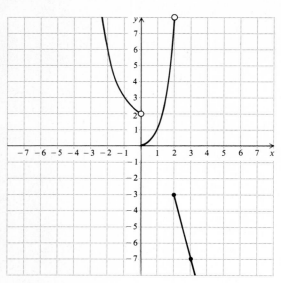

48. Graph $y = f(x)$. Then reflect that portion that lies below the x-axis, across the x-axis.

CHAPTER 4

Margin Exercises, Section 4.1

1. (a), (b)

2.

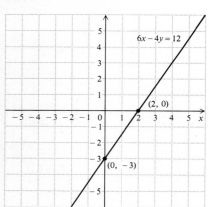

3.

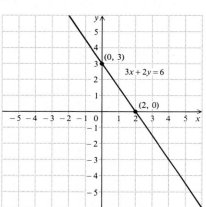

4.

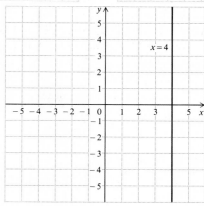

5.

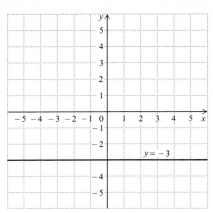

6. $m = 2$ **7.** $m = -2$ **8.** $m = 2$ **9.** $m = 6$ **10.** -1 **11.** $-\frac{1}{3}$ **12.** 0 **13.** m is undefined   **14.** 0

15. $y = -3x - \frac{23}{4}$ **16.** $y = \frac{x}{4} - 9$ **17.** $y = -\frac{x}{2} + \frac{5}{2}$ **18.** $y = -3x + 7$ **19.** $y = -\frac{10x}{3} + 4$ **20.** $m = -7; b = 11$

21. $m = 0; b = -4$ **22.** (a) $y = \frac{2}{3}x + 2$; (b) $m = \frac{2}{3}; b = 2$

Exercise Set 4.1, pp. 164–165

1. (a), (b), (d), (h)   **3.** Line through $(0, -8)$ and $(3, 0)$ **5.** Line through $(0, 3)$ and $(-4, 0)$

7. Horizontal line through $(0, -2)$ **9.** Vertical line through $(3, 0)$ **11.** $\frac{1}{8}$ **13.** $\frac{1}{2}$ **15.** $\frac{1}{\pi}$ **17.** $y = 4x - 10$

19. $y = 2x - 5$ **21.** $y = \frac{x}{2} + \frac{7}{2}$ **23.** $m = 2; b = 3$ **25.** $m = -3; b = 5$ **27.** $m = \frac{3}{4}; b = -3$ **29.** $m = 0; b = -\frac{10}{3}$

31. $y = 3.516x - 13.1602$ **33.** $y = 1.2222x + 1.0949$ **35.** $f(x) = mx$ **37.** $f(x) = x + b$

39. False   **41.** False   **43.** Yes   **45.** $\overline{AB}, \overline{DC}$: same slope; $\overline{BC}, \overline{AD}$: same slope   **47.** $F = \frac{9}{5}C + 32$

49. $P = mQ + b, m \neq 0$. Then we can solve for Q: $Q = \frac{P}{m} - \frac{b}{m}$.

51. Grade $= 4\%$; $y = 4\%x$

Margin Exercises, Section 4.2

1. No   **2.** Yes   **3.** Perpendicular   **4.** Neither   **5.** Parallel
6. Parallel: $y - 4 = -2(x - 3)$, or $y = -2x + 10$; perpendicular: $y - 4 = \frac{1}{2}(x - 3)$, or $y = \frac{1}{2}x + \frac{5}{2}$
7. Parallel: $x = 5$; perpendicular: $y = -4$ **8.** $\sqrt{149}$ **9.** $6\sqrt{2}$ **10.** 16 **11.** 8 **12.** Yes   **13.** No   **14.** $\left(\frac{3}{2}, -\frac{5}{2}\right)$
15. $(9, -5)$

Exercise Set 4.2, pp. 170–171

1. Neither   **3.** Perpendicular   **5.** $y = 3x + 3$ **7.** $x = 3$ **9.** $y = -3$ **11.** $y = -\frac{2}{5}x - \frac{31}{5}$ **13.** $y = 3$
15. $x = -3$ **17.** $y = 0.6114x + 3.4094$ **19.** 5 **21.** $3\sqrt{2}$ **23.** $\sqrt{a^2 + 64}$ **25.** $\sqrt{a^2 + b^2}$ **27.** $2\sqrt{a}$ **29.** 18.8061
31. Yes   **33.** $\left(-\frac{1}{2}, -1\right)$ **35.** $(a, 0)$ **37.** $(-0.4485, -0.2733)$ **39.** $y = -\frac{7}{3}x + \frac{22}{3}$
41. $(5, 0)$

Margin Exercises, Section 4.3

1. (a)

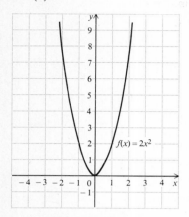

(b) upward; (c) y-axis, x = 0; (d) 0; (e) (0, 0)

2. (a)

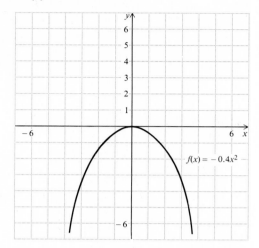

(b) downward; (c) y-axis, x = 0; (d) 0; (e) (0, 0)

3. (a) and (b)

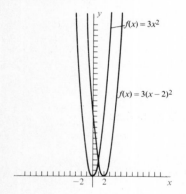

(c) (2, 0); (d) x = 2; (e) 0; (f) upward;
(g) horizontal translation to the right

4. (a) and (b)

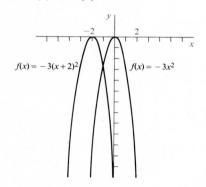

(c) (−2, 0); (d) x = −2; (e) 0; (f) downward;
(e) horizontal translation to the left

5.

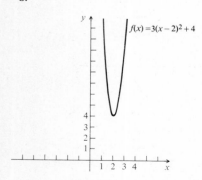

(a) (2, 4); (b) x = 2; (c) no; (d) yes, 4

6.

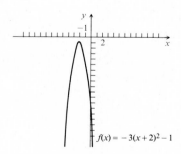

(a) (−2, −1); (b) x = −2; (c) yes, −1; (d) no

7. (a) $(5, \pi)$; (b) $x = 5$; (c) no; (d) yes, π **8.** (a) $(5, 0)$; (b) $x = 5$; (c) yes, 0; (d) no
9. (a) $(-\frac{1}{4}, -6)$; (b) $x = -\frac{1}{4}$; (c) no; (d) yes, -6 **10.** (a) $(-9, 3)$; (b) $x = -9$; (c) yes, 3; (d) no
11. $f(x) = (x - 2)^2 + 3$ **12.** $f(x) = 3(x + 4)^2 - 38$ **13.** (a) $(\frac{3}{2}, -14)$, $x = \frac{3}{2}$; (b) -14 is a minimum
14. $(-\frac{5}{2}, \frac{43}{2})$; $\frac{43}{2}$ is a maximum
15.   **16.**

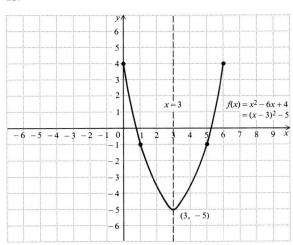

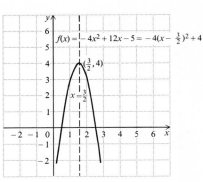

17. $(1 - \sqrt{6}, 0)$, $(1 + \sqrt{6}, 0)$ **18.** $(-1, 0)$, $(3, 0)$ **19.** $(-4, 0)$ **20.** None

Exercise Set 4.3, pp. 178–179

1. (a) $(0, 0)$; (b) $x = 0$; (c) 0 is a minimum   **3.** (a) $(9, 0)$; (b) $x = 9$; (c) 0 is a maximum
5. (a) $(1, -4)$; (b) $x = 1$; (c) -4 is a minimum   **7.** (a) $f(x) = -(x - 1)^2 + 4$; (b) $(1, 4)$; (c) 4 is a maximum
9. (a) $f(x) = (x + \frac{3}{2})^2 - \frac{9}{4}$; (b) $(-\frac{3}{2}, -\frac{9}{4})$; (c) $-\frac{9}{4}$ is a minimum
11. (a) $f(x) = -\frac{3}{4}(x - 4)^2 + 12$; (b) $(4, 12)$; (c) 12 is a maximum
13. (a) $f(x) = 3(x + \frac{1}{6})^2 - \frac{49}{12}$; (b) $(-\frac{1}{6}, -\frac{49}{12})$; (c) $-\frac{49}{12}$ is a minimum
15. $f(x) = -x^2 + 2x + 3 = -(x - 1)^2 + 4$ **17.** $f(x) = x^2 - 8x + 19 = (x - 4)^2 + 3$

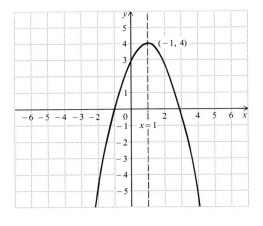

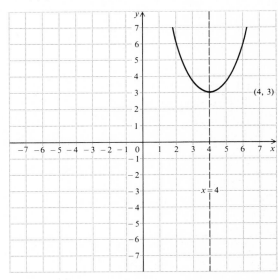

19. $f(x) = -\frac{1}{2}x^2 - 3x + \frac{1}{2} = -\frac{1}{2}(x + 3)^2 + 5$

21. $f(x) = 3x^2 - 24x + 50 = 3(x - 4)^2 + 2$

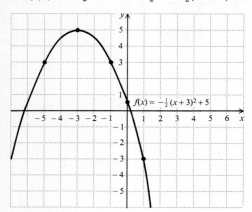

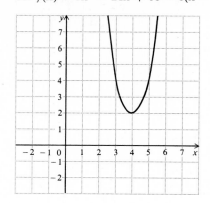

23. $(3, 0), (-1, 0)$ **25.** $(4 \pm \sqrt{11}, 0)$ **27.** None   **29.** $f(x) = a\left[x - \left(-\dfrac{b}{2a}\right)\right]^2 + \dfrac{4ac - b^2}{4a}$

31.   **33.** Minimum, -6.95 **35.** Minimum, -6.081; ± 2.466

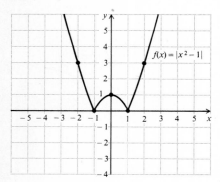

Margin Exercises, Section 4.4

1. (a) $S = \frac{14}{15}d + 9\frac{2}{3}$; (b) 85   **2.** (a) $P(x) = 60x - 15{,}000$; (b) $x = 250$ is the breakeven point.
3. Maximum height 12.4 m, at $t = \frac{2}{7}$. Reaches ground in 1.88 sec.

Exercise Set 4.4, pp. 185–186

1. (a) $E = 0.15t + 72$; (b) 76.5, 77.25   **3.** (a) $R = -0.01t + 10.43$; (b) 9.79, 9.73; (c) 2063
5. (a) $C(x) = 40x + 22{,}500$; (b) $R(x) = 85x$; (c) $P(x) = 45x - 22{,}500$; (d) profit = \$112,500; (e) $x = 500$; (f) $x > 500$;
(g) $x < 500$ **7.** (a) 1662.5 m after 15 sec; (b) after 33.4 sec   **9.** 10×10 **11.** 30
13. \$3.67

Margin Exercises, Section 4.5

1. $\{-3, 4\}$ **2.** $\{2, d, e\}$ **3.**    **4.** 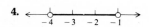   **5.**

6. Ø, nothing to graph   **7.** $\{1, 2, 3, 4, 5\}$ **8.** $\{-3, -4, 1, 2, 3, 4, 8, 9, 11\}$ **9.** $\{1, 2, a, b, c, d, e\}$

10.

11.

12.

13.   **14.** $4 < x < 8$ **15.** $-3 \leq x < 0$ **16.** $-5 \leq x \leq -2$ **17.** $-1 < x \leq -\frac{1}{4}$

18. $-\frac{1}{2} < x$ and $x < 1$ **19.** $-\frac{17}{3} \leq x$ and $x < -2$ **20.** $\frac{19}{4} \leq x$ and $x \leq \frac{37}{6}$

21. $\{x \mid -\frac{2}{3} < x \leq 4\} = \{x \mid -\frac{2}{3} < x\} \cap \{x \mid x \leq 4\}$ **22.** $\{x \mid -5 < x < 11\} = \{x \mid -5 < x\} \cap \{x \mid x < 11\}$

23. $\{x \mid -\frac{1}{3} \leq x \leq \frac{1}{6}\} = \{x \mid -\frac{1}{3} \leq x\} \cap \{x \mid x \leq \frac{1}{6}\}$ **24.** $\{x \mid x < -7 \text{ or } x > -1\} = \{x \mid x < -7\} \cup \{x \mid x > -1\}$

25. $\{x \mid x \leq -1 \text{ or } x > 4\} = \{x \mid x \leq -1\} \cup \{x \mid x > 4\}$ **26.** $\{x \mid x \geq \frac{5}{3} \text{ or } x \leq 1\} = \{x \mid x \geq \frac{5}{3}\} \cup \{x \mid x \leq 1\}$

27. $\{x \mid x < -\frac{11}{4} \text{ or } x \geq \frac{1}{4}\} = \{x \mid x < -\frac{11}{4}\} \cup \{x \mid x \geq \frac{1}{4}\}$

Exercise Set 4.5, p. 191

1. $\{3, 4, 5\}$ **3.** $\{0, 2, 4, 6, 8, 9\}$ **5.** $\{c\}$ **7.** Entire set of real numbers
9. $-\frac{1}{2}$ and all numbers between $-\frac{1}{2}$ and $\frac{1}{2}$
11.   **13.**

15. $\emptyset$ **17.** $\{x \mid -3 \leq x < 3\}$ **19.** $\{x \mid 8 \leq x \leq 10\}$ **21.** $\{-7\}$ **23.** $\{x \mid -\frac{3}{2} < x < 2\}$ **25.** $\{x \mid 1 < x \leq 5\}$
27. $\{x \mid -\frac{11}{3} < x \leq \frac{13}{3}\}$ **29.** $\{x \mid x \leq -2 \text{ or } x > 1\}$ **31.** $\{x \mid x \leq -\frac{7}{2} \text{ or } x \geq \frac{1}{2}\}$ **33.** $\{x \mid x < 9.6 \text{ or } x > 10.4\}$
35. $\{x \mid x \leq -\frac{57}{4} \text{ or } x \geq -\frac{55}{4}\}$ **37.** (1)   **39.** $\{x \mid x > -\frac{1}{5}\}$ **41.** $\{x \mid x \geq -\frac{3}{2}\}$ **43.** $\{w \mid 4.885 \text{ cm} < w < 53.67 \text{ cm}\}$
45. $\{S \mid 97\% \leq S \leq 100\%\}$; yes   **47.** $\{x \mid x > 2\}$

Margin Exercises, Section 4.6

1. $\{-5, 5\}$ **2.** $\{\frac{1}{4}, -\frac{1}{4}\}$ **3.** $\{x \mid -5 < x < 5\}$ **4.** $\{x \mid -\frac{1}{4} \leq x \leq \frac{1}{4}\}$ **5.** $\{x \mid x \leq -5 \text{ or } x \geq 5\}$
6. $\{x \mid x < -\frac{1}{4} \text{ or } x > \frac{1}{4}\}$ **7.** $\{-5, 3\}$ **8.** $\{x \mid -9 < x < -5\}$ **9.** $\{x \mid x < 1 \text{ or } x > 5\}$ **10.** $\{-1, \frac{11}{3}\}$
11. $\{x \mid -\frac{3}{10} < x < \frac{1}{10}\}$ **12.** $\{x \mid x < -1 \text{ or } x > \frac{11}{3}\}$

Exercise Set 4.6, pp. 194–195

1. $\{-7, 7\}$ **3.** $\{x \mid -7 < x < 7\}$ **5.** $\{x \mid x \leq -\pi \text{ or } x \geq \pi\}$ **7.** $\{-3, 5\}$ **9.** $\{x \mid -17 < x < 1\}$
11. $\{x \mid x \leq -17 \text{ or } x \geq 1\}$ **13.** $\{x \mid -\frac{1}{4} < x < \frac{3}{4}\}$ **15.** $\{-\frac{1}{3}, \frac{1}{3}\}$ **17.** $(-1, -\frac{1}{3})$ **19.** $\{x \mid -\frac{1}{3} < x < \frac{1}{3}\}$
21. $\{x \mid -6 \leq x \leq 3\}$ **23.** $\{x \mid x < 4.9 \text{ or } x > 5.1\}$ **25.** $\{x \mid -\frac{7}{3} \leq x \leq 1\}$
27. $\{x \mid -\frac{1}{2} \leq x \leq \frac{7}{2}\}$

29. $\{x \mid x < -8 \text{ or } x > 7\}$ **31.** $\{x \mid x < -\frac{7}{4} \text{ or } x > -\frac{3}{2}\}$ **33.** $\{x \mid \frac{3}{8} \le x \le \frac{9}{8}\}$ **35.** $\emptyset$ **37.** $\emptyset$
39. $\{x \mid 1.9234 < x < 2.1256\}$ **41.** $\{x \mid 0.98414 \le x \le 4.9808\}$ **43.** $\{2, \frac{4}{5}\}$ **45.** $\{-4, 4\}$ **47.** $\{x \mid x \le \frac{3}{2}\}$
49. $\{x \mid -\frac{9}{2} < x < \frac{11}{2}\}$ **51.** $\{x \mid x < -\frac{8}{3} \text{ or } x > -2\}$

Margin Exercises, Section 4.7

1. $\{x \mid -2 < x < 5\}$ **2.** $\{x \mid x > -1 + \sqrt{5} \text{ or } x < -1 - \sqrt{5}\}$ **3.** $\{x \mid x < -1 \text{ or } x > 4\}$
4. $\{x \mid -1 < x < 4\}$ **5.** $\{x \mid -3 < x < 0 \text{ or } x > 2\}$ **6.** $\{x \mid -11 \le x < -4\}$
7. $\{x \mid x < 5 \text{ or } x \ge 10\}$

Exercise Set 4.7, pp. 198–199

1. $\{x \mid -1 < x < 2\}$ **3.** $\{x \mid x \le -1 \text{ or } x \ge 1\}$ **5.** All real numbers   **7.** $\{x \mid 3 - \sqrt{5} < x < 3 + \sqrt{5}\}$
9. $\{x \mid -2 \le x \le 10\}$ **11.** $\{x \mid x < -2 \text{ or } x > 4\}$ **13.** $\{x \mid -3 < x < \frac{5}{4}\}$ **15.** $\left\{x \mid x < \dfrac{-1 - \sqrt{41}}{4} \text{ or } x > \dfrac{-1 + \sqrt{41}}{4}\right\}$
17. $\{x \mid -1 < x < 0 \text{ or } x > 1\}$ **19.** $\{x \mid x < -3 \text{ or } -2 < x < 1\}$ **21.** $\{x \mid x > 4\}$ **23.** $\{x \mid x < -\frac{2}{3} \text{ or } x > 3\}$
25. $\{x \mid \frac{3}{2} < x \le 4\}$ **27.** $\{x \mid x \le -\frac{5}{2} \text{ or } x > -2\}$ **29.** $\{x \mid x < -\frac{11}{7}\}$ **31.** $\{x \mid x > 1\}$ **33.** $\{x \mid x > 0 \text{ and } x \ne 2\}$
35. $\{x \mid x < 0 \text{ or } x \ge 1\}$ **37.** $\{x \mid x < -3 \text{ or } -2 < x < -1 \text{ or } x > 2\}$ **39.** $\{x \mid x < \frac{5}{3} \text{ or } x > 11\}$ **41.** $\emptyset$
43. $\{x \mid x < -\frac{1}{4} \text{ or } x \ge \frac{1}{2}\}$ **45.** $\{x \mid x \ne 0\}$ **47.** $\{x \mid -4 < x < -2 \text{ or } -1 < x < 1\}$
49. $\{h \mid h > -2 + 2\sqrt{6} \text{ cm}\}$ **51.** (a) 10, 35; (b) $\{x \mid 10 < x < 35\}$; (c) $\{x \mid x < 10 \text{ or } x > 35\}$
53. (a) $\{k \mid k > 2 \text{ or } k < -2\}$; (b) $\{k \mid -2 < k < 2\}$ **55.** $\{x \mid -1 \le x \le 1\}$
57. $\{x \mid x \le -3 \text{ or } x \ge 1\}$

Chapter 4 Review, pp. 199–200

1.

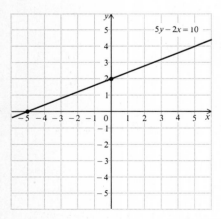

2. -2 **3.** -1 **4.** $y = 3x + 5$ **5.** $y = \frac{1}{3}x - \frac{1}{3}$ **6.** $\sqrt{34}$ **7.** $(\frac{1}{2}, \frac{11}{2})$
8. $y = -\frac{2}{3}x - \frac{1}{3}$ **9.** $y = \frac{3}{2}x - \frac{5}{2}$ **10.** Parallel   **11.** Neither   **12.** Perpendicular
13. (a) $f(x) = 3(x + 1)^2 - 2$; (b) $(-1, -2)$; (c) $x = -1$; (d) -2 is a minimum
14. (a) $f(x) = -2(x + \frac{3}{4})^2 + \frac{57}{8}$; (b) $(-\frac{3}{4}, \frac{57}{8})$; (c) $x = -\frac{3}{4}$; (d) $\frac{57}{8}$ is a maximum

15.

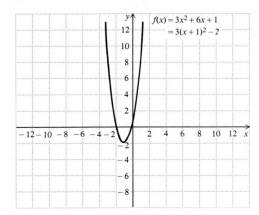

16. None

17. $\{x \mid 2 \leq x \leq 4\}$ **18.** $\{x \mid 1 < x < 11\}$ **19.** $\{x \mid x > 0 \text{ or } x < -\frac{4}{3}\}$ **20.** $\{x \mid -20 \leq x \leq 4\}$ **21.** $\{2, -7\}$
22. $\{x \mid -3 < x < 3\}$ **23.** $\{x \mid x > 2 \text{ or } x < -\frac{1}{2}\}$ **24.** $\{x \mid -4 < x < 1 \text{ or } x > 2\}$ **25.** $\{x \mid x < -\frac{14}{3} \text{ or } x > -3\}$
26. (a) $R = -0.4t + 260$; (b) 220 sec $\approx$ 3:40   **27.** 20×20 **28.** (a) 20; (b) $\{x \mid x > 20\}$; (c) $\{x \mid x < 20\}$
29. $\{x \mid -2 \leq x < \frac{1}{3}\}$ **30.** $x \mid x < -\frac{1}{2} \text{ or } x > \frac{1}{2}\}$ **31.** $\{x \mid x < 2\}$ **32.** $m \neq 0$ **33.** $\{x \mid \frac{1}{3} \leq x \leq 1\}$ **34.** $\{x \mid -1 < x < \frac{3}{7}\}$

CHAPTER 5

Margin Exercises, Section 5.1

1. Yes   **2.** No   **3.** $(1, 2)$ **4.** $(4, -2)$ **5.** $(-2, 5)$ **6.** $(-3, 2)$ **7.** $(-\frac{1}{3}, \frac{1}{2})$ **8.** $(2, -1)$ **9.** $\frac{11}{2}, \frac{9}{2}$ **10.** $\frac{11}{2}, -\frac{9}{2}$
11. 10 km/h, 2 km/h   **12.** 5 L, 3 L

Exercise Set 5.1, pp. 207–208

1. No   **3.** $(-1, 3)$ **5.** $(\frac{39}{11}, -\frac{1}{11})$ **7.** $(-4, -2)$ **9.** $(-3, 0)$ **11.** $(10, 8)$ **13.** $(1, 1)$ **15.** $-\frac{11}{2}, -\frac{9}{2}$
17. 20 km/h, 3 km/h   **19.** 12.5 L, 7.5 L   **21.** $(-12, 0)$ **23.** $(0.924, -0.833)$ **25.** $(-\frac{1}{4}, -\frac{1}{2})$
27. $\{(5, 3), (-5, 3), (5, -3), (-5, -3)\}$

Margin Exercises, Section 5.2

1. (a) No; (b) yes   **2.** $(2, \frac{1}{2}, -2)$ **3.** $f(x) = x^2 - 2x + 1$ **4.** $f(x) = \frac{5}{8}x^2 - 50x + 1150$; 510 accidents

Exercise Set 5.2, pp. 212–213

1. Yes   **3.** $(3, -2, 1)$ **5.** $(-3, 2, 1)$ **7.** $(\frac{1}{2}, \frac{2}{3}, -\frac{5}{6})$
9. $(-2, 4, -1, 1)$; solution listed in the order in which the variables occur in the equations   **11.** $y = 2x^2 + 3x - 1$
13. (a) $E = -4t^2 + 40t + 2$; (b) \$98   **15.** $(-1, \frac{1}{5}, -\frac{1}{2})$ **17.** A: 4 hr; B: 6 hr; C: 12 hr

Margin Exercises, Section 5.3

1. $\emptyset$ **2.** (a) Inconsistent; (b) consistent   **3.** No solution (inconsistent system)   **4.** (a) Inconsistent; (b) consistent
5. $\left(\frac{2y - 5}{3}, y\right)$ or $\left(x, \frac{3x + 5}{2}\right)$, $(1, 4)$, $(3, 7)$, $(0, \frac{5}{2})$, etc.   **6.** $\left(x, \frac{1 - 2x}{5}\right)$ or $\left(\frac{1 - 5y}{2}, y\right)$, $(-7, 3)$, $(3, -1)$, $(0, \frac{1}{5})$, etc.
7. $(-2z + 7, 3z - 5, z)$; $(7, -5, 0)$, $(5, -2, 1)$, $(3, 1, 2)$, etc.   **8.** $(-2z, 3z, z)$; $(-2, 3, 1)$, $(2, -3, -1)$, $(-4, 6, 2)$, etc.
9. $(0, 0, 0)$, only solution

Exercise Set 5.3, p. 218

1. $\left(\dfrac{y+5}{3}, y\right)$ or $(x, 3x - 5)$, $(0, -5)$, $(1, -2)$, $(-1, -8)$, etc.   **3.** $\emptyset$ **5.** $\left(\dfrac{5 - 2y}{3}, y\right)$ or $\left(x, \dfrac{5 - 3x}{2}\right)$

7. $\left(\dfrac{6y - 3}{4}, y\right)$ or $\left(x, \dfrac{4x + 3}{6}\right)$ **9.** $\emptyset$ **11.** $\left(\dfrac{10 + 11z}{9}, \dfrac{-11 + 5z}{9}, z\right)$; $(\tfrac{10}{9}, -\tfrac{11}{9}, 0)$, etc.   **13.** $(\tfrac{11}{9}z, \tfrac{5}{9}z, z)$: $(\tfrac{11}{9}, \tfrac{5}{9}, 1)$, etc.

15. $(4z - 5, -3z + 2, z)$: $(-1, -1, 1)$, etc.   **17.** $(0, 0, 0)$

19. Consistent: 1, 5, 7, 11, 13, 15, 17, the others are inconsistent; dependent: 1, 5, 7, 11, 13, 15, the others are independent

21. $\left(\dfrac{724y + 9160}{2013}, y\right)$ or $\left(x, \dfrac{2013x - 9160}{724}\right)$ **23.** (a) $\emptyset$; (b) inconsistent; (c) dependent

Margin Exercises, Section 5.4

1. $\left(-\tfrac{63}{29}, -\tfrac{114}{29}\right)$ **2.** $(-1, 2, 3)$

Exercise Set 5.4, pp. 221–222

1. $(\tfrac{3}{2}, \tfrac{5}{2})$ **3.** $(-1, 2, -2)$ **5.** $(\tfrac{1}{2}, \tfrac{3}{2})$ **7.** $(\tfrac{3}{2}, -4, 3)$ **9.** $(r - 2, 3 - 2r, r)$ **11.** $(-3, -2, -1, 1)$ **13.** 4 dimes, 30 nickels

15. 10 nickels, 4 dimes, 8 quarters   **17.** 5 lb of $4.05; 10 lb of $2.70   **19.** $30,000 at $12\tfrac{1}{2}\%$; $40,000 at 13%

Margin Exercises, Section 5.5

1. -13 **2.** -2 **3.** $-2x + 12$ **4.** $(3, 1)$ **5.** $\left(-\tfrac{10}{41}, -\tfrac{13}{41}\right)$ **6.** $\left(\dfrac{3\sqrt{2} + 4\pi}{2 + \pi^2}, \dfrac{4\sqrt{2} - 3\pi}{2 + \pi^2}\right)$ **7.** 93   **8.** 60   **9.** $x^3 - x^2$

10. $(1, 3, -2)$

Exercise Set 5.5, pp. 225–226

1. -11 **3.** $x^3 - 4x$ **5.** -109 **7.** $-x^4 + x^2 - 5x$ **9.** $\left(-\tfrac{25}{2}, -\tfrac{11}{2}\right)$ **11.** $\left(\dfrac{4\pi - 5\sqrt{3}}{3 + \pi^2}, \dfrac{4\sqrt{3} + 5\pi}{-3 - \pi^2}\right)$ **13.** $(\tfrac{3}{2}, \tfrac{13}{14}, \tfrac{33}{14})$

15. $(\tfrac{1}{2}, \tfrac{2}{3}, -\tfrac{5}{6})$ **17.** $2, -2$ **19.** $\{x \mid x \le -\sqrt{3} \text{ or } x \ge \sqrt{3}\}$ **21.** -34 **23.** 4   **25.** $\begin{vmatrix} L & -W \\ 2 & 2 \end{vmatrix}$ **27.** $\begin{vmatrix} a & b \\ -b & a \end{vmatrix}$

29. $\begin{vmatrix} 2\pi r & 2\pi r \\ -h & r \end{vmatrix}$

Margin Exercises, Section 5.6

1. No

2.

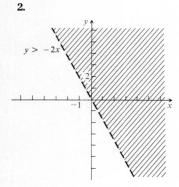

$y > -2x$

3.

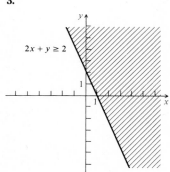

$2x + y \ge 2$

4.

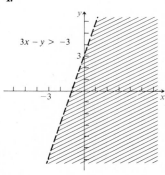

$3x - y > -3$

5.

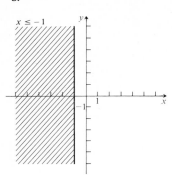

$x \le -1$

6.

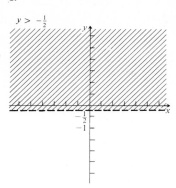

$y > -\frac{1}{2}$

7.

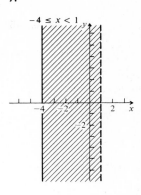

$-4 \le x < 1$

8.

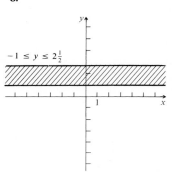

$-1 \le y \le 2\frac{1}{2}$

9.

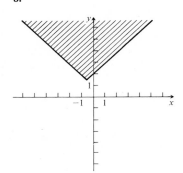

10. Vertices: $(0, 0)$, $(4, 0)$, $(4, \frac{5}{3})$, $(0, 3)$, $(\frac{12}{5}, 3)$

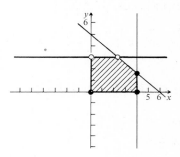

Exercise Set 5.6, p. 230

1.

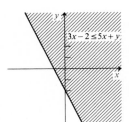

$y < x$

3.

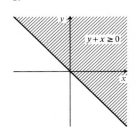

$y + x \ge 0$

5.

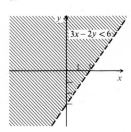

$3x - 2y < 6$

7.

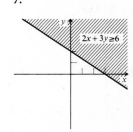

$2x + 3y \ge 6$

9.

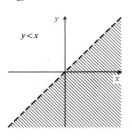

$3x - 2 \le 5x + y$

11.

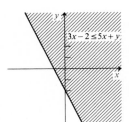

$x < -4$

13.

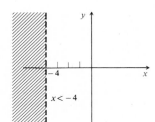

$0 \le x < 5\frac{1}{2}$

15.

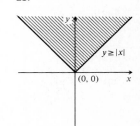

$y \ge |x|$

$(0, 0)$

17.

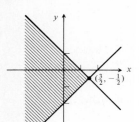

19.

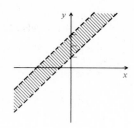

21.

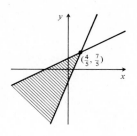

23.

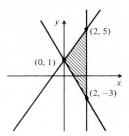

25.

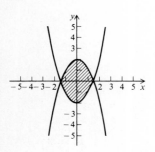

27.

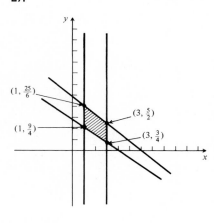

29.

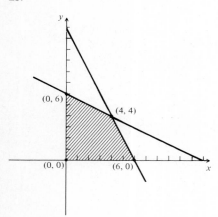

31.

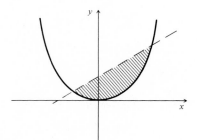

Margin Exercises, Section 5.7

1. Maximum 120, when $x = 3$, $y = 3$; minimum 34, when $x = 1$, $y = 0$
2. Maximum 45, when $x = 6$, $y = 0$; minimum 12, when $x = 0$, $y = 3$
3. Maximum $23.70, by selling 50 hot dogs and 40 hamburgers

Exercise Set 5.7, p. 234

1. Maximum 168, when $x = 0$, $y = 6$; minimum 0, when $x = 0$, $y = 0$
3. Maximum 152, when $x = 7$, $y = 0$; minimum 32, when $x = 0$, $y = 4$ 5. Maximum 102, 8 of A, 10 of B
7. $7000 bank X, $15,000 bank Y, maximum income $1395   9. Maximum $192, 2 knits, 4 worsteds
11. Minimum $460, 30 P1 airplanes, 10 P2 airplanes

Chapter 5 Review, pp. 235–236

1. $(-2, -2)$ **2.** $(-5, 4)$ **3.** $\varnothing$ **4.** $\varnothing$ **5.** $(0, 0, 0)$ **6.** $(13, 8, 2, -5)$
7. Consistent: 1, 2, 5, 6; the others are inconsistent   **8.** Dependent: none   **9.** 31 nickels, 44 dimes
10. $1600 at 10%, $3400 at 10.5%   **11.** $(1, 2)$ **12.** $(-3, 4, -2)$ **13.** $\left(\dfrac{z}{2}, -\dfrac{z}{2}, z\right)$; $(0, 0, 0)$, $(\frac{1}{2}, -\frac{1}{2}, 1)$, $(1, -1, 2)$, etc.

14. $(1, -2, 3, -4)$ **15.** $y = -x^2 - 2x + 3$ **16.** 10   **17.** -18 **18.** $2x + 12$ **19.** -6 **20.** -16.588 **21.** 0
22. $(3, -2)$ **23.** $(a, 0)$ **24.** $\left(\frac{3}{2}, \frac{13}{14}, \frac{33}{14}\right)$ **25.** $(0, 9)$, $(2, 5)$, $(5, 1)$, $(8, 0)$
26. Minimum $= 12$ at $(2, 0)$; maximum $= 60$ at $(10, 0)$ **27.** Type A: 0; type B: 10; maximum score $= 120$ pts
28. $10,000 at 12%, $12,000 at 13%, $18,000 at $14\frac{1}{2}$%   **29.** $\left(\frac{5}{18}, \frac{1}{7}\right)$ **30.** $(1, \frac{1}{2}, \frac{1}{3})$
31.   **32.**

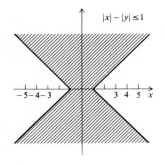

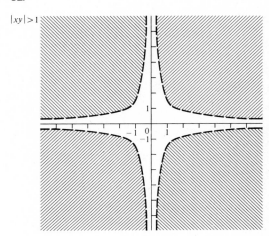

CHAPTER 6

Margin Exercises, Section 6.1

1. 3×2 **2.** 2×2 **3.** 3×3 **4.** 1×2 **5.** 2×1 **6.** 1×1 **7.** 2, 3, 6   **8.** $A + B = \begin{bmatrix} -2 & -6 \\ 13 & 0 \end{bmatrix} = B + A$

9. $\begin{bmatrix} 1 & 1 & -10 \\ 2 & 4 & -3 \end{bmatrix}$ **10.** $A + 0 = \begin{bmatrix} 4 & -3 \\ 5 & 8 \end{bmatrix} = 0 + A = A$ **11.** $\begin{bmatrix} -1 & 4 & -7 \\ -2 & -4 & 8 \end{bmatrix}$ **12.** $\begin{bmatrix} -6 & 6 \\ 1 & -4 \\ -7 & 5 \end{bmatrix}$ **13.** $\begin{bmatrix} -2 & 1 & -5 \\ -6 & -4 & 3 \end{bmatrix}$

14. $\begin{bmatrix} 0 & 0 & 0 \\ 0 & 0 & 0 \end{bmatrix}$ **15.** $\begin{bmatrix} -1 & 4 & -7 \\ -2 & -4 & 8 \end{bmatrix}$ **16.** $\begin{bmatrix} 5 & -10 & 5x \\ 20 & 5y & 5 \\ 0 & -25 & 5x^2 \end{bmatrix}$ **17.** $\begin{bmatrix} t & -t & 4t & tx \\ ty & 3t & -2t & ty \\ t & 4t & -5t & ty \end{bmatrix}$ **18.** $[-13]$ **19.** $\begin{bmatrix} 12 \\ 13 \\ 5 \\ 16 \end{bmatrix}$

20. $\begin{bmatrix} 0 & 26 \\ -8 & 3 \\ -13 & 33 \\ -7 & 32 \end{bmatrix}$ **21.** $AB = \begin{bmatrix} 2 & 8 & 6 \\ -29 & -34 & -7 \end{bmatrix}$; BA not possible   **22.** $[8 \;\; 5 \;\; 4]$ **23.** $AB = \begin{bmatrix} -2 & 32 \\ 4 & 16 \end{bmatrix}$, $BA = \begin{bmatrix} 8 & -13 \\ -16 & 6 \end{bmatrix}$

24. $AI = \begin{bmatrix} 3 & 2 \\ -1 & 5 \end{bmatrix} = IA = A$ **25.** $\begin{bmatrix} 3 & 4 & -2 \\ 2 & -2 & 5 \\ 6 & 7 & -1 \end{bmatrix} \begin{bmatrix} x \\ y \\ z \end{bmatrix} = \begin{bmatrix} 5 \\ 3 \\ 0 \end{bmatrix}$

Exercise Set 6.1, pp. 244–245

1. $\begin{bmatrix} -2 & 7 \\ 6 & 2 \end{bmatrix}$ **3.** $\begin{bmatrix} 1 & 3 \\ 2 & 6 \end{bmatrix}$ **5.** $\begin{bmatrix} 9 & 9 \\ -3 & -3 \end{bmatrix}$ **7.** $\begin{bmatrix} 11 & 13 \\ 5 & 3 \end{bmatrix}$ **9.** $\begin{bmatrix} -4 & 3 \\ -2 & -4 \end{bmatrix}$ **11.** $\begin{bmatrix} 17 & 9 \\ -2 & 1 \end{bmatrix}$ **13.** $\begin{bmatrix} 0 & 0 \\ 0 & 0 \end{bmatrix}$ **15.** $\begin{bmatrix} 1 & 2 \\ 4 & 3 \end{bmatrix}$ or $\mathbf{A}$

17. $\begin{bmatrix} -5 & 4 & 3 \\ 5 & -9 & 4 \\ 7 & -18 & 17 \end{bmatrix}$ **19.** $\begin{bmatrix} -2 & 9 & 6 \\ -3 & 3 & 4 \\ 2 & -2 & 1 \end{bmatrix}$ or $\mathbf{C}$ **21.** $[-16]$ **23.** $[2 \quad -19]$ **25.** $\begin{bmatrix} 3 & -2 & 4 \\ 2 & 1 & -5 \end{bmatrix}\begin{bmatrix} x \\ y \\ z \end{bmatrix} = \begin{bmatrix} 17 \\ 13 \end{bmatrix}$

27. $\begin{bmatrix} 1 & -1 & 2 & -4 \\ 2 & -1 & -1 & 1 \\ 1 & 4 & -3 & -1 \\ 3 & 5 & -7 & 2 \end{bmatrix}\begin{bmatrix} x \\ y \\ z \\ w \end{bmatrix} = \begin{bmatrix} 12 \\ 0 \\ 1 \\ 9 \end{bmatrix}$ **29.** $\begin{bmatrix} -40.19 & 37.94 & 142.24 \\ -36.78 & 16.63 & 119.62 \\ -1.659 & 14.97 & 12.65 \end{bmatrix}$

31. $(\mathbf{A} + \mathbf{B})(\mathbf{A} - \mathbf{B}) = \begin{bmatrix} -2 & 1 \\ 2 & -1 \end{bmatrix}$, $\mathbf{A}^2 - \mathbf{B}^2 = \begin{bmatrix} 0 & 3 \\ 0 & -3 \end{bmatrix}$

Margin Exercises, Section 6.2

1. $a_{11} = -8$, $a_{13} = 6$, $a_{22} = -6$, $a_{31} = -1$, $a_{32} = -3$

2. $M_{22} = \begin{vmatrix} -8 & 6 \\ -1 & 5 \end{vmatrix} = -34$, $M_{31} = \begin{vmatrix} 0 & 6 \\ -6 & 7 \end{vmatrix} = 36$, $M_{13} = \begin{vmatrix} 4 & -6 \\ -1 & -3 \end{vmatrix} = -18$ **3.** $A_{22} = -34$, $A_{32} = 80$, $A_{13} = -18$

4. $|\mathbf{A}| = 0 \cdot A_{12} + (-6)A_{22} + (-3)A_{32} = -36$ **5.** $|\mathbf{A}| = 6 \cdot A_{13} + 7A_{23} + 5A_{33} = -36$

Exercise Set 6.2, pp. 248–249

1. $a_{11} = 7$, $a_{32} = 2$, $a_{22} = 0$ **3.** $M_{11} = 6$, $M_{32} = -9$, $M_{22} = -29$ **5.** $A_{11} = 6$, $A_{32} = 9$, $A_{22} = -29$ **7.** $|\mathbf{A}| = -10$
9. $|\mathbf{A}| = -10$ **11.** $M_{41} = -14$, $M_{33} = 20$ **13.** $A_{24} = 15$, $A_{43} = 30$ **15.** $|\mathbf{A}| = 110$ **17.** -195

Margin Exercises, Section 6.3

1. 0   **2.** 0   **3.** (a) $|\mathbf{A}| = -18$, $|\mathbf{B}| = 18$; (b) The rows are interchanged.
4. (a) $|\mathbf{C}| = -10$, $|\mathbf{D}| = 10$; (b) The first and third columns are interchanged.   **5.** 0   **6.** $x = -4$ **7.** $x = 6$ **8.** 0
9. $\begin{vmatrix} -2 & 3 & 4 \\ -3 & 10 & 5 \\ 0 & 9 & 7 \end{vmatrix}$ **10.** 68   **11.** -195 **12.** $(b - a)(c - a)(b - c)$

Exercise Set 6.3, pp. 253–254

1. -70 **3.** -4 **5.** 9072   **7.** -153 **9.** 0   **11.** 0   **13.** $(x - y)(y - z)(x - z)$ **15.** $xyz(x - y)(y - z)(z - x)$

Margin Exercises, Section 6.4

1. (a) $\mathbf{A}$; (b) $\mathbf{A}$; (c) both equal $\mathbf{A}$ **2.** (a) $\begin{bmatrix} 1 & 0 & 0 \\ 0 & 1 & 0 \\ 0 & 0 & 1 \end{bmatrix} = \mathbf{I}$; (b) $\mathbf{I}$; (c) both equal $\mathbf{I}$

3. $\mathbf{A}^t = \begin{bmatrix} -8 & -4 & 6 \\ 1 & 0 & 7 \\ -2 & -1 & 8 \end{bmatrix}$; $\mathbf{B}^t = \begin{bmatrix} -7 \\ 9 \\ 10 \\ 4 \end{bmatrix}$; $\mathbf{C}^t = [-20 \quad 11]$; $\mathbf{D}^t = \begin{bmatrix} -4 & 1 & 0 \\ 5 & 0 & 1 \end{bmatrix}$ **4.** $\mathbf{A}^{-1} = \begin{bmatrix} 2 & -\frac{5}{2} \\ -1 & \frac{3}{2} \end{bmatrix}$ **5.** $\mathbf{A}^{-1} = \mathbf{A}$

6. $A^{-1} = \begin{bmatrix} \frac{3}{8} & \frac{1}{8} & -\frac{1}{4} \\ -\frac{1}{8} & -\frac{3}{8} & \frac{3}{4} \\ -\frac{1}{4} & \frac{1}{4} & \frac{1}{2} \end{bmatrix}$ **7.** $A^{-1} = \begin{bmatrix} -\frac{1}{2} & \frac{1}{2} & \frac{1}{2} \\ 1 & 0 & -1 \\ \frac{3}{2} & -\frac{1}{2} & -\frac{1}{2} \end{bmatrix}$ **8.** $A^{-1} = \begin{bmatrix} 2 & -\frac{5}{2} \\ -1 & \frac{3}{2} \end{bmatrix}$ **9.** (a) $\begin{bmatrix} 4 & -2 \\ 1 & 5 \end{bmatrix} \begin{bmatrix} x_1 \\ x_2 \end{bmatrix} = \begin{bmatrix} -1 \\ 1 \end{bmatrix}$;

(b) $A = \begin{bmatrix} 4 & -2 \\ 1 & 5 \end{bmatrix}$; (c) $A^{-1} = \frac{1}{22} \begin{bmatrix} 5 & 2 \\ -1 & 4 \end{bmatrix}$; (d) $x_1 = -\frac{3}{22}$, $x_2 = \frac{5}{22}$

Exercise Set 6.4, pp. 260–261

1. $A^{-1} = \begin{bmatrix} -3 & 2 \\ 5 & -3 \end{bmatrix}$ **3.** $A^{-1} = \begin{bmatrix} 2 & -3 \\ -7 & 11 \end{bmatrix}$ **5.** $A^{-1} = \begin{bmatrix} \frac{2}{11} & \frac{3}{11} \\ -\frac{1}{11} & \frac{4}{11} \end{bmatrix}$ **7.** $A^{-1} = \begin{bmatrix} \frac{3}{8} & -\frac{1}{4} & \frac{1}{8} \\ -\frac{1}{8} & \frac{3}{4} & -\frac{3}{8} \\ -\frac{1}{4} & \frac{1}{2} & \frac{1}{4} \end{bmatrix}$ **9.** $A^{-1} = \begin{bmatrix} \frac{1}{3} & 0 & \frac{1}{3} \\ -\frac{2}{5} & \frac{2}{5} & \frac{1}{5} \\ \frac{2}{15} & \frac{1}{5} & -\frac{1}{15} \end{bmatrix}$

11. A^{-1} does not exist.   **13.** $A^{-1} = \begin{bmatrix} 1 & -2 & 3 & 8 \\ 0 & 1 & -3 & 1 \\ 0 & 0 & 1 & -2 \\ 0 & 0 & 0 & -1 \end{bmatrix}$ **15.–27.** See Exercises 1–13.   **29.** $\left(-\frac{1}{39}, \frac{55}{39}\right)$ **31.** $(3, -3, -2)$

33. Find AI and IA and compare with A.   **35.** A^{-1} exists if and only if $xy \neq 0$. $A^{-1} = \begin{bmatrix} x^{-1} & 0 \\ 0 & y^{-1} \end{bmatrix}$.

37. A^{-1} exists if and only if $xyzw \neq 0$. $A^{-1} = \begin{bmatrix} \frac{1}{x} & -\frac{1}{xy} & -\frac{1}{xz} & -\frac{1}{xw} \\ 0 & \frac{1}{y} & 0 & 0 \\ 0 & 0 & \frac{1}{z} & 0 \\ 0 & 0 & 0 & \frac{1}{w} \end{bmatrix}$.

Chapter 6 Review, pp. 261–262

1. $\begin{bmatrix} 0 & -1 & 6 \\ 3 & 1 & -2 \\ -2 & 1 & -2 \end{bmatrix}$ **2.** $\begin{bmatrix} -3 & 3 & 0 \\ -6 & -9 & 6 \\ 6 & 0 & -3 \end{bmatrix}$ **3.** $\begin{bmatrix} -1 & 1 & 0 \\ -2 & -3 & 2 \\ 2 & 0 & -1 \end{bmatrix}$ **4.** $\begin{bmatrix} -2 & 2 & 6 \\ 1 & -8 & 18 \\ 2 & 1 & -15 \end{bmatrix}$ **5.** Not possible

6. $\begin{bmatrix} 2 & -1 & -6 \\ 1 & 5 & -2 \\ -2 & -1 & 4 \end{bmatrix}$ **7.** $\begin{bmatrix} 3 & 1 & 2 \\ 2 & 1 & 2 \\ 6 & 2 & 5 \end{bmatrix}$ **8.** B^{-1} does not exist.   **9.** $\begin{bmatrix} -\frac{1}{2} & 0 \\ \frac{1}{6} & \frac{1}{3} \end{bmatrix}$ **10.** $\begin{bmatrix} 0 & 0 & \frac{1}{4} \\ 0 & -\frac{1}{2} & 0 \\ \frac{1}{3} & 0 & 0 \end{bmatrix}$ **11.** $\begin{bmatrix} 1 & 0 & 0 & 0 \\ 0 & \frac{1}{9} & \frac{5}{18} & 0 \\ 0 & -\frac{1}{9} & \frac{2}{9} & 0 \\ 0 & 0 & 0 & 1 \end{bmatrix}$

12. $\begin{bmatrix} 3 & -2 & 4 \\ 1 & 5 & -3 \\ 2 & -3 & 7 \end{bmatrix} \begin{bmatrix} x \\ y \\ z \end{bmatrix} = \begin{bmatrix} 13 \\ 7 \\ -8 \end{bmatrix}$ **13.** -31 **14.** -1 **15.** 0 **16.** 120

17. $\begin{vmatrix} 5a & 5b & 5c \\ 3a & 3b & 3c \\ d & e & f \end{vmatrix} = 5(3) \begin{vmatrix} a & b & c \\ a & b & c \\ d & e & f \end{vmatrix} = 0$, since the first two rows are the same.   **18.** $(b - a)(c - b)(c - a)$

19. $(y - x)(z - x)(z - y)(yz + xz + yx)$ **20.** $(b - a)(c - a)(d - a)(c - b)(d - b)(d - c)$ **21.** If a matrix has all 0's below the main diagonal, then its determinant is the product of the elements on the main diagonal. *Proof:* Expand about the first column.

CHAPTER 7

Margin Exercises, Section 7.1

1. (a) increasing; (b) set of all real numbers; (c) set of all positive numbers; (d) 1; (e) 3.32
2. (a)–(d) all the same as Exercise 1; (e) 7.10; (f) 4^x

3.

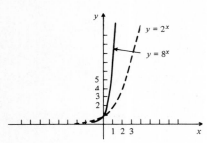

4.

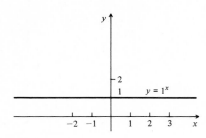

5.

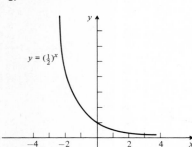

6.

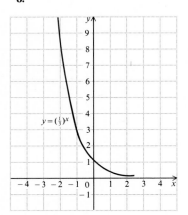

7. Domain: positive real numbers; range: all real numbers

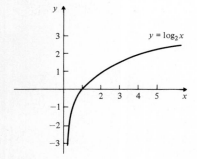

8. Domain: positive real numbers; range: all real numbers

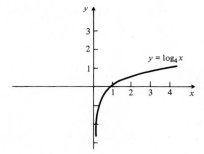

9. $\log_6 1 = 0$ **10.** $\log_{10} 0.001 = -3$ **11.** $\log_{16} 2 = \frac{1}{4}$ **12.** $\log_{6/5} \frac{25}{36} = -2$ **13.** $2^5 = 32$ **14.** $10^3 = 1000$
15. $10^{-2} = 0.01$ **16.** $(\sqrt{5})^2 = 5$ **17.** 10,000   **18.** 3   **19.** 4   **20.** 1   **21.** -2 **22.** 3   **23.** π **24.** 42   **25.** 37   **26.** M
27. 3.2

Exercise Set 7.1, p. 270

1. (a) and (b)   (c)

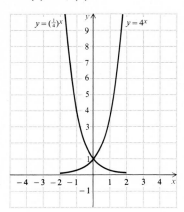

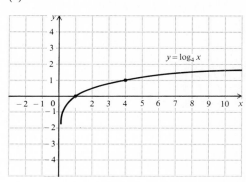

3.   **5.**

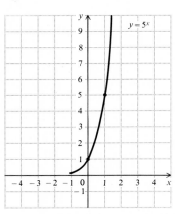

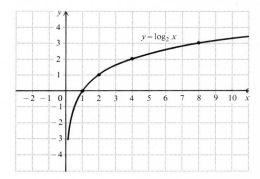

7. $2^5 = 32$ **9.** $10^{-2} = 0.01$ **11.** $6^1 = 6$ **13.** $\log_6 1 = 0$ **15.** $\log_{6/5} \frac{25}{36} = -2$ **17.** $\log_5 \frac{1}{25} = -2$ **19.** $\log_e 1.0833 = 0.08$
21. $10{,}000$ **23.** $\frac{1}{2}$ **25.** 4   **27.** $\frac{1}{2}$ **29.** 2   **31.** 1   **33.** -3 **35.** $4x$ **37.** $\sqrt{5}$
39.   **41.**

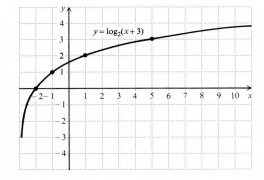

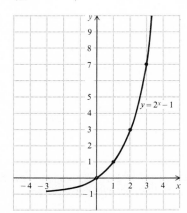

43.

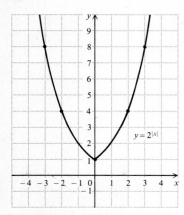

45.

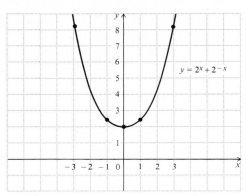

47. All real numbers **49.** $\{x \mid x \neq 0\}$ **51.** $\{x \mid x > \frac{4}{3}\}$ **53.** $\{x \mid x < -3 \text{ or } x > 3\}$ **55.** $\{x \mid x \leq 0\}$
57. $\{x \mid x \geq 16\}$ **59.** $\{x \mid x > -3\}$ **61.** π^5 **63.**

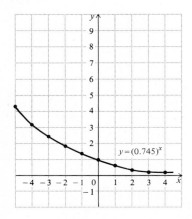

Margin Exercises, Section 7.2

1. $\log_a M + \log_a N$ **2.** $\log_5 25 + \log_5 5$ **3.** $\log_3 35$ **4.** $\log_a CABIN$ **5.** $5 \log_7 4$ **6.** $\frac{1}{2} \log_a 5$
7. (a) $\log_a M - \log_a N$; (b) $\log_c 1 - \log_c 4$ **8.** $\log_{10} 4 + \log_{10} \pi - \frac{1}{2} \log_{10} 23$ **9.** $\frac{1}{2}[3 \log_a z - \log_a x - \log_a y]$
10. $\log_a \dfrac{x^5 \sqrt[4]{z}}{y}$ **11.** (a) 0.954; (b) 0.1505; (c) 0.1003; (d) 0.176; (e) 1.585 **12.** 1 **13.** 0 **14.** 0 **15.** 1

Exercise Set 7.2, p. 274

1. $2 \log_a x + 3 \log_a y + \log_a z$ **3.** $\log_b x + 2 \log_b y - 3 \log_b z$ **5.** $\log_a 4$ **7.** $\log_a \dfrac{2x^4}{y^3}$ **9.** $\log_a \dfrac{\sqrt{a}}{x}$ or $\frac{1}{2} - \log_a x$
11. $\log_a (x^2 - xy + y^2)$ **13.** $\frac{1}{2}[\log_a (1 - x) + \log_a (1 + x)]$ **15.** 0.602 **17.** 1.699 **19.** 1.778 **21.** -0.088 **23.** 1.954
25. -0.046 **27.** False **29.** True **31.** False **33.** False **35.** $\frac{1}{2}$ **37.** $\sqrt{7}$ **39.** $-2, 0$ **41.** $\{x \mid x > 0\}$

Margin Exercises, Section 7.3

1. (a) 8; (b) 16 **2.** 8 **3.** 5 **4.** 2 **5.** 9 **6.** 0.4969 **7.** 0.9996 **8.** 0.6021 **9.** 5.74 **10.** 1.00 **11.** 3.62
12. $0.4609 + 2$ **13.** $0.4609 + (-4)$ **14.** 4.8312 **15.** 5.9504 **16.** 1.6618 **17.** 8.7846 **18.** $8.8932 - 10$ **19.** $6.0453 - 10$
20. $7.8976 - 10$ **21.** 64,100 **22.** 64,100 **23.** 8,560 **24.** 0.000425 **25.** 0.0105 **26.** 0.0601 **27.** 797,000

Exercise Set 7.3, pp. 280–281

1. 0.3909 **3.** 2.5403 **5.** 1.7202 **7.** 5.7952 **9.** 8.8463 − 10 **11.** 6.3345 − 10 **13.** 233 **15.** 0.018 **17.** 0.00000105
19. 25.2 **21.** 0.0973 **23.** 4.49 **25.** 0.0133 **27.** 190 **29.** 272 **31.** 8.77 **33.** 3.64 **35.** 25.7 **37.** 4.754264
39. −0.321371 **41.** 78,397,100 **43.** 0.000583

Margin Exercises, Section 7.4

1. 3.6592 **2.** 8.3779 − 10 **3.** 2856 **4.** 0.0005956

Exercise Set 7.4, p. 284

1. 1.6194 **3.** 0.4689 **5.** 2.8130 **7.** 9.1538 − 10 **9.** 7.6291 − 10 **11.** 9.2494 − 10 **13.** 2.7786 **15.** 9.8445 − 10
17. 224.5 **19.** 14.53 **21.** 70,030 **23.** 0.09245 **25.** 0.5343 **27.** 0.007295 **29.** 0.8268

Margin Exercises, Section 7.5

1. 2.8076 **2.** 2.9999 **3.** $x = \dfrac{\log (t + \sqrt{t^2 + 1})}{\log e}$ or $\log_e (t + \sqrt{t^2 + 1})$ **4.** 125 **5.** 8.75 **6.** 2 **7.** 10 yr **8.** 34 db
9. 60 db **10.** 7.8 **11.** (a) 68; (b) 54; (c) 40 **12.** $\log_2 7$ **13.** 3

Exercise Set 7.5, pp. 290–291

1. 5 **3.** $\frac{12}{5}$ **5.** $\frac{1}{2}$, −3 **7.** 2.7093 **9.** 10 **11.** 1 **13.** 5 **15.** 1; 100 **17.** 4 **19.** $x = \log_e (t + \sqrt{t^2 + 1})$
21. $x = \frac{1}{2} \log_5 \dfrac{t + 1}{1 - t}$ **23.** 11.9 years **25.** 65 db **27.** 140 db **29.** 6.7 **31.** $10^5 \cdot I_0$ **33.** (a) 82; (b) 68 **35.** 9 months
37. 4.2 **39.** 1; 10,000 **41.** ∅ **43.** −9; 9 **45.** $\frac{7}{4}$ **47.** $\log_x y - \log_x a$ **49.** $t = \dfrac{100(\log_e P - \log_e P_0)}{r}$
51. $t = -\dfrac{1}{k} \log \left[\dfrac{T - T_0}{T_1 - T_0} \right]$ **53.** $Q = a^b \cdot \sqrt[3]{y}$ **55.** 10; 100

Margin Exercises, Section 7.6

1. 2.48832 **2.** 2.59374246
3.

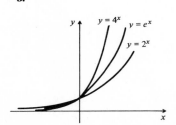

4.

5. 0.693147 **6.** 4.605170 **7.** −2.599375 **8.** −0.000100 **9.** 2.0956 **10.** 11.3059 **11.** −2.5096 **12.** 7.6009
13. −9.2103 **14.** $t \approx 4$ **15.** $t \approx 44$ **16.** (a) 0.08; (b) 170,524 **17.** 3.5 g **18.** 689 millibars **19.** 3 **20.** 4.7385
21. 0.4343 **22.** 6.937 **23.** −0.783 **24.** 2.8076

Exercise Set 7.6, pp. 298–300

1. and **3.**

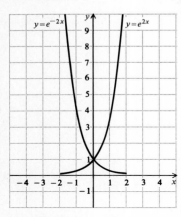

5.

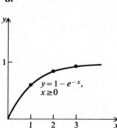

7. 0.6313   **9.** -3.9739 **11.** 0.7561   **13.** -1.5465 **15.** 8.4119   **17.** -7.4875 **19.** -2.5258 **21.** 13.7953   **23.** 4.6052
25. 4.0944   **27.** 2.3026   **29.** 140.67   **31.** $k = 0.0201$; $P = 1,243,000$ **33.** 0.4 gram   **35.** 1.2 days
37. (a) 3000 yr; (b) 5100 yr   **39.** 587 mb   **41.** (a) $k = 0.061$, $P = \$100e^{0.061t}$; (b) \$338.72; (c) 1978   **43.** 2.1610
45. -0.1544 **47.** 2.4849   **49.** $t = \dfrac{\ln P - \ln P_0}{k}$ **51.** By Theorem 7, $\ln x = \dfrac{\log x}{\log e} \approx \dfrac{\log x}{0.4343}$; by Table 2, $\approx 2.3026 \log x$
53. e^{π} **55.** 2; 2.25; 2.48832; 2.593742; 2.704814; 2.716924

Chapter 7 Review, pp. 300–301

1.

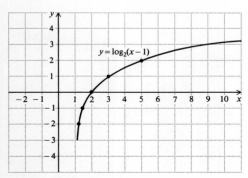

2.

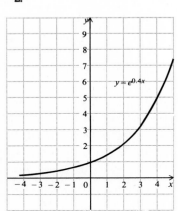

3.

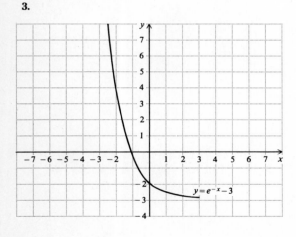

4. $8^{-2/3} = \frac{1}{4}$ **5.** $\log_7 x = 2.3$ **6.** 4   **7.** $\frac{1}{2}$ **8.** 3   **9.** $\log_b \dfrac{a^{1/2}c^{3/2}}{d^4}$ **10.** 1.255   **11.** 0.544   **12.** -0.602 **13.** 0.2385

14. $\frac{2}{3}\log M - \frac{1}{3}\log N$ **15.** 1.4200   **16.** $7.9063 - 10$ **17.** 73.9   **18.** 0.0276   **19.** 3.5   **20.** 0.0006934   **21.** 2.1861

22. 6.328   **23.** -3.0748 **24.** $x^2 + 1$ **25.** $\frac{1}{5}$ **26.** 9   **27.** $T = \dfrac{\log 2}{\log 1.13} \approx 5.7$ **28.** 30   **29.** 6.2 g   **30.** 1   **31.** 3   **32.** 3

33. 1   **34.** 64, $\frac{1}{64}$

35.

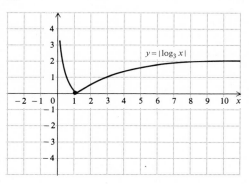

$y = |\log_3 x|$

36.

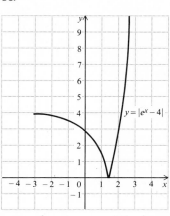

$y = |e^x - 4|$

37. $\{x \mid x > e^{6/5}\}$ **38.** $\{x \mid x \neq \frac{1}{4}\ln 10\}$

CHAPTER 8

Margin Exercises, Section 8.1

1. 43°   **2.** 7.5   **3.** 20 m   **4.** (a) $\sin\theta = \frac{12}{13}$, $\cos\theta = \frac{5}{13}$, $\tan\theta = \frac{12}{5}$, $\cot\theta = \frac{5}{12}$, $\sec\theta = \frac{13}{5}$, $\csc\theta = \frac{13}{12}$;
(b) $\sin\phi = \frac{5}{13}$, $\cos\phi = \frac{12}{13}$, $\tan\phi = \frac{5}{12}$, $\cot\phi = \frac{12}{5}$, $\sec\phi = \frac{13}{12}$, $\csc\phi = \frac{13}{5}$ **5.** 75.84 m   **6.** 25.5°

Exercise Set 8.1, pp. 308–310

1. 52°   **3.** 17.5   **5.** $\sin\theta = \frac{3}{5}$, $\cos\theta = \frac{4}{5}$, $\tan\theta = \frac{3}{4}$, $\cot\theta = \frac{4}{3}$, $\sec\theta = \frac{5}{4}$, $\csc\theta = \frac{5}{3}$
7. $\sin\theta = 0.8788$, $\cos\theta = 0.4771$, $\tan\theta = 1.8419$, $\cot\theta = 0.5429$, $\sec\theta = 2.0958$, $\csc\theta = 1.1379$
9. (a) 53.5°; (b) 16.8; (c) 22.7   **11.** (a) 52.5°; (b) 36.5 ft; (c) 46 ft
13. 3.3287   **15.** (a) $\sin 45° = \dfrac{1}{\sqrt{2}}$; (b) $\cos 45° = \dfrac{1}{\sqrt{2}}$; (c) $\tan 45° = 1$

Margin Exercises, Section 8.2

1. (a) $(3, 4)$; (b) $(-3, -4)$; (c) $(-3, 4)$; (d) yes, by symmetry   **2.** (a) $(-5, 2)$; (b) $(5, -2)$; (c) $(5, 2)$; (d) yes, by symmetry
3. (a) $\left(\dfrac{\sqrt{2}}{2}, -\dfrac{\sqrt{2}}{2}\right)$; (b) $\left(-\dfrac{\sqrt{2}}{2}, \dfrac{\sqrt{2}}{2}\right)$; (c) $\left(-\dfrac{\sqrt{2}}{2}, -\dfrac{\sqrt{2}}{2}\right)$; (d) yes, by symmetry   **4.** (a) $\dfrac{\pi}{2}$; (b) $\frac{3}{4}\pi$; (c) $\frac{3}{2}\pi$
5. (a) $\dfrac{\pi}{4}$; (b) $\dfrac{7\pi}{4}$; (c) $\dfrac{5\pi}{4}$; (d) $\dfrac{3\pi}{4}$ **6.** (a) $\dfrac{\pi}{6}$; (b) $\dfrac{5\pi}{6}$; (c) $\dfrac{4\pi}{3}$; (d) $\dfrac{11\pi}{6}$

7.

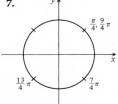

8.

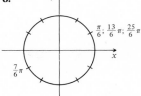

9.

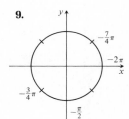

Exercise Set 8.2, pp. 314–315

1. (a) $(-5, -2)$; (b) $(5, 2)$; (c) $(5, -2)$; (d) yes, by symmetry   **3.** $M(0, 1)$; $N\left(-\dfrac{\sqrt{2}}{2}, \dfrac{\sqrt{2}}{2}\right)$; $P(0, -1)$; $Q\left(\dfrac{\sqrt{2}}{2}, -\dfrac{\sqrt{2}}{2}\right)$

5.

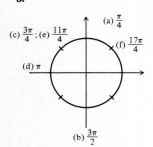

7.

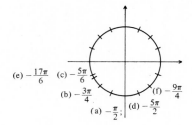

9.

11. $M: \dfrac{2}{3}\pi, -\dfrac{4}{3}\pi$; $N: \dfrac{5}{6}\pi, -\dfrac{7}{6}\pi$; $P: \dfrac{5}{4}\pi, -\dfrac{3}{4}\pi$; $Q: \dfrac{11}{6}\pi, -\dfrac{\pi}{6}$ **13.** $\left(\dfrac{\sqrt{3}}{2}, -\dfrac{1}{2}\right)$ **15.** $M: \dfrac{8\pi}{3}$; $N: \dfrac{17\pi}{6}$; $P: \dfrac{13\pi}{4}$; $Q: \dfrac{23\pi}{6}$

17. $\pm\dfrac{\sqrt{3}}{2}$ **19.** ±0.96649

Margin Exercises, Section 8.3

1. (a) 0; (b) -1; (c) $\dfrac{\sqrt{2}}{2}$; (d) $-\dfrac{\sqrt{2}}{2}$; (e) $\dfrac{\sqrt{3}}{2}$; (f) $-\dfrac{1}{2}$ **2.** Yes, 2π **3.** Odd   **4.** Yes   **5.** The set of all real numbers

6. The set of real numbers from -1 to 1, inclusive

7. (a) $(-a, -b)$; (b) $(-a, -b)$

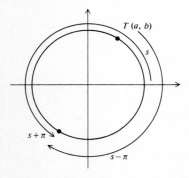

8. (a) -1; (b) 0; (c) $\dfrac{\sqrt{2}}{2}$; (d) $\dfrac{\sqrt{2}}{2}$; (e) $-\dfrac{\sqrt{3}}{2}$; (f) $\dfrac{1}{2}$

9. Yes, 2π **10.** Even   **11.** Yes   **12.** The set of all real numbers
13. The set of all real numbers from -1 to 1, inclusive

Exercise Set 8.3, pp. 321–323

1.

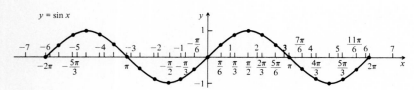

3. (a) See Exercise 1;
(b)

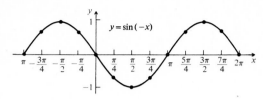

(c) same as (b); (d) same as (b)

5. (a) See Exercise 1;
(b)

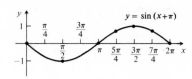

(c) same as (b); (d) the same

7. (a) See Exercise 2;
(b)

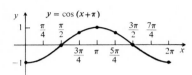

(c) same as (b); (d) same as (b)

9. $\cos x$ **11.** $-\sin x$ **13.** $-\cos x$ **15.** $\cos x$

17. $-\cos x$

19. (a) $\dfrac{\sqrt{2}}{2}$; (b) 0; (c) $\dfrac{1}{2}$; (d) $-\dfrac{\sqrt{2}}{2}$; (e) -1; (f) $\dfrac{1}{2}$; (g) $-\dfrac{\sqrt{2}}{2}$; (h) 0; (j) $\dfrac{\sqrt{2}}{2}$; (k) 1; (m) $-\dfrac{\sqrt{3}}{2}$

21. (a) $\dfrac{\sqrt{2}}{2}$; (b) -1; (c) $-\dfrac{\sqrt{2}}{2}$; (d) $\dfrac{\sqrt{3}}{2}$; (e) 0; (f) $-\dfrac{\sqrt{3}}{2}$; (g) $\dfrac{\sqrt{2}}{2}$; (h) -1; (j) $-\dfrac{\sqrt{2}}{2}$; (k) 0; (m) $\dfrac{1}{2}$

23. (a) $\dfrac{\pi}{2} + 2k\pi$, k any integer; (b) $\dfrac{3\pi}{2} + 2k\pi$, k any integer   **25.** $x = k\pi$, k any integer

27. $f \circ g(x) = \cos^2 x + 2 \cos x$, $g \circ f(x) = \cos (x^2 + 2x)$

29. (a) 0.8660; (b) 0.7071

Margin Exercises, Section 8.4

1. 0   **2.** -1 **3.** Does not exist   **4.** $\dfrac{3\pi}{2}$, $-\dfrac{\pi}{2}$, $-\dfrac{3\pi}{2}$, etc.   **5.** π

6. The set of all real numbers except $\dfrac{\pi}{2} + k\pi$, k any integer   **7.** The set of all real numbers   **8.** Odd

9. π, $-\pi$, 2π, -2π, etc.   **10.** π **11.** The set of all real numbers except $k\pi$, k any integer

12. The set of all real numbers   **13.** Positive: I, III; negative: II, IV   **14.** Odd   **15.** $\sqrt{2}$ **16.** -1

17. $\dfrac{\pi}{2}$, $-\dfrac{\pi}{2}$, $\dfrac{3\pi}{2}$, $-\dfrac{3\pi}{2}$, etc.   **18.** 2π **19.** The set of all real numbers except $k\pi$, k any integer

20. The set of real numbers 1 and greater, together with the set of real numbers -1 and less

21. Positive: I, II; negative: III, IV   **22.** Odd   **23.** Negative: sine, cosine, secant, cosecant. Others positive.

24. $\tan \dfrac{\pi}{11} = 0.29362$, $\cot \dfrac{\pi}{11} = 3.40571$, $\sec \dfrac{\pi}{11} = 1.04222$, $\csc \dfrac{\pi}{11} = 3.54950$

25. $\cot (-x) \equiv -\cot x$; $\cot (-x) \equiv \dfrac{1}{\tan (-x)} \equiv \dfrac{\cos (-x)}{\sin (-x)} \equiv \dfrac{\cos x}{-\sin x} \equiv -\cot x$; $\therefore \cot (-x) \equiv -\cot x$

26. $\tan (x - \pi) \equiv \dfrac{\sin (x - \pi)}{\cos (x - \pi)} \equiv \dfrac{-\sin x}{-\cos x} \equiv \dfrac{\sin x}{\cos x} \equiv \tan x$

27. $\csc (\pi - x) \equiv \dfrac{1}{\sin (\pi - x)} \equiv \dfrac{1}{\sin x} \equiv \csc x$

Exercise Set 8.4, pp. 329–330

1. 1   **3.** $\dfrac{\sqrt{3}}{3}$ **5.** $\sqrt{2}$ **7.** Does not exist   **9.** $-\sqrt{3}$

11.

	$\dfrac{\pi}{16}$	$\dfrac{\pi}{8}$	$\dfrac{\pi}{6}$	$\dfrac{\pi}{4}$	$\dfrac{3\pi}{8}$	$\dfrac{7\pi}{16}$
sin	0.19509	0.38268	0.50000	0.70711	0.92388	0.98079
cos	0.98079	0.92388	0.86603	0.70711	0.38268	0.19509
tan	0.19891	0.41421	0.57735	1.00000	2.41424	5.02737
cot	5.02737	2.41424	1.73206	1.00000	0.41421	0.19891
sec	1.01959	1.08239	1.15469	1.41421	2.61315	5.12584
csc	5.12584	2.61315	2.00000	1.41421	1.08239	1.01959

13.

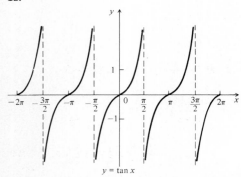

$y = \tan x$

15.

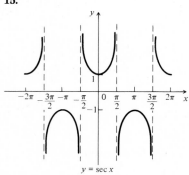

$y = \sec x$

17. cos, sec   **19.** sin, cos, sec, cosec   **21.** Positive: I, III; negative: II, IV   **23.** Positive: I, IV; negative: II, III

25. $\sec(-x) \equiv \dfrac{1}{\cos(-x)} \equiv \dfrac{1}{\cos x} \equiv \sec x$ **27.** $\cot(x + \pi) \equiv \dfrac{\cos(x + \pi)}{\sin(x + \pi)} \equiv \dfrac{-\cos x}{-\sin x} \equiv \dfrac{\cos x}{\sin x} \equiv \cot x$

29. $\sec(x + \pi) \equiv \dfrac{1}{\cos(x + \pi)} \equiv -\dfrac{1}{\cos x} \equiv -\sec x$

31. The graph of $\sec(x - \pi)$ is like that of $\sec x$, moved π units to the right. The graph of $-\sec x$ is that of $\sec x$ reflected across the x-axis. The graphs are identical.

33. If the graph of $\tan x$ were reflected across the y-axis and then translated to the right a distance of $\dfrac{\pi}{2}$, the graph of $\cot x$ would be obtained. There are other ways to describe the relation.

35. The sine and tangent functions; the cosine and cotangent functions.

37.

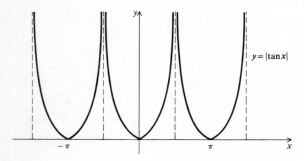

$y = |\tan x|$

Margin Exercises, Section 8.5

1. (a) $\dfrac{\sin^2 s}{\cos^2 s} + \dfrac{\cos^2 s}{\cos^2 s} \equiv \dfrac{1}{\cos^2 s}$, $\tan^2 s + 1 \equiv \sec^2 s$; (b) when $\cos^2 s = 0$ (or $\cos s = 0$); (c) yes; yes

2. $\sin^2 x + \cos^2 x \equiv 1$, $\sin^2 x \equiv 1 - \cos^2 x$ 3. $\sin^2 x \equiv 1 - \cos^2 x$, $|\sin x| \equiv \sqrt{1 - \cos^2 x}$, or $\sin x \equiv \pm \sqrt{1 - \cos^2 x}$

4. (a) and (b)   (c)   (d) the graphs of b and c are the same; (e) $\cos\left(x - \dfrac{\pi}{2}\right) \equiv \sin x$

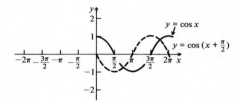 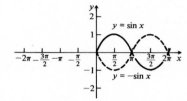

5. (a) and (b)   (c) and (d)   (e) the graphs are the same; (f) $\cos\left(x + \dfrac{\pi}{2}\right) \equiv -\sin x$

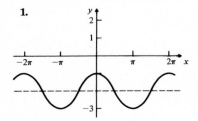 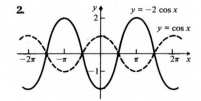

6. Since the cosine function is even, $\cos\left(\dfrac{\pi}{2} - x\right) \equiv \cos\left(x - \dfrac{\pi}{2}\right)$; but $\cos\left(x - \dfrac{\pi}{2}\right) \equiv \sin x$; $\therefore \cos\left(\dfrac{\pi}{2} - x\right) \equiv \sin x$

7. By definition of the cotangent function, $\cot\left(x - \dfrac{\pi}{2}\right) \equiv \dfrac{\cos\left(x - \dfrac{\pi}{2}\right)}{\sin\left(x - \dfrac{\pi}{2}\right)} \equiv \dfrac{\sin x}{-\cos x} \equiv \dfrac{-\sin x}{\cos x} \equiv -\tan x$;

$$\therefore \cot\left(x - \frac{\pi}{2}\right) \equiv -\tan x$$

Exercise Set 8.5, pp. 334–335

1. $\cot^2 x \equiv \csc^2 x - 1$, $\csc^2 x - \cot^2 x \equiv 1$ 3. (a) $\csc x \equiv \pm \sqrt{1 + \cot^2 x}$; (b) $\cot x \equiv \pm \sqrt{\csc^2 x - 1}$

5. $\tan\left(x - \dfrac{\pi}{2}\right) \equiv -\cot x$ 7. $\sec\left(\dfrac{\pi}{2} - x\right) \equiv \csc x$

9. $\sin\left(x \pm \dfrac{\pi}{2}\right) \equiv \pm\cos x$, $\cos\left(x \pm \dfrac{\pi}{2}\right) \equiv \mp\sin x$, $\tan\left(x \pm \dfrac{\pi}{2}\right) \equiv -\cot x$, $\cot\left(x \pm \dfrac{\pi}{2}\right) \equiv -\tan x$, $\sec\left(x \pm \dfrac{\pi}{2}\right) \equiv \mp\csc x$,

$$\csc\left(x \pm \frac{\pi}{2}\right) \equiv \pm\sec x$$

11. $\cos x = -0.9898$, $\tan x = -0.1440$, $\cot x = -6.9460$, $\sec x = -1.0103$, $\csc x = 7.0175$

13. $\sin x = 0.7987$, $\cos x = 0.6018$, $\tan x = 1.3273$, $\sec x = 1.6617$, $\csc x = 1.2520$ 15. $\begin{vmatrix} \sin x & \cos x \\ -\cos x & \sin x \end{vmatrix} = 1$

Margin Exercises, Section 8.6

1.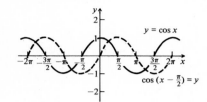

2.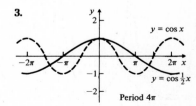

3.

4.

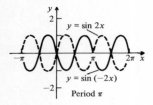

5.

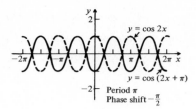

6.

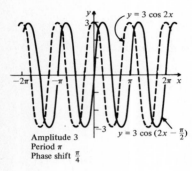

7.

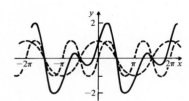

Exercise Set 8.6, pp. 339–340

1.

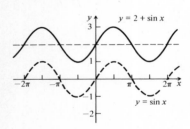

3.

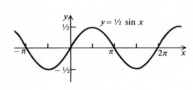

5.

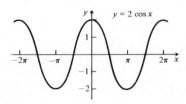

7.

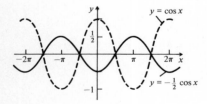

9.

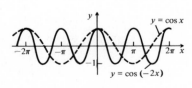

11.

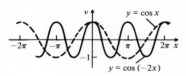

13.

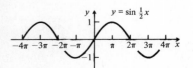

15.

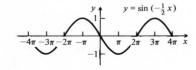

17.

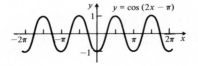

19.

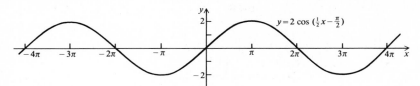

21.

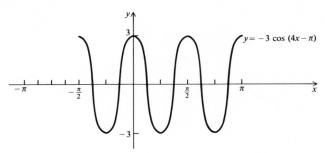

23.

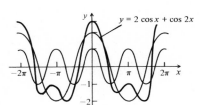

25.

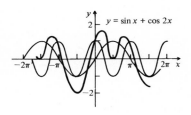

27.

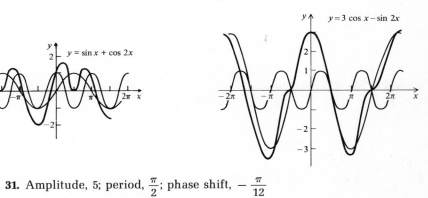

29. Amplitude, 3; period, $\frac{2\pi}{3}$; phase shift, $\frac{\pi}{6}$ **31.** Amplitude, 5; period, $\frac{\pi}{2}$; phase shift, $-\frac{\pi}{12}$

33. Amplitude, $\frac{1}{2}$; period, 1; phase shift, $-\frac{1}{2}$

35.

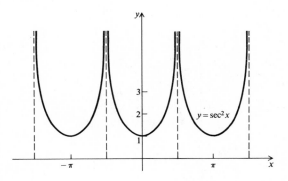

37.

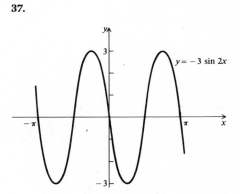

Margin Exercises, Section 8.7

1. $\cos x + 1$ **2.** $\sin x$ **3.** $\cos x \left(\dfrac{1}{\sin x} - 1 \right)$ or $\cos x (\csc x - 1)$ **4.** $\dfrac{1 + \sin x}{1 - \cot x}$ **5.** $\dfrac{5 \sin x - \cos x}{\sin^2 x - \cos^2 x}$

6. $\dfrac{2}{\cos^2 x(\cot x - 2)}$ **7.** $\tan x \sin x \sqrt{\sin x}$ **8.** $-\dfrac{\cos x}{\sqrt{3}\cos x}$ **9.** $3, -4$ **10.** $\dfrac{1 \pm \sqrt{97}}{8}$ **11.** $\cos x = \dfrac{-2}{3}, \cos x = \dfrac{1}{2}$

Exercise Set 8.7A, pp. 343–344

1. $\sin^2 x - \cos^2 x$ **3.** $\sin x - \sec x$ **5.** $\sin y + \cos y$ **7.** $\cot x - \tan x$ **9.** $1 - 2\sin y \cos y$ **11.** $2\tan x + \sec^2 x$
13. $\sin^2 y + \csc^2 y - 2$ **15.** $\cos^3 x - \sec^3 x$ **17.** $\cot^3 x - \tan^3 x$ **19.** $\sin^2 x$ **21.** $\cos x(\sin x + \cos x)$
23. $(\sin y - \cos y)(\sin y + \cos y)$ **25.** $\sin x(\sec x + 1)$ **27.** $(\sin x + \cos x)(\sin x - \cos x)$ **29.** $3(\cot y + 1)^2$
31. $(\csc^2 x + 5)(\csc x + 1)(\csc x - 1)$ **33.** $(\sin y + 3)(\sin^2 y - 3\sin y + 9)$ **35.** $(\sin y - \csc y)(\sin^2 y + 1 + \csc^2 y)$
37. $\sin x(\cos y + \tan y)$ **39.** $-\cos x(1 + \cot x)$

Exercise Set 8.7B, pp. 344–345

1. $\tan x$ **3.** $\dfrac{2\cos^2 x}{9\sin x}$ **5.** $\cos x - 1$ **7.** $\cos x + 1$ **9.** $\dfrac{2\tan x + 1}{3\sin x + 1}$ **11.** $\cos x + 1$ **13.** $\dfrac{\cos x - 1}{1 + \tan x}$ **15.** 1

17. $\cos x - 1$ **19.** $-\tan^2 y$ **21.** $\dfrac{1}{2\cos x}$ **23.** $\dfrac{3\tan x}{\cos x - \sin x}$ **25.** $\frac{1}{3}\cot y$ **27.** $\dfrac{1 - 2\sin y + 2\cos y}{\sin^2 y - \cos^2 y}$ **29.** -1

31. $\dfrac{5(\sin x - 3)}{3}$

Exercise Set 8.7C, p. 345

1. $\sin x \cos x$ **3.** $\sqrt{\sin y}(\sin y + \cos y)$ **5.** $\sin x + \cos x$ **7.** $1 - \sin y$ **9.** $\sin x(\sqrt{2} + \sqrt{\cos x})$ **11.** $\dfrac{\sqrt{\sin x \cos x}}{\cos x}$

13. $\dfrac{\sqrt{\cos x}}{\cot x}$ **15.** $\dfrac{\sqrt{2}\cot x}{2}$ **17.** $\dfrac{\cos x}{1 - \sin x}$ **19.** $3, -7$ **21.** $\frac{3}{4}, -\frac{1}{2}$ **23.** $-10, 1$ **25.** $2, -3$ **27.** $3 \pm \sqrt{13}$

29. $\dfrac{5 \pm \sqrt{73}}{12}$

Chapter 8 Review, pp. 346–347

1. $(-3, -2)$ **2.** $(-3, 2)$ **3.** $(3, 2)$
4.–7.

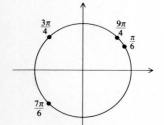

8., 12.

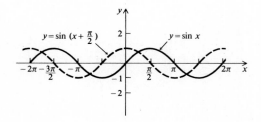

9. Set of all real numbers

10.

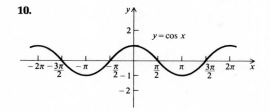

11. 2π　**12.** See graph for Exercise 8.
13.

	$\dfrac{\pi}{6}$	$\dfrac{\pi}{4}$	$\dfrac{\pi}{3}$	$\dfrac{\pi}{2}$	$\dfrac{3\pi}{4}$	$\dfrac{5\pi}{4}$
sin x	$\dfrac{1}{2}$	$\dfrac{\sqrt{2}}{2}$	$\dfrac{\sqrt{3}}{2}$	1	$\dfrac{\sqrt{2}}{2}$	$-\dfrac{\sqrt{2}}{2}$
cos x	$\dfrac{\sqrt{3}}{2}$	$\dfrac{\sqrt{2}}{2}$	$\dfrac{1}{2}$	0	$-\dfrac{\sqrt{2}}{2}$	$-\dfrac{\sqrt{2}}{2}$

14.

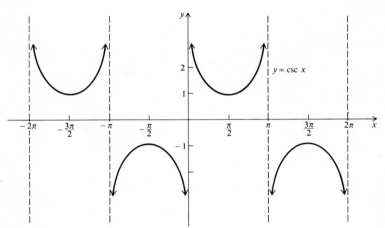

15. 2π　**16.** $x \geq 1$ and $x \leq -1$　**17.** I, IV

18. The points for s and $s - \pi$ are symmetric with respect to the origin. Cot $s = \dfrac{x}{y}$ and

$\cot (s - \pi) = \dfrac{-x}{-y} = \dfrac{x}{y}$, so $\cot s = \cot (s - \pi)$.

19. 1　**20.** $\csc^2 x$　**21.** $-\sin x$　**22.** $\sin x$　**23.** $-\cos x$

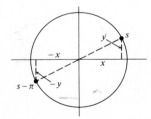

24. $\pm \sqrt{\sec^2 x - 1}$

25.

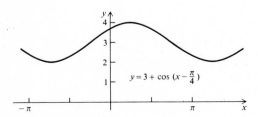

26. $\dfrac{\pi}{4}$　**27.** 2π

28.

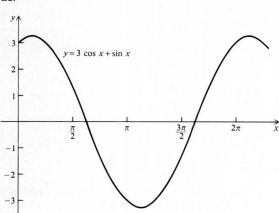

$y = 3 \cos x + \sin x$

29. $\dfrac{1}{\sin x}$ **30.** 1

31. $\dfrac{\sqrt{\sec x \tan x}}{\sec x}$ **32.** $\dfrac{1 \pm \sqrt{7}}{3}$ **33.** No, $\sin x = \frac{7}{5}$, but sines are never greater than 1.   **34.** All values

CHAPTER 9

Margin Exercises, Section 9.1

1. (a) I; (b) III; (c) IV; (d) III; (e) I; (f) IV; (g) I   **2.** (a) 360°; (b) 180°; (c) 90°; (d) 45°; (e) 60°; (f) 30°
3. (a) $\frac{5}{4}\pi$; (b) $\frac{7}{4}\pi$; (c) -4π **4.** (a) 1.26; (b) 5.23; (c) -5.50 **5.** (a) 240°; (b) 450°; (c) $-144°$
6. (a) III; (b) I; (c) IV; (d) I   **7.** 57.6 cm   **8.** 6 radians   **9.** $\frac{3}{2}$ radians

Exercise Set 9.1, pp. 353–354

1. I   **3.** III   **5.** I   **7.** II   **9.** $\dfrac{\pi}{6}$ **11.** $\dfrac{\pi}{3}$ **13.** $\dfrac{5\pi}{12}$ **15.** 0.2095π **17.** 1.1922π **19.** 2.093   **21.** 5.582   **23.** 3.489
25. 2.0550   **27.** 0.0236   **29.** 57.32°   **31.** 1440°   **33.** 135°   **35.** 74.69°   **37.** 135.6°   **39.** 1.1 radians, 63°
41. 5.233 radians   **43.** 16 m   **45.** $30° = \dfrac{\pi}{6}$, $60° = \dfrac{\pi}{3}$, $135° = \dfrac{3\pi}{4}$, $180° = \pi$, $225° = \dfrac{5\pi}{4}$, $270° = \dfrac{3\pi}{2}$, $315° = \dfrac{7\pi}{4}$
47. (a) 53.33; (b) 170; (c) 25; (d) 142.86   **49.** 111.7 km, 69.81 mi   **51.** $\frac{1}{30}$ radian

Margin Exercises, Section 9.2

1. 377 cm/sec   **2.** r in cm, ω in radians/sec   **3.** km/yr   **4.** 3.6 radians/sec   **5.** 70.4 radians, or 11.21 revolutions

Exercise Set 9.2, pp. 357–358

1. 3150 cm/min   **3.** 52.3 cm/sec   **5.** 1047 mph   **7.** 68.8 radians/sec   **9.** 10 mph   **11.** 75.4 ft   **13.** 377 radians
15. 2.093 radians/sec   **17.** (a) 395.35 rpm/sec, angular acc $= \dfrac{\Delta v}{t}$; (b) 41.4 radians/sec²

Margin Exercises, Section 9.3

1. $\sin \theta = -\frac{3}{5}$, $\cos \theta = -\frac{4}{5}$, $\tan \theta = \frac{3}{4}$, $\cot \theta = \frac{4}{3}$, $\sec \theta = -\frac{5}{4}$, $\csc \theta = -\frac{5}{3}$

2. $\sin 120° = \dfrac{\sqrt{3}}{2}$, $\cos 120° = -\dfrac{1}{2}$, $\tan 120° = -\sqrt{3}$, $\cot 120° = -\dfrac{1}{\sqrt{3}}$ or $-\dfrac{\sqrt{3}}{3}$, $\sec 120° = -2$, $\csc 120° = +\dfrac{2}{\sqrt{3}}$ or $+\dfrac{2\sqrt{3}}{3}$ **3.** $\sin(-135°) = -\dfrac{\sqrt{2}}{2}$, $\cos(-135°) = -\dfrac{\sqrt{2}}{2}$, $\tan(-135°) = 1$, $\cot(-135°) = 1$, $\sec(-135°) = -\sqrt{2}$, $\csc(-135°) = -\sqrt{2}$ **4.** $\sin 270° = -1$, $\cos 270° = 0$, $\tan 270°$: undefined, $\cot 270° = 0$, $\sec 270°$: undefined, $\csc 270° = -1$ **5.** 15.3 ft   **6.** $\sin 324° = -0.6283$, $\cos 324° = 0.8090$, $\tan 324° = -0.7265$, $\cot 324° = -1.3765$, $\sec 324° = 1.2361$, $\csc 324° = -1.5916$

Exercise Set 9.3, pp. 367–368

1. $\sin \theta = \frac{5}{13}$, $\cos \theta = -\frac{12}{13}$, $\tan \theta = -\frac{5}{12}$, $\cot \theta = -\frac{12}{5}$, $\sec \theta = -\frac{13}{12}$, $\csc \theta = \frac{13}{5}$
3. $\sin \theta = -\dfrac{3}{4}$, $\cos \theta = -\dfrac{\sqrt{7}}{4}$, $\tan \theta = \dfrac{3}{\sqrt{7}}$, $\cot \theta = \dfrac{\sqrt{7}}{3}$, $\sec \theta = -\dfrac{4}{\sqrt{7}}$, $\csc \theta = -\dfrac{4}{3}$
5. $\sin 30° = 0.5$, $\cos 30° = 0.866$, $\tan \theta = 0.577$, $\cot \theta = 1.732$, $\sec 30° = 1.155$, $\csc 30° = 2$ **7.** -1 **9.** Does not exist.
11. 1   **13.** -0.707 **15.** 0.5   **17.** 1.732   **19.** 1.414   **21.** 1   **23.** 0.5   **25.** 0.577   **27.** 0.707   **29.** -0.577
31.   **33.**

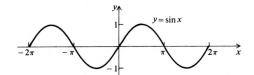   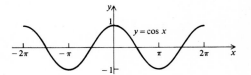

Margin Exercises, Section 9.4

1. $\sin \theta = -\dfrac{\sqrt{7}}{4}$, $\tan \theta = -\dfrac{\sqrt{7}}{3}$, $\cot \theta = -\dfrac{3}{\sqrt{7}}$, $\sec \theta = \dfrac{4}{3}$, $\csc \theta = -\dfrac{4}{\sqrt{7}}$
2. $\sin \theta = \dfrac{1}{\sqrt{10}}$, $\cos \theta = -\dfrac{3}{\sqrt{10}}$, $\tan \theta = -\dfrac{1}{3}$, $\sec \theta = -\dfrac{\sqrt{10}}{3}$, $\csc \theta = \sqrt{10}$
3. $\sin 15° = 0.2588$, $\cos 15° = 0.9659$, $\tan 15° = 0.2679$, $\cot 15° = 3.732$, $\sec 15° = 1.035$, $\csc 15° = 3.864$

Exercise Set 9.4, p. 370

1. $\cos \theta = -\dfrac{2\sqrt{2}}{3}$, $\tan \theta = \dfrac{\sqrt{2}}{4}$, $\cot \theta = 2\sqrt{2}$, $\sec \theta = -\dfrac{3\sqrt{2}}{4}$, $\csc \theta = -3$
3. $\sin \theta = -\frac{4}{5}$, $\tan \theta = -\frac{4}{3}$, $\cot \theta = -\frac{3}{4}$, $\sec \theta = \frac{5}{3}$, $\csc \theta = -\frac{5}{4}$
5. $\sin \theta = -\dfrac{\sqrt{5}}{5}$, $\cos \theta = \dfrac{2\sqrt{5}}{5}$, $\tan \theta = -\dfrac{1}{2}$, $\sec \theta = \dfrac{\sqrt{5}}{2}$, $\csc \theta = -\sqrt{5}$
7. $\sin 25° = 0.4226$, $\cos 25° = 0.9063$, $\tan 25° = 0.4663$, $\cot 25° = 2.145$, $\sec 25° = 1.103$, $\csc 25° = 2.366$
9. (a) $\cos \theta = -\dfrac{\sqrt{8}}{3}$, $\tan \theta = -\dfrac{1}{\sqrt{8}}$, $\cot \theta = -\sqrt{8}$, $\sec \theta = -\dfrac{3}{\sqrt{8}}$, $\csc \theta = 3$;
(b) $\sin(\pi + \theta) = -\dfrac{1}{3}$, $\cos(\pi + \theta) = \dfrac{\sqrt{8}}{3}$, $\tan(\pi + \theta) = -\dfrac{1}{\sqrt{8}}$, $\cot(\pi + \theta) = -\sqrt{8}$, $\sec(\pi + \theta) = \dfrac{3}{\sqrt{8}}$, $\csc(\pi + \theta) = -3$;
(c) $\sin(\pi - \theta) = \dfrac{1}{3}$, $\cos(\pi - \theta) = \dfrac{\sqrt{8}}{3}$, $\tan(\pi - \theta) = \dfrac{1}{\sqrt{8}}$, $\cot(\pi - \theta) = \sqrt{8}$, $\sec(\pi - \theta) = \dfrac{3}{\sqrt{8}}$, $\csc(\pi - \theta) = 3$;
(d) $\sin(2\pi - \theta) = -\dfrac{1}{3}$, $\cos(2\pi - \theta) = -\dfrac{\sqrt{8}}{3}$, $\tan(2\pi - \theta) = \dfrac{1}{\sqrt{8}}$, $\cot(2\pi - \theta) = \sqrt{8}$, $\sec(2\pi - \theta) = -\dfrac{3}{\sqrt{8}}$, $\csc(2\pi - \theta) = -3$ **11.** (a) $\cos 27° = 0.89101$, $\tan 27° = 0.50952$, $\cot 27° = 1.96262$, $\sec 27° = 1.12232$, $\csc 27° = 2.20269$;
(b) $\sin 63° = 0.89101$, $\cos 63° = 0.45399$, $\tan 63° = 1.96262$, $\cot 63° = 0.50952$, $\sec 63° = 2.20269$, $\csc 63° = 1.2232$
13. $\sin 128° = 0.78801$, $\cos 128° = -0.61566$, $\tan 128° = -1.27994$, $\cot 128° = -0.78129$, $\sec 128° = -1.62427$, $\csc 128° = 1.26902$

Margin Exercises, Section 9.5

1. 0.2644   **2.** 0.4699   **3.** 0.9026   **4.** 1.098   **5.** 0.0495 radian, or 2°50′; 0.0931 radian, or 5°20′   **6.** 37°27′   **7.** 43.917°
8. −0.6264   **9.** 1.901   **10.** 67°20′ or 1.1752 radians   **11.** 28°48′, or 0.5027 radian   **12.** −0.9182   **13.** 0.9853
14. 226°55′

Exercise Set 9.5, pp. 376–377

1. 0.2306   **3.** 0.5519   **5.** 0.4176   **7.** 0.7720   **9.** 46°23′   **11.** 67°50′   **13.** 45.75°   **15.** 76.88°   **17.** 0.4728   **19.** 0.5894
21. 0.2563   **23.** 0.2995   **25.** 0.7824   **27.** 13°40′   **29.** 68°10′   **31.** 69°17′   **33.** 0.4392   **35.** 0.8145   **37.** −0.9228
39. −1.590   **41.** 166°20′   **43.** 248°10′   **45.** 1.9324   **47.** 5.5245   **49.** 2°30′   **51.** 25°42′51″, 61.63944°   **53.** 0.48083
55. 865,000 miles

Margin Exercises, Section 9.6

1. $\cos \theta = 0.88$, $\tan \theta = -0.53$, $\cot \theta = -1.87$, $\sec \theta = 1.14$, $\csc \theta = -2.13$
2. $\sec \theta = -1.75$, $\cos \theta = -0.57$, $\sin \theta = -0.82$, $\csc \theta = -1.22$, $\cot \theta = 0.69$

3.

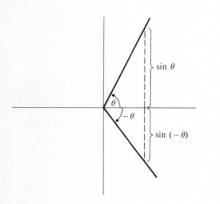

4.

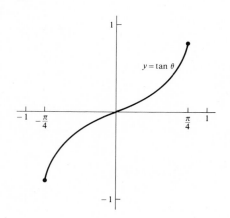

5. $\cot \left(\dfrac{\pi}{2} - x \right) \equiv \tan x$

Exercise Set 9.6, pp. 382–383

1. $\cos x = -0.9898$, $\tan x = -0.1440$, $\cot x = -6.9444$, $\sec x = -1.0103$, $\csc x = 7.0175$
3. $\sin x = 0.7987$, $\cos x = 0.6018$, $\tan x = 1.3273$, $\csc x = 1.2520$, $\sec x = 1.6618$ **5.** Odd
7. $\sin \theta = \dfrac{-y}{1}$, $\sin (-\theta) = \dfrac{y}{1}$

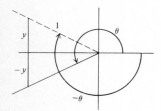

9. Same as Exercise 31 in Exercise Set 8.3   **11.** Domain: set of all real numbers; range: $[-1, 1]$; period: 2π
13. Domain: all real numbers except odd multiples of $\dfrac{\pi}{2}$; range: set of all real numbers; period: π

15. $\cos \left(\dfrac{\pi}{2} - x \right) \equiv \sin x$ **17.** $\tan \left(x - \dfrac{\pi}{2} \right) \equiv -\cot x$ **19.** $\sec \left(x - \dfrac{\pi}{2} \right) \equiv \csc x$

21.

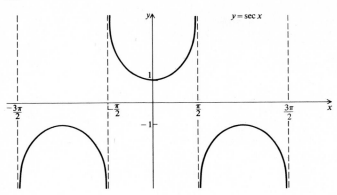

23.

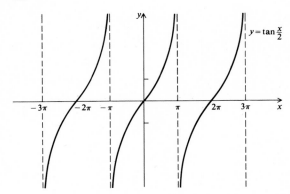

25. By similar triangles, $\dfrac{x}{\cos s} = \dfrac{y}{\sin s} = \dfrac{r}{1}$. Thus $\begin{cases} x = r\cos s \\ y = r\sin s \end{cases}$.

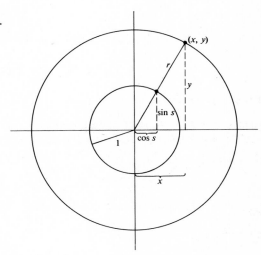

Chapter 9 Review, pp. 383–384

1. I, 0.483π, 1.52   **2.** II, 0.806π, 2.53   **3.** I, 0.167π, 0.525   **4.** IV, -0.167π, -0.525 **5.** 270°   **6.** 720°   **7.** $\dfrac{7\pi}{4}$ or 5.5 cm

8. 2.25, 129°   **9.** 1130 cm/min   **10.** 146,215 radians/hr

11. $\sin\theta = \dfrac{3}{\sqrt{13}}$, $\cos\theta = -\dfrac{2}{\sqrt{13}}$, $\tan\theta = -\dfrac{3}{2}$, $\cot\theta = -\dfrac{2}{3}$, $\sec\theta = -\dfrac{\sqrt{13}}{2}$, $\csc\theta = \dfrac{\sqrt{13}}{3}$

12.

	0°	30°	45°	60°	90°	270°
$\sin \theta$	0	$\dfrac{1}{2}$	$\dfrac{\sqrt{2}}{2}$	$\dfrac{\sqrt{3}}{2}$	1	-1
$\cos \theta$	1	$\dfrac{\sqrt{3}}{2}$	$\dfrac{\sqrt{2}}{2}$	$\dfrac{1}{2}$	0	0
$\tan \theta$	0	$\dfrac{\sqrt{3}}{3}$	1	$\sqrt{3}$	—	—
$\cot \theta$	—	$\sqrt{3}$	1	$\dfrac{\sqrt{3}}{3}$	0	0
$\sec \theta$	1	$\dfrac{2}{\sqrt{3}}$	$\sqrt{2}$	2	—	—
$\csc \theta$	—	2	$\sqrt{2}$	$\dfrac{2}{\sqrt{3}}$	1	-1

13. $\dfrac{\sqrt{2}}{2}$ **14.** 1 **15.** $\sqrt{3}$

16.

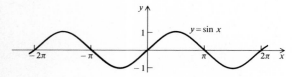

Domain: set of all real numbers; range: $[-1, 1]$; period: 2π

17. $-\dfrac{2}{3}$ **18.** $-\dfrac{\sqrt{5}}{3}$ **19.** $\dfrac{\sqrt{5}}{2}$ **20.** $-\dfrac{3}{\sqrt{5}}$ or $-\dfrac{3\sqrt{5}}{5}$ **21.** $-\dfrac{3}{2}$ **22.** 0.7314 **23.** 0.6820 **24.** 1.0724 **25.** 0.9325

26. 1.4663 **27.** 1.3673 **28.** 22°12′ **29.** 47.55° **30.** 0.9894 **31.** 0.5147 **32.** 1.402 **33.** 16°10′ **34.** 39°

35. $\sin x = 0.9898$, $\tan x = -0.1440$, $\cot x = -6.9460$, $\sec x = 7.0175$, $\csc x = -1.0103$

36. (a) cosine, secant; (b) sine, cosecant, tangent, cotangent

37.

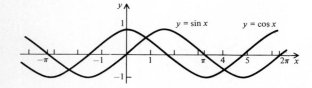

38. $\csc\left(x - \dfrac{\pi}{2}\right) \equiv \sec x$

39. Domain: Set of all reals; range: $[-3, 3]$; period: 4π

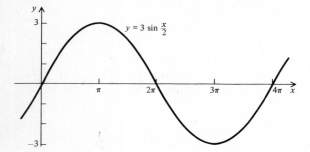

CHAPTER 10

Margin Exercises, Section 10.1

1. $\frac{1}{2}$ **2.** $\frac{-\sqrt{6} + \sqrt{2}}{4}$, or $\frac{\sqrt{2}}{4}(1 - \sqrt{3})$ **3.** $\cos \frac{7\pi}{12}$, or $\cos\left(-\frac{7\pi}{2}\right)$ **4.** $\cos 25°$ **5.** $\cos(\alpha + \beta)$

6. $\frac{\sqrt{2} + \sqrt{6}}{4}$, or $\frac{\sqrt{2}}{4}(1 + \sqrt{3})$ **7.** $\sin(\alpha - \beta)$

8. $\tan(\alpha - \beta) \equiv \dfrac{\sin(\alpha - \beta)}{\cos(\alpha - \beta)} \equiv \dfrac{\sin \alpha \cos \beta - \cos \alpha \sin \beta}{\cos \alpha \cos \beta + \sin \alpha \sin \beta} \cdot \dfrac{\dfrac{1}{\cos \alpha \cos \beta}}{\dfrac{1}{\cos \alpha \cos \beta}}$

$$\equiv \dfrac{\dfrac{\sin \alpha \cos \beta}{\cos \alpha \cos \beta} - \dfrac{\cos \alpha \sin \beta}{\cos \alpha \cos \beta}}{\dfrac{\cos \alpha \cos \beta}{\cos \alpha \cos \beta} + \dfrac{\sin \alpha \sin \beta}{\cos \alpha \cos \beta}} \equiv \dfrac{\tan \alpha - \tan \beta}{1 + \tan \alpha \tan \beta}$$

9. $\dfrac{3 + \sqrt{3}}{3 - \sqrt{3}}$ **10.** $\sin \dfrac{\pi}{6}$, or $\dfrac{1}{2}$

11. 30°

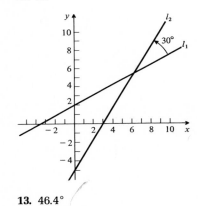

12. 150°

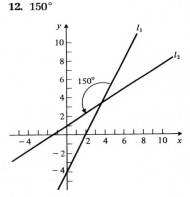

13. 46.4°

Exercise Set 10.1, pp. 392–393

1. 0.9659   **3.** 0.2588   **5.** 0.9659   **7.** 3.7322   **9.** 1   **11.** 0   **13.** Undefined   **15.** 0.8448   **17.** $\sin 59°$, or 0.8572

19. $\tan 52°$ or 1.280   **21.** $2 \sin \alpha \cos \beta$ **23.** $2 \cos \alpha \cos \beta$ **25.** $\cos u$ **27.** $\dfrac{2\pi}{3}$ **29.** 0°   **31.** 86°40'   **33.** 135°   **35.** $2 \sin \theta \cos \theta$

37. $\cot(\alpha + \beta) = \dfrac{\cos(\alpha + \beta)}{\sin(\alpha + \beta)} = \dfrac{\cos \alpha \cos \beta - \sin \alpha \sin \beta}{\sin \alpha \cos \beta + \cos \alpha \sin \beta} \cdot \dfrac{\dfrac{1}{\sin \alpha \sin \beta}}{\dfrac{1}{\sin \alpha \sin \beta}}$

$$= \dfrac{\dfrac{\cos \alpha \cos \beta}{\sin \alpha \sin \beta} - 1}{\dfrac{\sin \alpha \cos \beta}{\sin \alpha \sin \beta} + \dfrac{\cos \alpha \sin \beta}{\sin \alpha \sin \beta}} = \dfrac{\cot \alpha \cot \beta - 1}{\cot \alpha + \cot \beta}$$

39. $\cos x$ **41.** $\frac{1}{7}$ **43.** $\frac{1}{6}$ **45.** 173°, approximately   **47.** 22.8°

Margin Exercises, Section 10.2

1. $24/25$ **2.** II, $\cos 2\theta = -119/169$, $\tan 2\theta = -120/119$, $\sin 2\theta = 120/169$ **3.** $\cos\theta - 4\sin^2\theta\cos\theta$

4. $\sin^3 x \equiv \sin x\, \dfrac{1-\cos 2x}{2}$ **5.** $\dfrac{\sqrt{2+\sqrt{3}}}{2}$

6. $\dfrac{1}{2+\sqrt{3}}$, or $2-\sqrt{3}$

Exercise Set 10.2, pp. 399–400

1. $\sin 2\theta = \frac{24}{25}$, $\cos 2\theta = -\frac{7}{25}$, $\tan 2\theta = -\frac{24}{7}$, II **3.** $\cos 2\theta = \frac{7}{25}$, $\sin 2\theta = \frac{24}{25}$, $\tan 2\theta = \frac{24}{7}$, I
5. $\sin 2\theta = \frac{24}{25}$, $\cos 2\theta = -\frac{7}{25}$, $\tan 2\theta = -\frac{24}{7}$, II
7. $8\sin\theta\cos^3\theta - 4\sin\theta\cos\theta$, or $4\sin\theta\cos^3\theta - 4\sin^3\theta\cos\theta$, or $4\sin\theta\cos\theta - 8\sin^3\theta\cos\theta$
9. $\dfrac{3 - 4\cos 2\theta + \cos 4\theta}{8}$ **11.** $\dfrac{\sqrt{2+\sqrt{3}}}{2}$ **13.** $2+\sqrt{3}$ **15.** $\dfrac{\sqrt{2+\sqrt{2}}}{2}$ **17.** 0.6421
19. 0.9844 **21.** 0.1734 **23.** $\cos x$ **25.** $\cos x$ **27.** $\sin x$
29. $\cos x$ **31.** 1 **33.** 1 **35.** 8 **37.** $\sin 2x$
39. (i) $\frac{1}{2}[\sin(u+v) + \sin(u-v)] = \frac{1}{2}[\sin u \cos v + \cos u \sin v + \sin u \cos v - \cos u \sin v]$
$= \sin u \cdot \cos v$. The other formulas follow similarly.

Margin Exercises, Section 10.3

1.

$\dfrac{\cos^2 x}{\sin^2 x} - \cos^2 x$	$\cos^2 x \cdot \dfrac{\cos^2 x}{\sin^2 x}$
$\dfrac{\cos^2 x - \sin^2 x \cos^2 x}{\sin^2 x}$	$\dfrac{\cos^4 x}{\sin^2 x}$
$\dfrac{\cos^2 x (1 - \sin^2 x)}{\sin^2 x}$	
$\dfrac{\cos^4 x}{\sin^2 x}$	

2.

$\dfrac{2\sin\theta\cos\theta + \sin\theta}{(2\cos^2\theta - 1) + \cos\theta + 1}$	$\dfrac{\sin\theta}{\cos\theta}$
$\dfrac{\sin\theta(2\cos\theta + 1)}{2\cos^2\theta + \cos\theta}$	
$\dfrac{\sin\theta(2\cos\theta + 1)}{\cos\theta(2\cos\theta + 1)}$	
$\dfrac{\sin\theta}{\cos\theta}$	

3. $\sqrt{2}\sin\left(2x + \dfrac{\pi}{4}\right)$ **4.** $13\sin(x+b)$, where $\cos b = \frac{12}{13}$ and $\sin b = -\frac{5}{13}$ **5.** $\sqrt{3}\sin x + \cos x \equiv 2\sin\left(x + \dfrac{\pi}{6}\right)$

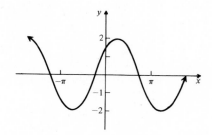

Exercise Set 10.3, pp. 404–405

1.

$$
\begin{array}{c|c}
\csc x - \cos x \cot x & \sin x \\
\hline
\dfrac{1}{\sin x} - \cos x \dfrac{\cos x}{\sin x} & \sin x \\
\dfrac{1 - \cos^2 x}{\sin x} & \\
\dfrac{\sin^2 x}{\sin x} & \\
\sin x &
\end{array}
$$

3.

$$
\begin{array}{c|c}
\dfrac{1 + \cos \theta}{\sin \theta} + \dfrac{\sin \theta}{\cos \theta} & \dfrac{\cos \theta + 1}{\sin \theta \cos \theta} \\
\hline
\dfrac{1 + \cos \theta}{\sin \theta} \cdot \dfrac{\cos \theta}{\cos \theta} + \dfrac{\sin \theta \cdot \sin \theta}{\cos \theta \sin \theta} & \dfrac{1 + \cos \theta}{\sin \theta \cos \theta} \\
\dfrac{\cos^2 \theta + \cos \theta + \sin^2 \theta}{\sin \theta \cos \theta} & \\
\dfrac{1 + \cos \theta}{\sin \theta \cos \theta} &
\end{array}
$$

5.

$$
\begin{array}{c|c}
\dfrac{1 - \sin x}{\cos x} & \dfrac{\cos x}{1 + \sin x} \\
\hline
\dfrac{1 - \sin x}{\cos x} \cdot \dfrac{\cos x}{\cos x} & \dfrac{\cos x}{1 + \sin x} \cdot \dfrac{1 - \sin x}{1 - \sin x} \\
\dfrac{\cos x - \sin x \cos x}{\cos^2 x} & \dfrac{\cos x - \sin x \cos x}{1 - \sin^2 x} \\
& \dfrac{\cos x - \sin x \cos x}{\cos^2 x}
\end{array}
$$

7.

$$
\begin{array}{c|c}
\dfrac{1 + \tan \theta}{1 + \cot \theta} & \dfrac{\sec \theta}{\csc \theta} \\
\hline
\dfrac{1 + \dfrac{\sin \theta}{\cos \theta}}{1 + \dfrac{\cos \theta}{\sin \theta}} & \dfrac{\sin \theta}{\cos \theta} \\
\dfrac{\dfrac{\cos \theta + \sin \theta}{\cos \theta}}{\dfrac{\sin \theta + \cos \theta}{\sin \theta}} & \tan \theta \\
\dfrac{\cos \theta + \sin \theta}{\cos \theta} \cdot \dfrac{\sin \theta}{\sin \theta + \cos \theta} & \\
\dfrac{\sin \theta}{\cos \theta} & \\
\tan \theta &
\end{array}
$$

9.

$$
\begin{array}{c|c}
\dfrac{\sin x + \cos x}{\sec x + \csc x} & \dfrac{\sin x}{\sec x} \\
\hline
\dfrac{\sin x + \cos x}{\dfrac{1}{\cos x} + \dfrac{1}{\sin x}} & \dfrac{\sin x}{\sec x} \\
\dfrac{\sin x + \cos x}{\dfrac{\sin x + \cos x}{\sin x \cos x}} & \\
(\sin x + \cos x) \cdot \dfrac{\sin x \cos x}{\sin x + \cos x} & \\
\sin x \cos x & \\
\dfrac{\sin x}{\sec x} &
\end{array}
$$

11.

$$
\begin{array}{c|c}
\dfrac{1 + \tan \theta}{1 - \tan \theta} + \dfrac{1 + \cot \theta}{1 - \cot \theta} & 0 \\
\hline
\dfrac{1 + \dfrac{\sin \theta}{\cos \theta}}{1 - \dfrac{\sin \theta}{\cos \theta}} + \dfrac{1 + \dfrac{\cos \theta}{\sin \theta}}{1 - \dfrac{\cos \theta}{\sin \theta}} & 0 \\
\dfrac{\dfrac{\cos \theta + \sin \theta}{\cos \theta}}{\dfrac{\cos \theta - \sin \theta}{\cos \theta}} + \dfrac{\dfrac{\sin \theta + \cos \theta}{\sin \theta}}{\dfrac{\sin \theta - \cos \theta}{\sin \theta}} & \\
\dfrac{\cos \theta + \sin \theta}{\cos \theta} \cdot \dfrac{\cos \theta}{\cos \theta - \sin \theta} + \dfrac{\sin \theta + \cos \theta}{\sin \theta} \cdot \dfrac{\sin \theta}{\sin \theta - \cos \theta} & \\
\dfrac{\cos \theta + \sin \theta}{\cos \theta - \sin \theta} + \dfrac{\sin \theta + \cos \theta}{\sin \theta - \cos \theta} & \\
\dfrac{\cos \theta + \sin \theta}{\cos \theta - \sin \theta} - \dfrac{\cos \theta + \sin \theta}{\cos \theta - \sin \theta} & \\
0 &
\end{array}
$$

13.

$$\dfrac{1 + \cos 2\theta}{\sin 2\theta} \qquad \cot\theta$$

$$\dfrac{1 + 2\cos^2\theta - 1}{2\sin\theta\cos\theta} \qquad \Bigg|\qquad \dfrac{\cos\theta}{\sin\theta}$$

$$\dfrac{\cos\theta}{\sin\theta}$$

15.

$$\sec 2\theta \qquad \dfrac{\sec^2\theta}{2 - \sec^2\theta}$$

$$\dfrac{1}{\cos 2\theta} \qquad \Bigg| \qquad \dfrac{1 + \tan^2\theta}{2 - 1 - \tan^2\theta}$$

$$\dfrac{1}{\cos^2\theta - \sin^2\theta} \qquad \dfrac{1 + \tan^2\theta}{1 - \tan^2\theta}$$

$$\dfrac{1 + \dfrac{\sin^2\theta}{\cos^2\theta}}{1 - \dfrac{\sin^2\theta}{\cos^2\theta}}$$

$$\dfrac{\cos^2\theta + \sin^2\theta}{\cos^2\theta - \sin^2\theta}$$

$$\dfrac{1}{\cos^2\theta - \sin^2\theta}$$

17.

$$\dfrac{\sin\alpha\cos\beta + \cos\alpha\sin\beta}{\cos\alpha\cos\beta} \qquad \Bigg| \qquad \dfrac{\sin\alpha}{\cos\alpha} + \dfrac{\sin\beta}{\cos\beta}$$

$$\dfrac{\sin\alpha\cos\beta + \cos\alpha\sin\beta}{\cos\alpha\cos\beta}$$

19.

$$1 - [\cos 5\theta \cos 3\theta + \sin 5\theta \sin 3\theta] \qquad \Bigg| \qquad 1 - \cos 2\theta$$

$$1 - \cos(5\theta - 3\theta)$$

$$1 - \cos 2\theta$$

21.

$$\dfrac{1}{2}\left[\dfrac{\dfrac{\sin\theta + \sin\theta\cos\theta}{\cos\theta}}{\dfrac{\sin\theta}{\cos\theta}}\right] \qquad \Bigg| \qquad \dfrac{1 + \cos\theta}{2}$$

$$\dfrac{1}{2}\left[\dfrac{\sin\theta(1 + \cos\theta)}{\cos\theta}\cdot\dfrac{\cos\theta}{\sin\theta}\right]$$

$$\dfrac{1 + \cos\theta}{2}$$

23.

$$(\cos^2 x - \sin^2 x)(\cos^2 x + \sin^2 x) \qquad \Bigg| \qquad \cos^2 x - \sin^2 x$$

$$\cos^2 x - \sin^2 x$$

25.

$$\tan(3\theta - \theta) \qquad \Bigg| \qquad \tan 2\theta$$

$$\tan 2\theta$$

27.

$$\dfrac{(\cos x - \sin x)(\cos^2 x + \cos x \sin x + \sin^2 x)}{\cos x - \sin x} \qquad \Bigg| \qquad \dfrac{2 + 2\sin x \cos x}{2}$$

$$1 + \cos x \sin x \qquad 1 + \sin x \cos x$$

29.

$$(\sin\alpha\cos\beta + \cos\alpha\sin\beta)(\sin\alpha\cos\beta - \cos\alpha\sin\beta) \qquad \Bigg| \qquad 1 - \cos^2\alpha - (1 - \cos^2\beta)$$

$$\sin^2\alpha\cos^2\beta - \cos^2\alpha\sin^2\beta \qquad \cos^2\beta - \cos^2\alpha$$

$$\cos^2\beta(1 - \cos^2\alpha) - \cos^2\alpha(1 - \cos^2\beta)$$

$$\cos^2\beta - \cos^2\beta\cos^2\alpha - \cos^2\alpha + \cos^2\alpha\cos^2\beta$$

$$\cos^2\beta - \cos^2\alpha$$

31.

$$\cos\alpha\cos\beta - \sin\alpha\sin\beta + \cos\alpha\cos\beta + \sin\alpha\sin\beta \qquad \Bigg| \qquad 2\cos\alpha\cos\beta$$

$$2\cos\alpha\cos\beta$$

33. $2 \sin \left(2x + \dfrac{\pi}{3}\right)$ **35.** $5 \sin (x + b)$, where $\cos b = \frac{4}{5}$, $\sin b = \frac{3}{5}$

37. $7.89 \sin (0.374x + 31.2°)$

39. $y = \sqrt{2} \sin \left(2x - \dfrac{\pi}{4}\right)$

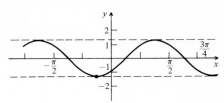

41. $\log (\cos x - \sin x) + \log (\cos x + \sin x) = \log [(\cos x - \sin x)(\cos x + \sin x)] = \log (\cos^2 x - \sin^2 x) = \log \cos 2x$

43. $\dfrac{1}{\omega C(\tan \theta + \tan \phi)} = \dfrac{1}{\omega C \left(\dfrac{\sin \theta}{\cos \theta} + \dfrac{\sin \phi}{\cos \phi}\right)} = \dfrac{1}{\omega C \left(\dfrac{\sin \theta \cos \phi + \sin \phi \cos \theta}{\cos \theta \cos \phi}\right)} = \dfrac{\cos \theta \cos \phi}{\omega C \sin (\theta + \phi)}$

45. By matrix multiplication, we obtain $\begin{bmatrix} \cos x \cos y - \sin x \sin y & \cos x \sin y + \sin x \cos y \\ -\sin x \cos y - \cos x \sin y & -\sin x \sin y + \cos x \cos y \end{bmatrix}$. The use of identities produces the desired result.

Margin Exercises, Section 10.4

1. Not a function   **2.** Not a function   **3.** $\dfrac{\pi}{4} + 2\pi k$, $-\dfrac{\pi}{4} + 2\pi k$, where k is an integer

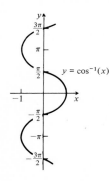

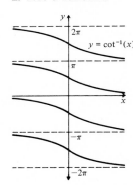

4. $\dfrac{\pi}{3} + 2\pi k$, $\dfrac{2\pi}{3} + 2k\pi$, where k is an integer   **5.** $25° + 360°k$, $155° + 360°k$ **6.** $\dfrac{3\pi}{4} + k\pi$ **7.** $\dfrac{\pi}{3}$

8. $\dfrac{3\pi}{4}$ **9.** $\dfrac{3\pi}{4}$ **10.** $-\dfrac{\pi}{4}$

Exercise Set 10.4, p. 410

1. $\dfrac{\pi}{4} + 2k\pi$, $\dfrac{3\pi}{4} + 2k\pi$ **3.** $\dfrac{\pi}{4} + 2k\pi$, $-\dfrac{\pi}{4} + 2k\pi$ **5.** $\dfrac{5\pi}{4} + 2k\pi$, $-\dfrac{\pi}{4} + 2k\pi$ **7.** $\dfrac{3\pi}{4} + 2k\pi$, $\dfrac{5\pi}{4} + 2k\pi$ **9.** $\dfrac{\pi}{3} + k\pi$

11. $\dfrac{\pi}{4} + k\pi$ **13.** $\dfrac{5\pi}{6} + k\pi$ **15.** $\dfrac{3\pi}{4} + k\pi$ **17.** $0 + 2k\pi$ **19.** $\dfrac{\pi}{2} + 2k\pi$ **21.** $23° + k \cdot 360°$, $157° + k \cdot 360°$

23. $39° + k \cdot 360°$, $141° + k \cdot 360°$ **25.** $36°58' + k \cdot 360°$, $323°02' + k \cdot 360°$ **27.** $21°25' + k \cdot 360°$, $338°35' + k \cdot 360°$

29. $20°10' + k \cdot 180°$ **31.** $38°20' + k \cdot 180°$ **33.** $31° + k \cdot 360°$, $329° + k \cdot 360°$

35. $9°10' + k \cdot 360°$, $170°50' + k \cdot 360°$ **37.** $\dfrac{\pi}{4}$ **39.** $\dfrac{\pi}{3}$ **41.** $-\dfrac{\pi}{3}$ **43.** $\dfrac{3\pi}{4}$ **45.** $-\dfrac{\pi}{6}$ **47.** $\dfrac{2\pi}{3}$ **49.** $13°30'$

51. $-39°30'$ **53.** $152°50'$ **55.** $-22°10'$ **57.** $169°30'$

Margin Exercises, Section 10.5

1. $-\dfrac{\pi}{2}$ **2.** $\dfrac{\pi}{6}$ **3.** $-\dfrac{\pi}{4}$ **4.** $\dfrac{3\pi}{4}$ **5.** $\dfrac{5\pi}{6}$ **6.** $\dfrac{\sqrt{2}}{2}$ **7.** $\dfrac{\sqrt{3}}{2}$ **8.** $\dfrac{\pi}{3}$ **9.** $\dfrac{\pi}{2}$ **10.** $\dfrac{3}{\sqrt{b^2+9}}$ **11.** $\dfrac{2}{\sqrt{5}}$ **12.** $\dfrac{t}{\sqrt{1-t^2}}$
13. 1 **14.** $\frac{1}{3}$

Exercise Set 10.5, pp. 414–415

1. 0.3 **3.** -4.2 **5.** $\dfrac{\pi}{3}$ **7.** $-\dfrac{\pi}{4}$ **9.** $\dfrac{\pi}{5}$ **11.** $-\dfrac{\pi}{3}$ **13.** $\dfrac{\sqrt{3}}{2}$ **15.** $\dfrac{1}{2}$ **17.** 1 **19.** $\dfrac{\pi}{6}$ **21.** $\dfrac{\pi}{3}$ **23.** $\dfrac{\pi}{2}$ **25.** $\dfrac{x}{\sqrt{x^2+4}}$
27. $\dfrac{\sqrt{x^2-9}}{3}$ **29.** $\dfrac{\sqrt{b^2-a^2}}{a}$ **31.** $\dfrac{3}{\sqrt{11}}$ **33.** $\dfrac{1}{3\sqrt{11}}$ **35.** $-\dfrac{1}{2\sqrt{6}}$ **37.** $\dfrac{1}{\sqrt{1+y^2}}$ **39.** $\dfrac{1}{\sqrt{1+t^2}}$ **41.** $\dfrac{\sqrt{1-y^2}}{y}$
43. $\sqrt{1-x^2}$ **45.** $\dfrac{1}{2}$ **47.** $\dfrac{\sqrt{2+\sqrt{3}}}{2}$ **49.** $\dfrac{24}{25}$ **51.** $\dfrac{119}{169}$ **53.** $\dfrac{3+4\sqrt{3}}{10}$ **55.** $-\dfrac{\sqrt{2}}{10}$ **57.** $xy+\sqrt{(1-x^2)(1-y^2)}$
59. $y\sqrt{1-x^2}-x\sqrt{1-y^2}$ **61.** 0.9861 **63.** $\theta = \text{Arctan}\,\dfrac{y+h}{x}-\text{Arctan}\,\dfrac{y}{x}$

Margin Exercises, Section 10.6

1. $60°+360°k$, $300°+360°k$, $\dfrac{\pi}{3}+2k\pi$, $\dfrac{5\pi}{3}+2k\pi$ **2.** $\dfrac{\pi}{6}$, $\dfrac{5\pi}{6}$, $\dfrac{7\pi}{6}$, $\dfrac{11\pi}{6}$; $30°$, $150°$, $210°$, $330°$ **3.** $\dfrac{\pi}{6}$, $\dfrac{5\pi}{6}$, $\dfrac{7\pi}{6}$, $\dfrac{11\pi}{6}$
4. $40°$, $320°$ **5.** $140°$, $220°$ **6.** $120°$, $240°$, $75°30'$, $284°30'$ **7.** $120°$, $240°$, $90°$, $270°$ **8.** $241°40'$, $118°20'$

Exercise Set 10.6, pp. 419–420

1. $\dfrac{\pi}{3}+2k\pi$, $\dfrac{2\pi}{3}+2k\pi$ **3.** $\dfrac{\pi}{4}+2k\pi$, $\dfrac{-\pi}{4}+2k\pi$ **5.** $20°10'+k\cdot360°$, $159°50'+k\cdot360°$ **7.** $236°40'$, $123°20'$ **9.** $\dfrac{4\pi}{3}$, $\dfrac{5\pi}{3}$
11. $123°40'$, $303°40'$ **13.** $\dfrac{\pi}{6}$, $\dfrac{5\pi}{6}$, $\dfrac{7\pi}{6}$, $\dfrac{11\pi}{6}$ **15.** $\dfrac{\pi}{6}$, $\dfrac{5\pi}{6}$, $\dfrac{7\pi}{6}$, $\dfrac{11\pi}{6}$ **17.** $\dfrac{\pi}{6}$, $\dfrac{5\pi}{6}$, $\dfrac{3\pi}{2}$ **19.** 0 **21.** 0, $\dfrac{\pi}{6}$, $\dfrac{5\pi}{6}$, π, $\dfrac{7\pi}{6}$, $\dfrac{11\pi}{6}$
23. $\dfrac{\pi}{6}$, $\dfrac{5\pi}{6}$ **25.** $109°30'$, $250°30'$, $120°$, $240°$ **27.** $\dfrac{\pi}{6}$, $\dfrac{5\pi}{6}$, π **29.** 0, π, $\dfrac{\pi}{2}$, $\dfrac{3\pi}{2}$ **31.** $60°$, $120°$, $240°$, $300°$ **33.** $60°$, $240°$
35. $30°$, $60°$, $120°$, $150°$, $210°$, $240°$, $300°$, $330°$ **37.** 0, π **39.** 0, π **41.** 0, ±4.8, ±14.3, ±17.1, etc.

Margin Exercises, Section 10.7

1. $\dfrac{\pi}{2}$, π **2.** $\dfrac{\pi}{2}$, π **3.** $\dfrac{\pi}{2}$, π **4.** $\dfrac{\pi}{2}$, $\dfrac{3\pi}{2}$, $\dfrac{\pi}{4}$, $\dfrac{3\pi}{4}$, $\dfrac{5\pi}{4}$, $\dfrac{7\pi}{4}$ **5.** $\dfrac{\pi}{2}$, $\dfrac{3\pi}{2}$, $\dfrac{7\pi}{6}$, $\dfrac{11\pi}{6}$ **6.** $\dfrac{\pi}{12}$, $\dfrac{5\pi}{12}$, $\dfrac{13\pi}{12}$, $\dfrac{17\pi}{12}$ **7.** $\dfrac{\pi}{3}$, π

Exercise Set 10.7, pp. 423–424

1. 0, π **3.** $\dfrac{3\pi}{4}$, $\dfrac{7\pi}{4}$ **5.** $\dfrac{\pi}{2}$, $\dfrac{3\pi}{2}$, $\dfrac{\pi}{6}$, $\dfrac{5\pi}{6}$ **7.** $\dfrac{\pi}{2}$, $\dfrac{3\pi}{2}$, $\dfrac{\pi}{4}$, $\dfrac{3\pi}{4}$, $\dfrac{5\pi}{4}$, $\dfrac{7\pi}{4}$ **9.** 0, $\dfrac{\pi}{2}$, π, $\dfrac{3\pi}{2}$ **11.** 0 **13.** 0, $\dfrac{\pi}{2}$, π, $\dfrac{3\pi}{2}$
15. $\dfrac{\pi}{6}$, $\dfrac{5\pi}{6}$, π **17.** $\dfrac{\pi}{6}$, $\dfrac{5\pi}{6}$, $\dfrac{7\pi}{6}$, $\dfrac{11\pi}{6}$ **19.** $63°30'$, $243°30'$, $101°20'$, $281°20'$ **21.** $\dfrac{\pi}{6}$, $\dfrac{\pi}{2}$, $\dfrac{5\pi}{6}$, $\dfrac{7\pi}{6}$, $\dfrac{3\pi}{2}$, $\dfrac{11\pi}{6}$ **23.** $\dfrac{2\pi}{3}$, $\dfrac{4\pi}{3}$
25. $\dfrac{\pi}{4}$, $\dfrac{7\pi}{4}$ **27.** $\dfrac{\pi}{12}$, $\dfrac{5\pi}{12}$, $\dfrac{13\pi}{12}$, $\dfrac{17\pi}{12}$ **29.** $\dfrac{\pi}{6}$, $\dfrac{3\pi}{2}$ **31.** 1 **33.** 1.15, 5.65, -0.65, etc.

Chapter 10 Review, p. 424

1. $\sin x$ **2.** $\dfrac{\tan 45° - \tan 30°}{1 + \tan 45° \tan 30°}$ **3.** $\cos(27° - 16°)$ or $\cos 11°$ **4.** $2 - \sqrt{3}$ **5.** $\frac{1}{2}\sqrt{2 - \sqrt{2}}$ **6.** $2 \cot \theta$

7.

$\tan 2\theta$	$\dfrac{2 \tan \theta}{1 - \tan^2 \theta}$
$\dfrac{\sin 2\theta}{\cos 2\theta}$	$\dfrac{2 \dfrac{\sin \theta}{\cos \theta}}{\dfrac{\cos^2 \theta}{\cos^2 \theta} - \dfrac{\sin^2 \theta}{\cos^2 \theta}}$
$\dfrac{2 \sin \theta \cos \theta}{\cos^2 \theta - \sin^2 \theta}$	$\dfrac{2 \sin \theta}{\cos \theta} \cdot \dfrac{\cos^2 \theta}{\cos^2 \theta - \sin^2 \theta}$
	$\dfrac{2 \sin \theta \cos \theta}{\cos^2 \theta - \sin^2 \theta}$

8. $\dfrac{\pi}{6} + 2k\pi, \dfrac{5\pi}{6} + 2k\pi$ **9.** $81° + k \cdot 180°$ **10.** $-45°$ or $-\dfrac{\pi}{4}$ **11.** $\dfrac{7}{8}$ **12.** $18°30'$ **13.** $0, \pi$

14. $\sqrt{40} \sin\left(3x + \text{Arcsin } \dfrac{\sqrt{10}}{10}\right)$

15.

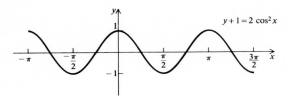

$y + 1 = 2 \cos^2 x$

16.

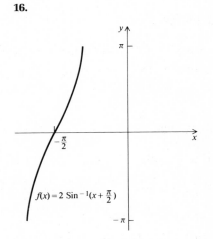

$f(x) = 2 \text{ Sin}^{-1}\left(x + \dfrac{\pi}{2}\right)$

CHAPTER 11

Margin Exercises, Section 11.1

1. $a = 38.43, b = 54.89$ **2.** $A = 56°19', B = 33°41', c = 7.211$ **3.** 35.7 ft   **4.** 240.4 ft   **5.** 136 km east, 63.4 km south
6. 22.1 km   **7.** 16 mi

Exercise Set 11.1, pp. 430–431

1. $B = 53°50', b = 37.2, c = 46.1$ **3.** $A = 77°20', a = 436, c = 447$ **5.** $B = 72°40', a = 4.24, c = 14.3$
7. $A = 66°50', b = 0.0148, c = 0.0375$ **9.** $B = 42°30', a = 35.6, b = 32.6$ **11.** $A = 7°40', a = 0.131, b = 0.973$
13. $c = 21.6, A = 33°40', B = 56°20'$ **15.** $b = 12.0, A = 53°10', B = 36°50'$ **17.** $a = 3.57, A = 63°20', B = 26°40'$
19. 47.9 ft   **21.** $1°40'$ **23.** $30°10'$ **25.** 3.52 mi   **27.** 109 km   **29.** 23.9 km   **31.** 7.92 km   **33.** 3.45 km

35.

$$A = \frac{1}{2}ab = \frac{1}{2}\frac{c^2}{c^2}ab = \frac{1}{2}c^2\frac{a}{c}\frac{b}{c}$$
$$= \frac{1}{2}c^2 \sin A \cos A = \frac{1}{4}c^2\, 2 \sin A \cos A$$
$$= \frac{1}{4}c^2 \sin 2A$$

Margin Exercises, Section 11.2

1. $C = 29°$, $c = 5.93$, $b = 11.1$ **2.** $\sin A = 2.8$; impossible   **3.** $B = 90°$, $C = 41°25'$, $c = 5$
4. (a) $A = 42°50'$, $C = 104°10'$, $c = 35.6$; (b) $A = 137°10'$, $C = 9°50'$, $c = 6.27$ **5.** $C = 18°$, $A = 124°$, $a = 26.9$

Exercise Set 11.2, p. 436

1. $C = 17°$, $a = 26.3$, $c = 10.5$ **3.** $A = 121°$, $a = 33.4$, $c = 14.0$ **5.** $B = 68°50'$, $a = 32.3$, $b = 32.3$
7. $B = 56°20'$, $C = 87°40'$, $c = 40.8$, and $B = 123°40'$, $C = 20°20'$, $c = 14.2$ **9.** $B = 19°$, $C = 44°40'$, $b = 6.25$
11. $A = 74°30'$, $B = 44°20'$, $a = 33.3$ **13.** 76.3 m   **15.** 50.8 ft   **17.** 1470 km   **19.** 10.6 km
21. $A = bh$, $h = a \sin \theta$, so $A = ab \sin \theta$

Margin Exercises, Section 11.3

1. $a = 40.5$, $B = 22°10'$, $C = 35°50'$ **2.** $A = 108°10'$, $B = 22°20'$, $C = 49°30'$

Exercise Set 11.3, pp. 439–440

1. $a = 14.9$, $B = 23°40'$, $C = 126°20'$ **3.** $a = 24.8$, $B = 20°40'$, $C = 26°20'$ **5.** $b = 74.8$, $A = 95°30'$, $C = 11°50'$
7. $A = 36°10'$, $B = 43°30'$, $C = 100°20'$ **9.** $A = 73°40'$, $B = 51°50'$, $C = 54°30'$ **11.** $A = 25°40'$, $B = 126°$; $C = 28°20'$
13. 28.2 nautical mi, S 55°20'E   **15.** 37 nautical mi   **17.** 59.4 ft   **19.** 68°, 68°, 44°   **21.** 42.6 ft
23. (a) 15.73 ft; (b) 120.8 ft²   **25.** 3424 yd²   **27.** 116.6 ft   **29.** $A = \frac{1}{2}a^2 \sin \theta$; when $\theta = 90°$

Margin Exercises, Section 11.4

1. 14.9 kg, 19°40'   **2.** 75°, 171 km/h

Exercise Set 11.4, p. 443

1. 57, 38°   **3.** 18.4, 37°   **5.** 20.9, 59°   **7.** 68.3, 18°   **9.** 13 kg, 67°   **11.** 655 kg, 21°   **13.** 21.6 ft/sec, 34°
15. 726 lb, 47°   **17.** 174 nautical mi, S 15°E   **19.** An angle of 12° upstream

Margin Exercises, Section 11.5

1. E: $50\sqrt{2}$, S: $50\sqrt{2}$ **2.** S: 36.9 lb; W: 25.8 lb   **3.** 18.9 km/h from S 32°E
4. (a) 4 up, 21 left; (b) 21.4, 10°50' with horizontal   **5.** $(12, -8)$ **6.** $(12, 4)$ **7.** $(8.54, 159°30')$ **8.** $(14.2, -4.88)$
9. (a) $(1, 7)$; (b) $(7, -3)$; (c) $(-23, 21)$; (d) 19.03

Exercise Set 11.5, p. 449

1. $(5, 16)$ **3.** $(4.8, 13.7)$ **5.** $(-662, -426)$ **7.** $(5, 53°10')$ **9.** $(18, 303°40')$ **11.** $(5, 233°10')$ **13.** $(18, 123°40')$
15. $(3.46, 2)$ **17.** $(-5.74, -8.19)$ **19.** $(17.3, -10)$ **21.** $(70.7, -70.7)$ **23.** Vertical 118, horizontal 92.4
25. S. 192 km/h, W. 161 km/h   **27.** 43 kg, S 35°30'W   **29.** (a) N. 28, W. 7; (b) 28.8, N 14°W   **31.** $(19, 36)$ **33.** $(14, 8)$
35. 18   **37.** 17.89   **39.** 35.78   **41. 0**

Margin Exercises, Section 11.6

1.

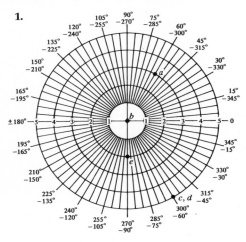

2. Many answers possible. A: $(4, 30°)$, $(4, 390°)$, $(-4, 210°)$, etc.; B: $(5, -60°)$, $(5, 300°)$; C: $(2, 150°)$, $(2, -210°)$; D: $(3, 225°)$, $(3, -135°)$; E: $(5, 60°)$, $(-5, -120°)$

3. (a) $(3\sqrt{2}, 45°)$; (b) $(4, 270°)$; (c) $(6, 120°)$; (d) $(4, 330°)$

4. (a) $\left(\dfrac{5}{2}\sqrt{3}, \dfrac{5}{2}\right)$; (b) $(5, 5\sqrt{3})$; (c) $\left(-\dfrac{5}{\sqrt{2}}, -\dfrac{5}{\sqrt{2}}\right)$; (d) $(4\sqrt{3}, 4)$

5. $2r\cos\theta + 5r\sin\theta = 9$

6. $r^2 + 8r\cos\theta = 0$ **7.** $x^2 + y^2 = 49$ **8.** $y = 5$ **9.** $x^2 + y^2 - 3x = 5y$ **10.**

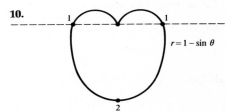

$r = 1 - \sin\theta$

Exercise Set 11.6, p. 453

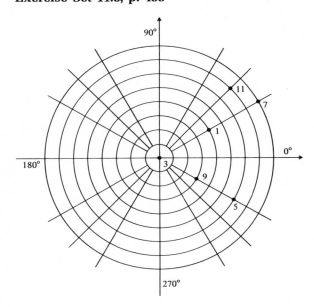

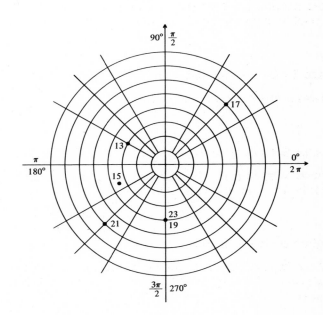

25. $(4\sqrt{2}, 45°)$ **27.** $(5, 90°)$ **29.** $(4, 0°)$ **31.** $(6, 60°)$ **33.** $(2, 30°)$ **35.** $(6, 30°)$

37. $\left(\dfrac{4}{\sqrt{2}}, \dfrac{4}{\sqrt{2}}\right)$ or $(2.83, 2.83)$ **39.** $(0, 0)$ **41.** $\left(\dfrac{-3}{\sqrt{2}}, \dfrac{-3}{\sqrt{2}}\right)$ **43.** $(3, -3\sqrt{3})$ **45.** $(5\sqrt{3}, 5)$

47. $(4.33, -2.5)$ **49.** $3r\cos\theta + 4r\sin\theta = 5$ **51.** $r\cos\theta = 5$ **53.** $r^2 = 36$ **55.** $r^2(\cos^2\theta - 4\sin^2\theta) = 4$

57. $x^2 + y^2 = 25$ **59.** $y = x$ **61.** $y = 2$ **63.** $x^2 + y^2 = 4x$ **65.** $x^2 - 4y = 4$ **67.** $x^2 - 2x + y^2 - 3y = 0$

69.   **71.**

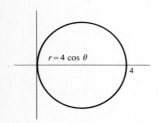

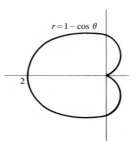

73.

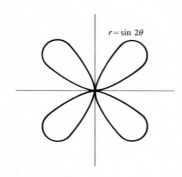

Margin Exercises, Section 11.7

1. Horizontal rope $300\sqrt{3}$ lb; other rope 600 lb
2. Parallel to incline 50 kg; perpendicular to incline $50\sqrt{3}$ or 86.6 kg
3. 500 kg on left, 866 kg on right

Exercise Set 11.7, pp. 456–458

1. Cable 224-lb tension, boom 166-lb compression
3. Horizontal rod 168-kg tension, other rod 261-kg compression
5. Lift 2472 lb, drag 1315 lb
7. 60 kg
9. 80°20′
11. Horizontal rope 400 kg, other rope 566 kg
13. 2000 kg in each
15. 2242 kg on left, 1830 kg on right

Chapter 11 Review, p. 458

1. $A = 58°10'$, $B = 31°50'$, $c = 4.55$ **2.** $A = 38°50'$, $b = 37.9$, $c = 48.6$ **3.** 1748 cm
4. $A = 34°10'$, $a = 0.621$, $c = 0.511$ **5.** 420 cm   **6.** 8.0   **7.** 13.95 ft   **8.** 13.72 cm^2 **9.** 75.8 mph; N 20°10′E
10. Parallel: 106 lb; perpendicular: 106 lb   **11.** $(-18.79, 6.84)$ **12.** $(\sqrt{13}, 123°40')$ **13.** $r + 2\cos\theta - 3\sin\theta = 0$
14. 23 lb down, 17 lb left; 28.6 lb, 53.5° downward from horizontal (to left)   **15.** (a) $(25, 0)$; (b) $\sqrt{50}$ **16.** 577 kg
17. $50.52°$, $129.48°$

CHAPTER 12

Margin Exercises, Section 12.1

1. $i\sqrt{6}$ **2.** $-i\sqrt{10}$ **3.** $2i$ **4.** $-5i$ **5.** $-\sqrt{10}$ **6.** $\sqrt{11}$ **7.** $i\sqrt{7}$ **8.** $7i$ **9.** $3i$ **10.** $(\sqrt{17} + 3)i$ **11.** i **12.** -1
13. $-i$ **14.** $12 + i$ **15.** $5 - i$ **16.** $2 + 14i$ **17.** 8   **18.** $-6 + 8i$ **19.** $3i$ **20.** $(x + 2i)(x - 2i)$ **21.** $(3 + yi)(3 - yi)$
22. Yes   **23.** $x = -1$, $y = 2$

Exercise Set 12.1, pp. 463–464

1. $i\sqrt{15}$ **3.** $4i$ **5.** $-2i\sqrt{3}$ **7.** $9i$ **9.** $i(\sqrt{7} - \sqrt{10})$ **11.** $-\sqrt{55}$ **13.** $2\sqrt{5}$ **15.** $\sqrt{\frac{5}{2}}i$ **17.** $-\frac{3}{2}i$ **19.** -2
21. $6 + 5i$ **23.** 8   **25.** $2 + 4i$ **27.** $-4 - i$ **29.** $-5 + 5i$ **31.** $7 - i$ **33.** $-6 + 12i$ **35.** $-5 + 12i$ **37.** i
39. $(2x + 5yi)(2x - 5yi)$ **41.** Yes   **43.** $x = -\frac{3}{2}$, $y = 7$ **45.** $-4 + 3i$
47. For example, $\sqrt{-1}\sqrt{-1} = i^2 = -1$, but $\sqrt{(-1)(-1)} = \sqrt{1} = 1$.   **49.** $\begin{bmatrix} -1 & -2 \\ 3 + 3i & 3 + 12i \end{bmatrix}$

Margin Exercises, Section 12.2

1. $7 - 2i$ **2.** $6 + 4i$ **3.** $5i$ **4.** $-3i$ **5.** -3 **6.** 8   **7.** $\frac{9}{13} + \frac{7}{13}i$ **8.** $\frac{4}{13} + \frac{7}{13}i$ **9.** $\frac{1}{3 + 4i}$, $\frac{3}{25} - \frac{4}{25}i$
10. Both are $7 + 3i$ **11.** Both are $-13 - 11i$ **12.** $\overline{z^3} = \overline{z \cdot z \cdot z} = \overline{z} \cdot \overline{z} \cdot \overline{z} = \overline{z}^3$ **13.** $5\overline{z}^3 + 4\overline{z}^2 - 2\overline{z} + 1$
14. $7\overline{z}^5 - 3\overline{z}^3 + 8\overline{z}^2 + \overline{z}$

Exercise Set 12.2, p. 468

1. $\frac{1}{2} + \frac{7}{2}i$ **3.** $\frac{1}{3} + \frac{2}{3}i\sqrt{2}$ **5.** $2 - 3i$ **7.** $\frac{1}{5} + \frac{2}{5}i$ **9.** $-\frac{1}{2} - \frac{i}{2}$ **11.** $\frac{28}{65} - \frac{29}{65}i$ **13.** $-\frac{1}{2} + \frac{3}{2}i$ **15.** $\frac{5}{2} + \frac{13}{2}i$ **17.** $\frac{4}{25} - \frac{3}{25}i$
19. $\frac{5}{29} + \frac{2}{29}i$ **21.** $-i$ **23.** $\frac{i}{4}$ **25.** $3\overline{z}^5 - 4\overline{z}^2 + 3\overline{z} - 5$ **27.** $4\overline{z}^7 - 3\overline{z}^5 + 4\overline{z}$ **29.** $z = 1$ **31.** a

33. $\frac{3 - i}{2 + i}$, or $1 - i$

Margin Exercises, Section 12.3

1. $2 + 5i$ **2.** $\dfrac{-1 + i \pm \sqrt{-18i}}{4}$ **3.** $\dfrac{-3 \pm 4i}{5}$ **4.** $x^2 - (1 + 2i)x + i - 1 = 0$ **5.** $x^3 - 2x^2 + x - 2 = 0$
6. $2 - i$, $-2 + i$

Exercise Set 12.3, pp. 471–472

1. $\dfrac{2}{5} + \dfrac{6}{5}i$ **3.** $\dfrac{8}{5} - \dfrac{9}{5}i$ **5.** $2 - i$ **7.** $\dfrac{11}{25} + \dfrac{2}{25}i$ **9.** $\dfrac{-1 + i \pm \sqrt{-6i}}{2}$ **11.** $-i, \dfrac{i}{2}$ **13.** $\dfrac{-1 - 2i \pm \sqrt{-15 + 16i}}{6}$

15. $1 \pm 2i$ **17.** $2 \pm 3i$ **19.** $-\dfrac{3}{2} \pm \dfrac{\sqrt{7}}{2}i$ **21.** $x^2 + 4 = 0$ **23.** $x^2 - 2x + 2 = 0$ **25.** $x^2 - 4x + 13 = 0$

27. $x^2 - 3x - ix + 3i = 0$ **29.** $x^3 - x^2 + 9x - 9 = 0$ **31.** $x^3 - 2x^2i - 3x^2 + 5ix + x - 2i + 2 = 0$

33. $x^3 + x - 2x^2i - 2i = 0$ **35.** $\sqrt{2} + \sqrt{2}i, -\sqrt{2} - \sqrt{2}i$ **37.** $2 + i, -2 - i$ **39.** $2, -1 \pm \sqrt{3}i$

41. $x = 2 + i, y = 1 - 3i$

Margin Exercises, Section 12.4

1.

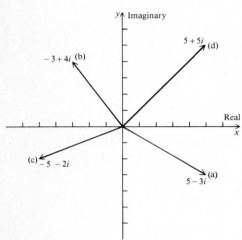

2.

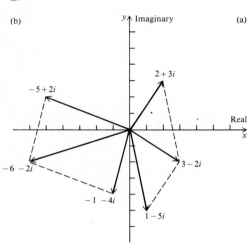

3. (a) 5; (b) 13   **4.** $1 - i$ **5.** $\sqrt{3} - i$ **6.** $\sqrt{2} \operatorname{cis} 315°$ **7.** $6 \operatorname{cis} 225°$ **8.** $20 \operatorname{cis} 55°$ **9.** $4 \operatorname{cis} \dfrac{5\pi}{4}$ **10.** $4 \operatorname{cis} 90°$

11. $2 \operatorname{cis} \dfrac{\pi}{4}$ **12.** $\sqrt{2} \operatorname{cis} 285°$

Exercise Set 12.4, pp. 477–478

1.

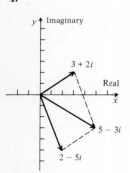

3.

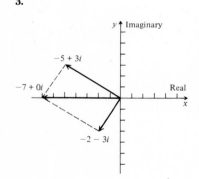

5.

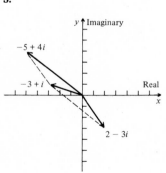

7.

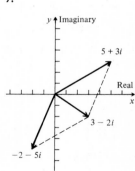

9. $\dfrac{3\sqrt{3}}{2} + \dfrac{3}{2}i$ **11.** $-10i$ **13.** $2 + 2i$ **15.** $-2 - 2i$ **17.** $\sqrt{2} \operatorname{cis} 315°$ **19.** $20 \operatorname{cis} 330°$ **21.** $5 \operatorname{cis} 180°$

23. $4 \operatorname{cis} 0°$, or 4   **25.** $8 \operatorname{cis} 120°$ **27.** $\operatorname{cis} 270°$, or $-i$ **29.** $2 \operatorname{cis} 270°$, or $-2i$

31. $z = a + bi, |z| = \sqrt{a^2 + b^2}; -z = -a - bi, |-z| = \sqrt{(-a)^2 + (-b)^2} = \sqrt{a^2 + b^2}, \therefore |z| = |-z|$

33. $|(a + bi)(a - bi)| = |a^2 + b^2| = a^2 + b^2$; $|(a + bi)^2| = |a^2 + 2abi - b^2| = |a^2 - b^2 + 2abi| = \sqrt{(a^2 - b^2)^2 + (2ab)^2} = \sqrt{a^4 + 2a^2b^2 + b^4} = a^2 + b^2$ **35.** $z \cdot w = (r_1 \operatorname{cis} \theta_1)(r_2 \operatorname{cis} \theta_2) = r_1 r_2 \operatorname{cis} (\theta_1 + \theta_2)$, $|z \cdot w| = \sqrt{[r_1 r_2 \cos (\theta_1 + \theta_2)]^2 + [r_1 r_2 \sin (\theta_1 + \theta_2)]^2} = \sqrt{(r_1 r_2)^2} = |r_1 r_2|$, $|z| = \sqrt{(r_1 \cos \theta_1)^2 + (r_1 \sin \theta_1)^2} = \sqrt{r_1^2} = |r_1|$, $|w| = \sqrt{(r_2 \cos \theta_2)^2 + (r_2 \sin \theta_2)^2} = \sqrt{r_2^2} = |r_2|$. Then $|z| \cdot |w| = |r_1| \cdot |r_2| = |r_1 r_2| = |z \cdot w|$

37.

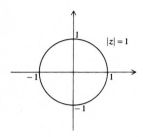

Margin Exercises, Section 12.5

1. $32 \operatorname{cis} 270°$, or $-32i$ **2.** $16 \operatorname{cis} 120°$, or $-8 + 8i\sqrt{3}$ **3.** (a) $1 + i$, $-1 - i$; (b) $\sqrt{5} + i\sqrt{5}$, $-\sqrt{5} - i\sqrt{5}$

4. $1 \operatorname{cis} 60°$, $1 \operatorname{cis} 180°$, $1 \operatorname{cis} 300°$; or $\dfrac{1}{2} + \dfrac{\sqrt{3}}{2}i$, -1, $\dfrac{1}{2} - \dfrac{\sqrt{3}}{2}i$

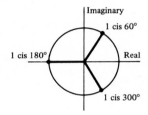

Exercise Set 12.5, pp. 480–481

1. $8 \operatorname{cis} \pi$ **3.** $64 \operatorname{cis} \pi$ **5.** $8 \operatorname{cis} 270°$ **7.** $-8 - 8\sqrt{3}i$ **9.** $-8 - 8\sqrt{3}i$ **11.** i **13.** 1

15. $\sqrt{2} \operatorname{cis} 60°$, $\sqrt{2} \operatorname{cis} 240°$; or $\dfrac{\sqrt{2}}{2} + \dfrac{\sqrt{6}}{2}i$, $\dfrac{-\sqrt{2}}{2} - \dfrac{\sqrt{6}}{2}i$ **17.** $\operatorname{cis} 30°$, $\operatorname{cis} 150°$, $\operatorname{cis} 270°$; or $\dfrac{\sqrt{3}}{2} + \dfrac{1}{2}i$, $\dfrac{-\sqrt{3}}{2} + \dfrac{1}{2}i$, $-i$

19. $2 \operatorname{cis} 0°$, $2 \operatorname{cis} 90°$, $2 \operatorname{cis} 180°$, $2 \operatorname{cis} 270°$; or 2, $2i$, -2, $-2i$

21. $-1.366 + 1.366i$, $0.366 - 0.366i$

Chapter 12 Review, p. 481

1. $14 + 2i$ **2.** $1 - 4i$ **3.** $2 - i$ **4.** $\dfrac{11}{10} + \dfrac{3}{10}i$ **5.** $x = 2$, $y = -4$ **6.** $3\bar{z}^3 + \bar{z} - 7$ **7.** $x^2 - 2x + 5 = 0$ **8.** $\dfrac{2}{5} \pm \dfrac{1}{5}i$

9. $\dfrac{(-3 \pm \sqrt{5})i}{2}$ **10.** $\sqrt{2} + \sqrt{2}i$, $-\sqrt{2} - \sqrt{2}i$

11.

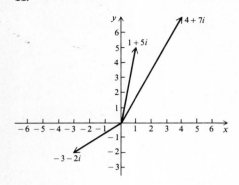

12. $-\sqrt{2} + \sqrt{2}i$ **13.** $\sqrt{2}\,\text{cis}\,45°$ **14.** $70\,\text{cis}\,50°$ **15.** $\sqrt[6]{2}\,\text{cis}\,15°$, $\sqrt[6]{2}\,\text{cis}\,135°$, $\sqrt[6]{2}\,\text{cis}\,255°$ **16.** $5 - 9i$ **17.** $1 + 2i$
18. $x = 2 - i$, $y = -1 - 3i$ **19.** $3, \frac{3}{2}(-1 \pm \sqrt{3}i)$
20.

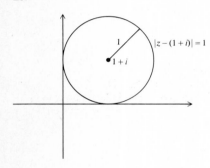

CHAPTER 13

Margin Exercises, Section 13.1

1. 5 **2.** 1 **3.** 2 **4.** 0 **5.** No degree **6.** 7 **7.** 2, -2 **8.** i, $-i$ **9.** (a) Yes; (b) no **10.** (a) Yes; (b) no; (c) no
11. $Q(x) = x^2 + 5x + 10$, $R(x) = 24$, $P(x) = (x - 3)(x^2 + 5x + 10) + 24$

Exercise Set 13.1, pp. 486–487

1. 4 **3.** 1 **5.** 2 **7.** 0 **9.** 2 yes, 3 no, -1 no **11.** (a) Yes; (b) no; (c) no
13. $Q(x) = x^2 + 8x + 15$, $R(x) = 0$, $P(x) = (x - 2)(x^2 + 8x + 15) + 0$
15. $Q(x) = x^2 + 9x + 26$, $R(x) = 48$, $P(x) = (x - 3)(x^2 + 9x + 26) + 48$
17. $Q(x) = x^2 - 2x + 4$, $R(x) = -16$, $P(x) = (x - 2)(x^2 - 2x + 4) - 16$
19. $Q(x) = x^2 + 5$, $R(x) = 0$, $P(x) = (x^2 + 4)(x^2 + 5) + 0$
21. $P(x) = (2x^2 - x + 1)\left(\dfrac{5}{2}x^5 + \dfrac{5}{4}x^4 - \dfrac{5}{8}x^3 + \dfrac{39}{16}x^2 + \dfrac{49}{32}x + \dfrac{35}{64}\right) + \dfrac{-63x - 227}{64}$
23. (a) -32; (b) -32; (c) -65; (d) -65

Margin Exercises, Section 13.2

1. $Q(x) = x^2 + 8x + 15$, $R(x) = 0$ **2.** $Q(x) = x^2 - 4x + 13$, $R(x) = -30$ **3.** $Q(x) = x^2 - x + 1$, $R(x) = 0$
4. (a) $P(10) = 73120$; (b) $P(-8) = -37292$ **5.** (a) Yes; (b) no; (c) yes **6.** No **7.** Yes
8. (a) Yes; (b) $x^2 + 8x + 15$; (c) $(x - 2)(x + 5)(x + 3)$

Exercise Set 13.2, p. 490

1. $Q(x) = 2x^3 + x^2 - 3x + 10$, $R(x) = -42$ **3.** $Q(x) = x^2 - 4x + 8$, $R(x) = -24$
5. $Q(x) = x^3 + x^2 + x + 1$, $R(x) = 0$ **7.** $Q(x) = 2x^3 + x^2 + \frac{7}{2}x + \frac{7}{4}$, $R(x) = -\frac{1}{8}$
9. $Q(x) = x^3 + x^2y + xy^2 + y^3$, $R(x) = 0$ **11.** $P(1) = 0$, $P(-2) = -60$, $P(3) = 0$
13. $P(20) = 5{,}935{,}988$, $P(-3) = -772$ **15.** -3 yes, 2 no   **17.** -3 no, $\frac{1}{2}$ no
19. $P(x) = (x - 1)(x + 2)(x + 3)$; $1, -2, -3$ **21.** $P(x) = (x - 2)(x - 5)(x + 1)$; $2, 5, -1$
23. $P(x) = (x - 2)(x - 3)(x + 4)$; $2, 3, -4$ **25.** $P(x) = (x - 1)(x - 2)(x - 3)(x + 5)$; $1, 2, 3, -5$
27. $-5 < x < 1$ or $x > 2$ **29.** $\frac{14}{3}$ **31.** $k = 0$

Margin Exercises, Section 13.3

1. 5, mult. 2; -6, mult. 1 **2.** -7, mult. 2; 3, mult. 1 **3.** -2, mult. 3; 3, mult. 1; -3, mult. 1
4. 4, mult. 2; 3, mult. 2 **5.** 1, -1, mult. 1 **6.** $x^3 - 6x^2 + 3x + 10$ **7.** $x^3 + (-1 + 5i)x^2 + (-2 - 5i)x - 10i$
8. $x^5 + 6x^4 + 12x^3 + 8x^2$ **9.** $x^4 + 2x^3 - 12x^2 + 14x - 5$ **10.** $7 + 2i$, $3 - \sqrt{5}$ **11.** $x^4 - 6x^3 + 11x^2 - 10x + 2$
12. $x^3 - 2x^2 + 4x - 8$ **13.** $-i$, -2, 1

Exercise Set 13.3, pp. 494–495

1. -3(m 2), 1(m 1)   **3.** 0(m 3), 1(m 2), -4(m 1)   **5.** $x^3 - 6x^2 - x + 30$ **7.** $x^3 + 3x^2 + 4x + 12$
9. $x^3 - \sqrt{3}x^2 - 2x + 2\sqrt{3}$; no   **11.** $-3 - 4i$, $4 + \sqrt{5}$ **13.** $x^3 - 4x^2 + 6x - 4$ **15.** $x^3 - 5x^2 + 16x - 80$
17. $x^4 + 4x^2 - 45$ **19.** i, 2, 3 **21.** $1 + 2i$, $1 - 2i$ **23.** 4, i, $-i$ **25.** i, $-i$, $1 + \sqrt{2}$, $1 - \sqrt{2}$ **27.** There is at least one
complex value for sin x satisfying the equation (not necessarily a value of x).

Margin Exercises, Section 13.4

1. (a) $3, -3, 1, -1$; (b) $2, -2, 1, -1$; (c) $\frac{3}{2}, -\frac{3}{2}, 3, -3, \frac{1}{2}, -\frac{1}{2}, 1, -1$; (d) $\frac{1}{2}, -3$; (e) $3 + \sqrt{10}$, $3 - \sqrt{10}$
2. (a) $2, -2, 1, -1, 4, -4, 7, -7, 14, -14, 28, -28$; (b) $1, -1$; (c) same as (a);
   (d) all coefficients positive; (e) -7; (f) $2i, -2i$
3. (a) All coefficients positive; (b) none   **4.** (a) None; (b) yes, $\dfrac{-3 \pm i\sqrt{3}}{2}$, quadratic formula
5. (a) $-\frac{1}{6}, -\frac{4}{3}, \frac{1}{6}, \frac{1}{3}$; (b) 6; (c) $6P(x) = 6x^4 - x^3 - 8x^2 + x + 2$; (d) $1, -1, -\frac{1}{2}, \frac{2}{3}$; (e) yes, $P(x) = 0$ and $6P(x) = 0$ are
   equivalent.

13

Exercise Set 13.4, pp. 498–499

1. $1, -1$ **3.** $\pm(1, \frac{1}{3}, \frac{1}{5}, \frac{1}{15}, 2, \frac{2}{3}, \frac{2}{5}, \frac{2}{15})$ **5.** -3, $\sqrt{2}$, $-\sqrt{2}$ **7.** $-\frac{1}{5}$, 1, $2i$, $-2i$ **9.** $-1, -2$, $3 + \sqrt{13}$, $3 - \sqrt{13}$
11. $1, -1, -3$ **13.** -2, $1 \pm i\sqrt{3}$ **15.** $\dfrac{1}{2}$, $\dfrac{1 \pm \sqrt{5}}{2}$ **17.** None   **19.** None   **21.** None   **23.** $-2, 1, 2$ **25.** 4 cm
27. 3 cm, $\dfrac{7 - \sqrt{33}}{2}$ cm

Margin Exercises, Section 13.5

1. 3 **2.** 2 **3.** Just 1 **4.** $5, 3$, or 1 **5.** 2 or 0 **6.** 2 or 0 **7.** 1 **8.** 0 **9.** Answers may vary, but 2 or 3 will do.
10. Answers may vary but 75 will do.   **11.** Answers may vary, but -4 will do.   **12.** Answers may vary, but -3
will do.   **13.** Answers may vary, but -4 will do.   **14.** Positive roots, 3 or 1; negative roots, 1; upper bound 6 (answer
may vary); lower bound -1 (answer may vary).   **15.** Positive roots, 1; negative roots, 1. Thus 2 nonreal roots; upper
bound 1 (answer may vary); lower bound -1 (in fact, -1 is a root—answer may vary).

Exercise Set 13.5, p. 507

1. 3 or 1   **3.** 0   **5.** 2 or 0   **7.** 0   **9.** 3 or 1   **11.** 2 or 0   **13.** 3 (other answers possible)   **15.** 4   **17.** -1 **19.** -3
21. 3 or 1 positive; 1 negative; upper bound, 2; lower bound, -3
23. 1 positive; 1 negative; 2 nonreal; upper bound, 2; lower bound, -2
25. 2 or 0 positive; 2 or 0 negative; upper bound, 4; lower bound, -3
27. 0 positive, 0 negative
29. Let $P(x) = x^n - 1$. There is one variation of sign, so there is just one positive root. Since n is even, $P(-x) = P(x)$. Hence $P(-x)$ has just one variation of sign, and there is just one negative root. Zero is not a root, so the total number of real roots is two.

Margin Exercises, Section 13.6

1.

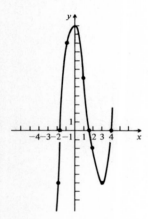

2. (a)

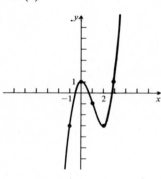

(b) 0.7, -0.5, 2.9

Exercise Set 13.6, p. 513

1.

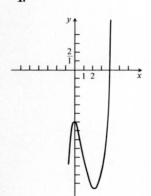

3.

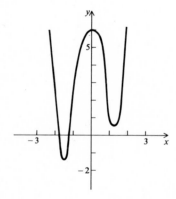

5.

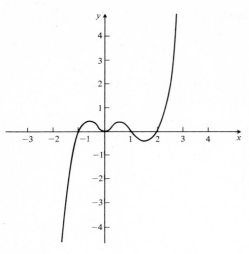

7.

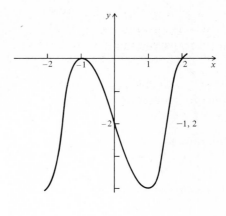

−1, 2

9.

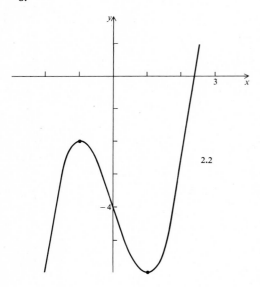

2.2

11.

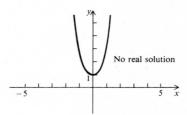

No real solution

13.

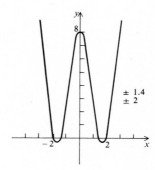

± 1.4
± 2

15.

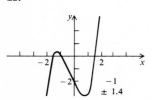

−1
± 1.4

17. 0.75   **19.** −1.27
21. 2.7

23. Consider, for example, $P(x) = 3x^4 - 5x^3 + 4x^2 - 5$.

a	3	-5	4	-5
		$3a$	$a(3a - 5)$	$a(a(3a - 5) + 4)$
	3	$3a - 5$	$a(3a - 5) + 4$	$\underbrace{a(a(3a - 5) + 4) - 5}_{R}$

The expression for R $\big($which is $P(a)\big)$ is the same expression obtained by factoring to find nested form. Generalizing completes the proof.

Margin Exercises, Section 13.7

1.

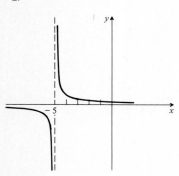

2.

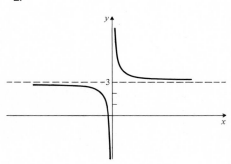

3.

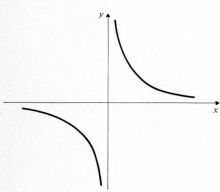

4.

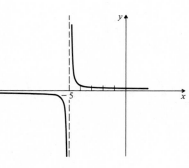

5.

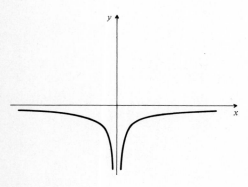

6.

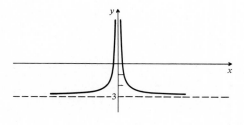

7.

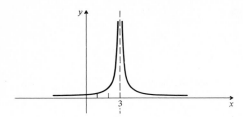

8. $x = 0$, $x = -2$, $x = 3$

9. (b) and (c)   **10.** $y = \frac{1}{2}$ **11.** $y = 3$ **12.** $y = 3x - 1$ **13.** $y = 5x$ **14.** 0, 3, −5   **15.** 0, 1, −3

16.

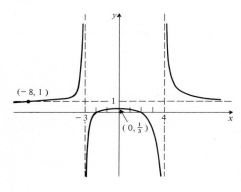

Exercise Set 13.7, pp. 521–522

1.

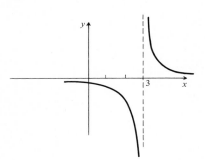

3.

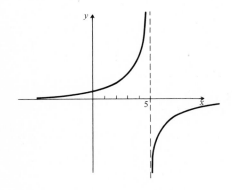

5.

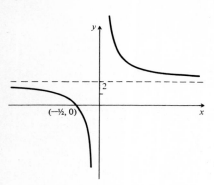

7.

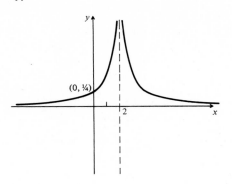

9.

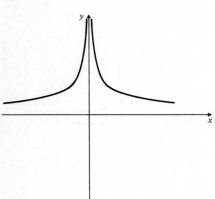

11.

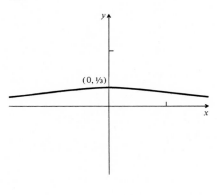

13.

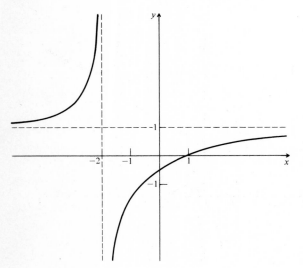

15.

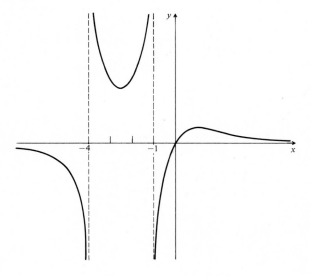

17.

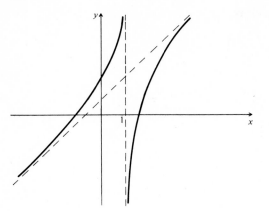

19.

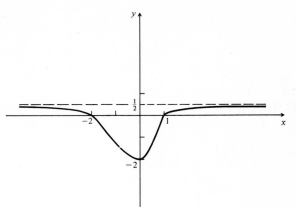

21.

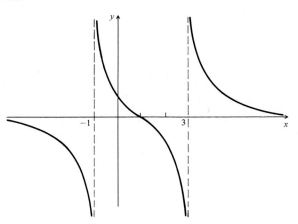

23.

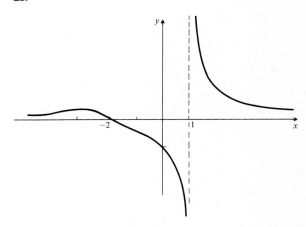

25.

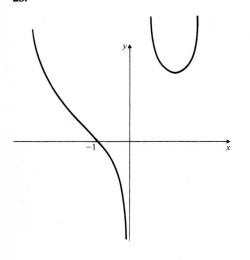

27.

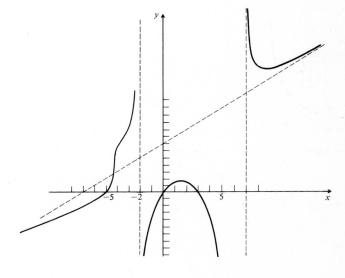

29.

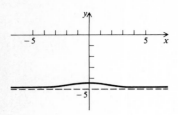

31.

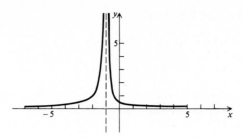

33.

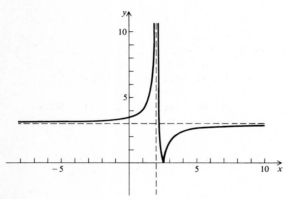

Chapter 13 Review, p. 522

1. 0   **2.** $Q = 2x^3 - 10x^2 + 27x - 59$, $R = 119$ **3.** $x^3 - 3x^2 + 2x$ **4.** $(x^2 - 1)(x - 2)^2(x + 3)^3$

5. $(x - 1)\left(x + \dfrac{1}{2} + i\dfrac{\sqrt{3}}{2}\right)\left(x + \dfrac{1}{2} - i\dfrac{\sqrt{3}}{2}\right)$ **6.** 88   **7.** $(x - 1)(x + 3)(x + 5)$; 1, -3, -5 **8.** No   **9.** 7   **10.** ± 3, $-3i$

11. 4   **12.** $-8 + 7i$, $10 - \sqrt{5}$ **13.** $\pm(1, 2, 3, 4, 6, 12, \frac{1}{2}, \frac{3}{2})$ **14.** 2 or none   **15.** 4, 2, or none   **16.** 2   **17.** -2 **18.** 1.41

19. -0.9, 1.3, 2.5

20.

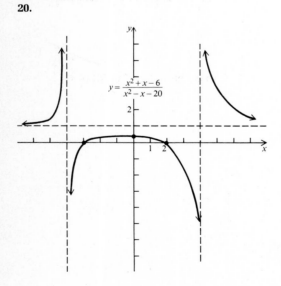

$$y = \frac{x^2 + x - 6}{x^2 - x - 20}$$

CHAPTER 14

Margin Exercises, Section 14.1

1. The union of the graphs of $y = -\frac{1}{2}x$ and $y = \frac{1}{2}x$
2. The union of the graphs of $y = 2$ and $y = -2$
3. Graph consists of a single line, $y = 3x$
4. Graph consists just of $(0, 0)$.
5. There is no real number solution, hence no graph.
6. $(x + 3)^2 + (y - 7)^2 = 25$ 7. $(-1, 3), 2$ 8. $(7, -2), 8$
9. $(x + 1)^2 + (y - 4)^2 = 41$

Exercise Set 14.1, pp. 530–532

1. The union of the graphs of $y = x$ and $y = -x$ 3. The union of the graphs of $y = -x$ and $y = \frac{3}{2}x$
5. The point $(0, 0)$ 7. $x^2 + y^2 = 49$ 9. $(-1, -3), 2$ 11. $(8, -3), 2\sqrt{10}$ 13. $(-4, 3), 2\sqrt{10}$ 15. $(2, 0), 2$
17. $(-4.123, 3.174), 10.071$ 19. $(0, 0), \frac{1}{3}$ 21. $x^2 + y^2 = 25$ 23. $(x - 2)^2 + (y - 4)^2 = 16$ 25. 2.68 ft, 37.32 ft
27. (a) No; (b) $y = \pm\sqrt{4 - x^2}$; (c) yes, domain $\{x \mid -2 \le x \le 2\}$,
range $\{y \mid 0 \le y \le 2\}$; (d) yes, domain $\{x \mid -2 \le x \le 2\}$, range $\{y \mid -2 \le y \le 0\}$ 29. Yes   31. No   33. Yes   35. No

Margin Exercises, Section 14.2

1. (a) Stretched; (b) shrunk
2. $V(-3, 0)(3, 0)(0, -1)(0, 1)$;
   $F(-2\sqrt{2}, 0)(2\sqrt{2}, 0)$

3. $V(-5, 0)(5, 0)(0, -3)(0, 3)$;
   $F(-4, 0)(4, 0)$

4. $V(-2, 0)(2, 0)(0, \sqrt{2})(0, -\sqrt{2})$;
   $F(-\sqrt{2}, 0)(\sqrt{2}, 0)$

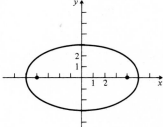

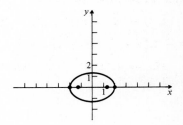

5. $C(0, 0)$; $V(-1, 0)(1, 0)(0, 3)(0, -3)$;
   $F(0, 2\sqrt{2})(0, -2\sqrt{2})$

6. $C(0, 0)$; $V(3, 0)(-3, 0)(0, 5)(0, -5)$; $F(0, 4)(0, -4)$

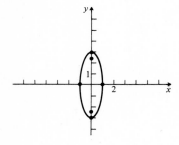

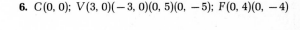
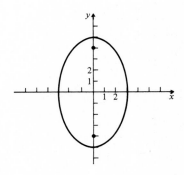

14

7. $C(0, 0)$; $V(\sqrt{2}, 0)(-\sqrt{2}, 0)(0, 2)(0, -2)$;
$F(0, \sqrt{2})(0, -\sqrt{2})$

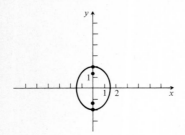

8. $C(-3, 2)$; $V(-2\frac{4}{5}, 2)(-3\frac{1}{5}, 2)(-3, 2\frac{1}{3})(-3, 1\frac{2}{3})$;
$F(-3, 2\frac{4}{15})(-3, 1\frac{11}{15})$

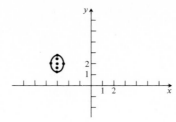

9. $C(2, -3)$; $V(2\frac{1}{3}, -3)(1\frac{2}{3}, -3)(2, -2\frac{4}{5})(2, -3\frac{1}{5})$; $F(2\frac{4}{15}, -3)(1\frac{11}{15}, -3)$

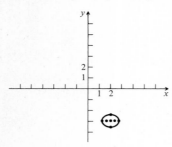

Exercise Set 14.2, pp. 537–538

1. V: $(2, 0)$, $(-2, 0)$, $(0, 1)$, $(0, -1)$; F: $(\sqrt{3}, 0)$, $(-\sqrt{3}, 0)$
3. V: $(-3, 0)$, $(3, 0)$, $(0, 4)$, $(0, -4)$; F: $(0, \sqrt{7})$, $(0, -\sqrt{7})$
5. V: $(-\sqrt{3}, 0)$, $(\sqrt{3}, 0)$, $(0, \sqrt{2})$, $(0, -\sqrt{2})$; F: $(-1, 0)$, $(1, 0)$
7. V: $(\pm\frac{1}{2}, 0)$, $(0, \pm\frac{1}{3})$; F: $\left(\frac{\pm\sqrt{5}}{6}, 0\right)$
9. $C(1, 2)$; V: $(3, 2)$, $(-1, 2)$, $(1, 3)$, $(1, 1)$; F: $(1 \pm \sqrt{3}, 2)$
11. $C(-3, 2)$; V: $(2, 2)$, $(-8, 2)$, $(-3, 6)$, $(-3, -2)$; F: $(0, 2)$, $(-6, 2)$
13. $C(-2, 1)$; V: $(-10, 1)$, $(6, 1)$, $(-2, 1 \pm 4\sqrt{3})$; F: $(-6, 1)$, $(2, 1)$
15. $C(2, -1)$; V: $(-1, -1)$, $(5, -1)$, $(2, 1)$, $(2, -3)$; F: $(2 - \sqrt{5}, -1)$, $(2 + \sqrt{5}, -1)$
17. $C(1, 1)$; V: $(0, 1)$, $(2, 1)$, $(1, 3)$, $(1, -1)$; F: $(1, 1 + \sqrt{3})$, $(1, 1 - \sqrt{3})$
19. $C(2.003125, -1.00513)$; V: $(5.0234302, -1.00515)$, $(-1.0171802, -1.00515)$, $(2.003125, -3.0186868)$, $(2.003125, 1.0083868)$
21. $\dfrac{x^2}{4} + \dfrac{y^2}{9} = 1$ **23.** $\dfrac{(x - 3)^2}{4} + \dfrac{(y - 1)^2}{25} = 1$ **25.** $\dfrac{(x + 2)^2}{\frac{1}{4}} + \dfrac{(y - 3)^2}{4} = 1$

27.

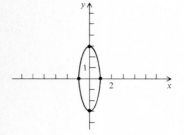

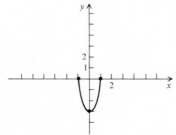

(a) No; (b) $y = \pm 3\sqrt{1 - x^2}$;
(c) yes, domain $\{x \,|\, -1 \le x \le 1\}$, range $\{y \,|\, 0 \le y \le 3\}$;
(d) yes, domain $\{x \,|\, -1 \le x \le 1\}$, range $\{y \,|\, -3 \le y \le 0\}$

31. $x = a\cos t$, so $\dfrac{x^2}{a^2} = \cos^2 t$; similarly, $\dfrac{y^2}{b^2} = \sin^2 t$. Adding gives us $\dfrac{x^2}{a^2} + \dfrac{y^2}{b^2} = 1$.

Margin Exercises, Section 14.3

1. $V(3, 0)(-3, 0)$; $F(\sqrt{13}, 0)(-\sqrt{13}, 0)$;
   A: $y = \frac{2}{3}x$, $y = -\frac{2}{3}x$

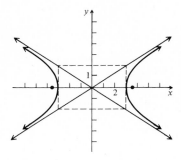

2. $V(4, 0)(-4, 0)$; $F(4\sqrt{2}, 0)(-4\sqrt{2}, 0)$;
   A: $y = x$, $y = -x$

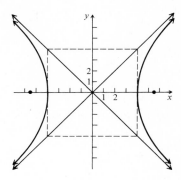

3. $V(0, 5)(0, -5)$; $F(0, \sqrt{34})(0, -\sqrt{34})$;
   A: $y = \frac{5}{3}x$, $y = -\frac{5}{3}x$

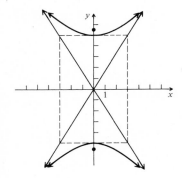

4. $V(0, 5)(0, -5)$; $F(0, 5\sqrt{2})(0, -5\sqrt{2})$;
   A: $y = x$, $y = -x$

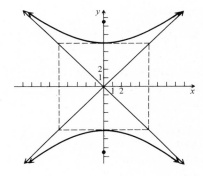

5. $C(1, -2)$; $V(6, -2)(-4, -2)$;
   $F(1 + \sqrt{29}, -2)(1 - \sqrt{29}, -2)$;
   A: $y = \frac{2}{5}x - \frac{12}{5}$, $y = -\frac{2}{5}x - \frac{8}{5}$

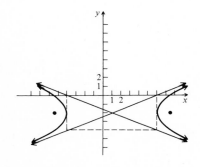

6. $C(-1, 2)$; $V(-1, 5)(-1, -1)$; $F(-1, 7)(-1, -3)$;
   A: $y = \frac{3}{4}x + \frac{11}{4}$, $y = -\frac{3}{4}x + \frac{5}{4}$

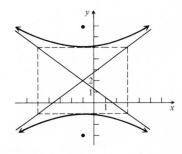

7.

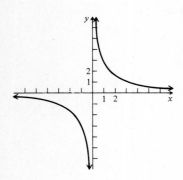

8.

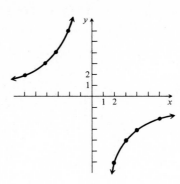

Exercise Set 14.3, p. 545

1. $C(0, 0)$; V: $(-3, 0)$, $(3, 0)$; F: $(-\sqrt{10}, 0)$, $(\sqrt{10}, 0)$; A: $y = \frac{1}{3}x$, $y = -\frac{1}{3}x$

3. $C(2, -5)$; V: $(-1, -5)$, $(5, -5)$; F: $(2 - \sqrt{10}, -5)$, $(2 + \sqrt{10}, -5)$; A: $y = -\dfrac{x}{3} - \dfrac{13}{3}$, $y = \dfrac{x}{3} - \dfrac{17}{3}$

5. $C(-1, -3)$; V: $(-1, -1)$, $(-1, -5)$; F: $(-1, -3 + 2\sqrt{5})$, $(-1, -3 - 2\sqrt{5})$; A: $y = \frac{1}{2}x - \frac{5}{2}$, $y = -\frac{1}{2}x - \frac{7}{2}$

7. $C(0, 0)$; V: $(-2, 0)$, $(2, 0)$; F: $(-\sqrt{5}, 0)$, $(\sqrt{5}, 0)$; A: $y = -\frac{1}{2}x$, $y = \frac{1}{2}x$

9. $C(0, 0)$; V: $(0, 1)$, $(0, -1)$; F: $(0, \sqrt{5})$, $(0, -\sqrt{5})$; A: $y = -\frac{1}{2}x$, $y = \frac{1}{2}x$

11. $C(0, 0)$; V: $(-\sqrt{2}, 0)$, $(\sqrt{2}, 0)$; F: $(-2, 0)$, $(2, 0)$; A: $y = \pm x$

13. $C(0, 0)$; V: $(-\frac{1}{2}, 0)$, $(\frac{1}{2}, 0)$; F: $\left(-\dfrac{\sqrt{2}}{2}, 0\right)$, $\left(\dfrac{\sqrt{2}}{2}, 0\right)$; A: $y = \pm x$

15. $C(1, -2)$; V: $(0, -2)$, $(2, -2)$; F: $(1 - \sqrt{2}, -2)$, $(1 + \sqrt{2}, -2)$; A: $y = -x - 1$, $y = x - 3$

17. $C(\frac{1}{3}, 3)$; V: $(-\frac{2}{3}, 3)$; $(\frac{4}{3}, 3)$; F: $(\frac{1}{3} - \sqrt{37}, 3)$, $(\frac{1}{3} + \sqrt{37}, 3)$; A: $y = 6x + 1$, $y = -6x + 5$ **19.** See text.   **21.** See text.

23. $C(1.023, -2.044)$; V: $(2.07, -2.044)$, $(-0.024, -2.044)$; A: $y = x - 3.067$, $y = -x - 1.021$ **25.** $\dfrac{x^2}{4} - \dfrac{y^2}{9} = 1$

27.

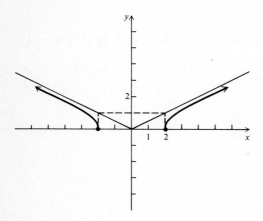

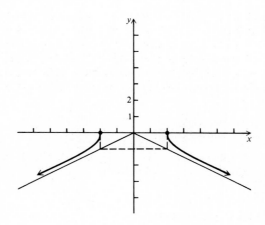

$$\left(\dfrac{x - h}{a}\right)\left(\dfrac{x - h}{a}\right) - \left(\dfrac{y - k}{b}\right)\left(\dfrac{y - k}{b}\right) = 1$$

$$\therefore \dfrac{(x - h)^2}{a^2} - \dfrac{(y - k)^2}{b^2} = 1$$

Margin Exercises, Section 14.4

1. $V(0, 0)$, $F(0, 2)$, D: $y = -2$ **2.** $V(0, 0)$, $F(0, \frac{1}{8})$, D: $y = -\frac{1}{8}$ **3.** $V(0, 0)$, $F(-\frac{3}{2}, 0)$, D: $x = \frac{3}{2}$

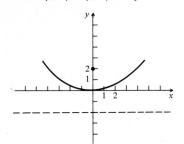 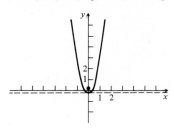 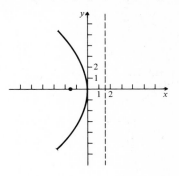

4. $y^2 = 12x$ **5.** $x^2 = 2y$ **6.** $y^2 = -24x$ **7.** $x^2 = -4y$
8. $V(-1, -\frac{1}{2})$, $F(-1, \frac{3}{2})$, D: $y = -\frac{5}{2}$ **9.** $V(2, -1)$, $F(1, -1)$, D: $x = 3$

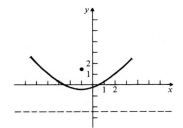 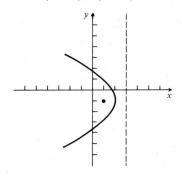

Exercise Set 14.4, pp. 550–551

1. $V(0, 0)$; $F(0, 2)$; D: $y = -2$ **3.** $V(0, 0)$; $F(-\frac{3}{2}, 0)$; D: $x = \frac{3}{2}$ **5.** $V(0, 0)$; $F(0, 1)$; D: $y = -1$
7. $V(0, 0)$; $F(0, \frac{1}{8})$; D: $y = -\frac{1}{8}$ **9.** $y^2 = 16x$ **11.** $y^2 = -4\sqrt{2}x$ **13.** $(y - 2)^2 = 14(x + \frac{1}{2})$
15. V: $(-2, 1)$; F: $(-2, -\frac{1}{2})$; D: $y = \frac{5}{2}$ **17.** V: $(-1, -3)$; F: $(-1, -\frac{7}{2})$; D: $x = -\frac{5}{2}$ **19.** V: $(0, -2)$; F: $(0, -1\frac{3}{4})$; D: $y = -2\frac{1}{4}$
21. V: $(-2, -1)$; F: $(-2, -\frac{3}{4})$; D: $y = -1\frac{1}{4}$ **23.** V: $(5\frac{3}{4}, \frac{1}{2})$; F: $(6, \frac{1}{2})$; D: $x = 5\frac{1}{2}$
25. V: $(0, 0)$; F: $(0, 2014.0625)$; D: $y = -2014.0625$ **27.** The graph of $x^2 - y^2 = 0$ is two lines; the others are, respectively, a hyperbola, a circle, and a parabola.   **29.** $(x + 1)^2 = -4(y - 2)$ **31.** $(y - k)^2 - (x - h)(4p) = 0$; $(y - k)^2 = 4p(x - h)$

Margin Exercises, Section 14.5

1. $(4, 3)$, $(-3, -4)$ **2.** $(4, 7)$, $(-1, 2)$ **3.** $(-4, 4)$, $(2, 1)$ **4.** $(4, 3)$, $(-3, -4)$ **5.** $(4, 7)$, $(-1, 2)$ **6.** $(-4, 4)$, $(2, 1)$
7. $(-\frac{5}{7}, \frac{22}{7})$, $(1, -2)$ **8.** 7, 11   **9.** 5 ft, 12 ft

Exercise Set 14.5, pp. 554–555

1. $(-4, -3)$, $(3, 4)$ **3.** $(0, -3)$, $(4, 5)$ **5.** $(3, 0)$, $(0, 2)$ **7.** $(-2, 1)$ **9.** $(3, 2)$, $(4, \frac{3}{2})$
11. $\left(\dfrac{5 + \sqrt{70}}{3}, \dfrac{-1 + \sqrt{70}}{3}\right)$, $\left(\dfrac{5 - \sqrt{70}}{3}, \dfrac{-1 - \sqrt{70}}{3}\right)$ **13.** $\left(\dfrac{15 + \sqrt{561}}{8}, \dfrac{11 - 3\sqrt{561}}{8}\right)$, $\left(\dfrac{15 - \sqrt{561}}{8}, \dfrac{11 + 3\sqrt{561}}{8}\right)$
15. 9, 5   **17.** 6 cm, 8 cm   **19.** 4 in., 5 in.   **21.** $(0.965, 4402.33)$, $(-0.965, -4402.33)$

23. $2(L + W) = P$, $L + W = \dfrac{P}{2}$, $LW = A$, $L = \dfrac{P}{2} - W$, $W\left(\dfrac{P}{2} - W\right) = A$, $W^2 - \dfrac{WP}{2} + A = 0$,

$W = \dfrac{\dfrac{P}{2} \pm \sqrt{\left(\dfrac{P}{2}\right)^2 - 4A}}{2} = \dfrac{P}{4} \pm \dfrac{\sqrt{P^2 - 16A}}{4} = \dfrac{1}{4}(P \pm \sqrt{P^2 - 16A})$ **25.** $(x - 2)^2 + (y - 3)^2 = 1$

Margin Exercises, Section 14.6

1. $(2, 0)$, $(-2, 0)$ **2.** $(4, 0)$, $(-4, 0)$ **3.** $(0, 2)$, $(0, -2)$ **4.** $(2, 3)$, $(2, -3)$, $(-2, 3)$, $(-2, -3)$
5. $(3, 2)$, $(-3, -2)$, $(2, 3)$, $(-2, -3)$ **6.** 1 ft, 2 ft

Exercise Set 14.6, pp. 558–559

1. $(-5, 0)$, $(4, 3)$, $(4, -3)$ **3.** $(3, 0)$, $(-3, 0)$ **5.** $(4, 3)$, $(-4, -3)$, $(3, 4)$, $(-3, -4)$ **7.** No solution
9. $(\sqrt{2}, \sqrt{14})$, $(-\sqrt{2}, \sqrt{14})$, $(\sqrt{2}, -\sqrt{14})$, $(-\sqrt{2}, -\sqrt{14})$ **11.** $(1, 2)$, $(-1, -2)$, $(2, 1)$, $(-2, -1)$
13. $(3, 2)$, $(-3, -2)$, $(2, 3)$, $(-2, -3)$ **15.** $\left(\dfrac{5 - 9\sqrt{15}}{20}, \dfrac{-45 + 3\sqrt{15}}{20}\right)$, $\left(\dfrac{5 + 9\sqrt{15}}{20}, \dfrac{-45 - 3\sqrt{15}}{20}\right)$
17. $(8.53, 2.53)$, $(8.53, -2.53)$, $(-8.53, 2.53)$, $(-8.53, -2.53)$
19. 13, 12 and -13, -12 **21.** 1 m, $\sqrt{3}$ m   **23.** 16 ft, 24 ft
25. $\left(x + \dfrac{5}{13}\right)^2 + \left(y - \dfrac{32}{13}\right)^2 = \dfrac{5365}{169}$

Chapter 14 Review, p. 559

1.

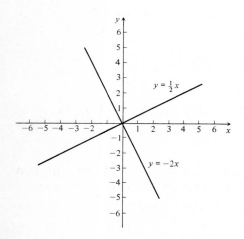

2. $(x + 2)^2 + (y - 6)^2 = 13$ **3.** $(x - 3)^2 + (y - 4)^2 = 25$
4. $C\left(\dfrac{3}{4}, \dfrac{5}{4}\right)$; $r = \dfrac{\sqrt{10}}{4}$ **5.** $(x - 2)^2 + (y - 4)^2 = 26$
6. $x^2 = -6y$ **7.** $F: (-3, 0)$; $V: (0, 0)$; $D: x = 3$

8.

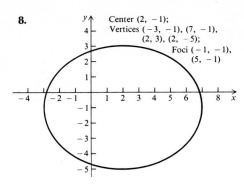

Center $(2, -1)$;
Vertices $(-3, -1)$, $(7, -1)$,
$(2, 3)$, $(2, -5)$;
Foci $(-1, -1)$,
$(5, -1)$

9. $\dfrac{x^2}{9} + \dfrac{y^2}{16} = 1$ **10.** $C(-2, \frac{1}{4})$; $V: (0, \frac{1}{4})$, $(-4, \frac{1}{4})$; $F: (-2 + \sqrt{6}, \frac{1}{4})$, $(-2 - \sqrt{6}, \frac{1}{4})$; $A: y - \frac{1}{4} = \pm \dfrac{\sqrt{2}}{2}(x + 2)$

11. $(-8\sqrt{2}, 8)$, $(8\sqrt{2}, 8)$ **12.** $\left(3, \dfrac{\sqrt{29}}{2}\right)$, $\left(-3, \dfrac{\sqrt{29}}{2}\right)$, $\left(3, \dfrac{-\sqrt{29}}{2}\right)$, $\left(-3, \dfrac{-\sqrt{29}}{2}\right)$ **13.** 7, 4   **14.** $\dfrac{16\sqrt{6}}{7} \approx 5.6$

CHAPTER 15

Margin Exercises, Section 15.1

1. (a) 1, 3, 5; (b) 67   **2.** (a) 1, $-\frac{1}{2}$, $\frac{1}{3}$, $-\frac{1}{4}$; (b) $-\frac{1}{48}$ **3.** $2n$ **4.** n **5.** n^3 **6.** $\dfrac{x^n}{n}$ **7.** 2^{n-1}

8. (a) 2, 2, 2, 2, 2; (b) -3, -9, -81, -6561, -43046721 **9.** $\frac{31}{32}$ **10.** $3 + 2\frac{1}{2} + 2\frac{1}{3}$ **11.** $5^0 + 5^1 + 5^2 + 5^3 + 5^4$

12. $8^3 + 9^3 + 10^3 + 11^3$ **13.** $\displaystyle\sum_{k=1}^{5} 2k$ **14.** $\displaystyle\sum_{k=1}^{4} k^3$ **15.** $\displaystyle\sum_{k=1}^{6} \dfrac{x^k}{k}$ **16.** $\displaystyle\sum_{k=1}^{4} (k + 1)$

Exercise Set 15.1, pp. 564–565

1. 4, 7, 10, 13; 31; 46   **3.** 1, $\frac{1}{2}$, $\frac{1}{3}$, $\frac{1}{4}$; $\frac{1}{10}$; $\frac{1}{15}$ **5.** $\frac{1}{2}$, $\frac{2}{3}$, $\frac{3}{4}$, $\frac{4}{5}$; $\frac{10}{11}$; $\frac{15}{16}$ **7.** $2n - 1$ **9.** $\dfrac{n+1}{n+2}$ **11.** $(\sqrt{3})^n$ **13.** $n - 1$

15. 4, $1\frac{1}{4}$, $1\frac{4}{5}$, $1\frac{5}{9}$ **17.** 64, 8, $2\sqrt{2}$, $\sqrt{2\sqrt{2}}$ **19.** 28   **21.** 30   **23.** $\frac{1}{2} + \frac{1}{4} + \frac{1}{6} + \frac{1}{8} + \frac{1}{10}$ **25.** $2^0 + 2^1 + 2^2 + 2^3 + 2^4 + 2^5$

27. $\log 7 + \log 8 + \log 9 + \log 10$ **29.** $\displaystyle\sum_{k=1}^{4} \dfrac{1}{k^2}$, $\frac{205}{144}$ **31.** $\displaystyle\sum_{k=1}^{6} (-1)^k 2^k$, 42   **33.** $\frac{3}{2}$, $\frac{3}{2}$, $\frac{3}{2}$, $\frac{3}{2}$, $\frac{3}{2}$ **35.** 1, 0, -1, 0, 1

37. π, 0, π, 0, π **39.** 2, 2.25, 2.37037, 2.441406, 2.488320, 2.521626, 14.071722
41. 2, 1.553774, 1.498834, 1.491398, 1.490379, 1.490238, 9.524623

Margin Exercises, Section 15.2

1. 50   **2.** 75   **3.** $a_1 = 7$, $d = 12$; 7, 19, 31, 43, ...   **4.** 20, 100   **5.** 225   **6.** 455   **7.** 1665   **8.** 120   **9.** 7, 10, 13

Exercise Set 15.2, pp. 570–571

1. $a_{12} = 46$ **3.** 27th   **5.** $a_{17} = 101$ **7.** $a_1 = 5$ **9.** $n = 28$ **11.** $a_1 = 8$; $d = -3$; 8, 5, 2, -1, -4 **13.** 670   **15.** 2500
17. 990   **19.** -264 **21.** 465   **23.** $5\frac{4}{5}$, $7\frac{3}{5}$, $9\frac{2}{5}$, $11\frac{1}{5}$ **25.** n^2 **27.** -10, -4, 2, 8   **29.** Sides are a, $a + d$, $a + 2d$, and
$a^2 + (a + d)^2 = (a + 2d)^2$. Solving, we get $d = \dfrac{a}{3}$. Thus the sides are a, $\dfrac{4a}{3}$, and $\dfrac{5a}{3}$ in the ratio 3:4:5.
31. 5200, 4687.5, 4175, 3662.5, 3150, 2637.5, 2125, 1612.5, 1100   **33.** 51,679.65

Margin Exercises, Section 15.3

1. 2^8, or 256   **2.** -9375 **3.** \$476.41   **4.** 11,718   **5.** $\dfrac{341}{256}$ **6.** 363   **7.** \$10,485.75, or an approximation of this

Exercise Set 15.3, pp. 574–576

1. $a_{10} = \frac{81}{64}$ **3.** 1275

5. $\frac{547}{18}$

7. $\frac{63}{32}$

9. \$933.12

11. $\frac{1}{256}$ ft

13. 5866.60

15. $\frac{a_n}{a_{n+1}} = r$, so $\frac{a_n^2}{a_{n+1}^2} = r^2$; hence $a_1^2, a_2^2, \ldots$, is geometric, with ratio r^2.

17. $\frac{a_n}{a_{n+1}} = r$; $\ln(a_n) - \ln(a_{n+1}) = \ln r$; hence $\ln a_1, \ln a_2, \ldots$, is arithmetic.

19. $G = pr$, $q = pr^2$; $\sqrt{pq} = \sqrt{p^2r^2} = pr = G$.

21. 10,485.76 inches

Margin Exercises, Section 15.4

1. No   **2.** No   **3.** Yes   **4.** $\frac{3}{2}$ **5.** 2

6. $3\frac{1}{5}$ **7.** $\frac{5}{11}$ **8.** $\frac{59}{11}$ **9.** \$3333.33

Exercise Set 15.4, pp. 579–580

1. No   **3.** Yes   **5.** Yes   **7.** Yes   **9.** 8   **11.** 125   **13.** 2   **15.** $\frac{160}{9}$ **17.** 12,500   **19.** $\frac{7}{9}$ **21.** $\frac{7}{33}$ **23.** $\frac{170}{33}$
25. $\$5.33 \times 10^{10}$ **27.** 3.33×10^6, $66\frac{2}{3}\%$

Margin Exercises, Section 15.5

1. $1^2 + 1 > 1 + 1$ or $2 > 2$; $2^2 + 1 > 2 + 1$ or $5 > 3$; $3^2 + 1 > 3 + 1$ or $10 > 4$; $4^2 + 1 > 4 + 1$ or $17 > 5$; $5^2 + 1 > 5 + 1$ or $26 > 6$

2. $1 = \dfrac{1(1 + 1)}{2}$ or $1 = 1$; $1 + 2 = \dfrac{2(2 + 1)}{2}$, or $1 + 2 = 3$; $1 + 2 + 3 = 6$; $1 + 2 + 3 + 4 = 10$; $1 + 2 + 3 + 4 + 5 = 15$

3. S_1: $x \leq x$, S_2: $x \leq x^2$. Both obviously true if $x > 1$. Thus basis step is complete. S_k: $x \leq x^k$, S_{k+1}: $x \leq x^{k+1}$. Assume S_k: $x \leq x^k$. We know by hypothesis that $1 < x$. Multiply the inequalities to get $x \cdot 1 \leq x^k \cdot x$, or $x < x^{k+1}$. We have arrived at S_{k+1}, hence have shown that $S_k \rightarrow S_{k+1}$ for all natural numbers k. We can now conclude that $x \leq x^n$ for all natural numbers n.

4. (a) $2 \cdot 1 = 1(1 + 1)$, $2 + 4 = 2(2 + 1)$; (b) $2 + 4 + \cdots + 2k = k(k + 1)$;
(c) $2 + 4 + \cdots + 2(k + 1) = (k + 1)[(k + 1) + 1]$; (d) $2 \cdot 1 \overset{?}{=} 1(1 + 1)$, $2 = 1 \cdot 2 = 2$; (e) Assume S_k as hypothesis:

$$2 + 4 + \cdots + 2k = k(k + 1).$$

Add $2(k + 1)$ on both sides:

$$2 + 4 + \cdots + 2k + 2(k + 1) = k(k + 1) + 2(k + 1)$$
$$= (k + 1)(k + 2) \qquad \text{Simplifying}$$

We now have

$$2 + 4 + \cdots + 2(k + 1) = (k + 1)(k + 2) \quad \text{or} \quad (k + 1)[(k + 1) + 1].$$

This is S_{k+1}. Hence $S_k \rightarrow S_{k+1}$ for all k. Finally, we conclude that $2 + 4 + \cdots + 2n = n(n + 1)$ for all natural numbers n.

5. S_1: $2 = \dfrac{1(3 + 1)}{2}$. True. S_k: $\displaystyle\sum_{p=1}^{k} (3p - 1) = \dfrac{k(3k + 1)}{2}$, or $2 + 5 + 8 + \cdots + 3k - 1 = \dfrac{k(3k + 1)}{2}$. S_{k+1}:

$\displaystyle\sum_{p=1}^{k+1} (3p - 1) = \dfrac{(k + 1)[3(k + 1) + 1]}{2}$, or $2 + 5 + \cdots + 3k - 1 + 3(k + 1) - 1 = \dfrac{(k + 1)[3(k + 1) + 1]}{2}$. Assume S_k.

Then add $[3(k + 1) - 1]$ on both sides:

$$2 + 5 + \cdots + [3(k + 1) - 1] = \dfrac{k(3k + 1)}{2} + [3(k + 1) - 1]$$

$$= \dfrac{k(3k + 1) + 2[3(k + 1) - 1]}{2}$$

$$= \dfrac{3k^2 + 7k + 4}{2} = \dfrac{(k + 1)(3k + 4)}{2}$$

$$= \dfrac{(k + 1)[3(k + 1) + 1]}{2}$$

We have arrived at

$$\sum_{p=1}^{k+1} (3p - 1) = \dfrac{(k + 1)[3(k + 1) - 1]}{2}.$$

This is S_{k+1}. So $S_k \to S_{k+1}$ for all k. We conclude that

$$\sum_{p=1}^{n} (3p - 1) = \dfrac{n(3n + 1)}{2}$$

for *all* natural numbers n.

Exercise Set 15.5, pp. 584–585

1. $1^2 < 1^3$, $2^2 < 2^3$, $3^2 < 3^3$, etc.   **3.** A polygon of 3 sides has $\dfrac{3(3 - 3)}{2}$ diagonals. A polygon of 4 sides has $\dfrac{4(4 - 3)}{2}$ diagonals, etc.

5. S_n: $1 + 2 + 3 + \cdots + n = \dfrac{n(n + 1)}{2}$

S_1: $1 = \dfrac{1(1 + 1)}{2}$

S_k: $1 + 2 + 3 + \cdots + k = \dfrac{k(k + 1)}{2}$

S_{k+1}: $1 + 2 + 3 + \cdots + k + (k + 1) = \dfrac{(k + 1)(k + 2)}{2}$

1. *Basis step:* S_1 true by substitution.
2. *Induction step:* Assume S_k. Deduce S_{k+1}.
   Starting with the left side of S_{k+1}, we have
   $\underbrace{1 + 2 + 3 + \cdots + k} + (k + 1)$

   $= \dfrac{k(k + 1)}{2} + (k + 1)$ (by S_k)

   $= \dfrac{k(k + 1) + 2(k + 1)}{2}$ (adding)

   $= \dfrac{(k + 1)(k + 2)}{2}$ (distributive law)

7. S_n: $1 + 5 + 9 + \cdots + (4n - 3) = n(2n - 1)$
S_1: $1 = 1(2 \cdot 1 - 1)$
S_k: $1 + 5 + 9 + \cdots + (4k - 3) = k(2k - 1)$
S_{k+1}: $1 + 5 + 9 + \cdots + (4k - 3) + [4(k + 1) - 3]$
$\qquad\qquad\qquad\qquad = (k + 1)[2(k + 1) - 1]$
$\qquad\qquad\qquad\qquad = (k + 1)(2k + 1)$

1. *Basis step:* S_1 true by substitution.
2. *Induction step:* Assume S_k. Deduce S_{k+1}.
   Starting with the left side of S_{k+1}, we have
   $\underbrace{1 + 5 + 9 + \cdots + (4k - 3)} + [4(k + 1) - 3]$
   $= k(2k - 1) + [4(k + 1) - 3]$ (by S_k)
   $= 2k^2 - k + 4k + 4 - 3$
   $= (k + 1)(2k + 1)$

9. S_n: $\dfrac{1}{1 \cdot 2} + \dfrac{1}{2 \cdot 3} + \cdots + \dfrac{1}{n(n+1)} = \dfrac{n}{n+1}$

S_1: $\dfrac{1}{1 \cdot 2} = \dfrac{1}{1+1}$

S_k: $\dfrac{1}{1 \cdot 2} + \dfrac{1}{2 \cdot 3} + \cdots + \dfrac{1}{k(k+1)} = \dfrac{k}{k+1}$

S_{k+1}: $\dfrac{1}{1 \cdot 2} + \dfrac{1}{2 \cdot 3} + \cdots + \dfrac{1}{k(k+1)} + \dfrac{1}{(k+1)(k+2)} = \dfrac{k+1}{k+2}$

13. S_1: $3^1 < 3^{1+1}$

S_k: $3^k < 3^{k+1}$

$3^k \cdot 3 < 3^{k+1} \cdot 3$

$3^{k+1} < 3^{(k+1)+1}$

17. S_1: $1 + \frac{1}{1} = 1 + 1$

S_k: $\left(1 + \dfrac{1}{1}\right) \cdots \left(1 + \dfrac{1}{k}\right) = k + 1$

Multiply by $\left(1 + \dfrac{1}{k+1}\right)$:

$\left(1 + \dfrac{1}{1}\right) \cdots \left(1 + \dfrac{1}{k+1}\right) = (k+1)\left(1 + \dfrac{1}{k+1}\right)$

$\qquad\qquad\qquad\qquad = (k+1)\left(\dfrac{k+1+1}{k+1}\right)$

$\qquad\qquad\qquad\qquad = (k+1) + 1$

21. *2. Induction step:* Assume S_k. Deduce S_{k+1}.

Starting with the left side of S_{k+1}, we have

$[r(\cos\theta + i\sin\theta)]^{k+1}$

$= [r(\cos\theta + i\sin\theta)]^k [r(\cos\theta + i\sin\theta)]$

$= r^k[\cos k\theta + i\sin k\theta][r(\cos\theta + i\sin\theta)]$

$= r^{k+1}(\cos k\theta + i\sin k\theta)(\cos\theta + i\sin\theta)$

$= r^{k+1}[\cos k\theta \cos\theta + i\sin k\theta \cos\theta + i\cos k\theta \sin\theta$
$\qquad\qquad\qquad\qquad\qquad\qquad - \sin k\theta \sin\theta]$

$= r^{k+1}[(\cos k\theta \cos\theta - \sin k\theta \sin\theta) + i(\sin k\theta \cos\theta$
$\qquad\qquad\qquad\qquad\qquad\qquad + \cos k\theta \sin\theta]$

$= r^{k+1}[\cos(k\theta + \theta) + i\sin(k\theta + \theta)]$

$= r^{k+1}[\cos(k+1)\theta + i\sin(k+1)\theta]$

23. *2. Induction step:* Assume S_k. Deduce S_{k+1}.

We start with the left side of S_{k+1}.

$\underbrace{\cos x \cdot \cos 2x \cdots \cos 2^{k-1}x} \cdot \cos 2^k x$

$= \dfrac{\sin 2^k x}{2^k \sin x} \cdot \cos 2^k x$

$= \dfrac{\sin 2^k x \cos 2^k x}{2^k \sin x}$

$= \dfrac{\frac{1}{2}\sin 2(2^k x)}{2^k \sin x} \qquad$ (since $\sin 2\theta = 2\sin\theta\cos\theta$)

$= \dfrac{\sin 2^{k+1} x}{2^{k+1} \sin x}$

11. *2. Induction step:* Assume S_k. Deduce S_{k+1}. Now

$\qquad k < k + 1 \qquad\qquad$ (by S_k)

$\qquad k + 1 < k + 1 + 1 \qquad$ (adding 1)

$\qquad \therefore\, k + 1 < k + 2$

15. S_1: $1^3 = \dfrac{1^2(1+1)^2}{4} = 1$

S_k: $1^3 + 2^3 + \cdots + k^3 = \dfrac{k^2(k+1)^2}{4}$

$1^3 + 2^3 + \cdots + (k+1)^3 = \dfrac{k^2(k+1)^2}{4} + (k+1)^3$

$\qquad\qquad\qquad\qquad = \dfrac{(k+1)^2}{4}\big(k^2 + 4(k+1)\big)$

$\qquad\qquad\qquad\qquad = \dfrac{(k+1)^2(k+2)^2}{4}$

19. S_1: $a_1 = \dfrac{a_1 - a_1 r}{1 - r} = \dfrac{a_1(1-r)}{1-r} = a_1$

S_k: $a_1 + \cdots + a_1 r^{k-1} = \dfrac{a_1 - a_1 r^k}{1 - r}$

Add $a_1 r^k$.

$a_1 + \cdots + a_1 r^k = \dfrac{a_1 - a_1 r^k}{1-r} + a_1 r^k \dfrac{1-r}{1-r}$

$\qquad\qquad\qquad = \dfrac{a_1 - a_1 r^k + a_1 r^k - a_1 r^{k+1}}{1-r}$

$\qquad\qquad\qquad = \dfrac{a_1 - a_1 r^{k+1}}{1-r}$

25. S_n: $\left(1 - \dfrac{1}{2^2}\right)\left(1 - \dfrac{1}{3^2}\right)\cdots\left(1 - \dfrac{1}{n^2}\right) = \dfrac{n+1}{2n}$

S_2: $1 - \dfrac{1}{2^2} = \dfrac{2+1}{2\cdot 2}$

S_k: $\left(1 - \dfrac{1}{2^2}\right)\left(1 - \dfrac{1}{3^2}\right)\cdots\left(1 - \dfrac{1}{k^2}\right) = \dfrac{k+1}{2k}$

S_{k+1}: $\left(1 - \dfrac{1}{2^2}\right)\left(1 - \dfrac{1}{3^2}\right)\cdots\left(1 - \dfrac{1}{k^2}\right)\left(1 - \dfrac{1}{(k+1)^2}\right) = \dfrac{k+2}{2(k+1)}$

1. *Basis step:* S_1 true by substitution.
2. *Induction step:* Assume S_k. Deduce S_{k+1}.
   Starting with the left side of S_{k+1} we have:

$$\underbrace{\left(1 - \frac{1}{2^2}\right)\left(1 - \frac{1}{3^2}\right)\cdots\left(1 - \frac{1}{k^2}\right)}\left(1 - \frac{1}{(k+1)^2}\right)$$

$$= \frac{k+1}{2k}\left(1 - \frac{1}{(k+1)^2}\right)$$

$$= \frac{k+1}{2k} - \frac{1}{2k(k+1)}$$

$$= \frac{(k+1)(k+1) - 1}{2k(k+1)}$$

$$= \frac{k^2 + 2k + 1 - 1}{2k(k+1)}$$

$$= \frac{k^2 + 2k}{2k(k+1)}$$

$$= \frac{k(k+2)}{2k(k+1)}$$

$$= \frac{k+2}{2(k+1)}$$

27. S_2: $\overline{z_1 + z_2} = \bar{z}_1 + \bar{z}_2$:

$$\overline{(a + bi) + (c + di)} = \overline{(a + c) + (b + d)i} = (a + c) - (b + d)i$$
$$\overline{a + bi} + \overline{(c + di)} = a - bi + c - di = (a + c) - (b + d)i.$$

S_k: $\overline{z_1 + z_2 + \cdots + z_k} = \bar{z}_1 + \bar{z}_2 + \cdots + \bar{z}_k.$

$$\overline{(z_1 + z_2 + \cdots + z_k) + z_{k+1}} = \overline{(z_1 + z_2 + \cdots + z_k)} + \overline{z_{k+1}} \qquad \text{(by } S_2\text{)}.$$
$$= \bar{z}_1 + \bar{z}_2 + \cdots + \bar{z}_k + \overline{z_{k+1}} \qquad \text{(by } S_k\text{)}.$$

29. S_1: i is either i or -1 or $-i$ or 1.
S_k: i^k is either i or -1 or $-i$ or 1.
$i^{k+1} = i^k \cdot i$ is then $i \cdot i = -1$ or $-1 \cdot i = -i$ or
$-i \cdot i = 1$ or $1 \cdot i = i$.

31. S_1: 2 is a factor of $1^2 + 1$.
S_k: 2 is a factor of $k^2 + k$.
$(k + 1)^2 + (k + 1) = k^2 + 2k + 1 + k + 1 = k^2 + k + 2(k + 1)$.
By S_k, 2 is a factor of $k^2 + k$, hence 2 is a factor of
the right-hand side, so is a factor of
$(k + 1)^2 + (k + 1)$.

Chapter 15 Review, p. 586

1. $3\frac{3}{4}$ **2.** $a + 4b$ **3.** 531 **4.** 465 **5.** 11 **6.** -4 **7.** $n = 6$, $S_n = -126$ **8.** $a_1 = 8$, $a_5 = \frac{1}{2}$

9. 0.27, 0.0027, 0.000027 **10.** $\frac{1}{2}$, $-\frac{1}{6}$, $\frac{1}{18}$ **11.** $\frac{211}{99}$ **12.** $5\frac{4}{5}$, $6\frac{3}{5}$, $7\frac{2}{5}$, $8\frac{1}{5}$ **13.** $3\sqrt{5}$ **14.** ≈ 40 ft   **15.** \$7.38, \$1365.10

16. $7,680,000$ **17.** $\dfrac{16\sqrt{2}}{\sqrt{2} - 1}$ in., or $32 + 16\sqrt{2}$ in.   **18.** $23\frac{1}{3}$ cm

19. S_n: $1 + 4 + 7 + \cdots + (3n - 2) = \dfrac{n(3n - 1)}{2}$

S_1: $1 = \dfrac{1(3 - 1)}{2}$

S_k: $1 + 4 + 7 + \cdots + (3k - 2) = \dfrac{k(3k - 1)}{2}$

S_{k+1}: $1 + 4 + 7 + \cdots + [3(k + 1) - 2]$

$\qquad = 1 + 4 + 7 + \cdots + (3k - 2) + (3k + 1)$

$\qquad = \dfrac{(k + 1)(3k + 2)}{2}$

1. *Basis step:* $1 = \dfrac{2}{2} = \dfrac{1(3 - 1)}{2}$ is true.

2. *Induction step:* Assume S_k.

$1 + 4 + 7 + \cdots + (3k - 2) + (3k + 1)$

$\qquad = \dfrac{k(3k - 1)}{2} + (3k + 1)$

$\qquad = \dfrac{k(3k - 1)}{2} + \dfrac{2(3k + 1)}{2}$

$\qquad = \dfrac{3k^2 - k + 6k + 2}{2}$

$\qquad = \dfrac{3k^2 + 5k + 2}{2}$

$\qquad = \dfrac{(k + 1)(3k + 2)}{2}$

21. S_n: $\left(1 - \dfrac{1}{2}\right)\left(1 - \dfrac{1}{3}\right) \cdots \left(1 - \dfrac{1}{n}\right) = \dfrac{1}{n}$

S_2: $\left(1 - \dfrac{1}{2}\right) = \dfrac{1}{2}$

S_k: $\left(1 - \dfrac{1}{2}\right)\left(1 - \dfrac{1}{3}\right) \cdots \left(1 - \dfrac{1}{k}\right) = \dfrac{1}{k}$

S_{k+1}: $\left(1 - \dfrac{1}{2}\right)\left(1 - \dfrac{1}{3}\right) \cdots \left(1 - \dfrac{1}{k}\right)\left(1 - \dfrac{1}{k + 1}\right) = \dfrac{1}{k + 1}$

1. *Basis step:* S_2 is true by substitution.

2. *Induction step:* Assume S_k. Deduce S_{k+1}.
   Starting with the left side of S_{k+1}, we have

$\underbrace{\left(1 - \dfrac{1}{2}\right)\left(1 - \dfrac{1}{3}\right) \cdots \left(1 - \dfrac{1}{k}\right)}\left(1 - \dfrac{1}{k + 1}\right)$

$= \dfrac{1}{k} \cdot \left(1 - \dfrac{1}{k + 1}\right) \qquad$ (by S_k)

$= \dfrac{1}{k} \cdot \left(\dfrac{k + 1 - 1}{k + 1}\right)$

$= \dfrac{1}{k} \cdot \dfrac{k}{k + 1}$

$= \dfrac{1}{k + 1} \qquad$ (simplifying)

22. S_1 fails.   **23.** $\dfrac{a_{k+1}}{a_k} = r_1$, $\dfrac{b_{k+1}}{b_k} = r_2$, so $\dfrac{a_{k+1}b_{k+1}}{a_k b_k} = r_1 r_2$ (constant)

24. $a_{k+1} - a_k = d$, so $\dfrac{c_{k+1}}{c_k} = \dfrac{b^{a_{k+1}}}{b^{a_k}} = b^{a_{k+1} - a_k} = b^d$ (constant)

25. (a) a_n is all positive or all negative; (b) always; (c) always; (d) never; (e) $a_n = 0$; (f) $a_n = 0$ **26.** $-2, 0, 2, 4$

27. 5, 51, 5203, 54, 142, 419   **28.** $\displaystyle\sum_{n=1}^{7} (n^2 - 1)$

20. S_1: $1 = \dfrac{3^1 - 1}{2}$; S_2: $1 + 3 = \dfrac{3^2 - 1}{2}$

S_k: $1 + 3 + 3^2 + \cdots + 3^{k-1} = \dfrac{3^k - 1}{2}$

$1 + 3 + \cdots + 3^{k-1} + 3^k$

$\qquad = \dfrac{3^k - 1}{2} + 3^k = \dfrac{3^k - 1}{2} + 3^k \cdot \dfrac{2}{2} = \dfrac{3 \cdot 3^k - 1}{2}$

$\qquad = \dfrac{3^{k+1} - 1}{2}$

CHAPTER 16

Margin Exercises, Section 16.1

1. $3 \cdot 2 \cdot 1$, or 6; $3 \cdot 3 \cdot 3$, or 27   **2.** $4 \cdot 2 \cdot 5$, or 40   **3.** $5 \cdot 4 \cdot 3 \cdot 2 \cdot 1$, or 120   **4.** $5 \cdot 4 \cdot 3 \cdot 2 \cdot 1$, or 120   **5.** $3 \cdot 2 \cdot 1$. or 6
6. $5 \cdot 4 \cdot 3 \cdot 2 \cdot 1$, or 120   **7.** $6 \cdot 5 \cdot 4 \cdot 3 \cdot 2 \cdot 1$, or 720   **8.** $4 \cdot 3 \cdot 2 \cdot 1$, or 24   **9.** $6 \cdot 5 \cdot 4 \cdot 3 \cdot 2 \cdot 1$, or 720
10. $8 \cdot 7 \cdot 6 \cdot 5 \cdot 4 \cdot 3 \cdot 2 \cdot 1$, or 40,320   **11.** 362,880   **12.** 40,320   **13.** 18!   **14.** (a) $10! = 10 \cdot 9!$; (b) $20! = 20 \cdot 19!$
15. $_7P_3 = 7 \cdot 6 \cdot 5 = 210$; $_7P_3 = \dfrac{7!}{4!} = \dfrac{7 \cdot 6 \cdot 5 \cdot 4 \cdot 3 \cdot 2 \cdot 1}{4 \cdot 3 \cdot 2 \cdot 1} = 7 \cdot 6 \cdot 5 = 210$
16. (a) $_{10}P_4 = 10 \cdot 9 \cdot 8 \cdot 7 = 5040$; (b) $_8P_2 = 8 \cdot 7 = 56$; (c) $_{11}P_5 = 11 \cdot 10 \cdot 9 \cdot 8 \cdot 7 = 55,440$; (d) $_nP_1 = n$;
(e) $_nP_2 = n(n-1) = n^2 - n$ **17.** $_{12}P_5 = 12 \cdot 11 \cdot 10 \cdot 9 \cdot 8 = 95,040$ **18.** $_{10}P_6 = 10 \cdot 9 \cdot 8 \cdot 7 \cdot 6 \cdot 5 = 151,200$
19. $_4P_4 \cdot {_3P_3} = 4! \cdot 3! = 144$ **20.** $\dfrac{11!}{1!\,4!\,4!\,2!} = 34,650$ **21.** $\dfrac{6!}{3!\,2!\,1!} = 60$ **22.** $\dfrac{8!}{3!\,3!\,2!} = 560$ **23.** $6! = 720$
24. $11! = 39,916,800$ **25.** $26^5 = 11,881,376$

Exercise Set 16.1, pp. 596–597

1. $4 \cdot 3 \cdot 2$, or 24   **3.** $_{10}P_7 = 10 \cdot 9 \cdot 8 \cdot 7 \cdot 6 \cdot 5 \cdot 4$, or 604,800   **5.** 120; 3125   **7.** 120; 24   **9.** $\dfrac{5!}{2!\,1!\,1!\,1!} = 5 \cdot 4 \cdot 3 = 60$
11. $9 \cdot 9 \cdot 8 \cdot 7 \cdot 6 \cdot 5 \cdot 4$, or 544,320   **13.** $\dfrac{9!}{2!\,3!\,4!} = 1260$ **15.** $12!$, or 479,001,600   **17.** (a) 120; (b) 3840
19. $52 \cdot 51 \cdot 50 \cdot 49 = 6,497,400$ **21.** $\dfrac{24!}{3!\,5!\,9!\,4!\,3!} = 16,491,024,950,400$
23. $4! = 24$, $8! \div 3! = 6720$, $\dfrac{13!}{2!\,2!\,2!\,2!\,2!} = 194,594,400$
25. $80 \cdot 26 \cdot 9999 = 20,797,920$ **27.** 11   **29.** 9

Margin Exercises, Section 16.2

1. (a) 6; (b) 1; (c) 4; (d) 1   **2.** (a) $5 \cdot 4 \cdot 3$, or 60; (b) *ABC, BCA, CAB, BAC, ACB, CBA, ABD, BDA, DAB, DBA, BAD,*
ADB, ABE, BEA, EAB, EBA, BAE, AEB, ACD, CDA, DAC, DCA, CAD, ADC, ACE, CEA, EAC, ECA, CAE, AEC, ADE,
DEA, EAD, EDA, EDB, DAE, AED, BCD, CDB, DBC, DCB, CBD, BDC, BCE, CEB, EBC, ECB, CBE, BEC, BDE, DEB,
EBD, DBE, BED, CDE, DEC, ECD, EDC, DCE, CED; (c) 10; (d) $\{A, B, C\}$, $\{A, C, D\}$, $\{B, C, D\}$, $\{D, C, E\}$, $\{A, C, E\}$,
$\{A, B, D\}$, $\{A, B, E\}$, $\{B, C, E\}$, $\{B, E, D\}$, $\{A, D, E\}$ **3.** (a) 45; (b) 45; (c) 21; (d) 1   **4.** 36, 36   **5.** $_{12}C_5 = 792$
6. $_{12}C_3 \cdot {_8C_2} = 6160$

Exercise Set 16.2, pp. 599–600

1. 126   **3.** 1225   **5.** $\dfrac{n(n-1)(n-2)}{3!}$ **7.** 8855   **9.** 210   **11.** $\binom{8}{2} = 28$, $\binom{8}{3} = 56$ **13.** $\binom{10}{7} \cdot \binom{5}{3} = 1200$ **15.** $\binom{58}{6} \cdot \binom{42}{4}$
17. $\binom{4}{3} \cdot \binom{48}{2} = 4512$ **19.** 2,598,960   **21.** $\binom{8}{3} = 56$
23. $\binom{5}{2}\binom{8}{2} = 280$ **25.** 5   **27.** 7

Margin Exercises, Section 16.3

1. $x^{10} - 5x^8 + 10x^6 - 10x^4 + 5x^2 - 1$ **2.** $16x^4 + 32x^3\dfrac{1}{y} + 24x^2\dfrac{1}{y^2} + 8x\dfrac{1}{y^3} + \dfrac{1}{y^4}$
3. $x^6 - 6\sqrt{2}x^5 + 30x^4 - 40\sqrt{2}x^3 + 60x^2 - 24\sqrt{2}x + 8$ **4.** $-1512x^5$ **5.** $5670x^4$ **6.** $8064y^5$ **7.** $243x^5$
8. $2^6 = 64$ **9.** 2^{50}

16

Exercise Set 16.3, pp. 603–604

1. $15a^4b^2$ **3.** $-745,472a^3$ **5.** $1120x^{12}y^2$ **7.** $-1,959,552u^5v^{10}$ **9.** $m^5 + 5m^4n + 10m^3n^2 + 10m^2n^3 + 5mn^4 + n^5$
11. $x^{10} - 15x^8y + 90x^6y^2 - 270x^4y^3 + 405x^2y^4 - 243y^5$ **13.** $x^{-8} + 4x^{-4} + 6 + 4x^4 + x^8$
15. $\binom{n}{0} - \binom{n}{1} + \binom{n}{2} - \binom{n}{3} + \cdots + (-1)^n\binom{n}{n}$ **17.** $140\sqrt{2}$ **19.** $9 - 12\sqrt{3}t + 18t^2 - 4\sqrt{3}t^3 + t^4$ **21.** 128
23. 2^{26}, or $67,108,864$ **25.** $-7 - 4\sqrt{2}i$ **27.** $\sin^7 t - 7\sin^5 t + 21\sin^3 t - 35\sin t + 35\csc t - 21\csc^3 t + 7\csc^5 t - \csc^7 t$
29. $\displaystyle\sum_{r=0}^{n}\binom{n}{r}(-1)^r a^{n-r}b^r$ **31.** -3 **33.** 5 **35.** $2^7 - 1$, or 127

Margin Exercises, Section 16.4

1. $\frac{1}{2}$ **2.** (a) $\frac{1}{13}$; (b) $\frac{1}{4}$; (c) $\frac{1}{2}$; (d) $\frac{2}{13}$ **3.** $\frac{6}{11}$ **4.** 0 **5.** 1 **6.** $\frac{11}{850}$ **7.** $\frac{15}{91}$ **8.** $\frac{6}{13}$ **9.** $\frac{1}{6}$

Exercise Set 16.4, pp. 608–610

1. 52 **3.** $\frac{1}{4}$ **5.** $\frac{1}{13}$ **7.** $\frac{1}{2}$ **9.** $\frac{2}{13}$ **11.** $\frac{5}{7}$ **13.** 0 **15.** $\frac{11}{4165}$ **17.** $\frac{28}{65}$ **19.** $\frac{1}{18}$ **21.** $\frac{1}{36}$ **23.** $\frac{30}{323}$ **25.** $\frac{9}{19}$ **27.** $\frac{1}{38}$ **29.** $\frac{1}{19}$
31. $2,598,960$ **33.** (a) 36; (b) 1.39×10^{-5}

Chapter 16 Review, p. 610

1. $6! = 720$ **2.** $9 \cdot 8 \cdot 7 \cdot 6 = 3024$ **3.** $\binom{15}{8} = 6435$ **4.** $24 \cdot 23 \cdot 22 = 12,144$ **5.** $\dfrac{9!}{1!\,4!\,2!\,2!} = 3780$ **6.** 36

7. (a) $_6P_5 = 720$; (b) $6^5 = 7776$; (c) $_5P_4 = 120$; (d) $_3P_2 = 6$ **8.** $6! = 720$ **9.** $220a^9x^3$ **10.** $\binom{18}{11}a^7x^{11}$, or $\dfrac{18!}{11!\,7!}a^7x^{11}$
11. $m^7 + 7m^6n + 21m^5n^2 + 35m^4n^3 + 35m^3n^4 + 21m^2n^5 + 7mn^6 + n^7$ **12.** $x^8 + 12x^6y + 54x^4y^2 + 108x^2y^3 + 81y^4$
13. $-6624 + 16,280i$ **14.** $\cos^8 t + 8\cos^6 t + 28\cos^4 t + 56\cos^2 t + 70 + 56\sec^2 t + 28\sec^4 t + 8\sec^6 t + \sec^8 t$
15. $\frac{1}{12}$, 0 **16.** $\frac{1}{4}$ **17.** $\frac{6}{5525}$ **18.** $\left(\log\dfrac{x}{y}\right)^{10}$ **19.** 36 **20.** 14

INDEX